**Textbook
of anatomy
and
physiology**

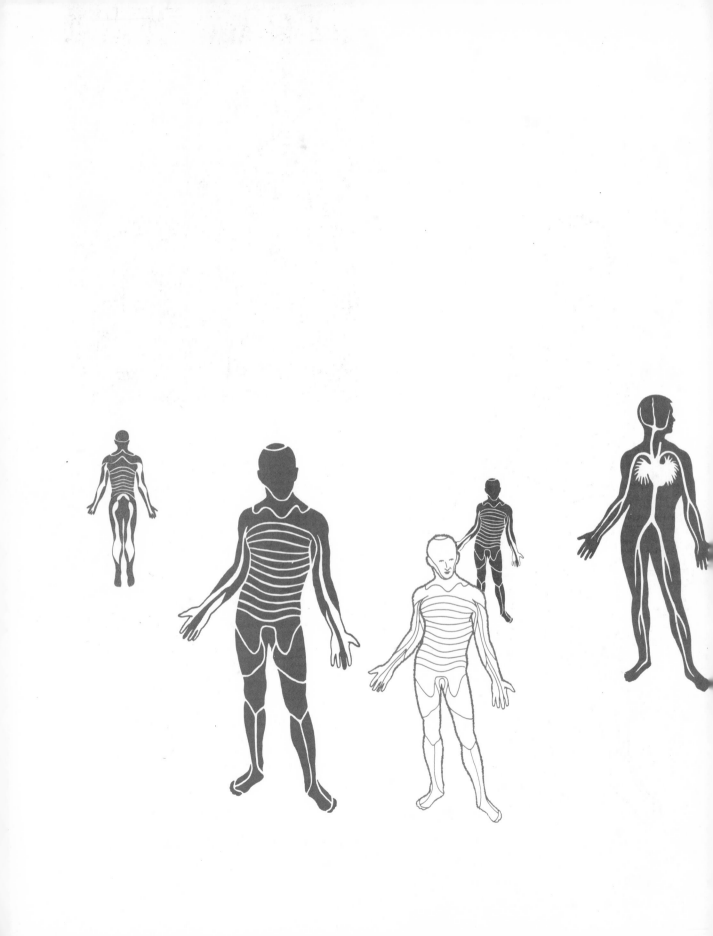

EIGHTH EDITION

Textbook
of anatomy
and
physiology

Catherine Parker Anthony, R.N., B.A., M.S.

*Formerly Assistant Professor of Nursing, Science Department,
and Assistant Instructor of Anatomy and Physiology,
Frances Payne Bolton School of Nursing,
Case Western Reserve University, Cleveland, Ohio;
formerly Instructor of Anatomy and Physiology,
Lutheran Hospital and St. Luke's Hospital,
Cleveland, Ohio*

with the collaboration of

Norma Jane Kolthoff, R.N., B.S., Ph.D.

*Professor of Nursing and Project Director for the School of Nursing
Institutional Nursing Research Development Program,
University of Wisconsin,
Madison, Wisconsin*

*With 320 figures (137 in color),
including 222 by Ernest W. Beck, and an
insert on human anatomy containing 15 full-color,
full-page color plates, with six in transparent Trans-Vision ®
(by Ernest W. Beck)*

The C. V. Mosby Company

St. Louis 1971

This book is affectionately dedicated to

Geoffrey, Scott, Rob, and Dawn

Our beloved third generation

Preface

One purpose dominated the revision of this eighth edition of *Textbook of anatomy and physiology* — to improve upon the seventh edition in every way we could envision. An earnest effort has been made to improve both the content of the book and its organization.

Since publication of the seventh edition, hundreds of comments have been received on the outstanding excellence of the illustrations contributed by Mr. Ernest W. Beck, one of our country's finest medical illustrators. More of Mr. Beck's drawings appear in this edition. If you are not familiar with his work, you may wish to examine Fig. 6-1, Fig. 8-19, and Fig. 13-4 as an introduction to it.

The basic format of the book remains unchanged, with each chapter beginning with a topical outline and closing with an outline summary and review questions. The review questions have been extensively revised. Because many of them now focus on likenesses, differences, and rela-tionships, it is hoped they will serve better to help you develop your discriminatory powers. Supplementary readings, abbreviations and prefixes, and a glossary are again included at the end of the book. Many new references have been substituted for outdated ones. You may like to consider using some of the new references as a basis for independent study. They almost certainly will increase your knowledge and appreciation of newer physiological research. They are also apt to sharpen your desire for "learning more" — surely a goal all education should seek.

More information about age changes in body structure and function appear in this edition than in former ones. Included also are discussions of the concepts of adaptation and maladaptation and their relationships to homeostasis and disease. And this eighth edition has an entirely new final chapter — one we hope you will find both interesting and valuable. It explains the concept of physiological stress and discusses its causes, the body's responses to it, and the relationships that seem to exist between it and psychological stress and disease. Extensive changes in content have been made in the introductory chapter and in the chapters on the cell and the circulatory system. The chapter on the nervous system has been reorganized. Lesser changes appear in most of the remaining chapters.

Dr. Norma Jane Kolthoff—nurse-scientist, superior teacher, and valued friend—contributed richly to the new chapter on stress and to the reorganization and rewriting of the chapter on the nervous system. I am deeply grateful to Dr. Kolthoff for these contributions as well as for many helpful suggestions, to Mr. Ernest W. Beck for his distinguished artwork, and to Mrs. Georgeanna Keefe for her loyal secretarial help. I also wish to acknowledge indebtedness and to express warmest appreciation to members of my family for their patience and faith.

Catherine Parker Anthony
Cleveland, Ohio

Contents

unit one
The body as a whole

1 Organization of the body, 2
2 Cells, 11
3 Tissues, 42
4 Membranes and glands, 53

unit two
The erect and moving body

5 The skeletal system, 62
6 The muscular system, 113

unit three
Integration and control of the body

7 The nervous system, 164
8 Sense organs, 239
9 The endocrine system, 263

unit four
Maintaining the metabolism of the body

10 The circulatory system, 292

11 The respiratory system, 362

12 The digestive system, 389

13 The urinary system, 446

unit five
Reproduction of the human being

14 The reproductive system, 466

unit six
Fluid, electrolyte, and acid-base balance

15 Fluid and electrolyte balance, 496

16 Acid-base balance, 512

unit seven
Stress

17 Stress, 524

Supplementary readings, 533

Abbreviations and prefixes, 537

Glossary, 539

Textbook
of anatomy
and
physiology

The body as a whole

1

Organization of the body

**Generalizations about
  body structure**

**Terms used in describing
  body structure**
Directional terms
Planes of body
Abdominal regions
Anatomical position

**Generalizations about
  body function**

*What a piece of work is man! how noble in
reason! how infinite in faculty! in form and
moving how express and admirable! in action
how like an angel! in apprehension how like a
god! the beauty of the world! the paragon of
animals!*

— Hamlet, in *Hamlet, Prince of Denmark,*
Act II, Scene II, by William Shakespeare

Anatomy and physiology are biological sciences; that is, they are branches of knowledge about living things. *Human anatomy* is the science of the structure and *human physiology* is the science of the function of the most wondrous of all structures—the human body.

Before attempting to explore details about the body, it seems feasible to examine some of its general characteristics, much as it is prudent to study the map of a strange city before embarking upon an excursion through it. This first chapter, therefore, will present some generalizations about the body's structure, define some terms used in describing its structure, and conclude with some generalizations about body function.

◾Generalizations about body structure

1 *The body is a large structural unit composed of four kinds of smaller units: cells, tissues, organs, and systems.* Smallest and most numerous of these are the cells. How many cells in the body? Far too many for our imaginations to visualize. One average-sized adult body, according to one authority,* consists of 100,000,000, 000,000 cells. In case you cannot translate

*Swanson, C. P.: The cell, ed. 3, Englewood Cliffs, N. J., 1969, Prentice-Hall, Inc.

this number—one with fourteen zeros after it—it is 100 trillions! or 100,000 billions! or 100 million millions!

Cells have long been recognized as the simplest units of living matter. In truth, however, they are far from simple. How extremely complex cells are, you can discover by reading the next chapter. A *cell* is the smallest unit that can maintain life and reproduce itself. (Viruses, although smaller living units than cells, cannot maintain life and reproduce unless they parasitize a living cell.)

Tissues are somewhat more complex units than cells. By definition, a *tissue* is an organization of a great many similar cells with varying amounts and kinds of nonliving, intercellular substance between them.

Organs are more complex units than tissues. An *organ* is an organization of several different kinds of tissues so arranged that together they can perform a special function. For example, the stomach is an organization of muscle, connective, epithelial, and nervous tissues. Muscle and connective tissues form its wall, epithelial and connective tissues form its lining, and nervous tissue extends throughout both its wall and lining.

Systems are the most complex of the component units of the body. A *system* is an organization of varying numbers and kinds of organs so arranged that together they can perform complex functions for the body. Nine major systems compose the human body: skeletal, muscular, nervous, endocrine, circulatory, respiratory, digestive, urinary, and reproductive. (For names and descriptions of the organs of a particular system, see the chapter on that system.)

2 *The human body contains a ventral cavity and a dorsal cavity.* Each of these consists of smaller cavities.

The *ventral cavity* consists of the *thoracic* or *chest cavity* and the *abdominopelvic cavity*. The thoracic cavity is separated from the abdominopelvic cavity by a musculomembranous partition, the diaphragm. The thoracic cavity has pleural, pericardial, and mediastinal subdivisions. The abdominopelvic cavity consists of an upper portion, the *abdominal cavity*, and a lower portion, the *pelvic cavity*.

The *dorsal cavity* consists of the cranial cavity and the spinal cavity. The *cranial cavity* lies in the skull and houses the brain. The *spinal cavity* lies in the spinal column and houses the spinal cord.

The body cavities contain the various internal organs or *viscera*. Located in the thoracic cavity are the lungs, heart, trachea, esophagus, thymus gland, and certain large blood and lymphatic vessels and nerves. The abdominal cavity contains the liver, gallbladder, stomach, pancreas, intestines, spleen, kidneys, and ureters. The bladder, certain reproductive organs (uterus, uterine tubes, and ovaries in the female; prostate gland, seminal vesicles, and part of the seminal ducts in the male), and part of the large intestine (namely, the sigmoid colon and rectum) lie in the pelvic cavity.

3 *The most outstanding characteristic of body structure is its organization.* Everyone of its component units—every system, every organ, every tissue, and even every cell—consists of smaller parts organized in a definite way. Organization is surely one of the most important characteristics of life since, as one scientist wrote, it "stands in direct opposition to the behavior of lifeless matter. The latter moves toward ever greater randomness. . . . A living organism, on the contrary, draws out from its chaotic environment particular substances and builds them into a system of ever greater and more organized complexity, thus steadily decreasing the randomness of matter. An organism is not an aggregate, but an integrate. . . . Life *is* organization."*

4 *Body structures change gradually over the years.* All of us have noticed some of these changes—for example, growth in children and wrinkles and muscle flabbiness in elderly people. But many structural age changes are less obvious. For instance, certain organs shrink in size in old age, and the number of cells in many tissues decreases markedly. By the age of 75, the brain weighs only about 56% as much as it did at the age of 30. And a man who lives to be 75 years old has only slightly more

than one-third the number of taste buds he had when he was 30.*

■ Terms used in describing body structure

Directional terms

superior or *cranial* toward the head end of the body; upper (example: the hand is part of the superior extremity).

inferior or *caudal* away from the head; lower (example: the foot is part of the inferior extremity).

anterior or *ventral* front (example, the kneecap is located on the anterior side of the leg).

posterior or *dorsal* back (example, the shoulder blades are located on the posterior side of the body).

medial or *mesial* toward the midline of the body (example, the great toe is located at the medial side of the foot).

lateral away from the midline of the body (example, the little toe is located at the lateral side of the foot).

proximal toward or nearest the trunk or the point of origin of a part (example, the elbow is located at the proximal end of the forearm).

distal away from or farthest from the trunk or the point of origin of a part (example, the hand is located at the distal end of the forearm).

Planes of body

sagittal a lengthwise plane running from front to back; divides the body or any of its parts into right and left sides (Fig. 1-1).

median sagittal plane through midline; divides the body or any of its parts into right and left halves.

coronal or *frontal* a lengthwise plane run-

---

*From Sinnott, E. W.: Two roads to truth, New York, 1953, The Viking Press, Inc., p. 131.

*Shock, N. W.: The physiology of aging, Sci. Amer. **206**:100–110 (Jan.), 1962.

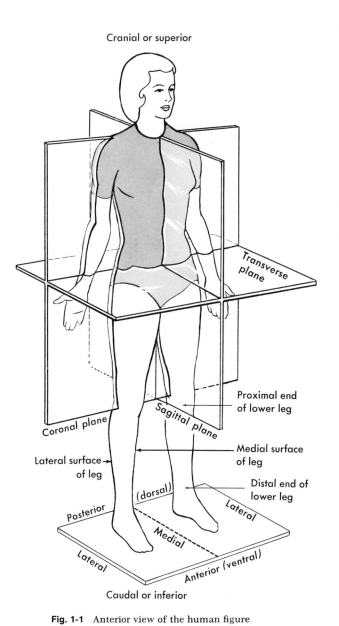

Fig. 1-1 Anterior view of the human figure demonstrating meanings of terms used in describing the body.

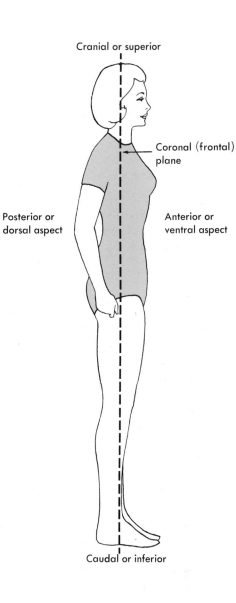

Fig. 1-2 Lateral view of the human figure demonstrating several terms used in describing the body.

ning from side to side; divides the body or any of its parts into anterior and posterior portions (Figs. 1-1 and 1-2).

transverse or *horizontal* a crosswise plane; divides the body or any of its parts into upper and lower parts (Fig. 1-1).

Abdominal regions

For convenience in locating abdominal organs, anatomists divide the abdomen into nine imaginary regions. See Fig. 1-3 for their names and boundaries.

Anatomical position

The term *anatomical position* means an erect position of the body, with arms at sides, palms turned forward (supinated).

■ Generalizations about body function

1 *Survival is the body's most important business—survival of itself and survival of the human species.* To achieve survival, the body must carry on ceaseless and almost numberless activities. Life is synonymous with activity; death is synonymous with cessation of activity.

2 *All body functions, in reality, are cell functions.* Some activities performed by a cell serve only itself. Self-serving cellular activities achieve the cell's own survival or its reproduction. In contrast, certain other cellular activities serve the body as a whole. These body-serving cellular functions achieve survival of the individual and of the human species. They make it possible for the body to stay alive and reproduce. Different kinds of cells, as we shall see, specialize in different kinds of body-serving functions. All kinds of cells, on the other hand, perform essentially the same self-serving functions (discussed in Chapter 2).

3 *Environment influences body functioning.* Many changes in the environment act as stimuli which induce responses by the body. Most responses are "adaptive." *Adaptive responses*, by definition, are those that tend to maintain healthy survival. They make it possible for the body to thrive as well as to survive. Collectively, adaptive responses perform the function of adaptation. Adaptation includes both the changes in the body induced by changes in the environment and, reciprocally, the changes in the environment produced by changes in the body. *Adaptation*, in other words, is the great complex of interactions between the body and its environment which works toward the well-being of the body. Adaptation is the successful coping of an organism with its environment. Successful adaptation means healthy survival. Inadequate adaptation (maladaptation) means disease. Failure of adaptation means death. Life, then, is not only organization and ceaseless activity, but it is also adaptation.

4 *Adaptive responses tend to maintain or restore homeostasis.* Homeostasis is a key word in physiology. It comes from two Greek words—*homoios*, meaning the same, and *stasis*, meaning standing. "Standing or staying the same," then, is the literal meaning of the word. Walter B. Cannon, a noted physiologist, suggested homeostasis as the special name for the steady states maintained by the body. He emphasized, however, that a steady state or homeostasis does not mean something set and immobile that stays exactly the same all the time. In his words, homeostasis "means a condition which may vary, but which is relatively constant."*

Homeostasis is a condition which changes and yet stays about the same—a condition which stays within the same narrow range. We might even use the word to describe a condition other than one in the body. For example, we might refer to homeostasis of the depth of water in a swimming pool (Fig. 1-4). Imagine that water ran into a pool at one end and out of it at the other. And suppose that the depth of the water changed from time to time, sometimes increasing, sometimes decreasing. But despite these changes, suppose that the depth always stayed within the same narrow range of 58 to 62 inches. When the water increased above 60 inches, something happened automatically to reverse this change and decrease the water so that its level moved back down toward

*From Cannon, W. B.: The wisdom of the body, rev. ed., New York, 1939, W. W. Norton & Co., Inc., p. 24.

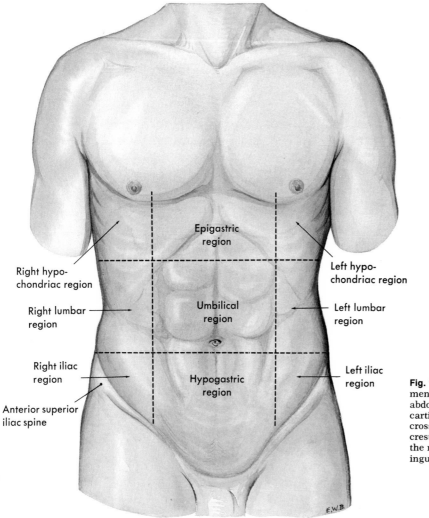

Epigastric
region

Right hypo-
chondriac region

Left hypo-
chondriac region

Right lumbar
region

Umbilical
region

Left lumbar
region

Right iliac
region

Left iliac
region

Anterior superior
iliac spine

Hypogastric
region

Fig. 1-3 The nine regions of the abdomen. The top horizontal line crosses the abdomen at the level of the ninth rib cartilages. The lower horizontal line crosses it at the level of the iliac crests. The vertical lines pass through the midpoints of the right and left inguinal (Poupart's) ligaments.

Fig. 1-4 A, Homeostasis as illustrated by constancy of a pool's water volume within a narrow range (indicated by double-headed arrow). When the volume of water entering the pool equals the volume leaving it, the pool volume remains constant at the level of the solid line, the midpoint of homeostasis for the pool's volume. **B,** The volume of water entering the pool has decreased. Because input is now less than output, the pool volume has decreased. **C,** The volume of water leaving the pool has decreased to compensate for the decreased volume entering it. Because output is now less than input, the pool volume has increased back to the original level; homeostasis of the pool volume is restored.

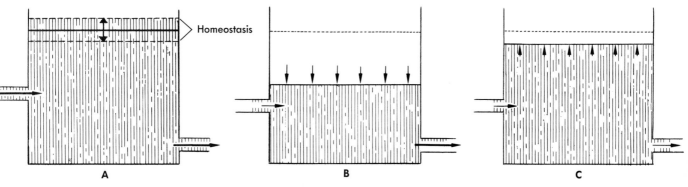

Homeostasis

A

B

C

the 60-inch mark. And when the depth of the water decreased below 60 inches, this change away from the balance point also automatically reversed itself. The water increased this time and again moved back toward the 60-inch level.

Human cells must have a relatively constant environment if they are to survive in a healthy condition. Homeostasis of the environment of cells, in other words, is an essential condition for their healthy survival. This requirement applies to many aspects of the cellular environment—its chemical composition, its osmotic pressure, its hydrogen-ion concentration, its temperature, etc. Even a small shift away from homeostasis of any of these factors makes cells unable to function normally. Because homeostasis is a requirement for healthy survival of cells, it necessarily follows that it is also a requirement for healthy survival of the body. Why? What microscopic living units compose the body?

Cellular environment, you must remember, is not synonymous with body environment. The body lives in the atmosphere surrounding it, the gaseous atmosphere of the external world. Body cells live in the atmosphere surrounding them, but this is the liquid atmosphere of an internal world.

The liquid environment around cells is called *extracellular fluid* for the obvious reason that it lies exterior to cells. Extracellular fluid occupies two main locations. It fills in the microscopic spaces between body cells, where it is called appropriately *intercellular fluid* (or *interstitial fluid*), and it flows through blood vessels, where it is called *blood plasma*.

More than a century ago, a great French physiologist, Claude Bernard, recognized that the extracellular fluid (or *milieu interne*, as he called it) constituted the internal environment, or, in other words, the cellular environment. He also noted that its physical and chemical composition

must be maintained relatively uniform if cells are to maintain their life and health. But it was many years later that Cannon named the relative constancy of the internal environment *homeostasis*.

The principle of homeostasis is one of the most fundamental of all physiological principles. It holds that the body must maintain relative constancy of its internal environment in order to survive. Stated another way, health and survival depend upon the body's maintaining or quickly restoring homeostasis.

All body structures play some role in maintaining the various steady states of the body. Homeostasis is virtually the job of every organ.

Homeostatic mechanisms are the devices by which the body maintains or restores homeostasis. In large part, they are also the devices by which the body accomplishes successful adaptation and thereby achieves healthy survival. All homeostatic mechanisms include adaptive responses. The study of homeostatic mechanisms, therefore, constitutes the main theme of physiology and of this book.

5 *Function is determined by structure.* Later chapters will cite innumerable examples to substantiate this cardinal principle about the relationship between structure and function.

6 *Body functions change gradually over the years.* What oldster has not noticed, for instance, that he cannot exercise as strenuously as he once did. Many functional changes go unnoticed by the individual but not undetected by laboratory tests. The latter have revealed, for example, that the volume of blood flowing to certain vital organs decreases as age increases. Only about 50% as much blood per minute flows through the kidneys of an average 75-year-old man as through those of an average 30-year-old man, and about 80% as much blood flows through the brain of an average 75-year-old man as through that of an average 30-year-old man.* Functional age changes will be noted in almost every chapter of this book.

• • •

Chapter 2 will describe cell structure and explain some of the intricacies of cell function.

*Shock, N. W.: The physiology of aging, Sci. Amer. **206**:100–110 (Jan.), 1962.

outline summary

Generalizations about body structure

1 Body composed of four kinds of smaller units
   a Cells—smallest units that can maintain life and reproduce
   b Tissues—organizations of many similar cells with nonliving intercellular substance between them
   c Organs—organizations of several different kinds of tissues so arranged that they can perform a special function
   d Systems—organizations of different kinds of organs so arranged that they can perform complex functions
2 Body contains ventral cavity and dorsal cavity, each with subdivisions
   a Ventral cavity
      1 Thoracic (or chest) cavity
         a Pleural portions—contain lungs
         b Pericardial portion—contains heart
         c Mediastinal portion—area between lungs; contains trachea, esophagus, thymus gland, and certain large blood vessels, lymphatic vessels, and nerves
      2 Abdominopelvic cavity
         a Abdominal portion—contains liver, gallbladder, stomach, pancreas, intestines, spleen, kidneys, and ureters
         b Pelvic portion—contains bladder, certain reproductive organs, and part of large intestine
   b Dorsal cavity
      1 Cranial cavity—contains brain
      2 Spinal cavity—contains spinal cord
3 Organization is most outstanding structural characteristic of body and all its parts
4 Body structures change gradually in variety of ways over years

Terms used in describing body structure

1 Directional terms
   a superior—upper
   b inferior—lower
   c anterior—front
   d posterior—back
   e medial—toward midline
   f lateral—away from midline
   g proximal—toward or nearest trunk or point of origin of part
   h distal—away from or farthest from trunk or point of origin
2 Planes of body
   a sagittal—lengthwise, dividing body or any of its parts into right and left portions
   b median—sagittal plane through midline
   c frontal or coronal—lengthwise, dividing body or any of its parts into anterior and posterior portions
   d transverse or horizontal—crosswise, dividing body or any of its parts into upper and lower sections
3 Abdominal regions—see Fig. 1-3
4 Anatomical position—erect, arms at sides, palms forward

Generalizations about body function

1 Survival, body's most important business—survival of individual and of species—depends upon ceaseless activity.
2 All body functions are cell functions; each cell performs functions to maintain its own life and specializes in some function that helps maintain life of body as whole.
3 Environment influences body functioning and vice versa; usually, body makes *adaptive responses* (those which tend to maintain healthy survival) to changes in its environment; great complex of interactions between body and its environment which tend to maintain healthy survival constitute broad function of *adaptation*.

4 Adaptive responses tend to maintain or restore *homeostasis*, that is, relative constancy of internal or cellular environment. *Principle of homeostasis*: healthy survival depends upon body's maintaining or quickly restoring homeostasis. *Homeostatic mechanisms*: devices which maintain or restore homeostasis and which are essential for successful adaptation and survival. All homeostatic mechanisms include adaptive responses.

5 Function is determined by structure.

6 Body functions change gradually over years.

review questions

1 Name and define briefly the four kinds of structural units that compose the body.

2 Name the two cavities on the dorsal surface of the body. What organs does each cavity contain?

3 Identify the divisions of the thoracic cavity and of the abdominopelvic cavity. Name the organs in each.

4 Define briefly each of the following terms: anatomical position; anatomy; anterior; distal; frontal; homeostatic mechanism; hypochondriac region; hypogastric region; iliac region; inferior; lateral; median plane; organ; physiology; posterior; proximal; sagittal plane; tissue; transverse plane.

5 What relationship exists between any structure and its function?

6 What relationship, if any, exists between an individual's age and the structure and function of his body?

7 What relationship, if any, exists between an individual's age and the number of cells composing his tissues?

Cells

During the past decade or two, research into the mysteries of the body's smallest units—its cells—has gone on at an almost feverish pace. Each year has seen the frontiers of this now vast field pushed farther and farther out. *Cytology*, the study of cell structure and function, has become a full-fledged member of the biological sciences. This chapter relates the main themes of this fascinating, though still unfinished, story. Some of the information presented is old and well established, some new and tenuous. We shall start with a discussion of cell structure, mentioning briefly the functions served by each cell part, and then consider the cell's self-serving functions in some detail.

Cell structure
Protoplasm
Cell membrane
Cytoplasm
  Endoplasmic reticulum
  Golgi apparatus
  Mitochondria
  Lysosomes
  Ribosomes
  Centrosome or
    centrosphere
Nucleus

Cell physiology
Movement of substances
  through cell membranes
  Diffusion
  Osmosis
  Active transport
    mechanisms
  Phagocytosis and
    pinocytosis
Cell metabolism
  Catabolism
    Glycolysis
    Citric acid cycle
Cell reproduction
  Mitosis

Cell structure

■ Protoplasm

Present-day knowledge about cells includes not only a great many facts about their microscopic structure, but also considerable information about their molecular structure. We know, for example, that *protoplasm* ("living substance," the complex chemical material that composes all living things) consists of no unique elements. Every element in protoplasm occurs also in nonliving matter. This is not true, however, of the compounds in protoplasm. Certain kinds of compounds—specifically, proteins (including enzymes), carbohydrates, lipids, and nucleic acids—occur naturally only in living matter, or matter that once was living. The most abundant elements in protoplasm are carbon, oxygen, hydrogen, and nitrogen. The most abundant compound is water.

The chemical composition of protoplasm varies considerably in different kinds of cells and even in one cell from time to time—a point worth mentioning because of the relationship between protoplasm's

11

molecular structure and its functions. Structure determines function! Does this basic principle already sound familiar to you? Here is just one of its many applications: Red blood cells contain hemoglobin, a protein whose molecular structure causes it to combine rapidly with oxygen under certain conditions and to dissociate from it rapidly under other conditions. In short, hemoglobin's structure determines red blood cells' function as oxygen carriers for the body.

Protoplasm's chemical structure endows it with the functional characteristics which distinguish it from nonliving matter. Living protoplasm exhibits *irritability* (the ability to respond to a stimulus), *conductivity* (the ability to transmit impulses), and *contractility* (the ability to contract or move) and carries on *metabolism* (the process of using foods to supply itself with energy and to synthesize various complex compounds). Protoplasm also has the ability to reproduce itself in all its complexities.

Not so many years ago, biologists knew only that cells consisted of an outer membrane, a nucleus, and a substance in-between that they called cytoplasm. Now they have learned, through the revealing powers of an imposing array of sophisticated tools and techniques, that a great many microscopic structures make up cells. The modern electron microscope has the power to magnify objects several hundred thousand times.* Thus magnified, an intercellular world teeming with tiny structures came into man's view. No one had even dreamed of their existence, and yet they number in the thousands. One of the remarkable techniques that has uncovered many secrets of both cell structure and

*Racker, E.: The membrane of the mitochondrion, Sci. Amer. **218**:32–39 (Feb.), 1968.

function is radioautography. It combines the use of radioactive atoms with a photographic technique.* Human cells, we now know, consist of the following parts:

Cell membrane
Cytoplasm
   Membranous organelles
     Endoplasmic reticulum
     Golgi apparatus
     Mitochondria
     Lysosomes
   Nonmembranous organelles
     Ribosomes
     Centrosome or centrosphere, including
      centrioles
Nucleus
   Nucleoli
   Chromosomes

Examine Plate XII of the color insert (p. 60) to see what the various cell structures look like under a light microscope. Then note in Fig. 2-1 how different they appear under the much greater magnification of an electron microscope.

■ Cell membrane

The most gossamer chiffon you can imagine is thick and heavy indeed compared to the membranes that form the boundaries of human cells. Their thinness truly does defy imagination. How can you possibly form a mental picture of anything that is only 3/10,000,000 of an inch thick!† A few tiny holes or *pores* penetrate this incredibly thin membrane. The scanty evidence available suggests that the pores have a diameter of perhaps 3 to 7 angstroms—very narrow doorways indeed for

*See supplementary readings for Chapter 2, references 4 and 9, if you would like to learn more about radioautography.
†*Thickness of a cell membrane* is 75 to 100 angstroms. (Since 1 angstrom, as shown on the facing page, equals roughly 1/250,000,000 of an inch, 75 angstroms equal about 3/10,000,000 of an inch.) *Diameter of a hydrogen atom* is 1 angstrom. (*Cont'd on opposite page*)

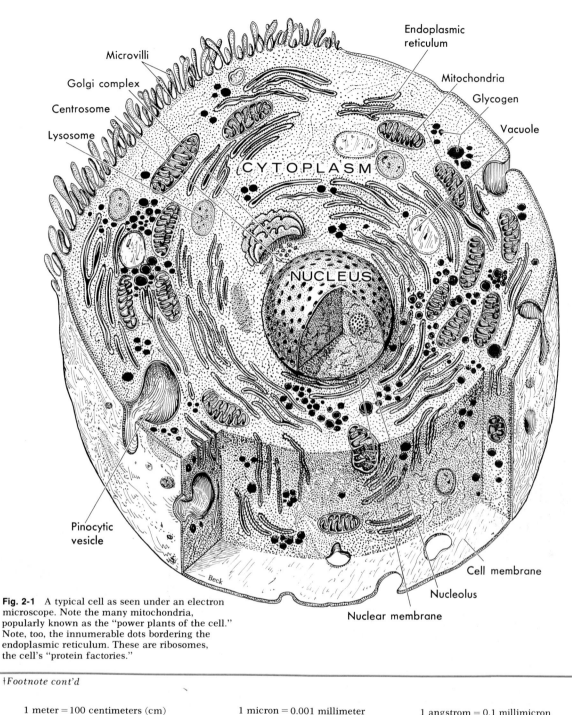

Microvilli

Golgi complex

Centrosome

Lysosome

Endoplasmic
reticulum

Mitochondria

Glycogen

Vacuole

CYTOPLASM

NUCLEUS

Pinocytic
vesicle

Beck

Cell membrane

Nucleolus

Nuclear membrane

Fig. 2-1 A typical cell as seen under an electron microscope. Note the many mitochondria, popularly known as the "power plants of the cell." Note, too, the innumerable dots bordering the endoplasmic reticulum. These are ribosomes, the cell's "protein factories."

---

†*Footnote cont'd*

1 meter = 100 centimeters (cm)
         1,000 millimeters (mm)
         39.37 inches

1 millimeter = 0.001 meter
         1,000 microns (μ)
         1,000,000 millimicrons (mμ)
         10,000,000 angstroms (Å)
         about 1/25 inch

1 micron = 0.001 millimeter
         1,000 millimicrons
         10,000 angstroms
         about 1/25,000 inch

1 millimicron = 0.001 micron
         10 angstroms
         about 1/25,000,000 inch

1 angstrom = 0.1 millimicron
         about 1/250,000,000 inch

1 inch = approximately
         2.5 centimeters
         25 millimeters
         25,000 microns
         25,000,000 millimicrons
         250,000,000 angstroms

molecules to pass through in entering or leaving cells!

Some cell membranes extend outward in hundreds of finger-shaped projections called *microvilli.* Note in Fig. 2-1 how greatly the microvilli shown there increase the surface area of the cell membrane. Obviously, more material per minute can move into a cell through a large surface than could move through it if it were smaller. Thus the structure of microvilli influences function. Their extensive surface area accelerates absorption into a cell. Not surprisingly, therefore, microvilli characterize the membranes of cells which specialize in the function of absorption — for example, certain cells lining the small intestine.

The molecular structure of cell membranes remains somewhat of a mystery. Protein and lipid molecules are known to compose them, but the organization or arrangement of these molecules has not yet been established. Probably the most persuasive data so far supports what we shall call the "triple-layer hypothesis" of cell membrane molecular structure. Fig. 2-2 illustrates this theory. Note that it shows protein molecules making up both inner and outer layers of the membrane and a double layer of lipid molecules forming its middle layer.

By virtue of its structure, the cell membrane performs two general functions: it serves as a boundary or barrier that maintains the cell's integrity, and it helps determine what substances can penetrate this barrier to enter or leave the cell. Despite its incredible delicacy, the cell membrane has sufficient firmness to maintain the arrangement of cell structures necessary for their carrying on the activities that maintain the cell's life. This sounds like a fantastic claim for a material only about 3/10,000,000 of an inch thick. Yet it is true. If a cell's membrane becomes torn, the cell loses its integrity, its organized wholeness. Its inner structure becomes disorganized as the cell's contents leak out. Result? The cell dies.

Summarizing, the cell membrane serves both as a barrier and at the same time as a gateway. As a barrier, it separates the living substance of the cell from its fluid environment. As a barrier, it keeps some things inside the cell and in their proper places and other things outside the cell. But while it is acting as a barrier, the cell membrane is also serving as a gateway — not an ordinary gateway, however, through which any and all substances may pass, but a highly selective gateway which allows only certain materials to move through it (discussed on pp. 20 to 29).

■Cytoplasm

Cytoplasm is the part of a cell between its membrane and its nucleus. In other words, it is all of a cell's protoplasm except its nucleus. Far from being the homogeneous substance once thought, it contains a half dozen or more different kinds of small structures. In many cells, some of these number in the thousands. Collectively, they are known as *organelles* — that is, the "little organs" of cells. Each organelle consists of molecules arranged or organized in such a way that they can perform some function essential for maintenance of the cell's life or for its reproduction. Membranes form the walls of four kinds of organelles: the endoplasmic reticulum, the Golgi apparatus, mitochondria, and lysosomes. We shall discuss these membranous cell structures first and then consider organelles that do not have membranous walls — that is, the ribosomes and the centrosome.

Endoplasmic reticulum

Endoplasm means the cytoplasm located toward the center of a cell. Reticulum

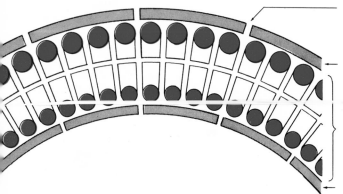

Membrane pore

Single layer of
protein molecules
(30 Å thick)

Double layer
of polarized
lipid molecules
(35 Å thick)

Single layer of
protein molecules
(30 Å thick)

Fig. 2-2 The "triple-layer hypothesis" of cell membrane (plasma membrane) molecular structure. Present evidence indicates that cell membranes range from 75 Å to 100 Å thick (approximately 3/10,000,000 or 4/10,000,000 of an inch).

means network. Therefore, the name endoplasmic reticulum (ER) means literally a network located deep inside the cytoplasm. And when first seen, it appeared to be just that. Later on, however, more highly magnified electron photomicrographs showed quite clearly that the endoplasmic reticulum has a much wider distribution than in the endoplasm alone. As you can observe in Fig. 2-1, it is scattered throughout the cytoplasm. The membranous walls of the endoplasmic reticulum have the triple-layered structure of the cell membrane.

There are two types of endoplasmic reticulum: rough and smooth. Innumerable small granules—ribosomes, by name—dot the outer surface of the membranous wall of the rough type and give it its "rough" appearance. Ribosomes are themselves organelles. The rough endoplasmic reticulum seems to consist, as shown in Fig. 2-1, of flat, curving sacs arranged in parallel rows. Actually, it is a system of connected sacs and canals. The canals wind tortuously through the cytoplasm, extending all the way from the nuclear membrane to the cell membrane.

The structural fact that the endoplasmic reticulum is an interconnected system of canals suggests that it might function as a miniature circulatory system for the cell. And in fact, proteins do move through the canals. The ribosomes attached to the rough endoplasmic reticulum synthesize proteins. These enter the canals and move

through them to the Golgi apparatus or complex. Thus, the rough endoplasmic reticulum functions both in protein synthesis and intracellular transportation.

No ribosomes border the membranous wall of the smooth endoplasmic reticulum —hence its smooth appearance and its name. Its functions are less well established and probably more varied than those of the rough type. In liver cells, for instance, it is thought that the smooth endoplasmic reticulum functions in lipid and cholesterol metabolism. On the other hand, in cells of the testes and adrenals, it apparently takes part in steroid hormone synthesis.

Golgi apparatus

The Golgi apparatus consists of tiny sacs stacked one upon the other and located near the nucleus. Note in Fig. 2-3 that the sacs look more and more distended in successive layers of the pile—as if some material were filling them up to the bulging point. And this actually is the case. Recently, several research teams have presented convincing evidence that the Golgi sacs synthesize large carbohydrate molecules and then combine them with proteins (brought to them through the canals of the endoplasmic reticulum) to form compounds called glycoproteins. As the amount of glycoproteins in the sacs increases, their shape changes from flat to "fat." They turn into perfect little spheres or globules. Then, one by one, they pinch away from the top of the stack. The Golgi apparatus, in other words, not only synthesizes carbohydrate and combines it with protein, but it even packages the product! Neat little globules of glycoprotein migrate outward away from the Golgi apparatus to and through the cell membrane. Once outside the cell, the globules break open, releasing their contents. The cell has secreted its product.

One might think that only in the secreting cells of glands would the Golgi apparatus function as just described. However, various other kinds of cells also make products for "export," products that move out of the cells that make them. In other words, various nonglandular cells secrete substances. Some examples—liver cells secrete blood proteins, plasma cells secrete antibodies, and connective tissue cells of certain types secrete substances used in bone and cartilage formation. The Golgi apparatus in all of these cells is believed to synthesize carbohydrate, to combine it with protein, and to package the product in globules for secretion. The Golgi apparatus no longer seems mysterious nor insignificant. Two scientists who have contributed greatly to our knowledge of this organelle make clear their high regard for it in the following words: "All in all, it looks as if the Golgi apparatus is a creative mechanism in the cell ranking in importance with the ribosome. Just as the ribosomes are responsible for the construction of proteins, so the Golgi apparatus seems to be the main agency for building a variety of large carbohydrates that serve many vital purposes."*

Mitochondria

Note the mitochondria shown in Fig. 2-1. Magnified thousands of times, as they are

*From Neutra, M., and Leblond, C. P.: The Golgi apparatus, Sci. Amer. **220**:100–107 (Feb.), 1969; copyright © 1969 by Scientific American, Inc.; all rights reserved.

there, they look like small, partitioned sausages—if you can imagine sausages only 15,000 angstroms long and a third as wide. (In case you can visualize inches better than angstroms, 15,000 angstroms equals about 3/50,000 of an inch.) Yet, like all organelles, and tiny as they are, mitochondria have a highly organized structure. The membranous wall of a mitochondrion consists not of one but of two delicate membranes. They form a sac within a sac. The mitochondrion's inner membrane folds into a number of extensions called *cristae*. Note in Fig. 2-4 how they jut into the interior of the mitochondrion like so many little partitions. With the very powerful magnification of an electron microscope, one can see small round particles attached to the cristae by short stalks and projecting inward from them. A single mitochondrion, it is said, may contain as many as 20,000 of these particles.* They may consist of enzyme molecules. Some uncertainty, however, still remains about this.

Both inner and outer membranes of the mitochondrion resemble the cell membrane in their molecular structure. Many of the proteins in the membranes of the cristae are known to be enzymes. Also, all evidence so far indicates that they are arranged most precisely in the order of their functioning. This is another example, but surely an impressive one, of the principle that organization is a foundation stone and a vital characteristic of life.

Enzymes in the mitochondrial inner membrane catalyze the chemical reactions by which cells provide themselves with most of the energy that does their many kinds of work and keeps both them and the body alive. Thus do mitochondria earn their now familiar title, the "power

*Swanson, C. P.: The cell, ed. 3, Englewood Cliffs, N. J., 1969, Prentice-Hall, Inc.

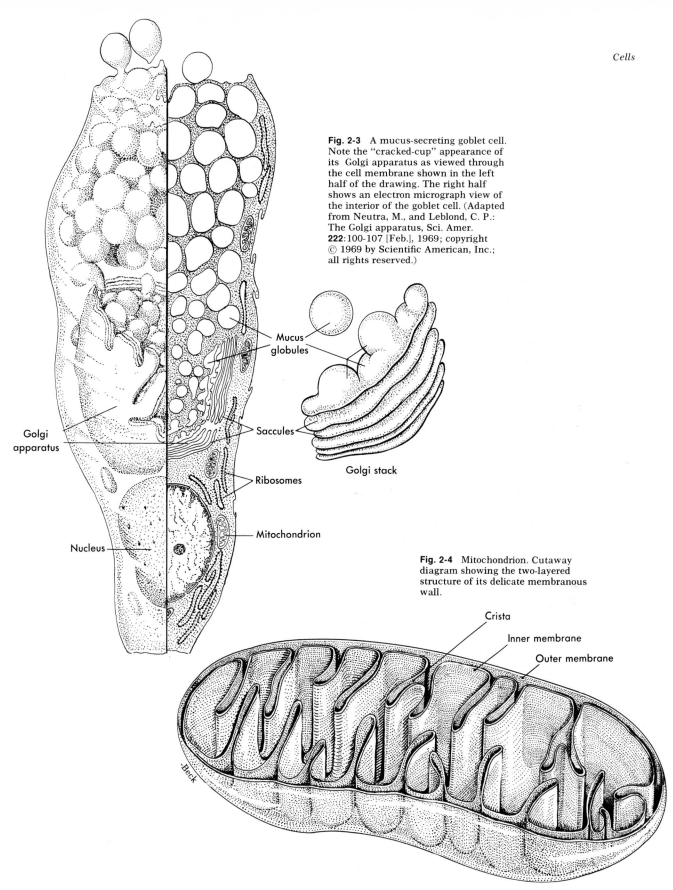

Fig. 2-3 A mucus-secreting goblet cell. Note the "cracked-cup" appearance of its Golgi apparatus as viewed through the cell membrane shown in the left half of the drawing. The right half shows an electron micrograph view of the interior of the goblet cell. (Adapted from Neutra, M., and Leblond, C. P.: The Golgi apparatus, Sci. Amer. **222**:100-107 [Feb.], 1969; copyright © 1969 by Scientific American, Inc.; all rights reserved.)

Mucus globules

Golgi apparatus

Saccules

Golgi stack

Ribosomes

Mitochondrion

Nucleus

Fig. 2-4 Mitochondrion. Cutaway diagram showing the two-layered structure of its delicate membranous wall.

Crista

Inner membrane

Outer membrane

plants" of cells (discussed on pp. 31 to 34).

From the knowledge that mitochondria generate most of the power for cellular work, one might deduce that the number of mitochondria in a cell would be directly related to its amount of activity. And this principle does seem to hold true. In general, the more work a cell does, the more mitochondria its cytoplasm contains. Liver cells, for example, do more work and have more mitochondria than sperm cells. Swanson estimates that a single liver cell may contain as many as 1,000 mitochondria, whereas a single sperm cell has only about 25 mitochondria.

Lysosomes

Lysosomes are membranous-walled organelles whose size and shape change with the stage of their activity. In their earliest, inactive stage, they look like mere granules. Later, as they become active, they take on the appearance of small vesicles or sacs and often contain tiny particles such as fragments of membranes or pigment granules (Fig. 2-1). The membranous wall of a lysosome presumably consists of protein and lipid molecules in the triple-layered arrangement characteristic of the cell membrane. The interior of the lysosome contains various kinds of enzymes capable of breaking down all the main components of cells. These enzymes can, and under some circumstances actually do, destroy cells by digesting them. The graphic nickname "suicide bags," therefore, seems appropriate for lysosomes. Little wonder that these powerful and dangerous substances are usually kept sealed up in lysosomes. However, lysosomal enzymes more often protect than destroy cells. Large molecules and larger particles (for example, bacteria) that find their way into cells enter lysosomes and their enzymes dispose of them by digesting them. "Digestive bags," therefore, and even "cellu-lar garbage disposals" are other nicknames for lysosomes. White blood cells contain a great many lysosomes. They serve as scavenger cells for the body, engulfing bacteria and destroying them in their lysosomes.

Ribosomes

Every cell contains thousands of ribosomes. The small spherical organelles you can see attached to the endoplasmic reticulum and scattered through the cytoplasm in Fig. 2-1 are ribosomes. Because ribosomes are too small to be seen with a light microscope, no one knew they existed until the electron microscope revealed them. Now scientists even know a good deal about their molecular structure. They know that they consist of approximately two-thirds ribonucleic acid (RNA) and one-third protein. They know, too, that at least three types of RNA molecules and perhaps fifty or more kinds of protein molecules make up the composition of ribosomes.[*]

The function of ribosomes is protein synthesis. Ribosomes are the molecular machines that make proteins. Or, to use a popular term, they are the cell's "protein factories." Ribosomes attached to the endoplasmic reticulum, as already mentioned, synthesize proteins for "export," whereas the ribosomes free in the cytoplasm make proteins for the cell's own domestic use. They make its structural proteins, in other words, and its enzymes. An interesting fact recently established is that one ribosome by itself cannot fabricate a protein. Groups of ribosomes, called *polysomes*, work together to make the complex molecules of proteins. Later on we shall tell more about this complicated process of protein synthesis.

[*]Nomura, R.: Ribosomes, Sci. Amer. **221**:28-35 (Oct.), 1969.

Centrosome or centrosphere

The names of this cytoplasmic organelle suggest its shape and location—a spherical area or body located near the center of the cell (Fig. 2-1). Actually, this means that the centrosome is located near the nucleus since the nucleus takes up the center space of most cells. With the light microscope and suitably prepared slides, one can see two dots (called *centrioles*) in the centrosome. The electron microscope, however, reveals that centrioles are more than just dots. Each one is a tiny cylinder whose wall consists of a definite number of groups of very fine tubules— nine groups of two or three tubules each to be specific.

Early in the process of cell division (mitosis), two pairs of centrioles can be seen. Curiously enough, one centriole of each pair lies at right angles to the other member of the pair. During mitosis, one pair of centrioles moves to one pole of the cell and the other pair goes to the other pole.

Centrioles play some part in the formation of the mitotic spindle fibers. They also control polarization of these fibers (see p. 34).

■ Nucleus

The nucleus is a spherical body located in the center of a cell. A double-layered, pore-containing membrane encloses the nucleus and separates it from the cytoplasm. In the nucleus are located a nucleolus (or several nucleoli) and a number of tiny structures of gigantic importance— the *chromosomes.*

Nucleoli consist mostly of ribonucleic acid (RNA) and proteins. In suitably prepared microscope slides, nucleoli appear as small, dense, and usually round bodies in the nucleus. No membrane encloses them. The best evidence so far indicates that nucleoli serve as the sites where ribosomal RNA combines with protein. This RNA-protein complex then migrates out of the nucleus into the cytoplasm. There it becomes organized into ribosomes— structures you will recognize as the protein synthesizers of cells. An interesting observation is that the more protein a cell synthesizes, the larger its nucleoli generally are. Cells of the pancreas, for example, make large amounts of protein and have large nucleoli.

The word chromosome is derived from two Greek words—*chroma*, meaning color, and *soma*, meaning body. And in a suitably stained cell, chromosomes do appear as "colored bodies" in the nucleus. They look like stubby little rods when a cell has just started the process of dividing to form two cells. But in between successive cell divisions, they appear as deeply stained granules and then are called *chromatin granules.* Chromosomes are composed partly of proteins but chiefly of a compound with a tongue-twisting name, deoxyribonucleic acid (usually shortened to DNA for obvious reasons). Later on in this chapter, we shall have a good deal to say about both the structure and function of— DNA—"the most golden molecule of all" as Watson called it.*

Cell physiology

Every cell carries on certain functions that maintain its own life. For example, all cells move substances through their membranes and metabolize foods. Also, all but a few types of cells reproduce themselves. In addition to these self-serving activities, every cell in the body also performs some special function that serves

---

*From Watson, J. D.: The double helix, New York, 1968, The New American Library Inc., p. 21.

the body as a whole. Muscle cells provide the function of movement, nerve cells contribute communication services, red blood cells transport oxygen, etc. In return, the body performs vital functions for all of its cells. It brings food and oxygen to them and removes waste from them, to mention only two examples. In short, a relationship of mutual interdependence exists between the body as a whole and its various parts. Optimum health of the body depends upon optimum health of each of its parts, down even to the smallest cell. Conversely, optimum health of each individual part depends upon optimum health of the body as a whole. In equation form, this physiological principle of mutual interdependence might be expressed as follows:

**body health ⇌ system health ⇌ organ
health ⇌ tissue health ⇌ cellular health**

Some parts of the body, of course, are more important for healthy survival than others. Obviously, the heart is far more important than the appendix. And the nerve cells which control respiration are infinitely more important for survival than muscle cells that move the little finger.

Cell physiology deals with all kinds of cell functions, but at this time we shall discuss only the movement of substances through cell membranes, cell metabolism, and cell reproduction.

■Movement of substances through cell membranes

Heavy traffic moves continuously in both directions through cell membranes. Streaming in and out of all cells, in endless procession, go molecules of water, foods, gases, wastes, and many kinds of ions. Several processes carry on this mass transportation. They are classified under two general headings as physical (or passive) and physiological (or active) processes.

Diffusion and osmosis qualify as the main physical or passive processes. The term physical process implies something which occurs in the physical (nonliving) world. But actually, substances can move through either living or nonliving membranes by diffusion and osmosis. This is because the energy that powers these processes is the kinetic energy provided by the random, never-ceasing movements of atoms, ions, and molecules and *not* from chemical reactions in cells. Since cells do not supply the energy that does the work of moving substances through their membranes by diffusion or osmosis, these physical processes are also called passive mechanisms.

Active transport, phagocytosis, and pinocytosis constitute the main physiological or active processes that move substances through cell membranes. These are active processes because cellular activity—specifically, chemical reactions carried on by living cells—supply the energy that powers them. Because this is true, the processes of active transport, phagocytosis, and pinocytosis can move substances through cell membranes only as long as cells are alive and functioning. When they die, these physiological processes cease. In contrast, the physical processes of diffusion and osmosis go on even after cellular death.

Diffusion

Diffusion means scattering or spreading. It occurs because small particles such as molecules and ions are forever on the go. They move continuously, rapidly, and at random. Here is a convincing way to observe the results of diffusion. Place a cube of sugar on the bottom of a cup of water. Immediately skim off a spoonful of liquid from the top and taste it. About ten minutes later skim off and taste another spoonful. The second sample will taste sweet; the first will not. The ten-minute interval will have given the sugar molecules time to move in all directions—up, out, down, in—

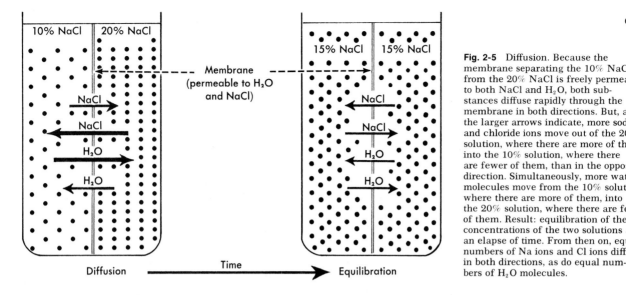

Fig. 2-5 Diffusion. Because the membrane separating the 10% NaCl from the 20% NaCl is freely permeable to both NaCl and H₂O, both substances diffuse rapidly through the membrane in both directions. But, as the larger arrows indicate, more sodium and chloride ions move out of the 20% solution, where there are more of them, into the 10% solution, where there are fewer of them, than in the opposite direction. Simultaneously, more water molecules move from the 10% solution, where there are more of them, into the 20% solution, where there are fewer of them. Result: equilibration of the concentrations of the two solutions after an elapse of time. From then on, equal numbers of Na ions and Cl ions diffuse in both directions, as do equal numbers of H₂O molecules.

to scatter or to diffuse, that is, throughout the water. More of them, however, will have diffused up and away from the cube than down and toward it. In other words, their *net diffusion* will have been away from areas of their greater concentration toward areas of their lesser concentration so that they eventually became evenly distributed throughout the water. Diffusion always results—if enough time elapses—in an even scattering of solute particles among solvent molecules. (Water is the solvent in all body fluids.)

Consider another example—two solutions, 20% and 10% sodium chloride (NaCl), separated by a membrane permeable to both NaCl and water (Fig. 2-5). Sodium chloride particles and water molecules racing in all directions through the solution collide with each other and with the membrane. Some inevitably hit the membrane pores from the 20% side and some from the 10% side. Just as inevitably, some bound through the pores in both directions. For a while more NaCl particles enter the pores from the 20% side, where they are more numerous or more concentrated. More NaCl particles therefore move through the membrane from the 20% solution into the 10% than diffuse through it in the opposite direction. Using different words for the same thought, *net*

diffusion of NaCl takes place from the solution where NaCl concentration is greater into the one where its concentration is lesser. Thus, net diffusion of NaCl occurs "down" the *NaCl concentration gradient*—that is, from the higher NaCl concentration down to the lower NaCl concentration.

During the time that net diffusion of NaCl is taking place between the 20% and 10% solutions, net diffusion of water is also going on. The direction of net diffusion of any substance is always down that substance's concentration gradient. Applying this principle, net NaCl diffusion occurs down the NaCl concentration gradient and net water diffusion occurs down the water concentration gradient. Since the greater concentration of water molecules lies in the more dilute 10% solution (where fewer water molecules have been displaced by NaCl molecules), more water molecules diffuse out of the 10% solution into the 20% solution than diffuse in the opposite direction. This net diffusion of water thus removes water from the more dilute solution and adds it to the more concentrated one, a process that obviously tends to equalize or equilibrate the concentrations of the two solutions. Simultaneously, the net diffusion of NaCl is removing NaCl from the more concentrated solution and adding it to the more dilute one, a process

21

which also tends to equilibrate the two solutions. Note that whereas net diffusion of NaCl and of water both go on at the same time, they go on in opposite directions.

Diffusion of NaCl and water continues even after equilibration has been achieved. But from that moment on, it is equal diffusion in both directions through the membrane and not net diffusion of either substance in either direction.

Dialysis is the separation of crystalloids from colloids by the diffusion of crystalloids through a membrane permeable to them but impermeable to colloids (Fig. 2-6). (Crystalloids or true solutes are solute particles with diameters less than 1 millimicron; for example, ions, glucose, oxygen, etc. Colloids are solute particles whose diameters range from about 1 to 100 millimicrons. Enzymes and all other proteins are colloids.)

The following paragraphs summarize some facts and principles worth remembering about diffusion. They have been culled from the diffusion examples just described.

1 *Diffusion* is the movement of solute and solvent particles in all directions through a solution or in both directions through a membrane.

2 *Net diffusion* is the movement of more particles of a substance in one direction than in the opposite direction.

3 Net diffusion of any substance occurs down its own concentration gradient, which means from the higher to the lower concentration of that substance. Following are two applications of this principle: (a) net diffusion of solute particles occurs from the more concentrated to the less concentrated solution; (b) net diffusion of water molecules, in contrast, occurs from the more dilute to the less dilute solution.

4 Net diffusion of the solute in one direction and of water in the opposite direction eventually results in *equilibration*. The concentrations of two solutions soon become equal when the membrane separating the solutions is freely permeable to both solute and water.

5 After two solutions have equilibrated, equal diffusion occurs in both directions through the membrane but net diffusion no longer occurs in either direction. (Equal diffusion means that the number of solute and water particles diffusing in one direction equals the number diffusing in the opposite direction. Net diffusion means more particles diffusing in one direction than in the other.)

6 Diffusion is a passive transport mech-

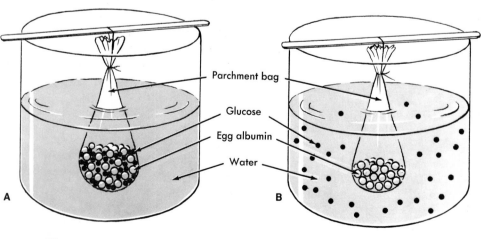

Parchment bag

Glucose

Egg albumin

Water

A   B

Fig. 2-6 Dialysis, the separation of crystalloids from colloids by means of a membrane permeable to crystalloids and impermeable to colloids. The parchment bag in **A**, which contains a solution of glucose and raw egg white, is permeable to glucose but not to albumin. Therefore, glucose moves out of the bag into the surrounding water while albumin stays inside the bag. Time elapses between **A** and **B**. **B** shows the result of dialysis—separation of the crystalloid, glucose, from the colloid, albumin.

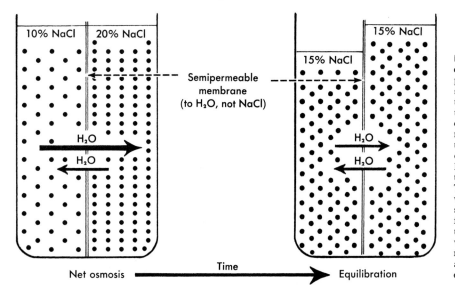

Fig. 2-7 Osmosis. Osmosis is the diffusion of water through a selectively permeable membrane. The membrane shown in this diagram is permeable to water but not to NaCl. Because there are relatively more water molecules in 10% NaCl than in 15% NaCl, more water molecules osmose from the more dilute into the more concentrated solution (as indicated by the larger arrow in the left-hand diagram) than osmose in the opposite direction. The *net* direction of osmosis, in other words, is toward the more concentrated solution. Net osmosis produces the following changes in these solutions: their concentrations equilibrate, the volume and pressure of the originally more concentrated solution increase, and the volume and pressure of the other solution decrease proportionately.

anism because cells are passive, not active and working in this process. Cellular chemical reactions do not supply the energy that moves diffusing particles. The continual random movements characteristic of all molecules and ions (molecular kinetic theory) furnish the energy for diffusion.

Think about these diffusion principles. Make sure that you understand them, for they have many applications in physiology. Our very lives, in fact, depend upon diffusion. Evidence? Oxygen, the "breath of life," enters cells by diffusion through their membranes.

The process known as *facilitated diffusion* resembles ordinary diffusion in certain respects. Both are passive processes and both move substances down their own concentration gradient. Facilitated diffusion differs from ordinary diffusion and resembles active transport in that a carrier substance combines with the particle being moved. Lipid-insoluble particles too large to go through membrane pores enter cells by facilitated diffusion. Also, glucose apparently enters red blood cells by this process.

Osmosis

Osmosis is the diffusion of water through a selectively permeable membrane (Fig. 2-7). As its name suggests, a selectively permeable membrane is one that is not equally permeable to all solute particles present. It permits some solutes to diffuse through it freely but hinders or prevents entirely the diffusion of others. Those solutes allowed to diffuse freely through the membrane obey the law of diffusion so eventually equilibrate across it. Their concentrations on both sides of the membrane become equal. But can particles not permitted to diffuse freely through a membrane also obey the law of diffusion? Can they, too, equilibrate across the membrane? The answer to both questions, as you can readily deduce, is "No." When a membrane hinders or prevents a substance from moving through it, the concentration of that substance necessarily remains higher on one side of the membrane than on the other. That solute cannot equilibrate across the membrane. In short, the selectively permeable membrane maintains a concentration gradient for the not freely diffusible solute.

Summarizing the preceding paragraph, a *selectively permeable membrane* may be defined either as one which does not permit free, unhampered diffusion of all the solutes present or as one which maintains at least one solute concentration gradient across itself. *Osmosis* is the diffusion of water through a selectively permeable membrane. Or, osmosis is the diffusion of water through a membrane which maintains at least one concentration gradient across itself.

Water osmoses through normal living cell membranes because they are selectively permeable. They maintain various solute concentration gradients. Most notably, healthy cell membranes maintain sodium and potassium concentration gradients between the interstitial fluid around them and the intracellular fluid within them. Interstitial fluid sodium concentration is many times higher than intracellular fluid sodium concentration. In contrast, intracellular potassium concentration is many times higher than interstitial fluid potassium concentration.

Perhaps the best way to explore the concept of osmosis further is by an example. Imagine that you have a 20% NaCl (table salt) solution separated from a 10% NaCl solution by a membrane. Assume that the membrane is impermeable to sodium and chloride ions but that it is freely permeable to water particles. Obviously then, sodium and chloride ions will not diffuse through this membrane. Water, on the other hand, will osmose through it in both directions—not in equal amounts, however.

To deduce the direction in which the greater volume of water will osmose (the direction, that is, of net osmosis), we must first decide which solution contains the greater concentration of water molecules. Pretty clearly, the more dilute of two solutions contains the greater concentration of water molecules. In this case, then, water concentration is greater on the 10% side of the membrane than it is on the 20% side. Next, we need to apply the principle already stated that "net diffusion of any substance occurs down the concentration gradient of that substance." Therefore, net osmosis (net water diffusion through the selectively permeable mem-

brane) occurs down the water concentration gradient. In our example, net osmosis takes place from the more dilute 10% salt solution into the more concentrated 20% solution. Thus, the solution which at first was more concentrated gains water by net osmosis and becomes more dilute. And the solution which at first was more dilute loses water and becomes more concentrated. Net osmosis, in other words, tends to make the concentrations of the two solutions equal. In briefest form, the principle is this: net osmosis occurs down a water concentration gradient and tends to equilibrate solutions separated by a selectively permeable membrane.

Net osmosis into the originally more concentrated of our two salt solutions

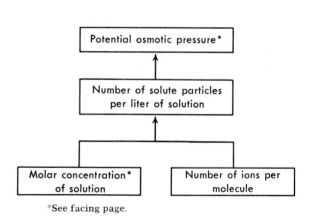

*See facing page.

Fig. 2-8 Diagram shows that the number of solute particles (ions and molecules) present in a liter of solution determines its potential osmotic pressure. Note, however, that the number of solute particles per liter is itself determined by two factors—the solution's molar concentration and the number of ions formed per molecule of electrolyte present in it.

increases the volume of this solution. Further, the increase in its volume causes an increase in its pressure, called osmotic pressure—a logical term since it is pressure caused by net osmosis. By definition, then, *osmotic pressure* is the pressure that develops in a solution as a result of net osmosis into that solution. An important principle stems from this definition—osmotic pressure develops in the solution which originally contains the higher concentration of the solute that does not diffuse freely through the membrane. For instance, in the example previously given, osmotic pressure would develop in the solution that originally had the 20% salt concentration.

Potential osmotic pressure is the maxi-

mum osmotic pressure that could develop in a solution if it were separated from distilled water by a selectively permeable membrane. (Actual osmotic pressure, on the other hand, is a pressure that already has developed, not just one that could develop.) What determines a solution's potential osmotic pressure? The answer is this: the number of solute particles in a unit volume of solution directly determines its potential osmotic pressure—the more solute particles per unit volume, the greater the potential osmotic pressure. The number of solute particles per unit volume (for example, per liter) of solution, in turn, is determined by the molar concentration of the solution and also, if the solute is an electrolyte, by the number of ions formed by

Formula:

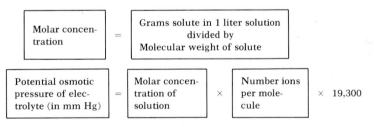

†Experimentation has shown that a 1.0 molar solution of any nonelectrolyte has a potential osmotic pressure of 19,304 mm Hg pressure (at body temperature, 37° C).

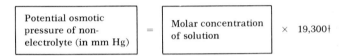

Example: Two solutions commonly used in hospitals are 0.85% NaCl and 5% glucose. What is the potential osmotic pressure of 0.85% NaCl at body temperature? (0.85% NaCl = 8.5 grams NaCl in 1 liter solution.)

Molecular weight of NaCl = 58 (NaCl yields 2 ions per molecule in solution)

Using the formula given for computing potential osmotic pressure of an electrolyte:

Problem: What is potential osmotic pressure of 5% glucose at body temperature? Molecular weight glucose = 180. Glucose does not ionize. It is a nonelectrolyte.

each molecule of solutes. This concept can be expressed by a diagram (Fig. 2-8) or by formula (see footnote to Fig. 2-8).

Since it is the number of solute particles per unit volume that directly determines a solution's potential osmotic pressure, one might at first thought jump to the conclusion that all solutions which have the same percent concentration also have the same potential osmotic pressures. Obviously, it is true that all solutions which contain the same percent concentration of the same solute do also have the same potential osmotic pressures. All 5% glucose solutions, for example, have a potential osmotic pressure at body temperature of about 5,300 mm Hg pressure. But all solutions with the same percent concentrations of different solutes do not have the same molar concentrations. Hence they do not have the same potential osmotic pressures. By applying the equation in the footnote, you will discover that 5% NaCl at body temperature has a potential osmotic pressure of nearly 33,000 mm Hg—quite different from 5% glucose's potential osmotic pressure of about 5,300 mm Hg.

Two solutions which have the same potential osmotic pressures are said to be *isosmotic* to each other. Because they have the same potential osmotic pressures, the same amount of water osmoses in both directions between them when they are separated by a selectively permeable membrane. No net osmosis, however, occurs in either direction between isosmotic solutions. Hence, no actual osmotic pressure develops in either solution. Their pressures remain the same. And because they do, isosmotic solutions are also called isotonic (Gr. *isos*, the same; *tonos*, tension or pressure).

By definition, *isotonic solutions* are those whose volumes and pressures will stay the same if the two solutions are separated by a membrane. For example,

0.85% NaCl is referred to as "isotonic saline," meaning that it is isotonic to the fluid inside human cells. Translated more fully, it means that if 0.85% NaCl is injected into human tissues or blood, no net osmosis occurs into or out of cells. Therefore, no change in intracellular volume or pressure takes place. Physicians, of course, understand this. They know that isotonic saline solution, given intramuscularly or intravenously, will not injure or kill cells by causing them either to lose water or to gain it. In more technical terms, they know that isotonic saline solution neither dehydrates nor hydrates human cells.

When two solutions have unequal potential osmotic pressures, their volumes and pressures will change if they are separated by a selectively permeable membrane. Net osmosis will occur into the solution which has the higher potential osmotic pressure. Net osmosis, as we have noted, always occurs down the water concentration gradient. Since the solution which has the lower potential osmotic pressure has the lower concentration of solute particles, it necessarily has the higher concentration of water molecules. Therefore, net osmosis occurs out of this solution into the solution with the higher potential osmotic pressure. This increases the volume of the solution with the higher potential osmotic pressure. And the increase in its volume produces an increase in its pressure. This increase in pressure is osmotic pressure—but actual osmotic pressure and not just a potential osmotic pressure.

In summary, when two solutions have unequal potential osmotic pressures and are separated by a selectively permeable or a semipermeable membrane, net osmosis occurs into the solution with the higher potential osmotic pressure. As a result, the volume of this solution increases and its potential osmotic pressure becomes an actual osmotic pressure.

When two solutions have unequal potential osmotic pressures, one of them is described as hypertonic and the other as hypotonic.

A *hypertonic solution* may be defined by any of the following statements: (1) a solution which has a higher potential osmotic pressure than another, (2) a solution into which net osmosis occurs when the solution is separated from another by a selectively permeable membrane, (3) a solution whose volume and pressure increase when the solution is separated from another by a selectively permeable membrane, and (4) a solution in which an actual osmotic pressure develops when it is separated from another solution by a selectively permeable membrane (Fig. 2-9).

A *hypotonic solution* has characteristics opposite to those of a hypertonic solution.

Example: The fluid inside human cells is hypertonic to distilled water and, conversely, distilled water is hypotonic to intracellular fluid. If, therefore, distilled water were injected into a vein, net osmosis would occur into blood cells. And eventually, if intracellular volume and pressure increased beyond a certain limit, blood cell membranes would rupture and the cells would die. When this happens to red blood cells, their hemoglobin leaks out and they are said to be hemolyzed. (*Hemolysis* means the destruction of red blood cells with the escape of hemoglobin from them into the surrounding medium.)

Active transport mechanisms

An active transport mechanism is a device that moves molecules or ions through cell membranes in an uphill direction, meaning up their concentration gradients or against their natural tendency. Their natural tendency is to diffuse down their concentration gradient. The law of diffusion requires that the net diffusion of any substance take place from the area of its

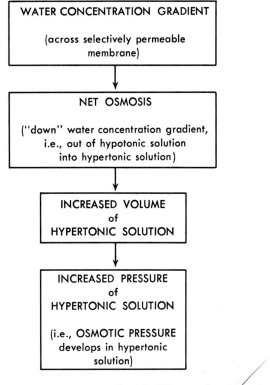

Fig. 2-9 Effects produced by existence of a water concentration gradient across a selectively permeable membrane.

higher concentration to that of its lower concentration. Moreover, as we all know, it is work to move anything in the opposite direction from that which it naturally tends to go (to move water uphill, for instance). And work demands energy expenditure. To do the work of active transport—one of the most important of all the kinds of cellular work—chemical reactions carried on by the living cell supply the energy. Some investigators[*] estimate that the body spends about one-fifth of all the energy its cells produce on active transport mechanisms.

A physical device which moves something against its natural tendency is called a pump. So it seems logical that we should think of active transport mechanisms as

*Ruch, T. C., and Patton, H. D.: Physiology and biophysics, ed. 19, Philadelphia, 1965, W. B. Saunders Co.

physiological or "biological pumps." Active transport mechanisms pump various substances through cell membranes, but probably the most important of these is sodium ions. Sodium ions diffuse down their concentration gradient from interstitial fluid into cells. But about as fast as they diffuse inward through a cell membrane, an active transport mechanism (commonly called the *sodium pump*) pumps them back out again. With energy supplied by cellular chemical reactions, the sodium pump moves sodium ions uphill against the sodium concentration gradient.* Normally, therefore, sodium does not equilibrate across living cell membranes. The sodium pump maintains a steep sodium concentration gradient across cell membranes. Of all the many processes that keep us alive, few are more important than active transport mechanisms. Sodium transport alone plays an essential part in several vital functions – in nerve impulse conduction, in the maintenance of water balance, and in the maintenance of acid-base balance, for example.

Sodium is not the only substance actively transported across cell membranes. Considerable evidence supports the hypothesis of a potassium transport mechanism that pumps potassium into cells and is coupled with sodium transport. Other ions may also be actively transported. Glucose was formerly thought to enter cells by active transport. But today it is known that glucose enters cells by facilitated transfusion.

Although very little is known about the chemistry of active transport, several theories have been proposed. Most of them assume the existence of carrier compounds, although none as yet has been identified. In simplest forms, the theories suggest the following steps as the essential parts of an active transport mechanism (Fig. 2-10):

1 On one surface of a cell membrane, a molecule of the substance to be transported (A) combines with a molecule of carrier compound (B) to form a new compound (AB).

$$A + B \longrightarrow AB$$

2 Molecule AB passes through the cell membrane into (or out of) the cell.

3 At the other surface of the cell membrane, AB dissociates, releasing the transported substance.

$$AB \longrightarrow A + B$$

4 Carrier compound molecule B moves back to the surface from which it came, ready to shuttle another molecule of A through the membrane.

Phagocytosis and pinocytosis

Phagocytosis and pinocytosis are also active mechanisms for moving substances through cell membranes but only in one direction – inward. More than sixty years ago, Elie Metchnikoff of the Pasteur Institute saw white blood cells engulf bacteria. It reminded him of eating. Therefore, from the Greek words for eating, cell, and action, he coined the word phagocytosis. Some thirty years later, in 1931, W. H. Lewis of Johns Hopkins University saw something similar in time-lapse photographs of some tissue culture cells. These cells, however, were engulfing droplets of fluid instead of solid particles. They seemed to be drinking rather than eating, so he named the process pinocytosis (from the Greek word for drinking).

Both phagocytosis and pinocytosis consist of the same essential steps. A segment

---

*One way of designating sodium concentration is as milligrams per liter. Expressed this way, interstitial fluid has a sodium concentration of about 3,200 mg per liter, whereas intracellular fluid has a sodium concentration of only about 345 mg per liter. Expressed in terms of milliequivalents per liter, interstitial fluid has a sodium concentration of about 139 mEq per liter and intracellular fluid, about 15 mEq per liter.

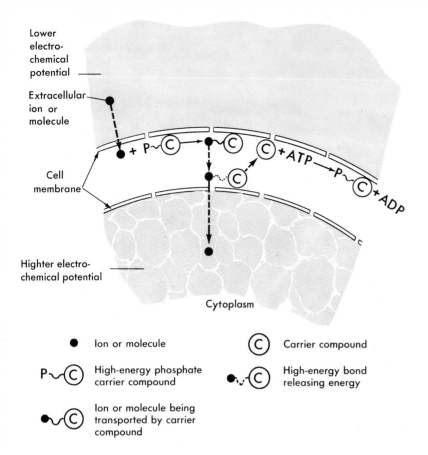

Lower electro-chemical potential

Extracellular ion or molecule

Cell membrane

Highter electro-chemical potential

Cytoplasm

● Ion or molecule

P~Ⓒ High-energy phosphate carrier compound

●~Ⓒ Ion or molecule being transported by carrier compound

Ⓒ Carrier compound

●~Ⓒ High-energy bond releasing energy

Fig. 2-10 Scheme to illustrate one hypothesis about active transport mechanisms. Diagram shows them consisting of the following steps: an ion or a molecule penetrates the outer surface of a cell membrane; it combines with a high-energy phosphate carrier compound; the compound thus formed moves through the membrane to its inner surface; its high-energy bond breaks down, releasing energy which propels the transported particle, but not the carrier compound, into the cytoplasm of the cell; the carrier compound moves back toward the outer part of the cell membrane and combines with ATP to again become a high-energy phosphate compound.

of cell membrane forms a small pocket around a bit of solid or liquid material outside the cell, pinches off from the rest of the membrane, and migrates inward as a closed vacuole or vesicle. Later, it releases its contents into the cell's cytoplasm.

Cell metabolism

The process of cell metabolism is as intricate as it is vital. It consists of two main processes called catabolism and anabolism. Each of these, in turn, consists of numerous chemical reactions. *Catabolism* is the process that supplies cells with energy in a form they can use for doing their work. *Anabolism* is the process by which cells synthesize complex compounds of many different kinds—hormones, the proteins called enzymes, structural proteins, and numerous other compounds. Anabolism is one of the several kinds of work which all cells perform. Catabolism supplies the energy that does the work of anabolism and all other kinds of cellular

work as well. For a brief account of the chemical and energy changes that make up the process of catabolism, read the following paragraphs. Anabolism is discussed in Chapter 12.

Catabolism

Catabolism, one of the two major processes of metabolism, consists of several processes, some of which are different for different kinds of foods. At this time, we shall discuss only the catabolism of the carbohydrate glucose and shall try to present only basic facts about this complex process since its detailed study belongs to the science of biochemistry.

Glucose catabolism consists of two successive processes: glycolysis and the citric acid cycle (also called tricarboxylic acid cycle, Krebs cycle, and cellular respiration).

Glycolysis. Glycolysis is the process which changes glucose to pyruvic acid and thereby releases some of the energy stored in glucose molecules. Glycolysis takes place in the cytoplasm of cells. It consists of a series of chemical changes catalyzed by enzymes* and accompanied by energy changes. Because the chemical changes of glycolysis do not use oxygen, glycolysis is referred to as the anaerobic phase of catabolism. Figs. 2-11 and 2-12 show the chemical changes of glycolysis in an abbreviated form. Stated very briefly, the chemical change produced by glycolysis is the breaking apart of 1 molecule of glucose to form 2 molecules of pyruvic acid.

The specific chemical changes, however, are not the most important facts to remember about glycolysis nor about the citric acid cycle. The most important fact is that the chemical changes of catabolism produce energy changes and that these energy changes supply all the energy which does all the work which keeps our cells and

*See Chapter 12 for a discussion of enzymes.

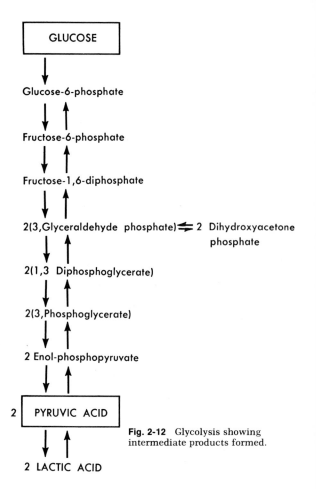

Fig. 2-12 Glycolysis showing intermediate products formed.

Fig. 2-11 Glycolysis. Glucose, a molecule containing 6 carbon atoms, is split by a series of chemical reactions into 2 molecules of pyruvic acid, each of which contains only 3 carbon atoms. Fig. 2-12 shows the series of intermediate reactions in brief form.

Fig. 2-13 Scheme to show that catabolism breaks larger molecules down into smaller ones. *Glycolysis* splits 1 molecule of glucose (6 carbon atoms) into 2 molecules of pyruvic acid (3 carbon atoms each). The *citric acid cycle* uses 6 molecules of oxygen to oxidize 2 molecules of pyruvic acid to 6 molecules of carbon dioxide (1 carbon atom each) and 6 molecules of water.

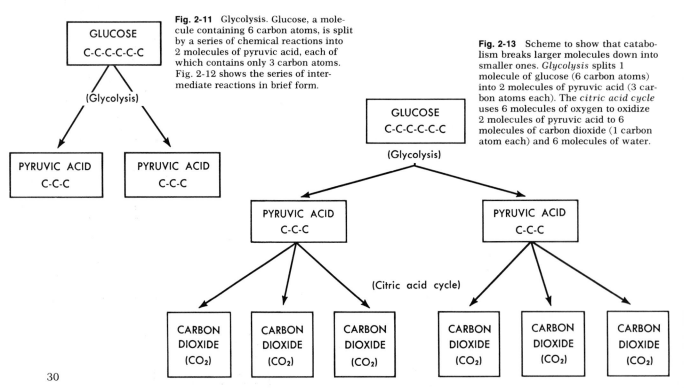

bodies alive. In essence, the energy changes produced by the chemical changes of catabolism are these (see Fig. 2-15):

1 Energy stored in the chemical bonds of food molecules is released.

2 More than half of the released energy is immediately recaptured—that is, put back into storage in chemical bonds (but in a different kind of bond in a different kind of molecule).

3 Somewhat less than half of the energy released from food molecules is transformed into heat energy. (Incidentally, this is the body's only way of producing heat—by catabolizing food molecules to transform part of their stored energy into heat energy.) In later paragraphs (pp. 32 to 34) more meaning will be given to the idea of energy changes during catabolism.

Citric acid cycle. The citric acid cycle (or *Krebs' cycle* as it is called in honor of the man who formulated it) is the second and final process in the catabolism of glu-

Fig. 2-14 The citric acid (Krebs') cycle. See text (pp. 31 to 33) for explanation.

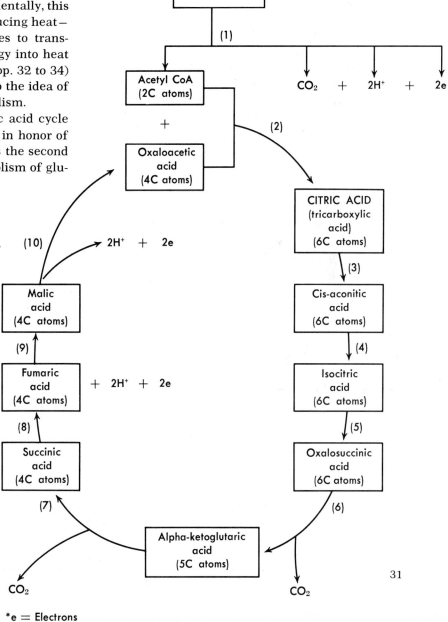

31

*e = Electrons

cose. It consists of a cyclical series of chemical reactions catalyzed by enzymes and accompanied by energy changes. Because some of these chemical reactions use oxygen, the citric acid cycle is referred to as the aerobic phase of catabolism. Figs. 2-13 and 2-14 show the chemical changes of the citric acid cycle in abbreviated forms. Put in fewest words, the chemical change produced by the citric acid cycle is that pyruvic acid is broken down and oxidized to form carbon dioxide and water. Six molecules of oxygen are used up in the process. Note well this fact. Why? Because this is the only way cells utilize oxygen—to carry on the citric acid cycle—and the reason they and you must have a continuous supply of oxygen to survive.

Enzymes which catalyze the chemical reactions of the citric acid cycle (Fig. 2-14) are protein molecules located in the thousands of particles attached to the outer membrane of each mitochondrion (p. 16). Investigators believe that even the placing of these enzyme molecules in the outer membrane particles is precise, that they are placed one after the other in the exact sequence of the chemical reactions they catalyze. Think of that! Even the molecules in a cell are organized so as to produce perfect timing of their functioning—a persuasive illustration, is it not, of the principle, "Organization is a dominant characteristic of life." Now let us examine some of the specific chemical changes of the citric acid cycle to see how they produce energy changes.

In Fig. 2-14, chemical reaction (1) removes 1 carbon dioxide molecule and 2 hydrogen atoms from each pyruvic acid molecule, or a total of 2 carbon dioxide molecules and 4 hydrogen atoms for each glucose molecule catabolized. The 4 hydrogen atoms immediately ionize to form 4 hydrogen ions and 4 electrons.

Molecules of the coenzyme DPN (diphos-

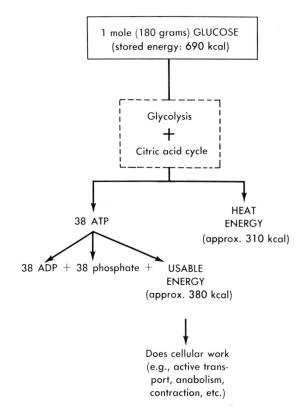

Fig. 2-15 Catabolism releases energy stored in chemical bonds of glucose molecules. 180 gm or 1 mole of glucose contains 690 kilocalories (kcal) of stored energy. Catabolism releases about 310 kilocalories of this energy as heat and puts about 380 kilocalories back in storage in high-energy bonds of 38 moles of ATP. Thus the efficiency of glucose catabolism as a mechanism for supplying cells with usable energy is about 55% (380/690).

phopyridine nucleotide) quickly accept these electrons,* shuttle them across the microscopic fluid-filled space between the outer and inner membranes of the mitochondrial wall, and pass them on to enzymes presumably located in particles attached to the inner membrane. There are thousands of these tiny particles. Evidence suggests that each one consists of a series or chain of several enzymes.

Because electrons are transferred from one enzyme to another in the chain, like so many buckets of water passed along a bucket brigade, these enzymes are referred to as "electron-transfer enzymes," and the

---

*Green, D. E.: The mitochondrion, Sci. Amer. **210**:67–74 (Jan.), 1964.

inner membrane particles which contain them are known as *"electron-transport particles."* The movement of electrons down the chain of electron-transfer enzymes releases energy. Somewhat more than half of this released energy is used to synthesize ATP (adenosine triphosphate) molecules. The rest of it is transformed to heat energy.

Finally, at the end of the electron-transfer chain, the electrons recombine with hydrogen ions to form hydrogen atoms which unite with oxygen to form water.

Briefly then, reaction (1) produces acetyl-CoA + CO_2 + H_2O + ATP + Heat. Acetyl-CoA immediately combines with oxaloacetic acid to form citric acid—reaction (2) in Fig. 2-14. Citric acid then undergoes a series of changes that eventually convert it to oxaloacetic acid, carbon dioxide, and water. Through the addition of oxygen during several reactions of the citric acid cycle, each pyruvic acid molecule entering the cycle is converted to 3 CO_2 molecules (reactions 1, 6, and 7) and 3 H_2O molecules (reactions 1, 8, and 10). Also, for each pyruvic acid molecule oxidized, 18 ATP molecules are synthesized, using energy released by electron transfer during reactions 1, 8, and 10. Note in Fig. 2-15 that the complete catabolism of 1 molecule of glucose yields 38 molecules of ATP. All but two of them are produced during the citric acid cycle.

•   •   •

Summarizing the changes of catabolism in equation form:

Glycolysis:

Glucose $\longrightarrow$ 2 pyruvic acid + 2 ATP + Heat

Citric acid cycle:

2 pyruvic acid + 6 O_2 $\longrightarrow$ 6 CO_2 + 6 H_2O + 36 ATP + Heat

We can even summarize the long series of chemical reactions that make up the process of glucose catabolism with one short equation:

$$C_6H_{12}O_6 + 6\,O_2 \longrightarrow 6\,CO_2 + 6\,H_2O +$$
(glucose) $\qquad\qquad$ 38 ATP + Heat

As you read the next few paragraphs, refer often to Fig. 2-15. It summarizes the energy changes accomplished by the chemical changes of glucose metabolism.

The synthesis of ATP is a phosphorylation reaction carried on during the citric acid cycle in the thousands of tiny particles attached to the inner membrane of each mitochondrion (the same structures in which electron-transfer occurs).

ADP + Phosphate + ENERGY → ATP
(adenosine $\qquad\qquad\qquad\qquad$ (adenosine
diphosphate) $\qquad\qquad\qquad\quad$ triphosphate)

The energy shown in the preceding equation (in other words, the energy used to do the work of synthesizing ATP molecules) is released from the stable chemical bonds in glucose molecules by the passage of electrons down the chains of electron-transfer enzymes. Immediately the process of ATP synthesis puts some of the released energy back in storage in chemical bonds, but in unstable (labile) high-energy bonds of ATP molecules.

As its name suggests, ATP or adenosine triphosphate contains three phosphate groups. Only the last two of these are attached by labile, high-energy bonds. A high-energy bond is one which can break apart with explosive rapidity to release active energy which performs some kind of work.

ATP $\longrightarrow$ ENERGY + ADP + Phosphate

All cellular work, it is now believed, is done by energy set free from high-energy phosphate bonds, mainly those in ATP molecules. Hence, ATP is one of the most important biological compounds known. It supplies energy directly to all kinds of energy-consuming mechanisms in all kinds of living organisms from one-celled plants to billion-celled animals, including man.

Since the citric acid cycle releases most of the energy from food molecules, and since this cycle of chemical reactions takes place inside mitochondria, these tiny structures are aptly described as the power plants of cells.

Albert Lehninger, the biochemist who discovered that citric acid cycle and electron-transfer enzymes are located in mitochondria, writes: "If the classical engineering science of energy transformation is humbled by what is now known about the power plants of the cell, so are the newer and more glamorous branches of engineering. The technology of electronics has achieved amazing success in packaging and miniaturizing the components of a computer. But these advances still fall far short of accomplishing the unbelievable miniaturization of complex energy-transducing components that has been perfected by organic evolution in each living cell."*

■Cell reproduction

Cells reproduce by a process called *cell division*. In other words, cells divide in order to multiply. One cell divides to form two cells. The process of cell division consists of two consecutive processes: mitosis (division of the nucleus) and cytokinesis (division of the cytoplasm). We shall discuss mitosis but not cytokinesis.

Mitosis

Mitosis, or nuclear division, consists of a succession of events plainly visible in suitably stained cells viewed with the light microscope. The events of mitosis occur in five phases: prophase, metaphase, anaphase, telophase, and interphase (formerly called resting stage). The events characteristic of each of these phases are summarized in Table 2-1 and illustrated in Fig. 2-16.

Chromosome replication, a process which usually occurs during the interphase, is actually *DNA replication*. Each DNA molecule within the cell makes a copy of itself—and almost always a perfect copy. Unquestionably, the process of DNA replication is one of the most unique and important of all biological phenomena. Justification for such an extravagant claim lies in these facts; a chromosome consists essentially

Table 2-1. Mitosis—nuclear division

Interphase	Prophase	Metaphase
1 Period when cell prepares for division and also grows in size 2 Chromosomes elongate and become too thin to be visible as such, but chromatin granules become visible 3 DNA of each chromosome replicates itself, forming two chromatids attached only at centromere (DNA replication discussed on p. 38)	1 Chromosomes shorten and thicken (due to coiling of DNA molecules which compose them); each chromosome consists of two *chromatids* attached at *centromere* 2 Centrioles move to opposite poles of cell; spindle fibers appear and, under control of centrioles, begin to orient between opposing poles	1 Chromosomes align across equator of spindle fibers; each pair of chromatids attached to spindle fiber at its centromere 2 Nucleoli and nuclear membrane disappear

*From Lehninger, A. L.: How cells transform energy, Sci. Amer. **205**:62–73 (Sept.); 1961; copyright © 1961 by Scientific American, Inc.; all rights reserved.

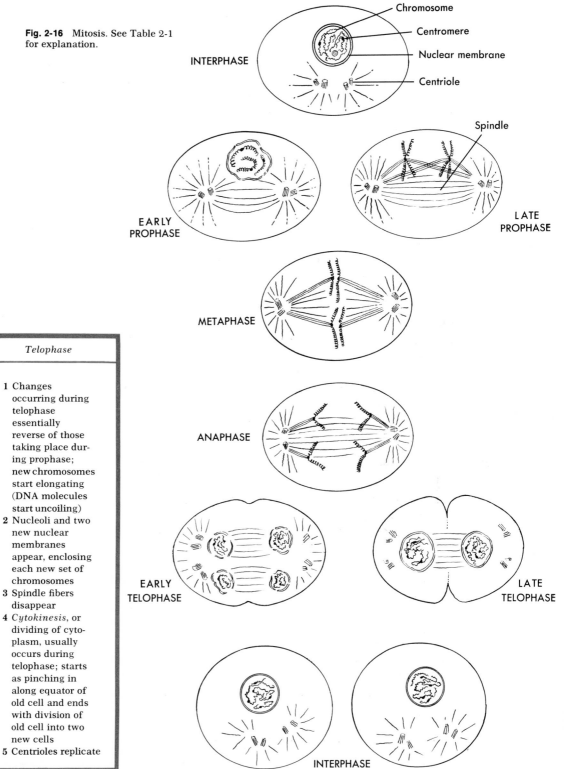

Fig. 2-16 Mitosis. See Table 2-1 for explanation.

INTERPHASE

Chromosome
Centromere
Nuclear membrane
Centriole

EARLY PROPHASE

LATE PROPHASE

Spindle

METAPHASE

ANAPHASE

EARLY TELOPHASE

LATE TELOPHASE

INTERPHASE

Anaphase	Telophase
1 Each centromere divides into two, thereby detaching two chromatids that compose each chromosome from each other 2 Divided centromeres start moving to opposite poles, each pulling its chromatid (now called a chromosome) along with it; there are now twice as many chromosomes as there were before mitosis started	1 Changes occurring during telophase essentially reverse of those taking place during prophase; new chromosomes start elongating (DNA molecules start uncoiling) 2 Nucleoli and two new nuclear membranes appear, enclosing each new set of chromosomes 3 Spindle fibers disappear 4 *Cytokinesis*, or dividing of cytoplasm, usually occurs during telophase; starts as pinching in along equator of old cell and ends with division of old cell into two new cells 5 Centrioles replicate

of a DNA molecule, a DNA molecule consists of a long line of genes, and genes determine all the potentialities of all new cells. In short, genes determine heredity. We shall, of course, discuss the structure and function of these extremely important structures. First, however, we shall present facts about DNA structure that you need to know if you are to gain an understanding of genes.

DNA is a giant among molecules. Not only does its size and the complexity of its shape exceed those of most molecules, the importance of its function surpasses that of any other molecule in the world. It is the main constituent of the nucleus in every living cell. (The cytoplasm of most cells also contains some DNA we now know.) Not quite twenty years ago, an American, James D. Watson, and two British scientists, Francis H. C. Crick and Maurice H. F. Wilkins, solved the puzzle of DNA's molecular structure. Watson and Crick published their findings in April, 1953. Nine years later the coveted Nobel Prize in Medicine was awarded all three scientists for their brilliant and significant work—almost surely the greatest biological discovery of our times. Watson tells how they accomplished this in *The double helix*, a book you will probably find hard to put down once you start it.*

Let us start our discussion of the DNA molecule by trying to visualize its shape. Picture to yourself an extremely long ladder made of a pliable material. Now see it twisting round and round on its axis and taking on the shape of a steep spiral staircase millions of turns long. This is the shape of the DNA molecule—namely, a double *helix* (Greek word for spiral).

As to its structure, the DNA molecule is a *polymer*. This means that it is a large

molecule made up of many smaller molecules joined together. DNA is a polymer of millions of pairs of nucleotides. A *nucleotide* is a compound formed by the combining of phosphoric acid with a sugar and a nitrogenous base. In the DNA molecule, each nucleotide consists of a phosphate group which attaches to the sugar deoxyribose which attaches to one of four bases —that is, either adenine or thymine or cytosine or guanine. (Deoxyribose is a sugar that is not sweet and one whose molecules contain only 5 carbon atoms.) Notice what the name deoxyribonucleic acid tells you—that this compound contains deoxyribose, that it occurs in nuclei, and that it is an acid.

Fig. 2-17 reveals additional and highly significant facts about DNA's molecular structure. First, observe which compounds form the sides of the DNA spiral staircase —a long line of phosphate and deoxyribose units joined alternately one after the other. Look next at the stair steps. Notice two facts about them: that two bases join (by means of a hydrogen bond) to form each step and that only two combinations of bases occur. The same two bases invariably pair off with each other in a DNA molecule—like teen-agers "going steady." Adenine always goes with thymine (or vice versa, thymine with adenine), and guanine always goes with cytosine (or vice versa). This fact about DNA's molecular structure is called *obligatory base-pairing*. Pay particular attention to it, for it is the open sesame for the door to understanding how a DNA molecule is able to duplicate itself. DNA duplication, or replication as it is usually called, is one of the most important of all biological phenomena because it is an essential and crucial part of the mechanism of heredity.

Another fact about DNA's molecular structure that has great functional importance is the sequence of its base pairs.

---

*Watson, J. D.: The double helix, New York, 1968, The New American Library Inc.

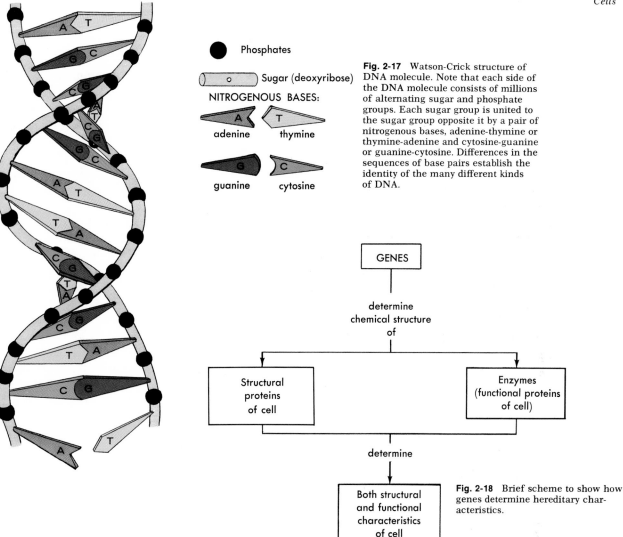

Phosphates

Sugar (deoxyribose)

NITROGENOUS BASES:

adenine

thymine

guanine

cytosine

Fig. 2-17 Watson-Crick structure of DNA molecule. Note that each side of the DNA molecule consists of millions of alternating sugar and phosphate groups. Each sugar group is united to the sugar group opposite it by a pair of nitrogenous bases, adenine-thymine or thymine-adenine and cytosine-guanine or guanine-cytosine. Differences in the sequences of base pairs establish the identity of the many different kinds of DNA.

GENES

determine
chemical structure
of

Structural
proteins
of cell

Enzymes
(functional proteins
of cell)

determine

Both structural
and functional
characteristics
of cell

Fig. 2-18 Brief scheme to show how genes determine hereditary characteristics.

Although the base pairs in all DNA molecules are the same, the sequence of these base pairs is not the same in all DNA molecules. For instance, the sequence of the base pairs composing the seventh, eighth, and ninth steps of one DNA molecule might be cytosine-guanine, adenine-thymine, and thymine-adenine. In another DNA molecule, the sequence of the base pairs making up these same steps might be entirely different, perhaps thymine-adenine, guanine-cytosine, and cytosine-guanine. Perhaps these strike you as picayune details. But nothing could be

further from the truth, since it is the sequence of the base pairs in the nucleotides composing DNA molecules that identifies each gene. Hence, it is the sequence of base pairs that determines all heredity traits.

A human *gene* is a segment of a DNA molecule. One gene consists of a chain of about one thousand pairs of nucleotides joined one after the other in a precise sequence. One gene controls the production of one protein (either a structural protein or a functional protein—that is, an enzyme). Each enzyme catalyses one

37

chemical reaction. Therefore, the millions of genes constituting a cell's DNA determine both the structural and the functional characteristics it inherits. The detailed story of how genes control protein synthesis although still unfinished, is far too long to tell here.* We shall, however, give a brief account of it in Chapter 12.

Some arithmetic about genes and DNA may interest you. Swanson says that the DNA of "each (human) chromosome contains an average of about 175 million nucleotide pairs."† You might then try to picture a human DNA molecule as a spiral staircase, microscopic in width, very, very steep, and 175 million steps long! If you divide 175 million (the average number of nucleotide pairs in one chromosome) by 1,000 (the number of nucleotide pairs in one gene), you find that one human chromosome consists of 175,000 genes. And if you multiply 175,000 genes per chromosome by 46 (the number of chromosomes in one human cell), you discover that approximately 8 million genes are present in one human cell!

Chromosome or *DNA replication* occurs during the interphase of mitosis. The tightly coiled DNA molecule uncoils except in small segments. Since the tight little coils remaining are denser than the thin elongated sections, they absorb more stain and show up as *chromatin granules*. The thin uncoiled sections, in contrast, absorb so little stain that they are invisible. As the DNA molecule uncoils, it comes apart in the midline, separating the two strands of the double helix. Then, along each of the two long, separated strands of nucleotides, a complementary strand forms. In other words, each half DNA

molecule duplicates itself to create a whole new DNA molecule. The next paragraph tells how.

Intracellular fluid contains many molecules of deoxyribose and nitrogenous bases and many phosphate ions. By the mechanism of "obligatory base-pairing," these substances become attached at their right places along each DNA strand. Interpreted, this means that new thymine (that is, from the intracellular fluid) attaches to "old" adenine in the original DNA strand. And conversely, new adenine attaches to old thymine. Also, new guanine joins old cytosine and, vice versa, new cytosine joins old guanine. Soon each of the two original DNA strands has a complete new complementary strand attached to it. So the original DNA molecule will have become two molecules. Once again, the miracle of DNA replication will have occurred and each original chromosome will have become two chromosomes. However, they are now called *chromatids*, not chromosomes, and the two chromatids formed from one original chromosome are joined together at one point by a structure called the *centromere*. Two new sets of genes now stand ready to control the destiny of the two new cells which will be formed by the next cell division.

By way of summary, a complete duplicate set of chromosomes (now called chromatids) is formed during the interphase of mitosis. During the anaphase, one set of chromatids moves to one pole of the cell and the other set moves to the opposite pole (and now chromatids become chromosomes). Then, finally, one set of chromosomes enters one new cell and the other set enters the other new cell and the parent cell divides. Hence, each of the two new cells contains a complete set of genes identical to those in the parent cell. And this means that both cells have the potentialities to become like their parent cell.

*See supplementary readings for Chapter 2, references 14, 19, and 21.
†From Swanson, C. P.: The cell, ed. 3, Englewood Cliffs, N. J., 1969, Prentice-Hall, Inc., p. 100.

Thus, the mechanism of mitosis makes possible heredity.

• • •

This chapter has presented information about the structure and self-serving functions of cells. The next one will relate facts and principles about the structure and body-serving functions of tissues—that is, organizations of cells.

outline summary

CELL STRUCTURE
See Fig. 2-1

Protoplasm
1 Definition—living matter; all cells composed of protoplasm
2 Composition
   a Main elements—carbon, hydrogen, oxygen, and nitrogen
   b Main compounds—water, inorganic salts, and compounds unique to living matter—i.e., proteins (including enzymes), carbohydrates, lipids, and nucleic acids
3 Functional properties—irritability, conductivity, contractility, metabolism, and reproduction

Cell membrane
1 Structure
   a About 3/10,000,000 of an inch thick according to present estimates
   b Composed of protein and lipid molecules; according to "triple-layer hypothesis," double layer of lipid molecules sandwiched between an inner and outer layer of protein molecules
   c Has tiny opening or pores; few in number; spaced far apart
2 Function
   a Maintains cell's integrity, its organization
   b Determines what substances can enter and leave cell

Cytoplasm
1 Definition—protoplasm located between cell membrane and nucleus
2 Contains thousands of organelles ("little organs")
   a Membranous organelles: endoplasmic reticulum, Golgi apparatus, mitochondria, and lysosomes
   b Nonmembranous organelles: ribosomes and centrosome
3 Endoplasmic reticulum
   a Structure—complicated network of canals and sacs extending through cytoplasm and opening at surface of cell; many ribosomes attached to membranes of rough endoplasmic reticulum but not to smooth
   b Functions—ribosomes attached to rough endoplasmic reticulum synthesize proteins; canals of reticulum serve cell as its inner circulatory system—e.g., proteins move through canals on way to Golgi apparatus
4 Golgi apparatus
   a Structure—membranous vesicles near nucleus
   b Function—synthesizes large carbohydrate molecules, combines them with proteins, and secretes product (glycoproteins)
5 Mitochondria
   a Structure—microscopic sacs; walls composed of inner and outer membranes separated by fluid; thousands of particles made up of enzyme molecules attach to both membranes
   b Function—"power plants" of cells; mitochondrial enzymes catalyze citric acid cycle, series of reactions that provide about 95% of cell's energy supply
6 Lysosomes
   a Structure—microscopic membranous sacs
   b Function—enzymes in lysosomes digest particles or large molecules which enter them; under some conditions, digest and thereby destroy cells
7 Ribosomes
   a Structure—microscopic spheres, large numbers of which attached to endoplasmic reticulum
   b Function—"protein factories"; ribosomes attached to endoplasmic reticulum synthesize proteins to be secreted by cell, and those lying free in cytoplasm make proteins for cell's own use—i.e., its structural proteins and enzymes; groups of ribosomes, called polysomes, not single ribosomes, synthesize proteins

8 Centrosome or centrosphere
  a Structure
    1 Centrosome is spherical body near center of cell – i.e., near nucleus
    2 Centrioles, located in centrosome, are tiny cylinders, walls of which are composed of fine tubules, nine groups of two or three tubules each
  b Function – centrioles control polarization of spindle fibers and play some part in their formation

Nucleus

1 Definition – spherical body in center of cell; enclosed by pore-containing membrane
2 Contains nucleoli and chromosomes
  a Nucleoli are small spherical bodies composed mostly of RNA and some protein; combining of ribosomal RNA with protein takes place in nucleoli
  b Chromosomes are deep-staining bodies in nucleus; composed of DNA and proteins

CELL PHYSIOLOGY
Movement of substances through cell membranes

1 By physical (or passive) processes
  a Energy which moves substances comes from random, never-ceasing movements of atoms, ions, and molecules, not from chemical reactions in cell
  b Diffusion and osmosis main physical processes which move substances through cell membrane or nonliving membranes
2 By physiological (or active) processes
  a Energy which moves substances comes from chemical reactions in living cell
  b Active transport mechanisms, phagocytosis, and pinocytosis types of physiological mechanisms that move substances through living cell membranes only

Diffusion

1 Movement of solute and solvent particles in all directions through solution or in both directions through membrane
  a Net diffusion of solute particles – down solute concentration gradient, i.e., from more to less concentrated solution
  b Net diffusion of water – down water concentration gradient, i.e., from less to more concentrated solution
2 Diffusion tends to produce equilibration of solutions on opposite sides of membrane but many exceptions to this rule when membrane living
3 Dialysis – separation of crystalloids from colloids by diffusion of crystalloids through membrane permeable to them but impermeable to colloids
4 Facilitated diffusion – diffusion in which carrier substance combines with particle being moved

Osmosis

1 In living systems, movement of water in both directions through membrane that maintains at least one solute concentration gradient across it
2 Net osmosis – more water osmoses in one direction through membrane than in opposite; net osmosis occurs down water concentration gradient (which is up solute concentration gradient and up potential osmotic pressure gradient); net osmosis tends to equilibrate two solutions separated by selectively permeable membrane
3 Osmotic pressure – pressure that develops in solution as result of net osmosis into it
4 Isotonic solution – one that has same potential osmotic pressure as solution it is isotonic to; no net osmosis between isotonic solutions
5 Hypertonic solution – has greater potential osmotic pressure and higher solute concentration but lower water concentration than solution to which it is hypertonic; net osmosis into hypertonic solution from hypotonic solution
6 Hypotonic solution – has lower potential osmotic pressure and lower solute concentration but higher water concentration than solution to which it is hypotonic; net osmosis out of hypotonic solution into hypertonic solution

Active transport mechanisms ("pumps")

Devices which move ions or molecules through cell membranes against their concentration gradient – i.e., direction opposite from net diffusion or net osmosis; energy supplied by cellular chemical reactions

Phagocytosis and pinocytosis

1 Phagocytosis – physiological process which moves solid particles into cell; segment of cell membrane forms pocket around particle outside cell, then pinches off from rest of membrane and migrates inward
2 Pinocytosis – physiological process which moves fluid into cell; process similar to phagocytosis

Cell metabolism
Catabolism

1 One of two major processes of metabolism
2 Consists of complex series of chemical reactions which take place inside cells and which yield energy, carbon dioxide, and water; about half of energy released from food molecules by catabolism put back in storage in unstable high-energy bonds of ATP molecules and rest transformed to heat; energy in high-energy bonds of ATP can be released as rapidly as needed for doing cellular work
3 Purpose – to continually provide cells with utilizable energy – i.e., energy supplied instantaneously to energy-consuming mechanisms which do cellular work
4 Glycolysis – series of anaerobic (nonoxygen-util-

izing) chemical reactions that convert 1 glucose molecule to 2 pyruvic acid molecules and yield small amount of high-energy ATP and of heat

5 Citric acid cycle – series of aerobic chemical reactions that utilize oxygen to oxidize 2 pyruvic acid molecules to 6 carbon dioxide molecules and 6 water molecules and yield about 95% of ATP and heat formed during catabolism

Anabolism

1 Other major process of metabolism
2 Synthesis of various compounds from simpler compounds; an important kind of cellular work that uses some of energy made available by catabolism
3 See Chapter 12

Cell reproduction

1 Process of cell division which consists of mitosis (division of nucleus) and cytokinesis (division of cytoplasm)
2 See Table 2-1 for summary of mitosis

review questions

1 Identify the following organelles with a brief statement about the structure and functions of each: endoplasmic reticulum; Golgi apparatus; mitochondria; lysosomes; ribosomes; centrosome.
2 One inch equals approximately how many angstroms? microns? millimeters? millimicrons?
3 One millimeter equals approximately what fractional part of an inch? about how many angstroms? about how many microns? about how many millimicrons?
4 Explain briefly what each of the following terms means: active transport; dialysis; diffusion; facilitated diffusion; net diffusion; osmosis; phagocytosis; pinocytosis.
5 How do active processes for moving substances through cell membranes differ from passive processes?
6 What characteristics identify a selectively permeable membrane?
7 What conditions are necessary for net osmosis to occur?
8 State the principle about the direction in which net osmosis occurs.
9 Differentiate between actual osmotic pressure and potential osmotic pressure.
10 What factor directly determines the potential osmotic pressure of a solution?
11 Explain the terms isotonic, hypotonic, and hypertonic.
12 State the principle about which solution develops an osmotic pressure, given appropriate conditions.
13 Define briefly: anabolism; ATP; catabolism; citric acid cycle; glycolysis; metabolism.
14 Which of the processes you defined in the preceding question yields almost 95% of the ATP formed during catabolism and is, therefore, the body's main energy-supplying process?

3

Tissues

Epithelial tissue
Simple squamous epithelium
Stratified squamous
   epithelium
Simple columnar epithelium

Muscle tissue

Connective tissue
Loose, ordinary connective
   tissue (areolar)
Adipose tissue
Dense fibrous tissue
Bone and cartilage
Hemopoietic tissue
Blood
Reticuloendothelial tissue

Nerve tissue

Tissues are organizations of cells. Non-living intercellular substances fill in any spaces between cells. Using appearance and functions as criteria for classification, there are four basic types of tissues—epithelial, muscle, connective, and nerve—and many subtypes (Table 3-1). Tissues differ in structure, and because structure determines function, they also differ in function. Not only do cell size, shape, and arrangement vary in different kinds of tissues, but so too does the amount and kind of intercellular substance present. Some tissues contain almost no intercellular material. Others consist predominantly of it. Some intercellular substance contains many fibers, some is unformed gel, and some is fluid, the interstitial fluid that bathes most living human cells.

■Epithelial tissue

Epithelial tissue performs the functions of protection, secretion, absorption, diffusion, and filtration. Only a fairly sturdy tissue can offer protection. Therefore, where this function is needed, epithelial tissue is stratified—that is, it consists of several layers of cells. In contrast, where substances need to move through a tissue, where they need to be absorbed, to filter, or to diffuse, here there is simple epithelial tissue which consists of a single layer of cells. All types of epithelial tissue are composed largely or entirely of cells. In other words, they contain little or no intercellular substance. They also contain no blood vessels. These are located in the connective tissue which epithelial tissue always overlies and adheres to firmly. Another common characteristic of all types of epithelial cells is that they undergo mitosis—a fact of practical importance since it means that old or destroyed epithelial cells can be replaced by new ones. There are several types of epithelial tissue. Three are described in the following paragraphs.

Simple squamous epithelium

Simple squamous epithelium consists of only one layer of flat, scalelike cells. Consequently, substances can readily diffuse or filter through this type of tissue. The microscopic air sacs of the lungs, for example, are composed of this kind of tissue, as are the linings of blood and lymphatic vessels and the surfaces of the pleura, pericardium, and peritoneum. (Blood and lymphatic vessel linings are called *endothelium*, and the surfaces of the pleura, pericardium, and peritoneum are called *mesothelium*. Some histologists classify these as connective tissue.)

Stratified squamous epithelium

Stratified squamous epithelium such as shown in Fig. 3-1 lines the mouth and esophagus. Its several layers of cells serve a protective function. The surface of the skin is composed of a special kind of stratified squamous epithelium (pp. 54 to 56).

Simple columnar epithelium

Simple columnar epithelium lines the stomach and intestines and parts of the respiratory tract. A single layer of cells composes this tissue, but two types of cells—goblet and columnar—may be present (Fig. 3-2). Goblet cells are the mucus-secreting specialists of the body. Columnar cells are its absorption specialists.

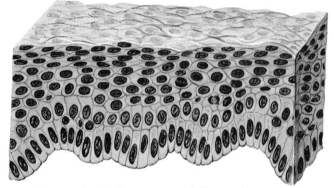

Fig. 3-1 Stratified squamous epithelium such as lines the mouth.

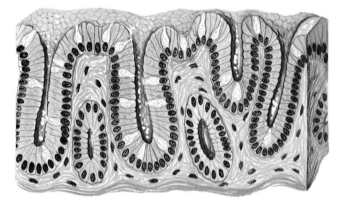

Fig. 3-2 Simple columnar epithelium with goblet cells such as lines intestines.

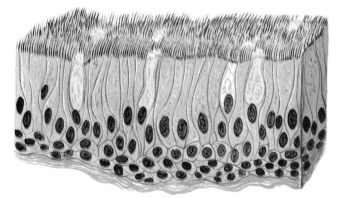

Fig. 3-3 Pseudostratified ciliated columnar epithelium with goblet cells. Stratified tissue consists of two or more layers of cells. Pseudostratified tissue appears, in certain sections, to meet this requirement but actually does not.

43

Table 3-1. Tissues

Tissue	Location	Function
EPITHELIAL *Simple squamous*	Alveoli of lungs	Diffusion of respiratory gases between alveolar air and blood
	Lining blood and lymphatic vessels (called endothelium; classed as connective tissue by some histologists)	Diffusion; filtration; osmosis
	Surface layer of pleura, pericardium, peritoneum (called mesothelium; classed as connective tissue by some histologists)	Diffusion; osmosis
Stratified squamous	Surface of lining of mouth and esophagus	Protection
	Surface of skin (epidermis)	Protection
Simple columnar	Surface layer of lining of stomach, intestines, and part of respiratory tract	Protection; secretion; absorption; moving of mucus (by ciliated columnar)
MUSCLE *Skeletal (striated voluntary)*	Muscles which attach to bones Extrinsic eyeball muscles Upper one third of esophagus	Movement of bones Eye movements First part of swallowing
Visceral (nonstriated involuntary or smooth)	In walls of tubular viscera of digestive, respiratory, and genitourinary tracts	Movement of substances along respective tracts
	In walls of blood vessels and large lymphatics	Change diameter of blood vessels, thereby aiding in regulation of blood pressure
	In ducts of glands	Movement of substances along ducts
	Intrinsic eye muscles (iris and ciliary body)	Change diameter of pupils and shape of lens
	Arrector muscles of hairs	Erection of hairs (gooseflesh)
Cardiac (striated involuntary)	Wall of heart	Contraction of heart
CONNECTIVE (most widely distributed of all tissues) *Loose, ordinary (areolar)*	Between other tissues and organs	Connection
	Superficial fascia	Connection

Table 3-1. Tissues—cont'd

Tissue	Location	Function
CONNECTIVE—cont'd		
Adipose (fat)	Under skin Padding at various points	Protection Insulation Support Reserve food
Dense fibrous	Tendons Ligaments Aponeuroses Deep fascia Dermis Scars Capsule of kidney, etc.	Furnishes flexible but strong connection
Bone	Skeleton	Support Protection
Cartilage Hyaline	Part of nasal septum Covering articular surfaces of bones Larynx Rings in trachea and bronchi	Furnishes firm but flexible support
Fibrous	Disks between vertebrae Symphysis pubis	
Elastic	External ear Eustachian tube	
Hemopoietic Myeloid (bone marrow)	Marrow spaces of bones	Formation of red blood cells, granular leukocytes, platelets; also reticuloendothelial cells and some other connective tissue cells
Lymphatic	Lymph nodes Spleen Tonsils and adenoids Thymus gland	Formation of lymphocytes and monocytes; also plasma cells and some other connective tissue cells
Blood	In blood vessels	Transportation Protection
Reticuloendothelial	Widely scattered, e.g., in lining of blood sinusoids of liver, spleen, and bone marrow; also in lining of lymph channels in lymph nodes	Phagocytosis
NERVE	Brain Spinal cord Nerves	Irritability and conduction

■Muscle tissue

The main specialty of muscle tissue is contraction. Because not all muscle tissue is alike—in location, microscopic appearance, and nervous control—these criteria are used to classify its types. Thus, using location as the criterion, there are three kinds of muscle tissue:

1 *skeletal muscle*—attached to bones
2 *visceral muscle*—in the walls of hollow internal structures such as blood vessels, intestines, uterus, and many others
3 *cardiac muscle*—composes the wall of the heart

With microscopic appearance as the basis of classification, there are only two types of muscle tissue: striated (named for cross striations seen in these cells) and nonstriated or smooth (no cross striations in cells).

On the basis of nervous control, there are also two kinds of muscle tissue: voluntary and involuntary.

Voluntary muscle receives nerve fibers from the cerebrospinal nervous system. Therefore, its contraction can be voluntarily controlled. Involuntary muscle, on the other hand, receives nerve fibers from the autonomic nervous system so that its contraction cannot be voluntarily controlled (except in a few rare individuals). Skeletal muscle is voluntary muscle. Visceral and cardiac muscles are involuntary muscles. Visceral and cardiac muscle are also automatic,* meaning that even without nervous stimulation they continue to contract. Skeletal muscle, in contrast, cannot contract automatically. Anything that cuts off its nerve impulses paralyzes it—that is, puts it immediately out of working

---

*For an account of an interesting experiment that demonstrates cardiac muscle automaticity, see Harary, I.: Heart cells in vitro, Sci. Amer. **206**:141–152 (May), 1962.

order. Poliomyelitis acts this way, for example. It damages nerve cells that conduct impulses to skeletal muscles so that they no longer conduct, and, deprived of stimulation, the muscles are paralyzed.

Finally, combining these classifications, we have the following:

1 *skeletal* or striated voluntary muscle
2 *cardiac* or striated involuntary muscle
3 *visceral* or nonstriated (smooth) involuntary muscle

Look at Fig. 3-4, and you can observe the following structural characteristics of skeletal muscle cells: They have many cross striations; each cell has many nuclei; the cells are long and narrow in shape—for example, often 1 1/2 or more inches in length but only between 1/100 and 1/10 of a millimeter (about 1/2500 to 1/250 of an inch) in diameter. Because such measurements give muscle cells a threadlike appearance, they are probably more often called muscle fibers than cells. The discussion of details about skeletal muscle structure appears in Chapter 6.

Smooth muscle cells are also long narrow fibers but not nearly so long as striated fibers. One can see the full length of a smooth muscle fiber in a microscopic field but only part of a striated fiber. (According to one estimate, the longest smooth muscle fibers measure about 500 microns and the longest striated fibers about 40,000 microns. Can you translate these measurements into millimeters and inches? If not, and if you are curious, see footnote on p. 13.) As Fig. 3-5 shows, smooth muscle fibers have only one nucleus per fiber and are nonstriated or smooth in appearance.

Under the light microscope, cardiac muscle fibers have cross striations and unique dark bands (intercalated disks) (Fig. 3-6). They also seem to be incomplete cells that branch into each other to form a big continuous mass of protoplasm known as a

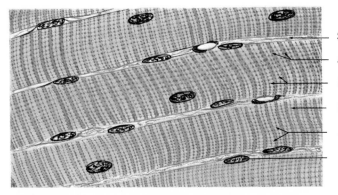

Sarcolemma

Anisotropic substance

Isotropic substance

Intermediate line

Myofibrils

Nucleus

Fig. 3-4 Skeletal or striated voluntary muscle tissue.

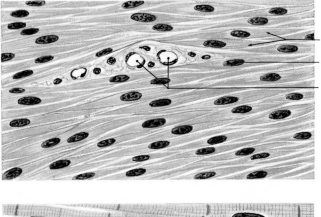

Smooth muscle cells

Nucleus

Blood capillaries

Fig. 3-5 Visceral or nonstriated (smooth) involuntary muscle tissue.

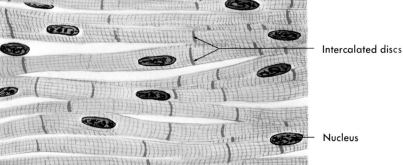

Intercalated discs

Nucleus

Fig. 3-6 Cardiac or striated involuntary muscle tissue.

syncytium. The electron microscope, however, has revealed that the intercalated disks are actually places where two cell membranes abut at the ends of adjacent cardiac fibers. Cardiac fibers do branch and anastomose but, contrary to previous belief, they do not form a syncytium. A complete cell membrane encloses each cardiac fiber—around its ends (at intercalated disks) as well as its sides.

■Connective tissue

Connective tissue is the most widespread and abundant tissue in the body. It connects and supports—connects tissues to each other, for example, and muscles to bones and bones to other bones. It forms a supporting framework for the body as a whole and for its organs individually. Connective tissue exists in more varied forms than the other three basic tissues. Delicate tissue paper webs, strong, tough cords, rigid bones – all are made of connective tissue.

One scheme of classification lists the following main types of connective tissue:

1 Loose, ordinary (areolar)
2 Adipose
3 Dense fibrous
4 Bone
5 Cartilage
6 Hemopoietic
7 Blood
8 Reticuloendothelial

Connective tissue consists predominantly of intercellular material and relatively few cells. Hence, the qualities of the intercellular material largely determine the qualities of each type of connective tissue. One connective tissue, for example, is a fluid because its intercellular material is a fluid. Some connective tissues have the consistency of a soft gel, some are firm but flexible, some are hard and rigid, some are tough, others are delicate – and in each case it is their intercellular substance that makes them so.

Intercellular substance may contain one or more of the following kinds of fibers – collagenous (or white), reticular, and elastic. They differ considerably in their properties. Collagenous fibers are tough and strong, reticular fibers are delicate, and elastic fibers are extendable and elastic. Collagenous or white fibers often occur in bundles – an arrangement that provides great tensile strength. Reticular fibers, in contrast, occur in networks and, although delicate, support small structures such as capillaries and nerve fibers. The protein collagen composes collagenous fibers. You know the hydrated form of collagen as gelatin. Of all the hundreds of different protein compounds in the body, collagen is the most abundant. According to one authority, it constitutes about 40% of all the protein

in the body.* Collagenous fibers, once formed, remain throughout life. The body cannot replace them with new fibers. Persuasive evidence indicates that the molecular structure of collagen changes with age and that this may be one of the most basic factors in the aging process.*

Like intercellular fibers, intercellular ground substance also varies in properties. The ground substance (matrix) in loose, ordinary connective tissue, for example, is a soft, viscous gel, whereas the matrix of bone is a very hard substance – as "hard as bone," in fact. A compound called hyaluronic acid gives the viscous quality to intercellular gels, but it can be converted to a watery consistency by the enzyme hyaluronidase. Physicians have made use of this latter fact for some time now. They frequently give a commercial preparation of hyaluronidase intramuscularly or subcutaneously with drugs or fluids. By decreasing the viscosity of intercellular material, the enzyme hastens diffusion and absorption and thereby lessens tissue tension and pain.

Loose, ordinary connective tissue (areolar)

First, a few words of explanation about the names of this kind of tissue. It is called loose because it is stretchable and ordinary because it is one of the most widely distributed of all tissues. It is common and ordinary, not special like some kinds of connective tissue (bone and cartilage, for example) that help form comparatively few structures. Areolar was the early name for the loose, ordinary connective tissue that connects many adjacent structures of the body. It acts like a glue spread between them but an elastic glue that permits movement. The word areolar means like a small space and refers to the bubbles

*Verzar, F.: The aging of collagen, Sci. Amer. **208**: 104–114 (April), 1963.

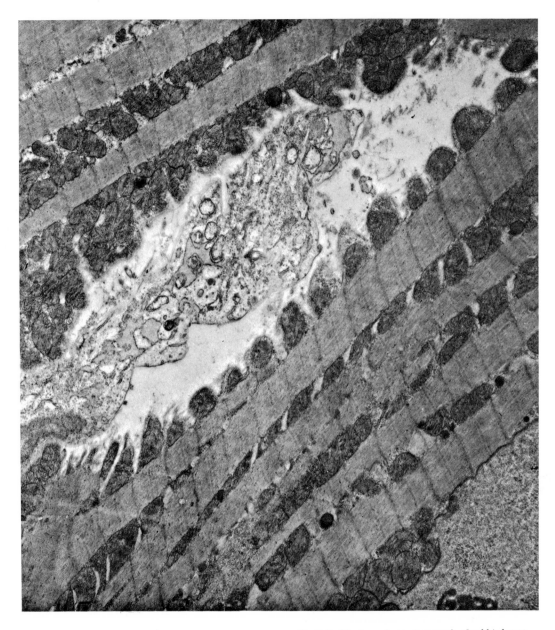

Fig. 3-7 Electron photomicrograph of rabbit heart muscle (×12,500). (Courtesy Department of Pathology, Case Western Reserve University, Cleveland, Ohio.)

that appear as areolar tissue is pulled apart during dissection.

Although intercellular substance is prominent in loose, ordinary connective tissue, cells also are numerous and varied (Fig. 3-8). Collagenous and elastic fibers are interwoven loosely and embedded in a soft viscous ground substance. Of the half dozen or so kinds of cells present, *fibroblasts* are the most common and *macrophages* are second. Fibroblasts synthesize intercellular substances of both types—that is, both fibers and gels. Macrophages carry on phagocytosis. *Plasma cells*, *fat cells*, *mast cells*, and some *white blood cells* (leukocytes) are also found in loose ordinary connective tissue but in smaller numbers than fibroblasts and macrophages. Plasma cells function as antibody producers.

Adipose tissue

Adipose tissue differs from loose, ordinary connective tissue mainly in that it contains predominantly fat cells and many fewer fibroblasts, macrophages, and mast cells (Fig. 3-9). Adipose tissue occurs mainly in certain areas of the body known as fat depots.

Dense fibrous tissue

Dense fibrous tissue consists mainly of bundles of fibers arranged in parallel rows in a fluid matrix. It contains relatively few fibroblast cells. It composes tendons and ligaments. Bundles of collagenous fibers endow tendons with great tensile strength and nonstretchability—desirable characteristics for these structures that anchor our muscles to bones. In ligaments, on the other hand, bundles of elastic fibers predominate. Hence, ligaments exhibit some degree of elasticity.

Bone and cartilage

Bone and cartilage are discussed in Chapter 5.

Hemopoietic tissue

Hemopoietic tissue is discussed in Chapter 10.

Blood

Blood consists of billions of cells afloat in a fluid intercellular substance called plasma. We shall discuss both the cells and the intercellular substance of this vital connective tissue in considerable detail in Chapter 10.

Reticuloendothelial tissue

Various types of connective tissue cells which carry on phagocytosis, although widely scattered throughout the body, are sometimes spoken of as reticuloendothelial tissue or as the *reticuloendothelial system*. They constitute an important part of the body's defense mechanism. Opinions differ somewhat as to which kinds of cells compose reticuloendothelial tissue. There does seem to be agreement, however, that among them are the following three types of phagocytic cells:

1 *reticuloendothelial cells*—located in the lining of blood sinusoids* in the liver, spleen, and bone marrow and in the lining of lymph channels in lymph nodes; another name for the reticuloendothelial cells of liver sinusoids is stellate cells of von Kuppfer—stellate because of their starlike shape and von Kuppfer for the man who first described them
2 *macrophages*—one of the commonest type cells in connective tissues; also called tissue histiocytes, resting wandering cells, and clasmatocytes
3 *microglia*—located in the central nervous system

Reticuloendothelial cells have a mark of

---

*Sinusoids are tiny blood vessels. They are analogous to capillaries in that they connect the arterial side of circulation to the venous side but differ in minor ways from capillaries.

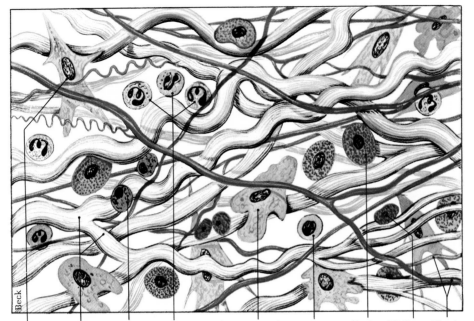

Fibrocyte | Collagenous | Plasma | Polymorphonuclear | Macrophage | Monocyte | Eosinophil | Mast | Elastic
(Fibroblast) | fibers | cell | leukocytes | | | | cell | fibers

Fig. 3-8 Areolar connective tissue. The bundles of fibers are collagenous fibers. The single (red) strand is a bundle of elastic fibers. Several fibroblasts are shown between the fibers. Also shown are macrophages, a plasma cell, a mast cell, and three types of white blood cells: polymorphonuclear leukocytes, eosinophils, and a monocyte.

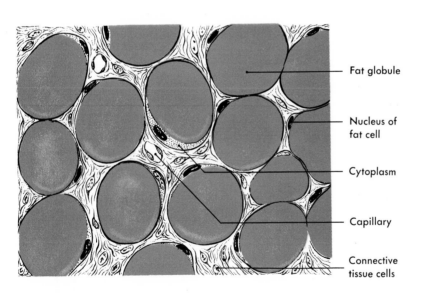

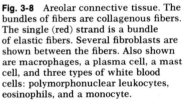

Fat globule

Nucleus of fat cell

Cytoplasm

Capillary

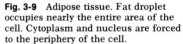

Connective tissue cells

Fig. 3-9 Adipose tissue. Fat droplet occupies nearly the entire area of the cell. Cytoplasm and nucleus are forced to the periphery of the cell.

distinction interesting to note—they may be either fixed or wandering cells. Many of them, especially macrophages, can come loose from their usual site and wander out into regions where their function of phagocytosis is needed, into inflamed or injured areas. Later, with their defense mission accomplished, they move back to their re-spective places and remain fixed there until a future need arises. Apropos of this dual nature of reticuloendothelial cells, did you notice one of the names macrophages go by? "Resting wandering cells."

■ Nerve tissue

Nerve tissue is discussed in Chapter 7.

outline summary

1 Definition – organization of cells with nonliving intercellular substances
2 Basic types
   a Epithelial
   b Muscle
   c Connective
   d Nerve

Epithelial tissue

1 General functions – protection, secretion, diffusion, filtration, and absorption
2 Main types
   a Simple squamous
   b Stratified squamous
   c Simple columnar

Simple squamous epithelium

1 Single layers of flat cells
2 Functions – diffusion and filtration

Stratified squamous epithelium

1 Several layers of cells
2 Function – protection

Simple columnar epithelium

1 Single layer of columnar and goblet-shaped cells and, in some places, ciliated cells
2 Functions – absorption, secretion, and moving mucus

Muscle tissue

1 General functions – contraction and therefore movement
2 Types
   a Skeletal; also called striated voluntary
   b Visceral; also called nonstriated or smooth involuntary
   c Cardiac; also called striated involuntary

Connective tissue

1 General functions – connection, support, and protection
2 Main types
   a Loose, ordinary connective (areolar)
   b Adipose
   c Dense fibrous
   d Cartilage
   e Bone
   f Hemopoietic
   g Blood
   h Reticuloendothelial
3 General characteristics – intercellular material predominates in most connective tissues and determines their physical characteristics; consists of fluid, gel, or solid matrix, with or without fibers (collagenous, reticular, and elastic)

Loose, ordinary connective tissue (areolar)

1 One of most widely distributed of all tissues; intercellular substance is prominent and consists of collagenous and elastic fibers loosely interwoven and embedded in soft viscous ground substance; several kinds of cells present, notably fibroblasts and macrophages, also mast cells, plasma cells, fat cells, and some white blood cells
2 Function – connection

Adipose tissue

1 Similar to loose, ordinary connective tissue but contains mainly fat cells
2 Functions – protection, insulation, support, and reserve food

Dense fibrous tissue

1 Fibrous intercellular substance (collagenous and elastic fibers); few fibroblasts cells
2 Function – furnishes flexible but strong connection

Reticuloendothelial tissue

1 Reticuloendothelial cells that line small lymph and blood channels in lymph nodes, liver, spleen, and bone marrow; function as phagocytes
2 Macrophages and microglia also classed as reticuloendothelial cells by many authors

review questions

1 Name the four basic types of tissue.
2 What are the main functions of each basic type of tissue?
3 What are the names of the subtypes of connective tissue?
4 Describe intercellular substance.
5 Explain the scientific basis for giving hyaluronidase with fluids or certain drugs that are injected.
6 Name several kinds of connective tissue cells.
7 What kind of cells produce intercellular substances?
8 Name three subtypes of muscle tissue. Give more than one name for each.
9 What special function do reticuloendothelial cells perform?
10 What are the main locations of reticuloendothelial cells?
11 Make a list of terms you have encountered for the first time in this chapter. Define each in your own words.
12 Recent evidence suggests that one of the most basic factors in the aging process is a molecular change in what intercellular protein?

4

Membranes and glands

Membranes

Membranes constitute a special class of organs in that they are merely thin sheets of tissues which cover or line various parts of the body. Of the numerous membranes in the body, four kinds are particularly important: mucous, serous, synovial, and cutaneous (skin). Other miscellaneous membranes (periosteum, fascia, dura mater, sclera, etc.) will be described from time to time.

■ Mucous membrane

Mucous membrane lines cavities or passageways of the body which open to the exterior, such as the lining of the mouth and entire digestive tract, the respiratory passages, and the genitourinary tract. It consists, as do serous, synovial, and cutaneous membranes, of a surface layer of epithelial tissue over a deeper layer of connective tissue. Mucous membrane performs the functions of protection, secretion, and absorption—protection, for example, against bacterial invasion, secretion of mucus, and absorption of water, salts, and other solutes.

■ Serous and synovial membranes

Serous and synovial membranes line cavities of the body which do not open to

Membranes
Mucous membrane
Serous and synovial
  membranes
Cutaneous membrane
  Epidermis and dermis
  Accessory organs of skin
    Hair
    Nails
    Skin glands
  Terms used in
    connection with skin

Glands

the exterior, otherwise known as closed cavities. Serous membrane that lines the thoracic cavity is called *pleura*, that which lines the abdominal cavity is called *peritoneum*, and that which lines the sac in which the heart lies is called *pericardium*.

Not only does serous membrane line the thoracic and abdominal cavities and the pericardial sac, but it also covers the organs lying in these spaces. The term *visceral layer* is applied to the part of the membrane which covers the organs, while that which lines the cavity is called *parietal layer* (Fig. 4-1). Between the two layers there is a potential space kept moist by a small amount of serous fluid. Think of the thoracic and abdominal cavities as

53

rooms in a house. The wallpaper then becomes comparable to the parietal layer of the serous membranes. Imagine the rooms filled with furniture, each piece wrapped tightly in muslin for protection. The muslin wrappings are comparable to the visceral serous membrane which covers each organ in the thoracic and abdominal cavities. Even though the articles of furniture in the rooms be stacked closely against each other and against the walls, still there is air between the pieces and between the pieces and the walls. This air might be likened to the small amount of lubricating serous fluid in the potential space between the visceral and parietal layers of serous membrane. When an organ moves against the body wall, as the lungs do in respiration, or when the heart beats in its serous sac, friction between the moving parts is prevented by the presence of the very smooth moist serous sheets lining the wall surface of the cavity and covering the organ surfaces. The mechanical principle that moving parts must have lubricated surfaces is thereby carried out in the body.

Synovial membrane lines joint cavities, tendon sheaths, and bursae. Its smooth moist surfaces protect against friction.

■ Cutaneous membrane

Vital, diverse, complex, extensive – these adjectives describe in part the body's largest and one of its most important organs – the skin. In terms of surface area, the skin is as large as the body itself – probably 2,500 to 3,000 square inches in most adults. Skin functions are crucial to survival. They are also diverse, including such different functions as protection, excretion, sensation, and playing a part in maintaining fluid and electrolyte balance and normal body temperature. The skin protects us against entry of unconquerable hordes of microorganisms and minimizes mechanical injury of underlying structures. It bars entry of excess sunlight and of most chemicals. Even water does not penetrate it under most circumstances. The skin protects against too much and too little heat loss. For example, if body temperature increases above certain limits, skin vessels dilate, more blood flows to the surface, and more heat may be lost by radiation. And at the same time sweat glands secrete more sweat, and more heat may be lost by evaporation.

Millions of microscopic nerve endings are distributed throughout the skin. These serve as antennas or receivers for the body, keeping it informed of changes in its environment – information vital at times to survival.

Epidermis and dermis

Two main layers compose the skin: an outer and thinner layer, the *epidermis*, and an inner, thicker layer, the *dermis* (Fig. 4-2). The epidermis consists of stratified squamous epithelial tissue and the dermis of fibrous connective tissue. Underlying the dermis is subcutaneous tissue or superficial fascia made of areolar and in many areas adipose tissue, too. The epidermis, in all parts of the body except the palms of the hands and soles of the feet, has four layers. In the skin of the palms and soles, there are five layers of epidermis. From the outside in, they are as follows:

1 *stratum corneum (horny layer)* – dead cells converted to a water-repellent protein called keratin that continually flakes off (desquamates)
2 *stratum lucidum* – so named because of the presence of a translucent compound (eleidin) from which keratin forms; this layer present only in thick skin of palms and soles
3 *stratum granulosum* – so named because of granules visible in cytoplasm of cells (cells die in this layer)
4 *stratum spinosum (prickle cell layer)*

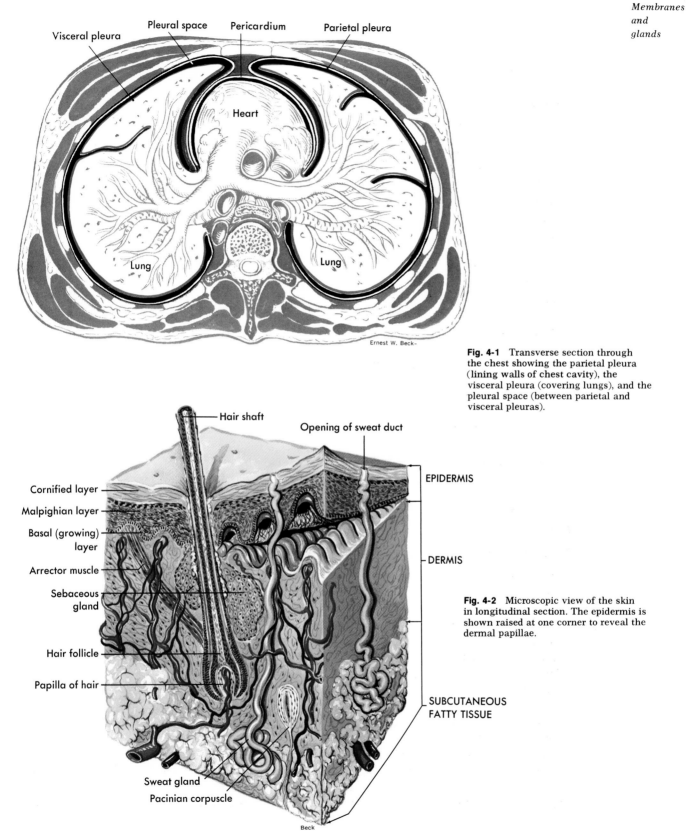

Ernest W. Beck

Fig. 4-1 Transverse section through the chest showing the parietal pleura (lining walls of chest cavity), the visceral pleura (covering lungs), and the pleural space (between parietal and visceral pleuras).

Visceral pleura
Pleural space
Pericardium
Parietal pleura
Heart
Lung
Lung

Hair shaft
Opening of sweat duct
EPIDERMIS
Cornified layer
Malpighian layer
Basal (growing) layer
DERMIS
Arrector muscle
Sebaceous gland

Fig. 4-2 Microscopic view of the skin in longitudinal section. The epidermis is shown raised at one corner to reveal the dermal papillae.

Hair follicle
Papilla of hair
SUBCUTANEOUS FATTY TISSUE
Sweat gland
Pacinian corpuscle

Beck

—several layers of irregularly shaped cells

5 *stratum germinativum* (or *basal layer*)—columnar-shaped cells, the only cells in the epidermis that undergo mitosis; new cells produced in this deepest stratum at the rate old keratinized cells lost from the stratum corneum; new cells continually push surfaceward from the stratum germinativum into each successive layer, only to die, become keratinized, and eventually flake off as did their predecessors.*

An interesting characteristic of the dermis or deep layer of the skin is its parallel ridges, suggestive on a miniature scale of the ridges of a contour plowed field. Epidermal ridges, the ones made famous by the art of fingerprinting, exist because the epidermis conforms to the underlying dermal ridges.

As everyone knows, human skin comes in a wide assortment of colors. Explanation for this fact lies in a constellation of factors that act and interact in complex ways (Fig. 4-3). The basic determinant, however, of skin color is the quantity of *melanin*—the main skin pigment—deposited in the epidermis. Heredity, first of all, determines this. Geneticists now tell us that four to six pairs of genes exert the primary control over the amount of melanin formed by special cells called *melanocytes*. Thus, heredity determines how dark or how light one's basic skin color will be. But other factors can modify the genetic effect. Sunlight is the obvious example. Prolonged exposure of the skin to sunlight causes melanocytes to increase melanin production and darken skin color. So, too, does an excess of adrenocorticotropic hormone

---

*Incidentally, this fact illustrates nicely the physiological principle that while life continues the body's work is never done; even at rest it is producing new cells to replace millions of old ones.

(ACTH) or an excess of melanocyte-stimulating hormone (MSH), two of the hormones secreted by the anterior pituitary gland.

An individual's basic skin color changes, as we have just observed, whenever its melanin content changes appreciably. But skin color can also change without any change in melanin. In this case, the change is usually temporary and most often stems from a change in the volume of blood flowing through skin capillaries. A marked increase in skin blood volume may cause the skin to take on a pink or red hue, and an appreciable decrease in skin blood volume may turn it frighteningly pale. In general, the sparser the pigments in the epidermis, the more transparent the skin is and, therefore, the more vivid the change in skin color will be with a change in skin blood volume. Conversely, the richer the pigmentation, the more opaque the skin is and the less skin color will change with a change in skin blood volume.

In some abnormal conditions, skin color changes because of an excess amount of unoxygenated hemoglobin in the skin capillary blood. If skin contains relatively little melanin, it will develop a bluish cast known as *cyanosis* when 100 milliliters of blood contains about 5 grams of unoxygenated hemoglobin. In general, the darker the skin pigmentation, the more unoxygenated hemoglobin must be present before cyanosis becomes visible.

Accessory organs of skin

The accessory organs of the skin consist of hair, nails, and microscopic glands.

Hair. Hair is distributed over the entire body except the palms and soles. The structure of a hair has several points of similarity to that of the epidermis. Just as the epidermis is formed by the cells of its deepest layer, reproducing and forcing the daughter cells, which become horny in

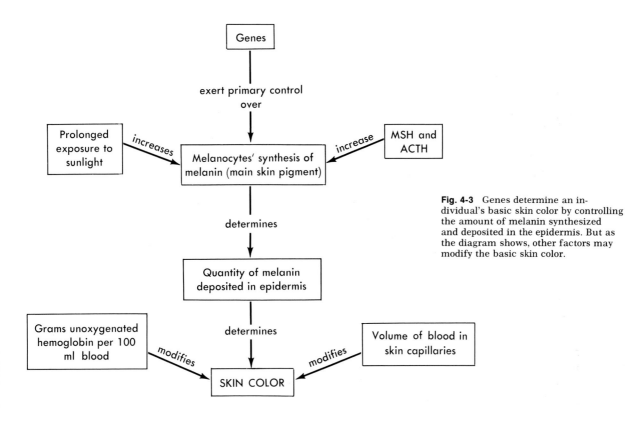

Fig. 4-3 Genes determine an individual's basic skin color by controlling the amount of melanin synthesized and deposited in the epidermis. But as the diagram shows, other factors may modify the basic skin color.

character, upward, so a hair is formed by a group of cells at its base multiplying and pushing upward and in so doing becoming keratinized. The part of the hair that is visible is the *shaft*, whereas that which is embedded in the dermis is the *root*. The root, together with its coverings (an outer connective tissue sheath and an inner epithelial coating which is a continuation of the stratum germinativum), forms the hair *follicle*. At the bottom of the follicle is a loop of capillaries enclosed in a connective tissue covering called the hair *papilla*. The cluster of epithelial cells lying over the papilla are the ones that reproduce and eventually form the hair shaft. As long as these cells remain alive, hair will regenerate even though it be cut or plucked or otherwise removed.

Each hair is kept soft and pliable by two or more *sebaceous glands* which secrete varying amounts of oily *sebum* into the follicle near the surface of the skin. Attached to the follicle, too, are small bundles of involuntary muscle known as the *arrector pili muscles*. These muscles are of interest because when they contract, the hair "stands on end," as it does in extreme fright or cold, for example. This mechanism is responsible also for gooseflesh. As the hair is pulled into an upright position, it raises the skin around it into the familiar little goose pimples.

Hair color is due to different amounts of melanin pigments in the outer layer (cortex) of the hair. White hair contains little or no melanin.

Some hair, notably that around the eyes and in the nose and ears, performs a protective function in that it keeps out some dust and insects. For the hair on the bulk of the skin, however, no function seems apparent.

Nails. The nails are epidermal cells that have been converted to keratin. They grow from epithelial cells lying under the white crescent (lunula) at the proximal end of each nail.

Skin glands. The skin glands include three kinds of microscopic glands—namely, sebaceous, sweat, and ceruminous.

Sebaceous glands secrete oil for the hair. Wherever hairs grow from the skin, there are sebaceous glands, at least two for each hair. The oil, or *sebum*, secreted by these tiny glands has value not only because it keeps the hair supple, but also because it keeps the skin soft and pliant. Moreover, it prevents excessive water evaporation from the skin and water absorption through the skin. And because fat is a poor conductor of heat, the sebum secreted onto the skin lessens the amount of heat lost from this large surface.

Sweat glands, though very small structures, are very important and very numerous—especially on the palms, soles, forehead, and axillae (armpits). Histologists estimate, for example, that a single square inch of skin on the palms of the hands contains about three thousand sweat glands.

Sweat secretion helps maintain homeostasis of fluid and electrolytes and of body temperature. For example, if too much heat is being produced, as in strenuous exercise, or if the environmental temperature is high, these glands secrete more sweat which, in evaporating, cools the body surface. Inasmuch as sweat contains some nitrogenous wastes, the sweat glands also function as excretory organs.

Ceruminous glands are thought to be modified sweat glands. They are located in the external ear canal. Instead of watery sweat, they secrete a waxy, pigmented substance, the *cerumen*.

Terms used in connection with skin

hypodermic beneath or under the skin.
subcutaneous same as hypodermic.
intracutaneous within the layers of the skin.

Fig. 4-4 Simple tubular gland, an exocrine gland with an unbranching duct.

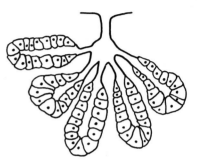

Fig. 4-5 Compound tubuloalveolar gland, an exocrine gland with a branching duct.

diaphoresis profuse perspiration.
pores minute openings of the sweat gland ducts on the surface of the skin; do not "open" or "close" since no muscle tissue enters into their formation; however, any agent which has an astringent action on the skin causes them to shrink or, in effect, to close.
furuncle a boil, an infection of a hair follicle.

Glands

Glands consist of epithelial cells specialized for synthesizing compounds which they secrete either into ducts or blood. *Exocrine glands* secrete into ducts and *endocrine glands* into blood capillaries. Exocrine glands are further classified in several ways. Sweat glands, for example, are *simple tubular exocrine glands* (Fig.

4-4). Interpreted, simple means that each gland has a single nonbranching duct, and tubular means that its secretory unit is tubular shaped. A salivary gland, in contrast, is a *compound tubuloalveolar gland* (Fig. 4-5) because it has a branching duct and some tubular and some flask-shaped secretory units.

outline summary

MEMBRANES

1 Definition—thin sheet of tissues that either covers or lines a part of body or divides an organ
2 Types—mucous, serous, synovial, cutaneous, and miscellaneous

Mucous membrane

1 Location—lines cavities and passages that open to exterior
2 Structure—surface layer of epithelial tissue over connective tissue
3 Functions—protection, secretion, and absorption

Serous membrane

1 Location—lines cavities that do not open to exterior
2 Functions—protection and secretion

Synovial membrane

1 Location—lines joint cavities, tendon sheaths, and bursae
2 Functions—protection and secretion

Cutaneous membrane

1 Functions
   a Protection against various factors—e.g., microorganisms, sunlight, and chemicals
   b Excretion of sweat
   c Sensations
   d Fluid and electrolyte balance—skin helps maintain this by secreting varying amounts of sweat
   e Normal body temperature—skin helps maintain this by varying amounts of blood flow through it and also by varying amounts of sweat secretion
2 Structure—two main layers: epidermis, outer, thinner layer of stratified squamous epithelium, and dermis, inner, thicker layer of connective tissue

Epidermis

1 Outer layer of stratified squamous epithelial cells
2 Surface cells dead, keratinized, and practically waterproof
3 Only deepest layer of cells undergoes mitosis to replace surface cells that continually desquamate
4 Melanin pigments mainly in deepest layer

Dermis

1 Dense fibrous connective tissue layer underlying epidermis
2 Dermis of palms and soles has numerous parallel ridges
3 Subcutaneous tissue (also called superficial fascia), underlying dermis, composed of areolar tissue or areolar and adipose tissues

Accessory organs of skin

1 Hair
   a Distribution—over entire body except palms and soles
   b Shaft—visible part of hair
   c Root—part of hair embedded in dermis
   d Follicle—root with coverings
   e Papilla—loop of capillaries enclosed in connective tissue covering
   f Germinal matrix—cluster of epithelial cells lying over papilla; these cells undergo mitosis to form hair; must be intact in order for hair to regenerate
   g Sebaceous glands and arrector pili muscles—attach to follicle; contraction of latter produces gooseflesh
   h Color—due to different amounts of melanin pigments in cortex of hair
2 Nails
   a Are epidermal cells converted to hard keratin
   b Grow from epithelial cells under lunula ("moons")
3 Skin glands
   a Sebaceous—secrete oil (sebum) that keeps hair and skin soft; helps prevent excess evaporation and absorption of water and excess heat loss

b Sweat—numerous throughout skin, especially on palms, soles, forehead, and axillae; important in heat regulation

c Ceruminous—thought to be modified sweat glands; located in external ear canal; secrete ear wax or cerumen

Terms used in connection with skin
See p. 58

GLANDS

1 Composed of epithelial cells specialized for synthesizing compounds which they secrete either into ducts or blood capillaries
2 Types
   a Exocrine glands—secrete into ducts
     1 Simple glands—have nonbranching ducts
     2 Tubular glands—secretory unit tubular-shaped
     3 Compound glands—have branching ducts
     4 Alveolar glands—secretory units flask-shaped
   b Endocrine glands—secrete into blood

review questions

1 What membranes line closed cavities? Cavities that open to the exterior?
2 What general functions do membranes serve?
3 What functions does the skin perform?
4 Describe the epidermis.
5 Describe the dermis.
6 What is keratin?
7 What is melanin and where is it found?
8 What is superficial fascia?
9 Name and describe the skin glands.
10 Define the following terms: alveolar gland; exocrine gland; endocrine gland; compound gland.
11 What organelle(s) would you expect to be prominent in secreting cells?

HUMAN ANATOMY

FULL-COLOR PLATES WITH SIX IN TRANSPARENT

"TRANS-VISION"® SHOWING STRUCTURES OF THE HUMAN TORSO

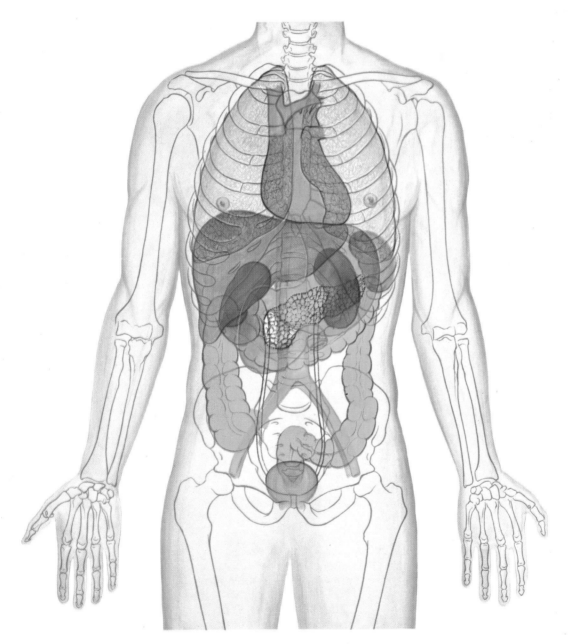

Plate I

ERNEST W. BECK, medical illustrator

in collaboration with
HARRY MONSEN , Ph.D.

Professor of Anatomy, College of Medicine, University of Illinois

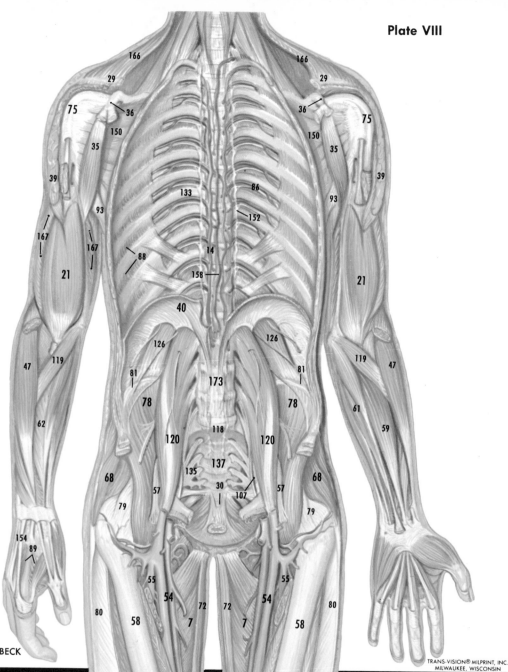

Plate VIII

ERNEST W. BECK

TRANS-VISION® MILPRINT, INC.
MILWAUKEE, WISCONSIN

ANTERIOR VIEW

7. Adductor magnus muscle
14. Azygos veins
21. Brachialis muscle
29. Clavicle
30. Coccyx
35. Coracobrachialis muscle
36. Coracoid process of the scapula
39. Deltoid muscle
40. Diaphragm
47. Extensor carpi radialis longus muscle
54. Femoral artery and vein

55. Femoral artery, deep
57. Femoral nerve
58. Femur
59. Flexor carpi radialis muscle
61. Flexor digitorum profundus muscle
62. Flexor digitorum superficialis muscle
68. Gluteus medius muscle
75. Humerus
78. Iliacus muscle
79. Iliofemoral ligament
80. Iliotibial tract
81. Ilium

86. Intercostal artery, vein and nerve
88. Intercostal muscle, internal
89. Interosseous muscles, dorsal
93. Latissimus dorsi muscle
107. Obturator nerve
118. Promontory
119. Pronator teres muscle
120. Psoas muscles (major and minor)
126. Quadratus lumborum muscle

133. Rib
135. Sacral nerves
137. Sacrum
150. Subscapularis muscle
152. Sympathetic (autonomic) nerve chain
154. Tendons of extensor muscles of hand
158. Thoracic duct
166. Trapezius muscle
167. Triceps brachii muscle
173. Vertebral column

Plate IX

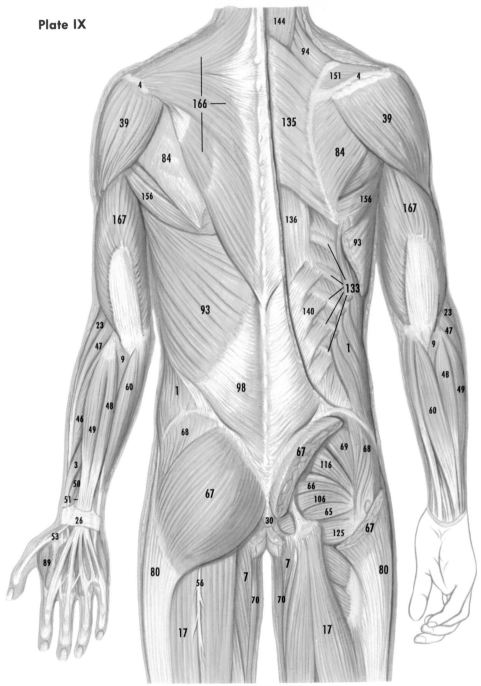

POSTERIOR VIEW

1. Abdominal oblique muscle, external
3. Abductor pollicis longus muscle
4. Acromion process of the scapula
7. Adductor magnus muscle
9. Anconeus muscle
17. Biceps femoris muscle
23. Brachioradialis muscle
26. Carpal ligament, dorsal
30. Coccyx
39. Deltoid muscle
46. Extensor carpi radialis brevis muscle

47. Extensor carpi radialis longus muscle
48. Extensor carpi ulnaris muscle
49. Extensor digitorum communis muscle
50. Extensor pollicis brevis muscle
51. Extensor pollicis longus muscle
56. Femoral cutaneous nerve, posterior
60. Flexor carpi ulnaris muscle
65. Gemellus inferior muscle

66. Gemellus superior muscle
67. Gluteus maximus muscle
68. Gluteus medius muscle
69. Gluteus minimus muscle
70. Gracilis muscle
80. Iliotibial tract
84. Infraspinatus muscle
89. Interosseous muscle, dorsal
93. Latissimus dorsi muscle
94. Levator scapulae muscle
98. Lumbodorsal fascia

106. Obturator internus muscle
116. Piriformis muscle
125. Quadratus femoris muscle
133. Ribs (VII-XII)
135. Rhomboideus muscle
136. Erector spinae muscle
140. Serratus posterior inferior muscle
144. Splenius capitis muscle
151. Supraspinatus muscle
156. Teres major muscle
166. Trapezius muscle
167. Triceps brachii muscle

Plate X

BONES AND SINUSES OF THE SKULL

Frontal bone

Parietal bone

Glabella

Supraorbital foramen

Temporal bone

Ethmoid bone

Sphenoid bone

Lacrimal bone

Zygomatic arch

Middle nasal concha

Infraorbital foramen

Inferior nasal concha

Maxilla

Mandible

Frontal sinus

Ethmoid sinuses

Sphenoid sinus
(behind ethmoid cells)

Maxillary sinus

Perpendicular plate
of the ethmoid bone

Mental foramen

HEMISECTION OF THE HEAD AND NECK

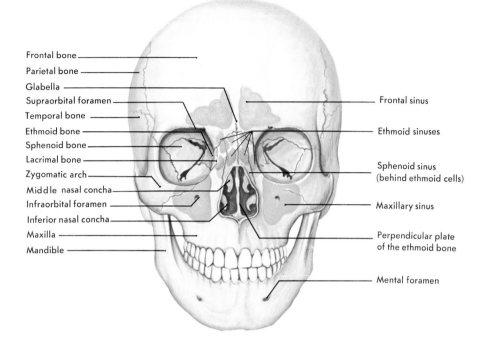

Rectus sinus

Frontal bone

Cerebrum

Corpus callosum

Frontal sinus

Pons

Pituitary gland

Sphenoid sinus

Cerebellum

Turbinates (nasal conchae)

Medulla oblongata

Maxilla

Tongue

Uvula

Tonsil (palatine)

Genioglossus muscle

Mandible

Epiglottis

Body of vertebra

Spinous process of cervical vertebra

Larynx (voice box)

Spinal cord

Plate XI

ANATOMY OF THE EAR

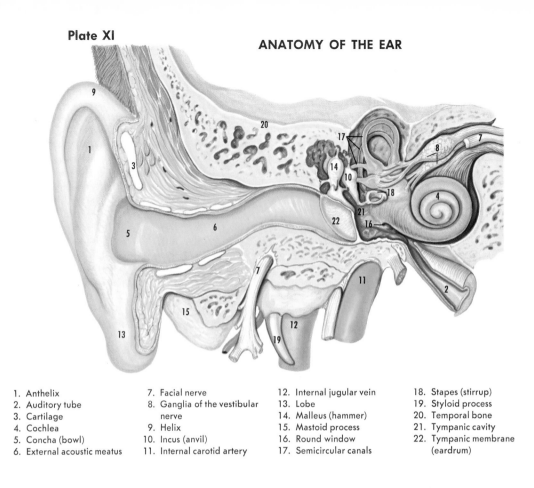

1. Anthelix	7. Facial nerve	12. Internal jugular vein	18. Stapes (stirrup)
2. Auditory tube	8. Ganglia of the vestibular	13. Lobe	19. Styloid process
3. Cartilage	nerve	14. Malleus (hammer)	20. Temporal bone
4. Cochlea	9. Helix	15. Mastoid process	21. Tympanic cavity
5. Concha (bowl)	10. Incus (anvil)	16. Round window	22. Tympanic membrane
6. External acoustic meatus	11. Internal carotid artery	17. Semicircular canals	(eardrum)

ANATOMY OF THE EYE

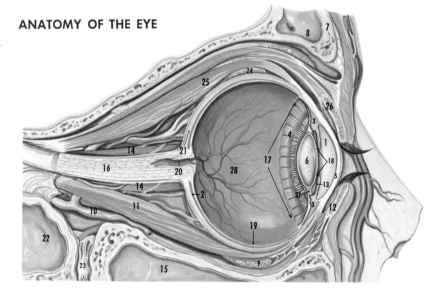

1. Aqueous chamber	8. Frontal sinus	15. Maxillary sinus	22. Sphenoid sinus
2. Choroid	9. Inferior oblique muscle	16. Optic nerve	23. Pterygopalatine ganglion
3. Ciliary muscle	10. Inferior ophthalmic vein	17. Ora serrata	24. Superior oblique muscle
4. Ciliary processes	11. Inferior rectus muscle	18. Pupil of the iris	25. Superior rectus muscle
5. Cornea	12. Inferior tarsus	19. Retina	26. Superior tarsus
6. Crystalline lens	13. Iris	20. Retinal artery and vein	27. Suspensory ligament
7. Frontal bone	14. Lateral rectus muscle	21. Sclera	28. Vitreous chamber

Plate XII

SCHEMATIC BODY CELL

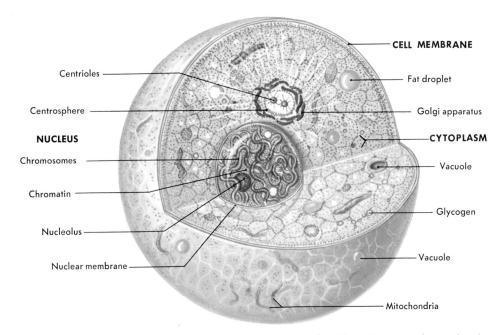

CELL MEMBRANE

Centrioles

Centrosphere

NUCLEUS

Chromosomes

Chromatin

Nucleolus

Nuclear membrane

Fat droplet

Golgi apparatus

CYTOPLASM

Vacuole

Glycogen

Vacuole

Mitochondria

Every living cell, regardless of its shape or size, has three main parts: the cell membrane, cytoplasm, and nucleus. Together they constitute protoplasm. Billions of such cells as shown above make up the tissues of our bodies.

TYPES OF CELLS

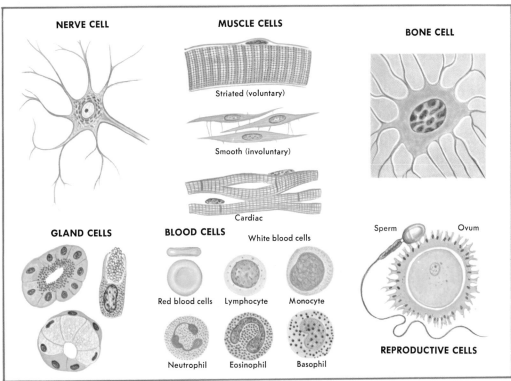

NERVE CELL

MUSCLE CELLS

Striated (voluntary)

Smooth (involuntary)

Cardiac

BONE CELL

GLAND CELLS

BLOOD CELLS

White blood cells

Red blood cells   Lymphocyte   Monocyte

Neutrophil   Eosinophil   Basophil

Sperm   Ovum

REPRODUCTIVE CELLS

Plate XIII

SKELETON

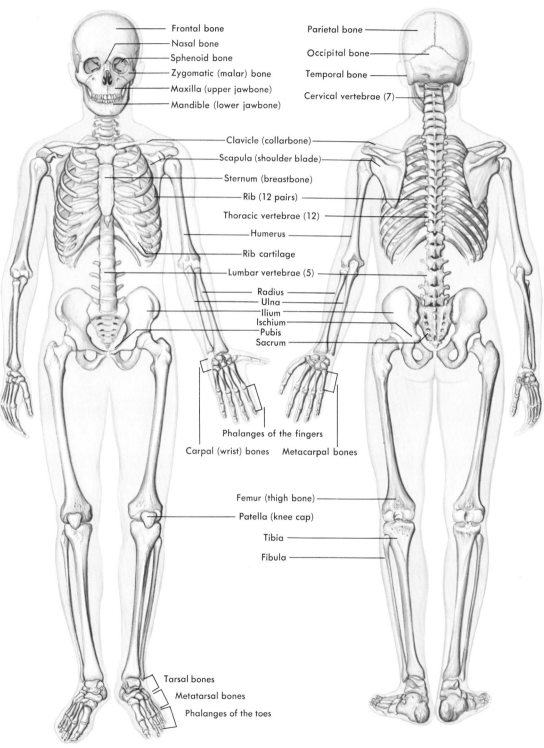

Frontal bone
Nasal bone
Sphenoid bone
Zygomatic (malar) bone
Maxilla (upper jawbone)
Mandible (lower jawbone)

Parietal bone
Occipital bone
Temporal bone
Cervical vertebrae (7)

Clavicle (collarbone)
Scapula (shoulder blade)
Sternum (breastbone)
Rib (12 pairs)
Thoracic vertebrae (12)
Humerus
Rib cartilage
Lumbar vertebrae (5)
Radius
Ulna
Ilium
Ischium
Pubis
Sacrum

Phalanges of the fingers
Carpal (wrist) bones
Metacarpal bones

Femur (thigh bone)
Patella (knee cap)
Tibia
Fibula

Tarsal bones
Metatarsal bones
Phalanges of the toes

Plate XIV

FEMALE PELVIC ORGANS

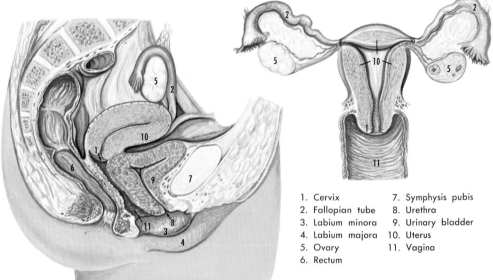

1. Cervix
2. Fallopian tube
3. Labium minora
4. Labium majora
5. Ovary
6. Rectum
7. Symphysis pubis
8. Urethra
9. Urinary bladder
10. Uterus
11. Vagina

MALE PELVIC ORGANS

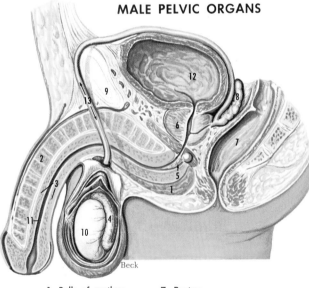

Beck

SECTION THROUGH PENIS

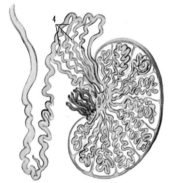

SCHEME OF DUCT ARRANGEMENT
IN THE TESTIS AND EPIDIDYMIS

1. Bulb of urethra
2. Corpus cavernosum
3. Corpus spongiosum
4. Epididymis
5. Duct of bulbourethral gland
6. Prostate gland
7. Rectum
8. Seminal vesicle
9. Symphysis pubis
10. Testis
11. Urethra
12. Urinary bladder
13. Vas deferens

Plate XV

LYMPHATIC ORGANS

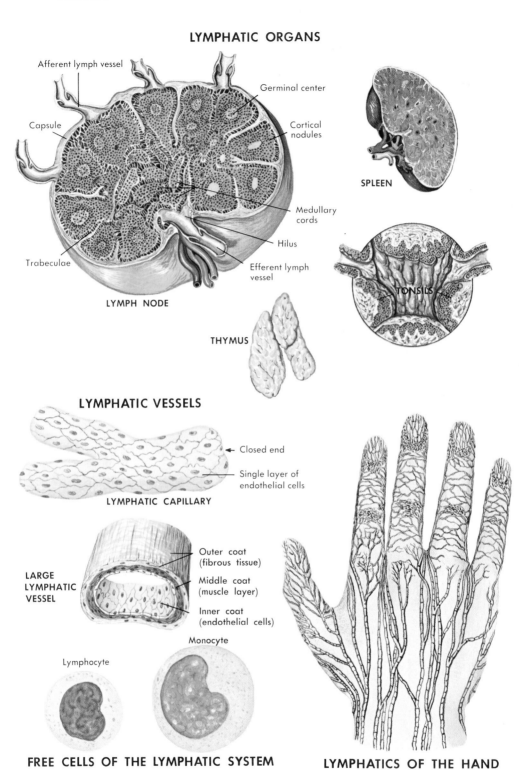

Afferent lymph vessel

Germinal center

Capsule

Cortical nodules

Medullary cords

Hilus

Trabeculae

Efferent lymph vessel

LYMPH NODE

SPLEEN

TONSILS

THYMUS

LYMPHATIC VESSELS

← Closed end

Single layer of endothelial cells

LYMPHATIC CAPILLARY

LARGE LYMPHATIC VESSEL

Outer coat (fibrous tissue)

Middle coat (muscle layer)

Inner coat (endothelial cells)

Monocyte

Lymphocyte

FREE CELLS OF THE LYMPHATIC SYSTEM

LYMPHATICS OF THE HAND

The erect and moving body

5

The skeletal system

Meaning

Functions

Bone and cartilage
Microscopic structure
  Bone
  Cartilage

Bones
Types
Structure
Formation and growth
Names and numbers
  Axial skeleton
    Skull
    Hyoid bone
    Vertebral column
    Sternum
    Ribs
  Appendicular skeleton
    Upper extremity
    Lower extremity
Markings
  Depressions and
    openings
  Projections or
    processes
  Identification
Differences between male
  and female skeletons
Age changes in skeleton

**Joints between bones
  (articulations)**
Kinds
Movements

Like all body structures, the skeletal and muscular systems play a part in the body's achievement of its overall goal of survival. These two systems work together to move the body and its parts. This is a function of tremendous importance not only for the enjoyment of life, but also for life itself, since without movement a favorable cellular environment cannot possibly be maintained.

The body must adjust to many changes in its external environment in order to maintain homeostasis. Sometimes it adjusts the external environment. Sometimes it adjusts itself. In either case, movements play a part. Suppose, for instance, that environmental temperature drops below the comfort zone. The body then needs to make some kind of adjustment in order to maintain homeostasis of its internal temperature. It may change the environmental temperature back to a comfortable level (by building a fire or setting up the room thermostat, for example), or it may change itself in some way to counteract the environmental change (for instance, shivering to produce more heat and surface blood vessel constriction to lose less heat).

chemical storage (calcium)
Hemopoiesis - manufacture of blood cells.

Whichever methods the body uses, movements are necessary. Movements require the coordinated activities of nearly all of the body's systems. In order to gain an understanding of how movements are accomplished, we shall start by investigating the two systems whose primary business is movement: the skeletal and muscular systems.

Meaning

The term skeletal system means all the bones of the body plus the joints formed by their attachments to each other. Predominant tissues of the system are two types of connective tissue: bone and cartilage.

Functions

The skeletal system serves the body by performing these functions: support, protection, and movement.

Support. Bones support the body much as steel girders support our modern buildings.

Protection. Hard, bony "boxes" protect delicate structures enclosed by them. For example, the skull protects the brain and the rib cage protects the lungs and heart.

Movement. Bones with their joints constitute levers. Muscles are anchored firmly to bones. As muscles contract, force is applied to the bony levers and movement necessarily results.

Bone and cartilage

Microscopic structure

BONE

Bone, like other tissues, consists of living cells and nonliving intercellular substance. And in bone, like other connective tissues, intercellular substance predominates over cells. But in bone the intercellular substance (matrix) is calcified. Calcium salts impregnate the cement substance of the matrix, a fact that explains the rigidity of bones and the familiar expression "as hard as bone." Embedded in the calcified matrix are many fibers of the body's most abundant protein, collagen.

Another unique feature of bone structure is the arrangement of its intercellular substance. Concentric cylindrical layers of calcified matrix (usually fewer than six of them) enclose a central longitudinal canal that contains a blood vessel. Each layer of bone matrix is called a *lamella*, the central canal is an *haversian canal*, and the entire

63

unit of canal and surrounding lamellae is an *haversian system.* According to Ham and Leeson,* most haversian canals contain a single large capillary, but some have a small arteriole and venule, and lymphatics. Bone cells (*osteocytes*) occupy minute spaces called *lacunae* between the lamellae. Microscopic canals (canaliculi), great numbers of them, radiate in all directions from the lacunae to connect them with haversian canals and provide routes for tissue fluid to reach bone cells. Each bone cell is said to lie not farther than 1/10 of a millimeter from an haversian canal (Fig. 5-1).

There are two types of bone based on the arrangement of lamellae—compact or dense and cancellous or spongy. In *compact bone*, adjacent haversian units fit closely together with the spaces between them filled in with interstitial lamellae. In *cancellous bone*, on the other hand, there are many open spaces between thin processes of bone (*trabeculae*) which are joined together somewhat like the beams of wood in a scaffold. Arrangement of trabeculae in different ways in different bones gives structural strength along the lines of strain on individual bones.

Bones are not the lifeless structures they seem to be. We tend to think of them as lifeless, perhaps because what we see when we look at a bone is its nonliving intercellular substance. But within this hard, lifeless material lie many living bone cells that must continually receive food and oxygen and that must be rid of their wastes. So blood supply to bone is important and abundant. For example, numerous blood vessels from the periosteum (bone covering) penetrate bone by way of *Volkmann's canals* to connect with blood vessels of an haversian canal. Also, one or more arteries supply the bone marrow in the internal medullary cavity of long bones.

CARTILAGE

Cartilage both resembles and differs from bone. Like bone, cartilage consists more of intercellular substance than of cells. Innumerable collagenous fibers reinforce the matrix of both tissues. But in cartilage the fibers are embedded in a firm gel instead of in a calcified cement substance as they are in bone. Hence, cartilage has the flexibility of a firm plastic material rather than the rigidity of bone. Another difference is this—no canal system and no blood vessels penetrate cartilage matrix. Cartilage is avascular and bone is abundantly vascular. Cartilage cells, like bone cells, lie in lacunae. However, because no canals and blood vessels interlace cartilage matrix, nutrients and oxygen can reach the scattered, isolated chondrocytes (cartilage cells) only by diffusion through the matrix gel, from capillaries in the fibrous covering of cartilage (perichondrium) or from synovial fluid in the case of articular cartilage.

Three types of cartilage are hyaline, fibrous, and elastic. They differ structurally mainly as to matrix fibers. Collagenous fibers are present in all three types but are most numerous in fibrocartilage. Hence it has the greatest tensile strength. Elastic cartilage matrix contains elastic fibers as well as collagenous fibers so has elasticity as well as firmness. Hyaline is the commonest type of cartilage. It resembles milk glass in appearance. In fact, its name is derived from the Greek word meaning glassy. A thin layer of hyaline cartilage covers articular surfaces of bones, where it helps to cushion jolts. Fibrocartilage discs between the vertebrae also serve this purpose. For other locations of cartilage, see Table 3-1, p. 45.

---

*Ham, A. W., and Leeson, T. S.: Histology, ed. 5, Philadelphia, 1965, J. B. Lippincott Co.

chondro cytes — cartilage cells

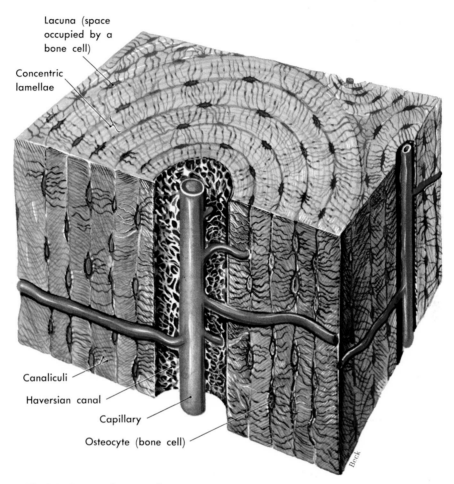

Lacuna (space occupied by a bone cell)

Concentric lamellae

Canaliculi

Haversian canal

Capillary

Osteocyte (bone cell)

Fig. 5-1 Section of compact bone showing details of a haversian system.

dermal—
cartilaginous—cartilage form first

■Bones

Types

There are four types of bones which are classified according to their shapes as follows:

1 *long bones*—femur, tibia, fibula, humerus, radius, ulna, and phalanges

2 *short bones*—carpals and tarsals (wrist and ankle bones)

3 *flat bones*—several cranial bones, such as frontal and parietal; also ribs and scapulae

4 *irregular bones*—vertebrae, sphenoid, ethmoid, sacrum, coccyx, and mandible

65

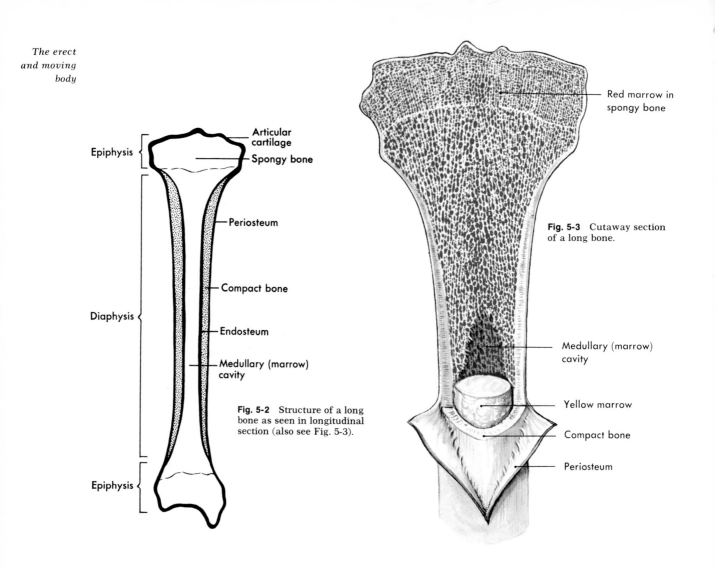

Fig. 5-2 Structure of a long bone as seen in longitudinal section (also see Fig. 5-3).

Fig. 5-3 Cutaway section of a long bone.

Labels for Fig. 5-2: Epiphysis; Articular cartilage; Spongy bone; Periosteum; Diaphysis; Compact bone; Endosteum; Medullary (marrow) cavity; Epiphysis

Labels for Fig. 5-3: Red marrow in spongy bone; Medullary (marrow) cavity; Yellow marrow; Compact bone; Periosteum

Structure

Each long bone of the body consists of the following parts: (1) diaphysis, (2) epiphyses, (3) articular cartilage, (4) periosteum, (5) medullary (or marrow) cavity, and (6) endosteum (Figs. 5-2 and 5-3).

The *diaphysis* is the main shaftlike portion of a long bone. Several structural features accommodate it to its function of providing strong support without cumbersome weight—the thick compact bone used as construction material, for example, and the hollow cylindrical shape which offers the dual advantages of greater strength with less weight compared with a solid cylinder of the same size.

The *epiphyses* are the extremities of long bones. Their somewhat bulbous shape provides generous space for muscle attachment near joints and makes for greater stability of the joint. Lightness despite size is achieved by construction of porous cancellous bone with only an outer layer of dense compact bone. Arrangement of the lamellae corresponding to the lines of stress gives added strength to the epiphyses. Marrow fills the cancellous spaces— red marrow in the proximal epiphyses of the humerus and femur and yellow marrow in other epiphyses in the adult.

The *articular cartilage* is the thin layer of hyaline cartilage covering the articular surface of each epiphysis. Resiliency of this material cushions jars and blows.

The *periosteum* is a dense white fibrous membrane that covers bone except at joint surfaces, where articular cartilage forms the covering. Many of the fibers of the periosteum penetrate the underlying bone to weld these two structures to each other. (The penetrating fibers are called Sharpey's fibers.) Muscle tendon fibers interlace with periosteal fibers to anchor muscles firmly to bone.

The inner layer of the periosteum of growing bones contains osteoblasts (bone-forming cells). Because of its bone-forming cells and blood vessels, the periosteum is necessary for bone growth and repair and for its nutrition and, therefore, for survival of its cells. In addition, it serves as the means for attaching muscle tendons and ligaments to the bone.

The *medullary* (or *marrow*) *cavity*, running the length of the diaphysis, contains yellow or fatty bone marrow in the adult.

The *endosteum* is the membrane that lines the medullary cavity and haversian canals. It is composed of cells that become active osteoblasts as needed.

Short bones consist of a core of cancellous bone encased in a thin layer of compact bone.

Flat bones consist of a layer of cancellous bone between two plates of compact bone. Cancellous bone of the skull bones (diploe), ribs, and sternum contains red marrow.*

*Note that red marrow is found in the adult in only a few locations, mainly in the diploe (spongy bone of the cranial bones), ribs, and sternum, in bodies of the vertebrae, and small amounts in the proximal epiphyses of the femurs and humeri. In newborn infants and children, red marrow occurs in many more bones. Because red bone marrow forms blood cells (process called hemopoiesis), it is one of the most important tissues of the body. Its very location, hidden within the bones, suggests its great importance. As the agent which coins millions of vital blood cells daily, it receives maximum protection from the body, the bones acting as a safe-deposit vault for it.

Irregular bones are similar in structure to short bones—that is, a thin layer of compact bone forms a casing over cancellous bone.

Formation and growth

The embryo skeleton, when first formed, consists of "bones" that are not really bones at all but hyaline cartilage or fibrous membrane structures shaped like bones. Gradually the process of *ossification* (or *osteogenesis*) replaces these prebone structures with bone. Ossification of any structure begins with the appearance in it of special bone-forming cells called *osteoblasts*. The Golgi apparatus in an osteoblast specializes in synthesizing and secreting carbohydrate compounds of the type named mucopolysaccharides and its endoplasmic reticulum makes and secretes collagen, a protein. In time, relatively large amounts of the mucopolysaccharide substance (or cement substance, as it is more commonly called) accumulates around each individual osteoblast and numerous bundles of collagenous fibers become embedded in it. Together, they constitute the organic intercellular substance of bone called the *bone matrix*. If you like nicknames, you might call bone matrix the reinforced concrete of the body. Bundles of collagenous fibers in bone increase its strength much as iron rods in reinforced concrete strengthen it.

About as fast as the organic bone matrix forms, inorganic compounds—complex calcium salts, not yet positively identified—begin depositing in it. This calcification of the matrix is the process that makes bone "hard as bone." To summarize briefly, ossification is a complex, still not completely understood mechanism consisting of two processes—synthesis of the organic bone matrix by osteoblasts, immediately followed by calcification of the matrix.

In long bones, endochondral ossification

starts in the diaphysis and in both epiphyses and proceeds toward each other. A layer of cartilage known as *epiphyseal cartilage* remains between the diaphyseal and epiphyseal centers of ossification. As long as bone growth continues, proliferation of epiphyseal cartilage cells brings about a thickening of the layer of cartilage from time to time. Ossification of this additional cartilage then follows—that is, osteoblasts synthesize organic bone matrix and the matrix undergoes calcification. As a result, the bone becomes longer. When epiphyseal cartilage cells stop multiplying and the cartilage has become completely ossified, bone growth has ended. This is the scientific fact that underlies the clinical practice of x-raying a child's wrist to determine whether he "will grow anymore." If the x-ray film reveals a layer of epiphyseal cartilage, the answer is "yes"; if not, it is "no." He will have attained his full height.

Bones grow in diameter by the combined action of two special kinds of cells: osteoblasts and osteoclasts. Osteoclasts enlarge the diameter of the medullary cavity by eating away the bone of its walls. At the same time, osteoblasts from the periosteum build new bone around the outside of the bone. By this dual process a bone with a larger diameter and larger medullary cavity has been produced from a smaller bone with a smaller medullary cavity.

Bone is not the lifeless, inert substance it appears to be but a living tissue that is continually being formed and destroyed. During childhood, bones grow in size because bone formation (ossification) goes on at a faster rate than does bone destruction. During the early and middle years of adulthood, the opposing processes balance each other and bones neither grow nor shrink (atrophy). In old age, bone often degenerates and is reabsorbed faster than

it is formed. As a result, it becomes porous and fragile, a condition known as *senile osteoporosis*. Physiologists cannot yet completely explain why this happens. They do know, however, that activity and estrogens have something to do with it. Both relate inversely to the incidence of osteoporosis. The less active individuals are, the more apt they are to develop osteoporosis. And in women, a decrease in blood estrogen concentration—for example, after the menopause—is very often accompanied by an increase in the incidence of osteoporosis. (Other hormones involved in bone calcification and decalcification—namely, parathyroid hormone and calcitonin—are discussed in Chapter 9.)

Names and numbers

The human skeleton consists of two main parts: the *axial skeleton*, composed of the bones which form the upright part or axis of the body (the skull, hyoid bone, vertebral column, sternum, and ribs), and the *appendicular skeleton*, made up of the bones which are attached to the axial skeleton as appendages (that is, the upper and lower extremities). The names and numbers of the bones in each division of the skeleton, with an identifying remark about each, are given in Table 5-1. When you are trying to learn these names, locate each bone on your own body. Feel its outline whenever possible. Locate each bone on a skeleton if one is available. Study Figs. 5-4 to 5-7.

AXIAL SKELETON

Skull. Twenty-eight irregularly shaped bones form the skull. Eleven of these are paired bones and six are single. All but one of the skull bones are so joined to each other as to be immovable. Only the lower jawbone (mandible) is movable. The skull consists of two major divisions: the cranium or brain case and the face.

Table 5-1. Bones of skeleton

Part of body	Name of bone	Number	Description
AXIAL SKELETON (80 bones)			Bones that form upright axis of body – skull, hyoid, vertebral column, sternum, and ribs
Skull (28 bones) Cranium (8 bones)			Cranium forms floor for brain to rest on and helmetlike covering over it
	Frontal	1	Forehead bone; also forms most of roof of orbits (eye sockets) and anterior part of cranial floor
	Parietal	2	Prominent, bulging bones behind frontal bone; form topsides of cranial cavity
	Temporal	2	Form lower sides of cranium and part of cranial floor; contain middle and inner ear structures
	Occipital	1	Forms posterior part of cranial floor and walls
	Sphenoid	1	Keystone of cranial floor; forms its midportion; resembles bat with wings outstretched and legs extended downward posteriorly; lies behind and slightly above nose and throat; forms part of floor and side walls of orbit
	Ethmoid	1	Complicated irregular bone that helps make up anterior portion of cranial floor, medial wall of orbits, upper parts of nasal septum, and side walls and part of nasal roof; lies anterior to sphenoid and posterior to nasal bones
Face (14 bones)	Nasal	2	Small bones forming upper part of bridge of nose
	Maxillary	2	Upper jaw bones; form part of floor of orbit, anterior part of roof of mouth, and floor of nose and part of sidewalls of nose
	Zygomatic (malar)	2	Cheekbones; form part of floor and sidewall of orbit
	Mandible	1	Lower jawbone; largest, strongest bone of face
	Lacrimal	2	Thin bones about size and shape of fingernail; posterior and lateral to nasal bones in medial wall of orbit; help form sidewall of nasal cavity; often missing in dry skull
	Palatine	2	Form posterior part of hard palate, floor, and part of sidewalls of nasal cavity and floor of orbit
	Inferior conchae (turbinates)	2	Thin scroll of bone forming kind of shell along inner surface of sidewall of nasal cavity; lies above roof of mouth
	Vomer	1	Forms lower and posterior part of nasal septum; shaped like ploughshare

Continued

Table 5-1. Bones of skeleton—cont'd

Part of body	Name of bone	Number	Description
AXIAL SKELETON —cont'd			
Skull—cont'd Ear bones (6 bones)	Malleus (hammer) Incus (anvil) Stapes (stirrup)	2 2 2	Tiny bones referred to as auditory ossicles in middle ear cavity in temporal bones; resemble, respectively, miniature hammer, anvil, and stirrup
Hyoid bone		1	U-shaped bone in neck between mandible and upper part of larynx; claims distinction as only bone in body not forming a joint with any other bone; suspended by ligaments from styloid processes of temporal bones
Vertebral column (26 bones)			Not actually a column but a flexible segmented rod shaped like an elongated letter S; forms axis of body; head balanced above, ribs and viscera suspended in front, and lower extremities attached below; encloses spinal cord
	Cervical vertebrae	7	First or upper seven vertebrae
	Thoracic vertebrae	12	Next twelve vertebrae; twelve pairs of ribs attached to these
	Lumbar vertebrae	5	Next five vertebrae
	Sacrum	1	Five separate vertebrae until about 25 years of age; then fused to form one wedge-shaped bone
	Coccyx	1	Four or five separate vertebrae in child but fused into one in adult
Sternum and ribs (25 bones)			Sternum, ribs, and thoracic vertebrae together form bony cage known as *thorax*; ribs attach posteriorly to vertebrae, slant downward anteriorly to attach to sternum (see description of false ribs below)
	Sternum	1	Breastbone; flat dagger-shaped bone
	True ribs	7 pairs	Upper seven pairs; fasten to sternum by costal cartilages
	False ribs	5 pairs	False ribs do not attach to sternum directly; upper three pairs of false ribs attach by means of costal cartilage of seventh ribs; last two pairs do not attach to sternum at all; therefore, called *"floating"*
APPENDICULAR SKELETON (126 bones)			Bones that are appended to axial skeleton: upper and lower extremities, including shoulder and hip girdles
Upper extremities (including shoulder girdle) (64 bones)	Clavicle	2	Collar bones; shoulder girdle joined to axial skeleton by articulation of clavicles with sternum; scapula does not form joint with axial skeleton

Table 5-1. Bones of skeleton—cont'd

Part of body	Name of bone	Number	Description
APPENDICULAR SKELETON—cont'd			
Upper extremities—cont'd	Scapula	2	Shoulder blades; scapulae and clavicles together comprise shoulder girdle
	Humerus	2	Long bone of upper arm
	Radius	2	Bone of thumb side of forearm
	Ulna	2	Bone of little finger side of forearm; longer than radius
	Carpals (scaphoid, lunate, triquetrum, pisiform, trapezium, trapezoid, capitate, and hamate)	16	Arranged in two rows at proximal end of hand (Figs. 5-31 and 5-32)
	Metacarpals	10	Long bones forming framework of palm of hand
	Phalanges	28	Miniature long bones of fingers, three in each finger, two in each thumb
Lower extremities (62 bones)	Ossa coxae or pelvic bones or innominate bones	2	Large hip bones; with sacrum and coccyx, these three bones form basinlike pelvic cavity; lower extremities attached to axial skeleton by pelvic bones
	Femur	2	Thigh bone; longest strongest bone of body
	Patella	2	Kneecap; largest sesamoid bone of body*; embedded in tendon of quadriceps femoris muscle
	Tibia	2	Shin bone
	Fibula	2	Long, slender bone of lateral side of lower leg
	Tarsals (calcaneus, talus, navicular, first, second, and third cuneiforms, cuboid)	14	Bones that form heel and proximal or posterior half of foot (Fig. 5-36)
	Metatarsals	10	Long bones of feet
	Phalanges	28	Miniature long bones of toes; two in each great toe, three in other toes
Total		206*	

* An inconstant number of small, flat, round bones known as *sesamoid bones* (because of their resemblance to sesame seeds) is found in various tendons in which considerable pressure develops. Because the number of these bones varies greatly between individuals, only two of them, the patellae, have been counted among the 206 bones of the body. Generally two of them can be found in each thumb (in flexor tendon near metacarpophalangeal and interphalangeal joints) and great toe plus several others in the upper and lower extremities. *Wormian bones*, the small islets of bone frequently found in some of the cranial sutures, have not been counted in this list of 206 bones either because of their variable occurrence.

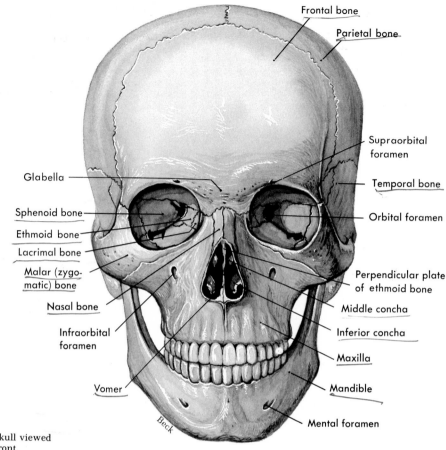

Frontal bone

Parietal bone

Supraorbital
foramen

Temporal bone

Orbital foramen

Perpendicular plate
of ethmoid bone

Middle concha

Inferior concha

Maxilla

Mandible

Mental foramen

Glabella

Sphenoid bone

Ethmoid bone

Lacrimal bone

Malar (zygo-
matic) bone

Nasal bone

Infraorbital
foramen

Vomer

Beck

Fig. 5-4 Skull viewed
from the front.

CRANIUM. The frontal, parietal, and occipital bones form the top of the cranium,
whereas the temporal bones and the great
wings of the sphenoid form its sides. These
same bones, plus the small cribriform
plate of the ethmoid bone, make up the
lower part of the cranium called the *cranial floor* or *base* (Fig. 5-8). Its midportion
is formed by the sphenoid bone which
serves as a keystone anchoring the frontal,
parietal, occipital, and ethmoid bones.

The *frontal bone* constitutes the skeletal
framework for the forehead. It contains
mucosa-lined air-filled spaces, the *frontal
sinuses*, and it forms the upper part of the

orbits. It unites with the two parietal bones
posteriorly in an immovable joint, the *coronal suture*. Several of the more prominent frontal bone markings are described
in Table 5-2, p. 94.

The two *parietal bones* give shape to the
bulging topsides of the cranium. They
form immovable joints with several bones:
the *lambdoidal suture* with the occipital
bone, the *squamous suture* with the temporal bone and part of the sphenoid, and
the *coronal suture*, mentioned before, with
the frontal bone.

The lower sides of the cranium and part
of its floor are fashioned from two *tem-*

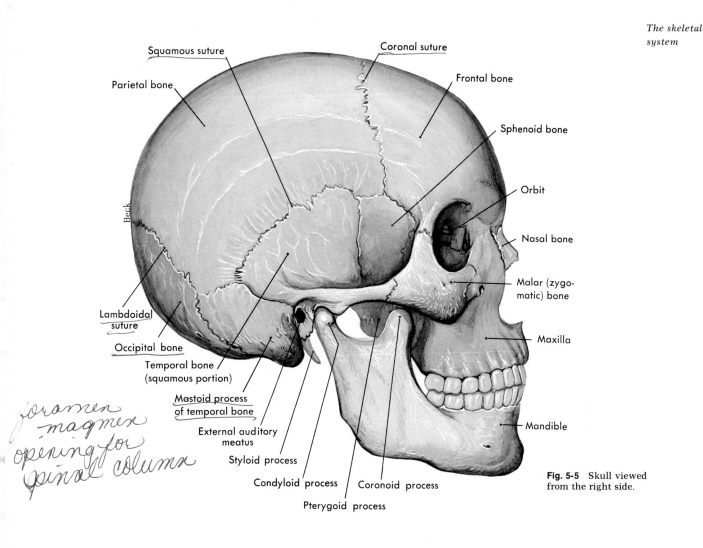

Squamous suture
Coronal suture
Parietal bone
Frontal bone
Sphenoid bone
Orbit
Beck
Nasal bone
Malar (zygo-matic) bone
Lambdoidal suture
Maxilla
Occipital bone
Temporal bone (squamous portion)
Mastoid process of temporal bone
Mandible
External auditory meatus
Styloid process
Condyloid process
Coronoid process
Pterygoid process

foramen magnum opening for spinal column

Fig. 5-5 Skull viewed from the right side.

poral bones. They house the middle and inner ear structures and contain the *mastoid air cells*, notable because of the occurrence of mastoiditis, an inflammation of the mucous lining of these spaces. A description of several other temporal bone markings is included in Table 5-2, p. 94.

The *occipital bone* makes the framework of the lower, posterior part of the skull. It forms immovable joints with three other cranial bones—the parietal, temporal, and sphenoid—and a movable joint with the first cervical vertebra. Consult Table 5-2, p. 95, for a description of some of its markings.

The *sphenoid bone* resembles a bat with its wings outstretched and legs extended downward posteriorly. It constitutes the center portion of the cranial floor and forms part of floor and sidewalls of the orbit. It contains fairly large mucosa-lined air-filled spaces, the *sphenoid sinuses*. Several prominent sphenoid markings are described in Table 5-2, p. 96 (also see Figs. 5-8 to 5-10).

The *ethmoid*, a complicated, irregular bone, lies anterior to the sphenoid but posterior to the nasal bones. It helps fashion the anterior part of the cranial floor, the medial walls of the orbits, the upper parts

7 true ribs attached to Sternum
5 other floating ribs (false)
(3 ~~not~~ indirectly connected)

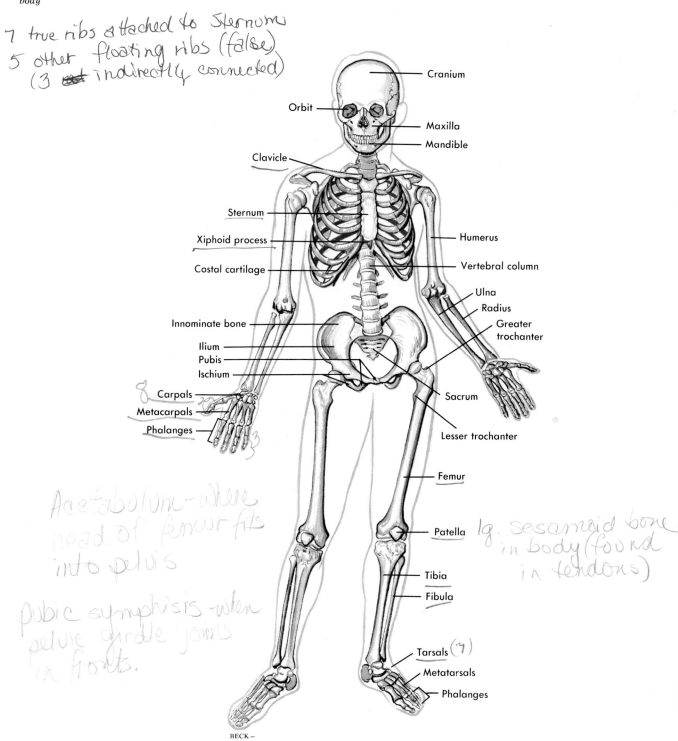

Cranium

Orbit

Maxilla

Mandible

Clavicle

Sternum

Xiphoid process

Costal cartilage

Humerus

Vertebral column

Ulna

Radius

Greater
trochanter

Innominate bone

Ilium

Pubis

Ischium

Sacrum

Carpals

Metacarpals

Phalanges

Lesser trochanter

Femur

Acetabulum—where
head of femur fits
into pelvis

Patella

lg. sesamoid bone
in body (found
in tendons)

pubic symphisis—when
pelvic girdle joins
in front.

Tibia

Fibula

Tarsals (7)

Metatarsals

Phalanges

BECK—

Fig. 5-6 Skeleton, anterior view.

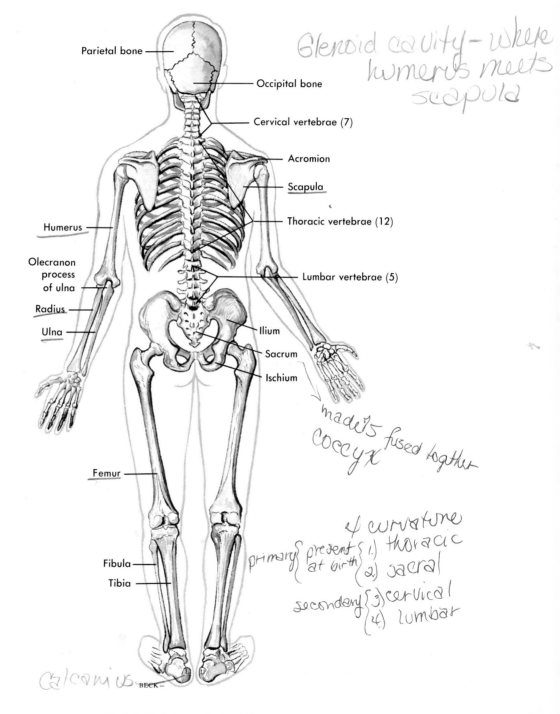

Glenoid cavity - where humerus meets scapula

Parietal bone

Occipital bone

Cervical vertebrae (7)

Acromion

Scapula

Thoracic vertebrae (12)

Humerus

Lumbar vertebrae (5)

Olecranon process of ulna

Radius

Ulna

Ilium

Sacrum

Ischium

made 5 fused together
Coccyx

Femur

4 curvature
primary { present at birth { 1.) thoracic
2.) sacral
secondary { 3.) cervical
4.) lumbar

Fibula

Tibia

Calcanius

BECK

Fig. 5-7   Skeleton, posterior view.

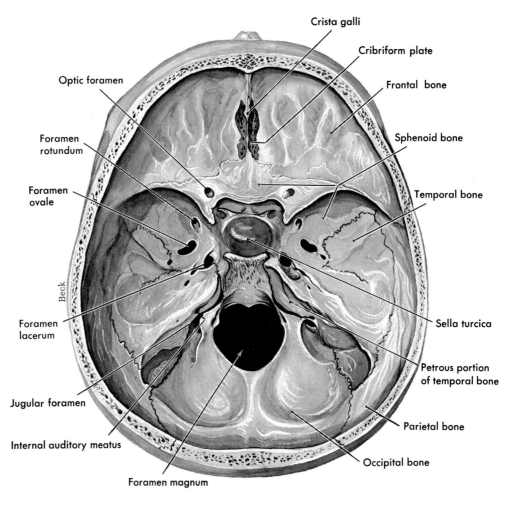

Fig. 5-8 Floor of the cranial cavity.

of the nasal septum and of the sidewalls of the nasal cavity, and the part of the nasal roof perforated by small foramina through which olfactory nerve branches reach the brain. The lateral masses of the ethmoid are honeycombed with sinus spaces. More ethmoid markings are described in Table 5-2 (also see Figs. 5-8, 5-10, and 5-14 to 5-16).

FACE BONES. Fourteen bones are commonly said to form the framework of the face, but actually more are involved since some of the cranial bones, particularly the frontal and ethmoid, also help shape the face.

Just as the sphenoid acts as the keystone in the architecture of the cranium, so the *maxillae* serve in this capacity for the face. With the exception of the mandible, all the face bones articulate with the maxillae, which also articulate with each other, in the midline. The maxillae form part of the floor of the orbits, part of the roof of the mouth, and part of the floor and sidewalls of the nose. Each maxilla contains a mucosa-lined space, the maxillary sinus or *antrum of Highmore*. This sinus is the largest of the paranasal sinuses. For other markings of the maxillae see Table 5-2.

Unlike the upper jaw, which is formed

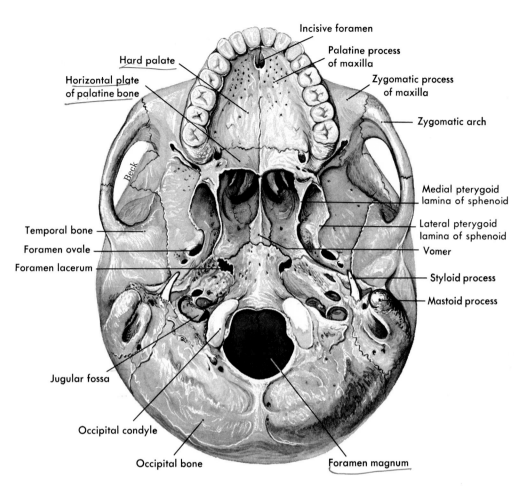

Incisive foramen

Hard palate

Horizontal plate of palatine bone

Palatine process of maxilla

Zygomatic process of maxilla

Zygomatic arch

Medial pterygoid lamina of sphenoid

Lateral pterygoid lamina of sphenoid

Vomer

Styloid process

Mastoid process

Temporal bone

Foramen ovale

Foramen lacerum

Jugular fossa

Occipital condyle

Occipital bone

Foramen magnum

Fig. 5-9 Skull viewed from below.

by a pair of bones, the lower jaw consists of a single bone, the *mandible*, due to fusion of its halves during infancy. It is the largest, strongest bone of the face. It articulates with the temporal bone in the only movable joint of the skull. Its major markings are identified in Table 5-2.

The cheek is shaped by the underlying *zygomatic* or malar bone. This bone also forms the outer margin of the orbit and, with the zygomatic process of the temporal bone, makes the zygomatic arch. It articulates with four other face bones: the maxillae and the temporal, frontal, and sphenoid bones.

Shape is given to the nose by the two *nasal bones*, which form the upper part of the bridge of the nose, and *cartilage*, the lower part. Though small bones, the nasal bones enter into five articulations: with the perpendicular plate of the ethmoid, the cartilaginous part of the nasal septum, the frontal bone, maxillae, and with each other.

An almost paper-thin bone, shaped and sized about like a fingernail, lies just posterior and lateral to each nasal bone. It helps form the sidewall of the nasal cavity and the medial wall of the orbit. Because it contains a groove for the nasolacrimal

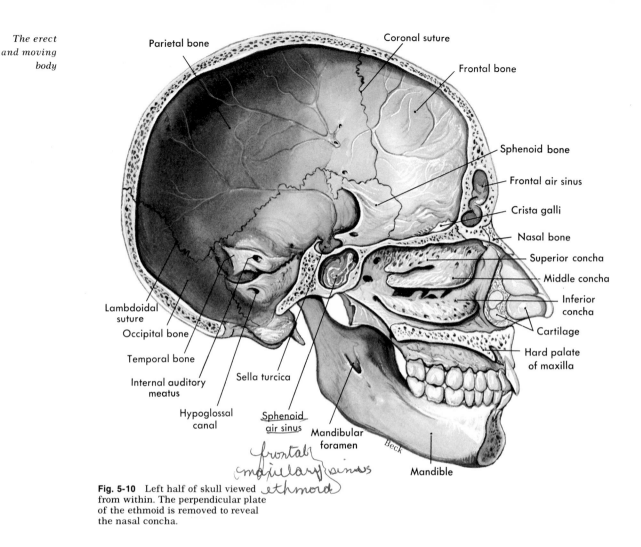

Parietal bone
Coronal suture
Frontal bone
Sphenoid bone
Frontal air sinus
Crista galli
Nasal bone
Superior concha
Middle concha
Inferior concha
Cartilage
Hard palate of maxilla
Lambdoidal suture
Occipital bone
Temporal bone
Internal auditory meatus
Hypoglossal canal
Sella turcica
Sphenoid air sinus
Mandibular foramen
Mandible
Beck

frontal maxillary sinus ethmoid

Fig. 5-10 Left half of skull viewed from within. The perpendicular plate of the ethmoid is removed to reveal the nasal concha.

duct, this bone is called the *lacrimal bone* (Fig. 5-4). It joins the maxilla, frontal bone, and ethmoid.

An irregular bone whose horizontal and vertical portions are joined in such a way as to make the bone roughly L shaped and which forms the framework for the inferior and lateral walls of the posterior part of the nasal cavity is known as the *palatine bone* because it also forms the posterior part of the hard palate. In addition, it has a small upper projection which helps form the floor of the orbit. The two palatine bones are united in the midline like two L's facing each other. They articulate also with the maxillae and the sphenoid.

The *inferior nasal concha* is a scroll-like

bone which forms a kind of ledge projecting into the nasal cavity from its lateral wall. In each nasal cavity there are three of these ledges, formed respectively by the superior and middle conchae (which are projections of the ethmoid) and the inferior concha (which is a separate bone). They are mucosa covered and divide each nasal cavity into three narrow, irregular channels, the *nasal meati*. The inferior nasal conchae form immovable joints with the ethmoid, lacrimal, maxillary, and palatine bones.

Two structures which enter into the formation of the nasal septum have already been mentioned, the perpendicular plate of the ethmoid bone and the septal cartilage.

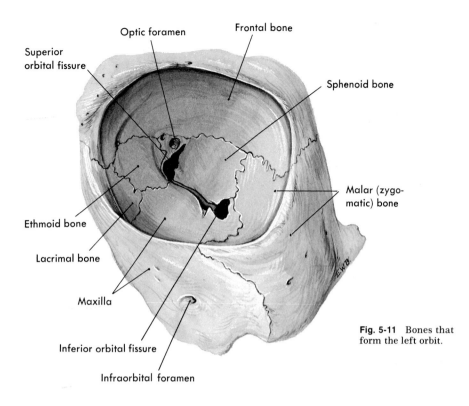

Superior orbital fissure

Optic foramen

Frontal bone

Sphenoid bone

Malar (zygomatic) bone

Ethmoid bone

Lacrimal bone

Maxilla

Inferior orbital fissure

Infraorbital foramen

Fig. 5-11 Bones that form the left orbit.

One other structure, the *vomer bone*, completes the septum posteriorly. It is usually described as being shaped like a ploughshare. It forms immovable joints with four bones: the sphenoid, ethmoid, palatine, and maxillae.

EAR BONES. See Table 5-1.

SPECIAL FEATURES. Sutures, fontanels, sinuses, orbits, nasal septum, and wormian bones are described in Table 5-2.

Hyoid bone. The hyoid bone is a single bone in the neck—a part of the axial skeleton. Its U shape may be felt just above the larynx and below the mandible where it is suspended from the styloid processes of the temporal bones. One of the pairs of extrinsic tongue muscles (hyoglossus) originates on the hyoid bone, and certain muscles of the floor of the mouth (mylohyoid, geniohyoid) insert on it (Fig. 5-17). The hyoid claims the distinction of being the only bone in the body that does not articulate with any other bone.

Vertebral column. The vertebral column constitutes the longitudinal axis of the skeleton. It is a flexible rather than a rigid column because it is segmented—that is, made up of twenty-six (typical in adult) separate bones called vertebrae, so joined to each other as to permit forward, backward, and sideways movement of the column. The head is balanced on top of this column, the ribs and viscera are suspended in front, the lower extremities are attached below, and the spinal cord is enclosed within. It is, indeed, the "backbone" of the body.

The seven *cervical vertebrae* constitute the skeletal framework of the neck (Fig. 5-18). The next twelve vertebrae are called *thoracic vertebrae* for the obvious reason that they lie behind the thoracic cavity. The next five spinal bones, the *lumbar vertebrae*, support the small of the back. Below the lumbar vertebrae lie the *sacrum* and *coccyx*. In the adult, the sacrum is a single bone which has resulted from the fusion of five separate vertebrae, and the coccyx is a single bone that has resulted from the fusion of four or five vertebrae.

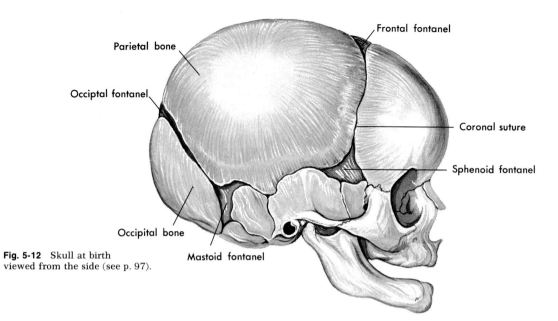

Parietal bone

Frontal fontanel

Occiptal fontanel

Coronal suture

Sphenoid fontanel

Occipital bone

Mastoid fontanel

Fig. 5-12 Skull at birth
viewed from the side (see p. 97).

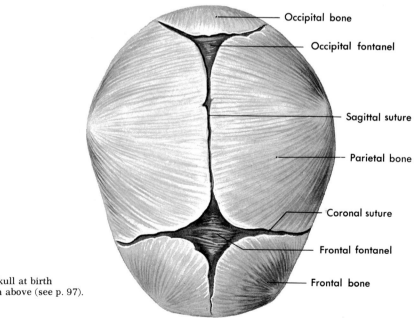

Occipital bone

Occipital fontanel

Sagittal suture

Parietal bone

Coronal suture

Frontal fontanel

Frontal bone

Fig. 5-13 Skull at birth
viewed from above (see p. 97).

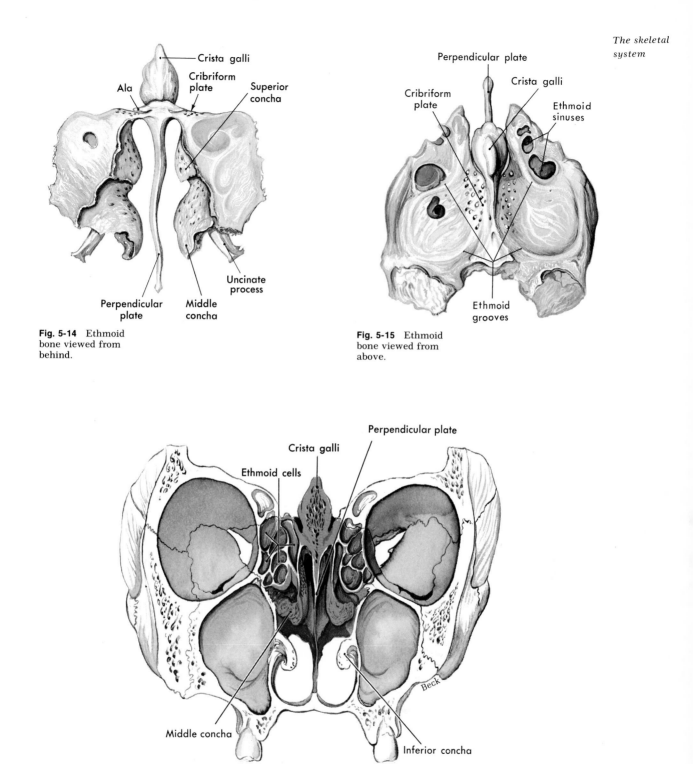

Fig. 5-14 Ethmoid bone viewed from behind.

Fig. 5-15 Ethmoid bone viewed from above.

Fig. 5-16 Skull in coronal section to reveal the ethmoid bone (shown in red) as seen from behind.

81

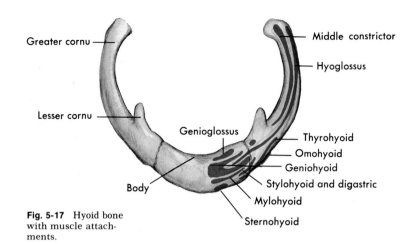

Greater cornu

Middle constrictor

Hyoglossus

Lesser cornu

Genioglossus

Thyrohyoid

Omohyoid

Geniohyoid

Stylohyoid and digastric

Mylohyoid

Sternohyoid

Body

Fig. 5-17 Hyoid bone with muscle attachments.

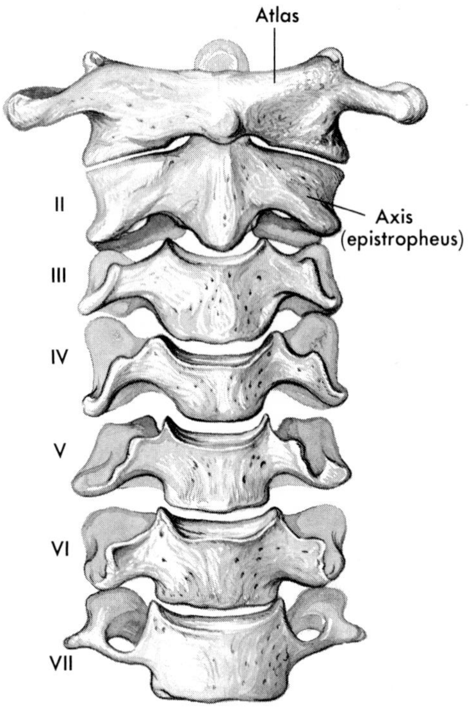

Atlas

II

Axis
(epistropheus)

III

IV

V

VI

VII

Fig. 5-18 The seven cervical vertebrae viewed from in front.

All the vertebrae resemble each other in certain features and differ in others. For example, all except the first cervical vertebra have a flat, rounded mass placed anteriorly and centrally, known as the *body*, plus a sharp or blunt *spinous process* projecting inferiorly in the posterior midline and two transverse processes projecting laterally (Figs. 5-19 to 5-23). All but the sacrum and coccyx have a central opening, the *vertebral foramen*. An upward projection (the *dens*) from the body of the second cervical vertebra furnishes an axis for rotating the head. A long, blunt spinous process which can be felt at the back of the base of the neck characterizes the seventh cervical vertebra. Each thoracic vertebra has articular facets for the ribs. More detailed descriptions of separate vertebrae are given in Table 5-2. The vertebral column as a whole articulates with the head, ribs, and hip bones, whereas the individual vertebrae articulate with each other in joints between their bodies and between their articular processes. For a description of intervertebral joints, see Table 5-4, p. 108.

In order to increase the carrying strength of the vertebral column and to make balance possible in the upright position, the vertebral column is curved. At birth there

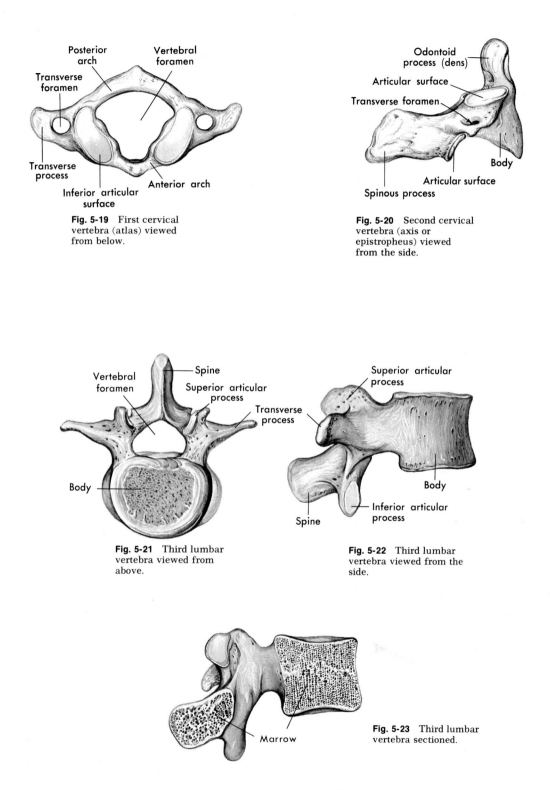

Fig. 5-19 First cervical vertebra (atlas) viewed from below.

Fig. 5-20 Second cervical vertebra (axis or epistropheus) viewed from the side.

Fig. 5-21 Third lumbar vertebra viewed from above.

Fig. 5-22 Third lumbar vertebra viewed from the side.

Fig. 5-23 Third lumbar vertebra sectioned.

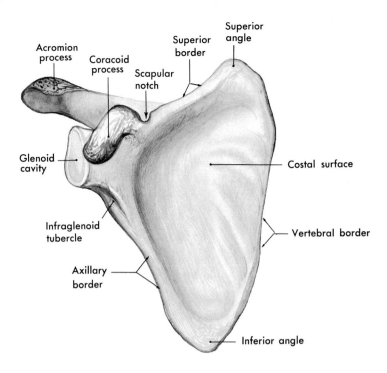

Acromion
process

Coracoid
process

Scapular
notch

Superior
border

Superior
angle

Glenoid
cavity

Costal surface

Infraglenoid
tubercle

Vertebral border

Axillary
border

Inferior angle

Fig. 5-24 Right scapula, anterior view.

is a continuous posterior convexity from
head to coccyx. Later, as the child learns to
sit and stand, secondary posterior concavi-
ties necessary for balance develop in the
cervical and lumbar regions. Not uncom-
monly, spinal curves deviate from the
normal. For example, the lumbar curve
frequently shows an exaggerated concavity (*lordosis*), whereas any of the regions
may have a lateral curvature (*scoliosis*).
The so-called hunchback is an exaggerated
convexity in the thoracic region (*kyphosis*).

Sternum. The *medial part* of the anterior chest wall is supported by the *sternum*,
a somewhat dagger-shaped bone consisting of three parts: the upper handle part
or *manubrium*, the middle blade part or
body, and a blunt cartilaginous lower tip,
the *xiphoid process*. The latter ossifies during adult life. The manubrium articulates
with the clavicle and first rib, whereas the
next nine ribs join the body of the sternum,
either directly or indirectly, by means of
the *costal cartilages*.

Ribs. *Twelve pairs of ribs*, together with
the vertebral column and sternum, form

the bony cage known as the *thorax*. Each
rib articulates with both the body and
transverse process of its corresponding
thoracic vertebra. In addition, the second
through the ninth ribs articulate with the
body of the vertebra above. From its vertebral attachment each rib curves outward,
then forward and downward, a mechanical
fact important for breathing. Anteriorly
each of the first seven ribs joins a costal
cartilage which attaches it to the sternum.
Each of the costal cartilages of the next
three ribs, however, joins the cartilage of
the rib above to be thus indirectly attached
to the sternum. Because the two costal
cartilages of the eleventh and twelfth ribs
do not attach even indirectly to the sternum, they are designated floating ribs.

APPENDICULAR SKELETON

Upper extremity. The upper extremity
consists of the bones of the shoulder girdle,
upper arm, lower arm, wrist, and hand.
Two bones, the *clavicle* and *scapula*, compose the *shoulder girdle*. Contrary to ap-
pearances, this girdle forms only one bony

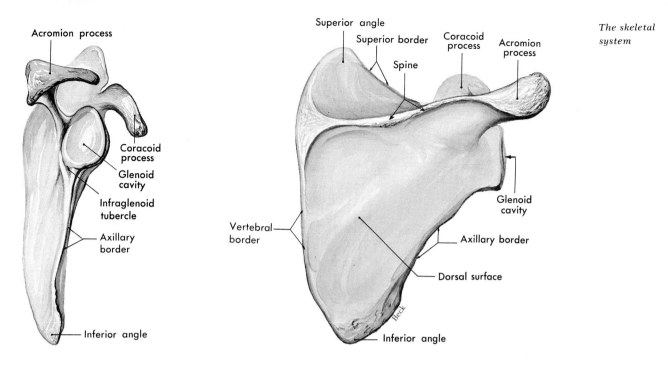

Fig. 5-25 Right scapula, lateral view.

Fig. 5-26 Right scapula, posterior view.

joint with the trunk: the sternoclavicular joint between the sternum and clavicle. At its outer end, the clavicle articulates with the scapula, which attaches to the ribs by muscles and tendons, not by a joint. All shoulder movements, therefore, involve the sternoclavicular joint. Various markings of the scapula are described in Table 5-2 (see also Figs. 5-24 to 5-26).

The *humerus* or upper arm bone, like other long bones, consists of a shaft or diaphysis and two ends or epiphyses (Figs. 5-27 and 5-28). The upper epiphysis bears several identifying structures: the head, anatomical neck, greater and lesser tubercles, intertubercular groove, and surgical neck. On the diaphysis are found the deltoid tuberosity and the radial groove. The distal epiphysis has four projections – the medial and lateral epicondyles, the capitulum, and the trochlea – and two depressions, the olecranon and coronoid fossae. For descriptions of all of these markings, see Table 5-2. The humerus articulates proximally with the scapula and distally with both the radius and ulna.

Two bones form the framework for the lower arm: the *radius* on the thumb side and the *ulna* on the little finger side. At the proximal end of the ulna, the olecranon process projects posteriorly and the coronoid process anteriorly. There are also two depressions: the semilunar notch on the anterior surface and the radial notch on the lateral surface. The distal end has two projections: a rounded head and a sharper styloid process. For more detailed identification of these markings, see Table 5-2. The ulna articulates proximally with the humerus and radius and distally with a fibrocartilaginous disc but not with any of the carpal bones.

The radius has three projections: two at its proximal end, the head and radial tuberosity, and one at its distal end, the styloid process (Figs. 5-29 and 5-30). There are two proximal articulations: one with the capitulum of the humerus and the other with the radial notch of the ulna. The three distal articulations are with the scaphoid and lunate carpal bones and with the head of the ulna.

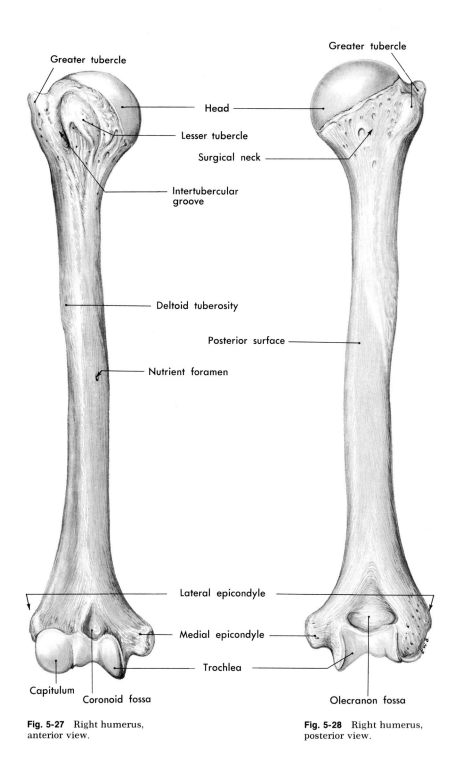

Greater tubercle

Greater tubercle

Head

Lesser tubercle

Surgical neck

Intertubercular
groove

Deltoid tuberosity

Posterior surface

Nutrient foramen

Lateral epicondyle

Medial epicondyle

Trochlea

Capitulum

Coronoid fossa

Olecranon fossa

Fig. 5-27 Right humerus,
anterior view.

Fig. 5-28 Right humerus,
posterior view.

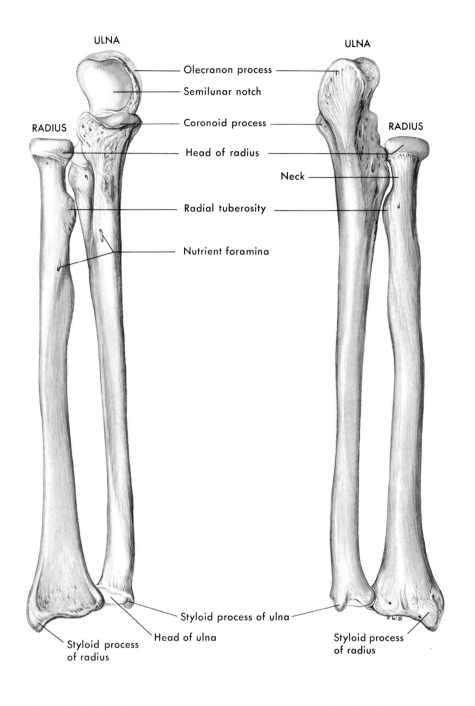

ULNA

Olecranon process

Semilunar notch

Coronoid process

RADIUS

Head of radius

Neck

Radial tuberosity

Nutrient foramina

Styloid process of ulna

Head of ulna

Styloid process of radius

ULNA

RADIUS

Styloid process of radius

Fig. 5-29 Right radius and ulna, anterior surfaces.

Fig. 5-30 Right radius and ulna, posterior surfaces.

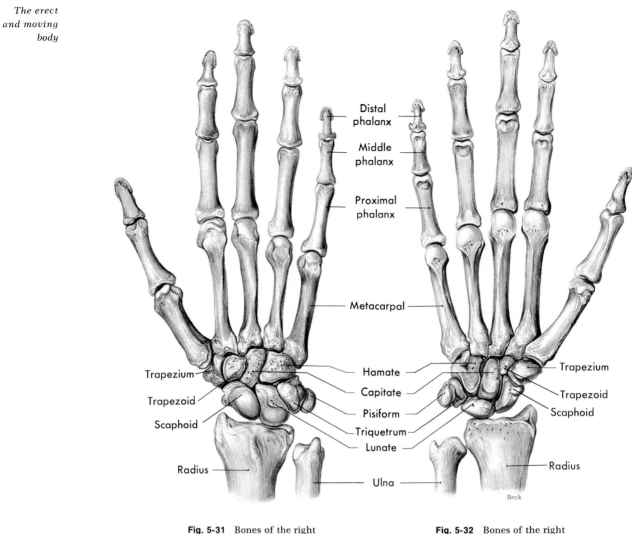

Fig. 5-31 Bones of the right hand and wrist, dorsal surface.

Fig. 5-32 Bones of the right hand and wrist, palmar surface.

The eight *carpal bones* (Figs. 5-31 and 5-32) form what most people think of as the upper part of the hand but what, anatomically speaking, is the wrist. Only one of these bones is evident from the outside, the *pisiform bone*, which projects posteriorly on the little finger side as a small, rounded elevation. Ligaments bind the carpals closely and firmly together in two rows of four each: proximal row (from little finger toward thumb)—pisiform, triquetrum, lunate, and scaphoid bones; distal row—hamate, capitate, trapezoid, and trapezium bones. The joints between the carpals and the joint between the carpals and radius permit wrist and hand movements.

Of the five *metacarpal bones* which form the framework of the hand, that of the thumb forms the most freely movable joint with the carpals. This fact has great significance. Because of the wide range of movement possible between the thumb metacarpal and the trapezium, particularly the ability to oppose the thumb to the fingers, the human hand has much greater dexterity than the forepaw of any animal and has enabled man to manipu-

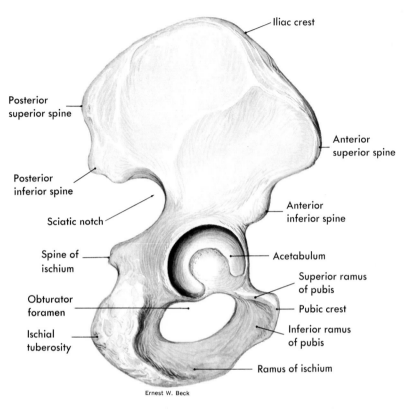

Iliac crest

Posterior superior spine

Anterior superior spine

Posterior inferior spine

Sciatic notch

Anterior inferior spine

Spine of ischium

Acetabulum

Superior ramus of pubis

Obturator foramen

Pubic crest

Ischial tuberosity

Inferior ramus of pubis

Ramus of ischium

Ernest W. Beck

Fig. 5-33 Right hip bone disarticulated from the skeleton viewed from the side with the bone turned so as to look directly into the acetabulum.

late his environment effectively. The heads of the metacarpals, prominent as the proximal knuckles of the hand, articulate with the phalanges.

Lower extremity. Bones of the hip, thigh, lower leg, ankle, and foot constitute the lower extremity. Strong ligaments bind the two hip bones (*os coxae* or *os innominatum*) to the sacrum posteriorly and to each other anteriorly to form the *pelvic girdle*, a stable, circular base which supports the trunk and attaches the lower extremities to it. In early life, each innominate bone is made up of three separate bones. Later on, they fuse into a single, massive, irregular bone which is broader than any other bone in the body. The largest and uppermost of the three bones is the *ilium*; the strongest, lowermost, the *ischium*; and the anteriormost, the *pubis*. Numerous markings are present on the three bones. These are identified in Table 5-2 (also see Fig. 5-33).

The two thigh bones or *femurs* have the distinction of being the longest and heaviest bones in the body. Several prominent markings characterize them. For example, three projections are conspicuous at each

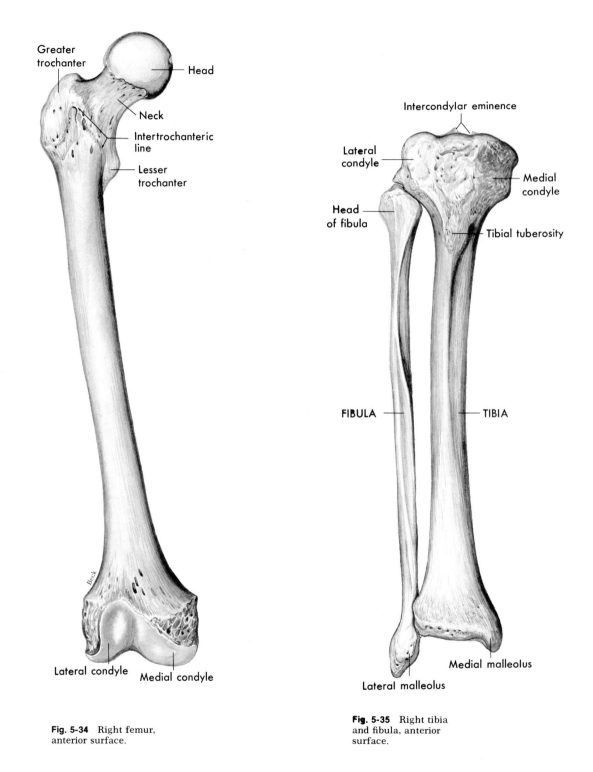

Fig. 5-34 Right femur, anterior surface.

Fig. 5-35 Right tibia and fibula, anterior surface.

epiphysis: the head and greater and lesser trochanters proximally and the medial and lateral condyles and adductor tubercle distally (Fig. 5-34). Both condyles and the greater trochanter may be felt externally. For a description of the various femur markings, see Table 5-2.

The largest sesamoid bone in the body, and the one which is almost universally present, is the *patella* or kneecap, located in the tendon of the quadriceps femoris muscle as a protection to the underlying knee joint. When the joint is extended, the patellar outline may be distinguished through the skin, but as the knee flexes, it sinks into the intercondylar notch of the femur and can no longer be delineated.

The *tibia* is the larger and stronger and the more medially and superficially located of the two lower leg bones, whereas the *fibula* is smaller and more laterally and deeply placed. The fibula articulates with the lateral condyle of the tibia. The tibia, in turn, articulates with the femur in the largest and one of the most stable joints of the body. Distally, the tibia articulates again with the fibula and also with the talus. The latter fits into a boxlike socket (ankle joint) formed by the medial and lateral malleoli, projections of the tibia and fibula, respectively. For other tibial markings, see Table 5-2 and Fig. 5-35.

Structure of the *foot* is similar to that of the hand with certain differences which adapt it for supporting weight. One example of this is the much greater solidity and the more limited mobility of the great toe compared to the thumb. Then, too, the foot bones are held together in such a way as to form springy lengthwise and crosswise arches. This is architecturally sound since arches are known to furnish more supporting strength per given amount of structural material than any other type of construction. Hence, the two-way arch construction makes a highly

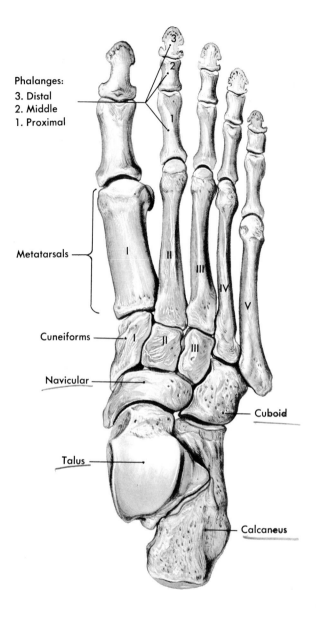

Fig. 5-36 Bones of right foot viewed from above. Tarsal bones consist of cuneiforms, navicular, talus, cuboid, and calcaneus.

Fig. 5-37 Arches of the
foot (see Table 5-2, p. 102).

Medial longitudinal arch

Transverse arch

Lateral longitudinal arch

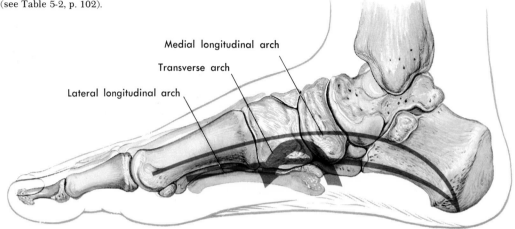

Fig. 5-38 Flatfoot results when there is a weakening
of tendons and ligaments attaching to the tarsal
bones. Downward pressure by the weight of the body
gradually flattens out the normal arch of bones.

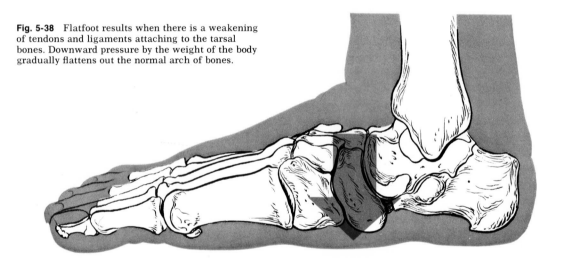

stable base. The longitudinal arch has an
inner or medial portion and an outer or
lateral portion, both of which are formed
by the placement of tarsals and metatarsals. Specifically, some of the tarsals (calcaneus, talus, navicular, and cuneiforms)
and the first three metatarsals form the
medial longitudinal arch, and the calcaneus and cuboid tarsals plus the fourth

and fifth metatarsals shape the lateral
longitudinal arch (Figs. 5-36 to 5-38). The
transverse arch results from the relative
placement of the distal row of tarsals and
the five metatarsals. (See Table 5-2 for
specific bones of different arches.) Strong
ligaments and leg muscle tendons normally hold the foot bones firmly in their
arched positions, but not infrequently

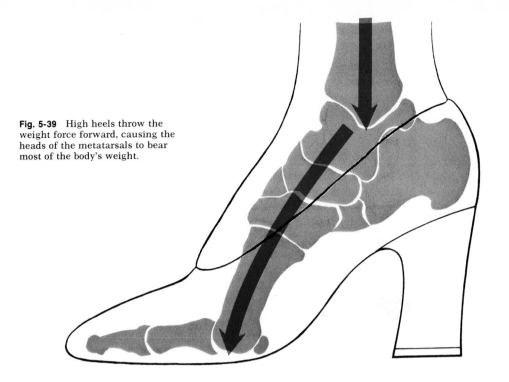

Fig. 5-39 High heels throw the weight force forward, causing the heads of the metatarsals to bear most of the body's weight.

these weaken, causing the arches to flatten, a condition aptly called fallen arches or flatfeet. Note that the tarsals and metatarsals play the major role in the functioning of the foot as a supporting structure, with the phalanges relatively unimportant. The reverse is true for the hand. Here, manipulation is the main function rather than support. Consequently, the phalanges are all important and the carpals and metacarpals are subsidiary.

Markings

Various points on bones are labeled according to the nature of their structure. This method of identifying definite parts of different bones proves helpful when locating other structures such as muscles, blood vessels, and nerves. Definitions of some of the common terms applied to bone markings follow.

DEPRESSIONS AND OPENINGS

1 *fossa* – a hollow or depression (example, mandibular fossa of the temporal bone)
2 *sinus* – a cavity or spongelike space in a bone (example, the frontal sinus)
3 *foramen* – a hole (example, foramen magnum of the occipital bone)
4 *meatus* – a tube-shaped opening (example, the external auditory meatus)

PROJECTIONS OR PROCESSES

Those projections or processes which fit into joints are as follows:

1 *condyle* – a rounded projection that enters into the formation of a joint (example, condyles of the femur)
2 *head* – a rounded projection beyond a narrow necklike portion (example, head of the femur)

Those projections to which muscles attach include the following:

1 *trochanter* – a very large projection; greater trochanter of femur
2 *crest* – a ridge (example, the iliac crest); a less prominent ridge is called a *line* (example, ileopectineal line)
3 *spinous process* or *spine* – a sharp projection (example, anterior superior iliac spine)
4 *tuberosity* – a large, rounded projection (example, ischial tuberosity)
5 *tubercle* – a small, rounded projection (example, rib tubercles)

Text continued on p. 103

Table 5-2. Indentification of bone markings*

Bone	Marking	Description
Frontal	*Supraorbital margin*	Arched ridge just below eyebrows
	Frontal sinuses	Cavities inside bone just above supraorbital margin; lined with mucosa; contain air
	Frontal tuberosities	Bulge above each orbit; most prominent part of forehead
	Superciliary arches	Ridges caused by projection of frontal sinuses; eyebrows lie over these ridges
	Supraorbital notch (sometimes foramen)	Notch or foramen in supraorbital margin slightly mesial to its midpoint; transmits supraorbital nerve and blood vessels
	Glabella	Smooth area between superciliary ridges and above nose
Temporal	*Mastoid process*	Protuberance just behind ear
	Mastoid air cells	Air-filled mucosa-lined spaces within mastoid process
	External auditory meatus (or *canal*)	Opening into ear and tube extending into temporal bone
	Zygomatic process	Projection which articulates with malar (or zygomatic) bone
	Internal auditory meatus	Fairly large opening on posterior surface of petrous portion of bone; transmits eighth cranial nerve to inner ear and seventh cranial nerve on its way to facial structures
	Squamous portion	Thin, flaring upper part of bone
	Mastoid portion	Rough-surfaced lower part of bone posterior to external auditory meatus
	Petrous portion	Wedge-shaped process that forms part of center section of cranial floor between sphenoid and occipital bones; name derived from Greek word for stone because of extreme hardness of this process; houses middle and inner ear structures
	Mandibular fossa	Oval-shaped depression anterior to external auditory meatus; forms socket for condyle of mandible
	Styloid process	Slender spike of bone extending downward and forward from undersurface of bone anterior to mastoid process; often broken off in dry skull; several neck muscles and ligaments attach to styloid process
	Stylomastoid foramen	Opening between styloid and mastoid processes where facial nerve emerges from cranial cavity
	Jugular fossa	Depression on undersurface of petrous portion; dilated beginning of internal jugular vein lodged here
	Jugular foramen	Opening in suture between petrous portion and occipital bone; transmits lateral sinus and ninth, tenth, and eleventh cranial nerves

*Italics suggest markings particularly useful to recognize.

Table 5-2. Identification of bone markings*—cont'd

Bone	Marking	Description
Temporal—cont'd	Carotid canal (or foramen)	Channel in petrous portion; best seen from undersurface of skull; transmits internal carotid artery
Occipital	*Foramen magnum*	Hole through which spinal cord enters cranial cavity
	Condyles	Convex, oval processes on either side of foramen magnum; articulate with depressions on first cervical vertebra
	External occipital protuberance	Prominent projection on posterior surface in midline short distance above foramen magnum; can be felt as definite bump
	Superior nuchal line	Curved ridge extending laterally from external occipital protuberance
	Inferior nuchal line	Less well-defined ridge paralleling superior nuchal line short distance below it
	Internal occipital protuberance	Projection in midline on inner surface of bone; grooves for lateral sinuses extend laterally from this process and one for sagittal sinus extends upward from it
Sphenoid	*Body*	Hollow, cubelike central portion
	Great wings	Lateral projections from body; form part of outer wall of orbit
	Lesser wings	Thin, triangular projections from upper part of sphenoid body; form posterior part of roof of orbit
	Sella turcica (or *Turk's saddle*)	Saddle-shaped depression on upper surface of sphenoid body; contains pituitary gland (Fig. 5-8)
	Sphenoid sinuses	Irregular air-filled mucosa-lined spaces within central part of sphenoid (Fig. 5-10)
	Pterygoid processes	Downward projections on either side where body and greater wing unite; comparable to extended legs of bat if entire bone is likened to this animal; form part of lateral nasal wall
	Optic foramen	Opening into orbit at root of lesser wing; transmits second cranial nerve
	Superior orbital fissure	Slitlike opening into orbit; lateral to optic foramen; transmits third, fourth, and part of fifth cranial nerves
	Foramen rotundum	Opening in greater wing that transmits maxillary division of fifth cranial nerve
	Foramen ovale	Opening in greater wing that transmits mandibular division of fifth cranial nerve
Ethmoid	*Horizontal (cribriform) plate*	Olfactory nerves pass through numerous holes in this plate

*Italics suggest markings particularly useful to recognize.

Continued

Table 5-2. Identification of bone markings*—cont'd

Bone	Marking	Description
Ethmoid—cont'd	*Crista galli*	See Figs. 5-14 and 5-16; meninges attach to this process
	Perpendicular plate	Forms upper part of nasal septum (Figs. 5-14 and 5-16)
	Ethmoid sinuses	Honeycombed, mucosa-lined air spaces within lateral masses of bone (Figs. 5-15 and 5-16)
	Superior and middle turbinates (conchae)	Help to form lateral walls of nose (Figs. 5-14 and 5-16)
	Lateral masses	Compose sides of bone; contain many air spaces (ethmoid cells or sinuses); inner surface forms superior and middle conchae
Mandible	Body	Main part of bone; forms chin
	Ramus	Process, one on either side, that projects upward from posterior part of body
	Condyle (or *head*)	Part of each ramus that articulates with mandibular fossa of temporal bone
	Neck	Constricted part just below condyles
	Alveolar process	Teeth set into this arch
	Mandibular foramen	Opening on inner surface of ramus; transmits nerves and vessels to lower teeth
	Mental foramen	Opening on outer surface below space between two bicuspids; transmits terminal branches of nerves and vessels which enter bone through mandibular foramen; dentists inject anesthetics through these foramina
	Coronoid process	Projection upward from anterior part of each ramus; temporal muscle inserts here
	Angle	Juncture of posterior and inferior margins of ramus
Maxilla	*Alveolar process*	Arch containing teeth
	Maxillary sinus or antrum of Highmore	Large air-filled mucosa-lined cavity within body of each maxilla; largest of sinuses
	Palatine process	Horizontal inward projection from alveolar process; forms anterior and larger part of hard palate
	Infraorbital foramen	Hole on external surface just below orbit; transmits vessels and nerves (Fig. 5-11)
	Lacrimal groove	Groove on inner surface; joined by similar groove on lacrimal bone to form canal which houses nasolacrimal duct
Palatine	Horizontal plate	Joined to palatine processes of maxillae to complete posterior part of hard palate
Special features of skull	*Sutures* 1 *Sagittal*	Immovable joints between skull bones 1 Line of articulation between two parietal bones

*Italics suggest markings particularly useful to recognize.

Table 5-2. Identification of bone markings*—cont'd

Bone	Marking	Discription
Special features of skull—cont'd	*Sutures*—cont'd **2** *Coronal* **3** *Lambdoidal*	**2** Joint between parietal bones and frontal bone **3** Joint between parietal bones and occipital bone
	Fontanels **1** *Anterior* (or *frontal*) **2** *Posterior* (or *occipital*) **3** Anterolateral (or *sphenoid*) **4** Posterolateral (or *mastoid*)	"Soft spots" where ossification incomplete at birth; allow some compression of skull during birth; also important in determining position of head before delivery; six such areas located at angles of parietal bones **1** At intersection of sagittal and coronal sutures (juncture of parietal bones and frontal bone); diamond shaped; largest of fontanels; usually closed by 1 1/2 years of age **2** At intersection of sagittal and lambdoidal sutures (juncture of parietal bones and occipital bone); triangular in shape; usually closed by second month **3** At juncture of frontal, parietal, temporal, and sphenoid bones **4** At juncture of parietal, occipital, and temporal bones; usually closed by second year
	Sinuses **1** *Air* (or *bony*) **2** *Blood*	**1** Spaces or cavities within bones; those which communicate with nose called *paranasal sinuses* (frontal, sphenoidal, ethmoidal, and maxillary); mastoid cells communicate with middle ear rather than nose, therefore not included among paranasal sinuses **2** Veins within cranial cavity (Figs. 10-20 and 10-21, p. 325)
	Orbits formed by **1** Frontal **2** Ethmoid **3** Sphenoid **4** Lacrimal **5** Maxillary **6** Zygomatic **7** Palatine	**1** Roof of orbit (Fig. 5-11) **2** Medial wall **3** Lateral wall **4** Medial wall **5** Floor **6** Lateral wall **7** Floor
	Nasal septum formed by **1** Perpendicular plate of ethmoid **2** Vomer bone **3** Cartilage	Partition in midline of nasal cavity; separates cavity into right and left halves **1** Forms upper part of septum **2** Forms lower, posterior part **3** Forms anterior part
	Wormian bones	Small islands of bones within suture
Vertebral column	General features	Anterior part of vertebrae (except first two cervical) consists of body; posterior part, of neural arch which, in turn, consists of two pedicles, two laminae, and seven processes projecting from laminae
	Thoracic vertebrae **1** *Body* **2** *Pedicles* **3** *Laminae*	**1** Main part; flat, round mass located anteriorly; supporting or weight-bearing part of vertebra **2** Short projections extending posteriorly from body **3** Posterior part of vertebra to which pedicles join and from which processes project

*Italics suggest markings particularly useful to recognize.

Continued

Table 5-2. Identification of bone markings*—cont'd

Bone	Marking	Description
Vertebral column—cont'd	4 *Neural arch*	4 Formed by pedicles and laminae; protects spinal cord posteriorly; together, neural arches form spinal cavity; congenital absence of one or more neural arches known as *spina bifida* (cord may protrude right through skin)
	5 *Spinous process*	5 Sharp process projecting inferiorly from laminae in midline
	6 *Transverse processes*	6 Right and left lateral projections from laminae
	7 *Superior articulating processes*	7 Project upward from laminae
	8 *Inferior articulating processes*	8 Project downward from laminae; articulate with superior articulating processes of vertebrae below
	9 *Spinal foramen*	9 Hole in center of vertebra formed by union of body, pedicles, and laminae; spinal foramina, when vertebrae superimposed one upon other, form spinal cavity which houses spinal cord
	Cervical vertebrae 1 General features	1 Foramen in each transverse process for transmission of vertebral artery, vein, and plexus of nerves; short bifurcated spinous processes except on seventh vertebrae where it is extra long and may be felt as protrusion when head bent forward; bodies of these vertebrae small, while spinal foramina large and triangular
	2 *Atlas*	2 First cervical vertebra; lacks body and spinous process; superior articulating processes concave ovals which act as rockerlike cradles for condyles of occipital bone; named atlas because supports head as Atlas was thought to have supported world (Fig. 5-19)
	3 *Axis* (epistropheus)	3 Second cervical vertebra; so named because atlas rotates about this bone in rotating movements of head; *dens*, or odontoid process, peglike projection upward from body of axis, forming pivot for rotation of atlas (Fig. 5-20)
	Lumbar vertebrae	Strong, massive; superior articulating processes directed inward instead of upward; inferior articulating processes, outward instead of downward; short, blunt spinous process (Figs. 5-21 to 5-23)
	Sacral promontory	Protuberance from anterior, upper border of sacrum into pelvis; of obstetrical importance because its size limits anteroposterior diameter of pelvic inlet
	Intervertebral foramina	Opening between vertebrae through which spinal nerves emerge
	Curves	Curves have great structural importance because increase carrying strength of vertebral column, make balance possible in upright position (if column were straight, weight of viscera would pull body forward), absorb jars from walking (straight column would transmit jars straight to head), and protect column from fracture
	1 *Primary*	1 Column curves at birth from head to sacrum with convexity posteriorly; after child stands, convexity persists only in *thoracic* and *sacral* regions which, therefore, are called primary curves
	2 *Secondary*	2 Concavities in *cervical* and *lumbar* regions; cervical concavity results from infant's attempts to hold head erect (3 to 4 months); lumbar concavity, from balancing efforts in learning to walk (10 to 18 months)
	3 *Abnormal*	3 *Kyphosis*, exaggerated convexity in thoracic region (hunchback); *lordosis*, exaggerated concavity in lumbar region, a very common condition; *scoliosis*, lateral curvature in any region

*Italics suggest markings particularly useful to recognize.

Table 5-2. Identification of bone markings*—cont'd

Bone	Marking	Description
Sternum	*Body*	Main central part of bone
	Manubrium	Flaring, upper part
	Xiphoid process	Projection of cartilage at lower border of bone
Ribs	*Head*	Projection at posterior end of rib; articulates with corresponding thoracic vertebra and one above; except last three pairs, which join corresponding vertebra only
	Neck	Constricted portion just below head
	Tubercle	Small knob just below neck; articulates with transverse process of corresponding thoracic vertebra; missing in lowest three ribs
	Body or shaft	Main part of rib
	Costal cartilage	Cartilage at sternal end of true ribs; attaches ribs (except floating ribs) to sternum
Scapula (Figs. 5-24 to 5-26)	*Borders* **1** Superior **2** Vertebral **3** Axillary	**1** Upper margin **2** Margin toward vertebral column **3** Lateral margin
	Spine	Sharp ridge running diagonally across posterior surface of shoulder blade
	Acromion process	Slightly flaring projection at lateral end of scapular spine; may be felt as tip of shoulder; articulates with clavicle
	Coracoid process	Projection on anterior surface from upper border of bone; may be felt in groove between deltoid and pectoralis major muscles, about 1 inch below clavicle
	Glenoid cavity	Arm socket
Humerus (Figs. 5-27 and 5-28)	*Head*	Smooth, hemispherical enlargement at proximal end of humerus
	Anatomical neck	Oblique groove just below head
	Greater tubercle	Rounded projection lateral to head on anterior surface
	Lesser tubercle	Prominent projection on anterior surface just below anatomical neck
	Intertubercular (bicipital) groove	Deep groove between greater and lesser tubercles; long tendon of biceps muscle lodges here
	Surgical neck	Region just below tubercles; so named because of its liability to fracture
	Deltoid tuberosity	V-shaped, rough area about midway down shaft where deltoid muscle inserts
	Radial groove	Groove running obliquely downward from deltoid tuberosity; lodges radial nerve

*Italics suggest markings particularly useful to recognize.

Continued

Table 5-2. Identification of bone markings*—cont'd

Bone	Marking	Description
Humerus— cont'd	*Epicondyles* (medial and lateral)	Rough projections at both sides of distal end
	Capitulum	Rounded knob below lateral epicondyle; articulates with radius; sometimes called radial head of humerus
	Trochlea	Projection with deep depression through center similar to shape of pulley; articulates with ulna
	Olecranon fossa	Depression on posterior surface just above trochlea; receives olecranon process of ulna when lower arm extends
	Coronoid fossa	Depression on anterior surface above trochlea; receives coronoid process of ulna in flexion of lower arm
Ulna (Figs. 5-29 and 5-30)	*Olecranon process*	Elbow
	Coronoid process	Projection on anterior surface of proximal end of ulna; trochlea of humerus fits snugly between olecranon and coronoid processes
	Semilunar notch	Curved notch between olecranon and coronoid, into which trochlea fits
	Radial notch	Curved notch lateral and inferior to semilunar notch; head of radius fits into this concavity
	Head	Rounded process at distal end; does not articulate with wrist bones but with fibrocartilaginous disc
	Styloid process	Sharp protuberance at distal end; can be seen from outside on posterior surface
Radius (Figs. 5-29 and 5-30)	*Head*	Disk-shaped process forming proximal end of radius; articulates with capitulum of humerus and with radial notch of ulna
	Radial tuberosity	Roughened projection on ulnar side, short distance below head; biceps muscle inserts here
	Styloid process	Protuberance at distal end on lateral surface (with forearm supinated as in anatomical position)
Os coxae (Fig. 5-33)	*Ilium*	Upper, flaring portion
	Ischium	Lower, posterior portion
	Pubic bone or pubis	Medial, anterior section
	Acetabulum	Hip socket; formed by union of ilium, ischium, and pubis
	Iliac crests	Upper, curving boundary of ilium
	Iliac spines 1 *Anterior superior* 2 Anterior inferior 3 Posterior superior 4 Posterior inferior	1 Prominent projection at anterior end of iliac crest; can be felt externally as "point" of hip 2 Less prominent projection short distance below anterior superior spine 3 At posterior end of iliac crest 4 Just below posterior superior spine
	Greater sciatic notch	Large notch on posterior surface of ilium just below posterior inferior spine

*Italics suggest markings particularly useful to recognize.

Table 5-2. Identification of bone markings*—cont'd

Bone	*Marking*	*Description*
Os coxae—cont'd	Gluteal lines	Three curved lines across outer surface of ilium—posterior, anterior, inferior, respectively
	Iliopectineal line	Rounded ridge extending from pubic tubercle upward and backward toward sacrum
	Iliac fossa	Large, smooth, concave inner surface of ilium above iliopectineal line
	Ischial tuberosity	Large, rough, quadrilateral process forming inferior part of ischium; in erect sitting position body rests on these tuberosities
	Ischial spine	Pointed projection just above tuberosity
	Symphysis pubis	Cartilaginous, amphiarthrotic joint between pubic bones
	Superior pubic ramus	Part of pubis lying between symphysis and acetabulum; forms upper part of obturator foramen
	Inferior pubic ramus	Part extending down from symphysis; unites with ischium
	Pubic arch	Angle formed by two inferior rami
	Pubic crest	Upper margin of superior ramus
	Pubic tubercle	Rounded process at end of crest
	Obturator foramen	Large hole in anterior surface of os coxa; formed by pubis and ischium; largest foramen in body
	Pelvic brim (or *inlet*)	Boundary of aperture leading into true pelvis; formed by pubic crests, iliopectineal lines, and sacral promontory; size and shape of this inlet has great obstetrical importance since if any of its diameters too small, infant skull cannot enter true pelvis for natural birth
	True (or *lesser*) *pelvis*	Space below pelvic brim; true "basin" with bone and muscle walls and muscle floor; pelvic organs located in this space
	False (or *greater*) *pelvis*	Broad, shallow space above pelvic brim, or pelvic inlet; name "false pelvis" is misleading since this space is actually part of abdominal cavity, not pelvic cavity
	Pelvic outlet	Irregular circumference marking lower limits of true pelvis; bounded by tip of coccyx and two ischial tuberosities
	Pelvic girdle (or bony pelvis)	Complete bony ring; composed of two hip bones (ossa coxae), sacrum, and coccyx; forms firm base by which trunk rests upon thighs and for attachment of lower extremities to axial skeleton
Femur (Fig. 5-34)	*Head*	Rounded, upper end of bone; fits into acetabulum
	Neck	Constricted portion just below head
	Greater trochanter	Protuberance located inferiorly and laterally to head

*Italics suggest markings particularly useful to recognize.

Continued

Table 5-2. Identification of bone markings*—cont'd

Bone	Marking	Description
Femur—cont'd	*Lesser trochanter*	Small protuberance located inferiorly and medially to greater trochanter
	Linea aspera	Prominent ridge extending lengthwise along concave posterior surface
	Gluteal tubercle	Rounded projection just below greater trochanter; rudimentary third trochanter
	Supracondylar ridges	Two ridges formed by division of linea aspera at its lower end; medial supracondylar ridge extends inward to inner condyle, lateral ridge to outer condyle
	Condyles	Large, rounded bulges at distal end of femur; one on medial and one on lateral surface
	Adductor tubercle	Small projection just above inner condyle; marks termination of medial supracondylar ridge
	Trochlea	Smooth depression between condyles on anterior surface; articulates with patella
	Intercondyloid notch	Deep depression between condyles on posterior surface; cruciate ligaments which help bind femur to tibia lodge in this notch
Tibia (Fig. 5-35)	*Condyles*	Bulging prominences at proximal end of tibia; upper surfaces concave for articulation with femur
	Intercondylar eminence	Upward projection on articular surface between condyles
	Crest	Sharp ridge on anterior surface
	Tibial tuberosity	Projection in midline on anterior surface
	Popliteal line	Ridge that spirals downward and inward on posterior surface of upper third of tibial shaft
	Medial malleolus	Rounded downward projection at distal end of tibia; forms prominence on inner surface of ankle
Fibula (Fig. 5-35)	*Lateral malleolus*	Rounded prominence at distal end of fibula; forms prominence on outer surface of ankle
Tarsals (Figs. 5-36 to 5-38)	*Calcaneus*	Heel bone
	Talus	Uppermost of tarsals; articulates with tibia and fibula; boxed in by medial and lateral malleoli
	Longitudinal arches 1 *Inner* 2 *Outer*	Tarsals and metatarsals so arranged as to form arch from front to back of foot 1 Formed by calcaneus, navicular, cuneiforms, and three medial metatarsals 2 Formed by calcaneus, cuboid, and two lateral metatarsals
	Transverse (or *metatarsal*) *arch*	Metatarsals and distal row of tarsals (cuneiforms and cuboid) so articulated as to form arch across foot; bones kept in two arched positions by means of powerful ligaments in sole of foot and by muscles and tendons

*Italics suggest markings particularly useful to recognize.

Differences between male and female skeletons

Both general and specific differences exist between male and female skeletons. The general difference is one of size and weight, the male skeleton being larger and heavier. The specific differences concern the shape of the pelvic bones and cavity. Whereas the male pelvis is deep and funnel shaped, with a narrow pubic arch (usually less than 90 degrees), the female pelvis is shallow, broad, and flaring, with a wider pubic arch (usually greater than 90 degrees). The childbearing function obviously explains the necessity for these and certain other modifications of the female pelvis.

Age changes in skeleton

Changes in the skeleton as a whole from infancy to adulthood are mainly changes in the size of the bones and in the proportionate sizes between different bones. Changes in individual bones from young adulthood to old age, on the other hand, are mainly a matter of changes in the texture and in the contour of the margins and bone markings. Some of the major changes in the skeleton from infancy to young adulthood are as follows:

1. The head becomes proportionately smaller. Whereas the infant head is approximately one-fourth the total height of the body, the adult head is only about one-eighth the total height.

2. The thorax changes shape, roughly speaking, from round to elliptical.

3. The pelvis becomes relatively larger and in the female relatively wider.

4. The legs become proportionately longer and the trunk proportionately shorter.

5. The vertebral column develops two curves not present at birth—the cervical curve when the infant starts lifting up his head (at about 3 months of age) and the lumbar curve when the child begins standing (toward the end of the first year). Both of these secondary curves are concave posteriorly, whereas the primary thoracic and sacral curves are convex posteriorly.

6. The cranium shows several modifications. It grows rapidly during early childhood, enlarging its capacity from approximately 350 ml at birth to approximately 1,500 ml (about adult size) by 6 years of age. The fontanels close by about 1 1/2 or 2 years of age, and the sutures begin to fuse in the 20's.

7. The facial bones also show several changes between infancy and adulthood. Unlike the cranial bones, their growth is slow during early childhood but rapid during the teens. Whereas the infant face compared with the entire skull bears the relationship of 1:8, the adult face bears the relationship of 1:2 to the adult skull. The sinuses are much larger in the adult. For example, at birth, only rudimentary maxillary and mastoid sinuses exist. The ethmoid and sphenoid sinuses start to appear at about 6 years of age and the frontal at about 7 years. All of the bony sinuses, but especially the frontal, grow rapidly during adolescence.

8. The epiphyses of the long bones are composed of cartilage at birth but become completely ossified (except for the thin layer of articular cartilage) by adulthood. Demonstration of epiphyseal cartilage on x-ray films indicates that skeletal growth has not ceased.

Changes in the skeleton continue to occur from adulthood to old age. Both bone margins and projections, for example, look different in old bones than in young. Instead of clean-cut, distinct margins, old bones characteristically have indistinct, shaggy-appearing margins (marginal lipping and spurs)—a regrettable change because the restricted movements of old age

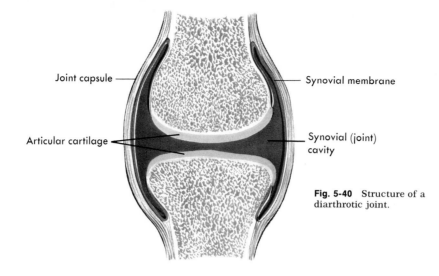

Joint capsule

Synovial membrane

Articular cartilage

Synovial (joint)
cavity

Fig. 5-40 Structure of a
diarthrotic joint.

stem partly from this piling up of bone
around joint margins. Also, an increase in
bone along various projections develops in
old age, making ridges and processes more
pronounced.

■ Joints between bones (articulations)

Bones are joined to one another in several ingenious ways which permit a great
variety of movement. Where free movement is essential, the articulating ends of
the bones are so shaped and the joint so
constructed as to permit and even facilitate unhampered motion. Where only slight
movement is desirable, bone shape and
joint structure make only slight movement
possible. Where no movement between the
bones is preferable, this, too, is accomplished by bone shape and joint structure.
How important normal joint structure and
function are for the productiveness and
enjoyment of life, probably most of us seldom consider. But disease often makes
this tragically clear. Joint structure in all
too many cases becomes so altered that
crippling immobility results—sometimes
only limitation of a single movement and
sometimes almost complete immobiliza-
tion. To inquire into joint structure and
action is, therefore, an essential part of
the study of anatomy and physiology. Basic

information includes understanding what
structural kinds of joints exist between
bones and what kinds of movements each
type permits.

Kinds

A confusing array of terms has grown up
around the subject of joint classification.
Different anatomists have used different
criteria for identifying joint types and have
muddled matters even more by using various names for the same kind of joint.

One of the simpler ways of classifying
joints is to divide them into two main structural types: diarthroses and synarthroses.

Diarthroses are joints in which a small
space, the joint cavity (Fig. 5-40), exists
between the articulating surfaces of the
two bones that form the joint. Because
there is this cavity with no tissue growing
between the articulating surfaces, the surfaces are free to move against one another.
And therefore functionally, diarthrotic
joints are classified as freely movable
joints. By far the majority of our joints
belong to this category.

Diarthroses share several characteristics
besides that of having a joint cavity. A
thin layer of hyaline cartilage covers the
joint surfaces of the articulating bones, a
sleevelike, fibrous capsule lined with
smooth slippery synovial membrane en-

cases the joint, and additional ligaments grow between the bones, lashing them firmly together. Crescent-shaped pieces of cartilage are found in some diarthrotic joints interposed between the articulating ends of the two bones. Examples are the semilunar cartilages of the knee joint and the glenoid cartilages of the shoulder joint.

Synarthroses are joints that do not have a joint cavity but instead have tissue (fibrous, cartilage, or bone) growing between their articulating surfaces and making them unable to move freely against one another. Defined functionally, synarthroses are joints that do not allow free movement. In fact, they allow little or no movement.

Diarthroses and synarthroses are divided into subtypes according to such characteristics as the shape of the joint surfaces of the united bones and the type of connective tissue between them. See Fig. 5-41 and Table 5-3 for a summary of the main kinds of joints with examples of each. A description of individual joints is given in Table 5-4.

Movements

Diarthrotic joints permit one or more of the following kinds of movements: (1) flexion, (2) extension, (3) abduction, (4) adduction, (5) rotation, (6) circumduction, and (7) special movements such as supination, pronation, inversion, eversion, protraction, and retraction.

Flexion. Flexion decreases the size of the angle between the anterior surfaces of articulated bones (exception: flexion of the knee and toe joints decreases the angle between the posterior surfaces of the articulated bones). Flexing movements are *bending* or *folding* movements. For example, bending the head forward is flexion of the joint between the occipital bone and the atlas, and bending the elbow is flexion of the elbow joint or of the lower arm. Flexing movements of the arms and legs may

be thought of as "withdrawing" movements.

Extension. Extension is the return from flexion. Whereas bending movements are flexions, *straightening* movements are extensions. Extension restores a part to its anatomical position from the flexed position. Continuation of extension beyond the anatomical position is called *hyperextension*. Examples include flexion of the head, bending it forward as in prayer, extension of the head, returning it to the upright anatomical position from the flexed position, and hyperextension of the head, stretching it backward from the upright position. Extension of the foot at the ankle joint is commonly referred to as *plantar flexion*, while flexion of the ankle joint is called *dorsal flexion*.

Abduction. Abduction moves the bone away from the median plane of the body. An example is moving the arms straight out to the sides.

Adduction. Adduction is the opposite of abduction. It moves the part toward the median plane of the body. Examples: bringing the arms back to the sides; adduction of the fingers means moving them toward the third finger; adduction of the toes is movement toward the second toe.

Rotation. Rotation is the pivoting or moving of a bone upon its own axis somewhat as a top turns on its axis. An example is holding the head in an upright position and turning it from one side to the other.

Circumduction. Circumduction causes the bone to describe the surface of a cone as it moves. The distal end of the bone describes a circle. It combines flexion, abduction, extension, and adduction in succession. Examples are dropping the head to one shoulder, then to the chest, to the other shoulder, and backward and describing a circle with the arms outstretched.

Special movements. *Supination* is a

Text continued on p. 110

Table 5-3. Joints

DIARTHROSES	SYNARTHROSES
1 Ball and socket *Other names:* spheroidal; endarthroses *Description:* ball-shaped head fits into concave socket *Movement:* widest range of all joints; triaxial *Examples:* shoulder joint and hip joint **2 Hinge** *Other name:* ginglymus *Description:* spool-shaped surface fits into concave surface *Movement:* in one plane about single axis (uniaxial); like hinged-door movement — viz., flexion and extension *Examples:* elbow, knee, ankle, and interphalangeal joints **3 Pivot** *Other name:* trochoid *Description:* arch-shaped surface rotates about rounded or peglike pivot *Movement:* rotation; uniaxial *Example:* between axis and atlas; between radius and ulna **4 Ellipsoidal** *Other names:* condyloid, ovoid *Description:* oval-shaped condyle fits into elliptical cavity *Movements:* in two planes at right angles to each other — specifically, flexion, extension, abduction, and adduction; biaxial *Example:* wrist joint (between radius and carpals) **5 Saddle** *Description:* saddle-shaped bone fits into socket that is concave-convex in opposite direction; modification of condyloid joint *Movements:* same kinds of movement as condyloid joint but freer; like rider in saddle; biaxial *Example:* thumb, between first metacarpal and trapezium **6 Gliding** *Other name:* arthrodia *Description:* articulating surfaces; usually flat *Movement:* gliding, a nonaxial movement *Example:* between carpal bones; between sacrum and ilium (sacroiliac joints)	**1 Cartilaginous** *Other name:* synchondrosis *Description:* cartilage grows between two articulating surfaces (e.g., cartilage disks between bodies of vertebrae) usually reinforced by ligaments *Movements:* bending and twisting or slight compression *Examples:* between bodies of vertebrae; between diaphysis and epiphysis of growing bones; replaced by bone in full-grown bones **2 Fibrous** *Other name:* syndesmosis *Description:* thin layer of fibrous tissue; continuous with periosteum; connects articulating bones *Movement:* none *Example:* sutures of skull; in older adults, fibrous connection replaced by bone

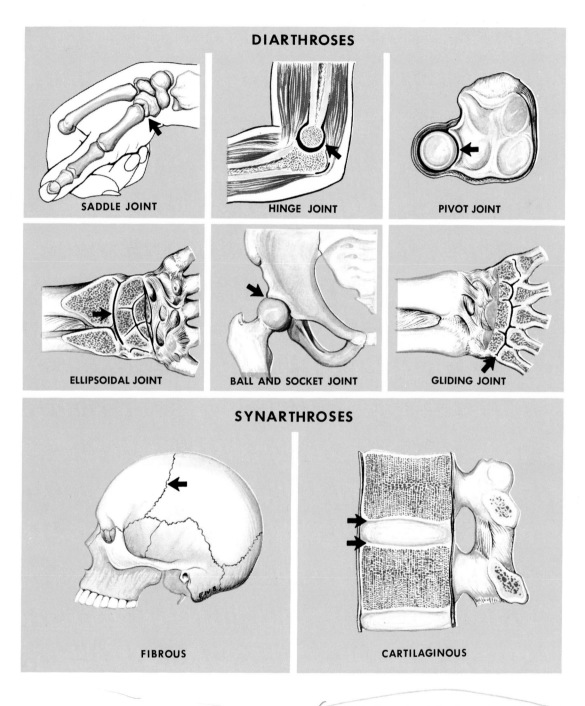

DIARTHROSES

SADDLE JOINT

HINGE JOINT

PIVOT JOINT

ELLIPSOIDAL JOINT

BALL AND SOCKET JOINT

GLIDING JOINT

SYNARTHROSES

FIBROUS

CARTILAGINOUS

Fig. 5-41 Examples of diarthroses (movable joints with a joint cavity) and synarthroses (joints without a joint cavity). Refer to Tables 5-3 and 5-4 for complete description and other examples.

Table 5-4. Description of individual joints

Name	Articulating bones	Type	Movements
Atlantoepistropheal	Anterior arch of atlas rotates about dens of axis (epistropheus)	Diarthrotic (pivot type)	Pivoting or partial rotation of head
Vertebral*	Between bodies of vertebrae	Synarthrotic, cartilaginous; amphiarthrotic by other system of classifying	Slight movement between any two vertebrae but considerable motility for column as whole
	Between articular processes	Diarthrotic (gliding)	
Clavicular Sternoclavicular	Medial end of clavicle with manubrium of sternum; only joint between upper extremity and trunk	Diarthrotic (gliding)	Gliding; weak joint that may be injured comparatively easily
Acromioclavicular	Distal end of clavicle with acromion of scapula	Diarthrotic (gliding)	Gliding; elevation, depression, protraction, and retraction
Thoracic	Heads of ribs with bodies of vertebrae	Diarthrotic (gliding)	Gliding
	Tubercles of ribs with transverse processes of vertebrae	Diarthrotic (gliding)	Gliding
Shoulder	Head of humerus in glenoid cavity of scapula	Diarthrotic (ball and socket type)	Flexion, extension, abduction, adduction, rotation, and circumduction of upper arm; one of most freely movable of joints
Elbow	Trochlea of humerus with semilunar notch of ulna; head of radius with capitulum of humerus	Diarthrotic (hinge type)	Flexion and extension
	Head of radius in radial notch of ulna	Diarthrotic (pivot type)	Supination and pronation of lower arm and hand; rotation of lower arm on upper as in using screwdriver
Wrist	Scaphoid, lunate, and triquetral bones articulate with radius and articular disc	Diarthrotic (condyloid)	Flexion, extension, abduction, and adduction of hand
Carpal	Between various carpals	Diarthrotic (gliding)	Gliding

*Vertebrae not easily dislocated; held securely together by following ligaments: anterior and posterior longitudinal ligaments—between anterior and posterior surfaces of bodies of vertebrae; supraspinous ligaments (called ligamentum nuchae in cervical region)—between tops of spinous processes; interspinous ligaments—between sides of spinous processes; ligamentum flavum—between laminae.

Table 5-4. Description of individual joints — cont'd

Name	Articulating bones	Type	Movements
Hand	Proximal end of first meta-carpal with trapezium	Diarthrotic (saddle)	Flexion, extension, abduction, adduction, and circumduction of thumb and opposition to fingers; motility of this joint accounts for dexterity of human hand compared with animal forepaw
	Distal end of metacarpals with proximal end of phalanges	Diarthrotic (hinge)	Flexion, extension, limited abduction, and adduction of fingers
	Between phalanges	Diarthrotic (hinge)	Flexion and extension of finger sections
Sacroiliac	Between sacrum and two ilia	Diarthrotic (gliding); joint cavity mostly obliterated after middle life	None or slight — e.g., during late months of pregnancy and during delivery
Symphysis pubis	Between two pubic bones	Synarthrotic (or amphiarthrotic), cartilaginous	Slight, particularly during pregnancy and delivery
Hip	Head of femur in acetabulum of os coxa	Diarthrotic (ball and socket)	Flexion, extension, abduction, adduction, rotation, and circumduction
Knee	Between distal end of femur and proximal end of tibia; largest joint in body	Diarthrotic (hinge type)	Flexion and extension; slight rotation of tibia
Tibiofibular	Head of fibula with lateral condyle of tibia	Diarthrotic (gliding type)	Gliding
Ankle	Distal ends of tibia and fibula with talus	Diarthrotic (hinge type)	Flexion (dorsiflexion) and extension (plantar flexion)
Foot	Between tarsals	Diarthrotic (gliding)	Gliding; inversion and eversion
	Between metatarsals and phalanges	Diarthrotic (hinge type)	Flexion, extension, slight abduction, and adduction
	Between phalanges	Diarthrotic (hinge type)	Flexion and extension

movement of the forearm which turns the palm forward as it is in the anatomical position. *Pronation* is turning the forearm so as to bring the back of the hand forward. *Inversion* is a special movement of the ankle which turns the sole of the foot inward, while *eversion* turns it outward. *Protraction* moves a part forward, such as sticking out the jaw. *Retraction* is a reverse of protraction.

outline summary

MEANING
All bones and their joints

FUNCTIONS
1 Furnishes supporting framework
2 Affords protection
3 Movement; bones constitute levers for muscle action

BONE AND CARTILAGE
Microscopic structure

Bone
1 Mainly calcified matrix – cement substance impregnated with calcium salts and reinforced by collagenous fibers
2 Lamellae – concentric cylindrical layers of calcified matrix enclosing haversian canal that contains blood vessel
3 Haversian system – canal and surrounding lamellae
4 Lacunae – microscopic spaces containing osteocytes (bone cells); lie between lamellae
5 Canaliculi – microscopic canals radiating in all directions from lacunae, connecting them with haversian canals; routes by which tissue fluid reaches osteocytes
6 Compact bone (or solid bone) – no empty spaces; lamellae fit closely together
7 Cancellous bone (or spongy bone) – many spaces in matrix which is arranged mainly in trabeculae rather than lamellae

Cartilage
1 Similar to that of bone with following exceptions:
a Cartilage matrix – firm gel; bone matrix – calcified cement substance
b Cartilage matrix – no canal system, no blood vessels; bone matrix – extensive canal network

BONES
Types
1 Long (femur)
2 Short (carpals)
3 Flat (parietal)
4 Irregular (vertebrae)

Structure
1 Long bones – see Figs. 5-1 to 5-3
2 Short bones – thin layer of compact bone encasing "core" of cancellous bone
3 Flat bones – layer of cancellous bone between two plates of compact bone
4 Irregular bones – thin layer of compact bone encasing cancellous bone

Formation and growth
1 Formation
a Skeleton preformed in hyaline cartilage and fibrous membranes; most cartilaginous or membranous structures changed into bone before birth but not complete until about 25 years of age
b Endochondral ossification – incompletely understood process which replaces hyaline cartilage "bones" with true bones
c Intramembranous ossification – process which replaces fibrous membrane "bones" with true bones
2 Growth
a In length – by continual thickening of epiphyseal cartilage followed by ossification
b In diameter – medullary cavity enlarged by osteoclasts destroying bone around it while new bone added around circumference by osteoblasts
3 Correlation with bone disease; osteoporosis – deficient synthesis of organic bone matrix by osteoblasts lacking stimulation of normal amounts of sex hormones in blood and of adequate muscular activity

Names and numbers
Total, 206 bones – Table 5-1

Axial skeleton (80 bones)

1 Skull (28 bones)
   a Cranium (8 bones) – frontal, parietal (2), temporal (2), occipital, sphenoid, and ethmoid
   b Face (14 bones) – nasal (2), maxillary (2), malar (2), mandible, lacrimal (2), palatine (2), inferior turbinates (2), and vomer
   c Ear ossicles (6 bones) – malleus (2), incus (2), and stapes (2)
2 Hyoid (1 bone)
3 Vertebral column (26 vertebrae)
   a Cervical (7)
   b Thoracic (12)
   c Lumbar (5)
   d Sacrum
   e Coccyx
4 Sternum and ribs (25 bones) – sternum, true ribs (7 pairs), false ribs (5 pairs, 2 pairs of which are floating)

Appendicular skeleton (126 bones)

1 Upper extremities (64 bones) – clavicle (2), scapula (2), humerus (2), ulna (2), radius (2), carpals (16), metacarpals (10), and phalanges (28)
2 Lower extremities (62 bones) – os coxa (2), femur (2), patella (2), tibia (2), fibula (2), tarsal (14), metatarsal (10), and phalanges (28)

Markings
Depressions and openings

1 Fossa – hollow or depression
2 Sinus – cavity of spongelike air space within bone
3 Foramen – hole
4 Meatus – tube-shaped opening

Projections or processes

1 Those which fit into joints
   a Condyle – rounded projection entering into formation of joint
   b Head – rounded projection beyond narrow neck
2 Those to which muscles attach
   a Trochanter – very large process
   b Crest – ridge
   c Spinous process or spine – sharp projection
   d Tuberosity – large, rounded projection
   e Tubercle – small rounded projection

Identification

See Table 5-2

Differences between male and female skeletons

1 Male skeleton larger and heavier
2 Male pelvis deep and funnel shaped with narrow pubic arch; female pelvis shallow, broad, and flaring with wider pubic arch and larger iliosacral notch

Age changes in skeleton

1 From infancy to young adulthood – absolute and relative sizes of bones change
2 From young adulthood to old age – texture of bones change as does contour of bone margins and markings

JOINTS BETWEEN BONES (ARTICULATIONS)
Kinds

1 Diarthroses
   a Characteristics
      1 Small space, joint cavity, present between articulating surfaces of two bones that form joint
      2 Thin layer of hyaline cartilage covers articular surfaces
      3 Fibrous, synovial-lined capsule encases joint
      4 Ligaments hold articulating bones firmly connected
      5 Free movement possible at all diarthrotic joints
   b Subtypes – see Table 5-3
2 Synarthroses
   a Characteristics
      1 No joint cavity
      2 Cartilage, fibrous tissue, or bone grows between articulating surfaces of bones
      3 Little or no movement possible at synarthrotic joints
   b Subtypes – see Table 5-3

Movements

1 Flexion – angle at joint decreases
2 Extension – angle at joint increases; return from flexion
3 Abduction – moving bone away from body's median plane
4 Adduction – moving bone back toward body's median plane
5 Rotation – pivoting bone upon its axis
6 Circumduction – describing surface of cone with moving parts
7 Special movements
   a Supination – movement of forearm which turns palm forward
   b Pronation – movement of forearm which turns back of hand forward
   c Inversion – ankle movement turning sole of foot inward
   d Eversion – ankle movement turning sole of foot outward
   e Protraction – moving part forward
   f Retraction – pulling part back; opposite of protraction

review questions

1 What general functions does the skeletal system perform?
2 Describe the microscopic structure of bone and cartilage.
3 Describe the structure of a long bone.
4 Describe the general plan of the skeleton.
5 Name the bones of the adult skeleton.
6 Describe the structural features of diarthrotic joints that facilitate movement.
7 Give examples of several types of diarthrotic joints.
8 Explain the functions of the periosteum.
9 What joint(s) unites the shoulder girdle with the trunk? The pelvic girdle with the trunk?
10 Name the several kinds of movements possible at joints. Define each movement named.
11 Do vertebrae become dislocated easily? Substantiate your answer.
12 Name the primary and secondary curves of the spine. Describe each.
13 Name the five pairs of bony sinuses in the skull.
14 Name the bones which fuse to form the coccyx.
15 What is the true pelvis? The false pelvis? Name the boundary line between the true and false pelves.
16 Through what opening does the spinal cord enter the cranial cavity?
17 Explain the basic steps in the process of ossification according to the concept described in the text.
18 Compare osteoblasts and osteoclasts as to function.
19 What two factors presumably relate to the high incidence of osteoporosis following the menopause?
20 Define or make an identifying statement about each of the following terms: condyle; crest; diaphysis; diarthroses; endosteum; epiphysis; foramen; fossa; haversian system; kyphosis; lordosis; medullary cavity; osteoblast; osteoclast; periosteum; rotation; scoliosis; sinus; spinous process; synarthroses; trochanter; trabeculae.

6

The muscular system

Man's survival depends in large part upon his ability to adjust to the changing conditions of his environment. Movements constitute the major part of this adjustment. Whereas most of the systems of the body play some role in accomplishing movement, it is the skeletal and muscular systems acting together which actually produce movements. We have investigated the architectural plan of the skeleton and have seen how its joint structures and firm supports make movement possible. However, bones and joints cannot move themselves. They must be moved by something. Muscle tissue, because of its contractility, extensibility, and elasticity, is admirably suited to this function. Our subject then for this chapter is skeletal muscles. (Cardiac muscle will be discussed in Chapter 10, and information about smooth muscle appears in several chapters.) In this chapter we shall try to discover how muscles move bones, and to do this we shall try to answer many other questions—how the structure of muscles adapts them to their function, how energy is made available for their work, how muscle activity contributes to the health and survival of the whole body—to mention only a few.

Properly speaking, the term muscular system means all the muscles of the body—those attached to the bones, those helping to make up the walls of numerous internal structures, and the muscle which com-

General functions

Skeletal muscle cells
Microscopic structure
Molecular structure
Functions
Energy sources for muscle
  contraction

Skeletal muscle organs
Structure
  Size, shape, and fiber
    arrangement
  Connective tissue
    components
  Nerve supply
  Age changes
Function
  Basic principles
  Hints on how to deduce
    actions

Names
  Reasons for names
  Muscles grouped
    according to location
  Muscles grouped
    according to function
Origins, insertions,
  functions, and
  innervations of
  representative muscles

Weak places in abdominal wall

Bursae
Definition
Locations
Function

Posture
Meaning
How maintained
Importance to body as
  whole

poses the wall of the heart. More commonly, however, the term refers to skeletal muscles only—those muscle masses which attach to bones and move them about, the masses of "red meat" of the body.

■ General functions

If you have any doubts about the importance of muscle function to normal life, you have only to observe a person with extensive paralysis—a victim of severe poliomyelitis, for example. Any of us possessed of normal powers of movement can little imagine life with this matchless power lost. But cardinal as it is, movement is not the only contribution muscles make to healthy survival. They also perform two other essential functions: maintenance of posture and production of a large portion of body heat.

Movement. Movement consists sometimes of locomotion, sometimes of movements of parts of the body, sometimes of changes in the size of openings, and sometimes of propulsion of substances through tubes, for example. Propulsion of blood through arteries by heart movements is one example of the latter. Passage of food through the digestive tract by contractions of the stomach and intestines is another. By means of locomotion, adjustments are made to the external environment. Desirable objects are approached, and undesirable or dangerous ones are repelled. By means of internal movements, vital adjustments and processes are accomplished. Consider the following examples: contraction and relaxation of the iris muscles which allow just the right amount of light to enter the eyes, contractions of the digestive tract muscles which promote digestion and elimination, and contractions of the heart which keep the blood circulating.

Posture. The continued partial contraction of many skeletal muscles makes possible standing, sitting, and other maintained positions of the body.

Heat production. Muscle cells produce heat by the same process as do all cells—namely, catabolism (see Fig. 2-15, p. 32). But because skeletal muscle cells are both highly active and numerous, they produce a major share of total body heat. Skeletal muscle contractions, therefore, constitute one of the most important parts of the mechanism for maintaining homeostasis of temperature.

■Skeletal muscle cells

Microscopic structure

Because they are so long and narrow, muscle cells are usually referred to as *muscle fibers* rather than cells. Also, several of the structures common to all cells go by different names in muscle fibers. For example, *sarcolemma* is the cell membrane of a muscle fiber. *Sarcoplasm* is its cytoplasm. Muscle cells contain a network of tubules and sacs known as the *sarcoplasmic reticulum*—a structure analogous, but not identical, to the endoplasmic reticulum of other cells. Muscle fibers contain certain structures found only in them. Most numerous of these are small fibers packed close together in the sarcoplasm and extending lengthwise through it. *Myofibrils* these are called. Alternating dark and light horizontal stripes characterize myofibrils. As you have probably already guessed, these are the cross striae that gave skeletal muscle its other name—striated muscle. The dark stripes are called *A bands* (for anisotropic), and the light stripes are called *I bands* (for isotropic). Note in Fig. 6-1 the rather curious fact that a light band—the *H zone*—crosses the middle of the dark A band and that a dark line—the *Z line*—passes through the center of the light I band. A myofibril is described as consisting of several segments or sarcomeres. A *sarcomere* is a section extending from one Z line of a myofibril to the next.

Another structure unique to skeletal muscle cells is the *T system*. This name derives from the fact that this system consists of tubules that extend transversely into the sarcoplasm. T system tubules enter the sarcoplasm at the levels of the Z lines—that is, in the middle of the light I bands. Because invaginations of the sarcolemma form the T system tubules, they open to the exterior of the muscle fiber.

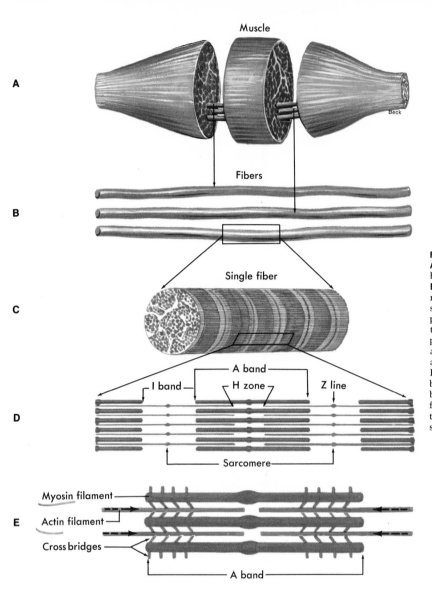

Fig. 6-1 Structure of skeletal muscle. **A,** Skeletal muscle organ, composed of bundles of muscle fibers—i.e., cells. **B,** Single fibers, enlarged. **C,** Greater magnification of single fiber showing smaller fibers—myofibrils—in its sarcoplasma. **D,** Myofibril magnified further to show thick and thin filaments composing it and producing its cross-striated appearance; dark A (anisotropic) bands alternate with light I (isotropic) bands; H zones, less dense midsections of A bands; Z lines, more dense midlines of I bands. **E,** Molecular structure of myofibril; thick filaments, myosin molecules; thin filaments, actin molecules. (Also see Fig. 6-3.)

A distinctive feature of the sarcoplasmic reticulum is that its tubules terminate in closed sacs at the ends of each sarcomere —that is, immediately above and below each Z line. Since T tubules constitute the Z lines, the sacs of the sarcoplasmic reticulum of one sarcomere lie just above a T tubule, whereas those of the next sarcomere lie just below it. This forms a triple-layered structure (a T tubule sandwiched between sacs of the sarcoplasmic reticulum) called a *triad*.

Molecular structure

Many of the same compounds compose skeletal muscle fibers as compose other cells—proteins, carbohydrates, lipids, and water, to name a few. Among the compounds present only in muscle are the proteins named actin, myosin, tropomyosin, and troponin. Myoglobin ("muscle hemoglobin") is a pigment present in skeletal muscle; it transports oxygen from blood capillaries to a muscle fiber's mitochondria.*

---

*Myoglobin, like hemoglobin, contains iron and gives a red color to cells which contain it. Muscle fibers deficient in myoglobin appear pale—the white meat of fowl, for example. Most of the fibers of red muscles, on the other hand, contain relatively large amounts of myoglobin.

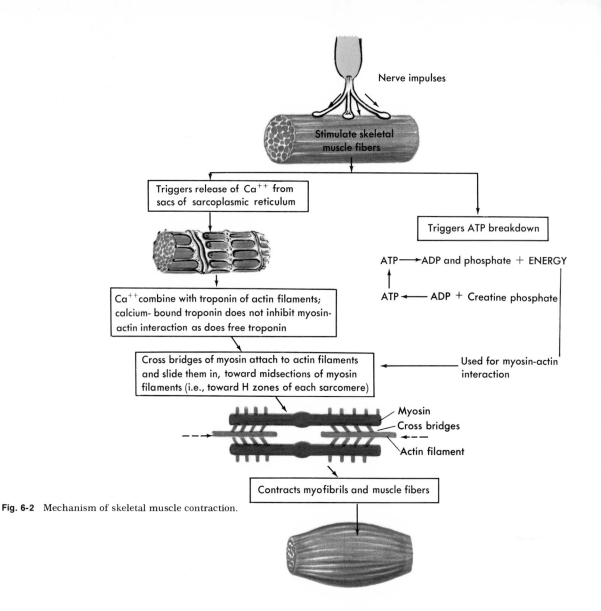

Fig. 6-2 Mechanism of skeletal muscle contraction.

The powerful magnification of electron microscopes has shown that myofibrils consist of still finer threads known as thick and thin filaments. Protein molecules compose them both. *Thin filaments* consist mainly of two chains of actin molecules twisted together. Associated with them, however, are two other proteins—namely troponin and tropomyosin. Myosin molecules compose the *thick filaments* of myofibrils. Now examine Fig. 6-1 to discover both the relative lengths of the thick myosin and the thin actin filaments and their placement in relation to each other. If you look carefully at the thin actin filaments, you will see that they extend from the mid-

section of one A band to the midsection of the next A band—in other words, from an H zone to an H zone but not through the H zones. Notice next that each thick myosin filament extends only the length of an A band. No myosin is present in the light-colored I bands, only actin filaments.

Summarizing then, A bands are the dark cross striae and I bands are the light cross striae of skeletal muscle fibers. Both ends of the A bands appear dark because their composition of both myosin and actin filaments gives them greater density. Their narrow midsections—the H zones—are less dense and, therefore, lighter in color because they consist of myosin only and no

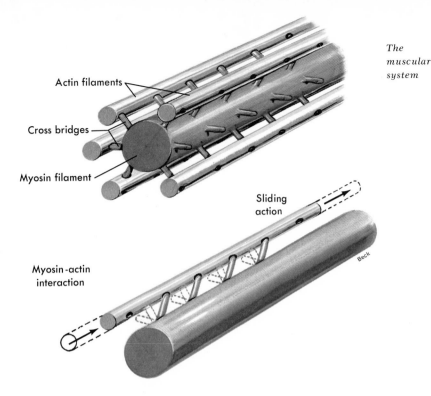

Actin filaments

Cross bridges

Myosin filament

Sliding action

Myosin-actin interaction

Beck

Fig. 6-3 Scheme to show how myosin interacts with actin to shorten muscle fibers. The cross bridges of a single myosin fiber are believed to point at 60° angles at six actin filaments. The cross bridges bond to the actin filaments and slide them toward the center of the sarcomere.

actin filaments. I bands are also less dense and lighter colored than A bands, but for a different reason—because they consist of actin only and no myosin.

Functions

Like other cells of the body, skeletal muscle cells perform both self-serving and body-serving functions. Muscle cells specialize in the body-serving function of contraction. But before we try to explain the mechanics of skeletal muscle contraction, we must call your attention to one more detail about the structure of a skeletal muscle fiber. Note in Fig. 6-1, *E*, the extensions or *cross bridges* projecting from both ends of the myosin filaments. When nerve impulses arrive at a skeletal muscle fiber, they initiate impulse conduction over its sarcolemma and inward via its T tubules. This triggers the release of calcium ions from the sacs of the sarcoplasmic reticulum (Fig. 6-2) into the sarcoplasm. Here, calcium ions combine with the troponin associated with the actin filaments of the myofibrils. In a resting muscle fiber, troponin prevents myosin from interacting with actin. But calcium-bound troponin

cannot exert this inhibiting action and therefore myosin-actin interaction takes place. Specifically, the cross bridges of myosin attach to the actin filaments and pull them toward the middle of the myosin filaments of each sarcomere. Fig. 6-3 shows how this sliding action of the actin filaments shortens the sarcomeres of the myofibrils. And obviously, the shortening of the sarcomeres shortens the myofibrils and this, in turn, shortens, that is, contracts the muscle fiber which the myofibrils compose.

Relaxation of a muscle fiber is now believed to be brought about by a reversal of the contraction mechanism just described. The calcium-troponin combinations separate, calcium ions reenter the sacs of the sarcoplasmic reticulum, and troponin, now no longer bound to calcium, inhibits myosin-actin interaction. In short, then, according to current theory, ". . . contraction is turned on by the release of calcium and turned off by its withdrawal."*

*From Hoyle, G.: How is muscle turned on and off? Sci. Amer. **222**:84-93 (April), 1970 (a fascinating account of recent experiments with individual fibers of a giant barnacle); copyright © 1970 by Scientific American, Inc.; all rights reserved.

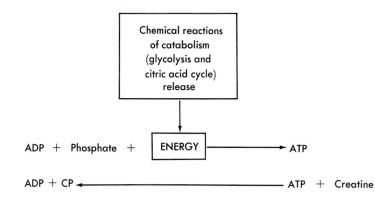

```
            ┌─────────────────────┐
            │  Chemical reactions │
            │    of catabolism    │
            │    (glycolysis and  │
            │    citric acid cycle)│
            │       release       │
            └─────────────────────┘
                      │
                      ▼
                 ┌─────────┐
ADP + Phosphate +│ ENERGY  │───────────────→ ATP
                 └─────────┘

ADP + CP ◄─────────────────────────── ATP + Creatine
```

Fig. 6-4 Noncontracting muscle cells use less ATP for their energy needs than they synthesize. They store some of this excess ATP and convert some of it to another high-energy compound, creatine phosphate. Thus, when muscle cells are not contracting, they are accumulating ATP and creatine phosphate as a reserve supply for future energy needs.

Muscle cells obey the all-or-none law when they contract. This means that they either contract with all the force possible under existing conditions or they do not contract at all. However, if conditions at the time of stimulation change, then the force of the cell's contraction changes. Suppose, for example, that a particular muscle fiber receives an adequate oxygen supply at one time and an inadequate supply at another time. It will contract more forcefully with an adequate than with a deficient oxygen supply.

Energy sources for muscle contraction

ATP breakdown presumably furnishes the energy for muscle contraction as well as for other kinds of cellular work. Nerve impulses arriving at a muscle cell trigger ATP breakdown. How they do this is not well understood. But in one way or another, stimulation acts to break ATP's terminal high-energy bonds. Result? ATP molecules split apart into adenosine diphosphate (ADP) and phosphate, and as ATP's terminal bonds are broken, energy is released from them and does the work of contracting the muscle fiber (see first equation in Fig. 6-2).

As you already know from the discussion on pp. 29 to 34, ATP is produced in all cells by catabolism. But muscle cells also have another and more rapid way of producing ATP—specifically, by combining ADP with creatine phosphate (CP; phosphocreatine)

as shown in the second equation in Fig. 6-2. CP is a high-energy phosphate compound synthesized only in muscle cells. They make CP when they are not contracting. At such times, muscle cells are synthesizing more ATP than they are breaking down to supply their low-energy needs of the moment. Thus, during these periods, they convert some of the excess ATP to high-energy creatine phosphate, which they also store (see second equation in Fig. 6-4). Then later, during periods of strenuous contraction, when the cell needs more ATP quickly, this reaction reverses itself and almost instantaneously provides more ATP, as indicated by the second equation in Fig. 6-2.

Other chemical changes also occur in muscle during strenuous exercise. Soon after vigorous activity begins, pyruvic acid starts to accumulate in muscle cells and is reduced to lactic acid. The reason is this: respiration and circulation cannot accelerate enough to supply the cells with as much oxygen as they need to oxidize the pyruvic acid via the citric acid cycle as rapidly as it forms by glycolysis. Most of the lactic acid (about four-fifths according to some investigators) diffuses out of the cells and is carried by the blood to the liver, there to be gradually resynthesized to glycogen or glucose. A smaller amount remains behind in the muscle cells. Here, after the strenuous bout of exercise ends and oxygen again becomes available, the lactic acid is oxidized back to pyruvic acid and on through the citric acid cycle to carbon dioxide and water.

The amount of oxygen required for oxidation of the lactic acid accumulated during strenuous exercise is referred to as the *oxygen debt*. It is a debt that has to be repaid before exercise can continue—a fact all of us have observed when we have stopped to catch our breath during some strenuous exertion. For example, if you were to run the 100-yard dash in 12 seconds, your body would need about 6 liters of oxygen to oxidize all the pyruvic acid formed by glycolysis during that short sprint. But the maximum amount of oxygen your body could actually use in that short period—about 4 liters per minute,* or 0.8 of a liter in 12 seconds—would fall far short of this amount. You would, therefore, have incurred an oxygen debt of several liters which would be repaid by your breathing rapidly and deeply for some time after your strenuous exertion. The maximum oxygen consumption possible varies considerably in different individuals, depending upon previous athletic training, age, general health, and other factors. In general, the more exercise a person habitually indulges in, the higher his maximum oxygen consumption. But also, the older an adult becomes, or the poorer his general health, the lower his oxygen consumption usually is.

▊ Skeletal muscle organs

Structure

SIZE, SHAPE, AND FIBER ARRANGEMENT

The structures called skeletal muscles are organs. They consist mainly of skeletal muscle tissue plus important connective and nervous tissue components. Skeletal

---

*Mountcastle, V. B., editor: Medical physiology, ed. 12, St. Louis, 1968, The C. V. Mosby Co.

muscles vary considerably in size, shape, and arrangement of fibers. They range from extremely tiny strands as, for example, the stapedius muscle of the middle ear, to large masses such as the muscles of the thigh. Some skeletal muscles are broad in shape and some narrow. Some are long and tapering and some short and blunt. Some are triangular, some quadrilateral, and some irregular. Some form flat sheets and others bulky masses.

Arrangement of fibers varies in different muscles. In some muscles, the fibers are parallel to the long axis of the muscle, in some they converge to a narrow attachment, and in some they are oblique and either pennate (like the feathers in an old-fashioned plume pen) or bipennate (double-feathered as in the rectus femoris). Fibers may even be curved, as in the sphincters of the face, for example. The direction of the fibers composing a muscle is significant because of its relationship to function. For instance, a muscle with the bipennate fiber arrangement can produce the strongest contraction.

CONNECTIVE TISSUE COMPONENTS

A fibrous connective tissue sheath (*epimysium*) envelops each muscle and extends into it as partitions between bundles of its fibers (*perimysium*) and between individual fibers (*endomysium*). Because all three of these structures are continuous with the fibrous structures that attach muscles to bones or other structures, muscles are most firmly harnessed to the structures they pull against during contraction. The epimysium, perimysium, and endomysium of a muscle, for example, may be continuous with fibrous tissue that extends from the muscle as a *tendon*, a strong tough cord continuous at its other end with the fibrous covering of bone (periosteum). Or the fibrous wrapping of a muscle may extend as a broad, flat sheet of connective tissue (*aponeurosis*) to attach it to adjacent structures, usually the fibrous wrappings of another muscle. So tough and strong are tendons and aponeuroses that they are not often torn, even by injuries forceful enough to break bones or tear muscles. They are, however, occasionally pulled away from bones.

Tube-shaped structures of fibrous connective tissue called *tendon sheaths* enclose certain tendons, notably those of the wrist and ankle. Like the bursae, tendon sheaths have a lining of synovial membrane. Its moist smooth surface enables the tendon to move easily, almost frictionlessly, in the tendon sheath.

You may recall that a continuous sheet of loose connective tissue known as the superficial fascia lies directly under the skin. Under this lies a layer of dense fibrous connective tissue, the *deep fascia*. Plate II (in transparent Trans-Vision®) of the color insert (p. 60) shows sections of the deep fascia very clearly. Extensions of the deep fascia form the epimysium, perimysium, and endomysium of muscles and their attachments to bones and other structures and also enclose viscera, glands, blood vessels, and nerves.

NERVE SUPPLY

A nerve cell that transmits impulses to a skeletal muscle is called a *somatic motoneuron*. One such neuron plus the muscle cells in which its axon terminates constitutes a *motor unit* (Fig. 6-5). The single axon fiber of a motor unit divides, upon entering the skeletal muscle, into a variable number of branches. Those of some motor units terminate in only a few muscle fibers, whereas others terminate in numerous fibers. Consequently, impulse conduction by one motor unit may stimulate only a half dozen or so muscle fibers to contract at one time, whereas conduction by another motor unit may activate a hun-

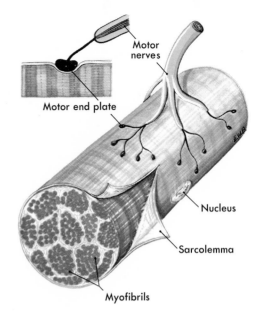

Fig. 6-5 A motor unit consists of one motoneuron and the muscle cells supplied by its axon branches. The diagram shows a motor axon ending in several unmyelinated branches. Each branch terminates in a motor end-plate embedded, as the insert shows, in a tiny trough on the surface of the muscle fiber.

dred or more fibers simultaneously. This fact bears a relationship to the function of the muscle as a whole. As a general rule, the fewer the number of fibers supplied by a skeletal muscle's individual motor units, the more precise the movements that muscle can produce. For example, in certain small muscles of the hand, each motor unit includes only a few muscle fibers, and these muscles produce precise finger movements. In contrast, motor units in large abdominal muscles that do not produce precise movements are reported to include more than a hundred muscle fibers each.

The area of contact between a nerve and muscle fiber is known as the *motor end plate* or *neuromuscular junction*. When nerve impulses reach the ends of the axon fibers in a skeletal muscle, small vesicles

in the axon terminals release a chemical
–acetylcholine–into the neuromuscular
junction. Diffusing swiftly across this microscopic trough, acetylcholine contacts
the sarcolemma of the adjacent muscle
fiber, stimulating the fiber to contract. In
addition to the many motor nerve endings, there are also many sensory nerve
endings in skeletal muscles. These will
be described in the chapter on the nervous
system.

AGE CHANGES

As a person grows old, his skeletal muscles undergo a process called *fibrosis*.
Gradually, some of the skeletal muscle
fibers degenerate, and fibrous connective
tissue replaces them. With this loss of
muscle fibers and increase in connective
tissue comes waning muscular strength,
a common finding in elderly people.* Other
factors probably also contribute to decreasing muscular strength in advanced age.

Function

Several methods of study have been used
to amass the present-day store of knowledge about how muscles function. They
vary from the traditional and relatively
simple procedures, such as observing and
palpating muscles in action, manipulating
dissected muscles to observe movements,
or deducing movements from knowledge of
muscle anatomy, to the newer more complicated method of electromyography (recording action potentials from contracting
muscles). As a result, there is now a somewhat overwhelming amount of knowledge
about muscle functions. So perhaps we
can thread our way through this maze of
detail more easily if we start with general
principles and then go on to the details
that seem to us most useful.

*Shock, N. W.: The physiology of aging, Sci. Amer.
206:100-110 (Jan.), 1962.

BASIC PRINCIPLES

1 *Skeletal muscles contract only if
stimulated.* They do not have the quality of
automaticity inherent in cardiac and visceral muscle. Although nerve impulses are
the natural stimuli for skeletal muscles,
electrical and some other artificial stimuli
can also activate them. A skeletal muscle
deprived of nerve impulses by whatever
cause is a functionless mass. One should,
therefore, think of a skeletal muscle and
its motor nerve as a physiological unit,
always functioning together, either useless
without the other.

2 *A skeletal muscle contraction may be
any one of several types.* It may be a tonic
contraction, an isotonic contraction, an
isometric contraction, a twitch contraction,
or a tetanic contraction. And there are also
other types of contraction–called treppe,
fibrillation, and convulsions–.

a A *tonic contraction* (*tone; tonus*) is
a continual, partial contraction. At any
one moment, a small number of the total
fibers in a muscle contract, producing a
tautness of the muscle rather than a recognizable contraction and movement. Different groups of fibers scattered throughout
the muscle contract in relays. Tonic contraction, or tone, is characteristic of the
muscles of normal individuals when they
are awake. It is particularly important for
maintaining posture. A striking illustration
of this fact is the following: when a person
loses consciousness, his muscles lose their
tone and he collapses in a heap, unable to
maintain a sitting or standing posture.
Muscles with less tone than normal are
described as flaccid muscles and those
with more than normal tone are called
spastic. Impulses over stretch reflex arcs
(Fig. 7-5, p. 173) maintain tone.

b An *isotonic contraction* (*iso*, same;
tonic, pressure or tension) is a contraction
in which the pressure or tension within a
muscle remains the same but in which the

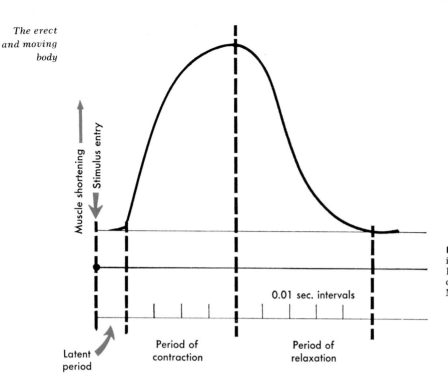

Muscle shortening

Stimulus entry

0.01 sec. intervals

Latent period

Period of contraction

Period of relaxation

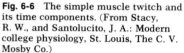

Fig. 6-6 The simple muscle twitch and its time components. (From Stacy, R. W., and Santolucito, J. A.: Modern college physiology, St. Louis, The C. V. Mosby Co.)

length of the muscle changes. It shortens, producing movement.

c An *isometric contraction* is a contraction in which muscle length remains the same but in which muscle tension increases. You can observe isometric contraction by pushing your arms against a wall and feeling the tension increase in your arm muscles. Isometric contractions "tighten" a muscle but they do not produce movements nor do work. Isotonic contractions, on the other hand, both produce movements and do work.

d A *twitch contraction* is a quick, jerky contraction in response to a single stimulus. Fig. 6-6 shows a record of such a contraction. It reveals that the muscle does not shorten at the instant of stimulation, but rather a fraction of a second later, and that it reaches a peak of shortening and then gradually resumes its former length. These three phases of contraction are spoken of, respectively, as the *latent period*, the *contraction phase*, and the *relaxation phase*. The entire twitch usually lasts less than 1/10 of a second. Twitch contractions rarely occur in the body.

e A *tetanic contraction (tetanus)* is a more sustained contraction than a twitch. It is produced by a series of stimuli bombarding the muscle in rapid succession. About 30 stimuli per second, for example, evoke a tetanic contraction by a frog gastrocnemius muscle, but the rate varies for different muscles and different conditions. Fig. 6-7 shows records of incomplete and complete tetanus. Normal movements are said to be produced by incomplete tetanic contractions.

f *Treppe (staircase phenomenon)* is a phenomenon in which increasingly stronger twitch contractions occur in response to constant strength stimuli repeated at the rate of about once or twice a second. In other words, a muscle contracts more forcefully after it has contracted a few times than when it first contracts — a principle made practical use of by athletes when they warm up but one not yet satisfactorily explained. Presumably, it relates partly to the rise in temperature of active muscles and partly to their accumulation of metabolic products. After the first few stimuli, muscle responds to

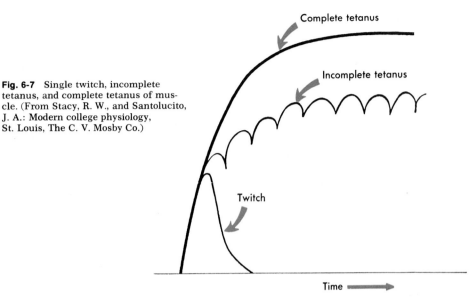

Fig. 6-7 Single twitch, incomplete tetanus, and complete tetanus of muscle. (From Stacy, R. W., and Santolucito, J. A.: Modern college physiology, St. Louis, The C. V. Mosby Co.)

a considerable number of successive stimuli with maximal contractions (Fig. 6-8). After these, it responds with less and less strong contractions. The relaxation phase becomes shorter and finally disappears entirely. In other words, the muscle stays partially contracted—an abnormal state of prolonged contraction called *contracture*.

Repeated stimulation of muscle eventually lessens its irritability and contractility and may result in muscle fatigue, a condition in which the muscle does not respond to the strongest stimuli. Complete muscle fatigue, however, very seldom occurs in the body, although it can be readily induced in an excised muscle.

g *Fibrillation* is an abnormal type of contraction in which individual fibers contract asynchronously, producing a flutter of the muscle but no effective movement. Fibrillation of the heart, for example, occurs fairly often.

h *Convulsions* are abnormal uncoordinated tetanic contractions of varying groups of muscles.

Fig. 6-8 Record of several successive muscle contractions, showing treppe occurring in the first few. (From Stacy, R. W., and Santolucito, J. A.: Modern college physiology, St. Louis, The C. V. Mosby Co.)

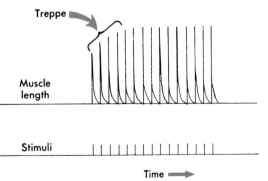

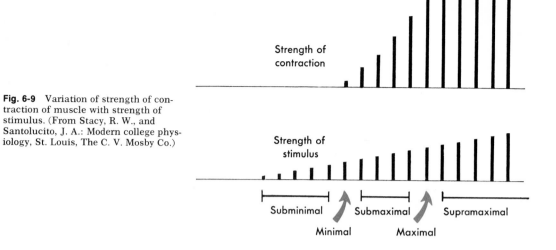

Strength of
contraction

Strength of
stimulus

Subminimal Submaximal Supramaximal

Minimal Maximal

Fig. 6-9 Variation of strength of contraction of muscle with strength of
stimulus. (From Stacy, R. W., and
Santolucito, J. A.: Modern college physiology, St. Louis, The C. V. Mosby Co.)

3 *Skeletal muscles contract according
to the graded strength principle* (Fig. 6-9)
—not according to the all-or-none principle,
as do the individual muscle cells composing them. In other words, skeletal muscles
contract with varying degrees of strength
at different times—a fact of practical importance. (How else, for example, could
we match the force of a movement to the
demands of a task?)

Several generalizations may help explain the fact of graded strength contractions. The strength of the contraction of a
skeletal muscle bears a direct relationship
to the initial length of its fibers, to their
metabolic condition, and to the number of
them contracting. If a muscle is moderately stretched at the moment when contraction begins, the force of its contraction
increases. This principle, established years
ago, applies experimentally to heart muscle also (Starling's law of the heart, discussed in Chapter 10). Outstanding among
metabolic conditions that influence contraction are oxygen and food supply.

With adequate amounts of these essentials, a muscle can contract with greater
force than possible with deficient amounts.
The greater the number of muscle fibers
contracting simultaneously, the stronger
the contraction of a muscle. How large
this number is depends upon how many

motor units are activated, and this, in turn,
depends upon the intensity and frequency
of stimulation. In general, the more intense and the more frequent a stimulus,
the more motor units and therefore the
more fibers are activated and the stronger
the contraction. Contraction strength also
relates to previous contraction, the warm-
up principle discussed on p. 122.

Another factor that influences the force
of contraction is the size of the load imposed on the muscle. Within certain limits,
the heavier the load, the stronger the contraction. Lift a pencil, for example, and
then a heavy book and you can feel your
arm muscles contract more strongly with
the book.

The factors that influence muscle contraction are summarized in Fig. 6-10.

4 *Skeletal muscles produce movements
by pulling on bones.* Most of our muscles
span at least one joint and attach to both
articulating bones. When they contract,
therefore, their shortening puts a pull on
both bones, and this pull moves one of the
bones at the joint—draws it toward the
other bone, much as a pull on marionette
strings moves a puppet's parts. (In case
you are wondering why both bones do not
move since both are pulled on by the contracting muscle, the reason is that one of
them is normally stabilized by contraction

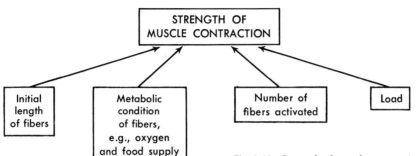

Fig. 6-10 Strength of muscle contraction.

of other muscles or by certain features of its own that make it less mobile.)

5 *Bones serve as levers, and joints serve as fulcrums of these levers.* (By definition, a *lever* is any rigid bar free to turn about a fixed point called its *fulcrum*.) A contracting muscle applies a pulling force on a bone lever at the point of the muscle's attachment to the bone. This causes the bone (referred to as the insertion bone) to move about its joint fulcrum. We have already noted that a skeletal muscle and its motor nerve act as a functional unit. Now we can add bones and joints to this unit and can describe the physiological unit for movement as a neuromusculoskeletal unit. Disease or injury of any one of these parts of the unit—of nerve or muscle or bone or joint—can, as you might surmise, cause abnormal movements or complete loss of movement. Poliomyelitis, for example, and multiple sclerosis and hemiplegia all involve the neural part of the unit. In contrast, muscular dystrophy affects the muscular part and arthritis the skeletal part.

6 *Muscles that move a part usually do not lie over that part.* In most cases, the body of a muscle lies proximal to the part moved. Thus, muscles that move the lower arm lie proximal to it—that is, in the upper arm. Applying the same principle, where

would you expect muscles that move the hand to be located? Those that move the lower leg? Those that move the upper arm?

7 *Skeletal muscles almost always act in groups rather than singly.* In other words, most movements are produced by the coordinated action of several muscles. Some of the muscles in the group contract while others relax. To identify each muscle's special function in the group, the following classification is used:

a *prime movers*—muscle or muscles whose contraction actually produces the movement

b *antagonists*—muscles which relax while the prime mover is contracting to produce movement (exception: contraction of the antagonist at the same time as the prime mover when some part of the body needs to be held rigid, such as the knee joint when standing*)

---

*Antagonistic muscles have opposite actions and opposite locations. If the flexor lies anterior to the part, the extensor will be found posterior to it. For example, the pectoralis major, the flexor of the upper arm, is located on the anterior aspect of the chest, while the latissimus dorsi, the extensor of the upper arm, is located on the posterior aspect of the chest. The antagonist of a flexor muscle is obviously an extensor muscle; that of an abductor muscle, an adductor muscle. Some frequently used antagonists are listed in Table 6-2.

c *synergists*—muscles which contract at the same time as the prime mover (may help the prime mover produce its movement or may stabilize a part—hold it steady—so that the prime mover produces a more effective movement)

HINTS ON HOW TO DEDUCE ACTIONS

To understand muscle actions, you need first to know certain anatomical facts such as which bones muscles attach to and which joints they pull across. Then if you relate these structural facts to functional principles (for instance, those discussed in the preceding paragraphs), you may find your study of muscles more interesting and less difficult than you anticipate. Some specific suggestions for deducing muscle actions follow.

1 Start by making yourself familiar with the names, shapes, and general locations of the larger muscles, using Table 6-1 as a guide.

2 Try to deduce which bones the two ends of a muscle attach to from your knowledge of the shape and general location of the muscle. For example, look carefully at the deltoid muscle as illustrated in Figs. 6-11 to 6-14. To what bones does it seem to attach? Check your deductions with Table 6-4.

3 Next, make a guess as to which bone moves when the muscle shortens. (The bone moved by a muscle's contraction is its *insertion* bone; the bone that remains relatively stationary is its *origin* bone.) In many cases, you can tell by trying to move one bone and then another which one is the insertion bone. In some cases, either bone may function as the insertion. Although not all muscle attachments can be deduced as readily as those of the deltoid, they can all be learned more easily by using this deduction method than by relying on rote memory alone.

4 Deduce a muscle's actions by applying the principle that its insertion moves toward its origin. Check your conclusions with the text. Here, as in steps 2 and 3, the method of deduction is intended merely as a guide and is not adequate by itself for determining muscle actions.

5 To deduce which muscle produces a given action (instead of which action a given muscle produces as in step 4), start by inferring the insertion bone (bone that moves during the action). The body and origin of the muscle will lie on one or more of the bones toward which the insertion moves—often a bone or bones proximal to the insertion bone. Couple these conclusions about origin and insertion with your knowledge of muscle names and locations to deduce the muscle that produces the action.

For example, if you wish to determine the prime mover for the action of raising the upper arms straight out to the sides, you infer that the muscle inserts on the humerus since this is the bone that moves. It moves toward the shoulder—that is, the clavicle and scapula—so that probably the muscle has its origin on these bones. Because you know that the deltoid muscle fulfills these conditions, you conclude, and rightly so, that it is the muscle that raises the upper arms sidewise.

6 Do not try to learn too many details about muscle origins, insertions, and actions. Remember, it is better to start by learning a few important facts thoroughly than to half learn a mass of relatively unimportant details. Remember, too, that trying to learn too many minute facts may well result in your not retaining even the main facts.

Names

REASONS FOR NAMES

Muscle names seem more logical and therefore easier to learn when one understands the reasons for the names. Each name describes one or more of the following features about the muscle.

1 *its action*—as flexor, extensor, adductor, etc.
2 *direction of its fibers*—as rectus or transversus
3 *its location*—as tibialis or femoris
4 *number of divisions composing a muscle*—as biceps, triceps, or quadriceps
5 *its shape*—as deltoid (triangular), trapezius, or quadratus
6 *its points of attachment*—as sternocleidomastoid

A good way to start the study of a muscle is by trying to find out what its name means.

MUSCLES GROUPED ACCORDING TO LOCATION

Just as names of people are learned by associating them with physical appearance, so the names of muscles should be learned by associating them with their appearance. As you learn each muscle name, study Figs. 6-11 to 6-34 to familiarize yourself with the muscle's size, shape, and general location. To help you in this task, the names of some of the major muscles are grouped according to their location in Table 6-1.

MUSCLES GROUPED ACCORDING TO FUNCTION

The following terms are used to designate muscles according to their main actions (see Table 6-2 for examples).

1 *flexors*—decrease the angle of a joint (between the anterior surfaces of the bones except in the knee and toe joints)
2 *extensors*—return the part from flexion to normal anatomical position; increase the angle of a joint

3 *abductors*—move the bone away from the midline
4 *adductors*—move the part toward the midline
5 *rotators*—cause a part to pivot upon its axis
6 *levators*—raise a part
7 *depressors*—lower a part
8 *sphincters*—reduce the size of an opening
9 *tensors*—tense a part—that is, make it more rigid
10 *supinators*—turn the hand palm upward
11 *pronators*—turn the hand palm downward

Origins, insertions, functions, and innervations of representative muscles

Basic information about many muscles is given in Tables 6-3 to 6-15. Each table has a description of a group of muscles that move one part of the body. Muscles which, in our judgment, are the most important for beginning students of anatomy to know are set in boldface type, and the origins and insertions so judged are set in italics. Remember that the actions listed for each muscle are those for which it is a prime mover. Actually, a single muscle contracting alone rarely accomplishes a given action. Instead, muscles act in groups as prime movers, synergists, and antagonists (p. 125) to bring about movements. As you study the muscles described in Tables 6-3 to 6-15, try to follow the hints for studying muscle actions given on p. 126.

■ Weak places in abdominal wall

There are several places in the abdominal wall where rupture (hernia) with protrusion of part of the intestine may occur. At these points the wall is weakened due to the presence of an interval or space in the abdominal aponeuroses. Any undue

Text continued on p. 156

Table 6-1. Muscles grouped according to location

Location	*Muscles*	*Figures illustrating*
Neck	Sternocleidomastoid	6-11
Back	Trapezius Latissimus dorsi	6-11 to 6-14 6-12
Chest	Pectoralis major Serratus anterior	6-11 6-11
Abdominal wall	External oblique	6-11, 6-12
Shoulder	Deltoid	6-11 to 6-14
Upper arm	Biceps brachii Triceps brachii Brachialis	6-13, 6-15 6-13, 6-14, 6-16 6-13, 6-18
Forearm	Brachioradialis Pronator teres	6-13, 6-14, 6-19 6-13, 6-17
Buttocks	Gluteus maximus Gluteus minimus Gluteus medius Tensor fasciae latae	6-21, 6-24 6-25 6-26 6-20

Thigh
  Anterior surface
	Quadriceps femoris group
	  Rectus femoris
	  Vastus lateralis
	  Vastus medialis
	  Vastus intermedius
		
		6-20, 6-22
		6-20, 6-22
		6-20, 6-22
		6-22

Medial surface	Gracilis Adductor group (brevis, longus, magnus)	6-20, 6-21, 6-23 6-20, 6-21, 6-23

Posterior surface	Hamstring group
	  Biceps femoris
	  Semitendinosus
	  Semimembranosus
		
		6-21, 6-27
		6-21, 6-27
		6-21, 6-27

Leg Anterior surface	Tibialis anterior	6-20
Posterior surface	Gastrocnemius Soleus	6-21 6-21
Pelvic floor	Levator ani Levator coccygeus Rectococcygeus	6-34 6-34 6-34

Table 6-2. Muscles grouped according to function

Part moved	Example of flexor	Example of extensor	Example of abductor	Example of adductor
Head	Sternocleidomastoid	Semispinalis capitis		
Upper arm	Pectoralis major	Trapezius Latissimus dorsi	Deltoid	Pectoralis major with latissimus dorsi
Forearm	With forearm supinated: biceps brachii With forearm pronated: brachialis With semisupination or semipronation: brachioradialis	Triceps brachii		
Hand	Flexor carpi radialis and ulnaris Palmaris longus	Extensor carpi radialis, longus, and brevis Extensor carpi ulnaris	Flexor carpi radialis	Flexor carpi ulnaris
Thigh	Iliopsoas Rectus femoris (of quadriceps femoris group)	Gluteus maximus	Gluteus medius and gluteus minimus	Adductor group
Leg	Hamstrings	Quadriceps femoris group		

Foot	Tibialis anterior	Gastrocnemius
		Soleus
			Evertors
			  Peroneus longus
			  Peroneus brevis
				Invertor
				  Tibialis anterior

Trunk	Iliopsoas Rectus abdominis	Sacrospinalis		

Table 6-3. Muscles that move shoulder*†

Muscle	*Origin*	*Insertion*	*Function*	*Innervation*
Trapezius	*Occipital bone* (protuberance)	*Scapula* (spine and acromion)	Raises or lowers shoulders and shrugs them	Spinal accessory, second, third, and fourth cervical nerves
	Vertebrae (cervical and thoracic)	*Clavicle*	Extends head when occiput acts as insertion	
Pectoralis minor	*Ribs* (second to fifth)	*Scapula* (coracoid)	Pulls shoulder down and forward	Medial and lateral anterior thoracic nerves
Serratus anterior	*Ribs* (upper eight or nine)	*Scapula* (anterior surface, vertebral border)	Pulls shoulder forward; abducts and rotates it upward	Long thoracic nerve

*When trying to learn the origins and insertion of the muscles listed, refer frequently to illustrations of each muscle and to the skeleton. Also, when possible, feel each muscle on your own body.
†Muscles judged to be most important for beginning students of anatomy to know are set in boldface type, and the origins and insertions so judged are set in italics.

Table 6-4. Muscles that move upper arm*†

Muscle	Origin	Insertion	Function	Innervation
Pectoralis major	*Clavicle* (medial half) *Sternum* *Costal cartilages of true ribs*	*Humerus* (greater tubercle)	Flexes upper arm Adducts upper arm anteriorly; draws it across chest	Medial and lateral anterior thoracic nerves
Latissimus dorsi	*Vertebrae* (spines of lower thoracic, lumbar, and sacral) *Ilium* (crest) Lumbodorsal fascia‡	*Humerus* (inter-tubercular groove)	Extends upper arm Adducts upper arm posteriorly	Thoracodorsal nerve
Deltoid	*Clavicle* *Scapula* (spine and acromion)	*Humerus* (lateral side about halfway down—deltoid tubercle)	Abducts upper arm Assists in flexion and extension of upper arm	Axillary nerve
Coracobrachialis	Scapula (coracoid process)	Humerus (middle third, medial surface)	Adduction; assists in flexion and medial rotation of arm	Musculocutaneous nerve
Supraspinatus	Scapula (supra-spinous fossa)	Humerus (greater tubercle)	Assists in abduct-ing arm	Suprascapular nerve
Teres major	Scapula (lower part, axillary border)	Humerus (upper part, anterior surface)	Assists in exten-sion, adduction, and medial rotation of arm	Lower subscapular nerve
Teres minor	Scapula (axillary border)	Humerus (greater tubercle)	Rotates arm out-ward	Axillary nerve
Infraspinatus	Scapula (infra-spinatus border)	Humerus (greater tubercle)	Rotates arm out-ward	Suprascapular nerve

*When trying to learn the origins and insertion of the muscles listed, refer frequently to illustrations of each muscle and to the skeleton. Also, when possible, feel each muscle on your own body.
†Muscles judged to be most important for beginning students of anatomy to know are set in boldface type, and the origins and insertions so judged are set in italics.
‡Lumbodorsal fascia—extension of aponeurosis of latissimus dorsi; fills in space between last rib and iliac crest.

Table 6-5. Muscles that move lower arm*†

Muscle	Origin	Insertion	Function	Innervation
Biceps brachii	*Scapula* (supra-glenoid tuberosity) *Scapula* (coracoid)	*Radius* (tubercle at proximal end)	Flexes supinated forearm Supinates forearm and hand	Musculocutaneous nerve
Brachialis	*Humerus* (distal half, anterior surface)	*Ulna* (front of coronoid process)	Flexes pronated forearm	Musculocutaneous nerve
Brachioradialis	Humerus (above lateral epicondyle)	Radius (styloid process)	Flexes semipro-nated or semi-supinated forearm; supinates forearm and hand	Radial nerve
Triceps brachii	*Scapula* (infra-glenoid tuberosity) *Humerus* (posterior surface—lateral head above radial groove; medial head, below)	*Ulna* (olecranon process)	Extends lower arm	Radial nerve
Pronator teres	Humerus (medial epicondyle) Ulna (coronoid process)	Radius (middle third of lateral surface)	Pronates and flexes forearm	Median nerve
Pronator quadratus	Ulna (distal fourth, anterior surface)	Radius (distal fourth, anterior surface)	Pronates forearm	Median nerve
Supinator	Humerus (lateral epicondyle) Ulna (proximal fifth)	Radius (proximal third)	Supinates forearm	Radial nerve

*When trying to learn the origins and insertion of the muscles listed, refer frequently to illustrations of each muscle and to the skeleton. Also, when possible, feel each muscle on your own body.
†Muscles judged to be most important for beginning students of anatomy to know are set in boldface type, and the origins and insertions so judged are set in italics.

Table 6-6. Muscles that move hand*

Muscle	Origin	Insertion	Function	Innervation
Flexor carpi radialis	Humerus (medial epicondyle)	Second metacarpal (base of)	Flexes hand Flexes forearm	Median nerve
Palmaris longus	Humerus (medial epicondyle)	Fascia of palm	Flexes hand	Median nerve
Flexor carpi ulnaris	Humerus (medial epicondyle) Ulna (proximal two-thirds)	Pisiform bone Third, fourth, and fifth metacarpals	Flexes hand Adducts hand	Ulnar nerve
Extensor carpi radialis longus	Humerus (ridge above lateral epicondyle)	Second metacarpal (base of)	Extends hand Abducts hand (moves toward thumb side when hand supinated)	Radial nerve
Extensor carpi radialis brevis	Humerus (lateral epicondyle)	Second, third metacarpals (bases of)	Extends hand	Radial nerve
Extensor carpi ulnaris	Humerus (lateral epicondyle) Ulna (proximal three-fourths)	Fifth metacarpal (base of)	Extends hand Adducts hand (move toward little finger side when hand supinated)	Radial nerve

*When trying to learn the origins and insertion of the muscles listed, refer frequently to illustrations of each muscle and to the skeleton. Also, when possible, feel each muscle on your own body.

Table 6-7. Muscles that move thigh*†

Muscle	Origin	Insertion	Function	Innervation
Iliopsoas (iliacus and psoas major)	*Ilium* (iliac fossa) *Vertebrae* (bodies of twelfth thoracic to fifth lumbar)	*Femur* (small trochanter)	Flexes thigh Flexes trunk (when femur acts as origin)	Femoral and second to fourth lumbar nerves
Rectus femoris	*Ilium* (anterior, inferior spine)	*Tibia* (by way of patellar tendon)	Flexes thigh Extends lower leg	Femoral nerve
Gluteal group				
Maximus	*Ilium* (crest and posterior surface) Sacrum and coccyx (posterior surface) Sacrotuberous ligament	*Femur* (gluteal tuberosity) *Iliotibial tract*‡	Extends thigh—rotates outward	Inferior gluteal nerve
Medius	*Ilium* (lateral surface)	*Femur* (greater trochanter)	Abducts thigh—rotates outward; stabilizes pelvis on femur	Superior gluteal nerve
Minimus	*Ilium* (lateral surface)	*Femur* (greater trochanter)	Abducts thigh; stabilizes pelvis on femur Rotates thigh medially	Superior gluteal nerve
Tensor fasciae latae	*Ilium* (anterior part of crest)	*Tibia* (by way of iliotibial tract)	Abducts thigh Tightens iliotibial tract‡	Superior gluteal nerve
Piriformis	Vertebrae (front of sacrum)	Femur (medial aspect of greater trochanter)	Rotates thigh outward Abducts thigh Extends thigh	First or second sacral nerves
Adductor group				
Brevis	*Pubic bone*	*Femur* (linea aspera)	Adducts thigh	Obturator nerve
Longus	*Pubic bone*	*Femur* (linea aspera)	Adducts thigh	Obturator nerve
Magnus	*Pubic bone*	*Femur* (linea aspera)	Adducts thigh	Obturator nerve
Gracilis	Pubic bone (just below symphysis)	*Tibia* (medial surface behind sartorius)	Adducts thigh and flexes and adducts leg	Obturator nerve

*When trying to learn the origins and insertion of the muscles listed, refer frequently to illustrations of each muscle and to the skeleton. Also, when possible, feel each muscle on your own body.

†Muscles judged to be most important for beginning students of anatomy to know are set in boldface type, and the origins and insertions so judged are set in italics.

‡The iliotibial tract is part of the fascia enveloping all the thigh muscles. It consists of a wide band of dense fibrous tissue attached to the iliac crest above and the lateral condyle of the tibia below. The upper part of the tract encloses the tensor fasciae latae muscle.

Table 6-8. Muscles that move lower leg*†

Muscle	Origin	Insertion	Function	Innervation
Quadriceps femoris group				
Rectus femoris	*Ilium* (anterior, inferior spine)	*Tibia* (by way of patellar tendon)	Flexes thigh Extends leg	Femoral nerve
Vastus lateralis	*Femur* (linea aspera)	*Same*	Extends leg	Femoral nerve
Vastus medialis	*Femur*	*Same*	Same	Femoral nerve
Vastus intermedius	*Femur* (anterior surface)	*Same*	Same	Femoral nerve
Sartorius	*Os innominatum* (anterior, superior iliac spines)	*Tibia* (medial surface of upper end of shaft)	Adducts and flexes leg Permits crossing of legs tailor fashion	Femoral nerve
Hamstring group				
Biceps femoris	*Ischium* (tuberosity)	*Fibula* (head of)	Flexes leg	Hamstring nerve (branch of sciatic nerve)
	Femur (linea aspera)	*Tibia* (lateral condyle)	Extends thigh	Hamstring nerve
Semitendinosus	*Ischium* (tuberosity)	*Tibia* (proximal end, medial surface)	Same	Hamstring nerve
Semimembranosus	*Same*	*Tibia* (medial condyle)	Same	Hamstring nerve

*When trying to learn the origins and insertion of the muscles listed, refer frequently to illustrations of each muscle and to the skeleton. Also, when possible, feel each muscle on your own body.
†Muscles judged to be most important for beginning students of anatomy to know are set in boldface type, and the origins and insertions so judged are set in italics.

Table 6-9. Muscles that move foot*†

Muscle	Origin	Insertion	Function	Innervation
Tibialis anterior	*Tibia* (lateral condyle of upper body)	*Tarsal* (first cuneiform) *Metatarsal* (base of first)	Flexes foot Inverts foot	Common and deep peroneal nerves
Gastrocnemius	*Femur* (condyles)	*Tarsal* (calcaneus by way of Achilles tendon)	Extends foot Flexes lower leg	Tibial nerve (branch of sciatic nerve)
Soleus	*Tibia* (underneath gastrocnemius) *Fibula*	*Same as gastrocnemius*	Extends foot (plantar flexion)	Tibial nerve
Peroneus longus	Tibia (lateral condyle) Fibula (head and shaft)	First cuneiform Base of first metatarsal	Extends foot (plantar flexion) Everts foot	Common peroneal nerve
Peroneus brevis	Fibula (lower two-thirds of lateral surface of shaft)	Fifth metatarsal (tubercle, dorsal surface)	Everts foot Flexes foot	Superficial peroneal nerve
Tibialis posterior	Tibia (posterior surface) Fibula (posterior surface)	Navicular bone Cuboid bone All three cuneiforms Second and fourth metatarsals	Extends foot (plantar flexion) Inverts foot	Tibial nerve
Peroneus tertius	Fibula (distal third)	Fourth and fifth metatarsals (bases of)	Flexes foot Everts foot	Deep peroneal nerve

*When trying to learn the origins and insertion of the muscles listed, refer frequently to illustrations of each muscle and to the skeleton. Also, when possible, feel each muscle on your own body.
†Muscles judged to be most important for beginning students of anatomy to know are set in boldface type, and the origins and insertions so judged are set in italics.

Table 6-10. Muscles that move head*†

Muscle	Origin	Insertion	Function	Innervation
Sternocleido-mastoid	*Sternum* *Clavicle*	*Temporal bone* *(mastoid process)*	Flexes head (prayer muscle) One muscle, alone, rotates head toward opposite side; spasm of this muscle alone or associated with trapezius called torticollis or wryneck	Accessory nerve
Semispinalis capitis	Vertebrae (transverse processes of upper six thoracic, articular processes of lower four cervical)	Occipital bone (between superior and inferior nuchal lines)	Extends head; bends it laterally	First five cervical nerves
Splenius capitis	Ligamentum nuchae Vertebrae (spinous processes of upper three or four thoracic)	Temporal bone (mastoid process) Occipital bone	Extends head Bends and rotates head toward same side as contracting muscle	Second, third, and fourth cervical nerves
Longissimus capitis	Vertebrae (transverse processes of upper six thoracic, articular processes of lower four cervical)	Temporal bone (mastoid process)	Extends head Bends and rotates head toward contracting side	

*When trying to learn the origins and insertion of the muscles listed, refer frequently to illustrations of each muscle and to the skeleton. Also, when possible, feel each muscle on your own body.
†Muscles judged to be most important for beginning students of anatomy to know are set in boldface type, and the origins and insertions so judged are set in italics.

Table 6-11. Muscles that move abdominal wall*†

Muscle	Origin	Insertion	Function	Innervation
External oblique	*Ribs* (lower eight)	*Ossa coxae* (iliac crest and pubis by way of inguinal ligament)‡ *Linea alba*§ by way of an aponeurosis‖	Compresses abdomen Important postural function of all abdominal muscles is to pull front of pelvis upward, thereby flattening lumbar curve of spine; when these muscles lose their tone, common figure faults of protruding abdomen and lordosis develop	Lower seven intercostal nerves and iliohypogastric nerves
Internal oblique	*Ossa coxae* (iliac crest and inguinal ligament) *Lumbodorsal fascia*	*Ribs* (lower three) *Pubic bone Linea alba*	Same as external oblique	Last three intercostals; iliohypogastric and ilioinguinal nerves
Transversalis	*Ribs* (lower six) *Ossa coxae* (iliac crest, inguinal ligament) *Lumbodorsal fascia*	*Pubic bone Linea alba*	Same as external oblique	Last five intercostals; iliohypogastric and ilioinguinal nerves
Rectus abdominis	*Ossa coxae* (pubic bone and symphysis pubis)	*Ribs* (costal cartilage of fifth, sixth, and seventh ribs) *Sternum* (xiphoid process)	Same as external oblique; because abdominal muscles compress abdominal cavity, they aid in straining, defecation, forced expiration, childbirth, etc.; abdominal muscles are antagonists of diaphragm, relaxing as it contracts and vice versa Flex trunk	Last six intercostal nerves

*When trying to learn the origins and insertion of the muscles listed, refer frequently to illustrations of each muscle and to the skeleton. Also, when possible, feel each muscle on your own body.

†Muscles judged to be most important for beginning students of anatomy to know are set in boldface type, and the origins and insertions so judged are set in italics.

‡*Inguinal ligament* (or Poupart's)—lower edge of aponeurosis of external oblique muscle, extending between the anterior superior iliac spine and the tubercle of the pubic bone. This edge is doubled under similarly to a hem on material. The inguinal ligament forms the upper boundary of the *femoral triangle*, a large triangular area in the thigh; its other boundaries are the adductor longus muscle mesially and the sartorius laterally.

§*Linea alba*—literally, a white line; extends from xiphoid process to symphysis pubis; formed by fibers of aponeuroses of the right abdominal muscles interlacing with fibers of aponeuroses of the left abdominal muscles; comparable to a seam up the midline of the abdominal wall, anchoring its various layers. During pregnancy the linea alba becomes pigmented and is known as the *linea niger*.

‖*Aponeurosis*—sheet of white fibrous tissue which attaches one muscle to another or attaches it to bone or other movable structures; e.g., the right external oblique muscle attaches to the left external oblique muscle by means of an aponeurosis.

Table 6-12. Muscles that move chest wall*†

Muscle	*Origin*	*Insertion*	*Function*	*Innervation*
External intercostals	Rib (lower border; forward fibers)	Rib (upper border of rib below origin)	Elevate ribs	Intercostal nerves
Internal intercostals	Rib (inner surface, lower border; backward fibers)	Rib (upper border of rib below origin)	Probably depress ribs	Intercostal nerves
Diaphragm	*Lower circumference of thorax (of rib cage)*	*Central tendon of diaphragm*	Enlarges thorax, causing inspiration	Phrenic nerves

*When trying to learn the origins and insertion of the muscles listed, refer frequently to illustrations of each muscle and to the skeleton. Also, when possible, feel each muscle on your own body.

†Muscles judged to be most important for beginning students of anatomy to know are set in boldface type, and the origins and insertions so judged are set in italics.

Table 6-13. Muscles of pelvic floor*

Muscle	*Origin*	*Insertion*	*Function*	*Innervation*
Levator ani	Pubis – posterior surface Ischium (spine)	Coccyx	Together form floor of pelvic cavity; support pelvic organs; if these muscles are badly torn at childbirth, or become too relaxed, uterus or bladder may prolapse, i.e., drop out	Pudendal nerve
Coccygeus (posterior continuation of levator ani)	Ischium (spine)	Coccyx Sacrum	Same as levator ani	Pudendal nerve

*When trying to learn the origins and insertion of the muscles listed, refer frequently to illustrations of each muscle and to the skeleton. Also, when possible, feel each muscle on your own body.

Table 6-14. Muscles that move trunk*†

Muscle	Origin	Insertion	Function	Innervation
Sacrospinalis (erector spinae)			Extend spine; maintain erect posture of trunk Acting singly, abduct and rotate trunk	Posterior rami of first cervical to fifth lumbar spinal nerves
Lateral portion:				
Iliocostalis lumborum	Iliac crest, sacrum (posterior surface), and lumbar vertebrae (spinous processes)	Ribs, lower six		
Iliocostalis dorsi	Ribs, lower six	Ribs, upper six		
Iliocostalis cervicis	Ribs, upper six	Vertebrae, fourth to sixth cervical		
Medial portion:				
Longissimus dorsi	Same as iliocostalis lumborum	Vertebrae, thoracic ribs		
Longissimus cervicis	Vertebrae, upper six thoracic	Vertebrae, second to sixth cervical		
Longissimus capitis	Vertebrae, upper six thoracic and last four cervical	Temporal bone, mastoid process		
Quadratus lumborum (forms part of posterior abdominal wall)	Ilium (posterior part of crest) Vertebrae (lower three lumbar)	Ribs (twelfth) Vertebrae (transverse processes of first four lumbar)	Both muscles together extend spine One muscle alone abducts trunk toward side of contracting muscle	First three or four lumbar nerves
Iliopsoas	See muscles that move thigh, p. 134		Flexes trunk	

*When trying to learn the origins and insertion of the muscles listed, refer frequently to illustrations of each muscle and to the skeleton. Also, when possible, feel each muscle on your own body.
†Muscles judged to be most important for beginning students of anatomy to know are set in boldface type.

Table 6-15. Muscles of facial expression and of mastication*

Muscle	Origin	Insertion	Function	Innervation
Epicranius (occipitofrontalis)	Occipital bone	Tissues of eyebrows	Raises eyebrows, wrinkling forehead, horizontally	Cranial nerve VII
Corrugator supercilii	Frontal bone (superciliary ridge)	Skin of eyebrow	Wrinkles forehead vertically	Cranial nerve VII
Orbicularis oculi	Encircles eyelid		Closes eye	Cranial nerve VII
Orbicularis oris	Encircles mouth		Draws lips together	Cranial nerve VII
Platysma	Fascia of upper part of deltoid and pectoralis major	Mandible—lower border Skin around corners of mouth	Draws corners of mouth down—pouting	Cranial nerve VII
Buccinator	Maxillae	Skin of sides of mouth	Permits smiling Blowing (e.g., as in playing a trumpet)	Cranial nerve VII
Muscles of mastication Masseter	Zygomatic arch	Mandible (external surface)	Closes jaw	Cranial nerve V
Temporal	Temporal bone	Mandible	Closes jaw	Cranial nerve V
Pterygoids (internal and external)	Undersurface of skull	Mandible (mesial surface)	Grate teeth	Cranial nerve V

*When trying to learn the origins and insertion of the muscles listed, refer frequently to illustrations of each muscle and to the skeleton. Also, when possible, feel each muscle on your own body.

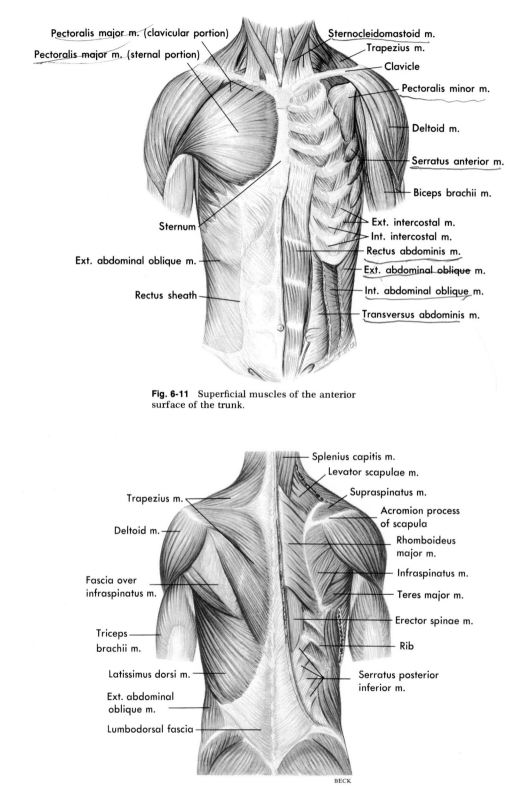

Pectoralis major m. (clavicular portion)

Pectoralis major m. (sternal portion)

Sternocleidomastoid m.

Trapezius m.

Clavicle

Pectoralis minor m.

Deltoid m.

Serratus anterior m.

Biceps brachii m.

Sternum

Ext. intercostal m.

Int. intercostal m.

Rectus abdominis m.

Ext. abdominal oblique m.

Ext. abdominal oblique m.

Int. abdominal oblique m.

Rectus sheath

Transversus abdominis m.

Fig. 6-11 Superficial muscles of the anterior
surface of the trunk.

Splenius capitis m.

Levator scapulae m.

Supraspinatus m.

Trapezius m.

Acromion process
of scapula

Deltoid m.

Rhomboideus
major m.

Fascia over
infraspinatus m.

Infraspinatus m.

Teres major m.

Triceps
brachii m.

Erector spinae m.

Rib

Latissimus dorsi m.

Serratus posterior
inferior m.

Ext. abdominal
oblique m.

Lumbodorsal fascia

BECK

Fig. 6-12 Superficial muscles of the posterior
surface of the trunk.

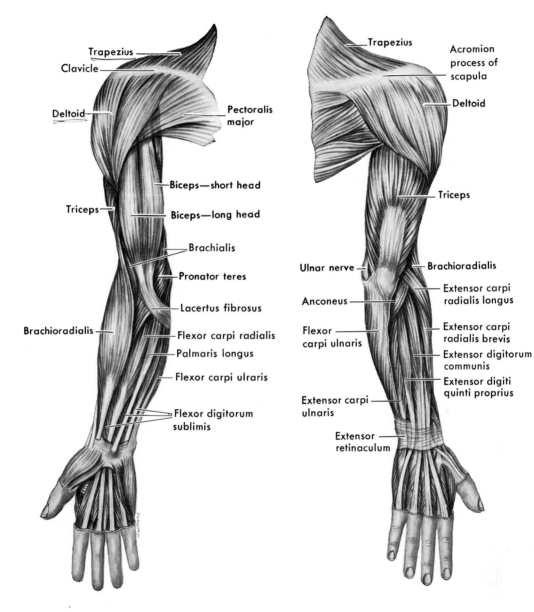

Fig. 6-13 Muscles of the flexor surface of the upper extremity.

Fig. 6-14 Muscles of the extensor surface of the upper extremity.

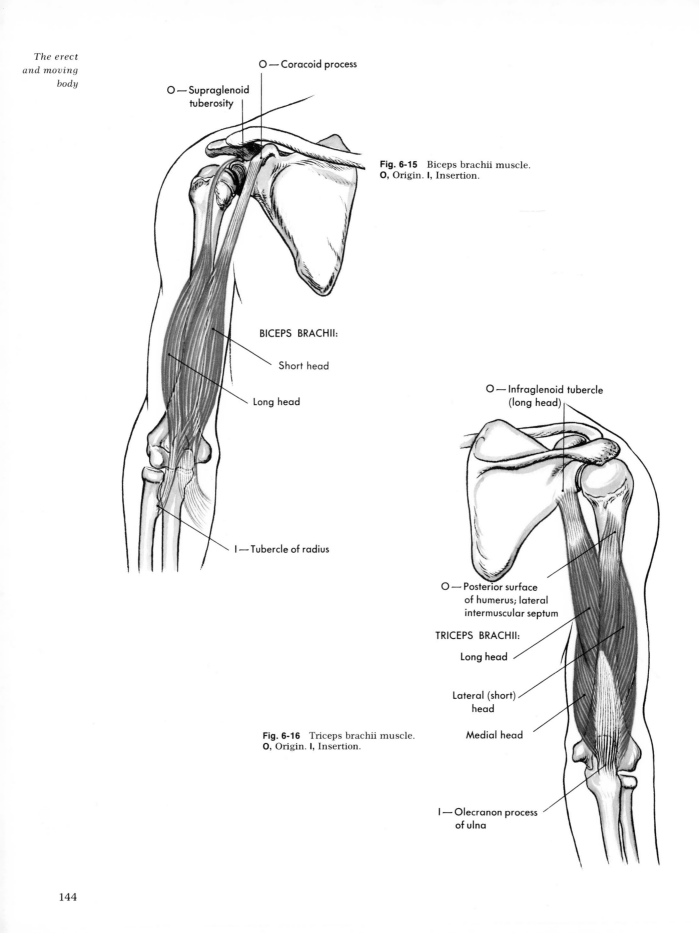

O — Coracoid process

O — Supraglenoid
tuberosity

Fig. 6-15 Biceps brachii muscle.
O, Origin. **I,** Insertion.

BICEPS BRACHII:

Short head

Long head

I — Tubercle of radius

O — Infraglenoid tubercle
(long head)

O — Posterior surface
of humerus; lateral
intermuscular septum

TRICEPS BRACHII:

Long head

Lateral (short)
head

Medial head

I — Olecranon process
of ulna

Fig. 6-16 Triceps brachii muscle.
O, Origin. **I,** Insertion.

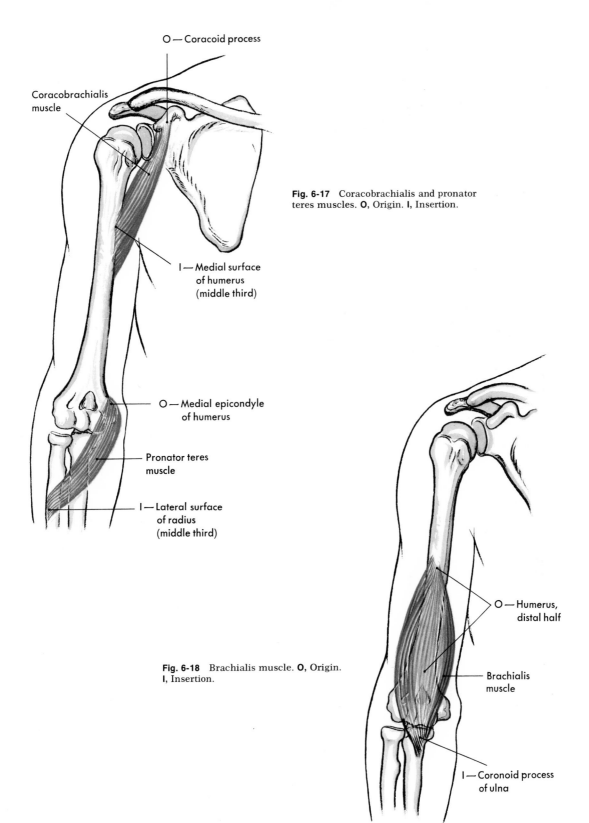

O — Coracoid process

Coracobrachialis muscle

Fig. 6-17 Coracobrachialis and pronator teres muscles. **O**, Origin. **I**, Insertion.

I — Medial surface of humerus (middle third)

O — Medial epicondyle of humerus

Pronator teres muscle

I — Lateral surface of radius (middle third)

O — Humerus, distal half

Brachialis muscle

I — Coronoid process of ulna

Fig. 6-18 Brachialis muscle. **O**, Origin. **I**, Insertion.

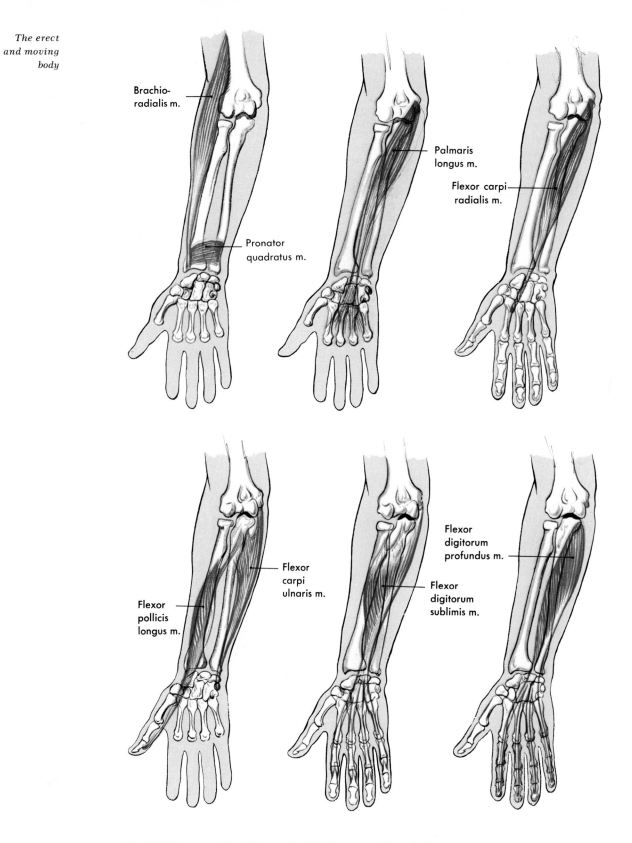

Fig. 6-19 Some muscles of the anterior (volar) aspect of the right forearm.

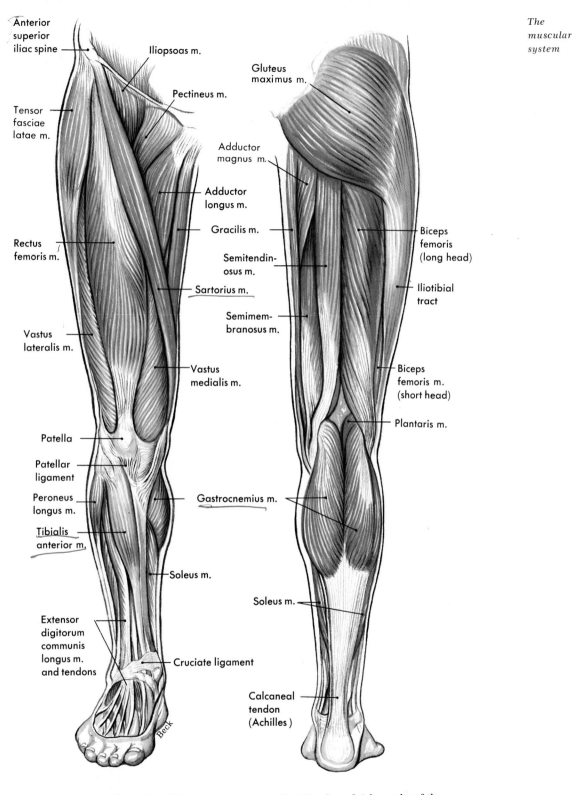

Anterior
superior
iliac spine

Iliopsoas m.

Pectineus m.

Tensor
fasciae
latae m.

Adductor
magnus m.

Adductor
longus m.

Rectus
femoris m.

Gracilis m.

Semitendin-
osus m.

Sartorius m.

Vastus
lateralis m.

Semimem-
branosus m.

Vastus
medialis m.

Patella

Patellar
ligament

Peroneus
longus m.

Gastrocnemius m.

Tibialis
anterior m.

Soleus m.

Extensor
digitorum
communis
longus m.
and tendons

Cruciate ligament

Gluteus
maximus m.

Biceps
femoris
(long head)

Iliotibial
tract

Biceps
femoris m.
(short head)

Plantaris m.

Soleus m.

Calcaneal
tendon
(Achilles)

Beck

Fig. 6-20 Superficial muscles of the
right thigh and leg, anterior view.

Fig. 6-21 Superficial muscles of the
right thigh and leg, posterior view.

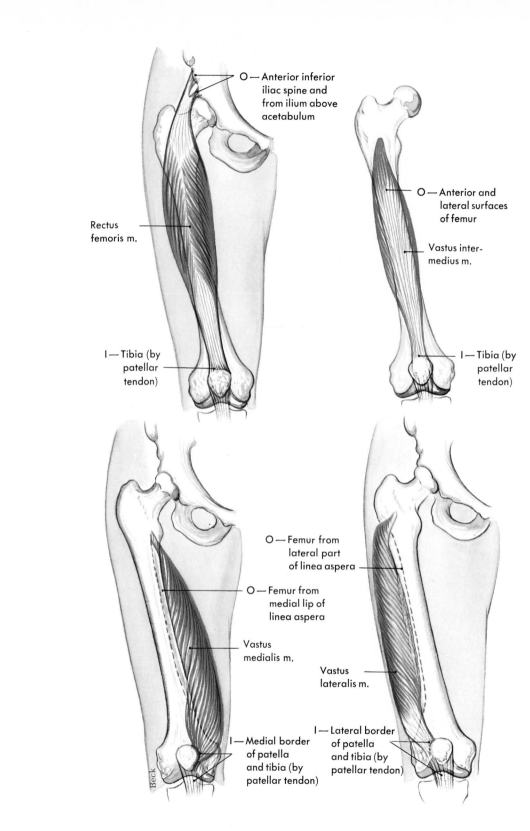

O — Anterior inferior
iliac spine and
from ilium above
acetabulum

Rectus
femoris m.

I — Tibia (by
patellar
tendon)

O — Anterior and
lateral surfaces
of femur

Vastus inter-
medius m.

I — Tibia (by
patellar
tendon)

O — Femur from
lateral part
of linea aspera

O — Femur from
medial lip of
linea aspera

Vastus
medialis m.

Vastus
lateralis m.

I — Medial border
of patella
and tibia (by
patellar tendon)

I — Lateral border
of patella
and tibia (by
patellar tendon)

Beck

Fig. 6-22 Quadriceps femoris group of thigh muscles: rectus femoris, vastus intermedius, vastus
medialis, and vastus lateralis.

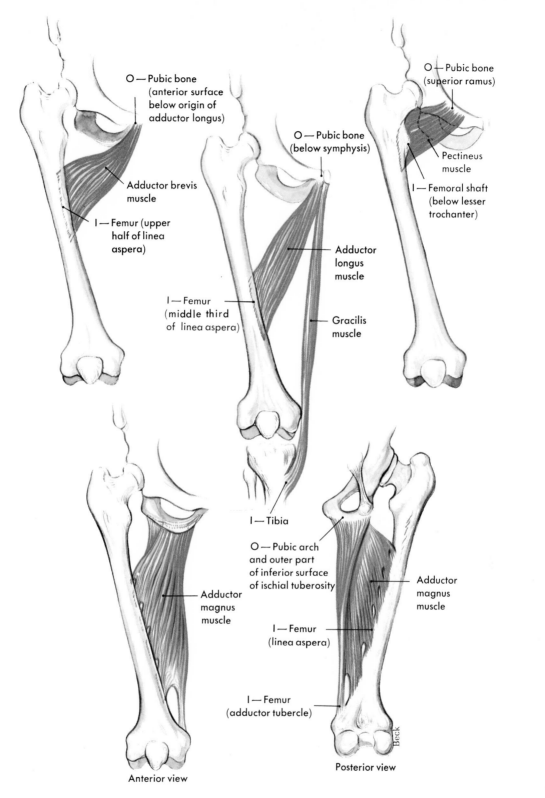

O — Pubic bone
(anterior surface
below origin of
adductor longus)

Adductor brevis
muscle

I — Femur (upper
half of linea
aspera)

O — Pubic bone
(below symphysis)

O — Pubic bone
(superior ramus)

Pectineus
muscle

I — Femoral shaft
(below lesser
trochanter)

Adductor
longus
muscle

Gracilis
muscle

I — Femur
(middle third
of linea aspera)

I — Tibia

O — Pubic arch
and outer part
of inferior surface
of ischial tuberosity

Adductor
magnus
muscle

Adductor
magnus
muscle

I — Femur
(linea aspera)

I — Femur
(adductor tubercle)

Anterior view

Posterior view

Fig. 6-23 Muscles which adduct the thigh.

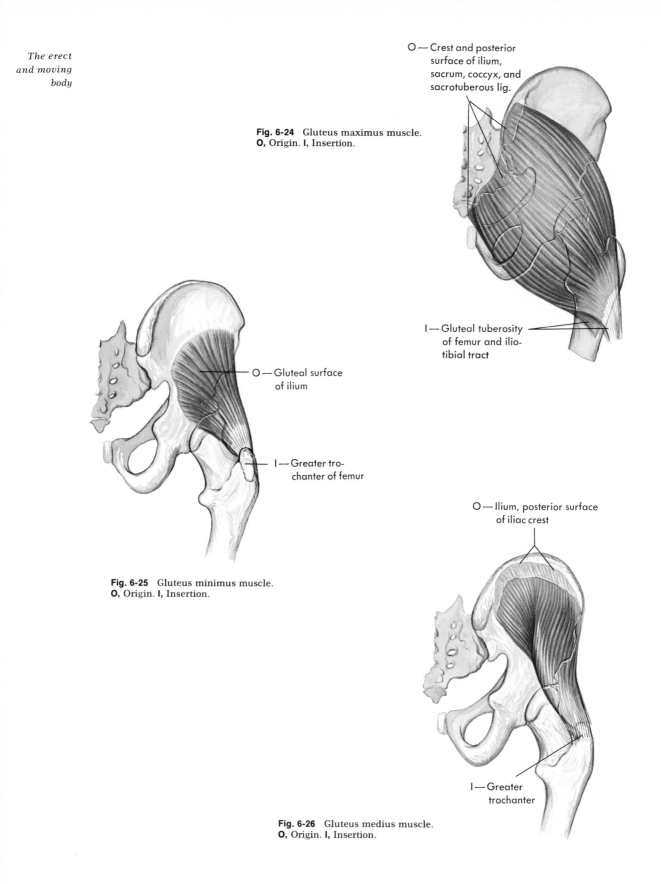

O — Crest and posterior
surface of ilium,
sacrum, coccyx, and
sacrotuberous lig.

Fig. 6-24 Gluteus maximus muscle.
O, Origin. **I,** Insertion.

I — Gluteal tuberosity
of femur and ilio-
tibial tract

O — Gluteal surface
of ilium

I — Greater tro-
chanter of femur

Fig. 6-25 Gluteus minimus muscle.
O, Origin. **I,** Insertion.

O — Ilium, posterior surface
of iliac crest

I — Greater
trochanter

Fig. 6-26 Gluteus medius muscle.
O, Origin. **I,** Insertion.

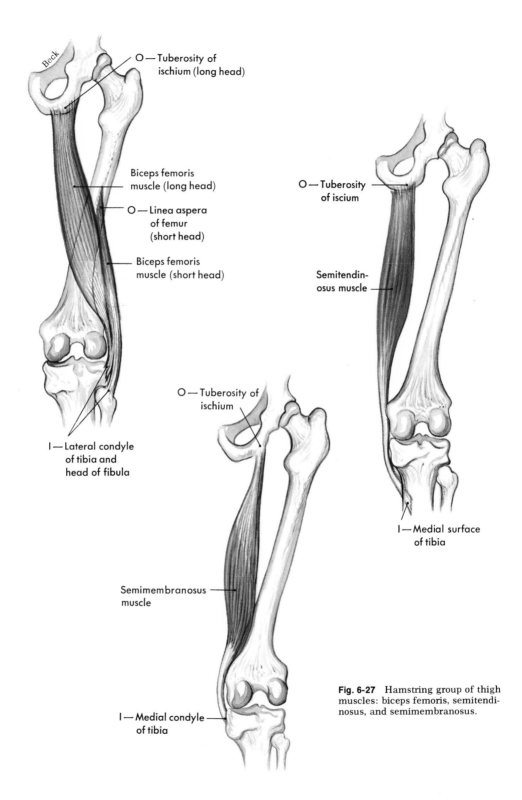

O — Tuberosity of ischium (long head)

Biceps femoris muscle (long head)

O — Linea aspera of femur (short head)

Biceps femoris muscle (short head)

I — Lateral condyle of tibia and head of fibula

Beck

O — Tuberosity of iscium

Semitendinosus muscle

I — Medial surface of tibia

O — Tuberosity of ischium

Semimembranosus muscle

I — Medial condyle of tibia

Fig. 6-27 Hamstring group of thigh muscles: biceps femoris, semitendinosus, and semimembranosus.

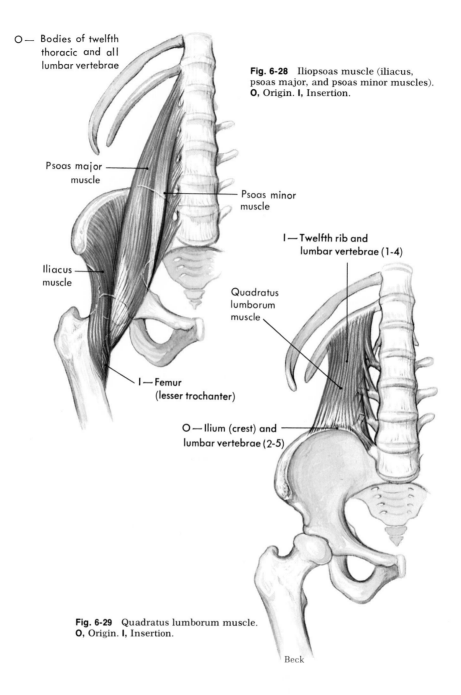

O— Bodies of twelfth
thoracic and all
lumbar vertebrae

Fig. 6-28 Iliopsoas muscle (iliacus,
psoas major, and psoas minor muscles).
O, Origin. **I,** Insertion.

Psoas major
muscle

Psoas minor
muscle

I—Twelfth rib and
lumbar vertebrae (1-4)

Quadratus
lumborum
muscle

Iliacus
muscle

I—Femur
(lesser trochanter)

O—Ilium (crest) and
lumbar vertebrae (2-5)

Fig. 6-29 Quadratus lumborum muscle.
O, Origin. **I,** Insertion.

Beck

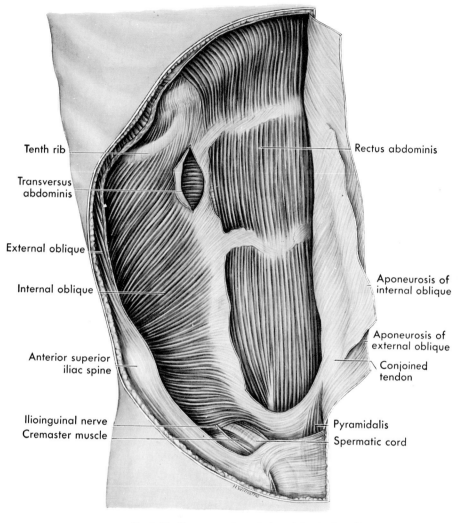

Tenth rib

Transversus abdominis

External oblique

Internal oblique

Anterior superior iliac spine

Ilioinguinal nerve
Cremaster muscle

Rectus abdominis

Aponeurosis of internal oblique

Aponeurosis of external oblique

Conjoined tendon

Pyramidalis
Spermatic cord

Fig. 6-30 Deep muscles of the abdominal wall. (From Francis, C. C, and Farrell, G. L.: Integrated anatomy and physiology, St. Louis, The C. V. Mosby Co.)

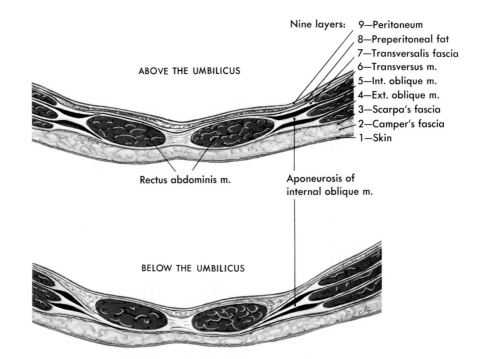

ABOVE THE UMBILICUS

Nine layers: 9—Peritoneum
8—Preperitoneal fat
7—Transversalis fascia
6—Transversus m.
5—Int. oblique m.
4—Ext. oblique m.
3—Scarpa's fascia
2—Camper's fascia
1—Skin

Rectus abdominis m.

Aponeurosis of
internal oblique m.

BELOW THE UMBILICUS

Fig. 6-31 Horizontal section of the anterolateral abdominal wall. The aponeurosis of the internal oblique muscle splits into two sections, one lying anterior and the other posterior to the rectus abdominis muscle, thereby forming an encasing sheath around this muscle.

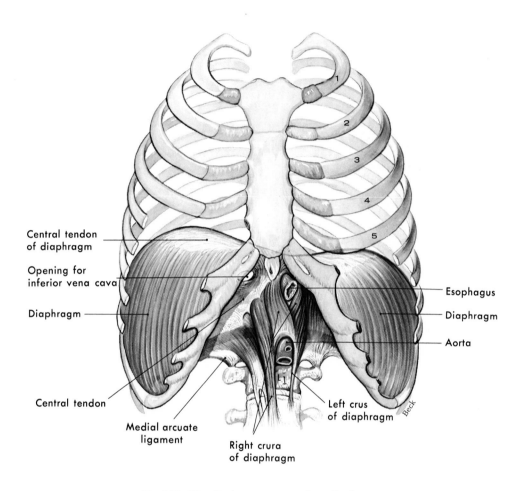

Central tendon
of diaphragm

Opening for
inferior vena cava

Diaphragm

Central tendon

Medial arcuate
ligament

Right crura
of diaphragm

Left crus
of diaphragm

Esophagus

Diaphragm

Aorta

Beck

Fig. 6-32 The diaphragm as seen from the front.
Note the openings in the vertebral portion for the
inferior vena cava, esophagus, and aorta.

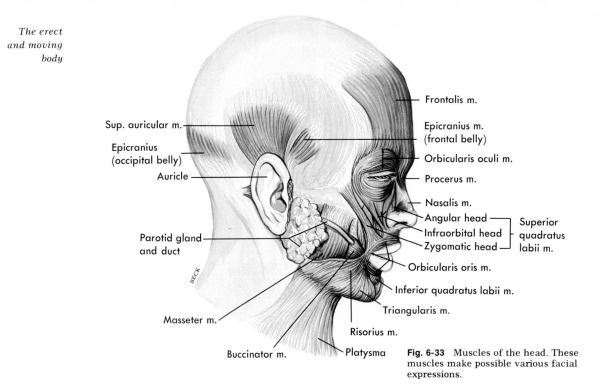

Fig. 6-33 Muscles of the head. These muscles make possible various facial expressions.

Labels on figure:
Sup. auricular m.
Epicranius (occipital belly)
Auricle
Parotid gland and duct
Masseter m.
Buccinator m.
Platysma
Risorius m.
Triangularis m.
Inferior quadratus labii m.
Orbicularis oris m.
Angular head
Infraorbital head
Zygomatic head
Superior quadratus labii m.
Nasalis m.
Procerus m.
Orbicularis oculi m.
Epicranius m. (frontal belly)
Frontalis m.
BECK

Text continued from p. 127

pressure on the abdominal viscera, therefore, can force a portion of the parietal peritoneum, and often a part of the intestine as well, through these nonreinforced places. The weak places are (1) the *inguinal canals*, (2) the *femoral rings*, and (3) the *umbilicus*. Hernia also occurs occasionally in the diaphragm and some other areas.

Piercing the aponeuroses of the abdominal muscles are two canals, the *inguinal canals*, one on the right and the other on the left. They lie above, but parallel to, the inguinal ligaments and are about 1 1/2 inches long. In the male the spermatic cords extend through the canals into the scrotum, whereas in the female the round ligaments of the uterus are in this location. The internal opening of each canal is a space in the aponeurosis of the transverse muscle known as the internal inguinal ring. The external openings or *external inguinal rings* are spaces in the aponeuro-

ses of the external oblique muscles. They are located inferiorly and mesially to the internal rings. The fact that they are larger in the male than in the female probably explains why external inguinal hernia occurs more often in men than in women.

The *femoral rings* are openings in the groin just below the inguinal ligaments, slightly lateral to the external inguinal ring and medial to the femoral veins. They have a diameter of about 1/2 inch and are usually somewhat larger in females, a fact which accounts for the greater prevalence of femoral hernia in women than in men.

■Bursae

Definition

Bursae are small connective tissue sacs lined with synovial membrane and containing synovial fluid.

Locations

Bursae are located wherever pressure is exerted over moving parts—for example,

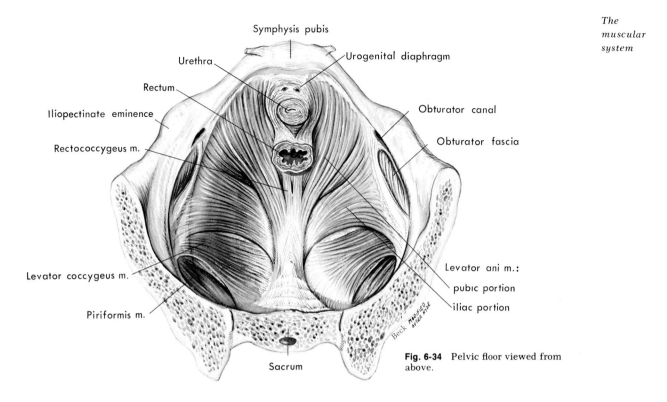

Symphysis pubis

Urethra

Rectum

Iliopectinate eminence

Rectococcygeus m.

Urogenital diaphragm

Obturator canal

Obturator fascia

Levator coccygeus m.

Piriformis m.

Levator ani m.:

pubic portion

iliac portion

Beck MODIFIED AFTER TYSE

Sacrum

Fig. 6-34 Pelvic floor viewed from above.

between skin and bone, between tendons and bone, or between muscles, or ligaments, and bone. Some bursae which fairly frequently become inflamed (bursitis) are as follows: the subacromial bursa, between the head of the humerus and the acromion process and the deltoid muscle; the olecranon bursa, between the olecranon process and the skin; the prepatellar bursa, between the patella and the skin. Inflammation of the prepatellar bursa is known as housemaid's knee, whereas olecranon bursitis is called student's elbow.

Function

Bursae act as cushions, relieving pressure between moving parts.

◼ Posture

We have already discussed the major role muscles play in movement and heat production. We shall now turn our attention to a third way in which muscles serve the

body as a whole—that of maintaining the posture of the body. Let us consider a few aspects of this important function.

Meaning

The term posture means simply position or alignment of body parts. "Good posture" means many things. It means body alignment which most favors function; it means position which requires the least muscular work to maintain, which puts the least strain on muscles, ligaments, and bones; it means keeping the body's center of gravity over its base. Good posture in the standing position, for example, means head and chest held high, chin, abdomen, and buttocks pulled in, knees bent slightly, and feet placed firmly on the ground about 6 inches apart.

How maintained

Since gravity pulls on the various parts of the body at all times, and since bones are too irregularly shaped to balance them-

157

selves upon each other, the only way the body can be held upright is for muscles to exert a continual pull on bones in the opposite direction from gravity. Gravity tends to pull the head and trunk forward and downward; muscles (head and truck extensors) must therefore pull backward and upward on them. Gravity pulls the lower jaw downward; muscles must pull upward on it, etc. Muscles exert this pull against gravity by virtue of their property of tonicity. Because tonicity is absent during sleep, muscle pull does not then counteract the pull of gravity. Hence, for example, we cannot sleep standing up.

Many structures other than muscles and bones play a part in the maintenance of posture. The nervous system is responsible for the existence of muscle tone and also regulates and coordinates the amount of pull exerted by the individual muscles. The respiratory, digestive, circulatory, excretory, and endocrine systems all contribute something toward the ability of muscles to maintain posture. This is one of many examples of the important principle that all body functions are interdependent.

Importance to body as whole

The importance of posture can perhaps be best evaluated by considering some of the effects of poor posture. Poor posture throws more work on muscles to counteract the pull of gravity and therefore leads to fatigue more quickly than good posture. Poor posture puts more strain on ligaments. It puts abnormal strains on bones and may eventually produce deformities. It interferes with various functions such as respirations, heart action, and digestion. It probably is not going too far to say that it even detracts from one's feeling of self-confidence and joy. In support of this last claim, consider our use of such expressions as "shoulders squared, head erect" to denote confidence and joy and "down-in-the-mouth," "long-faced," and "bowed down" to signify dejection and anxiety. The importance of posture to the body as a whole might be summed up in a single sentence: maximal health and good posture are reciprocally related—that is, each one depends upon the other.

outline summary

GENERAL FUNCTIONS
1 Movement—sometimes locomotion, sometimes movement within given area
2 Posture
3 Heat production

SKELETAL MUSCLE CELLS
Microscopic structure
1 Muscle cells usually called muscle fibers, term descriptive of their long, narrow shape
2 Sarcolemma—cell membrane of muscle fiber
3 Sarcoplasm—cytoplasm of muscle fiber
4 Sarcoplasmic reticulum—analogous but not identical to endoplasmic reticulum of cells other than muscle fibers
5 Myofibrils—numerous fine fibers packed close together in sarcoplasm
6 Cross striae
  a Dark stripes called A bands; light H zone runs across midsection of each dark A band
  b Light stripes called I bands; dark Z line extends across center of each light I band
7 Sarcomere—section of myofibril extending from one Z line to next; each myofibril consists of several sarcomeres
8 T system—transverse tubules that extend into sarcoplasm at levels of Z lines; formed by invaginations of sarcolemma
9 Triad—triple-layered structure consisting of T tubule sandwiched between sacs of sarcoplasmic reticulum

Molecular structure
1 Proteins
  a Twisted double chain of actin molecules (with some troponin and tropomyosin) composes thin filaments of myofibrils; actin molecules extend full length of each I band and half length of each adjacent A band

b Both myosin and actin filaments compose both ends of A bands, but H zones (light, narrow midsections of A bands) consist of myosin only

2 Pigment—myoglobin, iron-containing red pigment present in muscle cells; transports oxygen from blood capillaries to muscle fiber's mitochondria

Functions

1 Contraction—cross bridges of myosin molecules attach to actin filaments and pull them toward middle of myosin filaments of each sarcomere (see Fig. 6-3)

2 Muscle cells obey all-or-none law when they contract—i.e., they either contract with all force possible under existing conditions or do not contract at all

Energy sources for muscle contraction

1 From breakdown of ATP triggered by stimulation of muscle cell

2 Sources of ATP—more than 90% of ATP produced during oxygen-using citric acid cycle of catabolism; hence, without adequate oxygen supply, small amount of ATP formed during glycolysis is sufficient only for limited contractions; small amount of ATP also formed by ADP combining with high-energy creatine phosphate; creatine phosphate is synthesized and stored in muscle cells when they are not contracting

3 Oxygen debt—amount of oxygen required for oxidation (via citric acid cycle) of lactic acid that accumulates during strenuous exercise because respirations and circulation cannot deliver as much oxygen to muscle cells as they would need to oxidize lactic acid as fast as it forms

SKELETAL MUSCLE ORGANS

Structure

1 Size, shape, and fiber arrangement—wide variation in different muscles

2 Connective tissue components

  a Epimysium—fibrous connective tissue sheath that envelops each muscle

  b Perimysium—extensions of epimysium, partitioning each muscle into bundles of fibers

  c Endomysium—extensions of perimysium between individual muscle fibers

  d Tendon—strong, tough cord continuous at one end with fibrous wrappings (epimysium, etc.) of muscle and at other end with fibrous covering of bone (periosteum)

  e Aponeurosis—broad flat sheet of fibrous connective tissue continuous on one border with fibrous wrappings of muscle and at other border with fibrous coverings of some adjacent structure, usually another muscle

  f Tendon sheaths—tubes of fibrous connective tissue that enclose certain tendons, notably those of wrist and ankle; synovial membrane lines tendon sheaths

  g Deep fascia—layer of dense fibrous connective tissue underlying superficial fascia under skin; extensions of deep fascia form epimysium, etc. and also enclose viscera, glands, blood vessels, and nerves

3 Nerve supply

One motoneuron, together with skeletal muscle fibers it supplies, constitutes *motor unit;* number of muscle fibers per motor unit varies; in general, more precise movements produced by muscle in which motor units include fewer muscle fibers

4 Age changes

  a Fibrosis—with advancing years some skeletal muscle fibers degenerate and are replaced by fibrous connective tissue

  b Decreased muscular strength, due in part to fibrosis

Function

1 Basic principles

  a Skeletal muscles contract only if stimulated; natural stimulus, nerve impulses

  b Skeletal muscle contractions of several types:

   1 Tonic contraction (tone; tonus)—continual, partial contractions produced by simultaneous activation of small group of motor units, followed by relaxation of their fibers and activation of another group of motor units; all healthy muscles exhibit tone when individuals are awake

   2 Isotonic contraction—muscle shortens but its tension remains constant; isotonic contractions produce movements

   3 Isometric contraction—muscle length remains unchanged but tension within muscle increases; isometric contractions "tighten" muscles but do not produce movements

   4 Twitch contraction—quick, jerky contraction in response to single stimulus; consists of three phases—latent period, contraction phase, and relaxation phase; twitch contractions rare in normal body

   5 Tetanic contraction (tetanus)—sustained smooth contraction produced by series of stimuli bombarding muscle in rapid succession; normal movements said to be produced by incomplete tetanic contractions

   6 Treppe (staircase phenomenon)—series of increasingly stronger contractions in response to constant strength stimuli applied at rate of 1 or 2 per second; contracture—i.e., incomplete relaxation after repeated stimulation; fatigue—i.e., failure of muscle to contract in response to strongest stimuli after repeated stimulation; true muscle fatigue seldom occurs in body

    7 Fibrillation – abnormal contraction in which individual muscle fibers contract asynchronously, producing no effective movement

    8 Convulsions – uncoordinated tetanic contractions of varying groups of muscles

  c Skeletal muscles contract according to graded-strength principle in contrast to individual muscle cells that compose them which contract according to all-or-none law

  d Skeletal muscles produce movement by pulling on insertion bones across joints

  e Bones serve as levers and joints as fulcrums of these levers

  f Muscles that move part usually do not lie over that part but proximal to it

  g Skeletal muscles almost always act in groups rather than singly – i.e., most movements produced by coordinated action of several muscles

2 Hints of how to deduce actions

  a Deduce bones that muscle attaches to from illustrations of muscle

  b Make guess as to which bone moves (insertion)

  c Deduce movement muscle produces by applying principle that its insertion moves toward its origin

Names

1 Reasons for names – muscle names describe one or more of following features about muscle

  a Its action

  b Direction of fibers

  c Its location

  d Number of divisions composing it

  e Its shape

  f Its points of attachment

2 Muscles grouped according to location – see Table 6-1

3 Muscles grouped according to function – also see Table 6-2

  a Flexors – decrease angle of joint

  b Extensors – return part from flexion to normal anatomical position

  c Abductors – move bone away from midline of body

  d Adductors – move bone toward midline of body

  e Rotators – cause part to pivot upon its axis

  f Levators – raise part

  g Depressors – lower part

  h Sphincters – reduce size of opening

  i Tensors – tense part or make it more rigid

  j Supinators – turn hand palm upward

  k Pronators – turn hand palm downward

Origins, Insertions, functions, and innervations of representative muscles

See Tables 6-3 to 6-15

WEAK PLACES IN ABDOMINAL WALL

1 Inguinal rings – right and left internal; right and left external

2 Femoral rings – right and left

3 Umbilicus

BURSAE

1 Definition – small connective tissue sacs lined with synovial membrane and containing synovial fluid

2 Locations – wherever pressure exerted over moving parts

  a Between skin and bone

  b Between tendons and bone

  c Between muscles or ligaments and bone

  d Names of bursae that frequently become inflamed (bursitis)

    1 Subacromial – between deltoid muscle and head of humerus and acromion process

    2 Olecranon – between olecranon process and skin; inflammation called student's elbow

    3 Prepatellar – between patella and skin; inflammation called housemaid's knee

3 Function – act as cushion, relieving pressure between moving parts

POSTURE

1 Meaning – position or alignment of body parts

2 How maintained – by continual pull of muscles on bones in opposite direction from pull of gravity – i.e., posture maintained by continued partial contraction of muscles, or muscle tone; therefore, indirectly dependent on many other factors – e.g., normal nervous, respiratory, and circulatory systems, health in general

3 Importance to body as whole – essential for optimal functioning of most of body – e.g., respiration, circulation, digestion, joint action, etc.; briefly, maximal health dependent upon good posture, good posture dependent upon health

review questions

1 Differentiate between the three kinds of muscle tissue as to structure, location, and innervation.
2 Describe several physiological properties of muscle tissue.
3 What property is more highly developed in muscle than in any other tissue?
4 State a principle describing the usual relationship between a part moved and the location of muscles (insertion, body, and origin) moving the part.
5 Applying the principle stated in question 4, where would you expect muscles that move the head to be located? Name two or three muscles that fulfill these conditions.
6 Applying the principle stated in question 4, what part of the body do thigh muscles move? Name several muscles that fulfill these conditions.
7 What bone or bones serve as a lever in movements of the forearm? What structure constitutes the fulcrum for this lever?
8 Explain the meaning of the term neuromusculoskeletal unit.
9 Name the main muscles of the back, chest, abdomen, neck, shoulder, upper arm, lower arm, thigh, buttocks, leg, and pelvic floor.
10 Name the main muscles that flex, extend, abduct, and adduct the upper arm; that raise and lower the shoulder; that flex and extend the lower arm; that flex, extend, abduct, and adduct the thigh; that flex and extend the lower leg and thigh; that flex and extend the foot; that flex, extend, abduct, and adduct the head; that move the abdominal wall; that move the chest wall.

11 Discuss the chemical reactions thought to make available energy for muscle contraction.
12 What physiological reason can you give for athletes using a warming-up period before starting a game?
13 Why does an individual pant after strenuous exercise?
14 Curare preparations are often given during surgery. Would you expect this to make the patient's muscles more relaxed or more rigid? Why?
15 In general, where are bursae located? Give several specific locations.
16 Name several weak places in the abdominal wall where hernia may occur.
17 What and where are the inguinal canals? Of what clinical importance are they?
18 Good posture depends upon tonicity of the antigravity muscles, particularly of those which hold the head and trunk erect and the abdominal wall pulled in. Name several muscles that perform these functions.
19 Define the following terms: aponeurosis; bursa; contraction; contracture; elasticity; extensibility; fibrillation; insertion; motor unit; origin; oxygen debt; tetanus; tone; treppe; twitch.
20 Explain briefly the current theory about the role of calcium ions in muscle contraction and relaxation.

Integration
and control of
the body

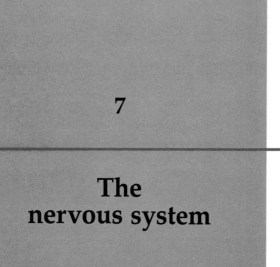

7

The nervous system

Cells
Neuroglia
  Types
  Structure and function
Neurons
  Structure
  Types
  Function

Nerve impulse (action potential)
Definition
Mechanism
Initiation
Conduction

Organs
Spinal cord
  Location
  Structure
  Functions
Brain
  Cerebrum
    Hemispheres, fissures, and lobes
    Cerebral cortex
    Cerebral tracts
    Basal ganglia
    Functions
  Diencephalon
    Thalamus
      Structure and location
      Functions
    Hypothalamus
      Structure and location
      Functions
  Cerebellum
    Structure and location
    Functions
  Medulla oblongata
    Structure and location
    Functions
  Pons varolii
    Structure and location
    Functions
  Midbrain
    Structure and location
    Functions

Cord and brain coverings
Cord and brain fluid spaces
Formation and circulation of cerebrospinal fluid
Spinal nerves
  Origin
  Distribution
  Microscopic structure
  Functions
Cranial nerves
  First (olfactory)
  Second (optic)
  Third (oculomotor)
  Fourth (trochlear)
  Fifth (trifacial or trigeminal)
  Sixth (abducens)
  Seventh (facial)
  Eighth (acoustic or auditory)
  Ninth (glossopharyngeal)
  Tenth (vagus or pneumogastric)
  Eleventh (accessory or spinal accessory)
  Twelfth (hypoglossal)

Sensory neural pathways

Arousal or alerting mechanism

Motor neural pathways to skeletal muscles

Reflexes
Definition
Some somatic reflexes of clinical importance

Autonomic nervous system
Definition
Divisions
Macroscopic structure
Microscopic structure
Some general principles
Higher autonomic centers

If you want to understand the body, you need to remind yourself frequently of some principles stated in the first chapter of this book—briefly, that the body is made up of millions of smaller structures that carry on a host of different activities. But, and this is the important point, together all of these diverse activities accomplish the one big, all-encompassing function of the body —survival. If you think about this for a moment, you realize that the only way many units can be made to function as a single unit (whether the units are cells, or people, or parts of a machine) is by organization—that is, controlling their numerous activities so as to coordinate and integrate them. And this necessitates communication between the units. In short, communication makes possible control, and control makes possible both organization and integration. Many familiar examples of this principle suggest themselves. A hospital, to name just one example, is such a unit. Quite obviously, the activities of the hundreds of individuals who make up a modern hospital must be organized, coordinated, and integrated, and communication must take place between the individuals. Complete chaos would prevail otherwise. A disorganized hospital could not survive functionally. In no time at all it would be utterly unable to carry on its one great function of giving care to sick people. Nerve impulses and chemicals—chiefly hormones, carbon dioxide, and certain ions — constitute the two kinds of "messages" or communications sent within the body. The nervous system and the endocrine system constitute its two main communication systems. The circulatory system serves as an assistant communication system by distributing hormones and other chemical messages.

Facts, theories, and questions about the nervous system are as abundant and complex as they are fascinating. We shall approach this large body of material by considering, in this chapter, the cells of the nervous system, the mechanism by which impulses are conducted along a single nerve cell, the structure and function of the various organs of the nervous system, the common pathways traveled by nerve impulses, the results of impulse conduction to certain muscles, and, finally, the autonomic nervous system. We shall discuss sense organs in Chapter 8 and the endocrine system in Chapter 9.

Cells

Two main kinds of cells compose nervous system structures — neurons and neuroglia. Neurons are the "specialists" of the nervous system; they specialize in impulse conduction, the function which makes possible all other nervous system functions. Neuroglia, on the other hand, perform the less specialized functions of support and protection. Neuroglia will be discussed first and then neurons.

■ Neuroglia

Types

Neuroglia are interesting cells — perhaps because their functions still remain somewhat of a mystery. Assuredly, however, some of them support neurons and anchor them to blood vessels, whereas others serve a defense or protective function. Neuroglia are important clinically because most tumors of the nervous system arise from them. Histologists identify these major types: astrocytes, oligodendroglia, and microglia (Fig. 7-1).

Structure and function

Astrocytes and oligodendroglia furnish support and connection. *Astrocytes* are star-shaped cells with numerous processes. Some of their processes form a thick network that twines around nerve cells. Other astrocyte processes attach by little "sucker feet" to adjacent blood vessels, thereby holding nerve cells close to their blood vessels. *Oligodendroglia* have fewer processes than astrocytes but are also interposed between neurons and their blood vessels.

Microglia serve two unrelated functions. They help support nerve tissue, and they protect against infection. Usually, they are small, stationary cells. In inflamed or degenerating brain tissue, however, microg-

165

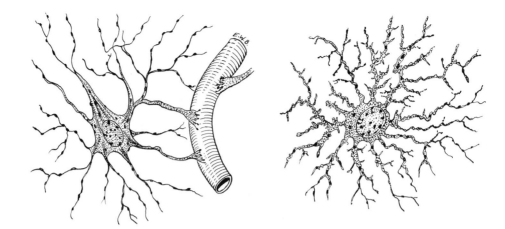

Different forms of astrocytes, with astrocyte at
left illustrated with footplates against blood vessel

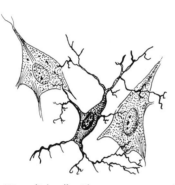

Microglial cell with processes extending
to two nerve cell bodies

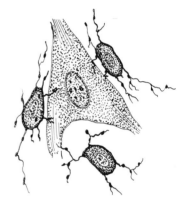

Oligodendroglial cells near
a nerve cell body

Fig. 7-1 Neuroglia, special connective tissue cells of the central nervous system. Astrocytes and oligodendroglia support neurons and connect them to blood vessels within the nerve tissue. Microglia show phagocytic activity when nerve tissue is injured or diseased so serve a protective function.

lia enlarge, move about, and carry on phagocytosis. In other words, they engulf and destroy microbes and cellular debris.

Whether support and protection are the sole functions of neuroglia is now being questioned. For example, some evidence indicates that they also play a part in forming the myelin sheath of neurons in the brain and spinal cord.

Neurons — basic unit of structure

Structure

All neurons consist of a cell body and at least two processes: one axon and one or more dendrites. These processes are threadlike extensions from the neuron's cell body. In many respects, the cell body, the largest part of a nerve cell, resembles other cells. It contains a nucleus, cytoplasm, and various organelles found in other cells—for example, mitochondria and a Golgi apparatus. Incidentally, Golgi first saw this apparatus in neurons. A neuron's cytoplasm extends from its cell body into its processes. A cell membrane encloses the entire neuron. It not only delimits the cell but, equally important, it works to maintain differences in chemical and electrical characteristics between the

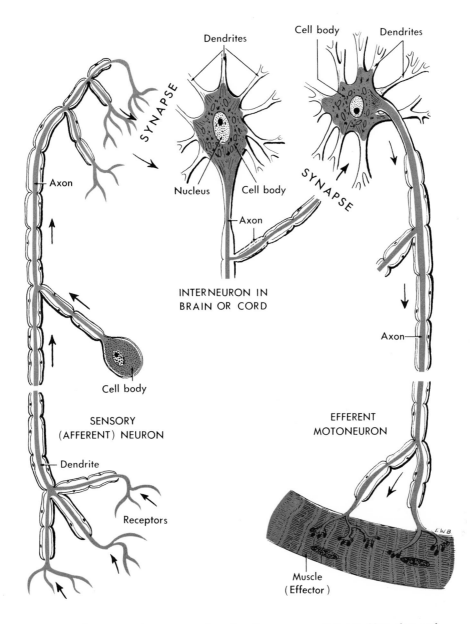

Dendrites

Cell body   Dendrites

SYNAPSE

Axon

Nucleus   Cell body

Axon

SYNAPSE

INTERNEURON IN
BRAIN OR CORD

Axon

Cell body

SENSORY
(AFFERENT) NEURON

EFFERENT
MOTONEURON

Dendrite

Receptors

Muscle
(Effector)

Fig. 7-2 Diagrammatic representation of a three-neuron reflex arc. Note that each neuron has three parts: a cell body and two types of extensions, dendrite(s) and an axon. Arrows indicate direction of impulse conduction. (See Fig. 7-3 for more details about coverings of processes.)

outside and inside of the neuron. Because the cell membrane regulates passage of ions into and out of the cell, it plays a crucial role in neuron functioning.

Certain structures—dendrites, axons, neurofibrils, Nissl bodies, myelin sheath, and neurilemma—are found only in neu-

rons. The following paragraphs describe them briefly.

Dendrites, as you can see in Fig. 7-2, branch extensively, like tiny trees. In fact, their name derives from the Greek word for tree. The distal ends of dendrites of sensory neurons are called *receptors* because they receive the stimuli that initiate conduction. Dendrites conduct impulses to the cell body of the neuron.

The *axon* of a neuron is a single process that extends out from the neuron cell body. Although a neuron has only one axon, it often has one or more side branches (*axon collaterals*). Moreover, axons terminate in many branched filaments, and, like dendrites, they vary considerably in length. Some are as much as 3 feet or more long. Others measure only a fraction of an inch. Axons vary also in diameter—a point of interest because it relates to velocity of impulse conduction. In general, fibers with a large diameter conduct more rapidly than those with a small diameter. A neuron's axon conducts impulses away from its cell body.

Neurofibrils are very fine fibers extending through dendrites, cell bodies, and axons. The electron microscope has revealed that bundles of neurofibrils interlace to form a network in neuron cytoplasm.

Nissl bodies consist of groups of flat, membranous sacs and numerous RNA granules scattered between them. In other words, Nissl bodies constitute the rough endoplasmic reticulum of a neuron. Since this organelle specializes in protein synthesis, one would expect Nissl bodies to perform the same function. But why should neurons, whose specialty is conduction, also specialize in protein synthesis? According to one speculation, they use the protein they make for maintaining and regenerating neuron processes. Radioautographic studies have shown that proteins synthesized in neuron cell bodies quickly migrate down their axons.

The *myelin sheath* is a segmented wrapping around a nerve fiber. Note in Fig. 7-3 the small gaps called *nodes of Ranvier* between segments of the sheath. The widely accepted "jelly roll hypothesis" of myelin formation holds that myelin consists of a double layer of the cell membranes of Schwann cells. *Schwann cells* are satellite cells located along the fibers of peripheral nerves. One Schwann cell winds itself in "jelly roll" fashion around each segment of the fiber (a segment is the section of fiber between two successive nodes of Ranvier).

The *neurilemma* is a continuous sheath that encloses the segmented myelin sheath of peripheral nerve fibers. Like the myelin sheath, the neurilemma is thought to also derive from Schwann cells. It plays an essential part in peripheral nerve fiber regeneration. Unfortunately, brain and spinal cord fibers do not have a neurilemma and, so far as we know, do not regenerate.* So if disease or injury causes them to degenerate, the destruction is permanent. This fact, as we shall see, may produce great loss of function.

Types

Neurons are classified according to two different systems—the direction in which they conduct impulses and the number of processes they have.

Classified according to the direction in which they conduct impulses, neurons are of three types: sensory, motor, and internuncial. *Sensory (afferent) neurons* transmit nerve impulses to spinal cord or brain. *Motoneurons* (*motor* or *efferent neurons*) transmit nerve impulses away from brain or spinal cord to or toward muscle or glandular tissue. *Interneurons* (*internuncial* or *intercalated neurons*) conduct impulses from sensory to motor neurons. They lie entirely within the central nervous system (brain and spinal cord).

---

*Recently, a few scientists have presented findings which suggest that regeneration of CNS neurons may, under some circumstances, be possible. See Culliton, B. J.: Spinal cord regeneration, Sci. News **98**:337-338 (Oct. 24), 1970.

Classified according to the number of their processes, neurons are of three types: multipolar, bipolar, and unipolar. *Multipolar neurons* have only one axon but several dendrites. Most of the neurons in the brain and spinal cord are multipolar. *Bipolar neurons* have only one axon and also only one dendrite. They are found in the retina of the eye and in the spiral ganglion of the inner ear, for example. *Unipolar neurons* originate in the embryo as bipolar neurons. But in the course of development, the two processes become fused into one for a short distance beyond the cell body. Then they separate into clearly distinguishable axon and dendrite. Generally, sensory neurons are unipolar. Fig. 7-2 illustrates all but one of the six classifications of neurons. Can you identify the missing one? Can you identify the ones that are shown?

Function

Neurons perform the specific function of conducting impulses and thereby contribute to the general functions of communication and integration. Nerve impulse conduction provides a means for rapid control of separate structures and for integrating the activities of many different structures.

The arrangement of neurons to form the organs of the nervous system is analogous to that of the wires and regulating devices of a telephone system. Also, the two systems—telephone and nervous—operate on similar basic principles. Let us say more specifically what we mean by this. In a telephone system, wires conduct messages from telephones to a central switching station. There, the messages may terminate, or they may be switched to other wires that conduct them out from the central station to other telephones. Also, messages may originate in a telephone system's central station and be relayed out from it. In the

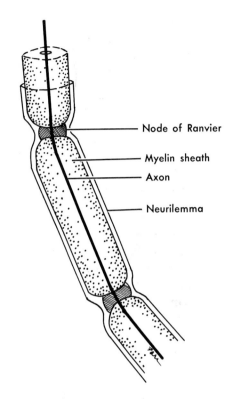

Fig. 7-3 Diagram of a nerve fiber and its coverings.

Node of Ranvier

Myelin sheath

Axon

Neurilemma

nervous system, nerve fibers are analogous to telephone wires and the central nervous system is analogous to a telephone central switching station. Nerve fibers conduct impulses from different parts of the body to the central nervous system (CNS; the brain and spinal cord). Some impulses terminate there. Others are switched in the central nervous system to other neurons and conducted by their fibers out from the CNS to different parts of the body. Also, some impulses presumably originate in the CNS and are sent out from it. Both telephone systems and nervous systems are, of course, communication systems. Both are extremely complex. The living system's intricacies of structure and function, however, far exceed those of the nonliving system.

All three parts of a neuron function in the conduction of impulses. The cell body also performs other functions—those necessary to maintain the neuron's life such as catabolism and anabolism. So vital is the neuron cell body to the life of its processes that they soon degenerate if the cell body dies or is destroyed. And if the cell body is separated from its processes by surgery or accident, it will survive but the processes will die.

Impulse conduction, a function performed by all neurons, is initiated by an adequate stimulus acting on a neuron's dendrites or cell body. Neurons in the brain, however, perhaps may initiate impulses without the intervention of a stimulus. And many brain neurons function in some way, still not understood, to produce mental activities—interpreting sensations, judging, willing, imagining, remembering, etc. How do neurons achieve such results? How can mere impulse conduction over certain neurons—marvelous as this mechanism is—how can it alone account for our many and varied and complex mental processes? After you have studied the paragraphs discussing the nerve impulse, pause long enough to wonder about this matter. Does it not seem to be an unfathomable secret? But before we attempt to explain the nerve impulse, we shall relate certain facts about the *nonconducting* or *"resting" neuron.* Hopefully, they will help you understand the discussion of the nerve impulse.

Hodgkin and co-workers, in the late 1930's, used recently developed methods to measure differences between the resting neuron's extracellular and intracellular characteristics. Among the differences they observed, two that eventually proved to have great physiological significance were these—the concentration of sodium

ions in extracellular fluid is much higher than it is in a nonconducting neuron's intracellular fluid, and the outer surface of its membrane is about 70 to 90 millivolts positive to its inner surface. Its membrane is *polarized*, to use a technical term, or a *potential difference* (a difference in electrical charges) exists across its membrane. The potential difference across a nonconducting neuron's membrane is commonly called the *resting potential.*

The complex mechanism that maintains the resting potential is not yet fully understood. And a complete explanation of what is known is beyond the scope of an introductory physiology textbook. We shall, therefore, attempt only a brief explanation, knowing that it omits much of today's knowledge about this crucially important mechanism and necessarily oversimplifies it.

The direct cause of the resting potential is the difference in concentrations of positively and negatively charged ions in extracellular and intracellular fluids. As you know, the law of diffusion dictates that the net diffusion of any particular ion occurs down its own concentration gradient. Another physical law requires that ions move down an electrical gradient—that is, from a more to a less electropositive area. Net diffusion of ions across the cell membrane, therefore, should equalize the ionic concentrations of extracellular and intracellular fluids and should equalize the electrical charges on both sides of the membrane. No potential difference would then exist across it. In the normal nonconducting cell, however, these equalizations do not develop. Activity of the living cell membrane prevents them from taking place. An active process within the membrane—still not completely understood—operates continuously to pump sodium ions out of the cell. This maintains a difference between extracellular and intracellular sodium concentrations. It keeps extracellular fluid sodium concentration many times higher than that of intracellular fluid. Moreover, by maintaining an ionic concentration difference between the two fluids, the sodium pump also maintains a difference in electrical potential across the cell membrane. It keeps the outer surface of a nonconducting cell membrane positive to its inner surface. The sodium pump, in other words, polarizes cell membranes.* To deduce how important living processes are in maintaining these differences across cell membranes, you need consider only one fact—that across the membranes of dead cells, no differences in ionic concentration nor in electrical potential exist.

Nerve impulse (action potential)

Definition

According to widely accepted present-day theory, a nerve impulse is a self-propagating wave of electrical negativity that travels along the surface of the neuron membrane. One can easily demonstrate the passage of a nerve impulse by measuring the voltage on the outer surface of a neuron's cell membrane. During the moment the impulse passes, the voltage becomes negative. Bernstein, in 1902, postulated the reason for the change in voltage. But more than thirty years passed before the experimental methods needed to prove or disprove his theory became available and were used by Hodgkin and associates. The following paragraphs give a step-by-step description of the mechanism of the nerve impulse. As you read them, refer often to Fig. 7-4.

*If you would like to delve further into the complex physiology of membrane polarization, see Mountcastle, V. B., editor: Medical physiology, ed. 12, St. Louis, 1968, The C. V. Mosby Co., pp. 1082-1127.

Mechanism

1 When an adequate stimulus is applied to a neuron, it greatly increases the membrane's permeability to sodium ions at the point of stimulation.

2 Sodium ions rush into the cell at the stimulated point. Therefore, at this point, the excess of positive ions outside rapidly dwindles to zero. Therefore, also, the membrane potential decreases to zero at this point. In other words, the stimulated point of the membrane is no longer polarized. It is *depolarized*. But only for an instant. As more sodium ions continue to stream into the cell, they almost instantaneously produce an excess of positive ions inside the cell and leave an excess of negative ions outside. In short, the influx of sodium ions reverses the resting potential and thereby changes it into an action potential. Whereas a typical resting potential is 70 millivolts with the outside of the membrane positive to the inside, the *action potential* is about 30 millivolts with the outside negative to the inside.

3 The stimulated negatively charged point of the membrane sets up a local current with the positive point adjacent to it, and this local current acts as a stimulus. Consequently, within a fraction of a second, the adjacent point on the membrane becomes depolarized and its potential reverses from positive to negative. The action potential has thus moved from the point originally stimulated to the adjacent point on the membrane. As the cycle goes on repeating itself over and over again in rapid succession, the action potential travels point by point out the full length of the nerve fiber, much as a wave of water moves in from sea to shore. Hence, the *action potential* (or its synonym, *nerve impulse*) is defined as a self-propagating wave of negativity that travels along the surface of a neuron's membrane.

4 By the time the action potential has

moved from one point on the membrane to the next (a matter of thousandths of a second), the first point has repolarized—its resting potential has been restored. *Repolarization* results from the fact that the increased permeability to sodium induced by stimulation lasts only momentarily. It is quickly replaced by increased permeability to potassium which, therefore, diffuses outward (because potassium concentration inside the cell is much greater than that outside the cell). Then the permeability of the membrane decreases and its pumping activity again becomes effective. Once again, it actively transports sodium ions out of the cell and potassium ions into it. Once again—and in much less time than it takes to tell it—more positive ions accumulate on the outer than on the inner surface of the cell membrane. The resting potential of the cell has, in other words, been restored. The outer surface of its membrane is again electrically positive to its inner surface.

Initiation

Normally, impulse conduction begins when a stimulus acts on receptors. (A *stimulus* is a change in the environment—for example, a change in pressure or temperature. *Receptors* are the distal ends of dendrites of sensory neurons.) When a stimulus acts on a receptor, the receptor membrane potential decreases below its resting level (explained in paragraphs 1 and 2 of the preceding discussion). If it decreases down to a certain critical level, known as the *threshold of stimulation*, it triggers off impulse conduction. Or, stated another way, it is a decrease in a neuron's membrane potential to its threshold of stimulation that initiates an action potential. Once initiated, the action potential then rapidly propagates itself the full length of the neuron. Whether or not a stimulus initiates conduction depends upon the intensity or "strength" of the stimulus. A stimulus just strong enough to decrease the receptor potential to its threshold is just strong enough to initiate impulse conduction and is called a *threshold stimulus* (or liminal stimulus). Any stimulus weaker than this is a *subthreshold stimulus* (or subliminal stimulus). It decreases the membrane potential but not down so low as the threshold level so does not trigger off impulse conduction. The effects of two or more subthreshold stimuli acting together, however, summate and can decrease the membrane potential to its threshold level and thus trigger a nerve impulse.

Conduction

Course (neural pathways). The usual route traveled by nerve impulses begins at receptors located outside of the central nervous system, extends from them to the central nervous system, then extends away from the central nervous system, and terminates in effectors. (*Effectors* are muscles and glands.) Such pathways followed by nerve impulses—from receptors to central nervous system to effectors—are called *reflex arcs*. Reflex arcs, of course, consist of neurons, the "conduction specialists" of the body. Each reflex arc is composed of neurons placed end to end. The contact point between the end of an axon of one neuron and the dendrite or cell body of another neuron is called a *synapse*.

The simplest reflex arcs are the two-neuron and three-neuron arcs. A *two-neuron* or *monosynaptic arc* consists of at least one sensory neuron, one motoneuron, and a synapse between them (Fig. 7-5). A *three-neuron arc* consists of at least one sensory neuron, one interneuron, one motoneuron, and two synapses—that is, a synapse between the sensory neuron and interneuron and another between the interneuron and motoneuron. Figs. 7-6 to 7-8

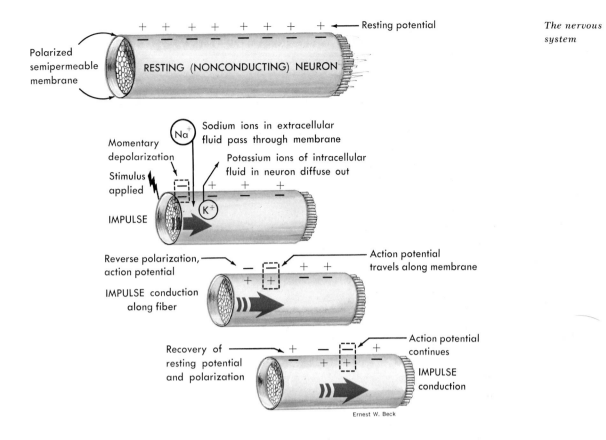

Fig. 7-4 Upper diagram represents polarized state of the membrane of a nerve fiber when it is not conducting impulses. Lower diagrams represent nerve impulse conduction—a self-propagating wave of negativity or action potential travels along membrane.

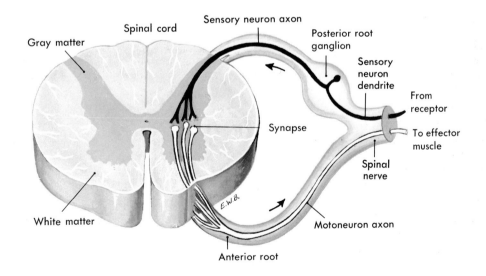

Fig. 7-5 A two-neuron ipsilateral reflex arc; also called a monosynaptic arc because impulses cross only one synapse in traversing it. In ipsilateral arcs, afferent impulses enter the cord and efferent impulses leave it on the same side. Conduction over such arcs produces stretch reflexes such as the knee jerk.

show different types of three-neuron arcs.

Besides simple two-neuron and three-neuron arcs, extremely complex arcs also exist. Moreover, all impulses that start in receptors do not invariably travel a complete reflex arc and terminate in effectors. Nor, according to current evidence, do all impulses that terminate in effectors invariably start in receptors. For example, many of them are thought to originate in the cerebral cortex.

Rate. One general principle is that the rate of impulse conduction varies directly with the diameter of an axon—the larger the diameter, the faster the conduction. Fibers with a large diameter are classed as A fibers, those with a small diameter are classed as C fibers, and those with a diameter of intermediate size are classed as B fibers. Read Table 7-1, in which the functions served by each class of fibers are summarized, and then try to answer this question: What "survival value" do you see in the fastest conducting fibers serving the functions they do?

A second principle about the rate of

impulse conduction holds that it relates directly to the thickness of myelin surrounding the fiber and to the distance between successive nodes of Ranvier. Heavily myelinated fibers with a greater distance between successive nodes conduct impulses many times faster than do so-called unmyelinated fibers with a lesser distance between nodes. ("Unmyelinated" fibers actually have a thin layer of myelin around them.)

Conduction across synapses. Many axon filaments terminate in little knobs called *synaptic knobs,* or *end feet,* or *end buttons* (Fig. 7-9). Each synaptic knob contacts either a dendrite or a cell body of another neuron to form a *synapse* between the two neurons. (Actually a space about a millionth of an inch wide, called the *synaptic cleft,* separates the synaptic knobs from the dendrites or cell body.) Numerous mitochondria and hundreds of tiny neurovesicles lie crowded together inside each synaptic knob. A *neurovesicle* contains a "transmitter substance," only a few thousand molecules per vesicle.

Table 7-1. Functions served by fibers classified according to diameter of axons

Classification	*Functions of sensory fibers*	*Functions and locations of motor fibers*
A fibers Fibers with large diameter; fastest conducting; about 100 meters per second or more than 3 miles per minute	Proprioception, touch, pressure, some heat, cold, some pain	Conduct impulses from cord to skeletal muscles causing them to contract
B fibers Fibers with diameter of intermediate size	Some pain	Conduct impulses from lower autonomic centers in cord or brainstem to autonomic ganglia
C fibers Fibers with small diameter; slowest conducting; about 1/2 meter per second or 1 mile per hour	Some pain, perhaps some touch, pressure, heat, cold	Conduct impulses from autonomic ganglia to smooth muscle, cardiac muscle, and glands activating them

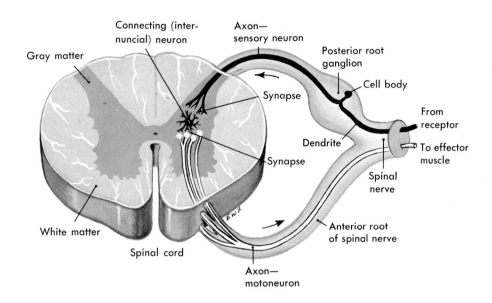

Fig. 7-6 Three-neuron ipsilateral reflex arc, consisting of a sensory neuron, a connecting (internuncial) neuron, and a motoneuron. Note the presence of two synapses in this arc—(1) between sensory neuron axon terminals and internuncial neuron dendrites and (2) between internuncial axon terminals and motoneuron dendrites and cell bodies (located in anterior gray matter). Nerve impulses traversing such arcs produce many spinal reflexes. Example: withdrawing the hand from a hot object. (See also Table 7-9.)

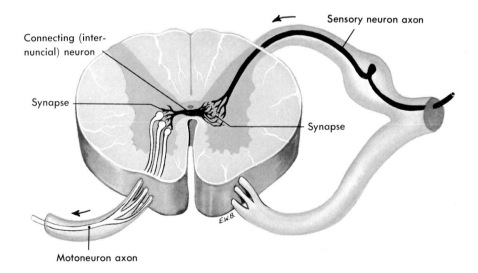

Fig. 7-7 Three-neuron contralateral reflex arc. Afferent impulses enter the cord on one side and efferent impulses leave it on the other side. Figs. 7-5 and 7-6 show ipsilateral arcs (impulses enter and leave the cord on the same side).

Once an action potential is initiated, it travels all the way down to the neuron's axon terminals, but there it ends. It does not cross over a synapse to another neuron. Instead, the action potential is generated anew in the next neuron, in the postsynaptic neuron, that is.

How is a nerve impulse generated anew in a postsynaptic neuron? Today's most widely held concept of synaptic conduction states that when the action potential reaches an axon's terminals it causes them to eject their transmitter substance (acetylcholine or possibly some other compound) into the synaptic cleft. In a flash, the chemical diffuses across this infinitesimal space and contacts the membrane of the postsynaptic neuron. Here it serves as a stimulus—that is, it lowers that neuron's membrane potential. But it may or may not trigger conduction by the postsynaptic neuron. Whether it does or not depends upon how much of it acts on the neuron. Sufficient transmitter substance must contact the postsynaptic neuron to lower its membrane potential to its threshold of stimulation level—generally somewhere between 60 and 50 millivolts (outside of membrane, positive to inside). The instant the potential reaches this critical level, it then decreases further with explosive rapidity all the way down through 0 to −30 millivolts. Briefly, then, a threshold amount of transmitter substance causes a postsynaptic neuron's potential to decrease from its resting level of, say, +70 to a threshold level of +55 and then instantaneously to −30 (the action potential or nerve impulse).

A subthreshold amount of transmitter substance also decreases a postsynaptic neuron's resting potential, but not down so far as the critical threshold level. A subthreshold amount of transmitter produces what is called an *excitatory postsynaptic potential* (EPSP)—that is, a low-

ered resting potential in the dendrite and cell body of the postsynaptic neuron. It does not initiate an action potential in this neuron. For example, assume that a neuron has a threshold of stimulation of +55 millivolts. A subthreshold amount of transmitter substance might decrease its resting potential from +70 to +65 millivolts. Since this is above the threshold level of +55 that triggers off impulse conduction, this neuron would be said to be *facilitated*, not stimulated. Its resting potential would have become an excitatory potential but not an action potential. In short, subthreshold amounts of transmitter substance from one or a few synaptic knobs produce an excitatory potential in the postsynaptic neuron. Subthreshold amounts of transmitter substance released simultaneously from many synaptic knobs, on the other hand, add together. Thus summated, they may constitute a threshold amount that triggers an action potential.

Acetylcholine is known to be the excitatory transmitter substance ejected at some synapses. Within seconds, an enzyme, *cholinesterase*, inactivates acetylcholine, thereby stopping synaptic conduction.

The synaptic knobs of some axons release a transmitter substance (still not identified) that inhibits rather than excites or stimulates postsynaptic neurons. This substance increases the neuron's membrane potential. In other words, it produces an *inhibitory postsynaptic potential* (IPSP)—perhaps a resting potential of +80 instead of the typical +70 millivolts.

Dozens of synaptic knobs from dozens of different axons synapse with any one postsynaptic neuron. Some of these knobs release excitatory transmitter. Some release inhibitory transmitter. And the algebraic sum of the two opposing chemicals determines their effect on the postsynaptic neuron. Together, they may merely facilitate it, or actually stimulate it, or, just the oppo-

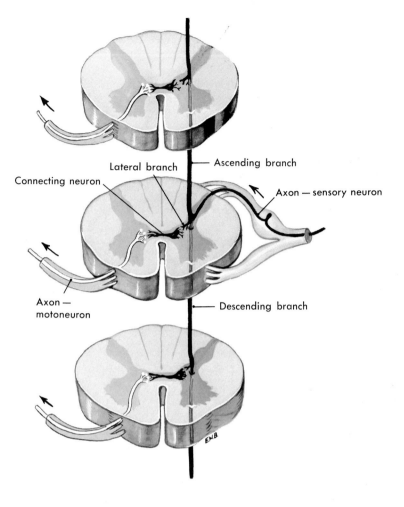

Lateral branch

Connecting neuron

Ascending branch

Axon — sensory neuron

Axon —
motoneuron

Descending branch

Fig. 7-8 Intersegmental reflex arcs, showing a sensory fiber splitting into ascending and descending branches that give rise to lateral branches that synapse with their respective connecting neurons. Such arrangements make possible the activation of more than one effector by impulses over a single sensory fiber and account for the phenomenon of divergence.

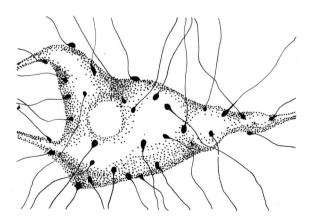

Fig. 7-9 Synaptic knobs (end feet or end buttons) on motoneuron cell body.

site, inhibit it. (Inhibition and facilitation are discussed further on pp. 219 to 221.)

Conduction across neuroeffector junctions. Neuroeffector junctions are contact places between a motoneuron's axon terminals and a muscle or a secreting cell's membrane. The chief transmitter substances released at these junctions are acetylcholine (ACh) and norepinephrine (NE). The axons of all somatic motoneurons terminate in skeletal muscle where they release acetylcholine. Axons of autonomic neurons, in contrast, terminate in either cardiac muscle or smooth muscle or glandular tissue. Some autonomic axons release acetylcholine, but others release norepinephrine. (Norepinephrine and epinephrine belong to the class of compounds called catecholamines.) Transmitter substances released at neuroeffector junctions stimulate the effector cell to contract, if it is a muscle cell, or to secrete if it is a glandular cell.

Organs

Although only a few organs—brain, spinal cord, and nerves—make up the nervous system, billions of cells compose these organs. Connective tissue cells called neuroglia (p. 165) hold in position arrangements of neuron cell bodies or of their processes. Nervous tissue is referred to as either white matter or gray matter. *White matter* consists mainly of myelinated nerve fibers (myelin, you may recall, is white). *Gray matter* consists mainly of neuron cell bodies.

Both nerves and tracts consist of white matter. They differ only in location. *Nerves* are bundles of nerve fibers located outside of the central nervous system (outside of the brain and spinal cord, that is). *Tracts* are bundles of nerve fibers located in the central nervous system. Later in this chap-

ter we shall discuss several brain and cord tracts in some detail.

Ganglia, nuclei, centers, and horns are structures that consist of neuron cell bodies. Therefore, which composes them—white matter or gray matter? *Ganglia* (singular, ganglion) consist of gray matter located outside the central nervous system. *Nuclei, centers,* and *horns* also consist of gray matter, but they are all located inside the central nervous system.

You can readily deduce the function of white matter from the fact that it consists of nerve fibers, the extended processes of nerve cells. White matter, therefore, conducts impulses over relatively long distances. Gray matter conducts only short distances. It consists of neuron cell bodies, as you know, and each one conducts impulses only from the cell's own dendrites to its axon. In addition to conduction, neuron cell bodies perform the many functions that all cells perform to keep themselves alive.

Only two organs—the brain and spinal cord—constitute the central nervous system (CNS). Hundreds of nerves and many ganglia make up the peripheral nervous system (PNS).

We shall discuss the spinal cord, the brain, their coverings, fluid spaces, and fluid, the spinal nerves, and the cranial nerves, in that order.

■ Spinal cord

Location

The spinal cord lies within the spinal cavity, extending from the foramen magnum to the lower border of the first lumbar vertebra (Fig. 7-10), a distance of 17 or 18 inches in the average body. The cord does not completely fill the spinal cavity —it also contains the meninges, spinal fluid, a cushion of adipose tissue, and blood vessels.

Structure

The spinal cord is an oval-shaped cylinder which tapers slightly from above downward and has two bulges, one in the cervical region and the other in the lumbar region. Two deep grooves, the *anterior* *median fissure* and the *posterior median* *sulcus*, just miss dividing the cord into separate symmetrical halves. The anterior fissure is the deeper and the wider of the two grooves—a useful fact to remember when you examine spinal cord diagrams.

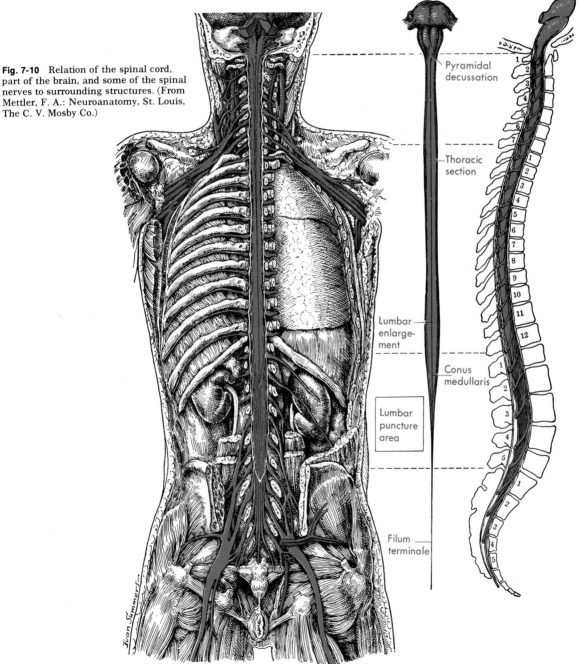

Fig. 7-10 Relation of the spinal cord, part of the brain, and some of the spinal nerves to surrounding structures. (From Mettler, F. A.: Neuroanatomy, St. Louis, The C. V. Mosby Co.)

Pyramidal decussation

Thoracic section

Lumbar enlargement

Conus medullaris

Lumbar puncture area

Filum terminale

It enables you to tell at a glance which part of the cord is anterior and which is posterior. Gray matter, shaped roughly like a three-dimensional letter H, composes the inner core of the cord. The limbs of the H are called the anterior, posterior, and lateral horns (or columns) of gray matter (Fig. 7-11). Horns consist primarily of cell bodies of interneurons and motoneurons. White matter surrounding the gray matter is subdivided on each half of the cord into three *columns* (or funiculi): the anterior, posterior, and lateral white columns. These consist of large bundles of nerve fibers arranged in tracts.

Tables 7-2 and 7-3 name important tracts of each white column and give essential information about them. Also refer to Fig. 7-12.

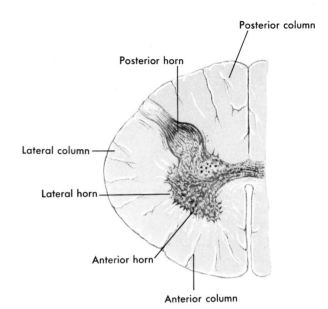

Fig. 7-11 Distribution of gray matter (horns) and white matter (columns) in a section of the spinal cord at the thoracic level.

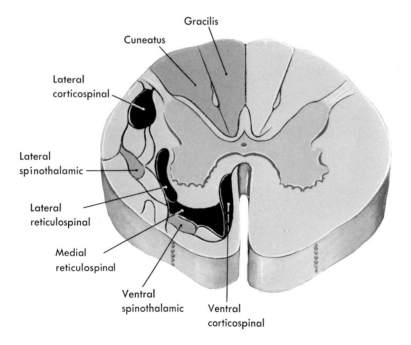

Fig. 7-12 Location in the spinal cord of some major projection tracts. Black areas, descending motor tracts. Shaded areas, ascending sensory tracts. (Also refer to Tables 7-2 and 7-3).

Table 7-2. Major ascending tracts of spinal cord

Name	Function	Location	Origin*	Termination†
Lateral spino-thalamic	Pain and temperature opposite side	Lateral white columns	Posterior gray column opposite side	Thalamus
Ventral spino-thalamic	Crude touch	Anterior white columns	Same	Same
Fasciculus gracilis and cuneatus	Conscious kinesthesia, sensations of vibration, stereognosis, deep touch and pressure, two-point discrimination	Posterior white columns	Spinal ganglia same side	Medulla
Spinocerebellar	Unconscious kinesthesia	Lateral white columns	Posterior gray column	Cerebellum

* Location of cell bodies of neurons from which axons of tract arise.
† Structure in which axons of tract terminate.

Table 7-3. Major descending tracts of spinal cord

Name	Function	Location	Origin*	Termination†
Lateral cortico-spinal (or crossed pyramidal)	Voluntary movement, contraction of individual or small groups of muscles, particularly those moving hands, fingers, feet, and toes of opposite side	Lateral white columns	Motor areas cerebral cortex (mainly areas 4 and 6) opposite side from tract location in cord	Intermediate or anterior gray columns
Ventral cortico-spinal (direct pyramidal)	Same as lateral corticospinal except mainly muscles of same side	Lateral white columns	Motor cortex but on same side as tract location in cord	Intermediate or anterior gray columns
Lateral reticulo-spinal	Mainly facilitatory influence on motoneurons to skeletal muscles	Lateral white columns	Reticular formation midbrain, pons, and medulla	Intermediate or anterior gray columns
Medial reticulo-spinal	Mainly inhibitory influence on motoneurons to skeletal muscles	Anterior white columns	Reticular formation medulla mainly	Intermediate or anterior gray columns

* Location of cell bodies of neurons from which axons of tract arise.
† Structure in which axons of tract terminate.

Functions

The spinal cord performs sensory, motor, and reflex functions. The following paragraphs describe these functions briefly.

Sensory and motor functions. Spinal cord tracts serve as two-way conduction paths between peripheral nerves and the brain. Ascending tracts conduct impulses up the cord to the brain. Descending tracts conduct impulses down the cord from the brain. Bundles of axons compose all tracts. Tracts are both structural and functional organizations of these nerve fibers. They are structural organizations in that all axons of any one tract originate from neuron cell bodies located in the same structure and all of the axons terminate in the same structure. For example, all fibers of the spinothalamic tract are axons which originate from neuron cell bodies located in the spinal cord and terminate in the thalamus. Tracts are functional organizations in that all the axons that compose one tract serve one general function. For instance, fibers of the lateral spinothalamic tract serve a sensory function. They transmit the impulses that produce our sensations of pain and of temperature.

Because so many different tracts make up the white columns of the cord, we shall mention only a few that seem most important in man. Locate each tract in Fig. 7-12. Consult Tables 7-2 and 7-3 for a brief summary of information about tracts. Four important ascending or sensory tracts and their functions, stated very briefly, are as follows:

1 *lateral spinothalamic tracts*—pain and temperature
2 *ventral spinothalamic tracts*—crude touch
3 *fasciculi gracilis and cuneatus*—discriminating touch and conscious kinesthesia
4 *spinocerebellar tracts*—unconscious kinesthesia

Further discussion of the sensory neural pathways may be found on pp. 211 to 215.

Four important descending or motor tracts and their functions in brief are as follows:

1 *lateral corticospinal tracts*—voluntary movement; contraction of individual or small groups of muscles, particularly those moving hands, fingers, feet, and toes on opposite side of body
2 *ventral corticospinal tracts*—same as preceding except mainly muscles of same side of body
3 *lateral reticulospinal tracts*—mainly facilitatory impulses to anterior horn motoneurons to skeletal muscles
4 *medial reticulospinal tracts*—mainly inhibitory impulses to anterior horn motoneurons to skeletal muscles

Further discussion of the motor neural pathways may be found on pp. 215 to 221.

Reflex functions. The spinal cord functions in all reflexes except those mediated by cranial nerves. Gray matter of the cord contains innumerable reflex centers. To mention just one example, gray matter of the second, third, and fourth lumbar segments contains the centers for the knee jerk reflex. The term reflex center means literally the center of a reflex arc or the place in the arc at which incoming sensory impulses become outgoing motor impulses. Some reflex centers are merely synapses between sensory and motoneurons, whereas others are interneurons interposed between sensory and motoneurons.

■Brain

The brain is one of the largest of adult organs. In most adults it weighs about 3 pounds but generally is smaller in women than in men and in older persons than in younger persons. Neurons of the brain undergo mitosis only during the prenatal period and the first few months of post-

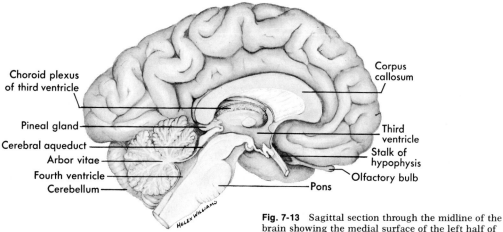

Choroid plexus
of third ventricle

Pineal gland

Cerebral aqueduct

Arbor vitae

Fourth ventricle

Cerebellum

Corpus
callosum

Third
ventricle

Stalk of
hypophysis

Olfactory bulb

Pons

HELEN WILLIAMS

Fig. 7-13 Sagittal section through the midline of the brain showing the medial surface of the left half of the brain (for more detailed structure, see Fig. 7-14). (From Francis, C. C, and Farrell, G. L.: Integrated anatomy and physiology, St. Louis, The C. V. Mosby Co.)

natal life. Although they grow in size after that, they do not increase in number. Malnutrition during those crucial months of neuron multiplication is reported* to hinder the process and result in fewer brain cells. The brain attains full size by about the eighteenth year but grows rapidly only during the first nine years or so.

The brain has six major divisions: cerebrum, diencephalon, cerebellum, medulla oblongata, pons varolii, and midbrain. The midbrain, pons, and medulla constitute the *brainstem* — an apt term since, viewed from the side, they look like a stem for the rest of the brain (see Fig. 7-14). Each division of the brain consists of gray matter (nuclei and centers) and white matter (tracts).

Cerebrum

The cerebrum is the largest and the most superiorly located division of the human brain.

Hemispheres, fissures, and lobes

Grooves, called *fissures* or *sulci,* dip into the surface of the cerebrum, dividing it

into two hemispheres and each hemisphere into five lobes. A deep groove, the *longitudinal fissure,* divides the cerebrum into two halves called hemispheres. They are, however, not completely separated from each other. A structure composed of white matter and known as the *corpus callosum* joins them medially on their inferior surfaces. Prominent fissures, in addition to the longitudinal fissure already named, include the central fissure (of Rolando), the lateral fissure (of Sylvius), and the parieto-occipital fissure. These prominent indentations subdivide each cerebral hemisphere into four *lobes,* each of which bears the name of the bone lying over it: frontal lobe, parietal lobe, temporal lobe, and occipital lobe. A fifth lobe, the insula (island of Reil), lies hidden from view in the lateral fissure. To see it, one must dissect the brain.

Cerebral cortex

Each hemisphere of the cerebrum consists of external gray matter, internal white matter, and islands of internal gray matter. The *cerebral cortex* is the thin surface layer of the cerebrum. Gray matter only 2 to 4 millimeters (roughly 1/12 to 1/6 of an inch) thick composes it. But despite its thinness, the cerebral cortex consists of six layers, and millions and millions of

*Malnutrition—effects on the brain: Sci. News **97**:70 (Jan. 17), 1970.

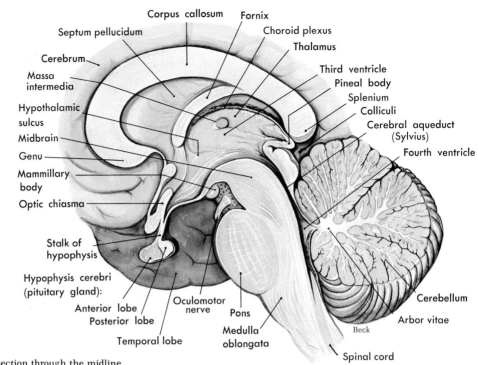

Fig. 7-14 Sagittal section through the midline
of the brain showing structures around
the third ventricle.

neuron cell bodies compose each layer. The
cerebral cortex overlays white matter
(mainly myelinated axons) which makes
up most of the interior of the cerebrum.
No doubt, when early anatomists observed
the outer darker layer, it reminded them of
tree bark, hence their choice of the name
cortex, which comes from Latin for bark.

Provided one uses a little imagination,
the surface of the cerebrum looks like a
group of small sausages. Each "sausage"
represents a convolution or *gyrus*. Between
adjacent gyri lie shallow grooves called
sulci or deeper crevices called *fissures*.

Cerebral tracts

Cerebral tracts lie interior to the cortex
and are composed of great numbers of
nerve fibers (axons). Tracts and nerves are
comparable structures. Both consist of
bundles of nerve fibers. However, whereas
tracts are bundles of axons located in the
brain and spinal cord, nerves are bundles
of dendrites or axons, or both, located outside the brain and cord.

Tracts that conduct impulses upward
are called sensory or *ascending projection
tracts,* and those that conduct downward
are referred to as motor or *descending projection tracts.* In one part of the interior
of the cerebrum, a group of sensory and
motor projection tracts forms a large irregular mass of white matter known as the
internal capsule. It lies between the thalamus on one side and the caudate and lenticular nuclei on the other (Figs. 7-29 and
7-31). Some tracts are short, extending
from one convolution to another in the
same hemisphere. These are called *association tracts.*

From the name of a specific tract, you
can tell from what structure to what structure its fibers conduct impulses, if you
remember only two principles. The first
part of a tract's name indicates the location of the dendrites and cell bodies of the
neurons whose axons compose the tract.
The last part of the tract's name indicates
the structure in which the tract fibers terminate. If you apply these principles to

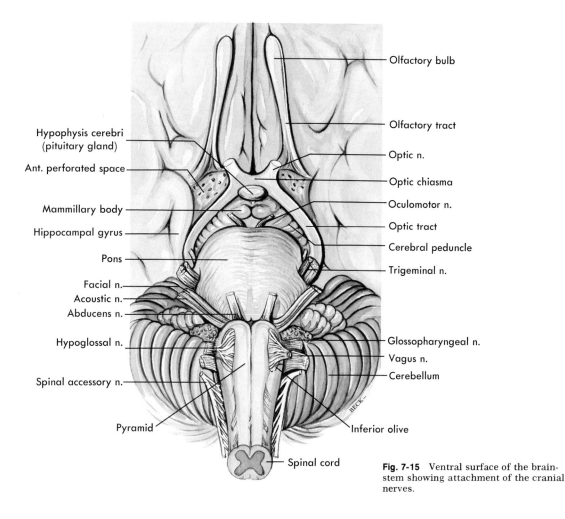

Olfactory bulb

Olfactory tract

Hypophysis cerebri (pituitary gland)

Optic n.

Ant. perforated space

Optic chiasma

Oculomotor n.

Mammillary body

Optic tract

Hippocampal gyrus

Cerebral peduncle

Pons

Trigeminal n.

Facial n.

Acoustic n.

Abducens n.

Hypoglossal n.

Glossopharyngeal n.

Vagus n.

Cerebellum

Spinal accessory n.

Pyramid

Inferior olive

Spinal cord

Fig. 7-15 Ventral surface of the brainstem showing attachment of the cranial nerves.

the corticospinal tract, where are the dendrites and cell bodies of the neurons whose axons compose the tract located? Where do the axons end? From what structure to what structure do corticospinal tract fibers conduct impulses? Are the right and left corticospinal tracts ascending or descending tracts?

Basal ganglia

Basal ganglia (or cerebral nuclei*) are islands of gray matter deep inside each

*When applied to the nervous system, the term *nucleus* means an area of gray matter in the brain or cord (composed mainly, as is all gray matter, of neuron cell bodies and dendrites). Such a cluster located outside the brain and cord is called a *ganglion*. So cerebral nuclei is a more accurate (but less common) name than basal ganglia.

cerebral hemisphere (Figs. 7-16 and 7-17). Names of the main basal ganglia are the caudate nucleus, the putamen, and the pallidum (globus pallidus). The putamen is a larger nucleus than the pallidum and lies lateral to it. Together, these two nuclei are called the lenticular nucleus. The caudate nucleus, the lenticular nucleus (made up of the putamen and pallidum), and the internal capsule together compose the *corpus striatum*. As its name suggests and as Fig. 7-17 shows, this region has a striped appearance. The stripes are narrow strips of gray matter that extend across the white internal capsule between the caudate and lenticular nuclei. Basal ganglia play an essential role in producing movements (see pp. 219 to 221).

Functions

Knowledge of cerebral function has accumulated in various ways: by studying symptoms of patients known to have brain lesions, by studying the effects of removing or destroying various cerebral areas in animals, by stimulating various cerebral areas in animals and human beings and recording results, and, more recently, by studying brain action potentials as recorded on electroencephalograms (EEG's). Another approach to investigating cerebral function has been to study the microscopic structure of the cerebrum. This revealed, for example, that several layers of neurons and neuroglia compose the cerebral cortex and that the number of layers and type and arrangement of cells differ in different areas. These findings led to the postulate that different areas of the cortex perform different functions—that cortical functions, in other words, are localized in different cortical areas. The current view differs somewhat. It holds that not one but many cortical areas take part in the performance of probably all cerebral functions. But, as we shall see, certain areas apparently do play a major role in the carrying out of certain functions.

What functions does the cerebral cortex perform? A general answer is this: the cerebral cortex performs all mental functions and many essential motor, sensory, and visceral functions. For example, conduction by cortical neurons occurs when we remember or imagine anything, when we experience sensations or emotions of any kind, and when we will to make any movements. Conduction by certain cortical neurons may even lead to visceral changes such as dilatation or constriction of facial blood vessels (blushing with embarrassment or turning pale with fear). Data presently available suggest that it is neurons in the frontal and temporal lobes that somehow produce our memories and our feelings of emotion. Neurons in the frontal lobe (precentral gyrus) initiate our willed movements, those in the parietal lobe (postcentral gyrus) are responsible for our general sensations, and those in certain gyri of the temporal and occipital lobes are responsible for our special senses of hearing, smelling, and vision. Neurons in widely scattered areas of the cortex perform visceral functions.

Sensations. Complex nervous mechanisms function to produce our sensations. Many gross structures are involved—nerves, ganglia, cord, medulla, pons, midbrain, thalamus, and cerebral cortex. Complex discriminative sensations depend upon the cerebral cortex, especially the *somesthetic area* or general sensory area and the visual and auditory areas (Fig. 7-18). These regions of the cortex do more than just register separate and simple sensations. They compare them and evaluate them. They integrate them into meaningful concepts. Suppose, for example, that someone blindfolded you and then put an ice cube in your hand. You would, of course, sense something cold touching your hand. But also, you would probably know that it was an ice cube because you would sense a total impression compounded of many sensations such as temperature, shape, size, weight, texture, and movement and position of your hand and arm.

Note the following locations in Fig. 7-18:

1 *general* or *somesthetic sensory area* —parietal lobe, postcentral gyrus (areas 3, 1, 2)
2 *primary auditory area* —temporal lobe, transverse gyrus along lateral fissure (areas 41, 42)
3 *primary visual area* —occipital lobe (area 17)
4 *primary olfactory area* —not precisely located but in temporal lobe (ventral or undersurface)

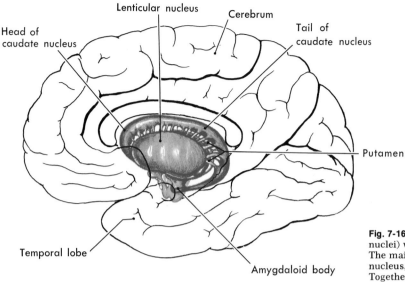

Head of
caudate nucleus

Lenticular nucleus

Cerebrum

Tail of
caudate nucleus

Putamen

Temporal lobe

Amygdaloid body

Fig. 7-16 Basal ganglia (or cerebral
nuclei) within a cerebral hemisphere.
The main basal ganglia are the caudate
nucleus, the putamen, and the pallidum.
Together, the putamen and pallidum
constitute the lenticular nucleus. (See
Fig. 7-17.)

5 *primary gustatory (taste) area*—not
precisely located; good evidence that
parietal lobe (lower part of postcentral convolution) is one taste area

Voluntary movements. Mechanisms
that control voluntary movements are extremely complex and are imperfectly
understood. It is known, however, that for
normal movements to take place, many
parts of the nervous system—including
certain areas of the cerebral cortex—must
function. The precentral gyrus (that is, the
most posterior gyrus of the frontal lobe;
area 4, Fig. 7-18) constitutes the *primary
motor area*. However, the gyrus immediately anterior to the precentral gyrus also
contains motor neurons. So, too, do many
other regions, including even the somatic
sensory area. Neurons in the precentral
gyrus are said to exert control over individual muscles, especially those that produce movements of distal joints (wrist,
hand, finger, ankle, foot, and toe movements). Neurons in the gyrus just anterior
to the precentral gyrus, on the other hand,
are thought to activate groups of muscles
simultaneously.

Mental functions. Included in the category of mental functions are many diverse
functions such as memory, foresight, emotional feelings, personality traits, speech
functions, and intelligence. These and
other mental and psychic functions depend
more on widespread cortical activity than
on localized regions. Memory, for example,
once considered a function of the temporal
lobes, now is viewed as a complex neural
process dependent on many parts.[*] Visual
memories, stored (in some as yet unknown
manner) in the occipital lobe, auditory
memories in the temporal lobe, and general sensory experiences in the parietal
lobe are all linked together and synthesized
into complex memories by numerous association tracts between lobes. Foresight and
personality traits depend largely on the
prefrontal lobes. Some of the most persuasive evidence of this lies in the changes
observed in patients who have undergone
a prefrontal lobectomy. Most noticeable
are their loss of initiative and planning
ability and their lack of restraint. A house-
wife might, for example, become almost
unable to plan a simple meal. And probably

*See Pribram, K. H.: The neurophysiology of remem-
bering, Sci. Amer. **220:**73-86 (Jan.), 1969.

she would become boastful and aggressive —personality changes indicative of a decreased ability to restrain behavior.

Speech functions consist of the use of language (speaking and writing) and the understanding of language (spoken and written). Today, these faculties are believed to depend on highly integrated cortical processes, with certain areas in the frontal, parietal, and temporal lobes called speech centers serving as the focal points for integration. Lesions in different ones of these areas are associated with different types of speech defects or *aphasias*. For example, with a lesion in *Broca's motor speech area* in the frontal lobe (Fig. 7-18), the individual becomes unable to express his ideas in spoken words, although he is not actually unable to speak. This condition is known as *motor aphasia*. With a lesion in the parietal lobe speech center, on the other hand, the individual has trouble finding the right names for things.

Diencephalon

The diencephalon is the part of the brain located between the cerebrum and the midbrain. It consists of structures around the third ventricle (one of the fluid spaces in the brain described on p. 196): the thalamus, epithalamus, subthalamus, and hypothalamus. We shall confine our discussion to the thalamus and hypothalamus.

Thalamus

Structure and location

The right thalamus is a rounded mass of gray matter about 1/2 of an inch wide and 1 1/2 inches long, bulging into the right lateral wall of the third ventricle. The left thalamus is a similar mass in the left lateral wall. Each thalamus consists of numerous nuclei. One of these, for example, is called the posterior ventral nucleus. Here, axons of the spinothalamic tracts terminate and synapse with neurons whose axons extend in thalamocortical tracts to the general sensory area of the cerebral cortex. Thus, the posterior ventral nucleus of the thalamus serves as a relay station for sensory impulses. In another nucleus of the thalamus, impulses from basal ganglia are relayed to the motor area of the cerebral cortex, and, vice versa, impulses from the motor area are relayed to basal ganglia. Still other thalamic nuclei relay impulses from the hypothalamus to various areas of the cortex and from these areas to the hypothalamus. Finally, some thalamic nuclei serve as centers for certain conscious and unconscious activities. (See functions **1a** and **2** below.)

Functions

The thalamus performs the following functions:

1 Plays two parts in the mechanism responsible for sensations:
   a Impulses from appropriate receptors, upon reaching the thalamus, produce conscious recognition of the cruder, less critical sensations of pain, temperature, and touch.
   b Neurons whose dendrites and cell bodies lie in certain nuclei of the thalamus relay all kinds of sensory impulses, except possibly olfactory, to the cerebrum.
2 Plays a part in the mechanism responsible for emotions by associating sensory impulses with feelings of pleasantness and unpleasantness
3 Plays a part in the arousal or alerting mechanism
4 Plays a part in mechanisms that produce complex reflex movements

Hypothalamus

Structure and location

The hypothalamus consists of several structures which lie beneath the thalamus

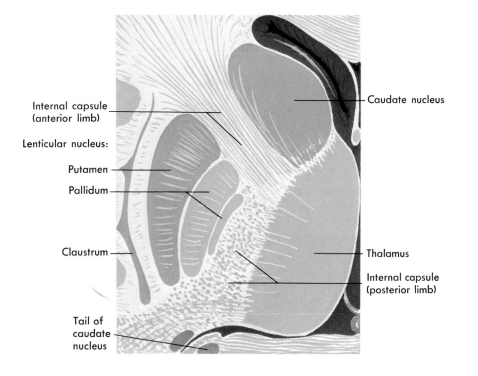

Internal capsule (anterior limb)

Lenticular nucleus:

Putamen

Pallidum

Claustrum

Tail of caudate nucleus

Caudate nucleus

Thalamus

Internal capsule (posterior limb)

Fig. 7-17 Basal ganglia and internal capsule as seen in a horizontal section through the cerebrum.

and form the floor and the lower part of the sidewall of the third ventricle. Prominent among the structures composing the hypothalamus are the supraoptic nuclei, the paraventricular nuclei, the stalk of the pituitary gland, the neurohypophysis (the posterior lobe of the pituitary gland), and the mamillary bodies. Identify as many of these as you can in Figs. 7-14 and 7-15. The supraoptic nuclei consist of gray matter located just above and on either side of the optic chiasma. The paraventricular nuclei of the hypothalamus are so named because of their location close to the wall of the third ventricle. The midportion of the hypothalamus consists of the stalk of the pituitary gland and the posterior lobe of the pituitary gland (neurohypophysis). The posterior part of the hypothalamus consists mainly of the mamillary bodies, in which are located the mamillary nuclei.

Functions

The hypothalamus is a small but functionally mighty area of the brain. It weighs little more than a quarter of an ounce, yet it performs many functions of the greatest importance both for survival and for the enjoyment of life. For instance, it functions as a link between the psyche (mind) and the soma (body). It also links the nervous system to the endocrine system. And if indications of recent experiments prove true, certain areas of the hypothalamus function as pleasure centers or reward centers for the primary drives such as eating, drinking, and mating. The following paragraphs give a brief summary of hypothalamic functions:

1 The hypothalamus functions as a higher autonomic center or, rather, as several higher autonomic centers. By this we mean that axons of neurons whose dendrites and cell bodies lie in nuclei of the

189

hypothalamus extend in tracts from the hypothalamus to both parasympathetic and sympathetic centers in the brainstem and cord. (Fig. 7-35 indicates these tracts in blue.) Thus, impulses from the hypothalamus can simultaneously or successively stimulate or inhibit few or many lower autonomic centers. In other words, the hypothalamus serves as a regulator and coordinator of autonomic activities. It helps control and integrate the responses made by visceral effectors all over the body.

2 The hypothalamus functions as the major relay station between the cerebral cortex and lower autonomic centers. Tracts conduct impulses from various centers in the cortex to the hypothalamus (also shown in blue in Fig. 7-35). Then, via numerous synapses in the hypothalamus, these impulses are relayed to other tracts that conduct them on down to autonomic centers in the brainstem and cord and also to spinal cord somatic centers (anterior horn motor neurons). Thus, the hypothalamus functions as the link between the cerebral cortex and lower centers—hence, between the psyche and the soma. It provides a crucial part of the route by which emotions can express themselves in changed bodily functions. It is the all-important relay station in the neural pathways which make possible the mind's influence over the body —sometimes, unfortunately, even to the profound degree of producing "psychosomatic disease."

3 Neurons in the supraoptic and paraventricular nuclei of the hypothalamus synthesize the hormones secreted by the posterior pituitary gland (neurohypophysis). They also help regulate the rate at which the posterior pituitary gland secretes these hormones. Therefore, the hypothalamus plays an essential role in maintaining water balance (see Fig. 15-9, p. 509).

4 The hypothalamus functions as an important part of the mechanism for controlling hormone secretion by the anterior pituitary gland—hence, it helps control the functioning of every cell in the body (see pp. 266 to 272).

5 The hypothalamus plays an essential role in maintaining the waking state. Presumably it functions as part of an arousal or alerting mechanism (p. 215). Clinical evidence of this is that somnolence characterizes some hypothalamic disorders.

6 The hypothalamus functions as a crucial part of the mechanism for regulating appetite and therefore the amount of food intake. Experimental and clinical findings seem to indicate the presence of a "feeding or appetite center" in the lateral part of the hypothalamus and a "satiety center" located medially. For example, an animal with an experimental lesion in the ventromedial nucleus of the hypothalamus will consume tremendous amounts of food. Similarly, a human being with a tumor in this region of the hypothalamus may eat insatiably and gain an enormous amount of weight.

7 The hypothalamus almost certainly helps control various reproductive functions (see p. 271).

8 The hypothalamus functions as a crucial part of the mechanism for maintaining normal body temperature. Hypothalamic neurons whose fibers connect with autonomic centers for vasoconstriction and dilatation and sweating and with somatic centers for shivering constitute heat-regulating centers. Marked elevation of body temperature frequently characterizes injuries or other abnormalities of the hypothalamus.

Cerebellum

Structure and location

The cerebellum, the second largest part of the brain, is located just below the posterior portion of the cerebrum and is par-

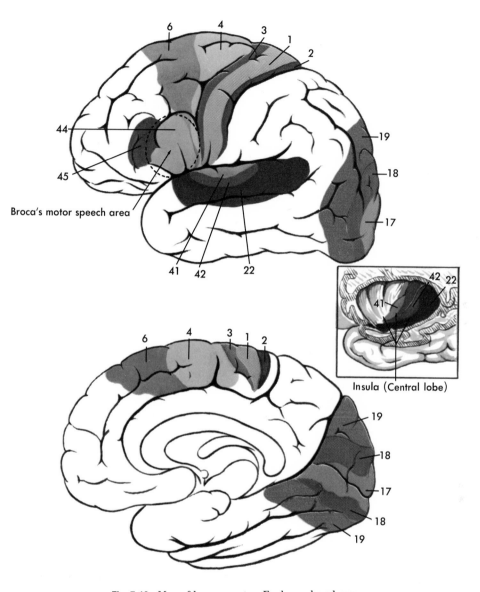

Fig. 7-18 Map of human cortex. Each numbered area shows different cellular structure. Some areas whose functions are best understood are the following: areas **3, 1,** and **2,** general somatic sensory areas; area **4,** primary motor area; area **6,** secondary motor area; area **17,** primary visual area; areas **18** and **19,** secondary visual areas; areas **41** and **42,** primary auditory areas; area **22,** secondary auditory area. Area **44** and part of area **45** constitute the approximate location of Broca's motor speech area. (Modified after Brodmann, K.: Feinere Anatomie des Grosshirns. In Handbuch der Neurologie, Berlin, 1910, Springer-Verlag.)

tially covered by it. A transverse fissure separates the cerebellum from the cerebrum. These two parts of the brain have several characteristics in common. For instance, gray matter makes up their outer portions and white matter predominates in their interiors. Turn to Fig. 7-14 to observe the *arbor vitae* – that is, the internal white matter of cerebellum. Note its distinctive pattern – similar to the veins of a leaf. Note, too, that the surfaces of both the cerebellum and the cerebrum have numerous grooves (sulci) and convolutions (gyri). The convolutions of the cerebellum, however, are much more slender and less prominent than those of the cerebrum. The cerebellum has two large lateral masses, the cerebellar hemispheres, and a central section called the vermis because in shape it resembles a worm coiled upon itself. (For a detailed description of the several subdivisions of the cerebellum, consult a textbook on neuroanatomy.)

The internal white matter of the cerebellum is composed of some short and some long tracts. The short association tracts connect the cerebellar cortex with nuclei located in the interior of the cerebellum. The longer tracts connect the cerebellum with other parts of the brain and with the spinal cord. These enter or leave the cerebellum by way of its three pairs of peduncles as follows:

1 *inferior cerebellar peduncles* (or *restiform bodies*) – composed chiefly of tracts into the cerebellum from the medulla and cord (notably, spinocerebellar, vestibulocerebellar, and reticulocerebellar tracts)

2 *middle cerebellar peduncles* (or *brachia pontis*) – composed almost entirely of tracts into the cerebellum from the pons (that is, pontocerebellar tracts)

3 *superior cerebellar peduncles* (or *brachia conjunctivum cerebelli*) – com-

posed principally of tracts from dentate nuclei through the red nucleus of the midbrain to the thalamus

An important pair of cerebellar nuclei are the *dentate nuclei*, one of which lies in each hemisphere. Tracts connect these nuclei with motor areas of the cerebral cortex (the dentatorubrothalamic tracts to the thalamus and thalamocortical tracts to the cortex). By means of these tracts, cerebellar impulses influence the motor cortex. Impulses also travel the reverse direction. Corticopontine and pontocerebellar tracts enable the motor cortex to influence the cerebellum.

Functions

The cerebellum performs three general functions, all of which have to do with the control of skeletal muscles. It acts with the cerebral cortex to produce skilled movements by coordinating the activities of groups of muscles. It controls skeletal muscles so as to maintain equilibrium. It helps control posture. It functions below the level of consciousness to make movements smooth instead of jerky, steady instead of trembling, and efficient and coordinated instead of ineffective, awkward, and uncoordinated (asynergic).

There have been many theories about cerebellar functions. One theory, based on comparative anatomy studies and substantiated by experimental methods, regards the cerebellum as three organs, each with a somewhat different function: synergic control of muscle action, excitation and inhibition of postural reflexes, and maintenance of equilibrium.

Synergic control of muscle action, which is ascribed to the neocerebellum (superior vermis and hemispheres), is closely associated with cerebral motor activity. Normal muscle action, you will recall, involves groups of muscles, the various members of which function together as a unit. In any given action, for example, the prime mover contracts, the antagonist relaxes but contracts weakly at the proper moment to act as a brake, checking the action of the prime mover, the synergists contract to assist the prime mover, and the fixation muscles of the neighboring joint contract. Through such harmonious coordinated group action, normal movements are smooth, steady, and precise as to force, rate, and extent. Achievement of such movements results from cerebellar activity added to cerebral activity. Impulses from the cerebrum start the action, but those from the cerebellum synergize or coordinate the contractions and relaxations of the various muscles once they have begun. Some physiologists consider this the main, if not the sole, function of the cerebellum.

One part of the cerebellum is thought to be concerned with both *exciting and inhibiting postural reflexes.*

Part of the cerebellum presumably discharges impulses important to the *maintenance of equilibrium.* Afferent impulses from the labyrinth of the ear reach the cerebellum. Here, connections are made with the proper efferent fibers for contraction of the necessary muscles for equilibrium.

Cerebellar disease (abscess, hemorrhage, tumors, trauma, etc.) produces certain characteristic symptoms, among which ataxia (muscle incoordination), hypotonia, tremors, and disturbances of gait and equilibrium predominate. As examples of ataxia may be mentioned overshooting a mark or stopping before reaching it when asked to touch a given point on the body (finger-to-nose test) and drawling, scanning, or singsong speech because of lack of coordinated actions of phonation and articulation muscles. Tremors are particularly pronounced toward the end of movements and with the exertion of effort. Distur-

bances of gait and equilibrium vary, depending upon the muscle groups involved, but the walk is often characterized by staggering or lurching and by a clumsy manner of raising the foot too high and bringing it down with a clap. Paralysis does not result from loss of cerebellar function.

Medulla oblongata

Structure and location

The medulla or bulb is the part of the brain that attaches to the spinal cord. It is in fact, an enlarged extension of the cord located just above the foramen magnum. It measures only slightly more than an inch in length and is separated from the pons above by a horizontal groove. It is composed mainly of white matter (projection tracts) and *reticular formation*, a term that means the interlacement of gray and white matter present in the cord, brainstem, and diencephalon. Nuclei in the reticular formation of the medulla include such important centers as respiratory and vasomotor centers.

On each side of the lower posterior part of the medulla are two prominent nuclei, the *nucleus gracilis* and the *nucleus cuneatus*. Here, afferent fibers from the posterior white columns (fasciculi gracilis and cuneatus) of the cord synapse with neurons whose axons extend to the thalamus and cerebellum.

The pyramids (Fig. 7-15) are two bulges of white matter located on the anterior surface of the medulla formed by fibers of the pyramidal projection tracts.

The olive (Fig. 7-15) is an oval projection appearing one on each side of the anterior surface of the medulla. It contains the *inferior olivary nucleus* and two accessory olivary nuclei. Fibers from the cells of these nuclei run through the inferior cerebellar peduncles (restiform bodies) into the cerebellum. Nuclei of the ninth to the twelfth cranial nerves are also located in the medulla.

Functions

Nuclei in the medulla contain a number of reflex centers. Some of these perform functions so necessary for survival that they are called the *vital centers.* They are the cardiac, vasomotor, and respiratory centers. As their names suggest, they serve as the centers for various reflexes controlling heart action, blood vessel diameter, and respirations. Because the medulla contains these centers, it is the most vital part of the entire brain—so vital, in fact, that injury or disease of the medulla often proves fatal. Blows at the base of the skull and bulbar poliomyelitis, for example, cause death if they interrupt impulse conduction by the vital respiratory centers.

The medulla functions in many nonvital reflexes. For example, it contains centers for vomiting, coughing, sneezing, hiccoughing, and swallowing.

All projection tracts between the cord and brain necessarily pass through the medulla. Hence, it functions in a great many sensory and motor mechanisms. Fibers of the crossed corticospinal tracts decussate—that is, cross from one side to the other in the pyramids of the medulla, an anatomical fact that explains why one side of the brain is said to control the other side of the body.

Pons varolii ~ bridge

Structure and location

Just above the medulla lies the pons, composed like the medulla of white matter and a few nuclei. Fibers that run transversely across the pons and through the brachia pontis (middle cerebellar peduncles) into the cerebellum make up the external white matter of the pons and give it its bridgelike appearance. The reticular formation extends into the pons from the medulla. One important reticular nucleus in the pons is called the *pneumotaxic center.* It functions in the control of respirations. Nuclei of the fifth to eighth cranial nerves are located in the upper part of the pons.

Functions

For functions of the pons varolii, see under midbrain below.

Midbrain

Structure and location

The midbrain lies below the inferior surface of the cerebrum and above the pons. It consists mainly of white matter with some internal gray matter around the cerebral aqueduct, the cavity within the midbrain. The *cerebral peduncles* form the ventral part of the midbrain, and the *corpora quadrigemina* or *colliculi* form the dorsal part. The cerebral peduncles are two ropelike masses of white matter that extend divergently from the pons to the undersurface of the cerebral hemispheres (Fig. 7-15). In other words, the cerebral peduncles are made up of tracts that constitute the main connection between the forebrain and hindbrain. Hence, the midbrain, by function as well as by location, is well named.

The *corpora quadrigemina* consist of four rounded eminences, the two superior and the two inferior colliculi (Fig. 7-14) which form the dorsal part of the midbrain. Certain auditory reflex centers lie in the inferior colliculi and visual centers in the superior colliculi.

An important nucleus in the midbrain reticular formation is the *red nucleus,* a large gray mass ventral to the superior colliculi. Fibers from the cerebellum and from the frontal lobe of the cerebral cortex end here, whereas fibers that extend into the rubrospinal tracts of the cord have their

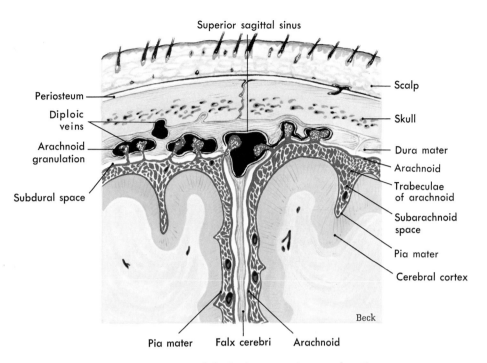

Superior sagittal sinus

Periosteum

Diploic veins

Arachnoid granulation

Subdural space

Scalp

Skull

Dura mater

Arachnoid

Trabeculae of arachnoid

Subarachnoid space

Pia mater

Cerebral cortex

Beck

Pia mater   Falx cerebri   Arachnoid

Fig. 7-19 Meninges of the brain as seen in coronal section through the skull.

cells of origin here. Nuclei of the third and fourth cranial nerves and the anterior part of the nucleus of the fifth cranial nerve are located deep in the midbrain. Also, as mentioned in the preceding paragraph, nuclei for certain auditory and visual reflexes lie in the colliculi of the midbrain.

Functions of pons and midbrain

The upper part of the brainstem serves as conduction pathways (projection tracts) between the cord and other parts of the brain. In addition, the pons functions as the reflex center for reflexes mediated by the fifth, sixth, seventh, and eighth cranial nerves (see Table 7-6 for functions). And because the pons contains the pneumotaxic centers, it helps regulate respirations (discussed on p. 384).

Like the pons, the midbrain also functions as a reflex center for certain cranial nerve reflexes—for example, pupillary re-

flexes and eye movements, mediated by the third and fourth cranial nerves, respectively.

■ Cord and brain coverings

Because the spinal cord and brain are both delicate and vital, nature has provided them with two protective coverings. The outer covering consists of bone: vertebrae encase the cord and cranial bones encase the brain. The inner covering consists of membranes, known as *meninges.* Three distinct layers compose the meninges: the dura mater, the arachnoid membrane, and the pia mater. Observe their respective locations in Figs. 7-19 and 7-27. The *dura mater,* made of strong white fibrous tissue, serves both as the outer layer of the meninges and also as the inner periosteum of the cranial bones. The *arachnoid membrane*, a delicate, cobwebby layer, lies between the dura mater and the pia mater or innermost layer of the meninges. The transparent *pia mater* adheres to the outer surface of the cord and brain and contains blood vessels.

Three extensions of the dura mater should be mentioned: the falx cerebri, falx cerebelli, and tentorium cerebelli. The *falx cerebri* projects downward into the longitudinal fissure to form a kind of partition between the two cerebral hemispheres. The *falx cerebelli* separates the two cerebellar hemispheres. The *tentorium cerebelli* separates the cerebellum from the occipital lobe of the cerebrum. It takes its name from the fact that it forms a tentlike covering over the cerebellum.

Between the dura mater and the arachnoid membrane is a small space called the *subdural space*, and between the arachnoid and the pia mater is another space, the *subarachnoid space*. Inflammation of the meninges is called *meningitis*. It most often involves the arachnoid and pia mater or the *leptomeninges*, as they are sometimes called.

The meninges of the cord continue on down inside the spinal cavity for some distance below the end of the spinal cord. The pia mater forms a slender filament known as the *filum terminale.* At the level of the third segment of the sacrum, the filum terminale blends with dura mater to form a fibrous cord that disappears in the periosteum of the coccyx. This extension of the meninges beyond the cord is convenient for performing lumbar punctures. It makes it possible to insert a needle between the third and fourth or fourth and fifth lumbar vertebrae into the subarachnoid space to withdraw cerebrospinal fluid without danger of injuring the spinal cord which ends more than an inch above that point. The fourth lumbar vertebrae can be easily located because it lies on a line with the iliac crest. Placing the patient on his side and arching his back by drawing the knees and chest together separates the vertebrae sufficiently to introduce the needle.

■ Cord and brain fluid spaces

In addition to the bony and membranous coverings, nature has further fortified the spinal cord and brain against injury by providing a cushion of fluid both around them and within them. The fluid is called *cerebrospinal fluid*, and the spaces containing it are as follows:

1 The subarachnoid space around the cord
2 The subarachnoid space around the brain
3 The central or spinal canal inside the cord
4 The ventricles and aqueduct inside the brain

The *ventricles* are cavities or spaces inside the brain. They are four in number. Two of them, the lateral (or first and second) ventricles, are located one in each cerebral hemisphere. Note in Fig. 7-20 the shape of these ventricles—roughly like the hemispheres themselves. The third ventricle is little more than a lengthwise slit in the cerebrum beneath the midportion of the corpus callosum and longitudinal fissure. The fourth ventricle is a diamond-shaped space between the cerebellum posteriorly and the medulla and pons anteriorly. Actually, it is an expansion of the central canal of the cord after the cord enters the cranial cavity and becomes enlarged to form the medulla.

■ Formation and circulation of cerebrospinal fluid

Cerebrospinal fluid is found in each ventricle. It is a clear fluid formed primarily by filtration out of the blood in networks of capillaries known as choroid plexuses. From each lateral ventricle the fluid seeps through an opening, the interventricular

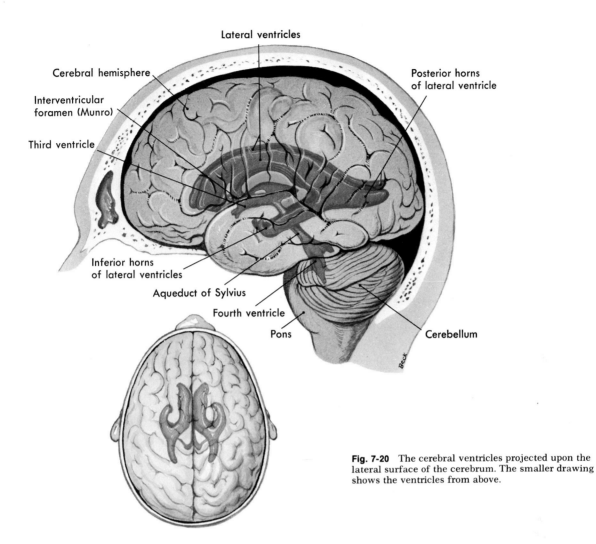

Lateral ventricles

Cerebral hemisphere

Interventricular foramen (Munro)

Third ventricle

Posterior horns of lateral ventricle

Inferior horns of lateral ventricles

Aqueduct of Sylvius

Fourth ventricle

Pons

Cerebellum

Fig. 7-20 The cerebral ventricles projected upon the lateral surface of the cerebrum. The smaller drawing shows the ventricles from above.

foramen (of Munro), into the third ventricle, thence through a narrow channel, the cerebral aqueduct (or aqueduct of Sylvius), into the fourth ventricle, from which it circulates into the central canal of the cord. Openings in the roof of the fourth ventricle (the foramen of Magendie and foramina of Luschka) permit the flow of fluid into the subarachnoid space around the cord and thence into the subarachnoid space around the brain. From the latter space it is gradually absorbed into the venous blood of the brain. Thus cerebrospinal fluid "circulates" from blood in the choroid plexuses, through

ventricles, central canal, and subarachnoid spaces, and back into the blood. Occasionally some condition interferes with this circuit. For example, a brain tumor may press against the cerebral aqueduct, shutting off the flow of fluid from the third to the fourth ventricle. In such an event, the fluid accumulates within the lateral and third ventricles because it continues to form even though its drainage is blocked. This condition is known as *internal hydrocephalus*. If the fluid accumulates in the subarachnoid space around the brain, *external hydrocephalus* results. Subarachnoid

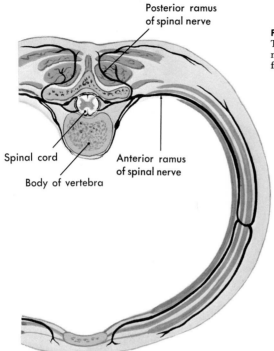

Posterior ramus
of spinal nerve

Spinal cord

Anterior ramus
of spinal nerve

Body of vertebra

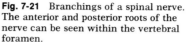

Fig. 7-21 Branchings of a spinal nerve. The anterior and posterior roots of the nerve can be seen within the vertebral foramen.

hemorrhage, for example, may lead to formation of blood clots which block drainage of the cerebrospinal fluid from the subarachnoid space. With decreased drainage, an increased amount of fluid, of course, remains in the space.

Withdrawal of some of the cerebrospinal fluid from the subarachnoid space in the lumbar region of the cord is known as a *lumbar puncture.*

The amount of cerebrospinal fluid in the average adult is about 120 milliliters.* Since approximately 550 milliliters are secreted daily, the fluid must continually circulate and be reabsorbed, or excessive amounts will soon accumulate.

*Mountcastle, V. B., editor: Medical physiology, ed. 12, St. Louis, 1968, The C. V. Mosby Co.

■ Spinal nerves

Origin

Thirty-one pairs of nerves have their origin on the spinal cord. Unlike the cranial nerves, they have no special names but are merely numbered according to the level of the spinal column at which they emerge from the spinal cavity. Thus, there are eight cervical, twelve thoracic, five lumbar, and five sacral pairs and one coccygeal pair of spinal nerves. The first cervical nerves emerge from the cord in the spaces above the first cervical vertebra (between it and the occipital bone). The rest of the cervical and all of the thoracic nerves pass out of the spinal cavity horizontally through the intervertebral foramina of their respective vertebrae. For example, the second cervical nerves emerge through the foramina above the second cervical vertebra.

Lumbar, sacral, and coccygeal nerves, on the other hand, have to descend from their point of origin at the lower end of the cord (which terminates at the level of the first lumbar vertebra) before reaching the intervertebral foramina of their respective vertebrae, through which they then emerge. This gives the lower end of the cord, with its attached spinal nerves, the appearance of a horse's tail. In fact, it bears the name *cauda equina* (Latin equivalent for horse's tail).

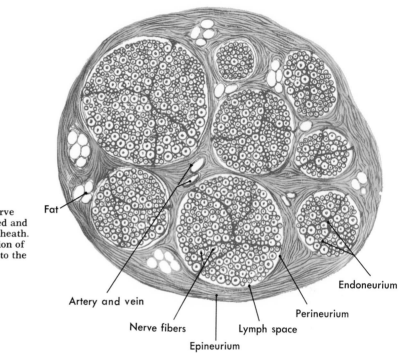

Fig. 7-22 Cross section of a nerve trunk. The fibers are medullated and are separated by a perineural sheath. The endoneurium is an extension of the perineurium which runs into the interior of the fasciculi.

Fat

Artery and vein

Nerve fibers

Epineurium

Lymph space

Perineurium

Endoneurium

Each spinal nerve, instead of attaching directly to the cord, attaches indirectly by means of two short roots, anterior and posterior. The posterior roots are readily recognized by the presence on them of a swelling, the posterior root ganglion or *spinal ganglion*. The roots lie within the spinal cavity with the ganglia in the intervertebral foramina (Fig. 7-21). Spinal ganglia are made up of cell bodies of sensory neurons. Fig. 7-22 shows the structure of a nerve in a cross-section view.

Distribution

After each spinal nerve emerges from the spinal cavity, it divides into anterior, posterior, and white rami. Anterior and posterior rami contain fibers belonging to the voluntary nervous system. White rami contain fibers of the involuntary or autonomic nervous system. The posterior rami subdivide into lesser nerves that extend into the muscles and skin of the posterior surface of the head, neck, and trunk.

The anterior rami (except those of the thoracic nerves) subdivide and supply fibers to skeletal muscles and skin of the extremities and of anterior and lateral surfaces. Subdivisions of the anterior rami form complex networks or plexuses (Table 7-4). For example, fibers from the lower four cervical and first thoracic nerves intermix in such a way as to form a fairly definite, although apparently hopelessly confused, pattern called the *brachial plexus* (Fig. 7-23). Emerging from this plexus are smaller nerves bearing names descriptive of their locations, such as the median

Table 7-4. Spinal nerves and peripheral branches

Spinal nerves	Plexuses formed from anterior rami	Spinal nerve branches from plexuses	Parts supplied
Cervical 1 2 3 4	Cervical plexus	Lesser occipital Great auricular Cutaneous nerve of neck Anterior supraclavicular Middle supraclavicular Posterior supraclavicular Branches to numerous neck muscles	Sensory to back of head, front of neck, and upper part of shoulder; motor to numerous neck muscles
		Suprascapular and dorsoscapular	Superficial muscles* of scapula
		Thoracic nerves, medial and lateral anterior	Pectoralis major and minor
		Long thoracic nerve	Serratus anterior
		Thoracodorsal	Latissimus dorsi
		Subscapular	Subscapular and teres major muscles
		Axillary (circumflex)	Deltoid and teres minor muscles and skin over deltoid
Cervical 5 6 7 8	Brachial plexus	Musculocutaneous	Muscles of front of arm (biceps brachii, coracobrachialis, and brachialis) and skin on outer side of forearm
Thoracic (or dorsal) 1		Ulnar	Flexor carpi ulnaris and part of flexor digitorum profundus; some of muscles of hand; sensory to medial side of hand, little finger, and medial half of fourth finger
2 3 4 5 6		Median	Rest of muscles of front of forearm and hand; sensory to skin of palmar surface of thumb, index, and middle fingers
7 8 9 10	No plexus formed; branches run directly to intercostal muscles and skin of thorax	Radial	Triceps muscle and muscles of back of forearm; sensory to skin of back of forearm and hand
11 12		Medial cutaneous	Sensory to inner surface of arm and forearm
		Phrenic (branches from cervical nerves before formation of plexus; most of its fibers from fourth cervical nerve)	Diaphragm

*Although nerves to muscles are considered motor, they do contain some sensory fibers that transmit proprioceptive impulses.

Continued on opposite page

nerve, the musculocutaneous nerve, and the ulnar nerve. These nerves (each containing fibers from more then one spinal nerve) divide further into smaller and smaller branches, resulting ultimately in the complete innervation of the hand and most of the arm. The brachial plexus is located in the shoulder region from the neck to the axilla. It is of clinical significance since it is sometimes stretched or torn at birth, causing paralysis and numbness of the baby's arm on that side. If untreated, it results in a withered arm. Branching from this plexus are several nerves to the skin and to voluntary muscles.

Table 7-4. Spinal nerves and peripheral branches — cont'd

Spinal nerves	Plexuses formed from anterior rami	Spinal nerve branches from plexuses	Parts supplied
Lumbar 1 2 3 4 5 Sacral 1 2 3 4 5 Coccygeal 1	Lumbosacral plexus	Iliohypogastric } Sometimes fused Ilioinguinal	Sensory to anterior abdominal wall
			Sensory to anterior abdominal wall and external genitalia; motor to muscles of abdominal wall
		Genitofemoral	Sensory to skin of external genitalia and inguinal region
		Lateral cutaneous of thigh	Sensory to outer side of thigh
		Femoral	Motor to quadriceps, sartorius, and iliacus muscles; sensory to front of thigh and to medial side of lower leg (saphenous nerve)
		Obturator	Motor to adductor muscles of thigh
		Tibial† (medial popliteal)	Motor to muscles of calf of leg; sensory to skin of calf of leg and sole of foot
		Common peroneal (lateral popliteal)	Motor to evertors and dorsiflexors of foot; sensory to lateral surface of leg and dorsal surface of foot
		Nerves to hamstring muscles	Motor to muscles of back of thigh
		Gluteal nerves, superior and inferior	Motor to buttocks muscles and tensor fasciae latae
		Posterior cutaneous nerve	Sensory to skin of buttocks, posterior surface of thigh, and leg
		Pudendal nerve	Motor to perineal muscles; sensory to skin of perineum

†Sensory fibers from the tibial and peroneal nerves unite to form the *medial cutaneous* (or sural) *nerve* that supplies the calf of the leg and the lateral surface of the foot. In the thigh, the tibial and common peroneal nerves are usually enclosed in a single sheath to form the *sciatic nerve*, the largest nerve in the body with its width of approximately 3/4 of an inch. About two-thirds of the way down the posterior part of the thigh, it divides into its component parts. Branches of the sciatic nerve extend into the hamstring muscles.

The right and left phrenic nerves whose fibers come from the third and fourth or fourth and fifth cervical spinal nerves before formation of the brachial plexus have considerable clinical interest since they supply the diaphragm muscle. If the neck is broken in a way that severs or crushes the cord above this level, nerve impulses from the brain can, of course, no longer reach the phrenic nerves, and therefore the diaphragm stops contracting. Unless artificial respiration of some kind is provided, the patient dies of respiratory paralysis as a result of the broken neck. Poliomyelitis that attacks the cord be-

tween the third and fifth cervical segments also paralyzes the phrenic nerve and, therefore, the diaphragm.

Another spinal nerve plexus is the *lumbar plexus,* formed by the intermingling of fibers from the first four lumbar nerves. This network of nerves is located in the lumbar region of the back in the psoas muscle. The large femoral nerve is one of several nerves emerging from the lumbar plexus. It divides into many branches which supply the thigh and leg.

Fibers from the fourth and fifth lumbar nerves and the first, second, and third sacral nerves form the *sacral plexus,* located in the pelvic cavity on the anterior surface of the piriformis muscle. Among other nerves which emerge from the sacral plexus are the tibial and common peroneal nerves which, in the thigh, form the largest nerve in the body, namely, the great sciatic nerve (Fig. 7-24). It pierces the buttocks and runs down the back of the thigh. Its many branches supply nearly all the skin

of the leg, the posterior thigh muscles, and the leg and foot muscles. Sciatica or neuralgia of the sciatic nerve is a fairly common and very painful condition.

At first glance, the distribution of spinal nerves does not appear to follow an ordered arrangement. But detailed mapping of the skin surface has revealed a close relationship between the source on the cord of

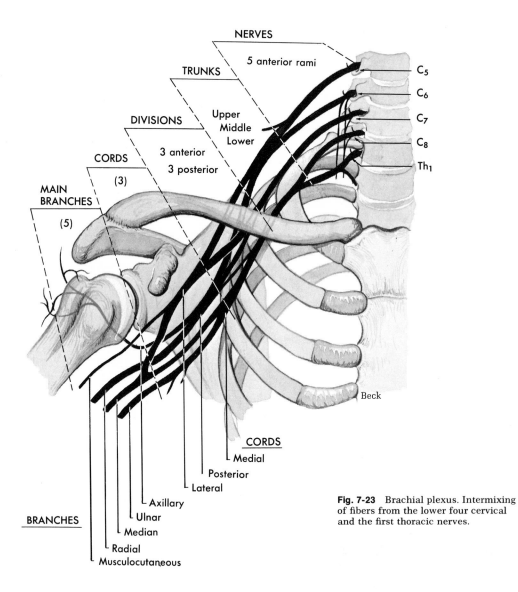

NERVES

5 anterior rami

TRUNKS

Upper
Middle
Lower

DIVISIONS

3 anterior
3 posterior

CORDS

(3)

MAIN
BRANCHES

(5)

C₅
C₆
C₇
C₈
Th₁

Beck

CORDS
└ Medial
└ Posterior
└ Lateral

BRANCHES

└ Axillary
└ Ulnar
└ Median
└ Radial
└ Musculocutaneous

Fig. 7-23 Brachial plexus. Intermixing of fibers from the lower four cervical and the first thoracic nerves.

each spinal nerve and the vertical position of the body it innervates (Figs. 7-25 and 7-26). Knowledge of the segmental arrangement of spinal nerves has proved useful to physicians. For instance, a neurologist can identify the site of spinal cord or nerve abnormality from the area of the body insensitive to a pinprick.

Microscopic structure

All spinal nerves are *mixed nerves*. As shown in Fig. 7-5, spinal nerves are composed of both sensory dendrites and motor axons. Note also in Fig. 7-5 that the sensory dendrite shown comes from a cell body located in the ganglion on the posterior root of the spinal nerve. The motor axon shown originates from a cell body located in the anterior gray column of the cord. These neurons whose cell bodies lie in the anterior gray columns and whose axons terminate in skeletal muscle are called *somatic motoneurons*. Spinal nerves also contain axons of postganglionic autonomic motoneurons (p. 224).

Functions

Because spinal nerves contain both sensory and motor fibers, they serve as two-way conduction paths between the periphery and the spinal cord. Hence spinal nerves function to make possible both sensations and movements. The sensory dendrites in spinal nerves constitute the first part of the neural pathway traveled by sensory impulses. And the motor axons in spinal nerves constitutes the last part of the neural pathway, the "*final common path*," traveled by motor impulses to skeletal muscles. Consequently, anything that interferes with the functioning of a spinal nerve produces both anesthesia and paralysis of the part innervated by that nerve. This principle does not hold, however, for all spinal nerve branches. Whereas some spinal nerve branches are mixed nerves,

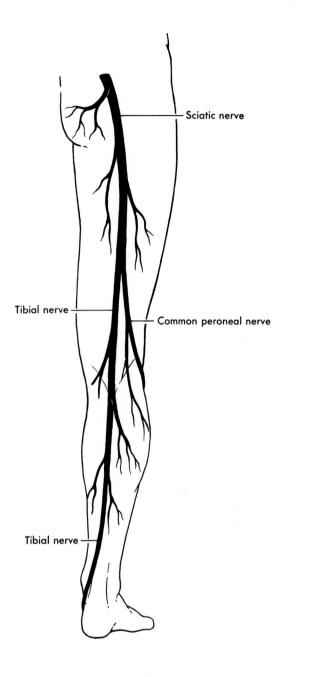

Sciatic nerve

Tibial nerve

Common peroneal nerve

Tibial nerve

Fig. 7-24 Main nerves of the lower extremity.

some are purely sensory and some are purely motor (Tables 7-4 and 7-5).

■ Cranial nerves

Twelve pairs of nerves arise from the undersurface of the brain, some from each division with the exception of the cerebellum. After leaving the cranial cavity by way of small foramina in the skull, they extend to their respective destinations. Both names and numbers identify the cranial nerves. Their names suggest their distribution or function. Their numbers indicate the order in which they emerge from front to back. Some cranial nerves, like all spinal nerves proper, consist of both afferent and efferent fibers—in short, they are mixed nerves. Some cranial nerves, on the other hand, consist of afferent fibers only and some of efferent fibers mainly. Cell bodies of the efferent fibers lie in the various nuclei of the brainstem. Cell bodies of the afferent fibers, with few exceptions, are located in ganglia (for example, the semilunar or gasserian ganglion of the fifth cranial nerve) outside the brainstem.

The first, second, and eighth cranial nerves are purely afferent. Information about cranial nerves is summarized in Tables 7-5 and 7-6.

First (olfactory)

The olfactory nerves are composed of axons of neurons whose dendrites and cell bodies lie in the nasal mucosa, high up along the septum and superior conchae (turbinates). Axons of these neurons form about twenty small fibers which pierce each cribriform plate and terminate in the olfactory bulbs, where they synapse with olfactory neurons II whose axons comprise the olfactory tracts. Summarizing:

Olfactory neurons I
Dendrites ⎫
Cell body ⎬ In nasal mucosa
Axons   In small fibers which extend through cribriform plate to olfactory bulb

Olfactory neurons II
Dendrites ⎫
Cell body ⎬ In olfactory bulb
Axons   In olfactory tracts

Table 7-5. Cranial nerves contrasted with spinal nerves

	Cranial nerves	*Spinal nerves*
Origin	Base of brain	Spinal cord
Distribution	Mainly to head and neck	Skin, skeletal muscles, joints, blood vessels, sweat glands, and mucosa except of head and neck
Structure	Some composed of sensory fibers only; some of both motor axons and sensory dendrites; some motor fibers belong to voluntary nervous system, some to autonomic	All of them composed of both sensory dendrites and motor axons; some of latter, somatic or voluntary, some autonomic
Function	Vision, hearing, sense of smell, sense of taste, eye movements, etc.	Sensations, movements, and sweat secretion

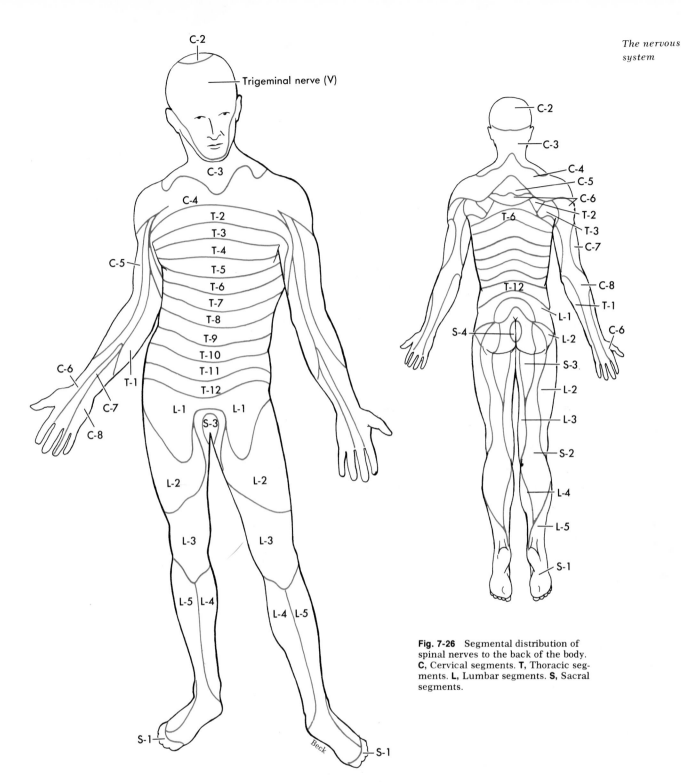

Fig. 7-25 Segmental distribution of spinal nerves to the front of the body. **C**, Cervical segments. **T**, Thoracic segments. **L**, Lumbar segments. **S**, Sacral segments.

Fig. 7-26 Segmental distribution of spinal nerves to the back of the body. **C**, Cervical segments. **T**, Thoracic segments. **L**, Lumbar segments. **S**, Sacral segments.

Table 7-6. Cranial nerves

Nerve*	Sensory fibers†			Motor fibers†		Functions‡
	Receptors	Cell bodies	Termination	Cell bodies	Termination	
I Olfactory	*Nasal mucosa*	*Nasal mucosa*	*Olfactory bulbs (new relay of neurons to olfactory cortex)*			*Sense of smell*
II Optic	*Retina*	*Retina*	*Nucleus in thalamus (lateral geniculate body); some fibers terminate in superior colliculus of midbrain*			*Vision*
III Oculomotor	*External eye muscles except superior oblique and lateral rectus*	?	?	**Midbrain (oculomotor nucleus and Edinger-Westphal nucleus)**	**External eye muscles except superior oblique and lateral rectus; fibers from E-W nucleus terminate in ciliary ganglion and thence to ciliary and iris muscles**	**Eye movements, regulation of size of pupil, accommodation,** *proprioception (muscle sense)*
IV Trochlear	*Superior oblique*	?	?	**Midbrain**	**Superior oblique muscle of eye**	**Eye movements,** *proprioception*
V Trigeminal	*Skin and mucosa of head, teeth*	*Gasserian ganglion*	*Pons (sensory nucleus)*	**Pons (motor nucleus)**	**Muscles of mastication**	**Sensations of head and face, chewing movements,** *muscle sense*
VI Abducens	*Lateral rectus*			**Pons**	**Lateral rectus muscle of eye**	**Abduction of eye,** *proprioception*
VII Facial	*Taste buds of anterior two-thirds of tongue*	*Geniculate ganglion*	*Medulla (nucleus solitarius)*	**Pons**	**Superficial muscles of face and scalp**	**Facial expressions, secretion of saliva,** *taste*
VIII Acoustic 1 Vestibular branch	*Semicircular canals and vestibule (utricle and saccule)*	*Vestibular ganglion*	*Pons and medulla (vestibular nuclei)*			*Balance or equilibrium sense*

	Organ of Corti in cochlear duct	Spiral ganglion	Pons and medulla (cochlear nuclei)			Hearing
2 Cochlear or auditory branch						
IX Glossopharyngeal	*Pharynx; taste buds and other receptors of posterior one-third of tongue*	*Jugular and petrous ganglia*	*Medulla (nucleus solitarius)*	**Medulla (nucleus ambiguus)**	**Muscles of pharynx**	*Taste and other sensations of tongue, swallowing movements, secretion of saliva, aid in reflex control of blood pressure and respirations*
	Carotid sinus and carotid body	*Same*	*Medulla (respiratory and vasomotor centers)*	**Medulla at junction of pons (nucleus salivatorius)**	**Otic ganglion and thence to parotid gland**	
X Vagus	*Pharynx, larynx, carotid body, and thoracic and abdominal viscera*	*Jugular and nodose ganglia*	*Medulla (nucleus, solitarius), pons (nucleus of fifth cranial nerve)*	**Medulla (dorsal motor nucleus)**	**Ganglia of vagal plexus and thence to muscles of pharynx, larynx, and thoracic and abdominal viscera**	*Sensations and movements of organs supplied; for example, slows heart, increases peristalsis, and contracts muscles for voice production*
XI Spinal accessory	?	?	?	**Medulla (dorsal motor nucleus of vagus and nucleus ambiguus)** / **Anterior gray column of first five or six cervical segments of spinal cord**	**Muscles of thoracic and abdominal viscera and pharynx and larynx** / **Trapezius and sternocleidomastoid muscle**	**Shoulder movements, turning movements of head, movements of viscera, voice production,** *proprioception?*
XII Hypoglossal	?	?	?	**Medulla (hypoglossal nucleus)**	**Muscles of tongue**	**Tongue movements,** *proprioception?*

*The first letters of the words in the following sentence are the first letters of the names of the cranial nerves. Many generations of anatomy students have used this sentence as an aid to memorizing these names. It is, "On Old Olympus Tiny Tops, A Finn And German Viewed Some Hops." (There are several slightly differing versions of this mnemonic.)

‡Italics indicate sensory fibers and functions. Boldface type indicates motor fibers and functions.

†An aid for remembering the general function of each cranial nerve is the following twelve-word saying: "Some say marry money but my brothers say bad business marry money." Words beginning with S indicate sensory function. Words beginning with M indicate motor function. Words beginning with B indicate both sensory and motor functions. For example, the first, second, and eighth words in the saying start with S, which indicates that the first, second, and eighth cranial nerves perform sensory functions.

Second (optic)

Axons from the third and innermost layer of neurons of the retina compose the second cranial nerves. After entering the cranial cavity through the optic foramina, the two optic nerves unite to form the *optic chiasma,* in which some of the fibers of each nerve cross to the opposite side and continue in the *optic tract* of that side. Thus each optic nerve contains only fibers from the retina of the same side, whereas each optic tract has fibers in it from both retinae, a fact of importance in interpreting certain visual disorders. Most of the optic tract fibers terminate in the thalamus (in the portion known as the lateral geniculate body). From here, a new relay of fibers runs to the visual area of the occipital lobe cortex. A few optic tract fibers terminate in the superior colliculi of the midbrain, where they synapse with motor fibers to the external eye muscles (third, fourth, and sixth cranial nerves).

Third (oculomotor)

Fibers of the third cranial nerve originate from cells in the oculomotor nucleus in the ventral part of the midbrain and extend to the various external eye muscles, with the exception of the superior oblique and the lateral rectus. Autonomic fibers whose cells lie in a nucleus of the midbrain are also contained in the oculomotor nerves. These fibers terminate in the ciliary ganglion, where they synapse with cells whose postganglionic fibers supply the intrinsic eye muscles (ciliary and iris). Still a third group of fibers are found in the third cranial nerves—namely, sensory fibers from proprioceptors in the eye muscles.

Fourth (trochlear)

Motor fibers of the fourth cranial nerve have their origin in cells in the midbrain, from whence they extend to the superior oblique muscles of the eye. Afferent fibers from proprioceptors in these muscles are also contained in the trochlear nerves.

Fifth (trifacial or trigeminal)

Three sensory branches (ophthalmic, maxillary, and mandibular nerves) carry afferent impulses from the skin and mucosa of the head and from the teeth to cell bodies in the semilunar or gasserian ganglion (a swelling on the nerve, lodged in the petrous portion of the temporal bone) (Fig. 7-28). Fibers extend from the ganglion to the main sensory nucleus of the fifth cranial nerve situated in the pons. A smaller motor root of the trigeminal nerve originates in the trifacial motor nucleus located in the pons just medial to the sensory nucleus. Fibers run from the motor root to the muscles of mastication by way of the mandibular nerve.

Neuralgia of the trifacial nerve, known as tic douloureux, is an extremely painful condition which can be relieved by removing the gasserian (or semilunar) ganglion, the large ganglion on the posterior root of the nerve (Fig. 7-28) containing the cell bodies of the nerve's afferent fibers. After such an operation, the patient's face, scalp, teeth, and conjunctiva on the side treated show anesthesia. Special care, such as wearing protective goggles and irrigating the eye frequently, is therefore prescribed. The patient is instructed also to visit his dentist regularly since he can go longer experience a toothache as a warning of diseased teeth.

Sixth (abducens)

The sixth cranial nerve is a motor nerve with fibers originating from a nucleus in the pons in the floor of the fourth ventricle and extending to the lateral rectus muscles of the eyes. It contains also some afferent

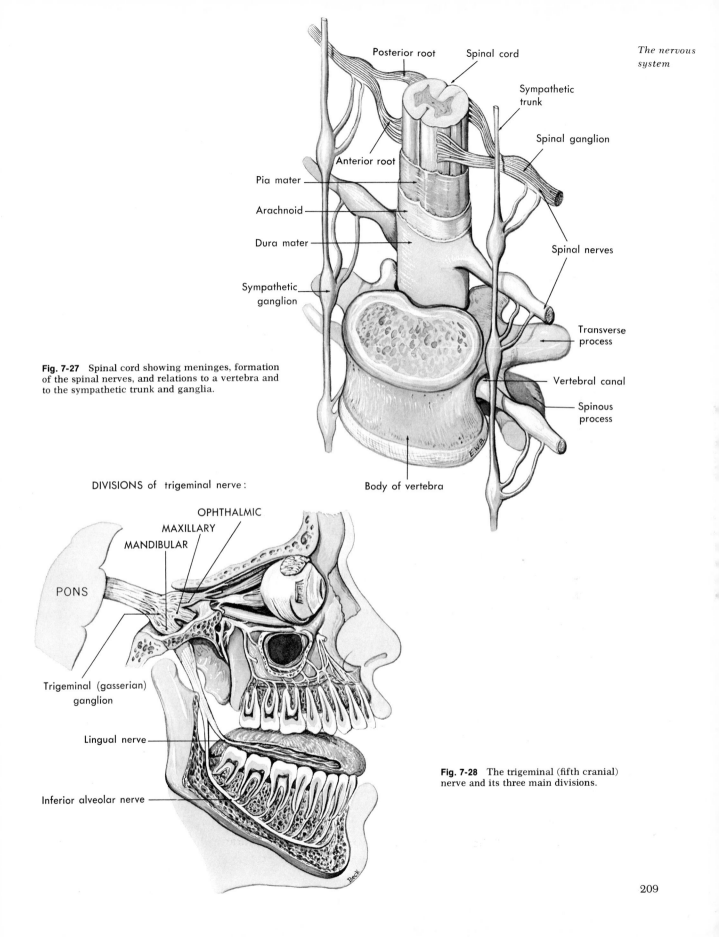

Posterior root

Spinal cord

Sympathetic trunk

Spinal ganglion

Anterior root

Pia mater

Arachnoid

Dura mater

Spinal nerves

Sympathetic ganglion

Transverse process

Vertebral canal

Spinous process

Fig. 7-27 Spinal cord showing meninges, formation of the spinal nerves, and relations to a vertebra and to the sympathetic trunk and ganglia.

Body of vertebra

DIVISIONS of trigeminal nerve:

OPHTHALMIC

MAXILLARY

MANDIBULAR

PONS

Trigeminal (gasserian) ganglion

Lingual nerve

Inferior alveolar nerve

Fig. 7-28 The trigeminal (fifth cranial) nerve and its three main divisions.

209

fibers from proprioceptors in the lateral rectus muscles.

Seventh (facial)

The motor fibers of the seventh cranial nerve arise from a nucleus in the lower part of the pons, whence they extend by way of several branches to the superficial muscles of the face and scalp and to the submaxillary and sublingual glands. Sensory fibers from the taste buds of the anterior two-thirds of the tongue run in the facial nerve to cell bodies in the geniculate ganglion, a small swelling on the facial nerve, where it passes through a canal in the temporal bone. From the ganglion, fibers extend to the nucleus solitarius in the medulla.

Eighth (acoustic or auditory)

The eighth cranial nerve has two distinct divisions: the vestibular nerve and the cochlear nerve. Both are sensory. Fibers from the semicircular canals run to the vestibular ganglion (in the internal auditory meatus), where their cell bodies are located and whence fibers extend to the vestibular nuclei in the pons and medulla. Together, these fibers constitute the *vestibular nerve*. Some of its fibers run to the cerebellum. The *cochlear nerve* consists of fibers which start in the organ of Corti in the cochlea, have their cell bodies in the spiral ganglion in the cochlea, and terminate in the cochlear nuclei located between the medulla and pons. The vestibular nerve transmits impulses which result in sensations of balance or imbalance. Conduction by the cochlear nerve results in sensations of hearing.

Ninth (glossopharyngeal)

Both sensory and motor fibers compose the ninth cranial nerve. This nerve supplies fibers not only to the tongue and pharynx, as its name implies, but also to other structures—for example, to the carotid sinus. The latter plays an important part in the control of blood pressure. Sensory fibers, with their receptors in the pharynx and posterior third of the tongue, have their cell bodies in the jugular (superior) and petrous (inferior) ganglia, located respectively in the jugular foramen and the petrous portion of the temporal bone. From these, fibers extend to the nucleus solitarius in the medulla. The motor fibers of the ninth cranial nerve originate in cells in the nucleus ambiguus in the medulla and run to muscles of the pharynx. There are also secretory fibers in this nerve, with cells of origin in the nucleus salivatorius (at the junction of the pons and medulla). These fibers run to the otic ganglion, whence postganglionic fibers extend to the parotid gland.

Tenth (vagus or pneumogastric)

The tenth cranial nerve is widely distributed and contains both sensory and motor fibers. Its sensory fibers supply the pharynx, larynx, trachea, heart, carotid body, lungs, bronchi, esophagus, stomach, small intestine, and gallbladder. Cell bodies for these sensory dendrites lie in the jugular and nodose ganglia, located respectively in the jugular foramen and just inferior to it on the trunk of the nerve. Centrally, the sensory axons terminate in the medulla (in the nucleus solitarius) and in the pons (in the nucleus of the trigeminal nerve). Motor fibers of the vagus originate in cells in the medulla (in the dorsal motor nucleus of the vagus) and extend to various autonomic ganglia in the vagal plexus, whence postganglionic fibers run to muscles of the pharynx, larynx, and thoracic and abdominal viscera.

Eleventh (accessory or spinal accessory)

The eleventh cranial nerve is a motor nerve. Part of its fibers originates in cells in the medulla (in the dorsal motor nucleus of the vagus and in the nucleus ambiguus) and pass by way of vagal branches to thoracic and abdominal viscera. The rest of the fibers have their cells of origin in the anterior gray column of the first five or six segments of the cervical spinal cord and extend through the spinal root of the accessory nerve to the trapezius and sternocleidomastoid muscles.

Twelfth (hypoglossal)

Motor fibers with cell bodies in the medulla (in the hypoglossal nucleus) compose the twelfth cranial nerve. They supply the muscles of the tongue. According to some anatomists, this nerve also contains sensory fibers from proprioceptors in the tongue.

* * *

The main facts about the distribution and function of each of the cranial nerve pairs are summarized in Table 7-6.

Severe head injuries often damage one or more of the cranial nerves, producing symptoms analogous to the functions of the nerve affected. For example, injury of the sixth cranial nerve causes the eye to turn in, due to paralysis of the abducting muscle of the eye, whereas injury of the eighth cranial nerve produces deafness. Injury to the facial nerve results in a poker-faced expression and a drooping of the corner of the mouth due to paralysis of the facial muscles.

An aneurysm of one of the middle cerebral arteries (Fig. 10-15, p. 321) with resulting pressure on the nearby oculomotor nerve (Fig. 7-15) is not uncommon. In such cases, the pupil on the same side remains dilated. In addition, the eye may turn outward and the lid may droop.

The second cranial nerve is particularly susceptible to atrophy, resulting in total blindness on the affected side.

Sensory neural pathways

Here are some important general principles about sensory neural pathways (also see Table 7-7):

1 The sensory neural pathway to the cerebral cortex consists of a chain of at least three sensory neurons. We shall call them sensory neurons I, II, and III.

a Sensory neuron I of the relay conducts from the periphery to the spinal cord, where it synapses with sensory neuron II. (If the receptors of sensory neuron I lie in regions supplied by cranial nerves, then its axon enters the brainstem instead of the cord.) Examine Fig. 7-29 and then try to answer question 40(a), p. 238. Check your answer with Table 7-7.

b Sensory neuron II conducts from the cord or brainstem up to the thalamus, where it synapses with sensory neuron III. Answer question 40(b), p. 238, after again examining Fig. 7-29.

c Sensory neuron III conducts from the thalamus to the general sensory area of the cerebral cortex (postcentral gyrus, parietal lobe). Look once more at Fig. 7-29 and then answer questions 40(c) and 43, p. 238.

2 Crude awareness of a sensation occurs when sensory impulses reach the thalamus. By crude awareness we mean that the individual is conscious both of the kind of sensation (for example, heat, cold, pain) he is experiencing and of whether it feels pleasant or unpleasant to him to a mild, moderate, or marked degree.

3 Full consciousness of sensations occurs when sensory impulses reach the cerebral cortex. Then the individual not only recognizes the kind of sensation present, but he also knows exactly where it is coming from. He can localize it accurately and can discriminate as to its intensity.

4 For the most part, sensory pathways to the cerebral cortex are crossed path-

Table 7-7. Sensory neural pathways

Neurons	Gross structures in which neuron parts are located
1 Pain and temperature	
Sensory neuron I	
Receptors	In skin, mucosa, muscles, tendons, viscera
Dendrite	In spinal nerve and branch of spinal nerve
Cell body	In spinal ganglion, on posterior root of spinal nerve
Axon	In posterior root of spinal nerve; terminates in posterior gray column of cord
Sensory neuron II	
Dendrite	Posterior gray column
Cell body	Posterior gray column
Axon	Decussates and ascends in lateral spinothalamic tract (Figs. 7-12 and 7-29); terminates in thalamus
Sensory neuron III	
Dendrite	Thalamus
Cell body	Thalamus
Axon	Thalamus via thalamocortical tract in internal capsule to general sensory area of cerebral cortex, i.e., postcentral gyrus in parietal lobe
2 Crude touch stimuli	
Sensory neuron I	Same as sensory neuron I for pain and temperature stimuli
Sensory neuron II	
Dendrite	Posterior gray column; same as sensory neuron II for pain and temperature
Cell body	Posterior gray column; same as sensory neuron II for pain and temperature
Axon	In ventral spinothalamic tract (Fig. 7-12) to thalamus
Sensory neuron III	Same as sensory neuron III for pain and temperature stimuli
3 Discriminating touch (two-point discrimination, vibrations) **deep touch,** and **pressure and conscious proprioception**	
Sensory neuron I	Same as sensory neuron I for pain, temperature, and crude touch stimuli, except that axon extends up cord in posterior white columns (fasciculi gracilis and cuneatus, Fig. 7-12) to nuclei gracilis or cuneatus in medulla instead of terminating in posterior gray columns of cord
Sensory neuron II	
Dendrite	In nuclei gracilis or cuneatus of medulla
Cell body	In nuclei gracilis or cuneatus of medulla
Axon	Decussates and ascends in medial lemniscus (broad band of fibers extending up through medulla and midbrain) and terminates in thalamus
Sensory neuron III	Same as sensory neuron III for pain, temperature, and crude touch stimuli

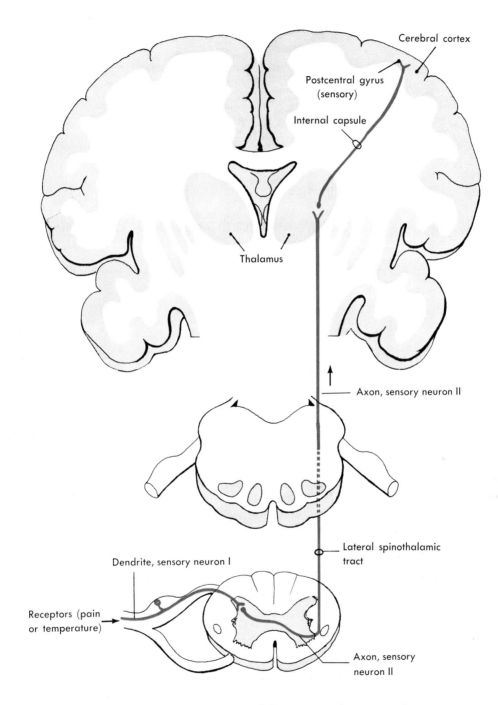

Fig. 7-29 The lateral spinothalamic tract relays sensory impulses from pain and temperature receptors.

DIVERGENCE CONVERGENCE

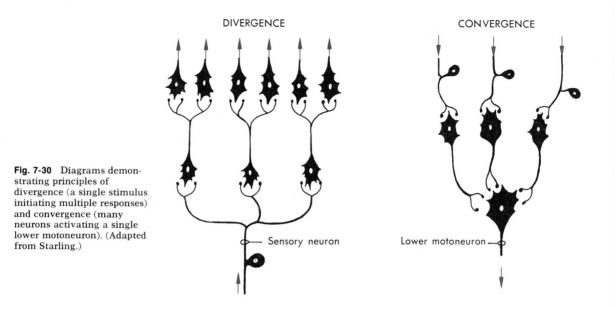

Fig. 7-30 Diagrams demonstrating principles of divergence (a single stimulus initiating multiple responses) and convergence (many neurons activating a single lower motoneuron). (Adapted from Starling.)

Sensory neuron

Lower motoneuron

ways. This means that each side of the brain registers sensations from the opposite side of the body. Usually it is the axon of sensory neuron II that decussates at some point in its ascent to the thalamus.

5 Most sensory neurons form synapses with many other neurons located at various levels of the cord and brain. This fact, sometimes referred to as the *principle of divergence* (Fig. 7-30), makes it possible for impulses initiated in any one receptor to be conducted to many different effectors almost simultaneously. Such an arrangement can have great survival value—somewhat as if a single line into a telephone exchange were to connect with a hospital, doctor's office, and police and fire departments all at the same time.

6 The neural pathway for sensations of *pain* and *temperature* is called the *lateral spinothalamic pathway.*

7 Two neural pathways conduct impulses that produce sensations of touch and pressure, namely, the medial lemniscal system and the ventral spinothalamic pathway. The *medial lemniscal system* consists of the tracts that make up the posterior white columns of the cord (the fasciculi cuneatus and gracilis) plus the *medial lemniscus,* a flat band of white fibers ex-

tending through the medulla, pons, and midbrain. (Derivation of the name lemniscus may interest you. It comes from the Greek word *lemniskos* meaning "woolen band." Apparently, to some early anatomist, the medial lemniscus looked like a band of woolen material running through the brainstem.)

The fibers of the medial lemniscus, like those of the spinothalamic tracts, are axons of sensory neurons II. They originate from cell bodies in the medulla, decussate, and then extend upward to terminate in the thalamus on the opposite side. The function of the medial lemniscal system is to transmit impulses that produce our more discriminating touch and pressure sensations. These include stereognosis (awareness of an object's size, shape, and texture), precise localization (sense of exact location of stimuli), and sense of vibrations.

The function of the *ventral spinothalamic pathway* is to transmit impulses that result in crude touch and pressure sensations—knowing when something touches the skin, for example.

8 The neural pathway for *conscious proprioception* or *kinesthesia* (sense of the position of body parts and of their movements) is also the medial lemniscal system. See Table 7-7 for a brief summary of the sensory neural pathways named in items **6, 7,** and **8.**

Arousal or alerting mechanism

According to an older theory, consciousness was the result of sufficient numbers of sensory impulses reaching the cerebral cortex, presumably via the great sensory pathways (mainly spinothalamic, auditory, and visual). Now, however, we know that these impulses alone can neither arouse the cortex to wakefulness nor maintain it in the conscious state. Instead, consciousness results from impulses reaching the cortex via the so-called *reticular activating system.* It is these impulses that arouse or alert the cortex and maintain consciousness and not those over the more direct great sensory paths.

The reticular activating system consists of nuclei in the brainstem reticular formation and tracts to and from it. Impulses continually feed into the brainstem reticular formation by direct spinoreticular tracts and by collaterals from spinothalamic, auditory, and visual tracts. Impulses also continually leave the reticular formation for the cortex. But presumably the pathways from the brainstem reticular formation to the cortex are long and multisynaptic routes. They probably include relays to the hypothalamus, thalamus, and possibly other parts of the brain too before the final relay to the cerebral cortex. At any rate, it is now an accepted concept that impulses over this system are the ones that arouse or alert the cortex and maintain consciousness.

Conversely, if anything blocks conduction by the reticular activating system, unconsciousness results. General anesthetics, for example, are thought to produce uncon-

sciousness by inhibiting conduction by the reticular activating system. And certain drugs, notably amphetamines and Adrenalin, produce the opposite effect. They stimulate the reticular activating system and thereby produce wakefulness.

In addition to its arousal function, the reticular formation also serves motor functions (see p. 219).

Motor neural pathways to skeletal muscles

Because neural pathways to skeletal muscles are extremely complex and because there are still so many unknowns about them, we shall confine our discussion to basic facts related to two major principles about these pathways: the principle of the final common path and the principle of convergence.

The *principle of the final common path* might be stated this way: motoneurons whose dendrites and cells lie in the anterior gray horns of the spinal cord constitute the final common path which all impulses to skeletal muscles must traverse. In other words, axons of anterior horn neurons (or lower motoneurons as they are called) are the only motor fibers terminating in skeletal muscles. Therefore, any condition which damages anterior horn neurons enough to make them unable to conduct causes paralysis of the associated muscles. The individual can then no longer contract these muscles at will. And reflex contraction, of course, also becomes impossible when impulses cannot travel over these final common paths to skeletal muscle cells. The poliomyelitis virus, probably the best known virus in the world today, produces paralysis by destroying anterior horn neurons.

The *principle of convergence,* briefly

stated, holds that axons of a great many neurons converge upon—that is, synapse with—each lower motoneuron (Fig. 7-30). Hence, many impulses from diverse sources continually bombard it, and together their combined or summated effect controls its functioning.

Motor pathways from the cerebral cortex down to anterior horn motoneurons are numerous and complex. Two methods are used to classify them—one based on the location of their fibers in the medulla and the other on their influence on the lower motoneurons. The first method divides them into pyramidal (Table 7-8) and extrapyramidal tracts, and the second classifies them as facilitatory and inhibitory tracts.

Pyramidal tracts are those whose fibers come together in the medulla to form the pyramids—hence their name. Because axons composing the pyramidal tracts originate from neuron cell bodies located in the cerebral cortex (presumably mainly from giant pyramidal cells of Betz in the primary motor area), pyramidal tracts are also called corticospinal tracts. Their fibers terminate at various levels of the cord in synapses with interneurons which, in turn, synapse with anterior horn neurons. (A few corticospinal fibers may synapse directly with the anterior horn neurons. Authorities disagree about this.) More than two-thirds of the fibers in the pyramidal tracts decussate in the medulla (Fig. 7-31) and extend down the cord in the lateral corticospinal tract on the opposite side. Other corticospinal fibers do not cross over from one side of the medulla to the other (Fig. 7-32). They extend down the cord in the ventral corticospinal tract located on the same side of the cord as the cerebral motor area from which they came. Hence, another name for ventral corticospinal tract is direct or uncrossed pyramidal tract, and another name for lateral corticospinal tract is crossed pyramidal tract. Impulses

Table 7-8. Pyramidal path from cerebral cortex

Microscopic structures	*Macroscopic structures in which neurons located*
Upper motoneuron (Betz cells)	
Dendrite	Motor area of cerebral cortex
Cell body	Motor area of cerebral cortex
Axon	Motor area of cerebral cortex; descends in corticospinal (pyramidal) tract through cerebrum and brainstem; decussates in medulla and continues descent in lateral corticospinal (crossed pyramidal) tract in lateral white column; or may descend uncrossed in ventral, or direct, pyramidal tract in anterior white column and either decussate or not prior to terminating in anterior gray column (Fig. 7-32)
Lower motoneuron	
Dendrite	Anterior gray column
Cell body	Anterior gray column
Axon	Anterior gray column to anterior root of spinal nerve to spinal nerve and branches; terminates in somatic effector—i.e., skeletal muscle

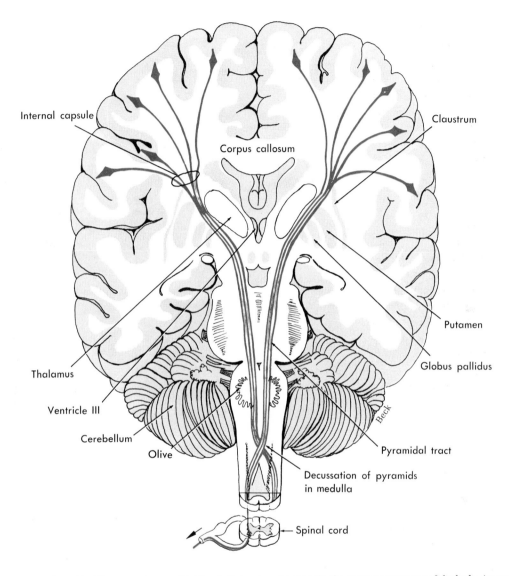

Internal capsule

Corpus callosum

Claustrum

Thalamus

Ventricle III

Cerebellum

Olive

Putamen

Globus pallidus

Pyramidal tract

Decussation of pyramids in medulla

Spinal cord

Fig. 7-31 The crossed pyramidal tracts (lateral corticospinal), the main motor tracts of the body. Axons that compose pyramidal tracts come from neuron cell bodies in the cerebral cortex (mainly in motor area 4). After they descend through the internal capsule of the cerebrum and the white matter of the brainstem, about three-fourths of the fibers decussate—cross over from one side to the other—in the medulla, as shown here. Then they continue downward in the lateral corticospinal tract on the opposite side of the cord. Each lateral corticospinal tract, therefore, conducts motor impulses from one side of the brain to skeletal muscles on the opposite side of the body.

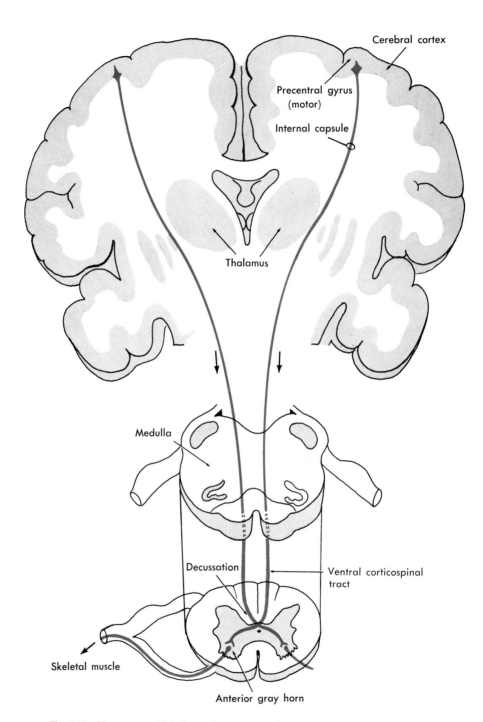

Fig. 7-32 Direct pyramidal (ventral corticospinal) tract fibers
do not decussate in medulla but most of them do decussate in
cord, as shown. A few direct pyramidal tract fibers do not decussate at all.

over these tracts stimulate anterior horn cells which, in turn, stimulate individual muscles (mainly of the hands and feet) to contract and produce small, discrete movements. In fact, stimulation of anterior horn cells by pyramidal tract impulses must occur in order for willed movements to occur. This means that paralysis results whenever pyramidal tract conduction is interrupted. For instance, the paralysis that so often follows cerebral vascular accidents comes from pyramidal neuron injury—sometimes of their cell bodies in the primary motor area, sometimes of their axons in the internal capsule (Fig. 7-31).

Extrapyramidal tracts are much more complex than pyramidal tracts. They consist of all pathways between the motor cortex and anterior horn cells except pyramidal tracts. Complicated incompletely worked out relays between the cortex, basal ganglia, thalamus, and brainstem form the upper portions of extrapyramidal tracts, and reticulospinal tracts are among the most important lower portions.

Fibers of the *reticulospinal tracts* originate from cell bodies in the reticular formation of the brainstem and terminate in gray matter of the spinal cord, where they synapse with interneurons that synapse with lower motoneurons. Some reticulospinal fibers constitute facilitatory tracts, while others form inhibitory tracts (see below). Extrapyramidal tracts conduct impulses important for muscle tone, automatic movements, and emotional expressions.

1 Impulses over some extrapyramidal fibers (for example, over facilitatory reticulospinal fibers) facilitate lower motoneurons and thereby tend to increase muscle tone. But impulses over other extrapyramidal fibers produce the opposite effect—they inhibit lower motoneurons and tend to decrease muscle tone.

2 Conduction by extrapyramidal tracts plays a crucial part in producing our larger, more automatic movements because they cause groups of muscles to contract in sequence or simultaneously. Such muscle action occurs, for example, in swimming and walking and, in fact, in all normal voluntary movements.

3 Conduction by extrapyramidal tracts plays an important part in our emotional expressions. For instance, most of us smile automatically at things that amuse us and frown at things that irritate us. And it is extrapyramidal impulses, not pyramidal, that produce the smiles or frowns.

Facilitatory tracts are so called because they conduct impulses that facilitate—that is, decrease—the resting potential (p. 170) of lower motoneurons. The effects of facilitatory impulses acting rapidly on any one neuron add up or summate. Each impulse, in other words, decreases the neuron's resting potential a little bit more. If sufficient numbers of impulses impinge rapidly enough on a neuron, its potential decreases to threshold level. And at that moment, the neuron starts conducting impulses. In short, it is stimulated. Thus, facilitatory impulses can either facilitate or stimulate a neuron. Facilitation—a decrease in a neuron's resting potential to some point above its threshold level—occurs when relatively few facilitatory impulses act on a neuron. Stimulation—that is, initiation of impulse conduction—occurs when a relatively large number of facilitatory impulses summate and decrease a neuron's membrane potential to its threshold of stimulation.

Inhibitory tracts conduct impulses that inhibit—that is, increase—the lower motoneuron's membrane potential above its resting level. Inhibition is thus the opposite of facilitation.

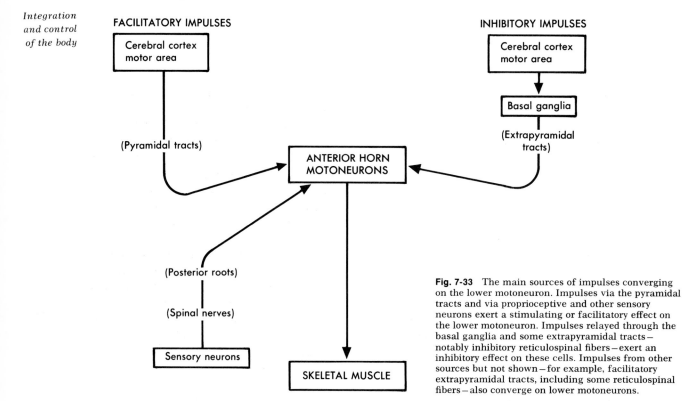

FACILITATORY IMPULSES

Cerebral cortex motor area

(Pyramidal tracts)

ANTERIOR HORN MOTONEURONS

(Posterior roots)

(Spinal nerves)

Sensory neurons

SKELETAL MUSCLE

INHIBITORY IMPULSES

Cerebral cortex motor area

Basal ganglia

(Extrapyramidal tracts)

Fig. 7-33 The main sources of impulses converging on the lower motoneuron. Impulses via the pyramidal tracts and via proprioceptive and other sensory neurons exert a stimulating or facilitatory effect on the lower motoneuron. Impulses relayed through the basal ganglia and some extrapyramidal tracts— notably inhibitory reticulospinal fibers—exert an inhibitory effect on these cells. Impulses from other sources but not shown—for example, facilitatory extrapyramidal tracts, including some reticulospinal fibers—also converge on lower motoneurons.

Fig. 7-33 shows some of the main sources of facilitatory and inhibitory impulses converging upon the lower motoneuron. Note that facilitatory impulses reach anterior horn cells via sensory neurons (whose axons lie in the posterior roots of spinal nerves) and via pyramidal tracts. In addition, some extrapyramidal tract fibers bring facilitatory impulses to anterior horn cells. Facilitatory extrapyramidal fibers are believed to come mainly from neuron cell bodies located in the so-called *facilitatory area* in the lateral reticular formation of the brainstem. According to recent evidence, impulses over these facilitatory reticulospinal fibers facilitate the lower motoneurons that supply extensor muscles. And at the same time, they reciprocally inhibit the lower motoneurons that supply flexor muscles. Hence, facilitatory reticulospinal impulses tend to increase the tone of extensor muscles and decrease the tone of flexor muscles.

Inhibitory impulses reach lower motoneurons mainly via inhibitory reticulospinal fibers that originate from cell bodies located in the *bulbar inhibitory area* in the medulla. They inhibit the lower motoneurons to extensor muscles (and reciprocally stimulate those to flexor muscles). Hence, inhibitory reticulospinal impulses tend to decrease extensor muscle tone and increase flexor muscle tone—opposite effects from facilitatory reticulospinal impulses.

Brainstem inhibitory and facilitatory areas receive impulses from various higher motor centers—notably from the cerebral cortex, cerebellum, and basal ganglia. But the pathways which transmit impulses from these centers to the brainstem centers are complex and poorly understood. Normally, the ratio of facilitatory and inhibitory impulses converging upon lower motoneurons is such as to maintain normal muscle tone. In other words, facilitatory impulses normally somewhat exceed inhibitory impulses. But disease sometimes alters this ratio. Parkinson's disease and "strokes," for example, may interrupt transmission by inhibitory extrapyramidal paths through basal ganglia to bulbar inhibitory centers. Facilitatory impulses then predominate, and excess muscle tone (rigidity or spasticity) develops. Injury of *upper motoneurons* (those whose axons lie in either pyramidal or extrapyramidal tracts) produces symptoms frequently referred to as "pyramidal signs," notably a spastic type of paralysis, exaggerated deep reflexes, and a positive Babinski reflex (see p. 223). Actually, pyramidal signs result from interruption of both pyramidal and extrapyramidal pathways. The paralysis stems from interruption of pyramidal tracts, whereas the spasticity (rigidity) and exaggerated reflexes come from interruption of inhibitory extrapyramidal pathways.

Injury of lower motoneurons produces different symptoms from upper motoneuron injury. Anterior horn cells or lower motoneurons, you will recall, constitute the final common path by which impulses reach skeletal muscles. This means that if they are injured, impulses can no longer reach the skeletal muscles they supply. This, in turn, means both that reflexes involving these muscles are no longer possible and that these muscles cannot be moved at the patient's will. As a result, they lose their normal tone and become soft and flabby (flaccid). In short, absence of reflexes and flaccid paralysis are the chief "lower motoneuron signs."

Reflexes

Definition

The action that results from a nerve impulse passing over a reflex arc is called a *reflex.* In other words, a reflex is a response to a stimulus. It may or may not be conscious. Usually the term is used to mean only involuntary responses rather than those directly willed (that is, involving cerebral cortex activity).

A reflex consists either of muscle contraction or glandular secretion. *Somatic reflexes* are contractions of skeletal muscles. Impulse conduction over somatic reflex arcs—arcs whose motoneurons are somatic motoneurons (that is, anterior horn neurons or lower motoneurons)—produces somatic reflexes. *Autonomic* (or *visceral*) *reflexes* consist either of contractions of smooth or cardiac muscle or secretion by glands; they are mediated by impulse conduction over autonomic reflex arcs, the motoneurons of which are autonomic neurons (discussed on pp. 224 to 228). The following paragraphs describe only somatic reflexes.

Some somatic reflexes of clinical importance

Clinical interest in reflexes stems from the fact that they deviate from normal in certain diseases. So the testing of reflexes is a valuable diagnostic aid. Physicians frequently test the following reflexes: knee jerk, ankle jerk, Babinski reflex, corneal reflex, and abdominal reflex.

The *knee jerk* or patellar reflex is an extension of the lower leg in response to tapping of the patellar tendon. The tap stretches both the tendon and its muscles, the quadriceps femoris, and thereby stimulates muscle spindles (receptors) in the muscle and initiates conduction over the following two-neuron reflex arc (Fig. 7-5):

1 *Sensory neurons*
   a Dendrites—in femoral and second, third, and fourth lumbar nerves
   b Cell bodies—second, third, and fourth ganglia
   c Axons—in posterior roots of second, third, and fourth lumbar nerves; terminate in spinal cord, anterior gray columns; synapse directly with lower motoneurons

Table 7-9. Correlation of microscopic and macroscopic structures of a three-neuron cord reflex arc*

Microscopic structures	Macroscopic structures in which neurons located
Sensory neuron	
Receptor	In skin or mucosa
Dendrite	In spinal nerve and branches
Cell body	In spinal ganglion on posterior root of spinal nerve
Axon	Posterior root of spinal nerve; terminates in posterior gray column of cord
Interneuron	
Dendrite	Posterior gray column
Cell body	Posterior gray column
Axon	Central gray matter of cord, extending into anterior gray column
Motoneuron	
Dendrite	Anterior gray column
Cell body	Anterior gray column
Axon	Anterior gray column, extending into anterior root of spinal nerve, spinal nerve, and its branches
Effector	In skeletal muscles

*See Figs. 7-6 and 7-7; Fig. 7-5 shows the two-neuron cord arc.

2 *Reflex center*—synapses in anterior gray column between axons of sensory neurons and dendrites and cell bodies of lower motoneurons

3 *Motoneurons*

  a Dendrites and cell bodies—in spinal cord anterior gray column

  b Axons—in anterior roots of second, third, and fourth lumbar spinal nerves and femoral nerves; terminate in quadriceps femoris muscle

The knee jerk can be classified in various ways as follows:

1 As a *spinal cord reflex*—because the center of the reflex arc (which transmits the impulses that activate the muscles which produce the knee jerk) lies in the spinal cord gray matter

2 As a *segmental reflex*—because impulses that mediate it enter and leave the same segment of the cord

3 As an *ipsilateral reflex*—because the impulses that mediate it come from and go to the same side of the body

4 As a *stretch reflex*, or *myotatic reflex* (Greek, *mys*, muscle; + *tasis*, stretching)—because of the kind of stimulation used to evoke it

5 As an *extensor reflex*—because produced by extensors of lower leg (muscles located on an anterior surface of thigh, which extend the lower leg)

6 As a *tendon reflex*—because tapping of a tendon is the stimulus that elicits it

7 *Deep reflex*—because of the deep location (in tendon and muscle) of the receptors stimulated to produce this reflex (*superficial reflexes*—those elicited by stimulation of receptors located in the skin or mucosa)

When a physician tests a patient's reflexes, he interprets the test results on the basis of what he knows about the reflex arcs which must function to produce normal reflexes.

To illustrate, suppose that a patient has been diagnosed as having poliomyelitis. In examining him, the physician finds that he cannot elicit the knee jerk when he taps the patient's patellar tendon. He knows that the poliomyelitis virus attacks anterior horn motoneurons. And he also knows the information previously related about which cord segments contain the reflex centers for the knee jerk. On the basis of this knowledge, therefore, he deduces that in this patient the poliomyelitis virus has damaged which segments of the spinal cord? Do you think that this patient's leg would be paralyzed, that he would be unable to move it voluntarily? What neurons would not be able to function which must function to produce voluntary contractions?

Ankle jerk or Achilles reflex is an extension (plantar flexion) of the foot in response to tapping of the Achilles tendon. Like the knee jerk, it is a tendon reflex and a deep reflex mediated by two-neuron spinal arcs with centers in the first and second sacral segments of the cord.

The *Babinski reflex* is an extension of the great toe, with or without fanning of the other toes, in response to stimulation of the outer margin of the sole of the foot. Normal babies, up until they are about 1 1/2 years old, show this positive Babinski reflex. By about this time, corticospinal fibers have become fully myelinated and the Babinski reflex becomes suppressed. Just why this is so is not clear. But at any rate it is, and a positive Babinski reflex after this age is abnormal. From then on, the normal response to stimulation of the outer edge of the sole is the *plantar reflex*. It consists of a curling under of all the toes (plantar flexion) plus a slight turning in and flexion of the anterior part of the foot. A positive Babinski reflex is one of the pyramidal signs (p. 221) and is interpreted to mean destruction of pyramidal tract (corticospinal) fibers.

The *corneal reflex* is winking in response to touching the cornea. It is mediated by reflex arcs with sensory fibers in the ophthalmic branch of the fifth cranial nerve,

centers in the pons, and motor fibers in the seventh cranial nerve.

The *abdominal reflex* is drawing in of the abdominal wall in response to stroking the side of the abdomen. It is mediated by arcs with sensory and motor fibers in the ninth to twelfth thoracic spinal nerves and centers in these segments of the cord and is classified as a superficial reflex. A decrease in this reflex or its absence occurs in lesions involving pyramidal tract upper motor neurons.

Autonomic nervous system
Definition

By definition, the autonomic nervous system consists microscopically of motoneurons only and only those motoneurons that conduct impulses from the central nervous system to visceral effectors. The term *visceral effectors* may be defined in two ways: in terms of tissues or of organs. In terms of tissues, visceral effectors consist of cardiac muscle, smooth muscle, and glandular epithelium. In terms of organs, visceral effectors consist of the heart, blood vessels, iris, ciliary muscles, hair muscles, various thoracic and abdominal organs, and the body's many glands. Note that all of these structures innervated by the autonomic nervous system are ones we think of as involuntary. They lie beyond our conscious control. They are our automatic parts. They function without our willing them to and, for the most part, without our even being conscious of them.

Even though the autonomic nervous system consists, by definition, of only motoneurons (specifically, those that conduct impulses from the brainstem or cord to visceral effectors), nevertheless sensory neurons also take part in autonomic functioning. The autonomic nervous system functions on the reflex arc principle just as the voluntary nervous system does. But any sensory neuron can theoretically function in both autonomic and somatic reflex arcs. For example, stimulation of cold receptors in the skin can initiate both an autonomic reflex (vasoconstriction of skin blood vessels) and a somatic reflex (shivering). In short, any one sensory neuron can function in both somatic and autonomic arcs, whereas any one motoneuron can function in either a somatic arc or an autonomic arc but not in both.

Divisions

Two anatomically and physiologically separate divisions compose the autonomic nervous system: the sympathetic (or thoracolumbar) division and the parasympathetic (or craniosacral) division. Both divisions consist macroscopically of ganglia and fibers and microscopically, as we have mentioned, of visceral motoneurons.

—stimulatory

—inhibitory

Macroscopic structure

Sympathetic ganglia lie lateral to the anterior surface of the spinal column. Because short fibers extend between the sympathetic ganglia, connecting them to each other, they look a little like two chains of beads (one chain on each side of the spinal column from the level of the second cervical vertebra to the coccyx) and are often referred to as the "sympathetic chain ganglia."*

Sympathetic ganglia — There are three cervical, ten or eleven thoracic, four lumbar, and four sacral ganglia in each sympathetic chain. A few sympathetic ganglia, notably the celiac, superior, and inferior mesenteric ganglia, are located a short distance from the cord and therefore are called *collateral ganglia.*
   Celiac ganglia (solar plexus) — two fairly large, flat ganglia located on either side of the celiac artery just below the diaphragm.
   Superior mesenteric ganglion — small ganglion located near the beginning of the superior mesenteric artery.
   Inferior mesenteric ganglion — small ganglion located close to the beginning of the inferior mesenteric artery.

Parasympathetic ganglia lie in or near visceral effectors, not near the spinal column, as do sympathetic ganglia. One example is the ciliary ganglion located in the posterior part of the orbit near the iris and ciliary muscle.

Microscopic structure

Microscopically, the autonomic nervous system consists of preganglionic neurons, postganglionic neurons, and synapses between them. As their names suggest, preganglionic neurons conduct impulses before they reach a ganglion and postganglionic neurons conduct after they reach a ganglion. For examples, look at the right side of Fig. 7-34. It shows both preganglionic and postganglionic neurons of the sympathetic system. Note the location of the preganglionic neuron's cell body in the lateral gray column of the cord. (In which column of the cord do somatic motoneuron cell bodies lie?) The axon of a preganglionic neuron terminates, as Fig. 7-34 shows, in an autonomic ganglion. Here, it synapses with the dendrites or cell body of a postganglionic neuron whose axon terminates in a visceral effector. Thus, it takes a relay of two neurons to conduct impulses from the central nervous system out to a visceral effector—a preganglionic neuron to conduct from either the cord or the brainstem to an autonomic ganglion and a postganglionic neuron to conduct from the ganglion to a visceral effector. How does this differ from conduction to somatic effectors from the central nervous system? Find the answer on the left side of Fig. 7-34 if you do not already know it. One neuron—the anterior horn neuron—and not a relay of two neurons conducts from the cord to somatic effectors.

Preganglionic neurons of sympathetic system. The preganglionic neurons of the sympathetic system (Figs. 7-34 and 7-35

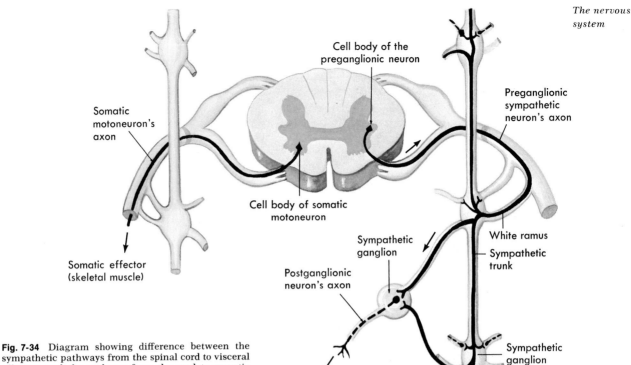

Cell body of the
preganglionic neuron

Somatic
motoneuron's
axon

Preganglionic
sympathetic
neuron's axon

Cell body of somatic
motoneuron

Sympathetic
ganglion

White ramus

Sympathetic
trunk

Postganglionic
neuron's axon

Somatic effector
(skeletal muscle)

Sympathetic
ganglion

Visceral effector
(smooth muscle, cardiac
muscle, glands)

Fig. 7-34 Diagram showing difference between the sympathetic pathways from the spinal cord to visceral effectors and the pathway from the cord to somatic effectors. A relay of two sympathetic neurons—preganglionic and postganglionic—conducts from the cord to visceral effectors. Note that, in contrast, only one somatic motoneuron (anterior horn neuron) conducts from the cord to somatic effectors with no intervening synapses. Parasympathetic impulses also travel over a relay of two neurons (parasympathetic preganglionic and postganglionic) to reach visceral effectors from the central nervous system. Note the location of the sympathetic preganglionic neuron's cell body and axon. Where are the sympathetic postganglionic neuron's cell body and axon located?

and Table 7-10) have their cell bodies in the lateral gray columns of the thoracic and first three or four lumbar segments of the cord. Axons from these cells extend through the anterior roots of the corresponding spinal nerves and through small side branches (the white rami) to the respective sympathetic ganglia. Here, some of them synapse with postganglionic neurons (each axon synapsing with several postganglionic neurons). Other preganglionic axons send branches up and down the sympathetic chain to terminate in ganglia above and below their point of origin. Still others extend through the sympathetic gan-

glia, out through the splanchnic nerves,* and terminate in the collateral ganglia.

But no matter which course a sympathetic preganglionic axon follows, it synapses with many postganglionic neurons, and these frequently terminate in widely separated organs. This anatomical fact explains a well-known physiological principle—sympathetic responses are usually widespread, involving many organs and not just one.

Postganglionic neurons of sympathetic system. The postganglionic neurons of the sympathetic system have their dendrites and cell bodies in the sympathetic

*The three splanchnic nerves constitute the main branches from the thoracic sympathetic trunk. Since their fibers synapse in the celiac and other collateral ganglia with postganglionic neurons to abdominal structures, they form the main routes for sympathetic stimulation of abdominal viscera.

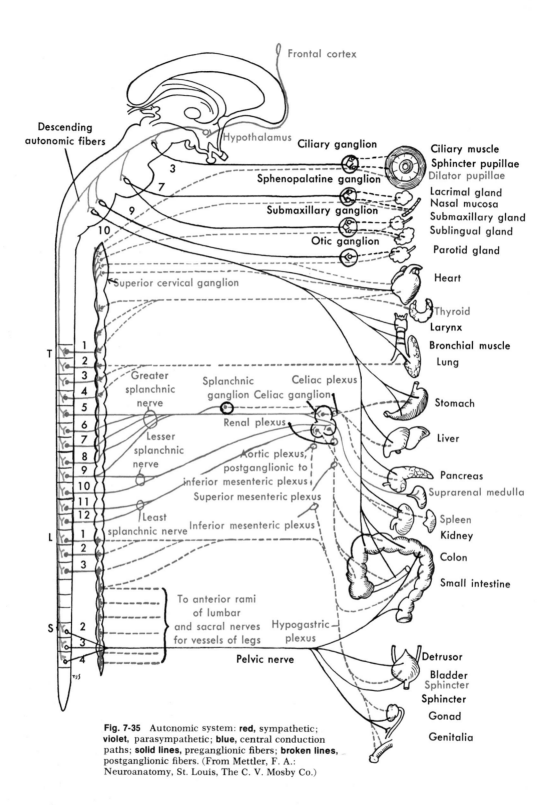

Fig. 7-35 Autonomic system: **red,** sympathetic;
violet, parasympathetic; **blue,** central conduction
paths; **solid lines,** preganglionic fibers; **broken lines,**
postganglionic fibers. (From Mettler, F. A.:
Neuroanatomy, St. Louis, The C. V. Mosby Co.)

Table 7-10. Locations of autonomic neurons

Autonomic neurons (motoneurons)	Macroscopic structures in which located
Sympathetic division	
Preganglionic neurons	
Dendrites and cell bodies	In lateral gray columns of thoracic and first four lumbar segments of spinal cord
Axons (preganglionic fibers)	In anterior roots of spinal nerves to spinal nerves (thoracic and first four lumbar) to white rami and thence in any of three following pathways: (1) through white rami to sympathetic ganglia, where they synapse with several postganglionic neurons; (2) through white rami to and through sympathetic ganglia, thence up or down sympathetic trunk before synapsing in a sympathetic ganglion with several postganglionic neurons; (3) through white rami to and through sympathetic ganglia, to and through splanchnic nerves to collateral ganglia (celiac, superior, and inferior mesenteric ganglia), where they synapse with several postganglionic neurons
Postganglionic neurons	
Dendrites and cell bodies	In sympathetic or collateral ganglia
Axons (postganglionic fibers)	In autonomic nerves that form various plexuses before supplying thoracic and abdominal viscera and blood vessels in these body cavities, or
	Through gray rami to spinal nerves to cutaneous blood vessels, sweat glands, and smooth muscles of hair follicles
Parasympathetic division	
Preganglionic neurons	
Dendrites and cell bodies	In midbrain, pons, or medulla (or lateral gray columns of sacral cord)
Axons (preganglionic fibers)	From midbrain to third cranial nerve to ciliary ganglion, or
	From pons to seventh cranial nerve to sphenopalatine ganglion or submaxillary ganglion, or
	From medulla to (1) ninth cranial nerve to otic ganglion or (2) tenth and eleventh cranial nerves to cardiac and celiac ganglia
Postganglionic neurons	
Dendrites and cell bodies	In various ganglia (ciliary, sphenopalatine, submaxillary, otic, cardiac, and celiac) located in or near organs
Axons (postganglionic fibers)	In short nerve filaments to various viscera, glands, blood vessels, and intrinsic eye muscles

chain ganglia or in collateral ganglia. Their axons are distributed by both spinal nerves and separate autonomic nerves. They reach the spinal nerves via small filaments (gray rami) that connect the sympathetic ganglia with the spinal nerves. They then travel in the spinal nerves to blood vessels, sweat glands, and arrector hair muscles all over the body.

The course of postganglionic axons through autonomic nerves is somewhat more complex. These nerves form complicated plexuses before fibers are finally distributed to their respective destinations. For example, postganglionic fibers from the celiac and superior mesenteric ganglia pass through the celiac plexus before reaching the abdominal viscera, those from the inferior mesenteric ganglion pass through the hypogastric plexus on their way to the lower abdominal and pelvic viscera, and those from the cervical ganglion pass through the cardiac nerves and the cardiac plexus at the base of the heart and are then distributed to the heart.

Preganglionic neurons of parasympathetic system. The preganglionic neurons of the parasympathetic system have their cell bodies in nuclei in the brainstem or in the lateral gray columns of the sacral cord. Their axons are contained in cranial nerves III, VII, IX, X, and XI and in some pelvic nerves. They extend a considerable distance before synapsing with postganglionic neurons. For example, axons arising from cell bodies in the vagus nuclei (located in the medulla) travel in the vagus nerve for a distance of a foot or more before reaching their terminal ganglia in the chest and abdomen (Fig. 7-35 and Table 7-10).

Postganglionic neurons of parasympathetic system. The postganglionic neurons of the parasympathetic system have their dendrites and cell bodies in the outlying parasympathetic ganglia and send short axons into the nearby structures. Each parasympathetic preganglionic neuron synapses, therefore, only with postganglionic neurons to a single structure. For this reason, parasympathetic stimulation frequently involves response by only one organ as contrasted with sympathetic responses which, as noted before, usually involve numerous organs.

Some general principles

Following are some general principles about the autonomic nervous system.

1 *Principle of autonomic functioning and homeostasis.* The autonomic nervous system regulates the functioning of visceral effectors so they tend to maintain or quickly restore homeostasis. In healthy individuals, homeostasis is maintained under all but the most stressful conditions.

2 *Principle of dual autonomic innervation.* Most visceral effectors receive both sympathetic and parasympathetic fibers. Examples are cardiac muscle, smooth muscle of the iris, ciliary body, bronchial tubes, and digestive tract, and some glands (Fig. 7-35).

3 *Principle of single autonomic innervation.* Some visceral effectors are believed to receive only sympathetic fibers. Examples are the adrenal medulla, sweat glands, and probably the smooth muscle of hairs and most blood vessels.

4 *Principle of autonomic chemical transmitters.* Terminals of autonomic axons, like those of all axons, release chemicals that transmit impulses across synapses and neuroeffector junctions.* Autonomic axons fall into two classifications based on the chemical transmitters

---

*Much of our present knowledge about chemical transmitters stems from years of research by Sweden's Dr. Ulf S. von Euler, the United States' Dr. Julius Axelrod, and England's Sir Bernard Katz. For their work, these eminent scientists shared the 1970 Nobel prize in medicine and physiology. Their most significant findings included the following:

1 Identification of norepinephrine (NE) as the major transmitter in some brain synapses (by Dr. von Euler)
2 Identification of the enzyme catechol-O-methyl transferase (COMT) and discovery that it inactivates norepinephrine (by Dr. Axelrod)
3 Discovery that conducting cholinergic fibers rapidly release numerous packets of acetylcholine into synapses (by Sir Bernard Katz)

The above knowledge led to other discoveries and to valuable applications. For example, researchers later learned that severe psychic depression occurs when a deficit of NE exists in certain brain synapses. And this finding led to the development of antidepressant drugs. Certain ones of these inhibit COMT. Because inhibited COMT does not inactivate NE, the amount of active NE in brain synapses increases and this relieves the individual's depression. The graphic names, "psychic energizers" and "mood elevators," refer to antidepressant drugs, now valuable weapons in medicine's arsenal for combating mental disease.

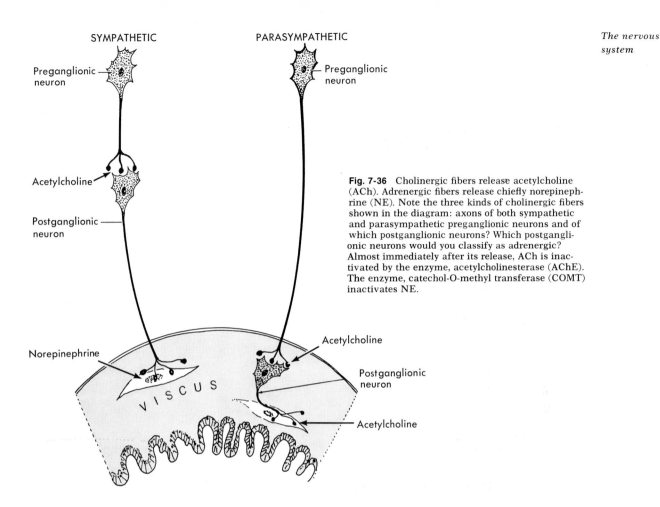

SYMPATHETIC   PARASYMPATHETIC

Preganglionic neuron

Preganglionic neuron

Acetylcholine

Postganglionic neuron

Norepinephrine

V I S C U S

Acetylcholine

Postganglionic neuron

Acetylcholine

Fig. 7-36 Cholinergic fibers release acetylcholine (ACh). Adrenergic fibers release chiefly norepinephrine (NE). Note the three kinds of cholinergic fibers shown in the diagram: axons of both sympathetic and parasympathetic preganglionic neurons and of which postganglionic neurons? Which postganglionic neurons would you classify as adrenergic? Almost immediately after its release, ACh is inactivated by the enzyme, acetylcholinesterase (AChE). The enzyme, catechol-O-methyl transferase (COMT) inactivates NE.

that they release. Some release acetylcholine so are classified as cholinergic fibers. Others release norepinephrine (and possibly some other substances) so are classified as adrenergic fibers.

The following are cholinergic autonomic fibers:

a All preganglionic axons

b Most, or perhaps all, parasympathetic postganglionic axons

c A few sympathetic postganglionic axons—namely, those to sweat glands, and also those to smooth muscle in the walls of certain blood vessels (that is, blood vessels in skeletal muscles and in the external genitalia)

The only adrenergic autonomic nerve fibers are sympathetic postganglionic axons. But not even all of these fibers are adrenergic. Sympathetic postganglionic axons to sweat glands, skeletal muscle blood vessels, and some skin vessels are cholinergic as mentioned in **c**.

5 *Principle of autonomic antagonism and summation.* Both sympathetic and parasympathetic impulses continually play on doubly innervated visceral effectors and tend to produce antagonistic effects on them. Stated differently, acetylcholine and its chief antagonist, norepinephrine, are continually released at the neuroeffector junctions of doubly innervated structures. Acetylcholine tends to make them respond in one way, while norepinephrine tends to make them respond in the opposite way. Hence, the algebraic sum of these opposing chemicals determines the actual response. Here is one example of this principle of autonomic antagonism in operation: parasympathetic fibers to the heart release acetylcholine which tends to slow the heart rate. At the same time, sympathetic fibers to the heart release norepinephrine, and this chemical tends to accelerate the heart rate. Therefore, at any one moment, the algebraic sum of these antagonists determines the actual rate of the heartbeat. If sympathetic impulses increase, the heart rate increases. If parasympathetic impulses increase, the heart rate decreases.

6 *Principle of parasympathetic dominance of digestive tract glands and smooth muscle.* Under normal conditions, parasympathetic impulses to the digestive glands and to the smooth muscle of the digestive tract dominate over sympathetic impulses to them. As shown in Table 7-11, parasympathetic impulses tend to increase digestive gland secretions and to stimulate the smooth muscle of the digestive tract. In short, parasympathetic impulses tend to promote digestion and peristalsis. They also tend to promote elimination (defecation and also urination).

7 *Principle of sympathetic dominance under stress conditions.* Under stress conditions, from either physical or emotional causes, sympathetic impulses to most visceral effectors increase greatly and cause them to respond in ways which enable the body to put forth its greatest physical effort, to expend its maximal amount of energy. Sympathetic dominance, to use Cannon's now classic phrase, prepares the body for "fight or flight." In fact, one of the very first steps in the body's complex defense mechanism against stress is a sudden and marked increase in sympathetic activity. Note the sympathetic effects listed in Table 7-11 and see if you can deduce the role each would play in making the body ready to fight or take flight.

Psychosomatic disorders illustrate an interesting principle—that emotions not outwardly expressed tend to be inwardly expressed. In other words, emotions not expressed through somatic effectors tend to be expressed through visceral effectors. You may have seen this principle operate in your own body. You have if, for example, you have ever been angry and gone for a walk to "work it off" and afterward felt "calmed down." Before your emotion expressed itself through your somatic effectors (skeletal muscle contractions during walking), it undoubtedly was expressing itself through your visceral effectors—your heart was probably beating faster than usual, your blood pressure may have increased, and your salivary glands probably were secreting less saliva than usual, perhaps so little that you were conscious of your mouth feeling dry.

Sympathetic impulses usually dominate the control of most visceral effectors in times of stress. But not always. Curiously enough, parasympathetic impulses frequently become excessive to some effectors at such times. For instance, one of the first symptoms of emotional stress in many individuals is that they feel hungry and want to eat more than usual. Presumably this is partly caused by increased parasympathetic impulses to the smooth muscle of the stomach. This stimulates increased gastric contractions which, in turn, may cause the feeling of hunger. The most famous disease of parasympathetic excess is peptic ulcer. In some individuals (presumably those born with certain genes), chronic stress leads to excessive parasympathetic stimulation of hydrochloric acid glands, with eventual development of peptic ulcer.

8 *Principle of nonautonomy.* The autonomic nervous system is not autonomous. It is neither anatomically nor physiologically independent of the rest of the nervous

Table 7-11. Autonomic functions

Visceral effectors	Parasympathetic (cholinergic) effects	Sympathetic (adrenergic or cholinergic) effects
Cardiac muscle	Slows heart rate; decreases strength of contraction	Accelerates rate; increases strength of contraction
Smooth muscle		
Of blood vessels in skin	No parasympathetic fibers	Adrenergic sympathetic fibers → stimulate → constrict skin vessels
Of blood vessels in skeletal muscles	No parasympathetic fibers	Adrenergic sympathetic fibers → stimulate → constrict skeletal muscle vessels
		Cholinergic sympathetic fibers → inhibit → dilate skeletal muscle vessels
Of blood vessels in brain, viscera, and genitalia	Parasympathetic fibers → inhibit → dilate vessels in brain, viscera, and genitalia	Adrenergic sympathetic fibers → stimulate → constrict vessels in brain and viscera
		Cholinergic sympathetic fibers → inhibit → dilate vessels in external genitalia
Of bronchi	Stimulates → bronchial constriction	Inhibits → bronchial dilatation
Of digestive tract	Stimulates → increased peristalsis	Inhibits → decreased peristalsis
Of anal sphincter	Inhibits → opens sphincter for defecation	Stimulates → closes sphincter
Of urinary bladder	Stimulates → contracts bladder	Inhibits → relaxes bladder
Of urinary sphincters	Inhibits → opens sphincter for urination	Stimulates → closes sphincter
Of eye (a) Iris	Stimulates circular fibers → constriction of pupil	Stimulates radial fibers → dilatation of pupil
(b) Ciliary	Stimulates → accommodation for near vision (bulging of lens)	Inhibits → accommodation for far vision (flattening of lens)
Of hairs (pilomotor muscles)	No parasympathetic fibers	Stimulates → "goose pimples"
Glands Sweat	No parasympathetic fibers	Cholinergic sympathetic fibers stimulate sweat glands
Digestive (salivary, gastric, etc.)	Stimulates secretion	Decreases secretion
Pancreas, including islets	Stimulates secretion	Decreases secretion
Liver	No parasympathetic fibers	Stimulates glycogenolysis which tends to increase blood sugar
Adrenal medulla	No parasympathetic fibers	Stimulates adrenalin secretion which tends to increase blood sugar, blood pressure, and heart rate and to produce many other sympathetic effects

system. It and all parts of the nervous system work together as a single functional unit. As evidence of the nonautonomy of the autonomic system, consider the following facts. All preganglionic neurons have their dendrites and cells located in the brain and cord gray matter, in the "lower autonomic centers." Moreover, fibers of numerous descending spinal cord tracts synapse with preganglionic neurons. And, therefore, impulses from various higher centers impinge upon and control the lower autonomic centers.

You may be wondering why the name autonomic system was ever chosen in the first place if the system is really not autonomous. Originally the term seemed appropriate. The autonomic system seemed to be self-regulating and independent of the rest of the nervous system. Common observations furnished abundant evidence of its independence from cerebral control, from direct control by the will, that is. But later, even this was found to be not entirely true. Some rare and startling exceptions were discovered. I have seen one such exception—a man who sat in a brightly lighted amphitheater in front of a class of medical students and made his pupils change from small, constricted dots (normal response to bright light) to widely dilated circles. This same man also willed gooseflesh to appear on his arms by contracting the smooth muscle of the hairs.

Higher autonomic centers

Reflex centers of the autonomic nervous system are classified as lower and higher centers depending upon their locations in the central nervous system. Lower autonomic centers lie in gray matter of the spinal cord or brainstem (medulla, pons, and midbrain). They consist essentially of synapses with dendrites and cell bodies of preganglionic neurons. Higher autonomic centers lie in various parts of the brain located above the brainstem—notably, in the hypothalamus and in the limbic system. Impulses from these centers are conducted to lower autonomic centers via various pathways. Hence, higher autonomic centers help control both sympathetic and parasympathetic functioning.

The term limbic comes from a word meaning border or fringe. And the *limbic system** consists mainly of cerebral cortex located on the medial surface of the brain and forming a border around the corpus callosum. The higher autonomic centers of the limbic system and hypothalamus function as important parts of the neural mechanisms that produce the responses characteristic of different emotions. As yet, however, these mechanisms are not clearly understood.

---

*Rhinencephalon (literally, "nose brain") is an older name for about the same structures now known as the limbic system.

---

outline summary

CELLS
Neuroglia
1 Types—astrocytes, oligodendroglia, and microglia
2 Structure and function
    a Astrocytes—star-shaped; numerous processes twine around neurons and attach them to blood vessels; support neurons and may carry on phagocytosis

    b Oligodendroglia—fewer processes than other two types of neuroglia; support neurons, connect them to blood vessels
    c Microglia—small cells but enlarge and move about in inflamed brain tissue; carry or phagocytosis

Neurons

1 Structure—see Figs. 7-2 and 7-3
2 Types
   a Classified according to direction of impulse conduction
     1 Sensory or afferent—conduct impulses to cord or brain
     2 Motoneurons or efferent—conduct impulses away from brain or cord to or toward muscle or glandular tissue
     3 Interneurons (internuncial or intercalated neurons)—conduct from sensory to motoneurons
   b Classified according to number of processes
     1 Multipolar—one axon and several dendrites
     2 Bipolar—one axon and one dendrite
     3 Unipolar—one process comes off neuron cell body but divides almost immediately into one axon and one dendrite
3 Function—respond to stimulation by conducting impulses
4 Nonconducting (resting) neuron
   a Membrane polarized—i.e., difference in electrical charge exists between inner and outer surfaces of resting neuron's cell membrane; specifically, outer surface of membrane about 70 to 90 mv positive to inner surface; polarization results from and depends upon operation of sodium pump
   b Potential difference (difference in electrical charge) existing across membrane of resting neuron called resting potential

NERVE IMPULSE (ACTION POTENTIAL)

1 Definition—self-propagating wave of electrical negativity that travels along surface of neuron membrane
2 Mechanism
   a Stimulus increases permeability of neuron membrane to sodium ions
   b Rapid inward diffusion of sodium ions causes outer surface of membrane to become negative to inner surface; this reversal of resting potential marks beginning of action potential or nerve impulse
3 Initiation of nerve impulse by threshold or stronger stimulus acting on dendrites or cell body of neuron
   a Threshold stimulus—just strong enough to decrease potential across neuron membrane from level of resting potential (e.g., outer surface of membrane 70 mv positive to inner) down to critical level (e.g., outer surface of membrane 30 mv negative to inner) that initiates acting potential
   b Subthreshold stimulus—decreases membrane potential below resting level but not down to threshold level

4 Conduction
   a Course (neural pathways)
     1 Generally, nerve impulse initiated in receptors (distal ends of sensory neurons), conducted over reflex arcs, and terminates in effectors (muscles and glands) causing reflex response
     2 Two-neuron or monosynaptic reflex arc—simplest arc possible; consists of at least one sensory neuron, one synapse, and one motoneuron (synapse is contact points between axon terminals of one neuron and dendrites or cell body of another neuron)
     3 Three-neuron arc—consists of at least one sensory neuron, synapse, interneuron, synapse, and motoneuron
     4 Complex, multisynaptic neural pathways also exist; many not clearly understood
   b Rate of conduction—see Table 7-1
   c Conduction across synapses
     1 When impulse reaches axon terminals, they eject chemical transmitter substance into synaptic cleft
     2 Chemical released by axon terminals of some neurons lowers postsynaptic neuron's resting potential; threshold or larger amounts of chemical excite neuron—i.e., stimulate impulse conduction; subthreshold amounts of chemical produce excitatory postsynaptic potential—i.e., facilitate neuron but do not stimulate it to conduct; acetylcholine is chemical released by some facilitatory axons
     3 Chemical released by terminals of some axons produces inhibitory postsynaptic potential—i.e., increases resting potential or inhibits postsynaptic neuron; inhibitory chemical unknown

ORGANS

1 Names—spinal cord, brain, and nerves
2 Definitions
   a White matter—bundles of myelinated nerve fibers
   b Gray matter—masses of neuron cell bodies mainly
   c Nerves—bundles of myelinated nerve fibers located outside central nervous system
   d Tracts—bundles of myelinated nerve fibers located in brain and cord
   e Ganglia (singular, ganglion)—macroscopic structures consisting of neuron cell bodies mainly; located outside central nervous system
   f Nuclei, centers, horns—consist mainly of neuron cell bodies (gray matter) located inside central nervous system
   g CNS (central nervous system)—brain and spinal cord
   h PNS (peripheral nervous system)—nerves and ganglia

Spinal cord

1 Location—in spinal cavity, from foramen magnum to first lumbar vertebra
2 Structure
  a Two deep grooves, anterior median fissure and posterior median sulcus, incompletely divide cord into right and left symmetrical halves; anterior median fissure deeper and wider than posterior groove
  b Core of gray matter shaped like three-dimensional letter H
  c White matter present in columns, anterior, lateral, and posterior, composed of numerous projection tracts
3 Functions
  a Sensory and motor conduction pathways between peripheral nerves and brain; for names and functions of tracts, see Fig. 7-2
  b Composite of reflex centers—for all spinal cord reflexes

Brain

Major parts of brain—cerebrum, diencephalon, cerebellum, medulla oblongata, pons varolii, and midbrain

Cerebrum

1 Hemispheres, fissures, and lobes—longitudinal fissure divides cerebrum into two hemispheres, connected only by corpus callosum; each cerebral hemisphere divided by fissures into five lobes: frontal, parietal, temporal, occipital, and island of Reil (insula)
2 Cerebral cortex—outer layer of gray matter arranged in ridges called convolutions or gyri
3 Cerebral tracts—bundles of axons compose white matter in interior of cerebrum; ascending projection tracts transmit impulses toward or to brain; descending projection tracts transmit impulses down from brain to cord; commissural tracts transmit from one hemisphere to other; association tracts transmit from one convolution to another in same hemisphere
4 Basal ganglia (or cerebral nuclei)—masses of gray matter embedded deep inside white matter in interior of cerebrum; caudate, putamen, and pallidum; putamen and pallidum constitute lenticular nucleus
5 Functions—in general, all conscious functions—e.g., analysis, integration, and interpretation of sensations, control of voluntary movements, use and understanding of language, and all other mental functions

Diencephalon
Thalamus

1 Structure and location—large rounded mass of gray matter in each cerebral hemisphere, lateral to third ventricle; composed of many nuclei

2 Functions
  a Conscious recognition of crude sensations of pain, temperature, and touch
  b Relays afferent impulses to cerebral cortex
  c Involved in emotional component of sensations; feelings of pleasantness or unpleasantness
  d Involved in alerting or arousal mechanism
  e Involved in production of complex reflex movements

Hypothalamus

1 Structure and location
  a Gray matter around optic chiasma, pituitary stalk, posterior lobe of pituitary gland, mamillary bodies, and adjacent regions; made up of many nuclei, notably supraoptic, paraventricular, and mamillary
  b Afferent tracts conduct impulses to hypothalamus from cerebral cortex, thalamus, and basal ganglia
  c Efferent tracts conduct from hypothalamus to thalamus and to autonomic centers in brainstem and cord
2 Functions
  a Higher autonomic center; helps control and integrate autonomic functions
  b Relay station between cerebral cortex and lower autonomic centers; crucial part of neural paths by which emotions influence bodily functions
  c Important part of mechanism for controlling anterior pituitary gland
  d Crucial part of mechanism for maintaining water balance
  e Essential part of arousal or alerting mechanism; essential for maintaining waking state
  f Crucial part of mechanism for regulating appetite and food intake
  g Probably helps control various reproductive functions
  h Crucial part of mechanism for maintaining normal body temperature

Cerebellum

1 Structure and location
  a Second largest part of human brain
  b Has two hemispheres and center section, vermis
  c Surface grooved with sulci
  d Slightly raised, slender convolutions
  e Internal white matter arranged in pattern like veins of leaf
  f Three pairs of tracts in cerebellum—inferior, middle, and superior cerebellar peduncles
  g Most important cerebellar nucleus—dentate nucleus
2 Functions
  a Synergic control of muscle action
  b Postural reflexes
  c Equilibrium

Medulla oblongata

1 Structure and location
  a Part of brain formed by enlargement of cord as it enters cranial cavity
  b Mainly white matter (projection tracts); also reticular formation (interlacement of gray and white matter, containing many nuclei)
  c Various autonomic centers in reticular formation, e.g., cardiac and vasomotor; also respiratory, vomiting, coughing, hiccoughing, sneezing, and swallowing centers
2 Functions
  a Helps control heartbeat, blood pressure, and respirations
  b Mediates reflexes of vomiting, coughing, hiccoughing, etc.
  c Conducts impulses between cord and brain

Pons varolii

1 Structure and location
  a Located just above medulla
  b White matter with few nuclei
2 Functions
  a Contains projection tracts between cord and various parts of brain
  b Centers for fifth to eighth cranial nerves

Midbrain

1 Location and structure
  a Located above pons, below cerebrum and diencephalon
  b White matter with few nuclei
  c Cerebral aqueduct within midbrain
  d Cerebral peduncles are two tracts composing ventral part of midbrain and connecting pons to cerebrum
  e Corpora quadrigemina—four rounded eminences (two superior and two inferior colliculi) on dorsal surface of midbrain
  f Important nucleus—red nucleus
2 Functions
  a Projection tracts function in sensations and movements
  b Centers for third and fourth cranial nerves, hence for pupillary reflexes and eye movements

Cord and brain coverings

1 Bony—vertebrae around cord; cranial bones around brain
2 Membranous—called meninges and consist of three layers
  a Dura mater—white fibrous tissue outer layer
  b Arachnoid membrane—cobwebby middle layer
  c Pia mater—transparent; adherent to outer surface of cord and brain; contains blood vessels; therefore, nutritive layer

Cord and brain fluid spaces

1 Subarachnoid space around cord
2 Subarachnoid space around brain
3 Central canal inside cord
4 Ventricles and cerebral aqueduct inside brain—four cavities within brain
  a First and second (lateral) ventricles—large cavities, one in each cerebral hemisphere
  b Third ventricle—vertical slit in cerebrum beneath corpus callosum and longitudinal fissure
  c Fourth ventricle—diamond-shaped space between cerebellum and medulla and pons; is expansion of central canal of cord

Formation and circulation of cerebrospinal fluid

1 Formed by plasma filtering from network of capillaries (choroid plexus) in each ventricle
2 Circulates from lateral ventricles to third ventricle, cerebral aqueduct, fourth ventricle, central canal of cord, subarachnoid space of cord and brain; venous sinuses

Spinal nerves

1 Thirty-one pairs
  a Eight cervical
  b Twelve thoracic
  c Five lumbar
  d Five sacral
  e One coccygeal
2 Origin—originate by anterior and posterior roots from cord, emerge through intervertebral foramina; spinal ganglion on each posterior root
3 Distribution—branches distributed to skin, mucosa, skeletal muscles; branches form plexuses, such as brachial plexus, from which nerves emerge to supply various parts
4 Microscopic structure—consist of sensory dendrites and motor axons—i.e., are mixed nerves; also contain autonomic postganglionic fibers; see Table 7-4 for peripheral branches
5 Functions—see Table 7-5

Cranial nerves

1 Twelve pairs
  a I (olfactory)
  b II (optic)
  c III (oculomotor)
  d IV (trochlear)
  e V (trifacial or trigeminal)
  f VI (abducens)
  g VII (facial)
  h VIII (acoustic or auditory)
  i IX (glossopharyngeal)
  j X (vagus or pneumogastric)
  k XI (accessory or spinal accessory)
  l XII (hypoglossal)
2 See Table 7-6 for distribution and function

SENSORY NEURAL PATHWAYS

1 Three-neuron relay conducts impulses from periphery to cerebral cortex
   a Sensory neuron I conducts from periphery to cord or to brainstem
   b Sensory neuron II conducts from cord or brainstem to thalamus
   c Sensory neuron III conducts from thalamus to general sensory area of cerebral cortex (areas 3, 1, 2)
2 Crude awareness of sensations occurs when impulses reach thalamus
3 Full consciousness of sensations with accurate localization and discrimination occurs when impulses reach cerebral cortex
4 Most sensory neuron II axons decussate; so one side of brain registers mainly sensations for opposite side of body
5 Principle of divergence applies to sensory impulse conduction; each sensory neuron synapses with more than one neuron; hence impulses over any sensory neuron diverge and may activate many effectors
6 Pain and temperature—lateral spinothalamic tracts to thalamus
7 Touch and pressure
   a Discriminating touch and pressure (stereognosis, precise localization, vibratory sense)—medial lemniscal system (fasciculi cuneatus and gracilis to medulla; medial lemniscus to thalamus)
   b Crude touch and pressure—ventral spinothalamic tracts
8 Conscious proprioception or kinesthesia—medial lemniscal system to thalamus
9 See also Table 7-7

AROUSAL OR ALERTING MECHANISM

Cerebral cortex stimulated by impulses from reticular formation which is stimulated by sensory impulses of all kinds reaching it

MOTOR NEURAL PATHWAYS TO SKELETAL MUSCLES

1 Principle of final common path—motoneurons in anterior gray horns of cord constitute final common path for impulses to skeletal muscles; are only neurons transmitting impulses into skeletal muscles
2 Principle of convergence—axons of many neurons synapse with each anterior horn motoneuron
3 Motor pathways from cerebral cortex to anterior horn cells classified according to way fibers enter cord:
   a Pyramidal (or corticospinal) tracts—dendrites and cells in cortex; axons descend through internal capsule, pyramids of medulla, and spinal cord and synapse directly with anterior horn cells or indirectly via internuncial neurons; impulses via pyramidal tracts essential for voluntary contractions of individual muscles to produce small discrete movements; also help maintain muscle tone
   b Extrapyramidal tracts—all pathways between motor cortex and anterior horn cells, except pyramidal tracts; upper extrapyramidal tracts relay impulses between cortex, basal ganglia, thalamus, and brainstem; reticulospinal tracts main lower extrapyramidal tracts; impulses via extrapyramidal tracts essential for large, automatic movements; also for facial expressions and movements accompanying many emotions
4 Motor pathways from cerebral cortex to anterior horn cells also classified according to influence on anterior horn cells:
   a Facilitatory tracts—have facilitating or stimulating effect on anterior horn cells; all pyramidal tracts and some extrapyramidal tracts facilitatory, notably facilitatory reticulospinal fibers
   b Inhibitory tracts—have inhibiting effect on anterior horn cells; inhibitory reticulospinal fibers main inhibitory tracts
5 Ratio of facilitatory and inhibitory impulses impinging on anterior horn cells controls activity; normally, slight predominance of facilitatory impulses maintains muscle tone

REFLEXES

1 Definition—action resulting from conduction over reflex arc; reflex is response (either muscle contraction or glandular secretion) to stimulus; term reflex usually used to mean only involuntary responses
2 Some reflexes of clinical importance (see also pp. 221 to 222)
   a Knee jerk
   b Ankle jerk
   c Babinski reflex
   d Corneal reflex
   e Abdominal reflex

AUTONOMIC NERVOUS SYSTEM

1 Definition—part of nervous system that sends efferent fibers to visceral effectors; specifically, to smooth muscle, cardiac muscle, and glands
2 Divisions—sympathetic (or thoracolumbar) and parasympathetic (or craniosacral)
3 Macroscopic structure
   a Sympathetic division consists of two chains of ganglia (one on either side of backbone) and fibers that connect ganglia with each other and with thoracic and lumbar segments of cord; other fibers extend from sympathetic ganglia out to visceral effectors
   b Parasympathetic division consists of ganglia located on or near viscera with fibers between ganglia and brainstem and between ganglia and sacral region of cord; also fibers from ganglia into viscera and glands

4 Microscopic structure
   a Cell bodies of preganglionic neurons of sympathetic system in lateral gray columns of thoracic and lumbar segments of cord; lower sympathetic centers also located here
   b Cell bodies of postganglionic neurons of sympathetic system in sympathetic chain ganglia or in collateral ganglia (celiac, superior, and inferior mesenteric)
   c Cells of preganglionic neurons of parasympathetic system in various nuclei of brainstem and in gray matter of sacral segments of cord; lower parasympathetic centers also located here
   d Cells of postganglionic neurons of parasympathetic system in ganglia on or near organs innervated
5 Some general principles
   a Autonomic functioning and homeostasis—autonomic system so regulates activities of visceral effectors as to maintain or quickly restore homeostasis in healthy individuals, except under highly stressful conditions
   b Dual autonomic innervation—both sympathetic and parasympathetic fibers supply most visceral effectors
   c Single autonomic innervation—sympathetic fibers but not parasympathetic fibers to adrenal medulla, sweat glands, and probably to smooth muscles of hairs and most blood vessels
   d Autonomic chemical transmitters—autonomic axon terminals release acetylcholine at synapses and either acetylcholine or norepinephrine at neuroeffector junctions; those which release acetylcholine called cholinergic fibers; those which release norepinephrine called adrenergic fibers
      1 All preganglionic axons are cholinergic fibers, as are most, or perhaps all, parasympathetic postganglionic axons and few sympathetic postganglionic axons (to sweat glands and to smooth muscle in walls of blood vessels in skeletal muscles and external genitalia)
      2 Only adrenergic autonomic fibers are sympathetic postganglionic axons but even few of these cholinergic (see preceding item)
   e Autonomic antagonism and summation—sympathetic and parasympathetic impulses tend to produce opposite effects; algebraic sum of two opposing tendencies determines response made by doubly innervated visceral effector
   f Parasympathetic dominance of digestive tract—normally, parasympathetic impulses to glands and smooth muscle of digestive tract dominate over sympathetic impulses to them; parasympathetic impulses promote digestive gland secretion, peristalsis, and defecation
   g Sympathetic dominance in stress—under stress conditions, sympathetic impulses to most visceral effectors increase greatly, causing them to respond in ways which enable body to expend maximal amount of energy; sympathetic dominance one of first defenses against stress, parasympathetic impulses, however, may become excessive to some effectors under stress conditions
   h Nonautonomy—autonomic nervous system not autonomous with relation to rest of nervous system; partly controlled by higher autonomic centers, notably in hypothalamus and limbic system
6 Higher autonomic centers
   a Located in various parts of brain above brainstem, notably in hypothalamus and limbic system
   b Help control both sympathetic and parasympathetic functioning; play part in expression of emotion

review questions

1 In a word or two, what general function does the nervous system perform?
2 What other system serves the same general function?
3 Compare neurons and neuroglia.
4 Make an identifying statement about each of the three types of neurons.
5 What function does the neurilemma serve?
6 What neurons do not have a neurilemma?
7 What microscopic structures compose gray matter? White matter?
8 Agree or disagree with the following statement and give your reasons: "Neurons located in the brain or cord, once destroyed, do not regenerate."
9 Briefly explain the following terms: resting potential; action potential; potential difference; depolarized.
10 What term in the preceding question is a synonym for nerve impulse?
11 What are receptors? Effectors?
12 What is a synapse?
13 Name the chemical transmitter released into synapses of autonomic ganglia. What structures release this transmitter—dendrites? axons? of preganglionic or of postganglionic neurons?
14 What function does cholinesterase serve?
15 Where is a lumbar puncture done? Why?
16 What general name is given to the membranous coverings of the brain and cord? What three layers compose this covering?
17 What are the cavities inside the brain called? How many are there? What do they contain?

18 Describe the circulation of cerebrospinal fluid.

19 What is the function of the cerebrospinal fluid?

20 What name is given to the outer portion of the cerebrum? Describe its appearance. What structure connects its hemispheres?

21 Compare nerves and tracts.

22 What are cerebral nuclei?

23 State at least one specific function performed by each of the lobes of the cerebral cortex: frontal, parietal, temporal, and occipital.

24 Explain the system for naming individual tracts.

25 Based on your explanation in answer to question 20, what is a spinothalamic tract?

26 Compare nucleus and ganglion.

27 What general functions does the cerebral cortex perform?

28 What is the diencephalon? What are its two main parts?

29 What general function does the thalamus perform?

30 What general functions does the hypothalamus perform?

31 What general functions does the cerebellum perform?

32 What general functions does the medulla perform?

33 What general functions does the spinal cord perform?

34 Name several ascending or sensory tracts in the spinal cord.

35 Name two descending or motor tracts in the spinal cord.

36 What is a reflex center?

37 Which cranial nerves transmit impulses that result in vision? In eye movements? Hearing? Taste sensations? Slowing of the heart?

38 What is the great sensory nerve of the head?

39 Digitalis, a drug said to have a stimulating effect on the vagus nerve, has been administered to a patient. What effect, if any, would you expect it to have on the patient's pulse rate?

40 Locate the dendrite, cell body, and axon of sensory neurons (a) I, (b) II, and (c) III.

41 Explain briefly the arousal or alerting mechanism.

42 Explain each of the following: lower motoneuron, upper motoneuron, and final common path.

43 Name the tracts that transmit impulses from each of the following types of receptors to the brain and state in which column of the cord each tract is located: pain and temperature receptors; crude touch receptors; proprioceptors (name two tracts for these); discriminating touch receptors (two tracts).

44 Compare pyramidal tract and extrapyramidal tract functions.

45 Explain briefly the general function of the autonomic nervous system; of the sympathetic system; of the parasympathetic system.

46 Sympathetic stimulation produces massive, widespread responses, whereas reactions to parasympathetic stimulation are often highly localized. What anatomical differences between the two systems explain this physiological difference?

47 What transmitter substance do axons of sympathetic postganglionic neurons release at visceral neuroeffector junctions?

48 Identify COMT by name, class of compound, and action.

49 Contrast these two neural paths: from the central nervous system to somatic effectors; from the central nervous system to visceral effectors.

50 Classify the following structures as somatic effectors, visceral effectors, or neither: adrenal glands, biceps femoris muscle, heart, iris, and skin.

51 Which of the following would indicate an increase in sympathetic impulses and which might indicate an increase in parasympathetic impulses to visceral effectors—constipation? dilated pupils? dry mouth? goose pimples? "I'm always hungry and eat too much when I am upset"? rapid heartbeat?

52 What is the limbic system, and what general function does it perform?

8

Sense organs

General remarks

**Somatic, visceral, and
    referred pain**

Eye
Anatomy
    Coats of eyeball
    Cavities and humors
    Muscles
    Accessory structures
        Eyebrows and eyelashes
        Eyelids
        Lacrimal apparatus
Physiology of vision
    Formation of retinal
        image
    Stimulation of retina
    Conduction to visual area

Auditory apparatus
Anatomy
    External ear
    Middle ear
    Inner ear
        Vestibule, utricle, and
            saccule
        Cochlea and cochlear
            duct
        Semicircular canals
Physiology
    Hearing
    Equilibrium

Olfactory sense organs

Gustatory sense organs

■ General remarks

The body has millions of sense organs. All of its receptors, the beginnings of dendrites of all its sensory neurons, are its sense organs. They serve two vital general functions—sensations and reflexes. All sensations and all reflexes result from stimulation of receptors. In short, receptors are the structures that detect changes in our external and internal environments and that initiate the responses necessary for adjusting the body to these changes so as to maintain or restore homeostasis. One more point—a matter more of interest than importance—we have more than just the "five senses," vision, hearing, taste, smell, and touch. For example, some of our other senses are warmth, cold, pain, and proprioception.

Receptors are located all over the body, inside as well as on its surfaces. Receptors have been classified according to their location as exteroceptors, visceroceptors, and proprioceptors. Exteroceptors are surface receptors. They are located in the skin, mucosa, eye, and ear. Visceroceptors and proprioceptors are both located internally. Visceroceptors are found, for example, in the walls of blood vessels, stomach, intestines, and various other organs. Proprioceptors are located in muscles, tendons, joints, and the internal ear.

Structurally, receptors differ considerably. Some, such as those in the eye and

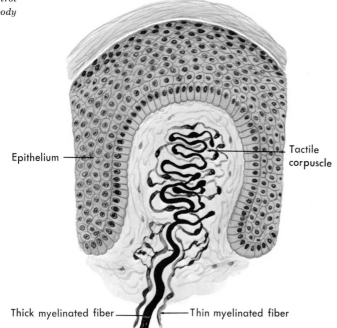

Epithelium —

Tactile
corpuscle

Thick myelinated fiber —

— Thin myelinated fiber

Fig. 8-1 A Meissner's corpuscle (tactile corpuscle) found in the connective tissue papillae of the skin. It is an example of an encapsulated specialized nerve ending found in the hairless portions of the skin.

Table 8-1. Receptors and sensations which they mediate

Sensations	Receptors
Touch	Meissner's corpuscles (Fig. 8-1) Merkel's disks Basketlike arrangements around bases of hairs
Pressure	Vater-Pacinian corpuscles
Heat	Corpuscles of Ruffini
Cold	Krause end bulbs
Pain	Naked nerve fibers
Proprioception	Neuromuscular spindles Neurotendinous spindles Ruffini endings in joint capsules

ear, constitute highly specialized sense organs. Others are simply naked or free nerve endings while other are encapsulated (Fig. 8-1).

According to the principle of specificity of receptors, specific kinds of receptors mediate specific sensations because they are sensitive to specific kinds of stimuli. Although generally accepted, this principle is now questioned by some investigators who think one kind of receptor may mediate more than one kind of sensation and that one kind of sensation may be mediated by more than one kind of receptor. Generally accepted ideas about which types of receptors mediate which sensations are included in Table 8-1.

■ Somatic, visceral, and referred pain

Because stimulation of pain receptors may give warning of potentially harmful environmental changes, pain receptors are also called *nociceptors* (L. *noceo*, to injure). Any type of stimulus, provided that it be sufficiently intense, seems to be adequate for stimulating nociceptors in the skin and mucosa. In contrast, only marked changes in pressure and certain chemicals can stimulate nociceptors located in the viscera. Sometimes this knowledge proves useful. For example, it enables a physician to cauterize the uterine cervix without giving an anesthetic but with assurance that the patient will not suffer pain from the intense heat. On the other hand, he knows that if the intestine becomes markedly distended (as it sometimes does following surgery), the patient will experience pain. So, too, will the individual whose heart becomes ischemic due to coronary occlusion. Presumably, the resulting cellular oxygen deficiency leads to the formation or accumulation of chemicals that stimulate nociceptors in the heart.

Two main types of pain are recognized: somatic and visceral.

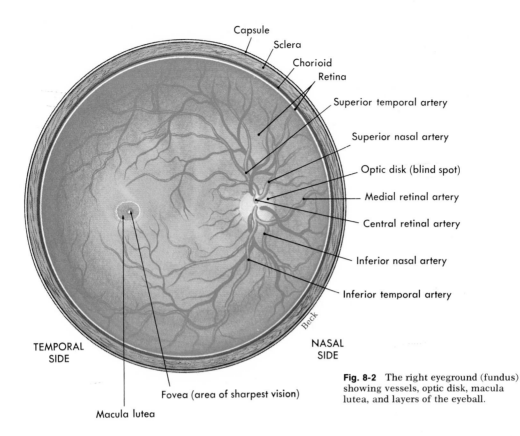

Capsule
Sclera
Chorioid
Retina
Superior temporal artery
Superior nasal artery
Optic disk (blind spot)
Medial retinal artery
Central retinal artery
Inferior nasal artery
Inferior temporal artery
Beck
TEMPORAL SIDE
NASAL SIDE
Fovea (area of sharpest vision)
Macula lutea

Fig. 8-2 The right eyeground (fundus) showing vessels, optic disk, macula lutea, and layers of the eyeball.

Somatic pain may be *superficial,* as when it arises from stimulation of skin receptors, or it may be *deep,* as when it results from stimulation of receptors in the skeletal muscles, fascia, tendons, and joints.

Visceral pain results from stimulation of receptors located in the viscera. Impulses are conducted from these receptors to the cord primarily by sensory fibers in sympathetic nerves and only rarely in parasympathetic nerves.

The cerebrum does not always interpret the source of pain accurately. Sometimes it erroneously refers the pain to a surface area instead of to the region in which the stimulated receptors actually are located. This phenomenon is called *referred pain.* It occurs only as a result of stimulation of pain receptors located in deep structures (notably, viscera, joints, and skeletal muscles)—never from stimulation of skin receptors. In other words, deep somatic

pain and visceral pain may be referred but not superficial somatic pain.

According to one theory, pain originating in the viscera and other deep structures is interpreted as coming from the skin area whose sensory fibers enter the same segment of the spinal cord as the sensory fibers from the deep structure. For example, sensory fibers from the heart enter the first to fourth thoracic segments. And so do sensory fibers from the skin areas over the heart and on the inner surface of the left arm. Pain originating in the heart, therefore, is referred to those skin areas. Several other theories have also been advanced to explain referred pain.

◼ Eye

Anatomy

COATS OF EYEBALL

Approximately five-sixths of the eyeball lies recessed in the orbit, protected by this bony socket. Only its small anterior sur-

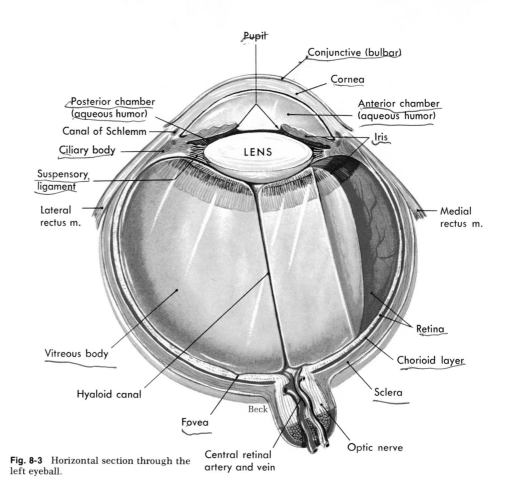

Fig. 8-3 Horizontal section through the left eyeball.

face is exposed. Three layers of tissues or coats compose the eyeball. From the outside in they are the sclera, the chorioid,* and the retina. Both the sclera and the chorioid coats consist of an anterior and a posterior portion.

Tough white fibrous tissue fashions the *sclera*. Deep within the anterior part of the sclera at its junction with the cornea lies a ring-shaped venous sinus, the *canal of Schlemm* (Figs. 8-6 to 8-8). The anterior portion of the sclera is called the *cornea* and lies over the colored part of the eye (iris). The cornea is transparent, whereas the rest of the sclera is white and opaque, a fact which explains why the visible

*The spelling *chorioid* is considered etymologically preferable to the widely used choroid.

anterior surface of the sclera is usually spoken of as the "whites" of the eyes. No blood vessels are found in the cornea, in the aqueous and vitreous humors, or in the lens.

The middle or *chorioid coat* of the eye contains a great many blood vessels and a large amount of pigment. Its anterior portion is modified into three separate structures: the ciliary body, the suspensory ligament, and the iris.

The *ciliary body* is formed by a thickening of the chorioid and fits like a collar into the area between the anterior margin of the retina and the posterior margin of the iris. The small *ciliary muscle*, composed of both radial and circular smooth muscle fibers, lies in the anterior part of the ciliary body. Attached to the ciliary body is the *suspensory ligament*, which

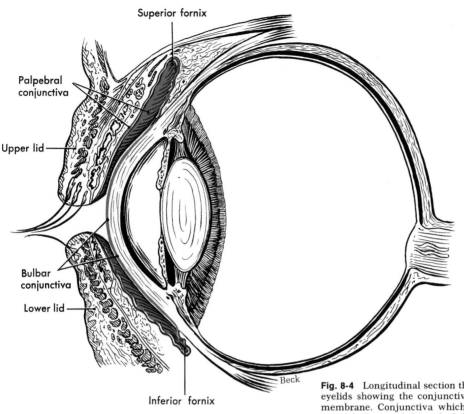

Superior fornix

Palpebral
conjunctiva

Upper lid

Bulbar
conjunctiva

Lower lid

Inferior fornix

Beck

Fig. 8-4 Longitudinal section through the eyeball and eyelids showing the conjunctiva, a lining of mucous membrane. Conjunctiva which covers the cornea is called bulbar, and that which lines the posterior surface of the eyelids is called palpebral.

blends with the elastic capsule of the *lens* and holds it suspended in place.

The *iris* or colored part of the eye consists of circular and radial smooth muscle fibers arranged so as to form a doughnut-shaped structure (the hole in the middle is called the *pupil*). The iris attaches to the ciliary body.

The *retina* is the incomplete innermost coat of the eyeball—incomplete in that it has no anterior portion. It consists mainly of nervous tissue and contains three layers of neurons (Fig. 8-5). Named in the order in which they conduct impulses, they are photoreceptor neurons, bipolar neurons, and ganglion neurons. The beginning of the dendrites of the photoreceptor neurons have been given names descriptive

of their shapes. Because some look like tiny rods and others like cones, they are called *rods* and *cones*, respectively. They constitute our visual receptors, structures highly specialized for stimulation by light rays (discussed on pp. 253 to 254). They differ as to numbers, distribution, and function. The estimated number of cones is 7,000,000 and that of rods, somewhere between ten and twenty times as many. Cones are most densely concentrated in the *fovea centralis*, a small depression in the center of a yellowish area, the *macula lutea*, found near the center of the retina. They become less and less dense from the fovea outward. Rods, on the other hand, are absent entirely from the fovea and macula and increase in density toward the periph-

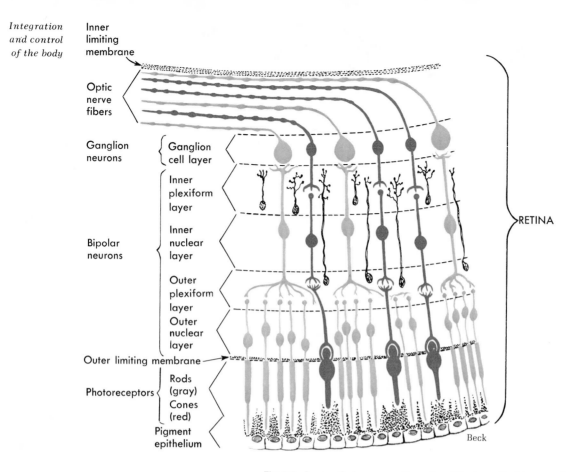

Fig. 8-5 Layers that compose the retina. The inner limiting membrane lies nearest the inside of the eyeball. It adheres to the vitreous humor. The pigment epithelium lies farthest from the inside of the eyeball. It adheres to the chorioid coat. Note relay of three neurons in the retina: photoreceptor, bipolar, and ganglion neurons (named in order of impulse transmission). Light rays pass through the vitreous humor and various layers of retina to stimulate rods and cones, the receptors of the photoreceptor neurons.

Table 8-2. Coats of the eyeball

Location	Posterior portion	Anterior portion	Characteristics
Outer coat (sclera)	Sclera proper	Cornea	Protective fibrous coat; cornea transparent; rest of coat white and opaque
Middle coat (chorioid)	Chorioid proper	Ciliary body; suspensory ligament; iris (pupil is hole in iris); lens suspended in suspensory ligament	Vascular, pigmented coat
Inner coat (retina)	Retina	No anterior portion	Nervous tissue; rods and cones (receptors for second cranial nerve) located in retina

ery of the retina. How these anatomical facts relate to rod and cone functions is revealed on p. 254.

All the axons of ganglion neurons extend back to a small circular area in the posterior part of the eyeball known as the *optic disk* or papilla. This part of the sclera contains perforations through which the fibers emerge from the eyeball as the *optic nerves*. The optic disk is also called the *blind spot* because light rays striking this area cannot be seen since it contains no rods or cones, only nerve fibers.

An outline summary of the coats of the eyeball is presented in Table 8-2.

CAVITIES AND HUMORS

The eyeball is not a solid sphere but contains a large interior cavity that is divided into two cavities, anterior and posterior.

The *anterior cavity* has two subdivisions known as the *anterior* and *posterior chambers*. The entire anterior cavity lies in front of the lens. The posterior chamber of the anterior cavity consists of the space directly posterior to the iris but anterior to the lens. And the anterior chamber of the anterior cavity is the small space anterior to the iris but posterior to the cornea. *Aqueous humor* fills both chambers of the anterior cavity. This substance is clear and watery and often leaks out when the eye is injured.

The *posterior cavity* of the eyeball is considerably larger than the anterior since it occupies all the space posterior to the lens, suspensory ligament, and ciliary body. It contains *vitreous humor*, a substance with a consistency comparable to soft gelatin. This semisolid material helps maintain sufficient intraocular pressure to prevent the eyeball from collapsing. (An obliterated artery, the hyaloid canal, runs through the vitreous humor between the lens and optic disk.)

An outline summary of the cavities of the eye is included in Table 8-3 (also see Fig. 8-3).

Still not established is the mechanism by which aqueous humor forms. It comes from blood in capillaries (located mainly in the ciliary body). Presumably, the capillaries actively secrete aqueous humor into the posterior chamber. But also passive filtration from capillary blood may contribute to aqueous humor formation. From the posterior chamber, aqueous humor moves between the iris and the lens and on through the pupil into the anterior chamber (Fig. 8-6). From here it drains into the canal of Schlemm and moves on into small veins. Normally, aqueous humor drains out of the anterior chamber at the same rate at which it enters the posterior chamber. So the amount of aqueous humor in the eye remains relatively constant. And so, too, does intraocular pressure. But sometimes something happens to upset this balance and intraocular pressure increases above the normal level of about

Table 8-3. Cavities of the eye

Cavity	Divisions	Location	Contents
Anterior	Anterior chamber Posterior chamber	Anterior to iris and posterior to cornea Posterior to iris and anterior to lens	Aqueous humor Aqueous humor
Posterior	None	Posterior to lens	Vitreous humor

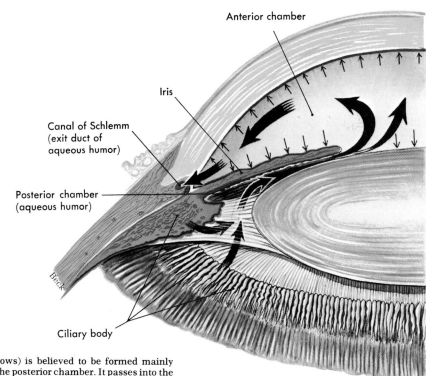

Fig. 8-6 Aqueous humor (heavy arrows) is believed to be formed mainly by secretion by the ciliary body into the posterior chamber. It passes into the anterior chamber through the pupil, from which it is drained away by the ring-shaped canal of Schlemm, and finally into the anterior ciliary veins. Small arrows indicate pressure of the aqueous humor.

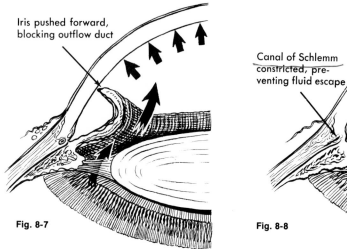

Fig. 8-7

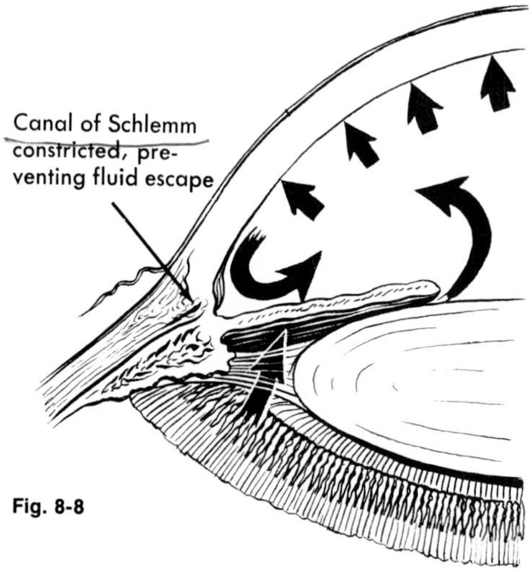

Fig. 8-8

Figs. 8-7 and 8-8 Acute glaucoma. When pressure of the aqueous humor in the anterior chamber becomes extreme, the iris is pushed forward, causing it to press upon and block the canal of Schlemm from draining the fluid. In most instances, proper eyedrop medication will dilate the duct, permitting excess fluid escape. When medication fails, a new outflow duct can be created surgically.

20 to 25 mm Hg pressure. The individual then has the eye disease known as glaucoma (Figs. 8-7 and 8-8). Either excess formation or, more often, decreased absorption is seen as an immediate cause of this condition, but underlying causes are unknown.

MUSCLES

Eye muscles are of two types: extrinsic and intrinsic.

Extrinsic muscles are those that attach to the outside of the eyeball and to the bones of the orbit. They move the eyeball in any desired direction and are, of course, voluntary muscles. Four of them are straight muscles and two are oblique. Their names describe their positions on the eyeball. They are the superior, inferior, mesial, and lateral rectus muscles and superior and inferior oblique muscles (Figs. 8-9 and 8-10).

Intrinsic eye muscles are those located within the eye. Their names are the iris and the ciliary muscles, and they are involuntary. Incidentally, the eye is the only organ in the body in which both voluntary and involuntary muscles are found. The iris regulates the size of the pupil. The ciliary muscle controls the shape of the lens. As the ciliary muscle contracts, it releases the suspensory ligament from the backward pull usually exerted upon it. And this allows the elastic lens, suspended in the ligament, to bulge or become more convex—a necessary accommodation for near vision (pp. 252 to 253). Some essential facts about eye muscles are summarized in Table 8-4.

ACCESSORY STRUCTURES

Accessory structures of the eye include the eyebrows, eyelashes, eyelids, and lacrimal apparatus.

Eyebrows and eyelashes. The eyebrows and eyelashes serve a cosmetic purpose and give some protection against the entrance of foreign objects into the eyes. Small glands located at the base of the lashes secrete a lubricating fluid. They are

Table 8-4. Eye muscles

	Extrinsic muscles	*Intrinsic muscles*
Names	Superior rectus Inferior rectus Lateral rectus Mesial rectus Superior oblique Inferior oblique	Iris Ciliary muscle
Kind of muscle	Voluntary (striated, skeletal)	Involuntary (smooth, visceral)
Location	Attached to eyeball and bones of orbit	Modified anterior portion of chorioid coat of eyeball; iris doughnut-shaped, sphincter muscle; pupil, hole in center of iris
Functions	Eye movements	Iris regulates size of pupil—therefore, amount of light entering eye; ciliary muscle controls shape of lens (accommodation)—therefore, its refractive power
Innervation	Somatic fibers of third, fourth, and sixth cranial nerves	Autonomic fibers of third and fourth cranial nerves

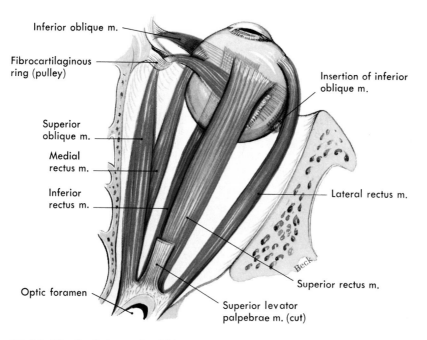

Inferior oblique m.

Fibrocartilaginous
ring (pulley)

Superior
oblique m.

Medial
rectus m.

Inferior
rectus m.

Optic foramen

Insertion of inferior
oblique m.

Lateral rectus m.

Superior rectus m.

Superior levator
palpebrae m. (cut)

Beck

Fig. 8-9 Muscles that move the right eye
as viewed from above.

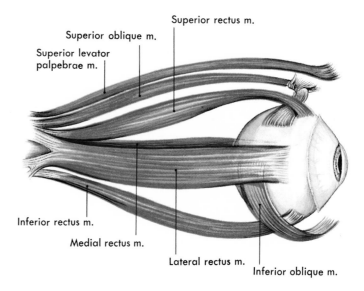

Superior rectus m.

Superior oblique m.

Superior levator
palpebrae m.

Inferior rectus m.

Medial rectus m.

Lateral rectus m.

Inferior oblique m.

Fig. 8-10 Muscles of the right orbit as
viewed from the side.

of interest because they frequently become infected, forming a *sty*.

Eyelids. The eyelids or palpebrae consist mainly of voluntary muscle and skin, with a border of thick connective tissue at the free edge of each lid known as the tarsal plate. One can feel the tarsal plate as a ridge when turning back the eyelid to remove a foreign object. Mucous membrane called conjunctiva lines each lid and continues over the surface of the eyeball, where it is modified to give transparency. Inflammation of the conjunctiva (conjunctivitis) is a fairly common infection. It is often called pinkeye because it produces a pinkish discoloration of the eye's surface.

The opening between the eyelids bears the technical name of palpebral fissure. The height of the fissure determines the apparent size of the eyes. In other words, if the eyelids are habitually held widely opened, the eyes appear large. Keeping the lids only partially open, on the other hand, gives the illusion of small eyes. Actually, there is very little difference in size between eyeballs of different adults. The upper and lower eyelids join, forming an angle or corner known as a *canthus*, the inner canthus being the mesial corner of the eye and the outer canthus the lateral corner.

Lacrimal apparatus. The lacrimal apparatus consists of the structures that secrete tears and drain them from the surface of the eyeball. They are the lacrimal glands, lacrimal ducts, lacrimal sacs, and nasolacrimal ducts (Fig. 8-11).

The *lacrimal glands*, comparable in size and shape to a small almond, are located in a depression of the frontal bone at the upper outer margin of each orbit. Approximately a dozen small ducts lead from each gland, draining the tears onto the conjunctiva at the upper outer corner of the eye.

The *lacrimal canals* are small channels, one above and the other below each *car-uncle* (small red body at inner canthus). They empty into the lacrimal sacs. The openings into the canals are called *punctae* and can be seen as two small dots at the inner canthus of the eye. The *lacrimal sacs* are located in a groove in the lacrimal bone. The *nasolacrimal ducts* are small tubes that extend from the lacrimal sac into the inferior meatus of the nose. All the tear ducts are lined with mucous membrane, an extension of the mucosa that lines the nose. When this membrane becomes inflamed and swollen, the nasolacrimal ducts become plugged, causing the tears to overflow from the eyes instead of draining into the nose as they do normally. Hence, when we have a common cold, "watering" eyes add to our discomforts.

Physiology of vision

In order for vision to occur, the following conditions must be fulfilled: an image must be formed on the retina to stimulate its receptors (rods and cones), and the resulting nerve impulses must be conducted to the visual areas of the cerebral cortex.

FORMATION OF RETINAL IMAGE

Four processes focus light rays so that they form a clear image on the retina: *refraction* of the light rays, *accommodation* of the lens, *constriction* of the pupil, and *convergence* of the eyes.

Refraction of light rays. Refraction means the deflection or bending of light rays. It is produced by light rays passing obliquely from one transparent medium into another of different optical density, and the more convex the surface of the medium, the greater its refractive power. The refracting media of the eye are the cornea, aqueous humor, lens, and vitreous humor. Light rays are bent or refracted at the anterior surface of the cornea as

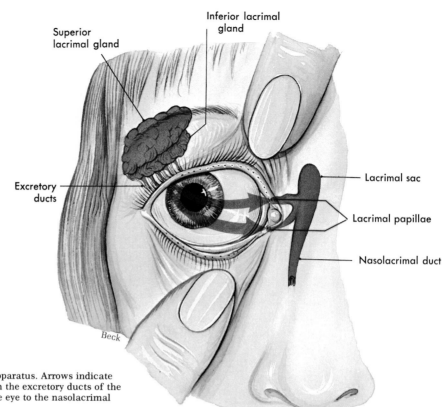

Superior
lacrimal gland

Inferior lacrimal
gland

Excretory
ducts

Lacrimal sac

Lacrimal papillae

Nasolacrimal duct

Beck

Fig. 8-11 The lacrimal apparatus. Arrows indicate
direction of drainage from the excretory ducts of the
lacrimal glands across the eye to the nasolacrimal
duct.

they pass from the rarer air into the denser
cornea, at the anterior surface of the lens
as they pass from the aqueous humor into
the denser lens, and at the posterior surface of the lens as they pass from the lens
into the rarer vitreous humor.

When an individual goes to an ophthal-
mologist for an eye examination, the doctor does a "refraction." In other words,
by various specially designed methods, he
measures the refractory or light-bending
power of that person's eyes.

In a normal (emmetropic) eye when it
is relaxed, the four refracting media together bend light rays sufficiently to bring
to a focus on the retina the parallel rays
reflected from an object 20 or more feet
away (Fig. 8-12). Of course, a normal eye
can also focus objects located much nearer
than 20 feet from the eye. This is accomplished by a mechanism known as accommodation (discussed on pp. 252 to 253).

Many eyes, however, show *errors of refraction* — that is, they are not able to focus
the rays on the retina under the stated conditions. Some common errors of refraction
are *nearsightedness* (myopia), *farsightedness* (hypermetropia), and *astigmatism.*

The nearsighted (myopic) eye sees distant objects as blurred images because it
focuses rays from the object at a point in
front of the retina (Fig. 8-13). According
to one theory, this occurs because the eyeball is too long and the distance too great
from the lens to the retina in the myopic
eye. Concave glasses, by lessening refraction, can give clear distant vision to nearsighted individuals. Presumably, opposite
conditions exist in the farsighted (hyperopic) eye (Fig. 8-14).

Astigmatism is a more complicated condition in which the curvature of the cornea
or of the lens is uneven, causing horizontal
and vertical rays to be focused at two dif-

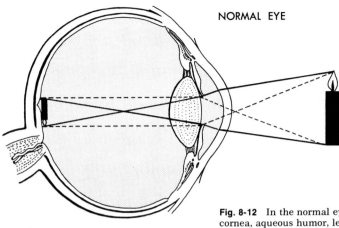

NORMAL EYE

Fig. 8-12 In the normal eye, light rays from an object are refracted by the cornea, aqueous humor, lens, and vitreous humor and are converged on the fovea of the retina, where an inverted image is clearly formed.

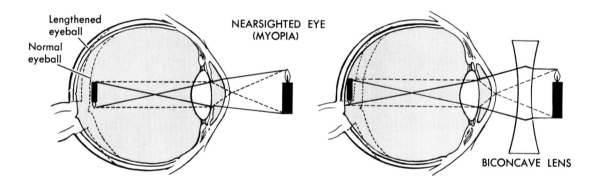

Lengthened eyeball

Normal eyeball

NEARSIGHTED EYE
(MYOPIA)

BICONCAVE LENS

Fig. 8-13 The nearsighted or myopic eye focuses the image in front of the retina. This may occur when the eyeball is too long or the lens is too thick. Correction is by concave lens.

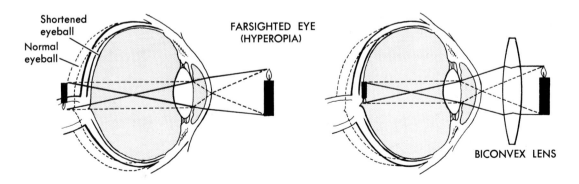

Shortened eyeball

Normal eyeball

FARSIGHTED EYE
(HYPEROPIA)

BICONVEX LENS

Fig. 8-14 The farsighted or hyperopic eye focuses the image behind the retina. This may occur when the eyeball is too short or the lens is too thin. Correction is by convex lens.

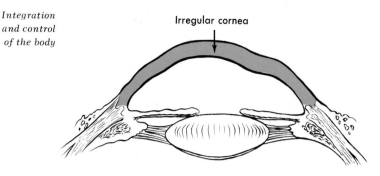

Irregular cornea

Fig. 8-15 Astigmatism results most frequently from an irregular cornea. The cornea may be only slightly flattened horizontally, vertically, or diagonally to produce distortion of vision. The compensating shape of the lens largely nullifies the irregularities of the cornea.

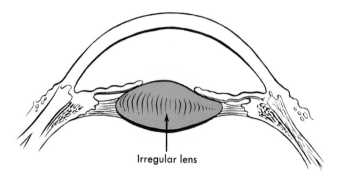

Irregular lens

Fig. 8-16 An irregular lens causes light to be bent in such a way that it does not focus the image on the sharpest area of vision on the retina. An astigmatic eye with this defect causes distorted or blurred vision.

ferent points on the retina (Figs. 8-15 and 8-16). Instead of the curvature of the cornea being a section of a sphere, it is more like that of a teaspoon with horizontal and vertical arcs uneven. Suitable glasses correct the refraction of such an eye.

Visual acuity or the ability to distinguish form and outline clearly is indicated by a fraction that compares the distance at which an individual sees an object (usually letters of a definite size and shape) clearly with the distance at which the normal eye would see the object. Thus, if he sees clearly at 20 feet an object that the normal eye would be able to see clearly at 20 feet, his visual acuity is said to be 20/20 or normal. But if he sees an object clearly at 20 feet which the normal eye sees clearly at 30 feet, then his visual acuity is 20/30 or two-thirds of normal.

As people grow older, they become far-sighted due to the lenses losing their elasticity and therefore their ability to bulge and to accommodate for near vision. This condition is called *presbyopia*.

Accommodation of lens. Accommodation for near vision necessitates three changes: increase in the curvature of the lens, constriction of the pupils, and convergence of the two eyes. Light rays from objects 20 or more feet away are practically parallel. The normal eye, as previously noted, refracts such rays sufficiently to focus them clearly on the retina. But light rays from nearer objects are divergent rather than parallel. So obviously they must be bent more acutely to bring them to a focus on the retina. Accommodation of the lens or, in other words, an increase in its curvature takes place to achieve this greater refraction. (It is a physical fact that the greater the convexity of a lens, the greater its refractive power.) Most observers accept Helmholtz' theory about the mechanism that produces accommodation of the lens. According to his theory, the ciliary muscle contracts, pulling the ciliary body and chorioid forward toward the lens. This releases the tension on the suspensory ligament and therefore on the lens which, being elastic, immediately bulges. For near vision, then, the ciliary muscle is contracted, and the lens is bulging, whereas for far vision the ciliary muscle is relaxed and the lens is comparatively flat. Continual use of the eyes for near work produces eyestrain because of the prolonged contraction of the ciliary muscle. Some of the strain can be avoided by looking into the distance at intervals while doing close work.

Constriction of pupil. The muscles of the iris play an important part in the formation of clear retinal images. Part of the accommodation mechanism consists of contraction of the circular fibers of the iris which constricts the pupil. This prevents divergent rays from the object from entering the eye through the periphery of the cornea and lens. Such peripheral rays could not be brought to a focus on the retina (due to spherical aberration of the lens) and therefore would cause a blurred image. Constriction of the pupil for near vision is called the *near reflex* of the pupil and occurs simultaneously with accommodation of the lens in near vision. The pupil constricts also in bright light (*photopupil reflex or pupillary light reflex*) to protect the retina from too intense or too sudden stimulation.

Convergence of eyes. Single binocular vision (seeing only one object instead of two when both eyes are used) occurs when light rays from an object fall on corresponding points of the two retinas. The foveas and all points lying equidistant and in the same direction from the foveas are corresponding points. Whenever the eyeballs move in unison, either with the visual axes parallel (for far objects) or converging upon a common point (for near objects), light rays strike corresponding points of the two retinas. Convergence is the movement of the two eyeballs inward so that their visual axes come together or converge at the object viewed. The nearer the object,

the greater the degree of convergence necessary to maintain single vision. A simple procedure serves to demonstrate the fact that single binocular vision results from stimulation of corresponding points of the two retinas. Gently press one eyeball out of line while viewing an object. Instead of one object, you will then see two. In order to achieve unified movement of the two eyeballs, a functional balance between the antagonistic extrinsic muscles must exist. If, for example, the right internal rectus muscle should contract more forcefully than its antagonist, the right external rectus, the right eye would be pulled in toward the nose instead of its visual axis being held parallel to that of the left eye in distant vision or converged upon the same point in near vision. Light rays from an object would then fall on noncorresponding points of the two retinas, and the object would be seen double (diplopia). Sometimes the individual can overcome the deviation of the visual axes by muscular effort (extra innervation of the weak muscle) and thereby achieve single vision, but only at the expense of muscular and nervous strain. The condition in which the imbalance of the eye muscles can be overcome by extra innervation of the weak muscle is called *heterophoria* (*esophoria* if the internal rectus is stronger and pulls the eye nasalward and *exophoria* if the external rectus is stronger and pulls the eye temporalward). *Strabismus* (cross-eye or squint) is an exaggerated esophoria which cannot be overcome by neuromuscular effort. An individual with strabismus usually does not have double vision, as you might expect, because he learns to suppress one of the images.

STIMULATION OF RETINA

Rods are known to contain *rhodopsin* (visual purple), a pigmented compound. As shown in Fig. 8-17, it forms by a protein

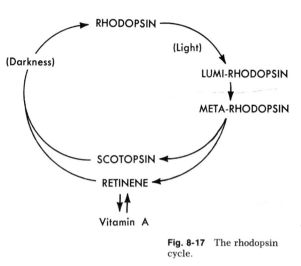

Fig. 8-17 The rhodopsin cycle.

scotopsin combining with retinene, a derivative of vitamin A. Rhodopsin is highly light-sensitive, so that when light rays strike a rod, its rhodopsin rapidly breaks down (also shown in Fig. 8-17). And in some way this chemical change initiates impulse conduction by the rod. Then, if the rod is exposed to darkness for a short time, rhodopsin reforms from the scotopsin and retinene and is ready to function again. *Cones* also contain photosensitive chemicals—iodopsin and possibly others. Just what they are, however, is not established. Presumably, the cone compounds are less sensitive to light than rhodopsin. Brighter light seems necessary for their breakdown. Cones, therefore, are considered to be the receptors responsible for daylight and color visions. Rods, on the other hand, are believed to be the receptors for night vision because their rhodopsin quickly becomes almost depleted in bright light due to its rapid breakdown but slow regeneration. This explains why you cannot see for a little while after you go from a bright light to darkness. But when rhodopsin has had time to reform, the rods again start functioning and dark adaptation has occurred. Or, as we say, we "can see in the dark once we get used to it." Night blindness

occurs in marked vitamin A deficiency. Why? Fig. 8-17 contains a clue. The fovea contains the greatest concentration of cones and is, therefore, the point of clearest vision in good light. For this reason, when we want to see an object clearly in the daytime, we look directly at it so as to focus the image on the fovea. But in dim light or darkness we see an object better if we look slightly to the side of it, thereby focusing the image nearer the periphery of the retina where rods are more plentiful.

CONDUCTION TO VISUAL AREA

Fibers that conduct impulses from the rods and cones reach the visual cortex in the occipital lobes via the optic nerves, optic chiasma, optic tracts, and optic radiations. Those fibers that originate in the medial half of each retina cross over in the optic chiasma and continue through the optic tract and radiation on the opposite side. This anatomical arrangement explains certain peculiar visual abnormalities seen occasionally in persons with brain tumors or other intracranial lesions.

■ Auditory apparatus
Anatomy

The ears, auditory nerves, and auditory areas of the temporal lobes of the cerebrum compose the auditory apparatus. Each ear consists of three parts: external ear, middle ear, and inner ear (Fig. 8-18).

EXTERNAL EAR

The external ear has two divisions: the flap or modified trumpet on the side of the head called the *auricle* or *pinna* and the tube leading from the auricle into the temporal bone and named the *external acoustic meatus (ear canal)*. This canal is about 1 1/4 inches long and takes, in general, an inward, forward, and downward direction, although the first portion of the tube slants

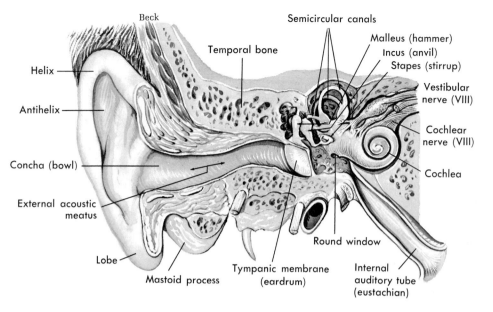

Beck

Helix

Antihelix

Concha (bowl)

External acoustic meatus

Lobe

Mastoid process

Temporal bone

Semicircular canals

Malleus (hammer)

Incus (anvil)

Stapes (stirrup)

Vestibular nerve (VIII)

Cochlear nerve (VIII)

Cochlea

Round window

Tympanic membrane (eardrum)

Internal auditory tube (eustachian)

Fig. 8-18 Components of the ear. The external ear consists of the auricle (pinna), the external acoustic meatus (ear canal), and the tympanic membrane (eardrum). The middle ear includes the malleus (hammer), incus (anvil), and stapes (stirrup) and the tympanic cavity. The inner ear contains the organs of balance and hearing.

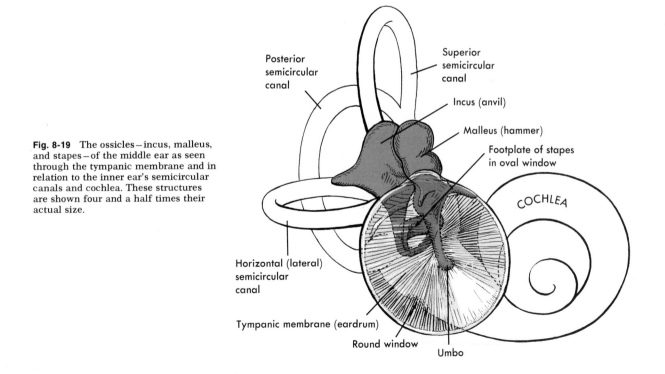

Posterior semicircular canal

Superior semicircular canal

Incus (anvil)

Malleus (hammer)

Footplate of stapes in oval window

COCHLEA

Horizontal (lateral) semicircular canal

Tympanic membrane (eardrum)

Round window

Umbo

Fig. 8-19 The ossicles—incus, malleus, and stapes—of the middle ear as seen through the tympanic membrane and in relation to the inner ear's semicircular canals and cochlea. These structures are shown four and a half times their actual size.

upward and then curves downward. Because of this curve in the auditory canal, the auricle should be pulled up and back to straighten the tube when medications are to be dropped into the ear. Modified sweat glands in the auditory canal secrete cerumen (waxlike substance) which occasionally becomes impacted and may cause pain and deafness. The tympanic membrane (eardrum) stretches across the inner end of the auditory canal, separating it from the middle ear.

MIDDLE EAR

The middle ear (tympanic cavity), a tiny epithelial-lined cavity hollowed out of the temporal bone, contains the three auditory ossicles: the malleus, incus, and stapes (Fig. 8-19). The names of these very small bones describe their shapes (hammer, anvil, and stirrup). The "handle" of the malleus is attached to the inner surface of the tympanic membrane, whereas the "head" attaches to the incus, which, in turn, attaches to the stapes. There are several openings into the middle ear cavity: one from the external auditory meatus, covered over with the tympanic membrane; two into the internal ear, the fenestra ovalis (oval window), into which the stapes fits, and the fenestra rotunda (round window), which is covered by a membrane; and one into the eustachian tube.

Posteriorly, the middle ear cavity is continuous with a number of mastoid cells in the temporal bone. The clinical importance of these openings is that they provide routes for infection to travel. Head colds, for example, especially in children, may lead to middle ear or mastoid infections via the nasopharynx–eustachian tube–middle ear–mastoid path.

The eustachian or auditory tube is composed partly of bone and partly of cartilage and fibrous tissue and is lined with mucosa. It extends downward, forward, and inward from the middle ear cavity to the nasopharynx (the part of the throat behind the nose).

Thus the eustachian tube provides the path by which throat infections may invade the middle ear. But the eustachian tube also serves a useful function. It makes possible equalization of pressure against inner and outer surfaces of the tympanic membrane and therefore prevents membrane rupture and the discomfort that marked pressure differences produce. The way the eustachian tube equalizes tympanic membrane pressures is this. When one swallows or yawns, air spreads rapidly through the open tube. Atmospheric pressure then presses against the inner surface of the tympanic membrane. And since atmospheric pressure is continually exerted against its outer surface, the pressures are equal. You might test this mechanism sometime when you are ascending or descending in an airplane—start chewing gum to increase your swallowing and observe whether this relieves the discomfort in your ears.

INNER EAR

The inner ear is also called the labyrinth because of its complicated shape. It consists of two main parts, a bony labyrinth and, inside this, a membranous labyrinth. The bony labyrinth consists of three parts: vestibule, cochlea, and semicircular canals. The membranous labyrinth consists of the utricle and saccule inside the vestibule, the cochlear duct inside the cochlea, and the membranous semicircular canals inside the bony ones (Fig. 8-20).

Vestibule, utricle, and saccule. The vestibule constitutes the central section of the bony labyrinth. Into it open both the oval and round windows from the middle ear as well as the three semicircular canals of the inner ear. The membranous utricle and saccule are suspended within the ves-

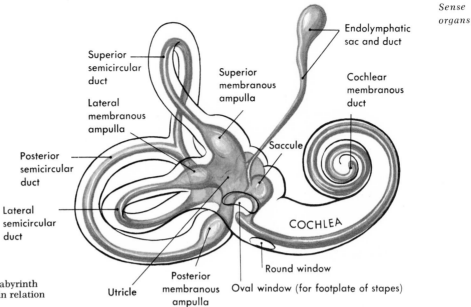

Fig. 8-20 The membranous labyrinth (red) of the inner ear shown in relation to the bony labyrinth.

Labels in figure: Superior semicircular duct; Endolymphatic sac and duct; Superior membranous ampulla; Cochlear membranous duct; Lateral membranous ampulla; Posterior semicircular duct; Saccule; Lateral semicircular duct; COCHLEA; Round window; Oval window (for footplate of stapes); Posterior membranous ampulla; Utricle

tibule. They are separated from the bony walls of the vestibule by fluid (perilymph), and both utricle and saccule contain a fluid called endolymph.

Located within the utricle (and also within the saccule) lies a small structure called the *macula*. It consists mainly of hair cells and a gelatinous membrane that contains *otoliths* (tiny ear "stones"—that is, small particles of calcium carbonate). A few delicate hairs protrude from the hair cells and are embedded in the gelatinous membrane. Receptors for the vestibular branch of the eighth cranial nerve contact the hair cells of the macula located in the utricle. Changing the position of the head causes a change in the amount of pressure on the gelatinous membrane and causes the otoliths to pull on the hair cells. This stimulates the adjacent receptors of the vestibular nerve. Its fibers conduct impulses to the brain which produce a sense of the position of the head and also a sensation of a change in the pull of gravity (for example, a sensation of acceleration). In addition, stimulation of the macula in the utricle evokes *righting reflexes*, muscular responses to restore the body and its

parts to their normal position when they have been displaced. (Impulses from proprioceptors and from the eyes also activate righting reflexes. And interruption of the vestibular or visual or proprioceptive impulses that initiate these reflexes may cause disturbances of equilibrium, nausea, vomiting, and other symptoms.)

Cochlea and cochlear duct. The word cochlea, which means snail, describes the outer appearance of this part of the bony labyrinth. When sectioned, the cochlea resembles a tube wound spirally around a cone-shaped core of bone, the *modiolus*. The modiolus houses the spiral ganglion which consists of cell bodies of the first sensory neurons in the auditory relay. Inside the cochlea lies the membranous *cochlear duct*. This structure is shaped like a tube but is a triangular rather than a round tube. It forms a shelf across the inside of the bony cochlea, dividing it into upper and lower sections all along its winding course (Figs. 8-21 and 8-22). The upper section (above the cochlear duct, that is) is called the *scala vestibuli*, whereas the lower section below the cochlear duct is the *scala tympani*. The roof of the coch-

257

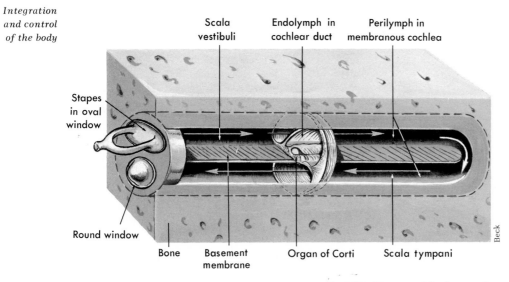

Stapes
in oval
window

Round window

Scala
vestibuli

Endolymph in
cochlear duct

Perilymph in
membranous cochlea

Bone

Basement
membrane

Organ of Corti

Scala tympani

Beck

Fig. 8-21 Diagram of the bony and membranous cochlea, uncoiled. Note
the end organ of Corti projecting into the endolymph contained in the
cochlear duct (membranous cochlea). The perilymph indicated above the
endolymph occupies the scala vestibuli. That in the lower compartment
lies in the scala tympani (see also Fig. 8-22).

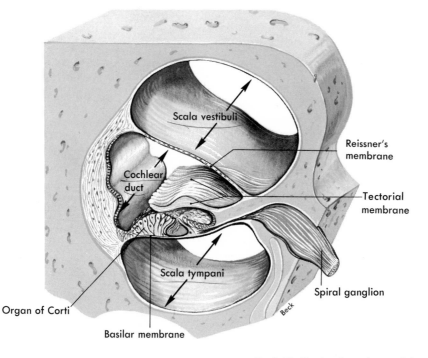

Scala vestibuli

Reissner's
membrane

Cochlear
duct

Tectorial
membrane

Scala tympani

Spiral ganglion

Organ of Corti

Basilar membrane

Beck

Fig. 8-22 Section through one of the coils of the cochlea. Perilymph
fills the scala vestibuli and scala tympani. Endolymph fills the cochlear
duct. A part of the spiral ganglion is shown (bulging stemlike structure at
right of diagram).

lear duct is known as *Reissner's membrane* or the vestibular membrane. *Basilar membrane* is the name given the floor of the cochlear duct. It is supported by bony and fibrous projections from the wall of the cochlea. Perilymph fills the scala vestibuli and scala tympani and endolymph the cochlear duct.

The hearing sense organ, the *organ of Corti*, rests on the basilar membrane throughout the whole length of the cochlear duct. The structure of the organ of Corti resembles that of the equilibrium sense organ (that is, the macula in the utricle). It consists of supporting cells plus the important *hair cells* which project into the endolymph and are topped by an adherent gelatinous membrane called the *tectorial membrane*. Dendrites of the sensory neurons whose cells lie in the spiral ganglion in the modiolus have their beginnings around the bases of the hair cells of the organ of Corti. Axons of these neurons extend in the cochlear nerve (a branch of the eighth cranial nerve) to the brain. They conduct impulses that produce the sensation of hearing.

Semicircular canals. Three semicircular canals, each in a plane approximately at right angles to the others, are found in each temporal bone. Within the bony semicircular canals and separated from them by perilymph are the membranous semicircular canals. Each contains endolymph and connects with the utricle, one of the membranous sacs inside the bony vestibule. Near its junction with the utricle the canal enlarges into an *ampulla*. Some of the receptors for the vestibular branch of the eighth cranial nerve lie in each ampulla. Like all receptors for both vestibular and auditory branches of this nerve, these receptors, too, lie in contact with hair cells in a supporting structure. Here in the ampulla, the hair cells and supporting structure together are named the *crista*

ampullaris, whereas in the utricle and saccule they are called the macula and in the cochlear duct, the organ of Corti. Sitting atop the crista is a gelatinous structure called the *cupula.*

The crista with its vestibular nerve endings presumably functions as the end organ for sensations of head movements, whereas the macula in the utricle serves as the end organ for sensations of head positions. Hence, both the crista and the macula of the utricle function as end organs for the sense of equilibrium.

The function of the macula in the saccule is not known but is postulated to play some part in the sense of hearing rather than in the sense of equilibrium.

Physiology
HEARING

Hearing results from stimulation of the auditory area of the cerebral cortex (temporal lobe, Fig. 7-18, p. 191). Before reaching this area of the brain, however, sound waves must be projected through air, bone, and fluid to stimulate nerve endings and set up impulse conduction over nerve fibers.

Sound waves in the air enter the external auditory canal, probably without much aid from the pinna in collecting and reflecting them because of its smallness in man. At the inner end of the canal they strike against the tympanic membrane, setting it in vibration. Vibrations of the tympanic membrane move the malleus, whose handle attaches to the membrane. The head of the malleus attaches to the incus, and the incus attaches to the stapes. So when the malleus vibrates, it moves the incus, which moves the stapes against the oval window into which it fits so precisely. At this point, fluid conduction of sound waves begins. To understand this, you will probably need to refer to Figs. 8-18 to 8-22 frequently as you read the next few sentences.

When the stapes moves against the oval window, pressure is exerted inward into the perilymph in the scala vestibuli of the cochlea. This starts a "ripple" in the perilymph which is transmitted through Reissner's membrane (the roof of the cochlear duct) to endolymph inside the duct and thence to the organ of Corti and to the basilar membrane that supports the organ of Corti and forms the floor of the cochlear duct. From the basilar membrane the ripple is next transmitted to and through the perilymph in the scala tympani and finally expends itself against the round window — on a much reduced scale, like an ocean wave expending itself as it breaks against the shore. Dendrites (of neurons whose cell bodies lie in the spiral ganglion and whose axons make up the cochlear nerve) terminate around the bases of the hair cells of the organ of Corti, and the tectorial membrane adheres to their upper surfaces. The movement of the hair cells against the adherent tectorial membrane somehow stimulates these dendrites and initiates impulse conduction by the cochlear nerve to the brainstem. Before reaching the auditory area of the temporal lobe, impulses pass through "relay stations" in nuclei in the medulla, pons, midbrain, and thalamus.

EQUILIBRIUM

In addition to hearing, the inner ear aids in the maintenance of equilibrium by making possible sensations of position and movements of the head (discussed previously on p. 257).

▪ Olfactory sense organs

The receptors for the fibers of the olfactory (first) cranial nerves lie in the mucosa of the upper part of the nasal cavity. Their location here explains the necessity for sniffing or drawing air forcefully up into the nose in order to smell delicate odors. The olfactory sense organ consists of hair cells and is relatively simple compared with the complex visual and auditory organs. Whereas the olfactory receptors are extremely sensitive — that is, are stimulated by even very slight odors, they are also easily fatigued — a fact which explains why odors that are at first very noticeable are not sensed at all after a short time.

▪ Gustatory sense organs

The receptors for the taste nerve fibers (in branches of the seventh and ninth cranial nerves) are known as *taste buds* or taste corpuscles. They are located in the papillae of the tongue. Not all taste receptors are stimulated by the same kinds of substances. Four different tastes are recognized, each resulting from stimulation of a different set of taste buds — for sweet, sour, bitter, and salt substances. All the other flavors experienced are a result of fusions of two or more of the four tastes named and as a result of stimulation of the olfactory receptors. In other words, the myriads of tastes recognized are not tastes alone but tastes plus odors. For this reason, a cold that interferes with the stimulation of the olfactory receptors by odors from foods in the mouth markedly dulls one's taste sensations.

The four kinds of taste corpuscles are not evenly distributed over the tongue. Most of those sensitive to bitter are located at the back of the tongue, those sensitive to sweet at the tip, and those sensitive to sour and to salt along the sides and tip. So, if you take a bitter medicine by placing it on the tip of your tongue and swallowing it quickly with water, you will experience less of the bitter taste than if you placed it on the back of your tongue, where there is a concentration of bitter-sensitive taste buds.

outline summary

GENERAL REMARKS
1 Millions of receptors constitute sense organs
2 For receptors and sensations they mediate, see Table 8-1

SOMATIC, VISCERAL, AND REFERRED PAIN
1 Somatic—results from stimulation of pain receptors (nociceptors) in skin or in deep structures (skeletal muscles, tendons, or joints)
2 Visceral—results from stimulation of pain receptors in viscera by pressure or chemical stimuli; conducted almost exclusively by sensory fibers in sympathetic nerves
3 Referred—pain interpreted as coming from skin area when it actually originates in deep structure

EYE
Anatomy

Coats of eyeball
1 Outer coat (sclera)
2 Middle coat (chorioid)
3 Inner coat (retina)
4 For outline summary, see Table 8-2

Cavities and humors
1 Anterior cavity
2 Posterior cavity
3 For outline summary, see Table 8-3

Muscles
1 Extrinsic
   a Attach to outside of eyeball and to bones of orbit
   b Voluntary muscles; move eyeball in desired directions
   c Four straight (rectus) muscles—superior, inferior, lateral, and mesial; two oblique muscles—superior and inferior
2 Intrinsic
   a Within eyeball; named iris and ciliary muscles
   b Involuntary muscles
   c Iris regulates size of pupil
   d Ciliary muscle controls shape of lens, making possible accommodation for near and far objects
3 For outline summary, see Table 8-4

Accessory structures
1 Eyebrows and eyelashes—protective and cosmetic
2 Eyelids
   a Lined with mucous membrane which continues over surface of eyeball; called conjunctiva
   b Opening between eyelids called palpebral fissure
   c Corners where upper and lower eyelids join called canthus, mesial and lateral

3 Lacrimal apparatus—lacrimal glands, lacrimal canals, lacrimal sacs, and nasolacrimal ducts

Physiology of vision
Fulfillment of following conditions results in conscious experience known as vision: formation of retinal image, stimulation of retina, and conduction to visual area

Formation of retinal image
1 Accomplished by four processes:
   a Refraction or bending of light rays as they pass through eye
   b Accommodation or bulging of lens—normally occurs if object viewed lies nearer than 20 feet from eye
   c Constriction of pupil; occurs simultaneously with accommodation for near objects and also in bright light
   d Convergence of eyes for near objects so light rays from object fall on corresponding points of two retinas; necessary for single binocular vision

Stimulation of retina
Accomplished by light rays producing photochemical change in rods and cones

Conduction to visual area
Fibers that conduct impulses from rods and cones reach visual cortex in occipital lobes via optic nerves, optic chiasma, optic tracts, and optic radiations

AUDITORY APPARATUS
Anatomy

External ear
1 Auricle or pinna
2 External acoustic meatus (ear canal)

Middle ear
1 Separated from external ear by tympanic membrane
2 Contains auditory ossicles (malleus, incus, and stapes) and openings from external acoustic meatus, internal ear, eustachian tube, and mastoid sinuses
3 Eustachian tube, collapsible tube, lined with mucosa, extending from nasopharynx to middle ear
   a Equalizes pressure on both sides of eardrum
   b Open when yawning or swallowing

Inner ear
1 Consists of bony and membranous portions, latter contained within former
2 Bony labyrinth has three divisions—vestibule, cochlea, and semicircular canals
3 Membranous cochlear duct contains receptors for cochlear branch of eighth cranial nerve (sense of hearing)

261

4 Utricle and membranous semicircular canals contain receptors for vestibular branch of eighth cranial nerve (sense of equilibrium)

Physiology

Hearing

Hearing results from stimulation of auditory area of temporal lobes by impulses over cochlear nerves, which are stimulated by sound waves being projected through air, bone, and fluid before reaching auditory receptors (organ of Corti)

Equilibrium

Stimulation of receptors in semicircular canals and utricle leads to sense of equilibrium; also initiates righting reflexes essential for balance

OLFACTORY SENSE ORGANS

1 Receptors for olfactory (first) cranial nerve located in nasal mucosa high along septum
2 Receptors very sensitive but easily fatigued

GUSTATORY SENSE ORGANS

1 Receptors for taste nerve fibers (in branches of seventh and ninth cranial nerves) called taste corpuscles or taste buds and located in papillae of tongue
2 Four different kinds of taste corpuscles or taste buds—those sensitive to sweet, salt, sour, and bitter
3 All other tastes result from fusion of two or more of these tastes or from olfactory stimulation

review questions

1 What two general functions do sense organs perform?
2 Explain briefly the principle of specificity of receptors.
3 Describe briefly one theory about the mechanism of referred pain.
4 Explain briefly the mechanism for accommodation for near vision.
5 Define briefly the term refraction. Name the refractory media of the eye.
6 Concave glasses are prescribed for nearsighted vision. Upon what principle is this based?
7 What is the name of the receptors for vision in dim light? For bright light?
8 Distinguish between exteroceptors, proprioceptors, and visceroceptors.
9 Describe the main features of middle ear structure.
10 Name the parts of the bony and membranous labyrinths and describe the relationship of membranous labyrinth parts to those of the bony labyrinth.
11 In what ear structure(s) is the hearing sense organ located? The equilibrium sense organs?
12 What is the name of the hearing sense organ? Of the equilibrium sense organs?

The endocrine system

■ Meaning

The endocrine system, according to the traditional view, consists of glands that release their secretions into the blood. They have no ducts, so another name for endocrine glands is ductless glands. Exocrine or duct glands, in contrast, release their secretions into ducts which convey the secretions into cavities or onto surfaces. Table 9-1 gives the names and locations of the main endocrine glands (see also Fig. 9-1). The table lists ten of them counting the pituitary, adrenal, and the ovary each as two separate glands, which functionally they are. Since the time when these glands were defined as *the* endocrine glands, other structures have also been shown to secrete regulatory chemicals into the blood. By definition, therefore, these structures, too—notably, certain cells of the stomach, small intestine, kidney, and brain —are endocrine glands. We have chosen to discuss them, however, with their respective systems instead of in this chapter.

The endocrine system and the nervous system both perform the same general functions for the body: communication, control, and integration. But they accom-

Meaning
**Pituitary body
(hypophysis cerebri)**
Location and component
glands
Anterior pituitary gland
(adenohypophysis)
Growth hormone
Prolactin
Tropic hormones
**Melanocyte-stimulating
hormone**
Control of secretion
Posterior pituitary gland
(neurohypophysis)
Antidiuretic hormone
Oxytocin
Control of secretion

Thyroid gland
Location and structure
Functions
Effects of hypersecretion
and hyposecretion

Parathyroid glands
Location and structure
Functions

Adrenal glands
Location and structure
Adrenal cortex
Glucocorticoids
Mineralocorticoids
Sex hormones
Control of secretion
Adrenal medulla
Hormones
Control of secretion

Islands of Langerhans

Ovaries
Estrogens
Progesterone

Testes

Thymus

**Pineal body (pineal
gland or epiphysis
cerebri)**

Placenta

plish these general functions through different kinds of mechanisms and with somewhat different types of results. The mechanism of the nervous system consists of nerve impulses conducted by neurons from one specific structure to another. In contrast, the mechanism of the endocrine

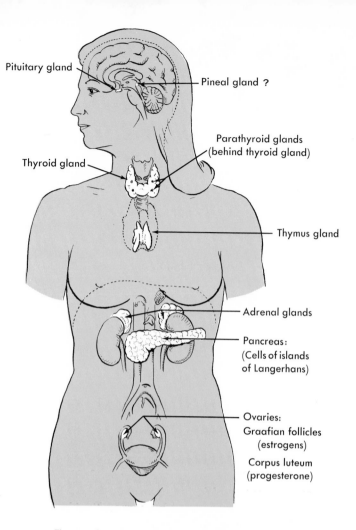

Pituitary gland

Pineal gland ?

Parathyroid glands
(behind thyroid gland)

Thyroid gland

Thymus gland

Adrenal glands

Pancreas:
(Cells of islands
of Langerhans)

Ovaries:
Graafian follicles
(estrogens)

Corpus luteum
(progesterone)

Fig. 9-1 Location of the endocrine glands in the female.
Pineal body postulated but not proved to be an endocrine
gland. Dotted line around thymus gland indicates maximum
size at puberty. Which glands appear in this figure but not
in Table 9-1?

Table 9-1. Names and locations of endocrine glands

Name	*Location*
Pituitary gland (hypophysis cerebri)	Cranial cavity
Anterior lobe (adenohypophysis)	
Posterior lobe (neurohypophysis)	
Thyroid gland	Neck
Parathyroid glands	Neck
Adrenal glands	Abdominal cavity (retroperitoneal)
Adrenal cortex	
Adrenal medulla	
Islands of Langerhans	Abdominal cavity (pancreas)
Ovaries	Pelvic cavity
Graafian follicle	
Corpus luteum	
Testes (interstitial cells)	Scrotum

system consists of secretions from endocrine gland cells entering the blood and circulating to all parts of the body. Endocrine gland secretions, as you undoubtedly know, are called *hormones*. Nerve impulses and hormones do not exert exactly the same kind of control, nor do they control precisely the same structures and functions. And this is fortunate. It makes for better timing and more precision of control. For instance, nerve impulses produce rapid, short-lasting responses, whereas hormones produce slower and generally longer-lasting responses. Nerve impulses control directly only two kinds of cells: muscle and gland cells. Some hormones, in contrast, exert control over all kinds of cells.

Endocrine gland cells synthesize hormones by the process of anabolism. They are either protein or steroid compounds—hence are relatively large molecules. Biochemists tell us that the growth hormone molecule, for instance, consists of 396 amino acids and has a molecular weight of 45,000! Compared with a carbon dioxide molecule, molecular weight 44, the growth hormone molecule is indeed a giant. (Carbon dioxide is one of several so-called "regulatory chemicals." These compounds resemble hormones in two respects—they are released into the blood from cells, and they exert profound influence on various structures and functions. They differ from hormones, however, in that they are neither proteins nor steroids but are much smaller, inorganic molecules or ions, and they are not products of anabolism. Instead, some regulatory chemicals, including carbon dioxide, are products of catabolism.)

Exaggerating the importance of endocrine glands is almost impossible. Hormones are the main regulators of metabolism (and, therefore, of growth and development), of reproduction, and of stress responses. They play roles of the utmost importance in maintaining homeostasis—fluid and electrolyte balance, acid-base balance, and energy balance, for example. Excesses or deficiencies of hormones make the difference between normalcy and all sorts of abnormalities such as idiocy, dwarfism, gigantism, and sterility—and even the difference between life and death in some instances.

We shall start our discussion of the endocrine system with a gland that is truly small but mighty. At its largest diameter, it measures only about 1/2 of an inch. By weight, it is even less impressive—only about 1/2 of a gram, 1/60 of an ounce! And yet, so crucial are its functions that the popular name "master gland" designates its anterior lobe. The anatomical name for the entire organ is hypophysis cerebri, but more often it is called simply the pituitary body or gland.

◾ Pituitary body (hypophysis cerebri)

Location and component glands

The pituitary body has a protected location. It lies in the sella turcica and is covered over by an extension of the dura mater known as the pituitary diaphragm. The deepest part of the sella turcica (saddle-shaped depression in the sphenoid bone) is called the pituitary fossa since the pituitary body lies in it. A stemlike portion, the pituitary stalk, juts up through the diaphragm and attaches the gland to the undersurface of the brain. More specifically, the stalk attaches the pituitary body to the hypothalamus. Although the pituitary looks like just one gland, it actually consists of two separate glands—the adenohypophysis or anterior pituitary gland and the neurohypophysis or posterior pituitary gland. These two glands develop from different embryonic structures, have different microscopic structure, and secrete different hormones. The adenohypophysis develops

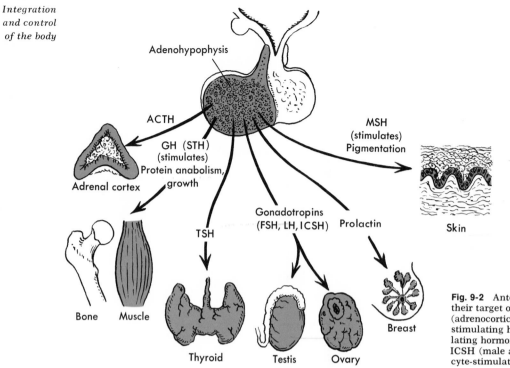

Fig. 9-2 Anterior pituitary hormones and their target organs. Tropic hormones: ACTH (adrenocorticotropic hormone); TSH (thyroid-stimulating hormone); FSH (follicle-stimulating hormone); LH (luteinizing hormone); ICSH (male analogue of LH); MSH (melanocyte-stimulating hormone).

as an upward projection from the embryo's pharynx, whereas the neurohypophysis develops as a downward projection from its brain. Microscopic differences between the two glands are suggested by their names— *adeno* means gland and *neuro* means nervous. In other words, the adenohypophysis has the microscopic structure of an endocrine gland, whereas the neurohypophysis has the structure of nervous tissue. As far as functions are concerned, the two glands are also dissimilar. The anterior lobe secretes seven hormones and the posterior lobe probably secretes only two.

Anterior pituitary gland (adenohypophysis)

The anterior pituitary gland consists chiefly of two main types of epithelial cells: chromophobes (so named because they resist staining) and chromophils (named for their ease of staining). Chromophils, in turn, are subdivided into two main types: acidophils (those that take acid stains) and basophils (those that take basic stains). Acidophils secrete growth

hormone (GH; also called somatotropin or STH) and prolactin. Basophils secrete the other five hormones of the anterior lobe: thyrotropin (TH; also called thyroid-stimulating hormone or TSH), adrenocorticotropin (ACTH), two gonadotropins (follicle-stimulating hormone [FSH] and luteinizing hormone [LH]), and melanocyte-stimulating hormone (MSH). Some investigators think it probable that a different type of cell secretes each of these different hormones.

Growth hormone

Growth hormone or somatotropin (Gr. *soma*, body; *trope*, turning) is thought to promote bodily growth indirectly by accelerating amino acid transport into cells—evidence, blood amino acid content decreases within hours after administration of growth hormone to a fasting animal. With the faster entrance of amino acid into cells, anabolism of amino acids to form tissue protein also accelerates. And this, in turn, tends to promote cellular growth. Growth hormone stimulates growth of both

Fig. 9-3 A pituitary giant and dwarf contrasted with normal-sized men. Excessive secretion of growth hormine* by the anterior lobe of the pituitary gland during the early years of life produces giants of this type, while deficient secretion of this substance produces well-formed dwarfs. (Courtesy Dr. Edmund E. Beard, Cleveland, Ohio.)

*The structure of the human growth hormone molecule was identified by Dr. C. H. Li, Professor of Biochemistry at the University of California, in 1966. It consists of a chain of 188 amino acids with one loop of 93 subunits and another of 6 subunits. In January of 1971, Dr. Li announced that his laboratory had succeeded in synthesizing human growth hormone.

Fig. 9-4 Acromegaly, a disease of adults caused by excess growth hormone secretion by the anterior pituitary gland. (Courtesy Dr. William McKendree Jefferies, Case Western Reserve University School of Medicine, Cleveland, Ohio.)

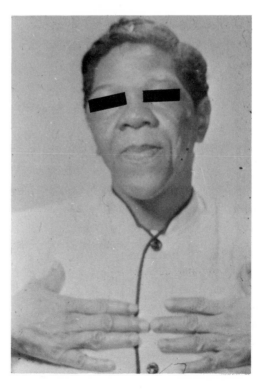

bone and soft tissues. If the anterior pituitary gland secretes an excess of growth hormone during the growth years (that is, before closure of the epiphyseal cartilages), bones grow more rapidly than normal and *gigantism* results (Fig. 9-3). But if oversecretion of growth hormone occurs after the individual is full grown, the condition known as *acromegaly* develops. Characteristic of this disease are enlargement of the bones of the hands, feet, jaws, and cheeks and an increase, too, in their overlying soft tissues (Fig. 9-4). Undersecretion of the growth hormone produces *dwarfism* when it occurs during the years of skeletal growth and *pituitary cachexia* (Simmonds' disease), a much rarer condition, when the deficiency develops during adult life. Premature aging with marked tissue atrophy characterizes the latter disease.

In addition to its stimulating effect on protein anabolism, growth hormone also influences fat and carbohydrate metabolism. For example, it tends to block fat deposition and to accelerate the mobilization of fats from adipose tissues and to increase their utilization (catabolism) by other tissues. Growth hormone, in other words, causes cells to shift from carbohydrate to fat utilization for energy. Did you notice that growth hormone exerts opposite effects on fat and protein metabolism? Growth hormone promotes *anabolism* of proteins but *catabolism* of fats.

The precise actions of growth hormone on carbohydrate metabolism are not known. But that it does influence carbohydrate metabolism is known. Growth hormone acts in some way to decrease carbohydrate utilization by cells. Hence, growth hormone tends to increase the blood concentration of glucose—it has a hyperglycemic effect. Insulin produces opposite effects. It increases carbohydrate utilization, thereby tending to de-

crease the blood concentration of glucose—to have a hypoglycemic effect. Whereas adequate amounts of insulin prevent diabetes mellitus, long-continued excess amounts of growth hormone produce diabetes. In short, growth hormone and insulin function as antagonists. Or, as more commonly stated, growth hormone has an anti-insulin effect, a fact of considerable clinical importance. Summarizing very briefly, growth hormone helps regulate metabolism in the following ways:

1 It promotes protein anabolism (synthesis of tissue proteins) so is essential for normal growth and for tissue repair and healing.

2 It promotes fat mobilization and catabolism.

3 It acts in some way to oppose insulin. This anti-insulin action decreases glucose utilization and seems to account for the hyperglycemic and diabetogenic effects of growth hormone.

Prolactin

The anterior pituitary cells called acidophils secrete two hormones: growth hormone and prolactin. Another name for prolactin—lactogenic hormone—suggests that it "generates" (that is, initiates) milk secretion. It stimulates the mammary glands to start secreting soon after delivery of a baby. And during pregnancy, it helps promote the breast development that makes possible milk secretion after pregnancy. Formerly prolactin was called *luteotropic hormone*.

Tropic hormones

The anterior pituitary cells known as basophils secrete four tropic hormones—hormones which have a stimulating effect on other endocrine glands. Names of the tropic hormones are thyrotropin, adrenocorticotropin, and the two gonadotropins known as follicle-stimulating hormone and luteinizing hormone. All tropic hormones perform the same general function. Each one stimulates one other endocrine gland—stimulates it both to grow and to secrete its hormone at a faster rate. It seems to affect only this one structure as if, like a bullet, it had been aimed at a target. In fact, "target gland" is the name given to the particular gland influenced by a particular hormone.

Individual tropic hormones perform the following functions:

1 *Thyrotropin* (thyroid-stimulating hormone, TSH) promotes and maintains growth and development of the thyroid gland and stimulates it to secrete thyroxin and tri-iodothyronine, its two hormones (together called thyroid hormone).

2 *Adrenocorticotropin* (ACTH) promotes and maintains normal growth and development of the adrenal cortex and stimulates it to secrete cortisol and other glucocorticoids.

3 *Follicle-stimulating hormone* (FSH) stimulates primary graafian follicles to start growing and to continue developing to maturity, that is, to the point of ovulation. FSH also stimulates follicle cells to secrete estrogens, one type of female sex hormones. In the male, FSH stimulates development of the seminiferous tubules and maintains spermatogenesis by them.

4 *Luteinizing hormone* (LH). The name of this hormone suggests one of its functions, that of stimulating formation of the corpus luteum. Prior to this, LH acts with FSH to bring about complete maturation of the follicle; LH then produces ovulation and stimulates formation of the corpus luteum in the ruptured follicle. Finally, LH in human females stimulates the corpus luteum to secrete progesterone and estrogens. The male pituitary gland also secretes LH, but it is called *intersti-tial cell–stimulating hormone* (ICSH) because it stimulates interstitial cells in the testes to develop and secrete testosterone.

●　　●　　●

During childhood, the anterior pituitary gland secretes insignificant amounts of gonadotropins. Then it steps up their production gradually a few years before puberty. And just before puberty begins, presumably its secretion of these hormones takes a sudden and marked spurt. As a result, the blood concentration of gonadotropins increases markedly. And this first high blood concentration of gonadotropins is the stimulus that brings on the first menstrual period and the many other changes that signal the beginning of puberty.

Melanocyte-stimulating hormone

In some species, the intermediate part of the adenohypophysis produces the melanocyte-stimulating hormone (MSH). Therefore, MSH was first named *intermedin*. However, in man, it is now thought that the anterior lobe of the pituitary gland produces most of the MSH and that both MSH and another anterior pituitary hormone (namely, ACTH) tend to produce increased pigmentation of the skin. Structurally, also, the two molecules resemble each other. Several of the same amino acids occur in the same sequence in both MSH and ACTH.

Control of secretion

Under ordinary circumstances, negative feedback mechanisms control the anterior pituitary gland's secretion of tropic hormones. A high blood level of a tropic hormone stimulates its target gland to increase its hormone secretion. This results in a high blood level of the target

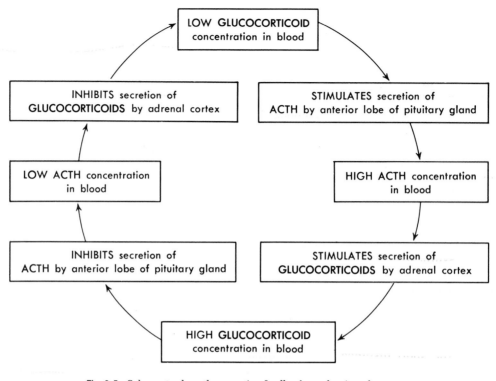

Fig. 9-5 Scheme to show the negative feedback mechanism that regulates secretion of ACTH by the anterior lobe of the pituitary gland and secretion of glucocorticoids by the adrenal cortex under ordinary conditions. This is a homeostatic mechanism since it tends to keep the blood levels of glucocorticoids stable within a narrow range.

gland's hormone which, "feeding back" via circulation to the anterior pituitary gland, inhibits its secretion of the tropic hormone. Fig. 9-5 will help you apply these principles of negative feedback to the specific mechanism that controls ACTH secretion by the pituitary gland. In studying this figure, probably the best place to start is with "HIGH ACTH concentration in blood" and then follow the arrows clockwise. Similar mechanisms regulate the anterior lobe's secretion of thyrotropin and the thyroid gland's secretion of thyroid hormone. Also, a similar mechanism controls the anterior lobe's secretion of FSH and the ovary's secretion of estrogens.

Clinical facts furnish interesting evidence about feedback control of hormone secretion by the anterior pituitary gland and its target glands. For instance, if a person has his pituitary gland removed (hypophysectomy) surgically or by radiation, he must be kept on hormone replacement therapy for the rest of his life. If not, he will develop thyroid, adrenocortical, and gonadotropic deficiencies—deficiencies, that is, of target gland hormones. Another well-known clinical fact is that estrogen deficiency develops in women between 40 and 50 years of age. By then, the ovaries seem to have tired of producing hormones and ovulating each month. Or,

releasing factors which stimulate "T" hormones.
hypothalamus: 1) corticotropin releasing factor
2) thyrotropin
3) somatotropin
4) gonadotropin

The
endocrine
system

at any rate, they no longer respond to FSH stimulation so estrogen deficiency develops, brings about the menopause, and persists after the menopause. What, therefore, would you deduce is true of the blood concentration of FSH after the menopause? What effect does a high concentration of a target gland hormone have on tropic hormone secretion by the anterior pituitary gland?

When the body is in a state of stress, neurosecretions from the hypothalamus regulate anterior pituitary secretion. Neurons in certain parts of the hypothalamus synthesize the chemicals called *neurosecretions*. They are released from the axons of these neurons into the blood in a complex of small veins known as the hypophyseal or *pituitary portal system*. Via this portal system, the neurosecretions travel the short distance from the hypothalamus down to the anterior lobe of the pituitary gland. There they stimulate its cells to secrete various hormones. For instance, one neurosecretion from the hypothalamus stimulates the anterior lobe to release ACTH into the blood. It has an appropriate name—corticotropin-releasing factor (CRF). Another hypothalamic neurosecretion, thyrotropin-releasing factor (TRF), stimulates the anterior pituitary gland to secrete thyrotropin. And almost surely the hypothalamus forms other neurosecretions too—for instance, a factor that controls the anterior pituitary gland's secretion of prolactin. At least, impressive evidence suggests that this may be true.

The hypothalamic–anterior pituitary mechanism described in the preceding paragraph operates as a kind of emergency device—it controls secretion by the anterior lobe in times of stress. It is not yet known definitely but perhaps under these conditions, changes in the blood concentration of various substances and presumably

nerve impulses from the cerebral cortex, limbic lobe, and other structures stimulate the hypothalamus to increase its output of neurosecretions into the pituitary portal system. Then, via these vessels, they quickly travel to the anterior lobe, where they stimulate it to increase its secretion of hormones.

In essence, what the hypothalamus does through its neurosecretions is to translate nerve impulses into hormone secretion by endocrine glands.* In other words, the hypothalamus links the nervous system to the endocrine system. It integrates the activities of these two great integrating systems—particularly, it seems, in times of stress. When healthy survival is threatened, the hypothalamus, via its neurosecretions into the pituitary portal system, seems to take over the command of the anterior lobe of the pituitary gland. By so doing, it indirectly controls all the endocrine glands influenced by the anterior lobe's tropic hormones—namely, the thyroid gland, the adrenal cortex, and the gonads. And, finally, by means of the hormones these target glands secrete, the hypothalamus can dictate the functioning of literally every cell in our bodies. These facts have tremendous implications, especially when coupled with the fact that tracts extend from the cerebral cortex to the hypothalamus. They mean that the cerebral cortex can do more than just receive impulses from all parts of the body and send out impulses to muscles and glands. They mean that the cerebral cortex —and therefore our thoughts and emotions —can, by way of the hypothalamus, influ-

*The term *neuroendocrine transducer* was coined to indicate a structure that translates nerve impulses into hormones. Known neuroendocrine transducers are the hypothalamus and the adrenal medulla. Some evidence seems to indicate that the pineal body may also qualify as one.

ence the functioning of all our billions of cells. In short, the brain has two-way contact with every tissue of the body. Not only can all tissues send information to the brain (via sensory conduction paths), but also the brain can send out information to all tissues—not just to muscles and glands as formerly supposed—via neurotransducer mechanisms. And this fact lends support to the now widely held circular theory of mind and body relations, the somatopsychosomatic theory. The state of the body influences mental processes, which, in turn influence the state of the body.

Posterior pituitary gland (neurohypophysis)

The posterior lobe of the pituitary gland secretes two hormones—one known as the antidiuretic hormone (ADH) and the other called oxytocin. But, strangely enough, cells of the posterior lobe do not themselves make these hormones. Neurons in the hypothalamus (supraoptic and paraventricular nuclei) synthesize them. From the cell bodies of these neurons the hormones pass down along axons (in the hypothalamohypophyseal tract) into the posterior lobe of the pituitary gland. Later, the posterior lobe secretes them into the blood.

Antidiuretic hormone

ADH acts on cells of the distal and collecting tubules of the kidney to make them more permeable to water. This causes faster reabsorption of water from tubular urine into blood. And this, in turn, automatically produces antidiuresis (smaller urine volume). Marked diuresis—that is, abnormally large urine volume—occurs if ADH secretion is inadequate, such as occurs in the disease *diabetes insipidus.* A preparation of ADH used to treat this condition is called vasopressin or Pitressin.

(The name vasopressin suggests the tendency of ADH to constrict blood vessels. However, as far as now known, this action of ADH has no physiological importance.)

Oxytocin

Oxytocin has two actions: it stimulates powerful contractions by the pregnant uterus and it causes milk ejection from the lactating breast. Under the influence of this hormone from the posterior lobe of the pituitary gland, alveoli (cells that synthesize milk) release the milk into the ducts of the breast. This is a highly important function of oxytocin because milk cannot be removed by suckling unless it has first been ejected into ducts. However, it was oxytocin's other action—its stimulating effect on contractions of the pregnant uterus—that inspired its name. The term means "swift childbirth" (Gr. *oxys,* swift; *tokos,* childbirth). Whether or not oxytocin takes part in initiating labor is still an unsettled question. Commercial preparations of oxytocin are given to stimulate uterine contractions after delivery of a baby in order to lessen the danger of hemorrhage.

Control of secretion

Details of the mechanism that controls the secretion of ADH by the posterior lobe of the pituitary gland have not been established. However, two factors play dominant roles: the osmotic pressure of the extracellular fluid and its total volume. Without going into a discussion of evidence or details, the general principles of control of ADH secretion are as follows:

1 An increase in the osmotic pressure of extracellular fluid stimulates ADH secretion. This leads to decreased urine output and tends, therefore, to increase the volume of extracellular fluid, which, in turn, tends to decrease its osmotic pressure back toward normal (see Fig. 15-9, p. 509).

Opposite effects result from a decrease in the osmotic pressure of extracellular fluid.

2 A decrease in the total volume of extracellular fluid acts in some way to stimulate ADH secretion and thereby to decrease urine output and increase extracellular volume back toward normal.

3 Stress acts in some way, presumably by the hypothalamus, to stimulate ADH secretion (see Fig. 15-9, p. 509). An interesting clinical application of this principle is the fact that after major surgery patients tend to show signs of water retention.

. . .

About all that is known about the mechanism that controls the secretion of oxytocin by the posterior lobe of the pituitary gland is that stimulation of the nipples by the baby's nursing initiates sensory impulses that eventually reach the supraoptic and paraventricular nuclei of the hypothalamus, stimulating them. They, in turn, stimulate the posterior lobe of the pituitary gland to increase its secretion of oxytocin (see Fig. 14-14, p. 482).

■Thyroid gland

Location and structure

Two fairly large lateral lobes and a connecting portion, the isthmus, constitute the thyroid gland (Fig. 9-6). It is located in the neck just below the larynx. The isthmus lies across the anterior surface of the upper part of the trachea. Thyroid tissue contains numerous small follicles. Colloid, composed largely of an iodine-containing protein known as thyroglobulin, fills these tiny sacs.

Functions

The thyroid gland secretes thyroid hormone and thyrocalcitonin. Actually, thyroid hormone consists of two hormones. Thy-

roxine is the name of the main one, and triiodothyronine is the name of the less abundant one. One molecule of thyroxine contains four atoms of iodine, and one molecule of triiodothyronine, as its name suggests, contains three iodine atoms. After synthesizing its hormones, the thyroid gland stores considerable amounts of them before secreting them. (Other endocrine glands do not store up their hormones.) As a preliminary to storage, thyroxine and triiodothyronine combine with a globulin in the thyroid cell to form a compound called thyroglobulin. Then thyroglobulin is stored in the colloid material in the follicles of the gland. Later, the two hormones are released from thyroglobulin and secreted into the blood as thyroxine and triiodothyronine. Almost immediately, however, they combine with a blood protein. They travel in the bloodstream in this protein-bound form. But in the tissue capillaries they are released from the protein and enter tissue cells as thyroxine and triiodothyronine.

The iodine in the thyroxine and triiodothyronine that is bound to protein in the blood is called protein-bound iodine (PBI). The amount of PBI can be measured by laboratory procedure and, in fact, is widely used as a test of thyroid functioning. (With a normal thyroid gland, the PBI is about 4 to 8 micrograms [μg] per 100 milliliters of plasma.)

The main physiological actions of thyroid hormone are to help regulate the metabolic rate and the processes of growth and tissue differentiation. Thyroid hormone increases the metabolic rate. How do we know this? Because oxygen consumption increases following thyroid administration. Like pituitary somatotropin, thyroid hormone stimulates growth, but unlike somatotropin, it also influences tissue differentiation and development. For example, cretins (individuals with thyroid deficiency)

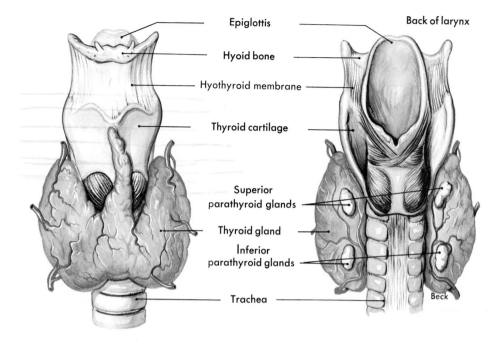

Epiglottis

Back of larynx

Hyoid bone

Hyothyroid membrane

Thyroid cartilage

Superior parathyroid glands

Thyroid gland

Inferior parathyroid glands

Trachea

Beck

Fig. 9-6 The thyroid and parathyroid glands. Note their relations to each other and to the larynx (voice box) and trachea.

not only are dwarfed but also are usually mentally retarded because the brain fails to develop normally. Their bones and many other tissues also show an abnormal pattern of development.

Convincing evidence indicates that the thyroid gland secretes *thyrocalcitonin.* (Formerly, the parathyroid gland was thought to secrete it.) Thyrocalcitonin acts quickly to decrease blood's calcium concentration. Apparently it produces this effect by inhibiting bone breakdown with calcium release into blood, or by promoting calcium deposition in bone, or by both actions. The function of thyrocalcitonin is to help maintain blood calcium homeostasis and prevent harmful hypercalcemia. Parathormone serves as an antagonist to thyrocalcitonin (p. 276).

Effects of hypersecretion and hyposecretion

Hypersecretion of thyroid hormone produces the disease *exophthalmic goiter* (Graves' disease, Basedow's disease, and several other names) (Fig. 9-7), characterized by an elevated PBI, an increased metabolism (+30 or more), an increased appetite, loss of weight, increased nervous irritability, and exophthalmos. Marked edema of the fatty tissue behind the eye, attributed to the high blood titer of thyrotropic hormone, produces the exophthalmos.

Hyposecretion during the formative years leads to malformed dwarfism or *cretinism,* a condition characterized by retarded mental, physical, and sexual development and lowered metabolic rate. Later in life, deficient thyroid secretion produces the disease *myxedema* (Fig. 9-8). This, characterized by decreased metabolic rate, in turn, leads to lessened mental and physical vigor and a gain in weight, loss of hair, and a thickening of the skin due to an accumulation of fluid in the subcutaneous tissues. Because of a high mucoprotein content, this fluid is viscous. Therefore, it gives firmness to the skin, and the skin does not pit when pressed, as it does in some other types of edema.

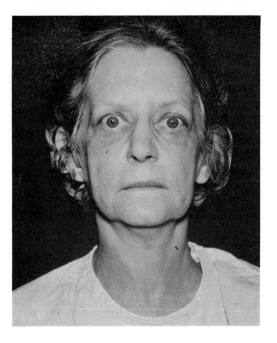

Fig. 9-7 Graves' disease caused by hypersecretion by the thyroid gland. (Courtesy Dr. William McKendree Jefferies, Western Reserve University School of Medicine, Cleveland, Ohio.)

Fig. 9-8 Myxedema, a condition produced by hyposecretion by the thyroid gland during the adult years. (Courtesy Dr. Edmund E. Beard, Cleveland, Ohio.)

■Parathyroid glands

Location and structure

The parathyroid glands are small round bodies attached to the posterior surfaces of the lateral lobes of the thyroid gland (Fig. 9-6). Usually there are four or five, but sometimes there are fewer and sometimes more of these glands.

Functions

The parathyroid glands secrete *parathormone.* Its chief function is to help maintain homeostasis of blood calcium concentration. To do this, it acts as follows:

1 Parathormone acts on intestine, bones, and kidney tubules to accelerate calcium absorption from them into the blood. Hence, parathormone tends to increase the blood concentration of calcium. Its primary action on bone is to stimulate bone breakdown or resorption. This releases calcium and phosphate, which diffuse into the blood.

2 Parathormone acts on kidney tubules to accelerate their excretion of phosphates from the blood into the urine. Note, therefore, that parathormone has opposite effects on the kidney's handling of calcium and phosphate. It accelerates calcium reabsorption but phosphate excretion by the tubules. Consequently, parathormone tends to increase the blood concentration of calcium and to decrease the blood concentration of phosphate.

•   •   •

The maintenance of calcium homeostasis is highly important for healthy survival. Normal neuromuscular irritability, blood clotting, cell membrane permeability, and also normal functioning of certain enzymes all depend upon the blood concentration of calcium being maintained at a normal level. Neuromuscular irritability is inversely related to blood calcium concentration. In other words, neuromuscular

irritability increases when the blood concentration of calcium decreases. Suppose, for example, that a parathormone deficiency develops and causes *hypocalcemia* (lower than normal blood concentration of calcium). The hypocalcemia increases neuromuscular irritability—sometimes so much that it produces muscle spasms and convulsions, a condition called tetany.

Parathormone excess produces hypercalcemia or a higher than normal blood concentration of calcium. Sometimes it causes a bone disease with the long name of osteitis fibrosa generalisata. Bone mass decreases (as a result of increased bone destruction followed by fibrous tissue replacement), decalcification occurs, and cystlike cavities appear in the bone.

■Adrenal glands

Location and structure

The adrenal glands are located atop the kidneys, fitting like a cap over these organs. The outer portion of the gland is called the *cortex* and the inner substance the *medulla*. Although the adrenal cortex and adrenal medulla are structural parts of one organ, they function as two separate endocrine glands.

■Adrenal cortex

Three different zones or layers of cells make up the adrenal cortex. Starting with the zone directly under the outer capsule of the gland, their names are zona glomerulosa, zona fasciculata, and zona reticularis. According to one theory based on good evidence, the zona glomerulosa secretes hormones called mineralocorticoids. Another type of corticoids, known as glucocorticoids, are secreted chiefly by the middle layer of cells. And the innermost layer secretes small amounts of both glucocorticoids and sex hormones. We shall now discuss briefly the functions of these three kinds of adrenal cortical hormones.

Glucocorticoids

The chief glucocorticoids secreted by the adrenal cortex are cortisol* (also called hydrocortisone and compound F) and corticosterone (also known as compound B). Glucocorticoids affect literally every cell in the body. Their general functions are to promote normal metabolism and to enable the body to resist stress – a good nickname for glucocorticoids might be "metabolic, stress-resisting hormones." Although the precise primary actions of glucocorticoids remain unknown, their most outstanding effects are as follows.

1 Glucocorticoids in adequate amounts promote normal protein metabolism. In excess amounts, they produce a "protein catabolic effect" – that is, a high blood concentration of glucocorticoids accelerates the breakdown of tissue proteins to form amino acids. This increased rate of tissue protein mobilization results in a net loss of tissue proteins – in other words, in a negative nitrogen balance. "Tissue wasting" this is sometimes graphically called. A deficient amount of glucocorticoids produces an opposite effect – it slows tissue protein mobilization.

2 Normal amounts of glucocorticoids promote normal carbohydrate metabolism.

---

*Steroid compounds have the following nucleus:

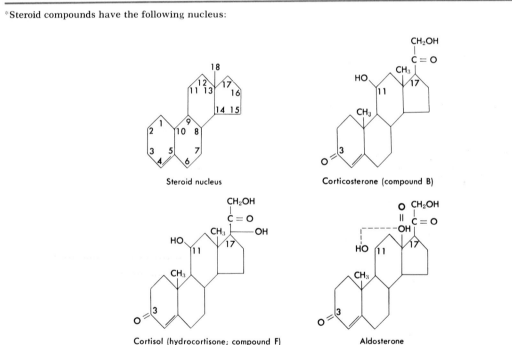

Steroid nucleus

Corticosterone (compound B)

Cortisol (hydrocortisone; compound F)

Aldosterone

Corticosterone (compound B) may be the parent substance of other corticoids.

Compound E or cortisone (chemical name, 17-hydroxy-11-dehydrocorticosterone, signifying that the molecule is the same as corticosterone with —OH instead of —H on C-17, and H on C-11).

DOC (11-desoxycorticosterone – corticosterone molecules without any oxygen at C-11).

Relation of molecular structure to function:
1 Oxygen at C-11 produces glucocorticoid effects described on pp. 277 to 278.
2 OH at C-17 (in addition to oxygen at C-11) enhances glucocorticoid effects.
3 Aldehyde group at C-18 (as in aldosterone) produces marked salt-retaining effect.

Glucocorticoids
|
(stimulate)
↓
Protein mobilization from tissues
into blood
↓
Increased amounts of amino acids
transported to liver
↓
Increased rate of gluconeogenesis
from amino acids
↓
Increased blood sugar

Fig. 9-9 Scheme showing postulated
mechanism by which corticoid control
of carbohydrate metabolism may
stem from corticoid control of protein
metabolism.

Increased glucocorticoids increase liver
glyconeogenesis or gluconeogenesis. Presumably, these effects result from the accelerated rate of tissue protein metabolism.
As larger amounts of mobilized amino
acids reach the liver, its cells synthesize
some of them into glycogen and some into
glucose. More glucose, therefore, moves
out of liver cells into blood and tends to
increase the blood glucose level—tends to
produce hyperglycemia, that is. Hence, an
excess amount of glucocorticoids is said
to have gluconeogenic and hyperglycemic
or diabetogenic effects. Since this is an
opposite effect from that of insulin, it is
also referred to as the anti-insulin effect
of glucocorticoids. A deficient amount of
glucocorticoids, on the other hand, results
in slower gluconeogenesis and tends to
produce hypoglycemia.

3 Normal amounts of glucocorticoids
promote normal fat metabolism. A high
blood concentration of glucocorticoids
causes faster mobilization of fats from
adipose tissue and tends to accelerate fat
catabolism. In other words, an increase in
glucocorticoids tends to cause a "shift"
to fat utilization from the usual carbohydrate utilization by cells for their energy.
But, also, the mobilized fats, like mobilized
amino acids, may be used for gluconeogenesis. Chronic excess of glucocorticoids,
as in Cushing's syndrome (Fig. 9-10), re-

sults in a redistribution of body fat. It apparently accelerates fat mobilization from
the extremities but promotes fat deposition
in the face ("moon face"), shoulders ("buf-
falo hump"), trunk, and abdomen.

4 Glucocorticoids sensitize blood vessels to vasopressor substances such as
norepinephrine. Therefore, glucocorticoids
help maintain the degree of blood vessel
constriction necessary for maintaining
normal blood pressure.

5 Glucocorticoids are essential to the
body's ability to cope with stress successfully. A high blood concentration of
glucocorticoids brings about many of the
changes that characteristically occur when
the body is in a condition of stress. Outstanding among the "stress responses" induced by glucocorticoids are the following:

a Lymphocytopenia and eosinopenia—
that is, a marked decrease in the number
of lymphocytes and eosinophils in the
blood (takes place within about two hours
after an increase in glucocorticoids occurs)

b Involution or decrease in the size of
lymphatic tissues, particularly the thymus
gland and lymph nodes

c Decreased antibody formation, therefore less immunity and less hypersensi-
tivity—the so-called anti-immunity and
antiallergic effects of glucocorticoids

d Decreased proliferation of fibroblasts
(connective tissue cells), therefore less
inflammation but greater tendency for an
infection to spread and also slower wound
healing

e Increased extracellular fluid volume

f Changes in metabolism noted in items
1, 2, 3 as effects of high glucocorticoid
concentration

Mineralocorticoids

Aldosterone and desoxycorticosterone
are classified functionally as mineralocorticoids because their main action concerns
mineral salt (electrolyte) metabolism. The

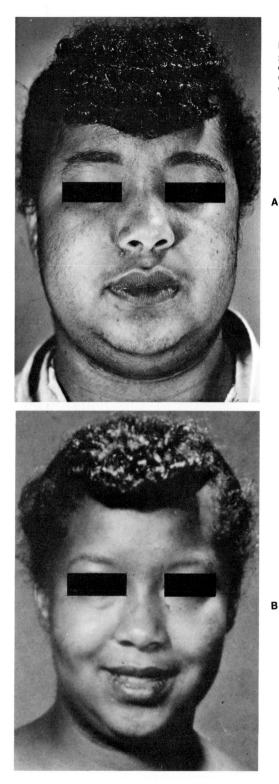

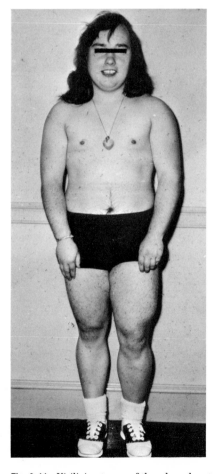

Fig. 9-10 Cushing's syndrome, the result of chronic excess glucocorticoids. **A**, Preoperatively. **B**, Six months post-operatively. (Courtesy Dr. William McKendree Jefferies, Case Western Reserve University School of Medicine, Cleveland, Ohio.)

Fig. 9-11 Virilizing tumor of the adrenal cortex. The tumor secretes excess androgens, thereby producing masculinizing effects. (Courtesy Dr. William McKendree Jefferies, Case Western Reserve University School of Medicine, Cleveland, Ohio.)

primary general function of mineralocorticoids seems to be to maintain homeostasis of the blood concentration of sodium. And, in doing this, these hormones also help maintain a normal ratio of blood sodium concentration to potassium concentration and normal volumes of extracellular and intracellular fluids. In short, mineralocorticoids play a crucial part in maintaining electrolyte and fluid balance and therefore in maintaining healthy survival. Aldosterone is some twenty to thirty times more powerful in these respects than desoxycorticosterone, so aldosterone is by far the more important mineralocorticoid.

Aldosterone acts on the distal renal tubule cells, stimulating them to increase their reabsorption of sodium ions from tubule urine back into blood. In exchange for each reabsorbed sodium ion, tubule cells excrete either a potassium or a hydrogen ion. Moreover, as each positive sodium ion is reabsorbed, a negative ion (bicarbonate or chloride) follows along, drawn by the attraction force between ions that bear opposite electrical charges. Also, the reabsorption of electrolytes causes net diffusion of water back into blood. Briefly, then, because of its primary sodium-reabsorbing effect on kidney tubules, aldosterone tends to produce sodium and water retention but potassium and hydrogen ion loss.

Sex hormones

The adrenal cortex in both sexes secretes small amounts of both male and female hormones—androgens, estrogens, and progesterone, respectively. In the amounts normally secreted, they promote normal development of bones and reproductive organs. But sometimes tumors of the adrenal cortex secrete excessive amounts of androgens. Because androgens produce masculinizing effects (for instance, growth of a beard in a woman), they are known as virilizing tumors (Fig. 9-11).

Control of secretion

Glucocorticoids (cortisol mainly). For discussion of mechanisms that control adrenal secretion of these hormones, see pp. 270 to 272.

Mineralocorticoids (aldosterone mainly). Many details of the apparently complex mechanism that controls aldosterone output by the adrenal cortex have not yet been established. But these two facts are known about it—that aldosterone secretion quickly increases (1) if the circulating blood volume and arterial blood pressure decrease markedly and (2) if the blood concentration of sodium decreases below a certain point. Probably these factors act in several ways to bring about increased aldosterone secretion. Fig. 9-12 shows one mechanism postulated to regulate aldosterone secretion. Fig. 15-1, p. 497, shows another.

According to the general consensus today, excess renin secretion, followed by excess angiotensin II formation, is one important cause of hypertension. Fig. 9-13 illustrates the mechanism involved.

Two clinical observations seem to illustrate this mechanism in operation—the fact that patients with malignant hypertension generally secrete large amounts of aldosterone and the fact that those with primary aldosteronism (excess aldosterone secretion) usually have hypertension. Excess production of angiotensin II would seem to cause both the malignant hypertension and the high aldosterone output. Interestingly enough, the trade name of a preparation of angiotensin II is Hypertensin.

Research findings of some investigators* suggest that the pineal body secretes a hormone which stimulates the zona glomerulosa of the adrenal cortex to secrete

---

*Dr. Gordon L. Farrell of Case Western Reserve University, Cleveland, Ohio, and others.

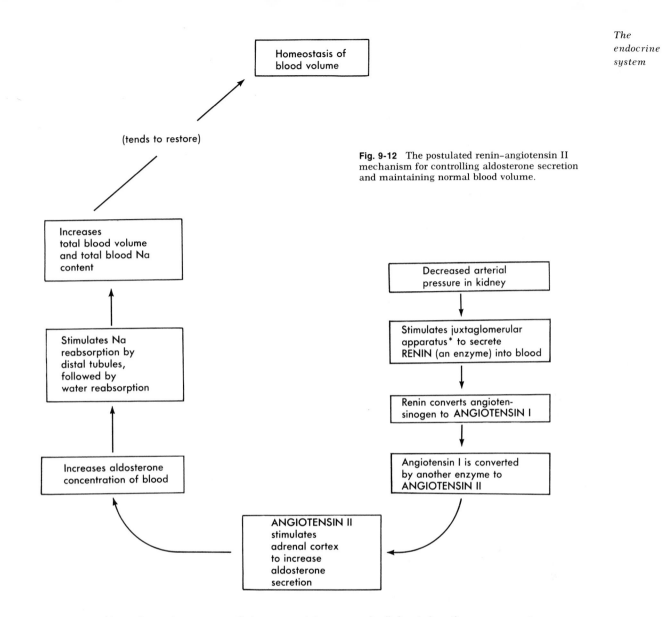

Fig. 9-12 The postulated renin–angiotensin II mechanism for controlling aldosterone secretion and maintaining normal blood volume.

*Juxtaglomerular apparatus (L. juxta, near to)—groups of cells located, as the name suggests, near glomeruli of kidney; lie in afferent arterioles just as they enter glomeruli.

aldosterone and which has therefore been named adrenoglomerulotropin.

■Adrenal medulla

Hormones

The adrenal medulla secretes the cate-cholamines, epinephrine and norepineph-rine. Like the adrenal corticoids, these hormones help the body meet stressful situations, but, unlike them, they are not essential for the maintenance of life. Epinephrine affects smooth and cardiac muscles and glands in a way similar to sympathetic stimulation of these structures. It serves to increase and prolong sympathetic effects.

Control of secretion

Increased epinephrine secretion by the adrenal medulla is one of the body's first

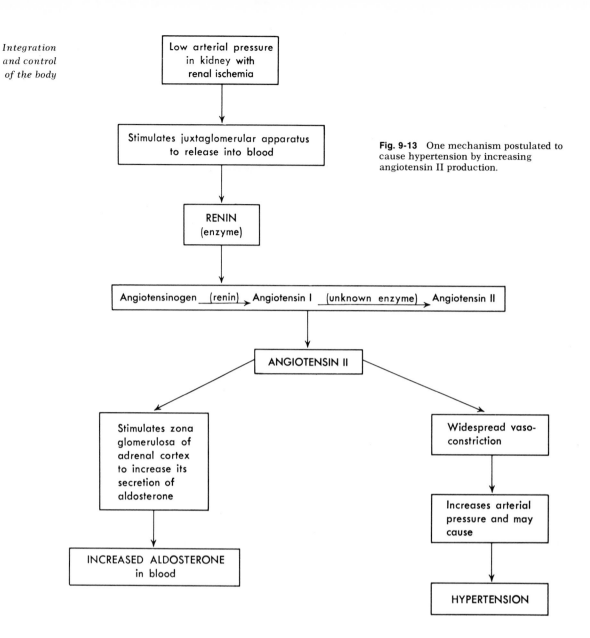

Fig. 9-13 One mechanism postulated to cause hypertension by increasing angiotensin II production.

responses to stress. It is thought to be brought about by impulses from the hypothalamus stimulating sympathetic preganglionic neurons which, in turn, stimulate the adrenal medulla to increase its output of epinephrine. Note that this is another example of the hypothalamus functioning in times of stress. And note, too, that the adrenal medulla is a neuroendocrine transducer—that is, a converter of nerve impulses into increased hormone secretion. We have already mentioned the hypothala-

mus as a neuroendocrine transducer. It brings about increased anterior pituitary secretion of ACTH (p. 271) and increased ADH secretion by the posterior lobe of the pituitary gland (p. 273) when a state of stress prevails.

■Islands of Langerhans

The beta cells of the islands of Langerhans secrete insulin, and the alpha cells secrete glucagon. Insulin's primary effects are to accelerate the transport of glucose, amino acids, and fatty acids through cell membranes into the intracellular fluid. Hence, insulin is an anabolic hormone in that it tends to promote glycogenesis, protein synthesis, and lipogenesis (deposition of fat in adipose tissue). But insulin also has a catabolic effect in that it tends to accelerate glucose utilization for energy (see Fig. 12-22, p. 420). Insulin, of course, tends to decrease blood glucose since it causes it to be transported into cells from the blood at a fast rate.

Glucagon serves as an antagonist to insulin. It accelerates liver glycogenolysis so tends to increase blood glucose. For control of islet secretion of insulin and glucagon, see Figs. 12-22 and 12-23, pp. 420 and 421.

■Ovaries

The ovaries produce two kinds of steroid female hormones: estrogens (chiefly estradiol and estrone) and progesterone.

Estrogens

Estrogens are secreted mainly by the graafian follicles and during pregnancy by the placenta. They carry on the following functions.

1 Estrogens accelerate proliferation of epithelial cells, particularly in organs of the female reproductive system. For example, during the early phase of each menstrual cycle, by stimulating epithelial cells to reproduce themselves more rapidly, estrogens promote regeneration of the endometrium (which sloughed off during the preceding menses) and cause it to thicken during the preovulatory phase.

Another example of estrogen's accelerating effect on epithelial cell mitosis is the thickening of the vaginal epithelium with cornification that occurs during the early part of each menstrual cycle.

2 Estrogens affect electrolyte and fluid metabolism in a way similar to the adrenal corticoids though to a lesser degree. They promote renal tubule reabsorption of sodium and water so tend to produce salt and water retention—particularly in the endometrium.

3 Estrogens affect protein metabolism. Specifically, they promote protein anabolism, an action that tends to promote growth. But also, estrogens exert an opposite influence on growth. They promote closure of the epiphyses and thus tend to terminate bone growth. Presumably, a girl's height is lessened if she reaches puberty (with its accompanying high rate of estrogen secretion) at an early age. For more about the effects of estrogen on bone, see p. 68.

4 Estrogens promote myometrial contractions.

5 Estrogens help control development and functioning of breasts. During adolescence and pregnancy, they especially promote development of ducts in the mammary glands. They inhibit secretion of lactogenic hormone by the anterior lobe of the pituitary gland. Hence, the high concentration of estrogens maintained during pregnancy inhibits lactation. Shedding of the placenta following delivery cuts off a major source of estrogens. Blood estrogen content drops rapidly and markedly. This decrease in the blood concentration of estrogens stimulates the secretion of prolactin which, in turn, initiates lactation.

6 A high blood concentration of estrogens inhibits secretion of FSH by the anterior lobe of the pituitary gland—a physiological fact of worldwide importance, for it constitutes the crucial part of the rationale which led to the development of contraceptive pills (p. 486).

7 Estrogens contribute to normal sexual desire (libido) but probably play a less crucial part than do various mental and emotional factors.

pancreas - alpha cells - [secretes insulin?]
beta cells - secretes insul

Progesterone

Progesterone is secreted by the corpus luteum and by the placenta. It has the following effects.

1 Progesterone acts on estrogen-primed endometrium, converting it to a secretory type of epithelium. An increasing blood concentration of progesterone causes an increasing secretion by endometrial glands and thereby prepares the uterine lining for implantation of a fertilized ovum. Incidentally, the name progesterone means a hormone that favors gestation—that is, pregnancy.

2 Progesterone acts on estrogen-primed breasts to promote development of the secreting cells to the stage where they can secrete milk when stimulated by prolactin following the delivery of the baby.

3 Progesterone decreases or inhibits contractions of the uterine smooth muscle (myometrium), another progestational effect.

4 Progesterone, like the corticoids, but to a lesser degree, promotes protein catabolism. It also is thought to have salt-retaining and water-retaining effects, particularly in the endometrium.

■ Testes

The interstitial cells of the testes secrete testosterone, a steroid hormone classed as an androgen—that is, a substance that promotes "maleness." It promotes development of the secondary sex organs and characteristics, maintains them in the adult state, and contributes to normal sexual behavior (also see p. 468). Urine of both men and women contains androsterone and possibly other androgens of the class of compounds known as 17-ketosteroids. Part of the urinary 17-ketosteroids derive from adrenal cortex secretions and part from testosterone.

■ Thymus

One of the body's best-kept secrets has been the function of the thymus. During all the years before 1961, there were no significant clues as to its role. Then a young Briton, Dr. Jacques F. A. P. Miller, removed the thymus glands from newborn mice. His findings proved startling and crucial. Almost like a chain reaction, further investigations followed and led at last to an uncovering of the thymus' long-held secret. It now seems clear that this small structure (it weighs at most only about an ounce) plays a critical part in the body's defenses against infections—in its vital immunity mechanism, that is.

In the mouse, and presumably in man as well, the thymus does two things. First, it serves as the original source of lymphocytes before birth. And then soon after birth, it starts secreting a hormone that enables lymphocytes to develop into plasma cells. Plasma cells, you will recall, synthesize antibodies against foreign proteins. Hence, the thymus functions to make possible immunity against microorganisms and other kinds of foreign proteins—for instance, against tissues transplanted from one individual to another. The thymus probably completes its essential work early in childhood. Its size is largest, relative to the rest of the body, when a child is about 2 years old. Its absolute size is largest at puberty. And from then on, it gradually atrophies until in great old age, it may be barely recognizable. The thymus is located in the mediastinal portion of the thoracic cavity and extends up to the lower edge of the thyroid gland in the neck.

■ Pineal body (pineal gland or epiphysis cerebri)

The pineal body has long been a mystery organ. Even now its function in the human being remains a matter of conjecture. Certain experiments* performed on rats, however, indicate that in these animals the pineal body is a neuroendocrine transducer that serves as a kind of "biological clock." It operates in this way—stimulation of the rat's retina by light initiates impulses in optic nerve fibers. Some of these impulses cross synapses to activate sympathetic neurons in the superior cervical ganglion. These sympathetic fibers then transmit impulses to the pineal gland, inhibiting its secretion of a hormone called melatonin. The resulting low blood concentration of melatonin has a stimulating effect on the rat's ovaries and accelerates its estrus cycle. Darkness initiates an opposite chain of events: increased melatonin synthesis, a resultant inhibition of the rat's ovaries, and a slowing of its estrus cycle. These findings and others have led to interesting speculations—that the pineal body may function in the human being to help regulate sex gland activity and the menstrual cycle and that perhaps it may even play a part in regulating various circadian rhythms such as adrenal cortical secretion. Whether or not these "educated guesses"

*See supplementary readings for Chapter 9, reference 3.

prove true remains to be shown by further investigations.

The pineal body is also postulated to secrete adrenoglomerulotropin (see p. 280) and serotonin in addition to melatonin. Serotonin plays an essential, but still not well understood, part in normal brain functioning.

The pineal body is shaped like a small cone about a centimeter long. It is located in the cranial cavity behind the midbrain and third ventricle and is attached to the roof of the latter. It degenerates at about 7 years of age and in the adult consists of fibrous tissue.

■ Placenta

The placenta functions as a temporary endocrine gland. During pregnancy, it produces chorionic gonadotropins—so-called because they are secreted by cells of the chorion, the outermost fetal membrane. In addition to gonadotropins, the placenta also produces estrogens and progesterone. The discovery more than fifty years ago by Aschheim and Zondek of the fact that large amounts of gonadotropic substances are excreted in the urine during pregnancy led to the development of the now well-known pregnancy tests.

---

outline summary

MEANING

1 Composed of glands that pour secretions into blood instead of into ducts
2 General functions and importance
  a Communication, control, integration; hormones main regulators of metabolism, reproduction, and responses to stress
  b Names and locations of endocrine glands—see Table 9-1

PITUITARY BODY (HYPOPHYSIS CEREBRI)

Consists of two endocrine glands: anterior lobe of pituitary gland (adenohypophysis) and posterior lobe of pituitary gland (neurohypophysis)

Anterior pituitary gland (adenohypophysis)
1 Cells called acidophils secrete growth hormone (STH) and prolactin
2 Basophils secrete thyrotropin (TSH), adrenocorticotropin (ACTH), follicle-stimulating hormone (FSH), luteinizing hormone (LH), and melanocyte-stimulating hormone (MSH)
3 Growth hormone (somatotropin or somatotropic hormone)
  a Accelerates protein anabolism so promotes growth
    1 Disorders caused by excess growth hormone: gigantism and acromegaly
    2 Disorders caused by deficient growth hormone: dwarfism and pituitary cachexia
  b Tends to inhibit fat deposition and to accelerate fat mobilization and catabolism

c Acts in some way to oppose insulin (anti-insulin, hyperglycemic, diabetogenic effects)
4 Prolactin—lactogenic hormone, also formerly called luteotropic hormone (LTH)
  a During pregnancy, promotes breast development
  b After delivery, initiates milk secretion
5 Tropic hormones
  a Thyrotropin—thyroid-stimulating hormone (TSH)
    1 Promotes growth and development of thyroid gland
    2 Stimulates thyroid gland to secrete thyroid hormone
  b Adrenocorticotropin (ACTH)
    1 Promotes growth and development of adrenal cortex
    2 Stimulates adrenal cortex to secrete cortisol and other glucocorticoids
  c Follicle-stimulating hormone (FSH)
    1 Stimulates primary graafian follicle to start growing and to develop to maturity
    2 Stimulates follicle cells to secrete estrogens
    3 In male, stimulates development of seminiferous tubules and maintains spermatogenesis by them
  d Luteinizing hormone (LH)
    1 Acts with FSH to cause complete maturation of follicle
    2 Brings about ovulation
    3 Stimulates formation of corpus luteum (luteinizing effect)
    4 Stimulates corpus luteum to secrete progesterone and estrogens
    5 In male, LH called interstitial cell–stimulating hormone (ICSH); stimulates interstitial cells in testis to develop and secrete testosterone
    6 Just before puberty, presumably sudden marked increase in secretion of gonadotropins (FSH, LH) initiates first menses
6 Melanocyte-stimulating hormone (MSH)—tends to produce increased pigmentation of skin
7 Control of secretion
  a Negative feedback mechanisms operate between target glands and anterior lobe of pituitary gland
    1 High blood concentration of target gland hormone inhibits secretion of tropic hormone by anterior lobe of pituitary gland
    2 High blood concentration of tropic hormone stimulates target gland secretion of its hormone
    3 Under ordinary conditions, negative feedback mechanisms control anterior pituitary secretion of ACTH, thyroid-stimulating hormone, and FSH and secretion of their respective target hormones

  b Neuroendocrine mechanisms dominate control of anterior pituitary secretion under stress conditions
    1 Hypothalamus releases neurosecretions into pituitary portal system; transported to anterior pituitary, where stimulate it to secrete certain hormones
    2 Stress acts in some way to stimulate the release of CRF, TRF, and perhaps other neurosecretions by hypothalamus—e.g., corticotropin-releasing factor (CRF) from hypothalamus stimulates secretion of ACTH by anterior lobe of pituitary gland; TRF stimulates pituitary secretion of thyroid hormone

Posterior pituitary gland (neurohypophysis)

1 Secretes hormones (antidiuretic hormone [ADH] and oxytocin) synthesized by neurons in hypothalamus
2 ADH (vasopressin or Pitressin) stimulates water reabsorption by distal and collecting tubules and stimulates smooth muscle of blood vessels and intestine
3 Oxytocin stimulates contractions of pregnant uterus and release of milk by lactating breast
4 Control of secretion
  a ADH secretion—details of mechanism controlling secretion of this hormone not established but general principles follows:
    1 Decreased ECF osmotic pressure leads to decreased ADH secretion
    2 Decreased ECF volume leads to increased ADH secretion
    3 Stress leads to increased ADH secretion
  b Oxytocin—details of mechanism controlling oxytocin secretion not established but is known that stimulation of nipples by suckling leads to increased oxytocin secretion

THYROID GLAND

1 Location and structure
  a Located in neck just below larynx
  b Two lateral lobes connected by isthmus
2 Functions
  a Stores iodine-containing protein, thyroglobulin
  b Secretes thyroxine and triiodothyronine—that stimulate rate of oxygen consumption (metabolic rate) of all cells and thereby help regulate physical and mental development, development of sexual maturity, and numerous other processes
  c Secretes thyrocalcitonin that decreases blood calcium concentration
3 Effects of hypersecretion and hyposecretion
  a Hypersecretion produces exophthalmic goiter
  b Hyposecretion in early life produces malformed dwarfism or cretinism; in later life, myxedema

PARATHYROID GLANDS

1 Location and structure
  a Attached to posterior surfaces of thyroid gland
  b Small round bodies, usually four or five in number
2 Functions—secrete parathormone which
  a Stimulates bone breakdown or resorption, releasing calcium and phosphate, thereby increasing calcium and phosphate absorption into blood from bone
  b Also accelerates calcium absorption from intestine and kidney tubules
  c Accelerates kidney tubule excretion of phosphates from blood into urine
3 Hypersecretion causes decrease in bone mass with replacement by fibrous tissue
4 Hyposecretion produces hypocalcemia and tetany and death in few hours

ADRENAL GLANDS

1 Location and structure
  a Located atop kidneys
  b Outer portion of gland called cortex and inner portion called medulla

Adrenal cortex

1 Zones or layers (from outside in)
  a Zona glomerulosa—secretes mineralocorticoids
  b Zona fasciculata—secretes glucocorticoids
  c Zona reticularis—secretes small amounts of glucocorticoids and sex hormones
2 Glucocorticoids (mainly cortisol, smaller amounts of corticosterone)
  a Promote normal protein metabolism; tend to accelerate tissue protein mobilization with resultant negative nitrogen balance
  b Promote normal carbohydrate metabolism; tend to accelerate liver gluconeogenesis, presumably from mobilized tissue proteins; hence, glucocorticoids tend to produce hyperglycemia
  c Promote normal fat metabolism; tend to accelerate fat mobilization and catabolism— i.e., tend to cause "shift" to fat utilization from usual carbohydrate utilization
  d Sensitize blood vessels to vasopressor substances such as norepinephrine
  e High blood concentration of glucocorticoids characteristic in stress; brings about many stress responses as follows:
    1 Lymphocytopenia and eosinopenia
    2 Involution of thymus gland and lymph nodes
    3 Decreased antibody formation, decreased immunity, and hypersensitivity
    4 Decreased connective tissue proliferation, less inflammation but greater tendency for infection to spread and slower wound healing
    5 Increased ECF volume
    6 Water diuresis
    7 Increased resistance to stress
    8 Increased tissue protein mobilization and catabolism, increased fat mobilization and catabolism, and increased liver gluconeogenesis
3 Mineralocorticoids (mainly aldosterone, also desoxycorticosterone)
  a Accelerate renal tubule reabsorption of sodium ions and excretion of potassium ions (or hydrogen ions)
  b Increased renal tubule reabsorption of bicarbonate ions (or chloride ions) and water result from increased sodium reabsorption
4 Sex hormones
  a Small amounts of both male and female hormones secreted in both sexes
  b Amounts normally secreted promote normal development of bones and reproductive organs
5 Control of secretion
  a Glucocorticoids
    1 Under ordinary conditions, negative feedback mechanism operates between anterior pituitary gland and adrenal cortex; high blood concentration of ACTH stimulates glucocorticoid secretion; high blood glucocorticoid concentration inhibits ACTH secretion
    2 Under stress conditions, increased CRF release by hypothalamus stimulates increased ACTH secretion and therefore increased glucocorticoid secretion
  b Mineralocorticoids
    1 Details of mechanism controlling aldosterone secretion not firmly established but known that decreased ECF volume and decreased arterial pressure lead to increased aldosterone secretion (see Fig. 9-12)
    2 Pineal body postulated to secrete hormone adrenoglomerulotropin, which stimulates aldosterone secretion

Adrenal medulla

1 Hormones—epinephrine mainly; some norepinephrine
2 Functions of epinephrine—affects visceral effectors (smooth muscle, cardiac muscle, and glands) in same way as does sympathetic stimulation of these structures; epinephrine from adrenal glands intensifies and prolongs sympathetic effects
3 Control of secretion—stress acts in some way to stimulate hypothalamus, which sends impulses to adrenal medulla via sympathetic neurons, stimulating medulla to increase its secretion of epinephrine

ISLANDS OF LANGERHANS

1 Hormones
  a Beta cells secrete insulin
  b Alpha cells secrete glucagon
2 Functions
  a Insulin
    1 Promotes glucose transport into cells, thereby increasing glucose utilization (catabolism) and glycogenesis and decreasing blood glucose
    2 Promotes fatty acid transport into cells and fat anabolism (lipogenesis or fat deposition) in them
    3 Promotes amino acid transport into cells and protein anabolism
  b Glucagon—accelerates liver glycogenolysis so tends to increase blood glucose; in short, is insulin antagonist

OVARIES

1 Hormones secreted—estrogens and progesterone
2 Functions
  a Estrogens
    1 Accelerate proliferation of epithelial cells, especially in female reproductive system— e.g., bring about regeneration of endometrium following menses and thickening of it before ovulation
    2 Mild stimulation of renal tubule reabsorption of sodium and water
    3 Promote protein anabolism and therefore growth but also promote epiphyseal closure so tend to terminate bone growth
    4 Promote myometrial contractions
    5 Promote development of mammary glands during adolescence and pregnancy; inhibit production of lactogenic hormone by anterior lobe of pituitary gland, thereby inhibiting lactation
    6 Inhibit production of FSH by anterior lobe of pituitary gland
    7 Probably contribute to normal sexual desire
  b Progesterone
    1 Converts estrogen-primed endometrium to secretory type epithelium, thereby causing increased secretion by endometrium
    2 Acts on estrogen-primed breasts, promoting development of secreting cells (to stage where can secrete milk when stimulated by lactogenic hormone)
    3 Inhibiting effect on myometrial contractions
    4 Mildly accelerating effect on protein catabolism
    5 Thought to promote salt and water retention, especially in endometrium

TESTES

1 Secrete testosterone, which
  a Accelerates protein anabolism so promotes growth, but also promotes epiphyseal closure so tends to terminate bone growth
  b Promotes development of secondary sex organs and characteristics and maintains them in adult condition
  c Contributes to normal sexual behavior

THYMUS

1 Postulated functions
  a Before birth, original lymphocytes formed in thymus
  b Soon after birth, thymus secretes hormone that enables lymphocytes to develop into plasma cells and synthesize antibodies; hence, thymus plays crucial role in immunity
2 Thymus gradually atrophies after puberty

PINEAL BODY (PINEAL GLAND OR EPIPHYSIS CEREBRI)

1 Functions not established; some evidence suggests that it may help regulate rhythmically recurring functions such as menstrual cycle and perhaps even circadian rhythms such as daily variations in amounts of corticoids secreted
2 Postulated to secrete adrenoglomerulotropin, melatonin, and serotonin

PLACENTA

1 Temporary endocrine glands
2 Secretes estrogens, progesterone, and chorionic gonadotropin
3 Helps maintain progestational state of endometrium

review questions

1 Name the endocrine glands and locate each one.
2 Name the hormone or hormones which help control each of the following: blood sugar level; blood calcium level; blood sodium level; blood potassium level. Explain mechanisms involved.
3 Name the hormone or hormones that help control each of the following: growth; development of secondary male characteristics and female characteristics; fluid and electrolyte balance; resistance to stress; functions of the adrenal cortex, thyroid gland, and ovaries; secretion of ACTH, TH, and FSH.
4 What hormone enhances and prolongs sympathetic effects?
5 What hormones help control protein metabolism? Explain.
6 What hormones help control fat metabolism? Explain.
7 What hormones help control carbohydrate metabolism? Explain.
8 Contrast the control of hormone secretion by the anterior pituitary gland under normal conditions and under stress conditions.
9 What conditions result from a deficiency of thyroid extract early in life? Later in life?
10 What disease results from too much thyroid secretion? From too much pituitary somatotropic hormone early in life? Too much later in life? Too little early in life?
11 Gigantism results from an oversupply of which endocrine secretion? Cretinism from a deficiency of which one? Acromegaly from an oversupply of which one? Myxedema from a deficiency of which one?
12 Which gland is called the "master gland"? Why?

Situation: A patient goes to his doctor because he feels "so terribly weak and tired all the time and has lost so much weight." After hospital admission and various tests, the doctor tells him he has "something wrong with his adrenal glands." Blood chemistry tests reveal low blood concentrations of sodium and chloride and high potassium.

13 Do you think this patient has a disease of the adrenal medulla or adrenal cortex? State reasons.
14 Is the condition hyperfunctioning or hypofunctioning of the gland? State reasons.
15 How would you explain the patient's chief complaints?
16 Might this patient have any type of water imbalance? If so, what and why? Explain mechanisms involved.
17 Eosinophil counts were done on this patient following injections of Adrenalin (commercial preparation of epinephrine). Do you think the counts would be normal or not? Explain why. Significance?
18 Blood sugar was measured after a twenty-four-hour fast. Would it be higher or lower than it would be in a normal individual under the same conditions? Why?

Situation: A patient who has a severe form of Cushing's disease is given insulin each day.

19 Should the nurse watch this patient for signs of hypoglycemia or hyperglycemia? Explain.
20 Would this patient be likely to be "insulin sensitive" or "insulin resistant"? Explain.
21 Explain the meaning of the statement "hormone functions are interrelated." Give examples to support or refute this statement.

Maintaining the metabolism of the body

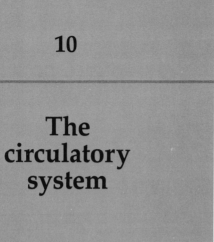

10

The circulatory system

Functions

Blood
Blood cells
  Erythrocytes
  Leukocytes
  Platelets
Blood types (or blood
  groups)
Blood plasma
Blood clotting
  Purpose
  Mechanism
  Factors that oppose
    clotting
  Factors that hasten
    clotting
  Pharmaceutical
    preparations that retard
    clotting
  Clot dissolution
  Clinical methods of
    hastening clotting

Heart
Covering
  Structure
  Function
Structure
  Wall
  Cavities
  Valves and openings
  Blood supply
  Conduction system
  Nerve supply
Physiology
  Function
  Cardiac cycle

Blood vessels
Kinds
Structure
Functions
Main blood vessels
  Systemic circulation
    Arteries
    Veins
  Portal circulation
  Fetal circulation

Circulation
Definitions
How to trace
Functions of control
  mechanisms
Principles
Local control of
  arterioles
Important factors
  influencing venous
  return to heart

Blood pressure
How arterial blood
  pressure measured
  clinically
Relation to arterial and
  venous bleeding

Velocity of blood

Pulse
Definition
Cause
Pulse wave
Where pulse can be felt
Venous pulse

Lymphatic system
Definition
Lymph and interstitial fluid
  (tissue fluid)
  Definition
Lymphatics
  Formation and
    distribution
  Structure
  Function
Lymph circulation
Lymph nodes
  Structure
  Location
  Functions

Spleen
Location
Structure
Functions

[handwritten: 3 basic kinds of cells]

[handwritten: function { nutritive, respiratory, excretory, protection, transportation, regulatory]

■ Functions

Transportation is the primary function of the circulatory system. Its secondary functions — the functions to which it contributes — are every function of every cell and every function of the body as a whole. This is a sweeping statement and is therefore suspect, but we shall try to support it with substantial evidence as this and the remaining chapters of the book unfold. For now, just a few examples: the circulatory system transports food and oxygen to all cells so plays a vital part in cellular metabolism and in all cellular functions. It transports water and electrolytes so makes vital contributions to the maintenance of homeostasis — homeostasis of fluid volume and pH and even of body temperature. It transports hormones and enzymes so takes part in the control and integration of countless functions. It transports antibodies so contributes heavily to the body's defense against microorganisms. Our discussion in this chapter will center around the following main topics: blood, heart, blood ves-sels, circulation, blood pressure, velocity of blood, pulse, lymphatic system, and spleen.

Blood

Blood is much more than the simple liquid it seems to be. Not only does it consist of a fluid, but also of cells — billions and billions of them. The fluid portion of blood — that is, the *plasma* — is one of the three major body fluids. (Interstitial fluid and intracellular fluid are the other two.)

How much blood does an adult body contain? We can answer that question only with some generalizations. The total blood volume varies markedly in different individuals. One of the chief variables influencing normal blood volume is the amount of body fat. Blood volume per kilogram of body weight varies inversely with the amount of excess body fat. This means that the less fat there is in your body, the more blood you have per kilogram of

[handwritten: 7.35 - 7.45 ph. 40-50% red blood cells]

293

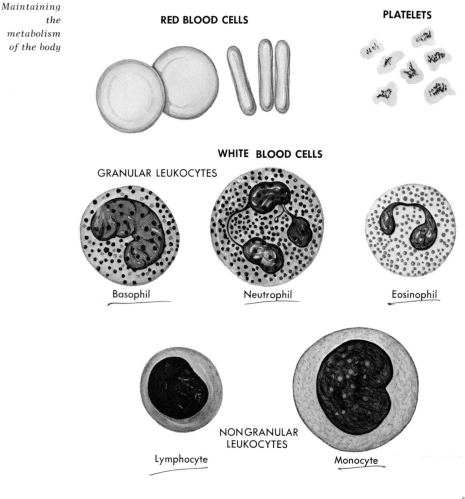

RED BLOOD CELLS

PLATELETS

WHITE BLOOD CELLS

GRANULAR LEUKOCYTES

Basophil

Neutrophil

Eosinophil

NONGRANULAR
LEUKOCYTES

Lymphocyte

Monocyte

Fig. 10-1 Human blood cells. There are close to 30 trillion red blood cells in the adult. Each cubic millimeter of blood contains from 4 1/2 to 5 1/2 million red blood cells and an average total of 7,500 white blood cells.

your body weight. According to Wennesland et al.,* healthy men average 71.4 milliliters of blood per kilogram of body weight. Approximately 55% of total blood volume is plasma volume, and the rest is blood cell volume. Now we shall take a rather detailed look at blood cells and then consider blood plasma.

■ Blood cells

Three main kinds of blood cells are recognized: red blood cells (erythrocytes), white blood cells (leukocytes), and plate-

*Wennesland, R., Brown, E., Hopper, J., Jr., Hodges, J. L., Jr., Guttentag, O. E., Scott, K. G., Tucker, I. N., and Bradley, B.: Red cell, plasma, and blood volume in healthy men measured by radiochromium (Cr⁵¹) cell tagging and hematocrit, J. Clin. Invest. **38:**1065-1077, 1959.

lets (thrombocytes). Leukocytes are further divided as shown in the following classifications of blood cells:

1 Red blood cells or erythrocytes
2 White blood cells or leukocytes
  a Granular leukocytes: basophils, neutrophils, and eosinophils
  b Nongranular leukocytes: lymphocytes and monocytes
3 Platelets (thrombocytes)

In another method of classification, blood cells are divided into two main types according to origin:

1 Myeloid cells (formed in myeloid tissue—that is, in red bone marrow)
  a Erythrocytes
  b Granular leukocytes: neutrophils, eosinophils, and basophils
  c Platelets

2 Lymphoid cells (or lymphatic cells; formed before birth in the thymus and afterward in lymphatic tissue mainly in lymph nodes and spleen)

    a Lymphocytes

    b Monocytes

Erythrocytes

Appearance and size

Facts about normal red cell size and shape hold more than academic interest. They are also clinically important. For example, an increase in red cell size characterizes one type of anemia. And a decrease in red cell size characterizes another type. Red cells are extremely small. More than 3,000 of them could be placed side by side in one inch since they measure only about 7 microns in diameter. A normal mature red cell has no nucleus. Just before the cell reaches maturity and enters the bloodstream from the bone marrow, the nucleus is extruded, with the result that the cell caves in on both sides. So, as you can see in Fig. 10-1, normal mature red cells are shaped like tiny biconcave disks.

Structure and functions

Red blood cell functions (specifically, the transport of oxygen and carbon dioxide) illustrate the familiar principle that structure determines function. Packed within one tiny red cell are an estimated 200 to 300 million molecules of the complex compound *hemoglobin.* One hemoglobin molecule consists of a protein molecule (globin) combined with four molecules of a pigmented compound (heme). Because each molecule of heme contains one atom of iron, one hemoglobin molecule contains four iron atoms. And this structural fact enables one hemoglobin molecule to unite with four oxygen molecules to form oxyhemoglobin (a reversible reaction). Hemoglobin can also combine with carbon dioxide to form carbaminohemoglobin (also reversible). But in this reaction, the structure of the globin part of the hemoglobin molecule rather than of its heme part makes the combining possible. Further discussion of oxygen and carbon dioxide transport appears in Chapter 11.

A man's blood usually contains more hemoglobin than a woman's. In most normal men, 100 milliliters of blood contains 13 to 16 grams of hemoglobin. The normal hemoglobin content of a woman's blood is a little less — specifically, in the range of 12 to 14 grams per 100 milliliters. Any adult who has a hemoglobin content of less than 12 (grams per 100 milliliters blood) is diagnosed as having *anemia.* An anemic person may or may not have an abnormally low red blood cell count. An adult whose blood contains more than about 17.5 grams hemoglobin per 100 milliliters is diagnosed as having *polycythemia.* His red blood cell count may or may not be higher than normal, although the word polycythemia means literally "many blood cells."

Formation (erythropoiesis)

Erythrocytes are formed in the red bone marrow* from nucleated cells known as hemocytoblasts or stem cells. These stem cells go through several stages of development (Fig. 10-2). First they develop into erythroblasts, then into reticulocytes, then into normoblasts, and finally they become mature, nonnucleated red blood cells or erythrocytes. Hemoglobin synthesis starts during the erythroblast stage. Extrusion

---

*Locations of red bone marrow are given on p. 67. Before birth, the liver and spleen produce red blood cells, but by birth and from that time on only the red bone marrow performs this function. An exception to this principle is that the liver and spleen may again produce red cells under some markedly abnormal conditions. With increased age comes decreased marrow productivity, a fact that may partially account for the anemia so common in old age.

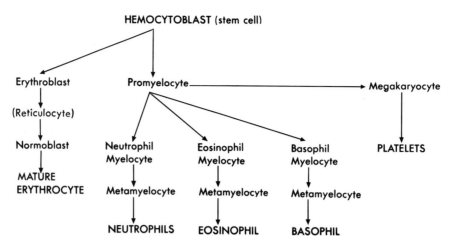

HEMOCYTOBLAST (stem cell)

Fig. 10-2 Stages of development of erythrocytes (red blood cells), granular leukocytes, and platelets in red bone marrow. Under ordinary conditions, when the need for red cells is not greater than normal, most erythroblasts develop directly into normoblasts without going through the reticulocyte stage.

of the nucleus occurs sometimes from erythroblasts but more often from normoblasts.

Frequently a physician needs information about the rate of erythropoiesis to help him make a diagnosis or prescribe treatment. A *reticulocyte count* gives this information. Approximately 1/2% to 1 1/2% of the red cells in normal blood are reticulocytes. A reticulocyte count of less than 1/2% of the red count usually indicates a slowdown in the process of red cell formation. Conversely, a reticulocyte count higher than 1 1/2% usually indicates an acceleration of red cell formation—as occurs, for example, following treatment of anemia.

Destruction

The life span of a red blood cell circulating in the bloodsteam is now believed to be about 120 days, based on studies using radioactive substances (isotopes) to "tag" red cells. Apparently as red cells grow older, their membranes become increasingly fragile and eventually rupture, causing the cell to break apart or fragment within the capillaries. Following this, reticuloendothelial cells in the liver, spleen, and bone marrow phagocytose the red cell fragments and break down their hemoglobin to yield an iron-containing pigment (hemosiderin) and bile pigments (bilirubin and biliverdin). Eventually, the bone marrow uses most of the iron over again for new red cell synthesis, and the liver excretes the bile pigments in the bile.

Erythrocyte homeostatic mechanism

Red cells are formed and destroyed at a breathtaking rate. Normally, every hour of every day of our adult lives, over 100 million red cells are formed to replace an equal number destroyed during that brief time. Obviously, some kind of homeostatic mechansim operates to balance the number of cells formed against the number destroyed, since, in health, the number of red cells remains relatively constant at about 4 1/2 to 5 1/2 million per cubic millimeter of blood. The exact mechanism responsible for this constancy is not known. It is known, however, that the rate of red cell production soon speeds up if either the

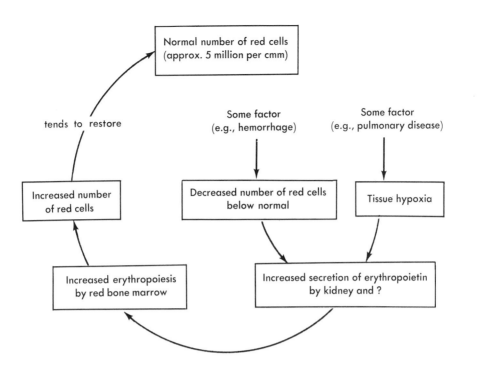

Fig. 10-3 Postulated red blood cell homeostatic mechanism—a negative feedback mechanism. A decrease in the number of circulating red blood cells "feeds back" to the red cell–forming structure (red bone marrow) to cause an increased rate of red cell formation which, in turn, tends to increase the number of red cells sufficiently to restore their normal number.

number of red cells decreases appreciably or if tissue hypoxia (oxygen deficiency) develops. Either of these conditions acts in some way to stimulate the kidneys (and perhaps some other structures) to increase their secretion of a hormone named *erythropoietin*. The resulting increased blood concentration of erythropoietin stimulates bone marrow to accelerate its production of red blood cells (Fig. 10-3). The name erythropoietin makes it easy to remember its function—erythropoietin stimulates erythropoiesis (process of red cell formation).

Note that for the red blood cell homeostatic mechanism to succeed in maintaining a normal number of red cells, the bone marrow must function adequately. To do this, the blood must supply it with adequate amounts of several substances with which to form the new red cells—iron and amino acids, for example, and also copper, vitamin B compounds, and

possibly cobalt to serve as catalysts. In addition, the gastric mucosa must provide some unidentified intrinsic factor necessary for absorption of vitamin B_{12} (called extrinsic factor because of its external source in foods; also called antianemic principle). The chain of reactions necessary for normal red cell production might be summarized as follows:

normal gastric mucosa
supplies
intrinsic factor
which
promotes absorption of vitamin B (extrinsic factor)
which
stimulates bone marrow
to produce
normal number of mature red blood cells

Failure to maintain homeostasis of red blood cells may result from interference at some point in the foregoing chain of reactions. For example, pernicious anemia develops when the gastric mucosa fails to produce sufficient intrinsic factor. Inadequate absorption of vitamin B_{12} then follows, and the bone marrow, deprived of stimulation by vitamin B_{12}, produces fewer but larger red cells than normal. Many of these cells are immature with overly fragile membranes, a fact which leads to their more rapid destruction.

A clinical example of failure of red cell homeostasis seen in recent years is anemia due to bone marrow injury by x-ray or gamma ray radiations.

Other factors may also cause marrow damage. To help diagnose this condition, a sample of marrow is removed from the sternum by means of a sternal puncture and is studied microscopically for abnormalities. When damaged marrow can no longer keep red cell production apace with destruction, red cell homeostasis is not maintained. Instead, the red cell count falls below normal. Anemia (fewer red cells than normal) develops whenever the rates of red cell formation and destruction become unequal. Therefore, either a decrease in red cell formation (as in pernicious anemia or bone marrow injury by radiation) or an increase in red cell destruction (as in infections and malignancies) can lead to anemia.

The number of red blood cells is determined by the *"red count"* or is estimated by the hematocrit. The *hematocrit* is the volume percentage of red cells in whole blood. To be more specific, a hematocrit of 47 means that in every 100 milliliters of whole blood there are 47 milliliters of blood cells and 53 milliliters of fluid (plasma). Normally, the average hematocrit for a man is about 47 (± 7, normal range) and for a woman about 42 (± 5).

Leukocytes

Appearance and size

Consult Fig. 10-1. Note particularly the differences in color of the cytoplasmic granules and in the shapes of the nuclei of the granular leukocytes. Neutrophils take their name from the fact that they stain with neutral dyes. And because their nuclei have two to five or more lobes, neutrophils are also called *polymorphonuclear leukocytes* or, to avoid that tongue twister, simply "polys." Eosinophils stain with acid dyes. Their nuclei have two oval lobes. Basophils stain with basic dyes. Their nuclei are roughly **S** shaped.

Lymphocytes and monocytes do not contain granules in their cytoplasm, a characteristic indicated by the classification of these cells as nongranular leukocytes.

Functions

White blood cells constitute part of the important defense mechanisms of the body. Granular leukocytes and monocytes carry on *phagocytosis,* a process in which they ingest and digest microorganisms and other foreign particles. Neutrophils and monocytes are most actively phagocytic; eosinophils are only moderately so. All leukocytes are motile cells. This characteristic enables them to move out of capillaries by squeezing through the intercellular spaces of the capillary wall (a process called *diapedesis*) and to migrate by ameboid movement toward microorganisms or other injurious particles that may have invaded the tissues. Neutrophils are highly motile, whereas lymphocytes, monocytes, and eosinophils are sluggishly so.

You may recall that reticuloendothelial cells also perform the function of phagocytosis. In general, however, they do this work within more localized areas than white cells. Reticuloendothelial cells, since they do not enter the bloodstream, cannot be transported by it to any part of the body, as white cells can. One more point—white blood cells carry on their function of phagocytosis in the tissues. Where do red cells perform their functions?

Lymphocytes play an important, though still not thoroughly understood, role in the development of immunity. Cogent evidence indicates that lymphocytes respond to the presence of antigens (usually foreign proteins introduced into the body) in the following two ways. Some of them multiply by mitosis and undergo some changes. They differentiate, that is, to become plasma cells. The plasma cells secrete antibodies into the blood which react, in the circulating blood, with the particular antigens that initiated the process that led to the formation of these particular antibodies. The reaction of an antibody with its antigen destroys that antigen or renders it harmless to the body —it "makes us immune," we say (to the particular antigens taking part in the antigen-antibody reaction).

As explained in the preceding paragraph, some lymphocytes differentiate into plasma cells which synthesize and secrete *circulating antibodies*—so-called for the obvious reason that they circulate in the bloodstream. Some lymphocytes, however, synthesize antibodies but do not secrete them into the blood. Instead, they remain in the cells that made them and, logically, are called *cellular antibodies*. The type of antigens that initiate cellular antibody formation are *cellular antigens*—that is, antigens present on or in cells. If, therefore, a tissue or an organ is transplanted from one person to another, its cellular antigens may initiate cellular-antibody formation by the recipient's lymphocytes. For example, lymphocytes may, and unfortunately often do, infiltrate the vicinity around a kidney transplant. There they proliferate and form cellular antibodies. These antibodies react with the kidney's cellular antigens and in the process destroy the foreign cells. In more picturesque language, "tissue-attacking lymphocytes reject" the transplanted kidney. *Tissue-at-*

tacking lymphocytes and *graft-rejection cells* are descriptive names for the same cells—that is, for lymphocytes that synthesize cellular antibodies. Recently, surgeons have given antilymphocyte serum (ALS) to patients receiving an organ transplant. Rationale? To try to decrease the number of infiltrating lymphocytes and the amount of cellular antibodies formed and, thereby, prevent rejection of the transplant. A team of University of Texas researchers has tried to prevent transplant rejection by a more dramatic method* (see p. 348).

Formation

All three types of granular leukocytes originate, as do erythrocytes, in myeloid tissue. In contrast, nongranular leukocytes (that is, lymphocytes and perhaps monocytes) derive from lymphatic tissue. The origin of monocytes is still uncertain. During the past several years, researchers have accumulated persuasive evidence that the thymus gland of the fetus synthesizes the body's original lymphocytes. From the thymus, lymphocytes enter the bloodstream, then circulate to the lymphatic tissues (mainly lymph nodes and spleen), and there take up residence. It is these lymphocytes in the lymphatic tissues that multiply to form new lymphocytes after birth. Many lymphocytes are found in bone marrow, but presumably they were formed in lymphatic tissues and carried to the bone marrow by the bloodstream.

Myeloid tissue (bone marrow) and lymphatic tissue together constitute the hem-

---

*Immune reaction: Alternative to drugs, Sci. News **96**:327 (Oct. 11), 1969. For information about new and persuasive evidence that lymphocytes may "recognize antigens and kill the tumors that carry them" and the exciting implications of this probability, see Culliton, B. J.: Antigens on the cell, Sci. News **95**:457-459 (May 10), 1969.

opoietic or blood cell-forming tissues of the body. Red bone marrow is myeloid tissue that is actually producing blood cells. Its red color comes from the red cells it contains. Yellow marrow, on the other hand, is yellow because it stores considerable fat. It is not active in the business of blood cell formation so long as it remains yellow. Sometimes, however, it becomes active and red in color when an extreme and prolonged need for red cell production occurs.

Granular leukocytes, like erythrocytes, pass through several recognized stages before becoming mature cells: hemocytoblast (or myeloblast), promyelocyte, myelocyte, metamyelocyte, and mature leukocyte (Fig. 10-2).

Destruction and life span

The life span of white blood cells is not known. Some evidence seems to indicate that granular leukocytes may live three days or less, whereas other evidence suggests that they may live about twelve days. Some of them are probably destroyed by phagocytosis and some by microorganisms. The most recent evidence about the life span of lymphocytes indicates that some, or perhaps all, lymphocytes live as long as 300 days. This refutes earlier evidence interpreted to mean that lymphocytes probably lived only twenty-four hours or so.

Numbers

A cubic millimeter of blood normally contains about 5,000 to 9,000 leukocytes, with definite percentages of each type (Table 10-1). Because these percentages change in certain abnormal conditions, they have clinical significance. In acute appendicitis, for example, the percentage of neutrophils increases and so, too, does the total white count. In fact, these characteristic changes may be the deciding points for surgery.

The procedure in which the different types of leukocytes are counted and their percentage of the total white count is computed is known as a *differential count*. In other words, a differential count is a percentage count of white cells. The different kinds of white cells and a normal differential count are listed in Table 10-1. A decrease in the number of white blood cells is *leukopenia*. An increase in the number of white cells is *leukocytosis*. (*Leukemia* is a malignant disease characterized by a marked increase in the number of white blood cells.)

Platelets

Appearance and size

To compare platelets with other blood cells as to appearance and size, see Fig. 10-1. Note that platelets look like small fragments of cells.

Functions

Platelets help set in operation the blood-clotting mechanism (p. 304).

Formation and life span

Platelets are formed in the red bone marrow presumably by fragmentation of very large cells known as megakaryocytes (Fig. 10-2). Presumably, platelets have a life span of a week or slightly more. A summary of the basic facts about blood cells is given in Table 10-2.

■ Blood types (or blood groups)

The term blood type refers to the type of antigens* present on red blood cell membranes. Antigens A, B, and Rh are the most important blood antigens as far as transfusions and newborn survival are con-

*Antigen—substance capable of stimulating formation of antibodies which can react with the antigen; e.g., to agglutinate or clump it.

Table 10-1. White blood cells

| | Differential count* | |
	Normal range (%)	Typical normal (%)
Class		
Those with nongranular cytoplasm and regular nucleus		
Lymphocytes (large and small)	20 to 25	25
Monocytes	3 to 8	6
Those with granular cytoplasm and irregular nuclei – leukocytes		
Eosinophils (acid staining)	2 to 5	3
Basophils (basic staining)	1/2 to 1	1
Neutrophils (neutral staining)	65 to 75	65
Total		100

*In any differential count, the sum of the percentages of the different kinds of leukocytes must, of course, total 100%.

Table 10-2. Blood cells

Cells	Number	Function	Formation (hemopoiesis)	Destruction
Red blood cells (erythrocytes)	4 1/2 to 5 1/2 million/cmm (total of approximately 30 trillion in adult body)	Transport oxygen and carbon dioxide	Red marrow of bones (myeloid tissue)	By fragmentation in circulating blood and by macrophages of spleen, liver, and red bone marrow; thought to live about 120 days in bloodstream
White blood cells (leukocytes)	5,000 to 9,000/cmm	Play important part in producing immunity – e.g., phagocytosis by neutrophils; lymphocytes form cellular antibodies; some lymphocytes become plasma cells, cells that form circulating antibodies	Granular leukocytes in red marrow; original lymphocytes formed in thymus gland of fetus; postnatally, lymphocytes formed in lymph nodes and other lymphatic tissues	Not known definitely; probably some destroyed by phagocytosis
Platelets (thrombocytes)	250,000 to 450,000/cmm; wide variation with different counting methods	Initiate blood clotting	Red marrow	Unknown

cerned. Many other antigens have also been identified, but they are less important clinically and are too complex to discuss here. Every person's blood belongs to one of the four AB blood groups and, in addition, is either Rh positive or Rh negative. Blood types are named according to the antigens present on red cell membranes. Here, then, are the four AB blood types:

1 *type A* — antigen A on red cells
2 *type B* — antigen B on red cells
3 *type AB* — both antigen A and antigen B on red cells
4 *type O* — neither antigen A nor antigen B on red cells

The term *Rh-positive blood* means that Rh antigen is present on its red cells. *Rh-negative blood*, on the other hand, is blood whose red cells have no Rh antigen present on them.

Blood plasma may or may not contain antibodies that can react with red cell antigens A, B, and Rh. An important principle about this is that plasma never contains antibodies against the antigens present on its own red blood cells — for obvious reasons. If it did, the antibody would react with the antigen and thereby destroy the red cells. But (and this is an equally important principle) plasma does contain antibodies against antigen A or antigen B if they are *not* present on its red cells. Applying these two principles: In type A blood, antigen A is present on its red cells; therefore, its plasma contains no anti-A antibodies but does contain anti-B antibodies. In type B blood, antigen B is present on its red cells; therefore, its plasma contains no anti-B antibodies but does contain anti-A antibodies. What antigens do you deduce are present on the red cells of type AB blood? What antibodies, if any, does its plasma contain?*

*Type AB blood — antigens A and B present on red cells; no anti-A nor anti-B antibodies in plasma.

No blood normally contains anti-Rh antibodies. However, anti-Rh antibodies can appear in the blood of an Rh-negative person provided Rh-positive red cells have at some time entered his bloodstream. One way this can happen is by giving an Rh-negative person a transfusion of Rh-positive blood. In a short time, his body makes anti-Rh antibodies, and these remain in his blood. There is one other way in which Rh-positive red cells can enter the bloodstream of an Rh-negative individual — but this can happen only to a woman. If she becomes pregnant, and if her mate is Rh positive, and if the fetus, too, is Rh positive, some of the red cells of the fetus may find their way into her blood from the fetal blood capillaries (via the placenta). These Rh-positive red cells then stimulate her body to form anti-Rh antibodies. Briefly, the only people who can ever have anti-Rh antibodies in their plasma are Rh-negative men or women who have been transfused with Rh-positive blood or Rh-negative women who have carried an Rh-positive fetus.

Anti-A, anti-B, and anti-Rh antibodies are agglutinins. *Agglutinins* are chemicals that agglutinate cells — that is, make them stick together in clumps. The danger in giving a blood transfusion is that antibodies present in the plasma of the person receiving the transfusion (the recipient's plasma) may agglutinate the donor's red blood cells. If, for example, type B blood were used for transfusing a person who had type A blood, the anti-B antibodies in the recipient's blood would agglutinate the donor's type B red cells. These clumped cells are potentially lethal. They can plug vital small vessels and cause the recipient's death.

Type O blood is referred to as *universal donor* blood. Not only can it be transfused safely into a person whose blood is also type O, but it can be given as well to one

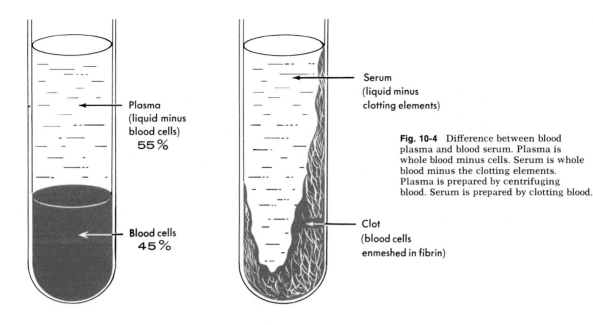

Plasma
(liquid minus
blood cells)
55%

Blood cells
45%

Serum
(liquid minus
clotting elements)

Clot
(blood cells
enmeshed in fibrin)

Fig. 10-4 Difference between blood plasma and blood serum. Plasma is whole blood minus cells. Serum is whole blood minus the clotting elements. Plasma is prepared by centrifuging blood. Serum is prepared by clotting blood.

sorption requires bile. If, therefore, the bile ducts become obstructed and bile cannot enter the intestine, a vitamin K deficiency develops. The liver cannot then produce prothrombin at its normal rate, and the blood's prothrombin concentration soon falls below normal. A prothrombin deficiency gives rise to a bleeding tendency. As a preoperative safeguard, therefore, patients with obstructive jaundice are generally given some kind of vitamin K preparation.

Factors that oppose clotting

Although blood clotting probably goes on continuously and concurrently with clot dissolution (fibrinolysis), several factors operate to oppose clot formation in intact vessels. Most important by far is the smooth nonwettable surface of the normal endothelial lining of blood vessels. Platelets do not adhere to it, consequently, they do not disintegrate and release platelet factors into the blood and, therefore, the blood clotting mechanism does not get started in normal blood vessels. As an additional deterrent to clotting, blood contains certain substances called *antithrombins*. The name suggests their function—they

oppose or inactivate thrombin. Thus, antithrombins prevent thrombin from converting fibrinogen to fibrin. *Heparin*, a natural constituent of blood, acts as an antithrombin. Its normal concentration in blood, however, is too low to have much effect in keeping blood fluid. Where it comes from is not definitely known. It was first prepared from liver (hence its name), but various other organs also contain heparin. Mast cells are known to contain considerable amounts of heparin, although they may not themselves synthesize it but only store it.

Factors that hasten clotting

Two conditions particularly favor thrombus formation: a rough spot in the endothelium (blood vessel lining) and abnormally slow blood flow. Atherosclerosis, for example, is associated with an increased tendency toward thrombosis because of endothelial rough spots in the form of plaques of accumulated cholesterol-lipid material. Immobility, on the other hand, may lead to thrombosis because blood flow slows down as movements decrease. Incidentally, this fact is one of the major reasons why physicians insist that bed patients must either move or be moved frequently. Presumably, sluggish blood flow allows thromboplastin to accumulate sufficiently to reach a concentration adequate for clotting.

Once started, a clot tends to grow. Platelets enmeshed in the fibrin threads disintegrate, releasing more thromboplastin which, in turn, causes more clotting, which enmeshes more platelets, and so on, in a vicious circle. Clot-retarding substances, available in recent years, have proved valuable for retarding this process.

Pharmaceutical preparations that retard clotting

The anticoagulant *Dicumarol* (bishydroxycoumarin) has become well known because of its clinical value in lessening thrombus and embolus formation. Dicumarol is postulated to block vitamin K's action and thereby decrease prothrombin synthesis by the liver. Blood to be used for transfusions is usually treated with a citrate compound. The latter combines with calcium ions. Therefore, citrate prevents coagulation by blocking thrombin formation from prothrombin.

Clot dissolution

Fibrinolysis is the physiological mechanism that dissolves clots. Newer evidence indicates that the two opposing processes of clot formation and fibrinolysis go on continuously. Dr. George Fulton of Boston University has presented one bit of dramatic evidence. He took micromovies that show tiny blood vessels rupturing under apparently normal circumstances and clots forming to plug them. Blood contains an enzyme, fibrinolysin, which catalyzes the hydrolysis of fibrin, causing it to dissolve. Many other factors, however, presumably also take part in clot dissolution — for instance, substances that activate profibrinolysin (inactive form of fibrinoly-

sin). Streptokinase, an enzyme from certain streptococci, can act this way and so can cause clot dissolution and even hemorrhage.

Clinical methods of hastening clotting

One way of treating excessive bleeding is to speed up the blood-clotting mechanism. The principle involved is apparent — to increase any of the substances essential for clotting. Application of this principle is accomplished in the following ways:

1 By applying a rough surface such as gauze, or by applying heat, or by gently squeezing the tissues around a cut vessel. Each of these procedures causes more platelets to disintegrate and release more platelet factors. And this, in turn, accelerates the first of the clotting reactions.

2 By applying purified thrombin (in the form of sprays or impregnated gelatin sponges which can be left in a wound). Which stage of the clotting mechanism does this accelerate?

3 By applying fibrin foam, films, etc.

Heart

The human heart is a four-chambered muscular organ, shaped and sized roughly like a man's closed fist. It lies in the mediastinum, with approximately two-thirds of its mass to the left of the midline of the body and one-third to the right.

The lower border of the heart, which forms a blunt point known as the *apex*, lies on the diaphragm, pointing toward the left. To count the apical beat, one must place a stethoscope directly over the apex — that is, in the space between the fifth and sixth ribs (fifth intercostal space) on a line with the midpoint of the left clavicle.

The upper border of the heart, that is, its base, lies just below the second rib. The

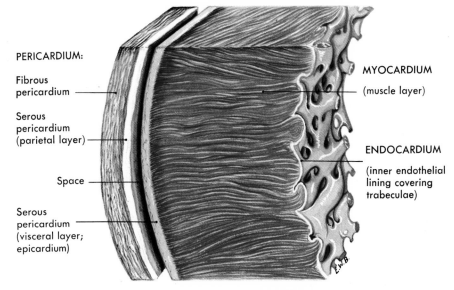

PERICARDIUM:

Fibrous pericardium

Serous pericardium (parietal layer)

Space

Serous pericardium (visceral layer; epicardium)

MYOCARDIUM

(muscle layer)

ENDOCARDIUM

(inner endothelial lining covering trabeculae)

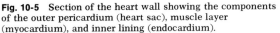

Fig. 10-5 Section of the heart wall showing the components of the outer pericardium (heart sac), muscle layer (myocardium), and inner lining (endocardium).

boundaries, which, of course, indicate its size, have considerable clinical importance since a marked increase in heart size accompanies certain types of heart disease. Therefore, when diagnosing heart disorders, the doctor charts the boundaries of the heart.

■Covering

Structure

The heart has its own special covering, a loose-fitting inextensible sac called the *pericardium*. The pericardium consists of two parts: a fibrous portion and a serous portion (Fig. 10-5). The sac itself is made of tough white fibrous tissue but is lined with smooth, moist serous membrane—the parietal layer of the serous pericardium. The same kind of membrane covers the entire outer surface of the heart. This covering layer is known as the visceral layer of the serous pericardium or as the *epicardium*. The fibrous sac attaches to the large blood vessels emerging from the top of the heart but not to the heart itself. Therefore, it fits loosely around the heart with a slight space between the visceral layer adhering to the heart and the parietal layer adhering to the inside of the fibrous

sac. This space is called the *pericardial space*. It contains a few drops of lubricating fluid secreted by the serous membrane and is called *pericardial fluid*.

The structure of the pericardium can be summarized in outline form as follows:

1 *Fibrous pericardium*—loose-fitting sac around the heart
2 *Serous pericardium*—consisting of two layers
   a *Parietal layer*—lining inside of the fibrous pericardium
   b *Visceral layer (epicardium)*—adhering to the outside of the heart; between visceral and parietal layers is a potential space, the pericardial space, which contains a few drops of pericardial fluid

Function

The fibrous pericardial sac with its smooth, well-lubricated lining provides protection against friction. The heart moves easily in this loose-fitting jacket with no danger of irritation from friction between the two surfaces so long as the serous pericardium remains normal. If, however, it becomes inflamed (pericarditis) and too much pericardial fluid or

myocardium
endocardium

307

fibrin or pus develops in the pericardial space, the visceral and parietal layers may stick together. And this, as you might guess, hampers heart action. In such cases, it sometimes becomes necessary to remove the fibrous pericardium with its lining of parietal serous membrane in order for the heart to continue functioning. This operation, a spectacular procedure, is called a *pericardectomy.*

◼ Structure

Wall

Three distinct layers of tissue make up the heart wall (Fig. 10-5). The bulk of the wall consists of specially constructed muscle tissue known as cardiac muscle or the *myocardium.* Covering the myocardium on the outside and adherent to it is the visceral layer of the *serous pericardium* (or *epicardium*) already described. Lining the interior of the myocardial wall is a delicate layer of endothelial* tissue known as the *endocardium.* On its inner surface, the myocardium is raised into ridgelike projections, the papillary muscles.

Cavities

The interior of the heart is divided into four chambers, two upper and two lower. The upper cavities are named *atria*† and the lower ones *ventricles* (Fig. 10-6). Of these, the ventricles are considerably larger and thicker walled than the atria because they carry a heavier pumping burden than the atria. Also, the left ventricle has thicker walls than the right for

---

*Endothelial tissue resembles simple squamous epithelial tissue in that it consists of a single layer of flat cells. It differs from epithelial tissue in that it arises from the mesoderm layer of the embryo, whereas epithelial tissue arises from the ectoderm.

†The atria are sometimes called auricles. Strictly speaking, the latter term means the earlike flaps protruding from the atria, although the two terms are often used synonymously.

the same reason. It has to pump blood through all the vessels of the body, except those to and from the lungs, whereas the right ventricle sends blood only through the lungs.

Valves and openings

The heart valves are mechanical devices that permit the flow of blood in one direction only. Four sets of valves are of importance to the normal functioning of the heart (Fig. 10-6). Two of these, the cuspid (atrioventricular) valves, are located in the heart, guarding the openings between the atria and ventricles (atrioventricular orifices). The other two, the semilunar valves, are located inside the pulmonary artery and the great aorta just as they arise from the right and left ventricles, respectively.

The cuspid valve guarding the right atrioventricular orifice consists of three flaps of endocardium anchored to the papillary muscles of the right ventricle by several cordlike structures called *chordae tendineae.* Because this valve has three flaps, it is appropriately named the *tricuspid valve.* The valve that guards the left atrioventricular orifice is similar in structure to the tricuspid except that it has only two flaps and is, therefore, called the *bicuspid* or, more commonly, the *mitral valve.* (An easy way to remember which valve is on the right and which is on the left is this: the names whose first letters come nearest together in the alphabet go together—thus, L and M for *l*eft side, *m*itral valve, and R and T for *r*ight side, *t*ricuspid.)

The construction of both cuspid valves allows blood to flow from the atria into the ventricles but prevents it from flowing back up into the atria from the ventricles. Ventricular contraction forces the blood in the ventricles hard against the cuspid valves, closing them and thereby ensur-

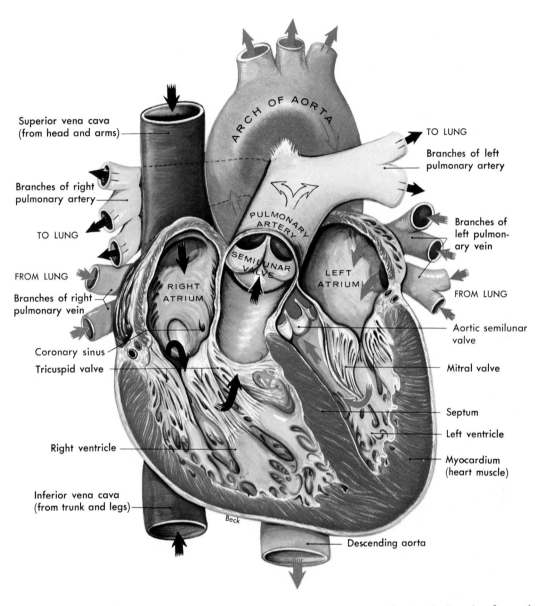

Superior vena cava
(from head and arms)

Branches of right
pulmonary artery

TO LUNG

FROM LUNG

Branches of right
pulmonary vein

Coronary sinus

Tricuspid valve

Right ventricle

Inferior vena cava
(from trunk and legs)

ARCH OF AORTA

TO LUNG

Branches of left
pulmonary artery

PULMONARY
ARTERY

SEMILUNAR
VALVE

RIGHT
ATRIUM

LEFT
ATRIUM

Branches of
left pulmon-
ary vein

FROM LUNG

Aortic semilunar
valve

Mitral valve

Septum

Left ventricle

Myocardium
(heart muscle)

Beck

Descending aorta

Fig. 10-6 Frontal section of the heart showing the four chambers, valves, openings, and major vessels. Arrows indicate direction of blood flow. Black arrows represent unoxygenated blood and red arrows oxygenated blood. The two branches of the right pulmonary vein extend from the right lung behind the heart to enter the left atrium.

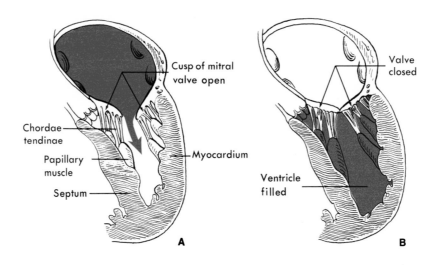

Fig. 10-7   Action of the cuspid (atrioventricular) valves. **A,** When the valves are open, blood passes freely from the atria to the ventricles. **B,** Filling of the ventricles closes the valves and prevents a back flow of blood into the atria when the ventricles contract.

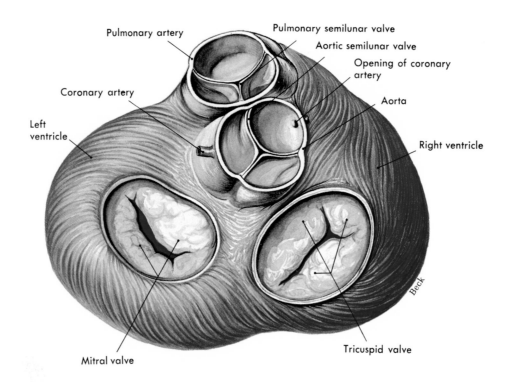

Fig. 10-8   The valves of the heart viewed from above. The atria are removed to show the mitral and tricuspid valves.

Table 10-3. Coronary arteries

Right coronary artery	Left coronary artery
Divides into two main branches:	Divides into two main branches:
1 Posterior descending artery–sends branches to both ventricles 2 Marginal artery–sends branches to right ventricle and right atrium	1 Anterior descending artery–sends branches to both ventricles 2 Circumflex artery–sends branches to left ventricle and left atrium

ing the movement of the blood upward into the pulmonary artery and aorta as the ventricles contract (see Fig. 10-7).

The *semilunar valves* consist of half-moon-shaped flaps growing out from the lining of the pulmonary artery and great aorta. When these valves are closed, as in Fig. 10-8, blood fills the spaces between the flaps and the vessel wall. Each flap then looks like a tiny filled bucket. Inflowing blood smooths the flaps against the blood vessel wall, collapsing the buckets and thereby opening the valves.

Like the cuspid valves, the semilunar valves, by preventing backflow of blood, cause it to flow forward in places where there would otherwise be considerable backflow. Whereas the cuspid valves prevent blood from flowing back up into the atria from the ventricles, the semilunar valves prevent it from flowing back down into the ventricles from the aorta and pulmonary artery.

Any one of the four valves may lose its ability to close tightly. Such a condition is known as *valvular insufficiency* or, because it permits blood to "leak" back into the part of the heart from which it came, "leakage of the heart." *Mitral stenosis* is an abnormality in which the left atrioventricular orifice becomes narrowed by scar tissue that forms as a result of disease. This hinders the passage of blood from the

left atrium to the left ventricle and leads to circulatory failure. But fortunately, the marvel of open-heart surgery has made possible the correction of many valvular defects.

Blood supply

Myocardial cells receive blood by way of two small vessels, the right and left coronary arteries. Since the openings into these vitally important vessels lie behind flaps of the aortic semilunar valve, they come off of the aorta at its very beginning and are its first branches. Both right and left coronary arteries have two main branches, as shown in Table 10-3.

More than a half million Americans die every year from coronary disease and another three and a half million or more are estimated to suffer some degree of incapacitation from this great killer.[*] Knowledge about the distribution of coronary artery branches, therefore, has the utmost practical importance. Here, then, are some principles about the heart's own blood supply that seem worth noting. Both ventricles receive their blood supply from branches of both the right and left coronary arteries. Each atrium, in contrast, receives blood only from a small branch of the corresponding coronary artery (Table 10-3). The most abundant blood supply of all goes to the myocardium of the left ventricle–an appropriate fact since the left ventricle does the most work so needs the most oxygen and nutrients delivered to it. The right coronary artery is dominant in about 50% of all hearts and the left coronary artery in about 20%, and in about 30% neither right nor left coronary artery dominates.

Another fact about the heart's own blood supply–and one of life-and-death impor-

---

[*]Effler, D. B.: Surgery for coronary disease, Sci. Amer. **219:**36-43 (Oct.), 1968.

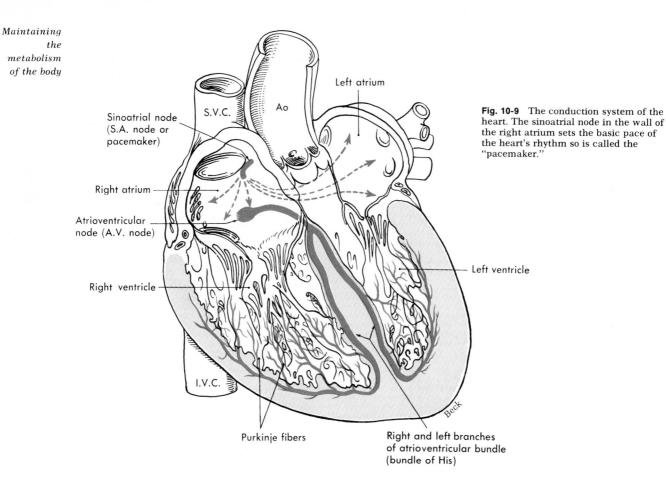

Sinoatrial node
(S.A. node or
pacemaker)

Right atrium

Atrioventricular
node (A.V. node)

Right ventricle

S.V.C.

Ao

Left atrium

Left ventricle

I.V.C.

Purkinje fibers

Right and left branches
of atrioventricular bundle
(bundle of His)

Fig. 10-9 The conduction system of the heart. The sinoatrial node in the wall of the right atrium sets the basic pace of the heart's rhythm so is called the "pacemaker."

tances—is the fact that only a few anastomoses exist between the larger branches of the coronary arteries. An *anastomosis* consists of one or more branches from the proximal part of an artery to a more distal part of itself or of another artery. Thus, anastomoses provide detours which arterial blood can travel if the main route becomes obstructed. In short, they provide collateral circulation to a part. This explains why the scarcity of anastomoses between larger coronary arteries looms so large as a threat to life. If, for example, a blood clot plugs one of the larger coronary artery branches, as it frequently does in coronary thrombosis or embolism, too little or no blood can reach some of the heart muscle cells. They become ischemic, in other words. Deprived of oxygen, they release too little energy for their own survival. Myocardial

infarction (death of ischemic heart muscle cells) soon results. There is another anatomical fact, however, which brightens the picture somewhat—many anastomoses do exist between very small arterial vessels in the heart, and, given time, new ones develop and provide collateral circulation to ischemic areas. In recent years, several surgical procedures have been devised to aid this process.*

Conduction system

Four structures—the sinoatrial node, atrioventricular node, atrioventricular bundle, and Purkinje fibers—compose the conduction system of the heart (Fig. 10-9). Each of these structures consists of cardiac muscle modified enough in struc-

*Effler, D. B.: Surgery for coronary disease, Sci. Amer. **219**:36-43 (Oct.), 1968.

ture to differ in function from ordinary cardiac muscle. The main specialty of ordinary cardiac muscle is contraction. In this, it is like all muscle and, like all muscle, ordinary cardiac muscle can also conduct impulses. But conduction alone is the specialty of the modified cardiac muscle that composes the conduction system structures.

Sinoatrial node. The sinoatrial node (S.A. node, Keith-Flack node, or pacemaker) consists of hundreds of cells located in the right atrial wall near the opening of the superior vena cava (Fig. 10-9). Although sinoatrial node cells resemble heart muscle cells, they differ somewhat from them in structure and they serve quite a different function. Sinoatrial node cells possess an intrinsic rhythm. This means that without any stimulation by nerve impulses, these cells will contract hour after hour at a constant rate characteristic for each species. As they contract, impulses spread from them throughout the myocardium and initiate a heartbeat. Thus, the sinoatrial node sets the basic pace for the heart rate and is appropriately called the pacemaker of the heart. The rate set by sinoatrial nodal activity is not, however, an unalterable one. Various factors can and do change the rate of the heartbeat. Two major modifiers of sinoatrial node activity —and, therefore, of the heart rate—are the ratio of sympathetic and parasympathetic impulses conducted to the node per minute and the blood concentrations of certain hormones (notably epinephrine and thyroid hormone).

Today, everyone has heard about artificial pacemakers, devices that electrically stimulate the heart at a set rhythm. They do an excellent job of maintaining a steady heart rate and of keeping many individuals with damaged hearts alive for many years. Nevertheless, they must be judged inferior to the heart's own natural pacemaker. Why? Because they cannot speed up the heartbeat (as is necessary, for example, to make strenuous physical activity possible), nor can they slow it down again when the need has passed. The sinoatrial node, influenced as it is by autonomic impulses and hormones, can produce these changes. Artificial pacemakers can only keep the heart beating at a steady pace.

Atrioventricular node. The atrioventricular node (A.V. node or node of Tawara), a small mass of special cardiac muscle tissue, lies in the right atrium along the lower part of the interatrial septum (Fig. 10-9).

Atrioventricular bundle and Purkinje fibers. The atrioventricular bundle (bundle of His) is a bundle of special cardiac muscle fibers which originate in the atrioventricular node and which extend by two branches down the two sides of the interventricular septum. From there, they continue as the Purkinje fibers (Fig. 10-9). The latter extend out to the papillary muscles and lateral walls of the ventricles.

Impulse conduction through the heart normally starts in the sinoatrial node and spreads through atrial muscle fibers in all directions, causing atrial contraction. When impulses reach the atrioventricular node, it relays them by way of the bundle of His and Purkinje fibers to the ventricles, causing their contraction. Impulse conduction generates tiny electrical currents in the heart that spread through surrounding tissues to the surface of the body. This fact has great clinical importance. Why? Because from the skin, visible records of heart conduction can be made with the electrocardiograph or oscillograph.

Nerve supply

Both divisions of the autonomic nervous system send fibers to the heart. Sympathetic fibers (contained in the middle, superior, and inferior cardiac nerves) and parasympathetic fibers (in branches of the vagus) combine to form *cardiac plexuses* located close to the arch of the aorta. From the cardiac plexuses, fibers accompany the right and left coronary arteries to enter the heart. Here, most of the fibers terminate in the sinoatrial node, but some end in the atrioventricular node and in the atrial myocardium. Sympathetic nerves to the heart are also called accelerator nerves. Vagus fibers to the heart serve as inhibitory or depressor nerves.

■ Physiology

Function

The function of the heart is to pump blood in sufficient amounts to meet the varying needs of the cells of the body for the substances it transports. Mechanisms that accomplish this function of pumping different volumes of blood per minute under different conditions are discussed on pp. 330 to 345.

Cardiac cycle

The term cardiac cycle means a complete heartbeat consisting of contraction (systole) and relaxation (diastole) of both atria plus contraction and relaxation of both ventricles. The two atria contract simultaneously. Then, as they relax, the two ventricles contract and relax, instead of the entire heart contracting as a unit. This gives a kind of milking action to the movements of the heart. The atria remain relaxed during part of the ventricular relaxation and then start the cycle over again. The events occurring during the cycle are described in Table 10-4. Note the following facts:

1 The contracting force of the atria completes the emptying of blood out of the atria into the ventricles. Cuspid valves are necessarily open during this phase, the ventricles relaxed, filling with blood, and the semilunar valves closed so that blood does not flow on out into the pulmonary artery or aorta.

2 The atria relax, and blood enters them from the veins during the first part of their diastole and starts draining out into the ventricles during the latter part of it. The cuspid valves are closed during the first part of the diastole (while the ventricles are contracting, squeezing blood through the open semilunar valves into the pulmonary artery and aorta) but open as the ventricles relax, the semilunar valves close, and the ventricles start to fill with blood from the atria. About what percent of the time are the atria relaxed or resting? The ventricles? Consult Table 10-4 to find the answers.

Heart sounds during cycle. The heart makes certain typical sounds during each cycle that are described as sounding like lubb-dupp through a stethoscope. The first or systolic sound is believed to be due to the contraction of the ventricles and to vibrations of the closing cuspid valves. It is longer and lower than the second or diastolic sound, which is thought to be caused by vibrations of the closing semilunar valves.

Both of these sounds have clinical significance since they give information about the valves of the heart. Any variation from normal in the sounds indicates imperfect functioning of the valves. *Heart murmur* is one type of abnormal sound frequently heard and may signify incomplete closing of the valves (valvular insufficiency) or stenosis of them.

Blood vessels

Kinds

There are three kinds of blood vessels: arteries, veins, and capillaries. By definition an *artery* is a vessel that carries blood away from the heart. All arteries except the pulmonary artery and its branches carry oxygenated blood. Small arteries are called *arterioles*.

A *vein*, on the other hand, is a vessel that carries blood toward the heart. All of the veins except the pulmonary vein contain deoxygenated blood. Small veins are called *venules*. Both arteries and veins are macroscopic structures.

Capillaries are microscopic vessels that carry blood from small arteries to small veins—that is, from arterioles to venules. They represented the "missing link" in the proof of circulation for many years—from the time William Harvey first declared that blood circulated from the heart through arteries to veins and back to the heart until the time that microscopes made it possible to find these connecting vessels between arteries and veins. Many people rejected Harvey's theory of circulation on the basis that there was no possible way for blood to get from arteries to veins. The discovery of the capillaries formed the final proof that the blood actually does circulate from the heart into arteries to arterioles, to capillaries, to venules, to veins, and back to the heart.

Structure

Consult Figs. 10-10 and 10-11 and Table 10-5 for structure of the blood vessels.

Functions

The capillaries, though seemingly the most insignificant of the three kinds of blood vessels because of their diminutive size, nevertheless are the most important vessels functionally. Since the prime function of blood is to transport essential materials to and from the cells and since the actual delivery and collection of these substances take place in the capillaries, the capillaries must be regarded as the most important blood vessels functionally. Arteries serve merely as "distributors," carrying the blood to the arterioles. Arterioles, too, serve as distributors, carrying blood from arteries to capillaries. But in addition, arterioles perform another function, one that is of great importance for maintaining normal blood pressure and circulation. They serve as resistance vessels (discussed on p. 335). Veins function both as collectors and as reservoir vessels. Not only do they return blood from the capillaries to the heart, but they also can accommodate varying amounts of blood. This reservoir function of veins, which we will discuss later, plays an important part in maintaining normal circulation. The heart acts as a "pump," keeping the blood moving through this circuit of vessels—arteries, arterioles, capillaries, venules, and veins. In short, the entire circulatory mechanism pivots around one essential, that of keeping the capillaries supplied with an amount of blood adequate to the changing needs of the cells. All the factors governing circulation operate to this one end.

Although capillaries are very tiny (on the average, only 1 millimeter long or about 1/25 of an inch), they are so numerous as to be incomprehensible. Someone has calculated that if these microscopic tubes were joined end to end, they would extend 62,000 miles, in spite of the fact that it takes twenty-five of them to reach a single inch! According to one estimate, 1 cubic inch of muscle tissue contains over 1 1/2 million of these important little vessels. None of the billions of cells composing the body lie very far removed from a capillary. The reason for this lavish distribution of capillaries is, of course, apparent in view of their function of keeping the cells supplied with needed materials and rid of injurious wastes.

Text continued on p. 320

Table 10-4. The cardiac cycle (in tenths of seconds; 0.8 second for complete cycle)

Sec.	*0.1*	*0.2*	*0.3*	*0.4*	*0.5*	*0.6*	*0.7*	*0.8*
Atria	Contract – blood squeezed into ventricles	Relax ——————————————————————————————→ Blood enters from venae cavae and pulmonary veins and drains into ventricles ————————————————————————————→						
Cuspid valves	Open	Closed ——————————→ Open ——————————————————————————————→						
Ventricles	Relaxed – filling with blood from atria	Contract Blood emptying into pulmonary artery and aorta		Relax Filling with →blood from atria ——————————————————→				
Semilunar valves	Closed	Open ——————————————→ Closed ——————————————————————→						

Table 10-5. Structure of blood vessels

		Arteries	*Veins*	*Capillaries*
Coats		Outer coat (tunica adventitia or externa) of white fibrous tissue; causes artery to stand open instead of collapsing when cut (see Fig. 10-10) Lining (tunica intima) of endothelium Muscle coat (tunica media) of smooth muscle, elastic, and some white fibrous tissues; this coat permits constriction and dilatation	Same three coats but thinner and fewer elastic fibers; veins collapse when cut; semilunar valves present at intervals	Only lining coat present; therefore walls only one cell thick

Blood supply

Endothelial lining cells supplied by blood flowing through vessels; exchange of oxygen, etc., between cells of middle coat and blood by diffusion; outer coat supplied by tiny vessels known as *vasa vasorum* or "vessels of vessels"

Nerve supply

Smooth muscle cells of tunica media innervated by autonomic fibers

Abnormalities

Arteriosclerosis – hardening of walls of arteries
Aneurysm – saclike dilatation of artery wall
Varicose veins – stretching of walls, particularly around semilunar valves
Phlebitis – inflammation of vein; "milk leg," phlebitis of femoral vein of women after childbirth

aneurysm - blood-filled sac-like dilation of the walls of the artery.

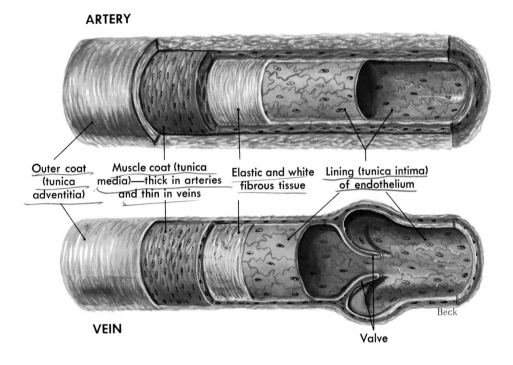

ARTERY

Outer coat (tunica adventitia) | Muscle coat (tunica media)—thick in arteries and thin in veins | Elastic and white fibrous tissue | Lining (tunica intima) of endothelium

VEIN

Beck

Valve

Fig. 10-10 Schematic drawings of an artery and vein showing comparative thicknesses of the three coats: outer coat (tunica adventitia), muscle coat (tunica media), and lining of endothelium (tunica intima). Note that the muscle and outer coats are much thinner in veins than in arteries and that veins have valves.

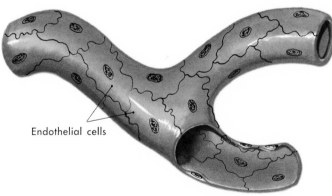

Endothelial cells

Fig. 10-11 The walls of capillaries consist of only a single layer of endothelial cells. These thin, flattened cells permit the rapid movement of substances between blood and interstitial fluid. Note that capillaries have no smooth muscle layer, elastic fibers, or surrounding adventitia.

Table 10-6. Main arteries

Artery	Branches (only largest ones named)
Ascending aorta	Coronary arteries (two, to myocardium)
Aortic arch	Innominate artery (or brachiocephalic) Left subclavian Left common carotid
Innominate artery	Right subclavian Right common carotid
Subclavian (right and left)	Vertebral* Axillary (continuation of subclavian)
Axillary	Brachial (continuation of axillary)
Brachial	Radial Ulnar
Radial and ulnar	Palmar arches (superficial and deep arterial arches in hand formed by anastomosis of branches of radial and ulnar arteries; numerous branches to hand and fingers)
Common carotid (right and left)	Internal carotid (brain, eye, forehead, and nose)* External carotid (thyroid, tongue, tonsils, ear, etc.)
Descending thoracic aorta	Visceral branches to pericardium, bronchi, esophagus, mediastinum Parietal branches to chest muscles, mammary glands, and diaphragm
Descending abdominal aorta	Visceral branches: 1 Celiac axis (or artery), which branches into gastric, hepatic, and splenic arteries (stomach, liver, and spleen) 2 Right and left suprarenal arteries (suprarenal glands) 3 Superior mesenteric artery (small intestine) 4 Right and left renal arteries (kidneys) 5 Right and left spermatic (or ovarian) arteries (testes or ovaries) 6 Inferior mesenteric artery (large intestine) Parietal branches to lower surface of diaphragm, muscles and skin of back, spinal cord, and meninges Right and left common iliac arteries – abdominal aorta terminates in these vessels in an inverted Y formation
Right and left common iliac arteries	Internal iliac or hypogastric (pelvic wall and viscera) External iliac (to leg)
External iliac (right and left)	Femoral (continuation of external iliac after it leaves abdominal cavity)
Femoral	Popliteal (continuation of femoral)
Popliteal	Anterior tibial Posterior tibial
Anterior and posterior tibial	Plantar arch (arterial arch in sole of foot formed by anastomosis of terminal branches of anterior and posterior tibial arteries; small arteries lead from arch to toes)

*The right and left vertebral arteries extend from their origin as branches of the subclavian arteries up the neck, through foramina in the transverse processes of the cervical vertebrae, and through the foramen magnum into the cranial cavity and unite on the undersurface of the brainstem to form the *basilar artery,* which shortly branches into the right and left *posterior cerebral arteries.* The internal carotid arteries enter the cranial cavity in the midpart of the cranial floor, where they become known as the *anterior cerebral arteries.* Small vessels, the *communicating arteries,* join the anterior and posterior cerebral arteries in such a way as to form an arterial circle (the *circle of Willis*) at the base of the brain, a good example of arterial anastomosis (Fig. 10-15).

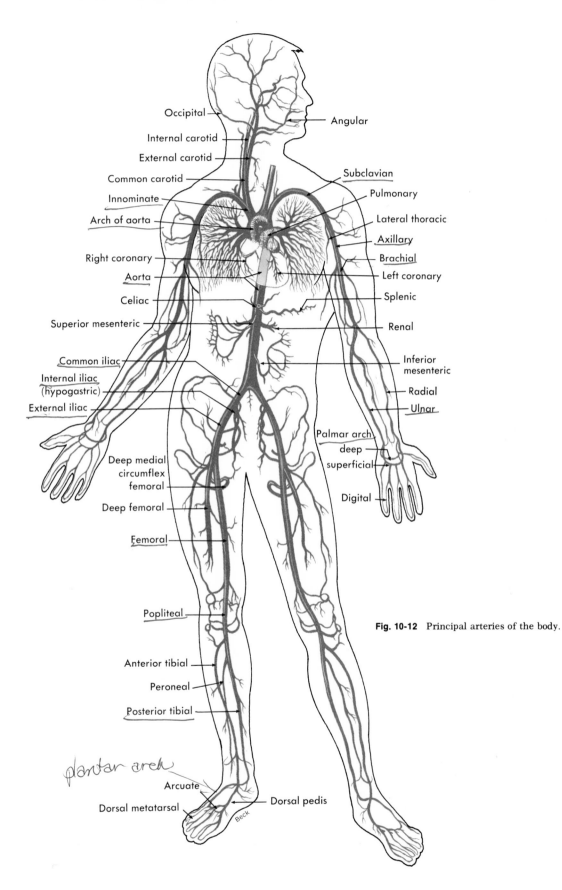

Occipital
Angular
Internal carotid
External carotid
Common carotid
Subclavian
Innominate
Pulmonary
Arch of aorta
Lateral thoracic
Axillary
Right coronary
Brachial
Aorta
Left coronary
Celiac
Splenic
Superior mesenteric
Renal
Common iliac
Inferior mesenteric
Internal iliac (hypogastric)
Radial
External iliac
Ulnar
Palmar arch
deep
superficial
Deep medial circumflex femoral
Digital
Deep femoral
Femoral
Popliteal
Anterior tibial
Peroneal
Posterior tibial
plantar arch
Arcuate
Dorsal metatarsal
Dorsal pedis

Beck

Fig. 10-12 Principal arteries of the body.

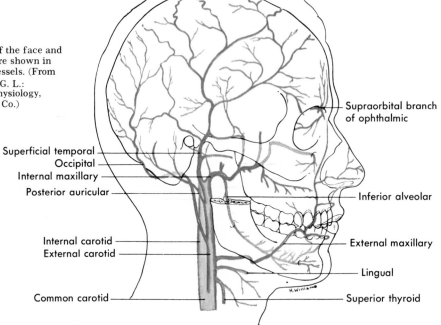

Fig. 10-13 Main arteries of the face and head. Superficial vessels are shown in brighter color than deep vessels. (From Francis, C. C, and Farrell, G. L.: Integrated anatomy and physiology, St. Louis, The C. V. Mosby Co.)

Supraorbital branch of ophthalmic

Superficial temporal
Occipital
Internal maxillary
Posterior auricular

Inferior alveolar

Internal carotid
External carotid

External maxillary

Lingual

Common carotid

Superior thyroid

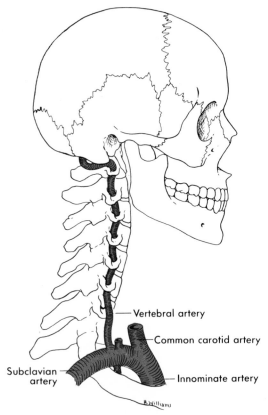

Vertebral artery

Common carotid artery

Subclavian artery

Innominate artery

Fig. 10-14 Location of the vertebral artery.

■ Main blood vessels

Systemic circulation

ARTERIES

Locate the arteries listed in Table 10-6 (see also Figs. 10-12 to 10-15). You may find it easier to learn the names of blood vessels and the relation of the vessels to each other from diagrams than from descriptions.

As you learn the names of the main arteries, keep in mind that these are only the major pipelines distributing blood from the heart to the various organs and that in each organ the main artery resembles a tree trunk in that it gives off numerous branches that continue to branch and rebranch, forming ever smaller vessels (arterioles), which also branch, forming microscopic vessels, the capillaries. In other words, most arteries eventually ramify into capillaries.

A few arteries open into other branches of the same or other arteries. Such a communication is termed an *arterial anastomosis*. Anatomoses, we have already noted, fulfill an important protective func-

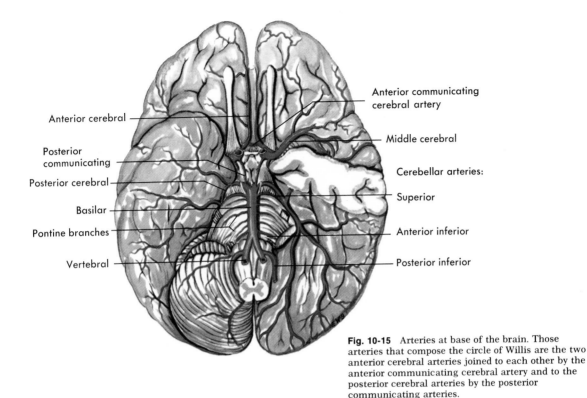

Anterior cerebral

Posterior communicating

Posterior cerebral

Basilar

Pontine branches

Vertebral

Anterior communicating cerebral artery

Middle cerebral

Cerebellar arteries:

Superior

Anterior inferior

Posterior inferior

Fig. 10-15 Arteries at base of the brain. Those arteries that compose the circle of Willis are the two anterior cerebral arteries joined to each other by the anterior communicating cerebral artery and to the posterior cerebral arteries by the posterior communicating arteries.

tion in that they provide detour routes for blood to travel in the event of obstruction of a main artery. Examples of arterial anastomoses are the circle of Willis at the base of the brain and the palmar and plantar arches. Other examples are found around several joints as well as in other locations.

VEINS

Several facts should be borne in mind while learning the names of veins.

1 Veins are the ultimate extensions of capillaries, just as capillaries are the eventual extensions of arteries. Whereas arteries branch into vessels of decreasing size to form arterioles and eventually capillaries, capillaries unite into vessels of increasing size to form venules and eventually veins.

2 Many of the main arteries have corresponding veins bearing the same name and located alongside or near the arteries. These veins, like the arteries, lie in deep, well-protected areas, for the most part

close along the bones—example: femoral artery and femoral vein, both located along the femur bone.

3 Veins found in the deep parts of the body are called *deep veins* in contradistinction to *superficial veins*, which lie near the surface. The latter are the veins that can be seen through the skin.

4 The large veins of the cranial cavity, formed by the dura mater, are not called veins but *sinuses*. They should not be confused with the bony sinuses of the skull.

•   •   •

The following list identifies the major veins. Locate each one as named on Figs. 10-16 to 10-21.

Veins of upper extremities (Figs. 10-16 and 10-17)

Deep

Palmar (volar) arch (also superficial)
Radial (partially deep, partially superficial)
Ulnar (partially deep, partially superficial)
Brachial
Axillary (continuation of brachial)
Subclavian (continuation of axillary)

Superficial

Veins of hand from dorsal and volar venous arches which, together with complicated network of superficial veins of lower arm, finally pour their blood into two large veins—cephalic (thumb side) and basilic (little finger side); these two veins empty into deep axillary vein

Veins of lower extremities (Figs. 10-16, 10-18, and 10-19)

Deep

Plantar arch
Anterior tibial
Posterior tibial
Popliteal
Femoral
External iliac

Superficial

Dorsal venous arch of foot
Great (or internal or long) saphenous
Small (or external or short) saphenous
(Great saphenous terminates in femoral vein in groin; small saphenous terminates in popliteal vein)

Veins of head and neck (Figs. 10-20 and 10-21)

Deep (in cranial cavity)

Longitudinal (or sagittal) sinus
Inferior sagittal and straight sinus
Numerous small sinuses
Right and left transverse (or lateral) sinuses
Internal jugular veins, right and left (in neck); continuations of transverse sinuses
Innominate veins, right and left; formed by union of subclavian and internal jugulars

Superficial

External jugular veins, right and left (in neck); receive blood from small superficial veins of face, scalp, and neck; terminate in subclavian veins (small emissary veins connect veins of scalp and face with blood sinuses of cranial cavity, a fact of clinical interest as a possible avenue for infections to enter cranial cavity)

Veins of abdominal organs (Fig. 10-16)

Spermatic (or ovarian)
Renal   } Drain into inferior
Hepatic   } vena cava
Suprarenal

Left spermatic and left suprarenal veins usually drain into left renal vein instead of into inferior vena cava; for return of blood from abdominal digestive organs, see discussion of portal circulation (next column); also Fig. 10-22

Veins of thoracic organs

Several small veins, such as bronchial, esophageal, pericardial, etc., return blood from chest organs (except lungs) directly into superior vena cava or into azygos vein; azygos vein lies to right of spinal column and extends from inferior vena cava (at level of first or second lumbar vertebra) through diaphragm to

terminal part of superior vena cava; hemiazygos vein lies to left of spinal column, extending from lumbar level of inferior vena cava through diaphragm to terminate in azygos vein; accessory hemiazygos vein connects some of superior intercostal veins with azygos or hemiazygos vein

Correlations. Middle ear infections sometimes cause infection of the transverse sinuses with the formation of a thrombus. In such cases, the internal jugular vein may be ligated to prevent the development of a fatal cardiac or pulmonary embolism.

Intravenous injections are most often given into the median basilic vein at the bend of the elbow. Blood that is to be used for various laboratory tests is also usually removed from this vein. In an infant, however, the longitudinal sinus is more often punctured (through the anterior fontanel) because the superficial arm veins are too tiny for the insertion of a needle.

Portal circulation

Veins from the spleen, stomach, pancreas, gallbladder, and intestines do not pour their blood directly into the inferior vena cava as do the veins from other abdominal organs. Instead, they send their blood to the liver by means of the portal vein. Here, the blood mingles with the arterial blood in the capillaries and is eventually drained from the liver by the hepatic veins that join the inferior vena cava. The reason for this detouring of the blood through the liver before it returns to the heart will be discussed in the chapter on the digestive system.

Fig. 10-22 shows the plan of the portal system. In most individuals, the portal vein is formed by the union of the splenic and superior mesenteric veins, but blood from the gastric, pancreatic, and inferior mesenteric veins drains into the splenic vein before it merges with the superior mesenteric vein.

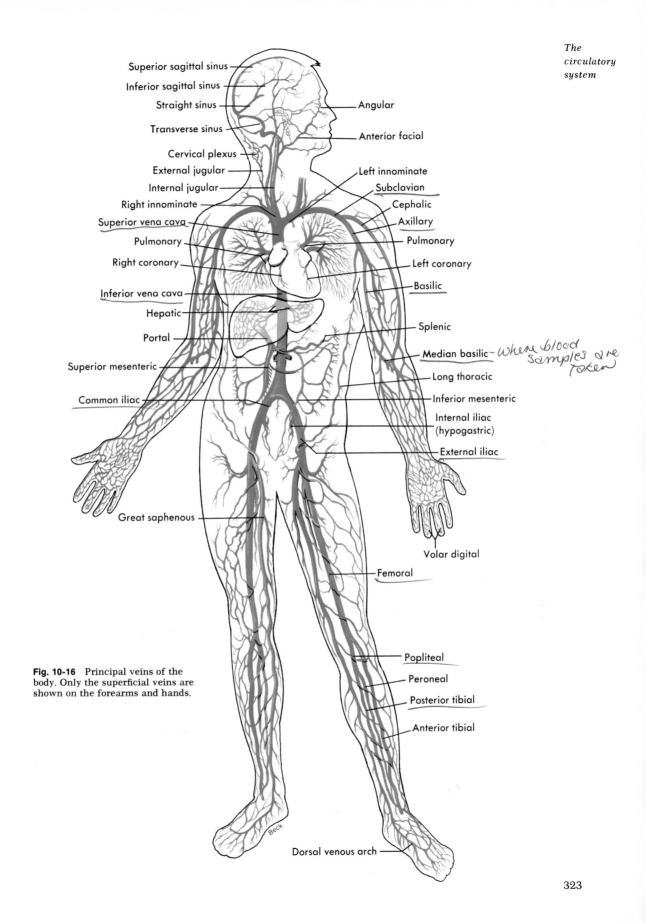

Superior sagittal sinus
Inferior sagittal sinus
Straight sinus
Transverse sinus
Cervical plexus
External jugular
Internal jugular
Right innominate
Superior vena cava
Pulmonary
Right coronary
Inferior vena cava
Hepatic
Portal
Superior mesenteric
Common iliac

Angular
Anterior facial
Left innominate
Subclavian
Cephalic
Axillary
Pulmonary
Left coronary
Basilic
Splenic
Median basilic — *Where blood samples are taken*
Long thoracic
Inferior mesenteric
Internal iliac (hypogastric)
External iliac

Great saphenous
Volar digital
Femoral
Popliteal
Peroneal
Posterior tibial
Anterior tibial
Dorsal venous arch

Fig. 10-16 Principal veins of the body. Only the superficial veins are shown on the forearms and hands.

Beck

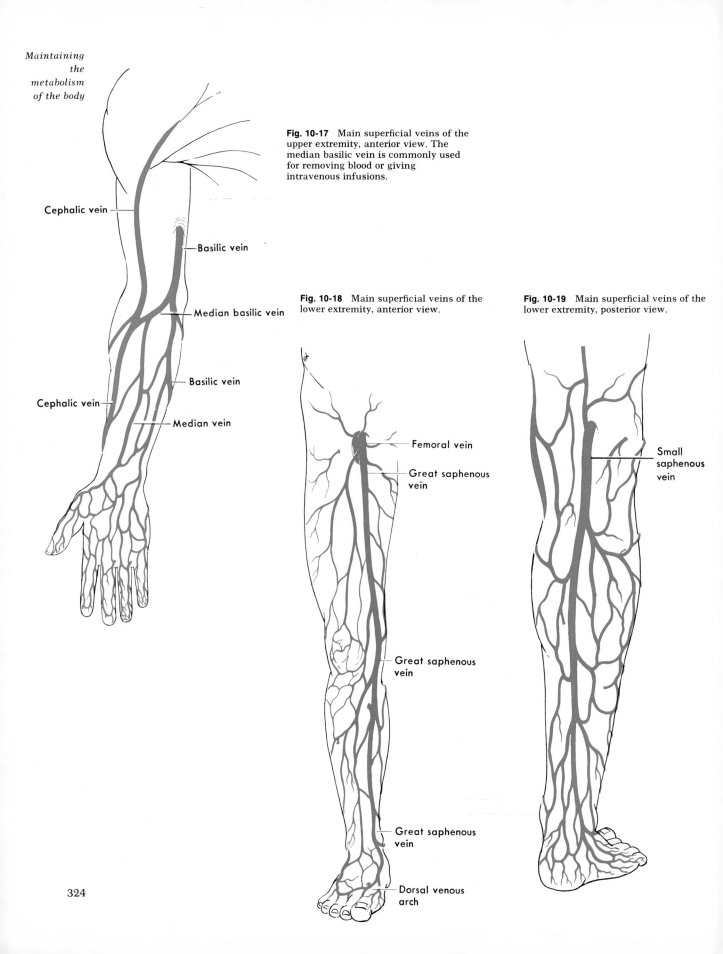

Cephalic vein

Basilic vein

Fig. 10-17 Main superficial veins of the upper extremity, anterior view. The median basilic vein is commonly used for removing blood or giving intravenous infusions.

Median basilic vein

Basilic vein

Cephalic vein

Median vein

Fig. 10-18 Main superficial veins of the lower extremity, anterior view.

Fig. 10-19 Main superficial veins of the lower extremity, posterior view.

Femoral vein

Great saphenous vein

Small saphenous vein

Great saphenous vein

Great saphenous vein

Dorsal venous arch

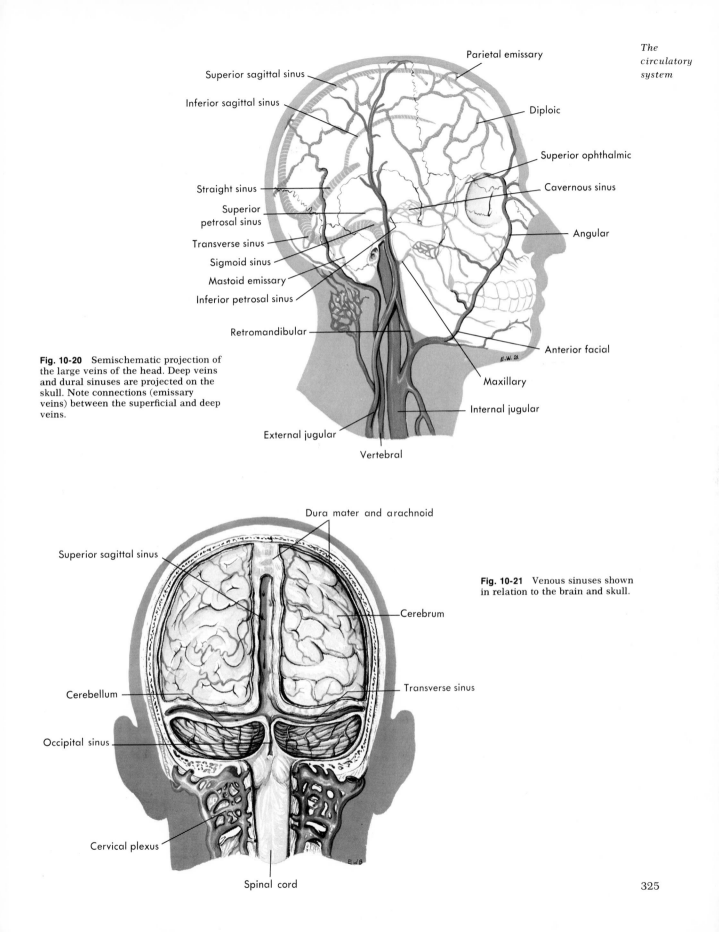

Fig. 10-20 Semischematic projection of the large veins of the head. Deep veins and dural sinuses are projected on the skull. Note connections (emissary veins) between the superficial and deep veins.

Parietal emissary

Superior sagittal sinus

Inferior sagittal sinus

Diploic

Superior ophthalmic

Cavernous sinus

Straight sinus

Superior petrosal sinus

Angular

Transverse sinus

Sigmoid sinus

Mastoid emissary

Inferior petrosal sinus

Retromandibular

Anterior facial

Maxillary

Internal jugular

External jugular

Vertebral

Dura mater and arachnoid

Superior sagittal sinus

Fig. 10-21 Venous sinuses shown in relation to the brain and skull.

Cerebrum

Cerebellum

Transverse sinus

Occipital sinus

Cervical plexus

Spinal cord

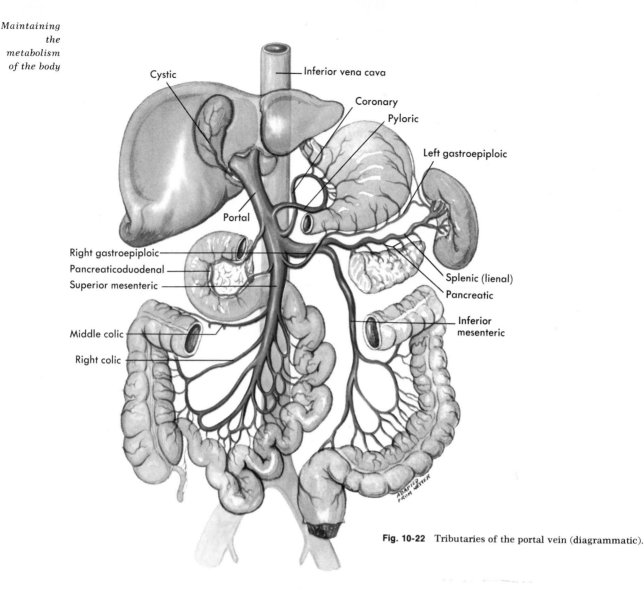

Cystic
Inferior vena cava
Coronary
Pyloric
Left gastroepiploic
Portal
Right gastroepiploic
Pancreaticoduodenal
Superior mesenteric
Splenic (lienal)
Pancreatic
Middle colic
Inferior mesenteric
Right colic

Fig. 10-22 Tributaries of the portal vein (diagrammatic).

If either portal circulation or venous return from the liver is interfered with (as they often are in certain types of liver or heart disease), then venous drainage from most of the other abdominal organs is necessarily obstructed also. The accompanying increased capillary pressure accounts at least in part for the occurrence of ascites ("dropsy" of abdominal cavity) under these conditions.

Fetal circulation

Circulation in the body before birth necessarily differs from circulation after birth for one main reason—because fetal blood secures oxygen and food from maternal blood instead of from its own lungs and digestive organs, respectively. Obviously, then, there must be additional blood vessels in the fetus to carry the fetal blood into close approximation with the maternal blood and to return it to the fetal body. These structures are the two *umbilical arteries*, the *umbilical vein*, and the *ductus venosus*. Also, there must be some structure to function as the lungs and digestive organs do postnatally—that is, a place where an interchange of gases, foods, and wastes between the fetal and maternal blood can take place. This structure is the *placenta*. The exchange of substances occurs without any actual mixing of maternal and fetal bloods since each flows in its own capillaries.

In addition to the placenta and umbilical vessels, three structures located within the fetus' own body play an important part in fetal circulation. One of them (ductus venosus) serves as a detour by which most of the blood returning from the placenta bypasses the fetal liver. The other two (foramen ovale and ductus arteriosus) provide detours by which blood bypasses the lungs. A brief description of each of the six structures necessary for fetal circulation follows (also see Fig. 10-23).

1 The *two umbilical arteries* are extensions of the internal iliac (hypogastric) arteries and carry fetal blood to the placenta.

2 The *placenta* is a structure attached to the uterine wall. Exchange of oxygen and other substances between maternal and fetal blood takes place in the placenta.

3 The *umbilical vein* returns oxygenated blood from the placenta, enters the fetal body through the umbilicus, extends up to undersurface of the liver where it gives off two or three branches to the liver, and then continues on as the ductus venosus. Two umbilical arteries and the umbilical vein together constitute the *umbilical cord* and are shed at birth along with the placenta.

4 The *ductus venosus* is a continuation of the umbilical vein along the undersurface of the liver and drains into the inferior vena cava. Most of the blood returning from the placenta bypasses the liver. Only a relatively small amount enters the liver by way of the branches from the umbilical vein into the liver.

3 shunts by placenta

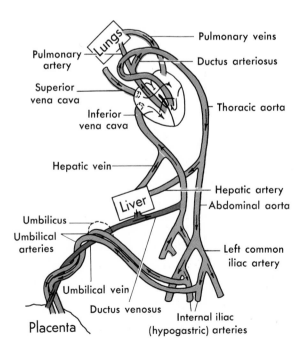

Fig. 10-23 Scheme to show the plan of fetal circulation. Note the following essential features: (1) two umbilical arteries, extensions of the internal iliac arteries, carry blood to (2) the placenta, which is attached to the uterine wall; (3) one umbilical vein returns blood, rich in oxygen and food, from the placenta; (4) the ductus venosus, a small vessel which connects the umbilical vein with the inferior vena cava; (5) the foramen ovale, an opening in the septum between the right and left atria; and (6) the ductus arteriosus, a small vessel which connects the pulmonary artery with the thoracic aorta.

5 The *foramen ovale* is an opening in the septum between the right and left atria. A valve at the opening of the inferior vena cava into the right atrium directs most of the blood through the foramen ovale into the left atrium so that it bypasses the fetal lungs. A small percentage of the blood leaves the right atrium for the right ventricle and pulmonary artery. But even most of this does not flow on into the lungs. Still another detour, the ductus arteriosus, diverts it.

6 The *ductus arteriosus* is a small vessel connecting the pulmonary artery with the descending thoracic aorta. It therefore enables another portion of the blood to detour into the systemic circulation without going through the lungs.

·   ·   ·

Almost all fetal blood is a mixture of oxygenated and deoxygenated blood. Examine Fig. 10-23 carefully to determine why this is so. What happens to the oxygenated blood returned from the placenta via the umbilical vein? It flows into what vessel?

Since the six structures that serve fetal circulation are no longer needed after birth, several changes take place. As soon as the umbilical cord is cut, the two umbilical arteries, the placenta, and the umbilical vein obviously no longer function. The placenta is shed from the mother's body as the afterbirth with part of the umbilical vessels attached. The sections of these vessels remaining in the infant's body eventually become fibrous cords which remain throughout life (the umbilical vein becomes the round ligament of the liver). The ductus venosus, no longer needed to bypass blood around the liver, eventually becomes the ligamentum venosum of the liver. The foramen ovale normally becomes functionally closed soon after a newborn

baby takes his first breath and full circulation through his lungs becomes established. Complete structural closure, however, requires longer. According to Gray, the foramen ovale "gradually decreases in size during the first month, but a small opening usually persists until the last third of the first year and often later."* Eventually, the foramen ovale becomes a mere depression (fossa ovalis) in the wall of the right atrial septum. About the ductus arteriosus, Gray writes that it "begins to contract immediately after respiration is established, and its lumen slowly becomes obliterated."* Eventually, it also turns into a fibrous cord.

Circulation

Definitions

The term circulation of blood suggests its meaning—namely, blood flow through vessels arranged to form a circuit or circular pattern. Blood flow from the heart (left ventricle) through all blood vessels except those of the lungs and back to the heart (to the right atrium) is spoken of as *systemic circulation*. The left ventricle pumps blood into the ascending aorta. From here it flows into arteries that carry it into the various tissues and organs of the body. Within each structure blood moves from arteries to arterioles to capillaries. Here, the vital two-way exchange of substances occurs between blood and cells. Blood flows next out of each organ by way of its venules and then its veins to drain eventually into the inferior or superior vena cava. These two great veins of the body return venous blood to the heart to

---

*From Goss, C. M., editor: Gray's Anatomy of the human body, ed. 28, Philadelphia, 1966, Lea & Febiger, pp. 541 and 542.

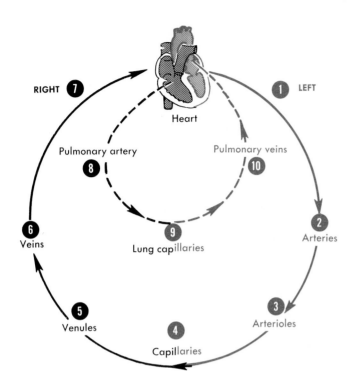

RIGHT **7** **1** LEFT

Heart

Pulmonary artery Pulmonary veins

8 **10**

6 **9** **2**

Veins Lung capillaries Arteries

5 **3**

Venules Arterioles

4

Capillaries

Fig. 10-24 Relationship of systemic and pulmonary circulation. As indicated by the numbers, blood circulates from the left side (ventricle) of the heart to arteries, to arterioles, to capillaries, to venules, to veins, to the right side of the heart (atrium to ventricle), to the lungs, and back to the left side of the heart, thereby completing a circuit. Refer to this diagram when tracing the circulation of blood to or from any part of the body.

the right atrium to complete systemic circulation. But the blood has not quite come full circle back to its starting point, the left ventricle. To do this and start on its way again, it must first flow through another circuit, the _pulmonary circulation._ Venous blood moves from the right atrium to the right ventricle to the pulmonary artery to lung arterioles and capillaries. Here, exchange of gases between blood and air takes place, converting venous blood to arterial blood. This oxygenated blood then flows on through lung venules into four pulmonary veins and returns to the left atrium of the heart. From the left atrium it enters the left ventricle to be pumped again through the systemic circulation.

How to trace

In order to enumerate the vessels through which blood flows in reaching a designated part of the body or in re-turning to the heart from a part, one must remember the following:

1 That blood always flows in this direction—from *left ventricle* of heart to *arteries*, to *arterioles*, to *capillaries* of each body part, to *venules*, to *veins*, to *right atrium, right ventricle, pulmonary artery, lung capillaries, pulmonary veins, left atrium,* and back to left ventricle (Fig. 10-24)

2 That when blood is in capillaries of abdominal digestive organs, it must flow through portal system before returning to heart

3 Names of main arteries and veins of body

For example, suppose glucose were instilled into the rectum. To reach the cells of the right little finger, the vessels through which it would pass after absorption from the intestinal mucosa into capillaries would be as follows: *capillaries into venules of large intestine into inferior* mesenteric *vein,* splenic vein, portal vein, capillaries of liver, hepatic veins, inferior vena cava, *right atrium* of heart, *right ventricle, pulmonary artery, lung capillaries, pulmonary veins, left atrium, left ventricle, ascending aorta,* aortic arch, innominate artery, right subclavian artery, right axillary artery, right brachial artery, right ulnar artery, arteries of palmar arch, arterioles, and *capillaries* of right little finger.

Note: The structures italicized show the direction of blood flow as described in points **1** and **2** and illustrated in Fig. 10-24. Follow this course of circulation first on Fig. 10-22 and then on Figs. 10-16 and 10-12. Try to answer question **39** of the review questions at the end of this chapter, using the plan outlined.

Functions of control mechanisms

Circulation is, of course, a vital function. It constitutes the only means by which cells can receive materials needed for energy and growth and can have their wastes removed. Not only is circulation necessary, but circulation of different volumes of blood per minute is also essential for healthy survival. More active cells need more blood per minute than less active cells. The reason underlying this principle is obvious. The more work cells do, the more energy they use and the more substances they need to supply this energy. Only arterial blood can deliver these energy suppliers (oxygen and foods). So the more active any part of the body is, the greater the volume of blood circulated to it per minute must be. And this requires that circulation control mechanisms accomplish two functions: maintain circulation (keep blood flowing, that is) and vary circulation (that is, change the volume of blood circulating per minute and the volume circulating to different tissues as their activity changes). At the right time, the

right amount of blood must be shifted from the right tissues to the right tissues—must be transferred, in other words, from the more active tissues to the less active ones.

To achieve these two ends, a great many factors must operate together. Incidentally, this is an important physiological principle that you have no doubt observed by now—that every body function depends upon many other functions. A constellation of separate processes or mechanisms act as a single integrated mechanism. Together, they perform some one large function. For example, many mechanisms together accomplish the large function we call circulation. To try to make the complexities of circulation mechanisms a little more understandable, we shall use a question and answer method for our discussion of the principles of circulation.

Principles

Why does blood circulate? What makes it keep moving over and over again, as long as life lasts, from the left side of the heart through the systemic vessels to the right side of the heart, to the lungs, and back to the left heart?

Answer: Blood flows for the same reason that any fluid flows—whether it be water in a river or in a garden hose or in hospital tubing or blood in vessels. A fluid flows because a pressure gradient exists between different parts of its bed. This primary fluid flow principle derives from Newton's first and second laws of motion. In essence, these laws state the following principles:

1 That a fluid does not flow when the pressure is the same in all parts of it
2 That a fluid flows only when its pressure is higher in one area than in another, and it flows always from the higher pressure area toward the lower pressure area

In brief, then, the answer to our first question and the primary principle about

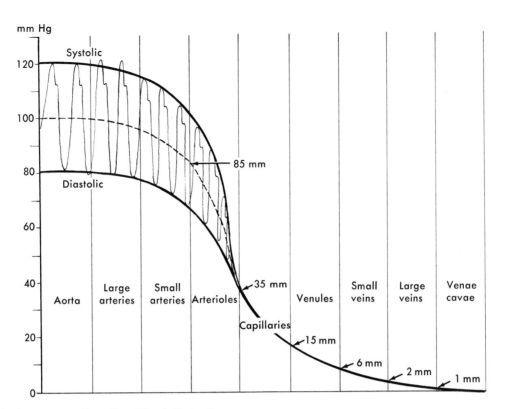

Fig. 10-25 Blood pressure gradient. Dotted line indicates the average or mean systolic pressure in arteries.

circulation is this: blood circulates because a blood pressure gradient exists within the circulatory system.

What is blood pressure gradient?

Answer: Blood pressure gradient is blood pressure difference. For example, if blood pressure in the aorta averages 100 mm Hg and pressure in a vein is 5 mm Hg, a blood pressure gradient of 95 mm Hg exists between the two areas. A blood pressure gradient might be thought of as a blood pressure hill down which blood flows. The symbol $(P_1 - P_2)$ is often used to stand for a pressure gradient, with P_1 the symbol for the higher pressure and P_2 the symbol for the lower pressure.

Suppose, for instance, that the average capillary pressure in the capillaries of your arm muscles is 25 mm Hg and the average pressure in the arterioles from which

these capillaries branch is 60 mm Hg. Which is P_1? P_2? What is the blood pressure gradient? In which direction would blood necessarily flow?

Fig. 10-25 shows a typical blood pressure gradient. What is the gradient for systemic circulation taken as a whole? P_1 in this gradient is the average or mean systolic pressure in the aorta and P_2 is the pressure in the venae cavae at their junction with the right atrium.

What factors determine arterial blood pressure?

Answer: The primary determinant is the volume of blood in the arteries. A direct relationship exists between arterial blood volume and arterial pressure. This means that an increase in arterial blood volume tends to increase arterial pressure and, conversely, a decrease in arterial volume

331

tends to decrease arterial pressure. Many factors together indirectly determine arterial pressure through their influence on arterial volume. Two of the most important are cardiac minute output (CMO) and peripheral resistance. A change in either tends to change the volume of blood within the arteries and thereby to change the blood pressure in the same direction. More specifically, anything that increases cardiac minute output tends to increase arterial blood volume and thereby to increase arterial blood pressure. Anything that decreases cardiac minute output tends to decrease arterial blood volume and pressure. Anything that increases peripheral resistance tends to increase arterial blood volume and pressure, and anything that decreases peripheral resistance tends to decrease arterial volume and pressure. Now let us turn our attention to these terms cardiac minute output and peripheral re-

sistance. Examine Fig. 10-26. Be sure to read its legend.

Cardiac minute output means what you would guess—the volume of blood pumped out of the left ventricle into the aorta each minute. How big this volume is, of course, depends both upon the number of heart contractions per minute and upon the amount of blood pumped per contraction. Contraction of the heart is called *systole*. Therefore, the volume of blood pumped by one contraction is known as *systolic discharge*. *Stroke volume* means the same thing, the amount of blood pumped by one stroke (contraction) of the ventricle. Stroke volume reflects the force or strength of ventricular contraction—the stronger the contraction, the greater the stroke volume tends to be. Cardiac minute output can be computed by the following simple equation:

$$\text{Stroke volume} \times \text{Heart rate} = \text{CMO}$$

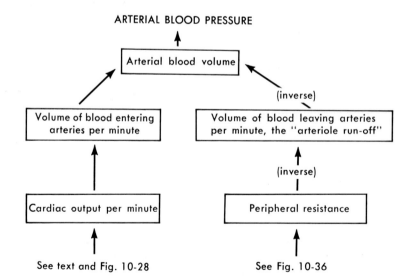

Fig. 10-26 Scheme to show how cardiac minute output and peripheral resistance affect arterial blood pressure. If cardiac minute output increases, the amount of blood entering the arteries increases and tends to increase the volume of blood in the arteries. The resulting increase in arterial volume increases arterial blood pressure. If peripheral resistance increases, it tends to decrease the amount of blood leaving the arteries, which tends to increase the amount of blood left in them. The increase in arterial volume increases arterial blood pressure.

From this equation we can derive the following principles. Anything that changes either the rate of the heartbeat or its stroke volume tends to change cardiac minute output, arterial blood volume, and blood pressure in the same direction. In other words, anything that makes the heart beat faster or anything that makes it beat stronger (increases its stroke volume) tends to increase cardiac minute output and, therefore, arterial blood volume and pressure. Conversely, anything that causes the heart to beat more slowly or more weakly tends to decrease cardiac minute output, arterial volume, and blood pressure. But do not overlook the word *tend* in the preceding sentences. A change in heart rate or in stroke volume does not always change the heart's output, or the amount of blood in the arteries, or the blood pressure. To see whether this is true, do the following simple arithmetic, using the formula just given for computing CMO. Assume a normal rate of 72 beats per minute and a normal stroke volume of 70 milliliters. Next, suppose the rate drops to 60 and the stroke volume increases to 100. Does the decrease in heart rate actually cause a decrease in cardiac minute output in this case? Clearly not—the cardiac minute output increases. Do you think it is valid, however, to say that a slower rate *tends* to decrease the heart's minute output? By itself, without any change in any other factor, would not a slowing of the heartbeat cause cardiac minute volume, arterial volume, and blood pressure to fall?

Peripheral resistance, another factor that helps determine arterial blood pressure, is resistance to blood flow imposed by the force of friction between blood and the walls of its vessels. Friction develops partly because of a characteristic of blood —its viscosity or stickiness—and partly from the small diameter of arterioles and capillaries. Peripheral resistance helps determine arterial pressure by controlling the rate of "arteriole runoff," the amount of blood that runs out of the arteries into the arterioles. The greater the resistance, the less the arteriole runoff tends to be— and, therefore, the more blood left in the arteries and the higher the arterial pressure tends to be.

Summarized, arterial blood pressure is determined directly by arterial blood volume which is determined by many factors but especially by the heart's output and peripheral resistance (Fig. 10-26).

What factors regulate the stroke volume of the heart?

Answer: Mechanical, neural, and chemical factors regulate the strength of the heartbeat and therefore its stroke volume. The mechanical factor that helps determine stroke volume is the length of myocardial fibers at the beginning of ventricular contraction.

Many years ago Starling described a principle later made famous as Starling's law of the heart. In this principle he stated the factor he had observed as the main regulator of heartbeat strength in experiments performed on denervated animal hearts. Starling's law of the heart, in essence, is this: within limits, the longer or more stretched the heart fibers at the beginning of contraction, the stronger is their contraction.

The factor determining how stretched the animal hearts were at the beginning of contractions was, as you might deduce, the amount of blood in the hearts at the end of diastole. The more blood returned to the hearts per minute, the more stretched were their fibers, the stronger were their contractions, and the larger was the volume of blood they ejected with each contraction. If, however, too much blood stretched the hearts too far, beyond a certain critical point, they seemed to lose their elasticity. They then contracted less vigorously—

much as a rubber band, stretched too much, rebounds with less force.

Physiologists have long accepted as fact that Starling's law of the heart operates in animals under experimental conditions. But they have questioned its validity and importance in the intact human body. Today, the prevailing opinion seems to be that Starling's law of the heart operates as a major regulator of stroke volume under ordinary conditions, but there is still disagreement as to its role during exercise.* Operation of Starling's law of the heart ensures that increased amounts of blood returned to the heart will be pumped out of it. It automatically adjusts cardiac output to venous return under usual conditions.

What factors regulate the heart rate?
Answer: <u>Pressoreflexes</u> constitute the dominant heart rate control mechanism, although various other factors also influence heart rate. *Chemoreflexes*

Cardiac pressoreflexes. Pressure-sensitive cells called baroreceptors (or pressoreceptors) (Fig. 10-27) are located in the aortic arch and the carotid sinus. (The <u>carotid sinus</u> is a small dilatation at the beginning of the internal carotid artery just above the bifurcation of the common carotid artery to form the internal and external carotid arteries.) The sinus lies just under the sternocleidomastoid muscle at the level of the upper margin of the thyroid cartilage. Sensory fibers extend from the aortic baroreceptors in the vagus (tenth cranial) nerve to terminate in the medulla in its cardiac and vasomotor centers. Sensory fibers from carotid sinus baroreceptors, on the other hand, run through the carotid sinus nerve (of Hering) and on through the glossopharyngeal (or ninth cranial) nerve to the cardiac and vasomotor centers.

*Mountcastle, V. B., editor: Medical physiology, ed. 12, St. Louis, 1968, The C. V. Mosby Co.

If blood pressure within the aorta or carotid sinus increases suddenly, it stimulates the aortic or carotid baroreceptors. As shown in Fig. 10-27, this leads to stimulation of the cardioinhibitory centers and reciprocal inhibition of the accelerator centers, which in turn leads to more impulses per second over parasympathetic fibers in the vagus nerve and fewer impulses over the sympathetic fibers in the cardioaccelerator nerves to the heart. As a result, reflex slowing of the heart occurs.

On the other hand, a decrease in aortic or carotid blood pressure usually initiates reflex acceleration of the heart. The lower blood pressure decreases baroreceptors' stimulation. Hence, the cardioinhibitory center receives fewer stimulating impulses and the cardioaccelerator center fewer inhibitory impulses. Net result? The heart beats faster.

Baroreceptors located in the right atrium of the heart may respond to changes in right atrial pressure. An increase in this pressure results in reflex acceleration of the heart, and a decrease in right atrial pressure produces reflex slowing.

Almost fifty years ago the noted physiologist Bainbridge demonstrated reflex heart acceleration in dogs following injection of saline solution or blood intravenously but did not establish the mechanism involved. But since then some have postulated that this Bainbridge reflex (accelerated heartbeat following increased venous return) is initiated by stimulation of baroreceptors in the vena cava or right atrium. Others deny that such a reflex exists at all.

Miscellaneous factors that influence heart rate. Included in this category are such important factors as emotions, exercise, hormones, blood temperature, and stimulation of various exteroceptors. Anxiety, fear, and anger often make the heart beat faster. Grief, in contrast, tends to slow it. Presumably, emotions produce changes

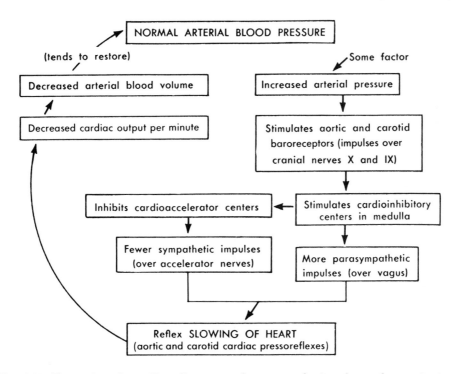

Fig. 10-27 The aortic and carotid cardiac pressoreflexes, a mechanism that tends to maintain or restore homeostasis of arterial blood pressure by regulating the rate of the heartbeat. Note that by this mechanism an increase in arterial blood pressure leads to reflex slowing of the heart and tends to lower blood pressure. The converse is also true. A decrease in blood pressure leads to reflex acceleration of the heart and tends to raise the pressure upward toward normal.

in the heart rate through the influence of impulses from the cerebrum via the hypothalamus to cardiac centers in the medulla and cord.

In exercise, the heart normally accelerates. The mechanism is not definitely known. But it is thought to include impulses from the cerebrum through the hypothalamus to cardiac centers. Epinephrine is the hormone most noted as a cardiac accelerator.

Increased blood temperature or stimulation of skin heat receptors tends to increase the heart rate, and decreased blood temperature or stimulation of skin cold receptors tends to slow it. Sudden intense stimulation of pain receptors also tends to decrease the heart rate. Fig. 10-28 shows the major factors that control the rate of the heartbeat.

What factors determine peripheral resistance?

Answer: Peripheral resistance exists mainly because of blood viscosity and the small diameter of arterioles.

What determines blood viscosity?

Answer: Viscosity is the characteristic of a fluid that results from attraction forces between its molecules or other small particles and that causes it to resist flowing. Blood viscosity stems mainly from the red cells but also partly from the protein molecules present in blood. An increase in either blood protein concentration or in the red cell count tends to increase viscosity, and a decrease in either tends to decrease it. Under normal circumstances, blood viscosity changes very little. But under certain abnormal conditions, such as marked anemia or hemorrhage, a decrease in blood

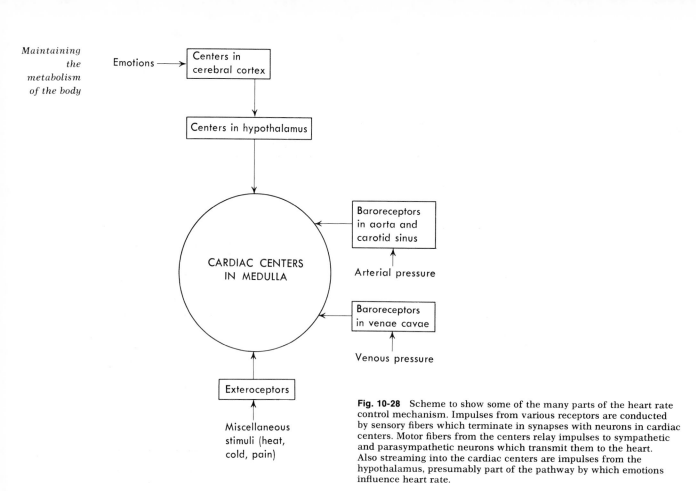

Fig. 10-28 Scheme to show some of the many parts of the heart rate control mechanism. Impulses from various receptors are conducted by sensory fibers which terminate in synapses with neurons in cardiac centers. Motor fibers from the centers relay impulses to sympathetic and parasympathetic neurons which transmit them to the heart. Also streaming into the cardiac centers are impulses from the hypothalamus, presumably part of the pathway by which emotions influence heart rate.

viscosity may be the crucial factor lowering peripheral resistance and arterial pressure even to the point of circulatory failure.

What factors regulate arteriole diameter?

Answer: Factors that control arteriole diameter might be said to constitute the vasomotor control mechanism. Like most physiological mechanisms, it consists of many parts (Figs. 10-29 to 10-34). A change in either arterial blood pressure or in arterial blood's oxygen or carbon dioxide content sets vasomotor control mechanisms in operation. A change in arterial blood pressure initiates a *vasomotor pressoreflex*. A change in arterial oxygen or carbon dioxide content acts in two ways to bring about a change in arteriole diameter—by stimulating chemoreceptors and thereby initiating a *vasomotor chemo-*

reflex and by stimulating the medulla's vasomotor centers directly and thereby initiating what we shall call the *medullary ischemic reflex.*

Vasomotor pressoreflexes (Fig. 10-29). A sudden increase in arterial blood pressure stimulates aortic and carotid baroreceptors —the same ones that initiate cardiac reflexes. Not only does this lead to stimulation of cardioinhibitory centers but also to inhibition of vasoconstrictor centers. More impulses per second go out to the heart over parasympathetic vagal fibers and fewer over sympathetic fibers to blood vessels. As a result, the heartbeat slows, and arterioles and the venules of the "blood reservoirs" dilate. Since sympathetic vasoconstrictor impulses predominate at normal arterial pressures, inhibition of these is considered the major mechanism of vasodilatation.

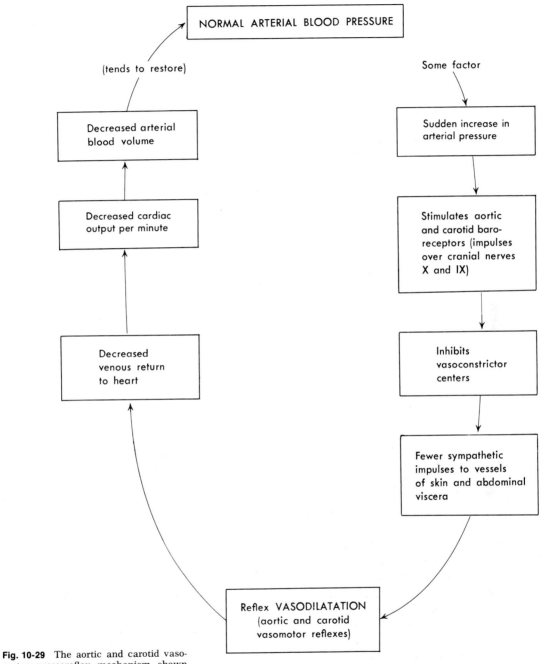

NORMAL ARTERIAL BLOOD PRESSURE

(tends to restore)

Some factor

Decreased arterial blood volume

Sudden increase in arterial pressure

Decreased cardiac output per minute

Stimulates aortic and carotid baro-receptors (impulses over cranial nerves X and IX)

Decreased venous return to heart

Inhibits vasoconstrictor centers

Fewer sympathetic impulses to vessels of skin and abdominal viscera

Reflex VASODILATATION (aortic and carotid vasomotor reflexes)

Fig. 10-29 The aortic and carotid vasomotor pressoreflex mechanism shown here is set in operation when some factor causes a sudden increase in arterial blood pressure. This mechanism and the aortic and carotid cardiac pressoreflexes (Fig. 10-27) operate simultaneously to maintain or restore homeostasis of arterial blood pressure.

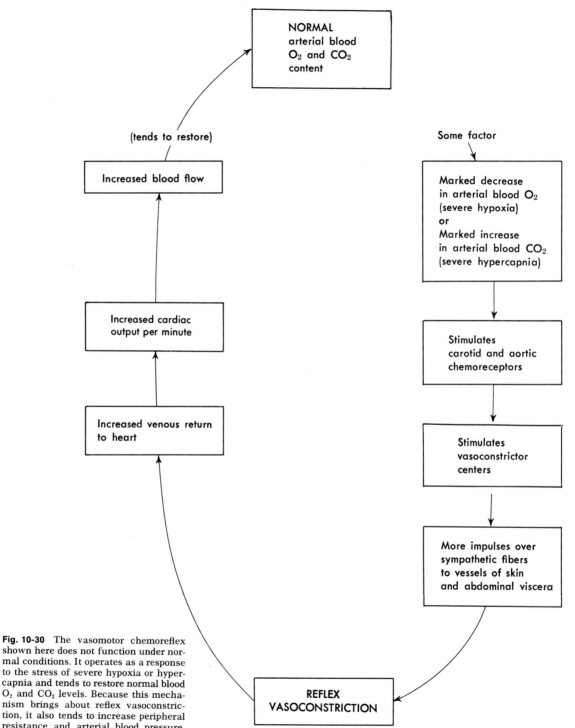

Fig. 10-30 The vasomotor chemoreflex shown here does not function under normal conditions. It operates as a response to the stress of severe hypoxia or hypercapnia and tends to restore normal blood O_2 and CO_2 levels. Because this mechanism brings about reflex vasoconstriction, it also tends to increase peripheral resistance and arterial blood pressure. (Also see Fig. 10-31.)

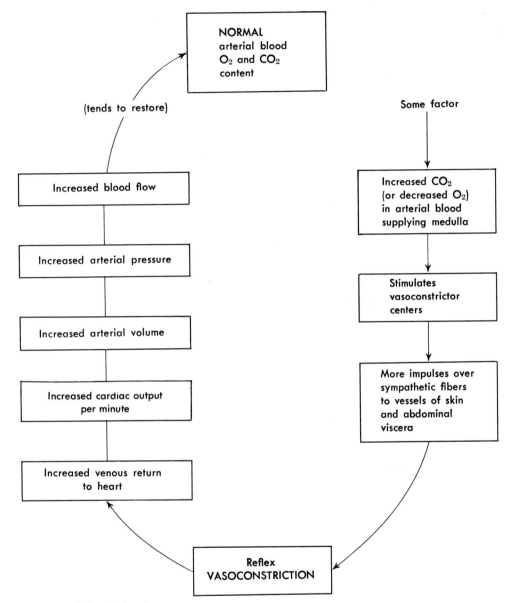

Fig. 10-31 The medullary ischemic reflex, a homeostatic mechanism that tends to restore homeostasis of blood oxygen and carbon dioxide content.

The main blood reservoirs are the venous plexuses and sinuses in the skin and abdominal organs (especially in the liver and spleen). In other words, blood reservoirs are the venous networks in most parts of the body — all but those in the skeletal muscles, heart, and brain. The term reservoir is apt, since these veins serve as storage depots for blood. It can quickly be moved out of them and "shifted" to heart and skeletal muscles when increased activity demands. The reflex vasoconstrictor mechanism that accomplishes this operates as follows.

A decrease in arterial pressure causes the aortic and carotid baroreceptors to send more impulses to the medulla's vasoconstrictor centers, thereby stimulating them. These centers then send more impulses via sympathetic fibers to the smooth muscle in the arterioles, venules, and veins of the blood reservoirs, causing their constriction. This squeezes more blood out of them, increasing the amount of venous return to the heart. Eventually, this extra blood is redistributed to more active structures such as skeletal muscles and heart because their arterioles become dilated due largely to the operation of a local mechanism (next column). Thus, the vasoconstrictor pressoreflex and the local vasodilating mechanism together serve as an important device for shifting blood from reservoirs to structures that need it more. It is an especially valuable mechanism during exercise.

Vasomotor chemoreflexes (Fig. 10-30). Chemoreceptors located in the aortic and carotid bodies are particularly sensitive to a deficiency of blood oxygen (hypoxia) and somewhat less sensitive to excess blood carbon dioxide (hypercapnia) and to decreased arterial blood pH. When one or more of these conditions stimulates the chemoreceptors, their fibers transmit more impulses to the medulla's vasoconstrictor centers, and vasoconstriction of arterioles and venous reservoirs soon follows. This stress mechanism functions as an emergency device when severe hypoxia or hypercapnia endangers survival.

The medullary ischemic reflex (Fig. 10-31). The medullary ischemic reflex mechanism is said to exert the most powerful control of all on small blood vessels. When the blood supply to the medulla becomes inadequate (ischemia), its neurons suffer from both oxygen deficiency and carbon dioxide excess. But, presumably, it is the latter, the hypercapnia, that intensely and directly stimulates the vasoconstrictor centers to bring about marked arteriole and venous constriction. If the oxygen supply to the medulla decreases below a certain level, its neurons, of course, cannot function and the medullary ischemic reflex cannot operate.

Vasomotor control by higher brain centers (Fig. 10-32). Impulses from centers in the cerebral cortex and in the hypothalamus are believed to be transmitted to the vasomotor centers in the medulla and to thereby help control vasoconstriction and dilatation. One evidence supporting this view, for example, is the fact that vasoconstriction and a rise in arterial blood pressure characteristically accompany emotions of intense fear or anger. Also, laboratory experiments on animals in which stimulation of the posterior or lateral parts of the hypothalamus leads to vasoconstriction support the belief that higher brain centers influence the vasomotor centers in the medulla.

Local control of arterioles

Some kind of local mechanism operates to produce vasodilatation in localized areas. Although the mechanism is not clear, it is known to function in times of increased tissue activity. For example, it probably accounts for the increased blood flow into

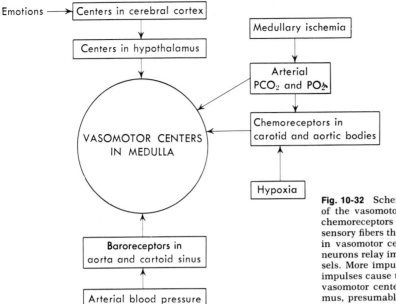

Emotions → Centers in cerebral cortex

Centers in hypothalamus

VASOMOTOR CENTERS IN MEDULLA

Medullary ischemia

Arterial PCO_2 and PO_2

Chemoreceptors in carotid and aortic bodies

Hypoxia

Baroreceptors in aorta and cartoid sinus

Arterial blood pressure

Fig. 10-32 Scheme to show some of the many parts of the vasomotor control mechanism. Impulses from chemoreceptors and baroreceptors are conducted by sensory fibers that terminate in synapses with neurons in vasomotor centers. From the centers sympathetic neurons relay impulses to smooth muscle in blood vessels. More impulses cause vessels to constrict. Fewer impulses cause them to dilate. Also streaming into the vasomotor centers are impulses from the hypothalamus, presumably part of the pathway by which emotions influence blood pressure.

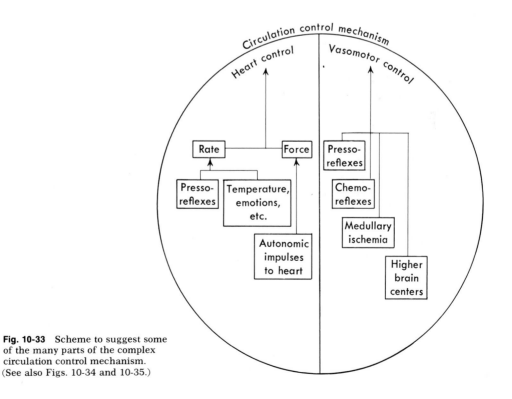

Circulation control mechanism

Heart control

Vasomotor control

Rate — Force

Presso-reflexes — Temperature, emotions, etc.

Autonomic impulses to heart

Presso-reflexes

Chemo-reflexes

Medullary ischemia

Higher brain centers

Fig. 10-33 Scheme to suggest some of the many parts of the complex circulation control mechanism. (See also Figs. 10-34 and 10-35.)

skeletal muscles during exercise. It also operates in ischemic tissues, serving as a homeostatic mechanism that tends to restore normal blood flow. Norepinephrine, histamine, lactic acid, and other locally produced substances have been suggested as the stimuli that activate the local vasodilator mechanism. Local vasodilation is also referred to as *reactive hyperemia*.

What factors determine the volume of blood flow per minute? Or, stated differently, what makes circulation increase or decrease from time to time?

Answer: The answer to this question lies within the answers to all of the preceding questions about circulation. In most abbreviated form, it might be stated this way: the amount of blood flowing through the body per minute is determined by the magnitude of both the blood pressure gradient and the peripheral resistance (Figs. 10-34 to 10-36).

A nineteenth century physiologist and physicist, Poiseuille, described the relationship between these three factors—pressure gradient, resistance, and volume of fluid flow per minute—with a mathematical equation known as *Poiseuille's law*. Because the central venous pressure at the opening of the venae cavae into the right atrium is zero, the systemic blood pressure gradient (mean arterial pressure—central venous pressure) normally equals the mean or average arterial blood pressure. Therefore, we can state Poiseuille's law as applied to circulation in the simplest terms as follows: the volume of blood circulated per minute is directly related to arterial pressure and inversely related to resistance or:

$$\text{Volume of blood circulated per minute} = \frac{\text{Arterial pressure}}{\text{Resistance}}$$

The preceding statement and equation need qualifying with regard to the in-fluence of peripheral resistance on circulation. For instance, according to the equation, an increase in peripheral resistance would tend to decrease blood flow. (Why? Increasing peripheral resistance increases the denominator of the fraction in the preceding equation. And increasing the denominator of any fraction necessarily does what to its value?) Increased peripheral resistance, however, has a secondary action that acts against its primary tendency to decrease blood flow. An increase in peripheral resistance hinders or decreases arteriole runoff. And this, of course, tends to increase the volume of blood left in the arteries so tends to increase arterial pressure. Note also that increasing arterial pressure tends to increase the value of the fraction in Poiseuille's equation. Therefore, it tends to increase circulation. In short, to say unequivocally what the effect of an increased peripheral resistance will be on circulation is impossible. It depends also upon arterial blood pressure—whether it increases, decreases, or stays the same when peripheral resistance increases. The clinical condition arteriosclerosis with hypertension (high blood pressure) illustrates this point. Both peripheral resistance and arterial pressure are increased in this condition. If resistance increases more than arterial pressure, circulation—that is, volume of blood flow per minute—decreases. But if arterial pressure increases proportionately to resistance, circulation remains normal.

Important factors influencing venous return to heart

Two important factors that promote the return of venous blood to the heart are respirations and skeletal muscle contractions. Both produce their facilitating effect on venous return by increasing the pressure gradient between peripheral veins and venae cavae.

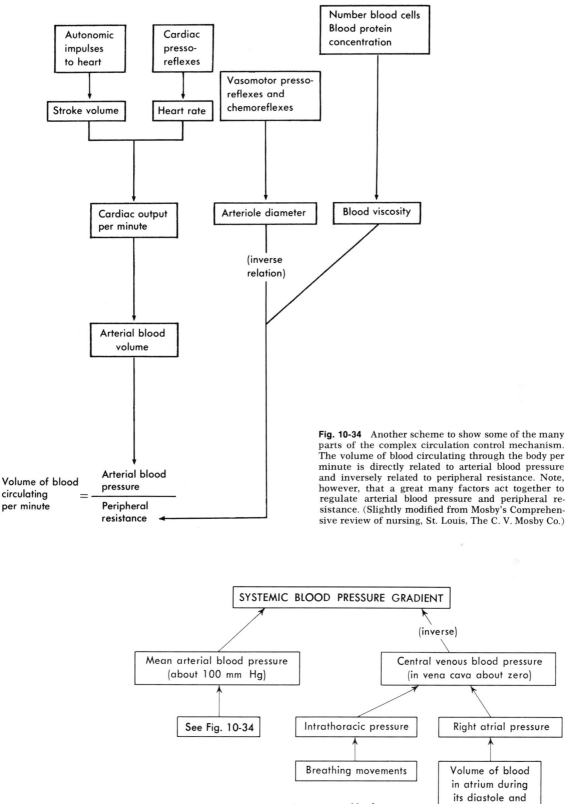

Fig. 10-34 Another scheme to show some of the many parts of the complex circulation control mechanism. The volume of blood circulating through the body per minute is directly related to arterial blood pressure and inversely related to peripheral resistance. Note, however, that a great many factors act together to regulate arterial blood pressure and peripheral resistance. (Slightly modified from Mosby's Comprehensive review of nursing, St. Louis, The C. V. Mosby Co.)

Fig. 10-35 Factors that determine the systemic blood pressure gradient.

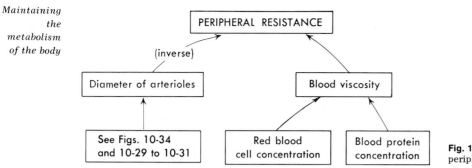

PERIPHERAL RESISTANCE

(inverse)

Diameter of arterioles

Blood viscosity

See Figs. 10-34
and 10-29 to 10-31

Red blood
cell concentration

Blood protein
concentration

Fig. 10-36 The main determinants of peripheral resistance.

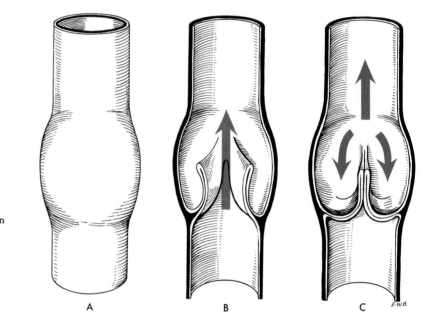

A   B   C

Fig. 10-37 Diagram showing the action of venous valves. **A,** External view of vein showing dilation at site of the valve. **B,** Interior of vein with the semilunar flaps in open position, permitting flow of blood through the valve. **C,** Valve flaps approximating each other, occluding the cavity and preventing the backflow of blood.

The process of breathing increases the pressure gradient between peripheral and central veins by decreasing central venous pressure and also by increasing peripheral venous pressure. Each time the diaphragm contracts, the thoracic cavity necessarily becomes larger and the abdominal cavity smaller. Therefore, the pressures in the thoracic cavity, in the thoracic portion of the vena cava, and in the atria decrease, and those in the abdominal cavity and the abdominal veins increase. Deeper respirations intensify these effects and therefore tend to increase venous return to the heart more than do normal respirations. This is part of the reason why the principle is true that increased respirations and increased circulation tend to go hand in hand.

Skeletal muscle contractions serve as "booster pumps" for the heart. They promote venous return in the following way. As each skeletal muscle contracts, it squeezes the soft veins scattered through its interior, thereby milking the blood in them upward or toward the heart. And the closing of the semilunar valves present in veins prevents blood from falling back as the muscle relaxes. Their flaps catch the blood as gravity pulls backward on it (Fig. 10-37). The net effect of skeletal muscle contraction plus venous valvular action, therefore, is to move venous blood toward the heart, to increase the venous return.

The value of skeletal muscle contractions in moving blood through veins is illustrated by a common experience. Who has not noticed how much more uncomfortable and tiring standing still is than walking? After several minutes of standing quietly, the feet and legs feel "full" and swollen. Blood has accumulated in the veins because the skeletal muscles are not contracting and squeezing it upward. The repeated contractions of the muscles when walking, on the other hand, keep the blood moving in the veins and prevent the discomfort of distended veins.

Blood pressure

How arterial blood pressure measured clinically

Blood pressure is measured with the aid of an apparatus known as a sphygmomanometer which makes it possible to measure the amount of air pressure equal to the blood pressure in an artery. The measurement is made in terms of how many millimeters high the air pressure raises a column of mercury in a glass tube.

The sphygmomanometer consists of a rubber cuff attached by a rubber tube to a compressible bulb and by another tube to a column of mercury that is marked off in millimeters. The cuff is wrapped around the arm over the brachial artery, and air is pumped into the cuff by means of the bulb. In this way air pressure is exerted against the outside of the artery. Air is added until the air pressure exceeds the blood pressure within the artery or, in other words, until it compresses the artery. At this time no pulse can be heard through a stethoscope placed over the brachial artery at the bend of the elbow along the inner margin of the biceps muscle. By slowly releasing the air in the cuff, the air pressure is decreased until it approximately equals the blood pressure within the artery. At this point, the vessel opens slightly and a small spurt of blood comes through, producing the first sound, one with a rather sharp, taplike quality. This is followed by increasingly louder sounds which suddenly change. They become more muffled, then disappear altogether. The nurse must train herself to hear these different sounds and simultaneously to read the column of mercury since the first taplike sound represents the *systolic blood pressure*. Systolic pressure is the force with which the blood is pushing against the artery walls when the ventricles are contracting. The lowest point at which the sounds can be heard, just before they disappear, is approximately equal to the *diastolic pressure* or the force of the blood when the ventricles are relaxed. Systolic pressure gives valuable information about the force of the left ventricular contraction, and diastolic pressure gives valuable information about the resistance of the blood vessels. Clinically, diastolic pressure is considered more important than systolic pressure because it indicates the pressure or strain to which blood vessel walls are constantly subjected. It also reflects the condition of the peripheral vessels since diastolic pressure rises or falls with the peripheral resistance. If, for instance, arteries are sclerosed, peripheral resistance and diastolic pressure both increase.

Blood in the arteries of the average adult exerts a pressure equal to that required to raise a column of mercury about 120 mm (or a column of water over 5 feet) high in a glass tube during systole of the ventricles and 80 mm high during their diastole. For the sake of brevity, this is expressed as a blood pressure of 120 over 80 (120/80). The first or upper figure represents systolic pressure and the second diastolic pressure. From the figures just given, we observe that blood pressure fluctuates considerably during each heartbeat. During ventricular systole, the force is great enough to raise the mercury column 40 mm higher than during ventricular diastole. This difference between systolic and diastolic pressure is called *pulse pressure*. It characteristically increases in arteriosclerosis due mainly to increased systolic pressure. Pulse pressure increases even more markedly in aortic valve insufficiency due both to a rise in systolic and a fall in diastolic pressure.

Relation to arterial and venous bleeding

Because blood exerts a comparatively high pressure in arteries and a very low pressure in veins, it gushes forth with considerable force from a cut artery but seeps in a slow, steady stream from a vein. As we have just seen, each ventricular contraction raises arterial blood pressure to the systolic level, and each ventricular relaxation lowers it to the diastolic level. As the ventricles contract, then, the blood spurts forth forcefully due to increased pressure in the artery, but as the ventricles relax, the flow ebbs to almost nothing due to the fall in pressure. In other words, blood escapes from an artery in spurts because of the alternate raising and lowering of arterial blood pressure but flows slowly and steadily from a vein because of the low, practically constant pressure. A uniform instead of a pulsating pressure exists in the capillaries and veins. Why? Because the arterial walls, being elastic, continue to squeeze the blood forward while the ventricles are in diastole. Therefore, blood enters capillaries and veins under a steady pressure.

Velocity of blood

The speed with which blood flows (that is, distance per minute) through its vessels is governed in part by the physical principle that when a liquid flows from an area of one cross-section size to an area of larger size, its velocity slows in the area with the larger cross section. For example, a narrow river whose bed widens flows more slowly through the wider section than through the narrow. In terms of blood vascular system, the total cross-section area of all arterioles together is greater than that of the arteries. Therefore, blood flows more slowly through arterioles than through arteries. Likewise, the total cross-section area of all capillaries together is greater than that of all arterioles and, therefore, capillary flow is slower than arteriole. Venule cross-section area, on the other hand, is smaller than capillary cross-section area. Therefore, the blood velocity increases in venules and again in veins, which have a still smaller cross-section area. In short, the most rapid blood flow takes place in arteries and the slowest in capillaries. Can you think of a valuable effect stemming from the fact that blood flows most slowly through the capillaries?

Pulse

Definition

Pulse is defined as the alternate expansion and recoil of an artery.

Cause

Two factors are responsible for the existence of a pulse which can be felt:

1 Intermittent injections of blood from the heart into the aorta, which alternately increase and decrease the pressure in that vessel. If blood poured steadily out of the

heart into the aorta, the pressure there would remain constant and there would be no pulse.

2 The elasticity of the arterial walls, which makes it possible for them to expand with each injection of blood and then recoil. If the vessels were fashioned from rigid material such as glass, there would still be an alternate raising and lowering of pressure within them with each systole and diastole of the ventricles, but the walls could not expand and recoil and, therefore, no pulse could be felt.

Pulse wave

Each ventricular systole starts a new pulse which proceeds as a wave of expansion throughout the arteries and is known as the pulse wave. It gradually dissipates as it travels, disappearing entirely in the capillaries. The pulse felt in the radial artery at the wrist does not coincide with the contraction of the ventricles. It follows each contraction by an appreciable interval (the length of time required for the pulse wave to travel from the aorta to the radial artery). The farther from the heart the pulse is taken, therefore, the longer that interval is.

Any nurse has only to think of the number of times she has counted pulses to become aware of the diagnostic importance of the pulse. It reveals important information about the cardiovascular system, about heart action, blood vessels, and circulation.

Where pulse can be felt

In general, the pulse can be felt wherever an artery lies near the surface and over a bone or other firm background. Some of the specific locations where the pulse is most easily felt are as follows:

1 *radial artery* – at wrist
2 *temporal artery* – in front of ear or above and to outer side of eye

3 *common carotid artery* – along anterior edge of sternocleidomastoid muscle at level of lower margin of thyroid cartilage
4 *facial artery* – at lower margin of lower jawbone on a line with corners of mouth and in a groove in mandible about one-third of way forward from angle
5 *brachial artery* – at bend of elbow along inner margin of biceps muscle
6 *posterior tibial artery* – behind the medial malleolus (inner "ankle bone")
7 *dorsalis pedis artery* – on the dorsum (upper surface) of the foot

Note: The so-called pressure points or points at which pressure may be applied to stop arterial bleeding and the points where the pulse may be felt are related. To be more specific, both are found where an artery lies near the surface and near a bone which can act as a firm background for pressure. There are six important pressure points:

1 *temporal artery* – in front of ear
2 *facial artery* – same place as pulse is taken
3 *common carotid artery* – point where pulse is taken, with pressure back against spinal column
4 *subclavian artery* – behind mesial third of clavicle, pressing against first rib
5 *brachial artery* – few inches above elbow on inside of arm, pressing against humerus
6 *femoral artery* – in middle of groin, where artery passes over pelvic bone; pulse can also be felt here

In trying to stop arterial bleeding by pressure, one must always remember to apply the pressure at the pressure point that lies between the bleeding part and the heart. Why? Because blood flows from the heart through the arteries to the part. Pressure between the heart and bleeding point, therefore, cuts off the source of the blood flow to that point.

Venous pulse

A pulse exists in the large veins only. It is most prominent in the veins near the heart, due to changes in venous blood pressure brought about by alternate contraction and relaxation of the atria of the heart. Venous pulse does not have as great clinical significance as arterial pulse so is less often measured.

Lymphatic system — *absorption of fat.*

Definition

The lymphatic system is actually part of the circulatory system since it consists of a moving fluid (lymph and tissue fluid) derived from the blood and a group of vessels (lymphatics) which return the lymph to the blood.

■ Lymph and interstitial fluid (tissue fluid)

Definition

Lymph is the clear, watery-appearing fluid found in the lymphatic vessels. Interstitial fluid is also a clear, watery fluid, but it occupies the microscopic spaces between cells. In some tissues it is part of a semi-fluid ground substance. In others it is the bound water in a gelatinous ground substance. Interstitial fluid and blood plasma together constitute the extracellular fluid or, in the words of Claude Bernard, the "internal environment of the body"—the fluid environment of cells in contrast to the atmosphere or external environment of the body. Both lymph and interstitial fluid closely resemble blood plasma in composition. The main difference is that they contain a lower percentage of proteins than does plasma.

■ Lymphatics

Formation and distribution

Lymphatic vessels originate as microscopic blind-end vessels called *lymphatic capillaries* (see Plate XV of color insert). These tiny vessels are located in the intercellular spaces and are widely distributed throughout the body. As twigs of a tree join to form branches and branches join to form larger branches, and large branches join to form the tree trunk, so do lymphatic capillaries merge, forming slightly larger lymphatics that join other lymphatics to form still larger vessels, which merge to form the main lymphatic trunks: the *right lymphatic ducts* and the *thoracic duct.* Lymph from the entire body, except the upper right quadrant (Fig. 10-39), drains eventually into the thoracic duct, which drains into the left subclavian vein at the point where it joins the left internal jugular vein.* Lymph from the upper right quadrant of the body empties into the right lymphatic duct (or, more commonly, into three collecting ducts) and thence into the right subclavian vein. Since most of the lymph of the body returns to the bloodstream by way of the thoracic duct, this vessel is considerably larger than the other main lymph channels, the right lymphatic ducts, but is much smaller than the large veins, which it resembles in structure. It has a diameter about the size of a goose

---

*An interesting application of knowledge about the thoracic duct and lymphocyte function has recently been made by a group of researchers at the University of Texas (Immune reaction: Alternative to drugs, Sci. News **96:**327 [Oct. 11], 1969). Before a kidney transplant was to be done, they surgically opened the thoracic duct and drew off lymph. They then filtered it to remove lymphocytes and other cells and returned the cell-free lymph to the patient. Since lymphocytes synthesize antibodies that destroy foreign cells, this procedure greatly reduced the chances of the patient's body rejecting the transplanted kidney.

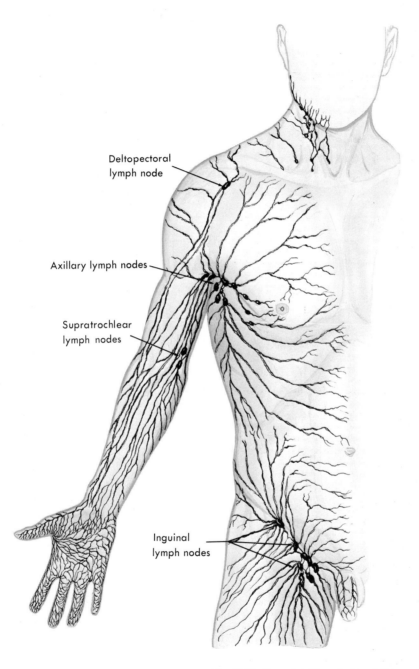

Deltopectoral
lymph node

Axillary lymph nodes

Supratrochlear
lymph nodes

Inguinal
lymph nodes

Fig. 10-38 Superficial lymphatics of the
upper extremity, anterior surface, and
superficial lymphatics and lymph nodes
of the front of the trunk.

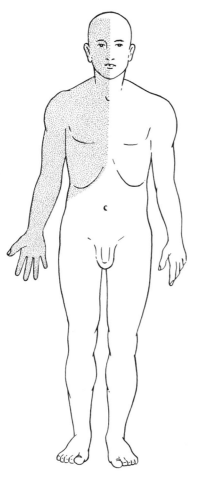

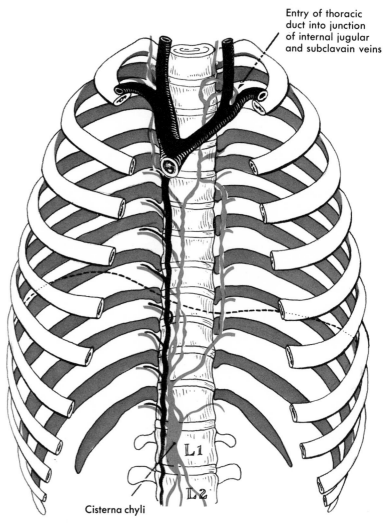

Fig. 10-39 Lymph drainage. The right lymphatic ducts drain lymph from the part of the body indicated by the stippled area. Lymph from all the rest of the body enters the general circulation by way of the thoracic duct.

Entry of thoracic duct into junction of internal jugular and subclavain veins

Cisterna chyli

Fig. 10-40 Position of the cisterna chyli and the thoracic duct and its tributaries and the entry of the duct into the junction of the internal jugular and subclavian veins.

quill and a length of from 15 to 18 inches. It originates as a dilated structure, the *cisterna chyli*, in the lumbar region of the abdominal cavity and ascends by a flexuous course to the root of the neck, where it joins the subclavian vein as just described (see Fig. 10-40). The presence of semilunar valves at frequent intervals along the duct gives it a somewhat varicose appearance.

Structure

Lymphatics resemble veins in structure with these exceptions:
1 Lymphatics have thinner walls
2 Lymphatics contain more valves
3 Lymphatics contain lymph nodes (or glands) located at certain intervals along their course

Lymphatics originating in the villi of the small intestine are called *lacteals*, and the milky fluid found in them after digestion is *chyle*.

Function

The function of lymphatics is to return water and proteins from the interstitial fluid to blood from which they came. Proteins can return to blood only via lymphatics. This fact has great clinical importance. For instance, if anything blocks lymphatic return, blood protein concentration and blood osmotic pressure soon fall below normal and fluid imbalance results (discussed in Chapter 15).

■ Lymph circulation

Water and solutes continually filter out of capillary blood into the interstitial fluid. To balance this outflow, fluid continually reenters blood from the interstitial fluid. Experimental studies have shown that only about 40% of the fluid that filters out of blood capillaries returns to them by osmosis. The remaining 60% returns to the blood by way of the lymphatics. Also, newer evidence has disproved the old idea

that healthy capillaries do not "leak" proteins. In truth, each day about 50% of the total blood proteins leak out of the capillaries into the tissue fluid and return to the blood by way of the lymphatic vessels. For more details about fluid exchange between blood and interstitial fluid, see Chapter 15. From lymphatic capillaries, lymph flows through progressively larger lymphatic vessels to eventually reenter blood at the junction of the internal jugular and subclavian veins (Fig. 10-40).

Although there is no pump connected with the lymphatic vessels to force lymph onward as the heart does blood, still lymph moves slowly and steadily along in its vessels. And this occurs despite the fact that most of the flow is uphill. What mechanisms establish the pressure gradient required by the basic law of fluid flow? Two of the same mechanisms that contribute to the blood pressure gradient also establish a lymph pressure gradient. These are breathing movements and skeletal muscle contractions.

The mechanism of inspiration, due to the descent of the diaphragm, causes intra-abdominal pressure to increase as intrathoracic pressure decreases. And this simultaneously causes pressure to increase in the abdominal portion of the thoracic duct and to decrease in the thoracic portion. In other words, the process of inspiring establishes a pressure gradient in the thoracic duct that causes lymph to flow upward through it.

In addition, contracting skeletal muscles exert pressure on the lymphatics to push the lymph forward because valves within the lymphatics prevent backflow.

■ Lymph nodes

Structure

Lymph nodes or glands, as some people call them, are oval-shaped or bean-shaped

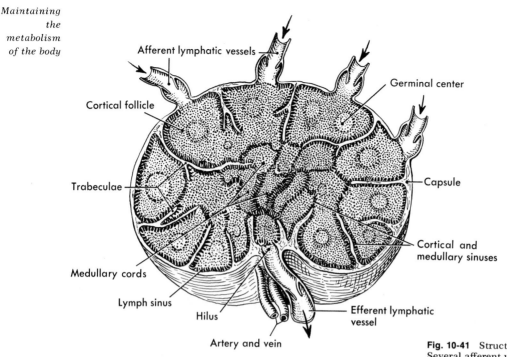

Afferent lymphatic vessels

Germinal center

Cortical follicle

Trabeculae

Capsule

Cortical and medullary sinuses

Medullary cords

Lymph sinus

Hilus

Efferent lymphatic vessel

Artery and vein

Fig. 10-41 Structure of a lymph node. Several afferent valved lymphatics bring lymph to the node. An efferent lymphatic leaves the node at the hilus. Note that the artery and vein enter and leave at the hilus.

structures. Some are as small as a pinhead and others as large as a lima bean. As shown in Fig. 10-41, lymph moves into the nodes via several afferent lymphatic vessels. Here it moves slowly through sinus channels lined with phagocytic reticuloendothelial cells and emerges usually by one efferent vessel. Lymphatic tissue, densely packed with lymphocytes, composes the substance of the node.

Location

With the exception of comparatively few single nodes, most of the lymph nodes occur in groups or clusters in certain areas. The group locations of greatest clinical importance are as follows:

1 *submental and submaxillary groups* in the floor of the mouth—lymph from the nose, lips, and teeth drains through these nodes.

2 *superficial cervical glands* in the neck along the sternocleidomastoid muscle

—these nodes drain lymph from the head (which has already passed through other nodes) and neck.

3 *superficial cubital or supratrochlear nodes* located just above the bend of the elbow—lymph from the forearm passes through these nodes.

4 *axillary nodes* (twenty to thirty large nodes clustered deep within the underarm and upper chest regions)—lymph from the arm and upper part of the thoracic wall, including the breast, drains through these nodes.

5 *inguinal nodes* in the groin—lymph from the leg and external genitals drains through these.

Functions

Lymph nodes perform two unrelated functions, defense and hemopoiesis.

1 *Defense functions: filtration and phagocytosis.* The structure of the sinus channels within lymph nodes slows the

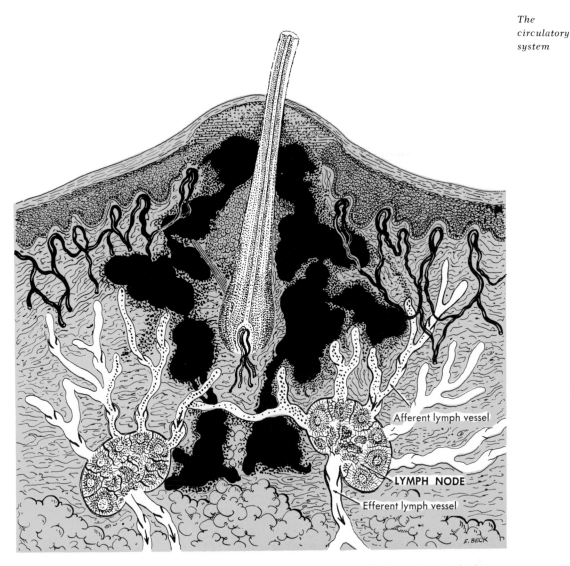

Fig. 10-42 Diagrammatic representation of a skin section in which an infection surrounds a hair follicle. The black areas represent dead and dying cells (pus). Black dots around the black areas represent bacteria. Leukocytes phagocytose many bacteria in tissue spaces. Others may enter the lymph nodes by way of afferent lymphatics. The nodes filter out those bacteria and usually destroy them all by phagocytosis.

lymph flow through them. This gives the reticuloendothelial cells that line the channels time to remove microorganisms and other injurious particles—cancer cells and soot, for example—from the lymph and phagocytose them (see Fig. 10-42). Sometimes, however, such hordes of microorganisms enter the nodes that the phagocytes cannot destroy enough of them to prevent their injuring the node. An infection of the node, adenitis, then results. Also, because cancer cells often break away from a malignant tumor and enter lymphatics, they travel to the lymph nodes where they may set up new growths. This may leave too few channels for lymph to return to the blood. For example, if tumors block axillary node channels, fluid accumulates in the interstitial spaces of the arm, causing the arm to become markedly swollen.

2 *Hemopoiesis*. The lymphatic tissue of lymph nodes forms lymphocytes and monocytes, the nongranular white blood cells, and plasma cells. Some lymphocytes, you will recall, synthesize cellular or tissue-rejecting antibodies. Some become plasma cells and produce circulating antibodies (p. 299).

Spleen — *reserve supply of blood*

Location

The spleen is located in the left hypochondrium directly below the diaphragm, above the left kidney and descending colon, and behind the fundus of the stomach.

Structure

As Fig. 10-43 shows, the spleen is roughly ovoid in shape. Its size varies greatly in different individuals and in the same individual at different times. For example, it hypertrophies during infectious diseases and atrophies in old age. Within the spleen are numerous areas of lymphatic tissue and many venous sinuses.

Functions

The spleen has long puzzled physiologists who have ascribed many and sundry

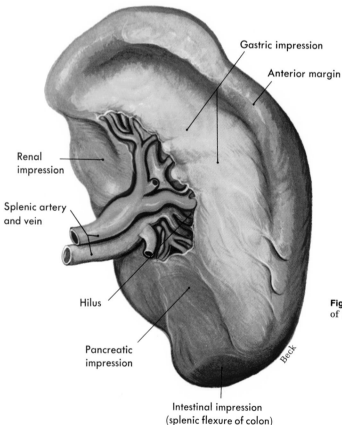

Gastric impression

Anterior margin

Renal impression

Splenic artery and vein

Hilus

Pancreatic impression

Intestinal impression (splenic flexure of colon)

Fig. 10-43 Spleen, medial aspect. Arrangement of the vessels at the hilus is highly variable.

functions to it. According to present-day knowledge, it performs several functions: defense, hemopoiesis, and red blood cell and platelet destruction; it also serves as a reservoir for blood.

1 *Defense.* As blood passes through the sinusoids of the spleen, reticuloendothelial cells (macrophages) lining these venous spaces remove microorganisms from the blood and destroy them by phagocytosis. Therefore, the spleen plays a part in the body's defense against microorganisms.

2 *Hemopoiesis.* Nongranular leukocytes —that is, monocytes and lymphocytes— and plasma cells are formed in the spleen. Before birth, red blood cells are also formed in the spleen, but after birth the spleen is said to form red cells only in extreme hemolytic anemia.

3 *Red blood cell and platelet destruction.* Macrophages lining the spleen's sinusoids remove worn out red cells and imperfect platelets from the blood and destroy them by phagocytosis. They also break apart the hemoglobin molecules from the destroyed red cells and salvage their iron and globin content by returning

them to the blood stream for storage in bone marrow and liver.

4 *Blood reservoir.* The pulp of the spleen and its venous sinuses store considerable blood. Its normal volume of about 350 milliliters is said to decrease about 200 milliliters in less than a minute's time following sympathetic stimulation that produces marked constriction of its smooth capsule. This occurs, for example, as a response to the stress imposed by hemorrhage.

      ●   ●   ●

Although the spleen's functions make it a most useful organ, it is not a vital one. Dr. Charles Austin Doan in 1933 took the daring step of performing the first splenectomy. He removed the spleen from a 4-year-old girl who was dying of hemolytic anemia. Presumably, he justified his radical treatment on the basis of what was then merely conjecture—that is, that the spleen destroys red blood cells. The child recovered and Dr. Doan's operation proved to be a landmark. It created a great upsurge of interest in the spleen and led to many investigations of it.

outline summary

FUNCTIONS

1 Primary function—transportation of various substances to and from body cells
2 Secondary functions—contributes to all bodily functions, for example:
   a Cellular metabolism
   b Homeostasis of fluid volume
   c Homeostasis of pH
   d Homeostasis of temperature
   e Defense against microorganisms

BLOOD

Whole blood consists of approximately 55% fluid (blood plasma) and 45% blood cells

Blood cells

Three main kinds—red blood cells (erythrocytes), white blood cells (leukocytes), platelets (thrombocytes); see Fig. 10-1
1 Erythrocytes (red blood cells)
   a Size and appearance—biconcave disks about 7 microns in diameter
   b Structure and functions—millions of molecules of hemoglobin inside each red cell makes possible red cell functions of oxygen and carbon dioxide transport
   c Formation (erythropoiesis)—by myeloid tissue (red bone marrow); see Fig. 10-2 for stages of red cell development
   d Destruction—by fragmentation in capillaries; reticuloendothelial cells phagocytose red cell fragments and break down hemo-

globin to yield iron containing pigment and bile pigments; bone marrow reuses most of iron for new red cell synthesis; liver excretes bile pigments; life span of red cells about 120 days according to studies made with radioactive isotopes

    e Erythrocyte (red blood cell) homeostatic mechanism—stimulus thought to be tissue hypoxia; for description of mechanisms, see Fig. 10-3; clinical applications (pernicious anemia, vitamin B_{12}, bone marrow damage, reticulocyte count, red count, hematocrit, and sternal punctures) discussed on p. 298

    f Functions—see Table 10-2

2 Leukocytes (white blood cells)

    a Appearance and size—vary; some relatively large cells (e.g., monocytes) and some small cells (e.g., small lymphocytes); nuclei vary from spherical to **S** shaped to lobulated; cytoplasm of neutrophils, eosinophils, and basophils contain granules that take neutral, acid, and basic stains, respectively

    b Functions—defense or protection since carry on phagocytosis

    c Formation—in myeloid tissue (red bone marrow) for granular leukocytes; nongranular leukocytes (lymphocytes and monocytes) in lymphatic tissue—i.e., mainly in lymph nodes and spleen

    d Destruction and life span—some destroyed by phagocytosis; life span unknown

    e Numbers—about 5,000 to 10,000 leukocytes per cubic millimeter of blood; for differential count, see Table 10-1

    f Functions—see Table 10-2

3 Platelets

    a Appearance and size—platelets are small fragments of cells

    b Functions—disintegrating platelets release factor that combines with blood proteins and calcium ions to form prothrombinase, which initiates blood clotting

    c Formation and life span—formed in red bone marrow by fragmentation of large cells (megakaryocytes, Fig. 10-2); life span not known—probably only few days

    d Functions—see Table 10-2

Blood types (or blood groups)

1 Names—indicate type of antigen on or in red cell membranes

2 Every person's blood belongs to one of the four AB blood groups (type A, type B, type AB, or type O) and, in addition, is either Rh positive or Rh negative

3 Plasma does not normally contain antibodies against antigens present on its red cells but does normally contain antibodies against A or B antigens not present on its red cells

4 No blood normally contains anti-Rh antibodies; anti-Rh antibodies appear in blood only of an Rh-negative person and only after Rh-positive red cells have entered his bloodstream (e.g., by transfusion or by carrying an Rh-positive fetus)

5 Universal donor: type O; universal recipient: type AB

Blood plasma

1 Liquid part of blood or whole blood minus its cells; constitutes about 55% of total blood volume

2 Composition

    a About 90% water and 10% solutes

    b Most solutes crystalloids but some colloids

    c Most solutes electrolytes but some non-electrolytes

    d Solutes include foods, wastes, gases, hormones, enzymes, vitamins, and antibodies and other proteins

3 Plasma proteins

    a Types—albumins, globulins, and fibrinogen

    b Total plasma proteins—6 grams to 7 grams per 100 milliliters of plasma

    c Albumins—about 55% of total plasma proteins

    d Globulins—about 38% of total plasma proteins

    e Fibrinogen—about 7% of total plasma proteins

    f Functions—contribute to blood viscosity, osmotic pressure, and volume; prothrombin (an albumin) and fibrinogen play key roles in blood clotting; modified gamma globulins are circulating antibodies, essential for immunity

    g All plasma proteins except circulating antibodies are synthesized in liver cells

Blood clotting

1 Purpose—to plug up ruptured vessels and thus prevent fatal hemorrhage

2 Mechanism—complicated process, still incompletely understood

    a Triggered by appearance of rough spot in blood vessel lining

    b Clumps of platelets adhere to rough spot; their membranes rupture, releasing "platelet factors" which initiate a series of rapidly occurring chemical reactions

    c Last two clotting reactions convert prothrombin to thrombin, an enzyme which catalyzes conversion of soluble fibrinogen to insoluble fibrin

    d Clot consists of tangled mass of fibrin threads and blood cells; blood serum is fluid left after clot forms

3 Factors that oppose clotting

    a Smoothness of endothelium that lines blood vessels prevents platelets' adherence and consequent disintegration; some platelet disintegration continuously despite this preventive

b Blood normally contains certain anticoagulants—e.g., antithrombins, substances which inactivate thrombin so that it cannot catalyze fibrin formation

4 Factors that hasten clotting

a Endothelial "rough" spots (e.g., lipoid plaques in atherosclerosis)

b Sluggish blood flow

5 Pharmaceutical preparations that retard clotting—commercial heparin, Dicumarol, citrates (for transfusion blood)

6 Clot dissolution—process called fibrinolysis; presumably this and opposing process (clot formation) go on all time

7 Clinical methods of hastening clotting

a Apply rough surfaces to wound to stimulate platelets and tissues to liberate more thromboplastin

b Apply purified thrombin

c Apply fibrin foam, film, etc.

HEART

1 Four-chambered muscular organ

2 Lies in mediastinum with apex on diaphragm, two-thirds of its bulk to left of midline of body and one-third to right

3 Apical beat may be counted by placing stethoscope in fifth intercostal space on line with left midclavicular point

Covering

1 Structure

a Loose-fitting, inextensible sac (fibrous pericardium) around heart, lined with serous pericardium (parietal layer), which also covers outer surface of heart (visceral layer or epicardium)

b Small space between parietal and visceral layers of serous pericardium contains few drops of pericardial fluid

2 Function—protection against friction

Structure

Wall

1 Myocardium—name of muscular wall

2 Endocardium—lining

3 Pericardium—covering

Cavities

1 Upper two—atria

2 Lower two—ventricles

Valves and openings

1 Openings between atria and ventricles—atrioventricular orifices, guarded by cuspid valves, tricuspid on right and mitral or bicuspid on left; valves consist of three parts: flaps, chordae tendineae, and papillary muscle

2 Opening from right ventricle into pulmonary artery guarded by semilunar valves

3 Opening from left ventricle into great aorta guarded by semilunar valves

Blood supply

From coronary arteries; branch from ascending aorta behind semilunar valves

1 Left ventricle receives blood via both major branches of left coronary artery and from one branch of right coronary artery

2 Right ventricle receives blood via both major branches of right coronary artery and from one branch of left coronary artery

3 Each atrium receives blood only from one branch of its respective coronary artery

4 Usually only few anastomoses between larger branches of coronary arteries, so that occlusion of one of these produces areas of myocardial infarction; if not fatal, anastomoses between smaller vessels grow and provide collateral circulation

Conduction system

1 Sinoatrial node (S.A. node; pacemaker of heart)—small mass of modified cardiac muscle at junction of superior vena cava and right atrium; inherent rhythmicity of S.A. node impulses set basic rate of heartbeat; ratio of sympathetic/parasympathetic impulses to node per minute and blood concentrations of epinephrine and thyroid hormone act on node to modify its activity and alter heart rate

2 Atrioventricular node (A.V. node)—small mass of modified cardiac muscle in septum between two atria

3 Atrioventricular bundle (bundle of His)—special cardiac muscle fibers originating in A.V. node and extending down interventricular septum

4 Purkinje fibers—extension of bundle of His fibers out into walls of ventricles

Nerve supply

1 Sympathetic fibers (in cardiac nerves) and parasympathetic fibers (in vagus) form cardiac plexuses

2 Fibers from plexuses terminate mainly in S.A. node

3 Sympathetic impulses tend to accelerate and strengthen heartbeat

4 Vagal impulses slow heartbeat

Physiology

1 Primary functions of circulatory system—maintain blood flow and vary rate of flow according to energy needs of cells

2 Cardiac cycle

a Nature—consists of systole and diastole of atria and of ventricles; atria contract and as they relax, ventricles contract b Time required for cycle – about 4/5 second or from 70 to 80 times per minute

c Events of cycle

1 Atria contracted – cuspid valves open; ventricles relaxed; semilunar valves closed

2 Atria relaxed – cuspid valves closed during first part of atrial diastole while ventricles are contracted and then open as ventricles relax; semilunar valves open during ventricular contraction

d Heart sounds during cycle – lubb due to contraction of ventricles and closure of cuspid valves; dupp due to closure of semilunar valves

BLOOD VESSELS

1 Kinds

a Arteries – vessels that carry blood away from heart; all except pulmonary artery carry oxygenated blood

b Veins – vessels that carry blood toward heart; all except pulmonary veins carry deoxygenated blood

c Capillaries – microscopic vessels that carry blood from small arteries (arterioles) to small veins (venules)

2 Structure – see Table 10-5

3 Functions

a Arteries and arterioles – carry blood away from heart to capillaries

b Capillaries – deliver materials to cells (by way of tissue fluid) and collect substances from them; vital function of entire circulatory system

c Veins and venules – carry blood from capillaries back to heart

Main blood vessels

Systemic circulation

1 Arteries – see pp. 314 and 320, Fig. 10-12, and Table 10-5

2 Veins – see pp. 315 and 321 to 326, Fig. 10-16, and Table 10-5

Portal circulation

See p. 322 and Fig. 10-22

Fetal circulation

See p. 326 and Fig. 10-23

CIRCULATION

1 Definitions

a Circulation – blood flow through closed circuit of vessels

b Systemic circulation – blood flow from left ventricle into aorta, other arteries, arterioles, capillaries, venules, and veins, to right atrium of heart

c Pulmonary circulation – blood flow from right ventricle to pulmonary artery to lung arterioles, capillaries, and venules, to pulmonary veins, to left atrium

2 Functions of control mechanisms

a Maintain circulation

b Vary circulation; increase blood flow per minute when activity increases and decreases blood flow when activity decreases

3 Principles

a Blood circulates because blood pressure gradient exists within its vessels; systemic blood pressure gradient (mean arterial pressure minus central venous pressure) equals about 100 mm Hg

b Arterial blood pressure determined primarily by volume of blood in arteries; other factors remaining constant – greater arterial blood volume, greater arterial blood pressure

c Arterial blood volume determined mainly by cardiac minute output and peripheral resistance – directly related to cardiac output and inversely related to resistance

d Cardiac minute output determined by heart's rate of contraction and its systolic discharge and directly related to both factors

e Heart's systolic discharge regulated mainly by ratio of sympathetic-parasympathetic impulses

f Heart rate regulated by pressoreflexes and by many miscellaneous factors; increased arterial pressure in aorta or carotid sinus tends to produce reflex slowing of heart, whereas increased right atrial pressure tends to produce reflex cardiac acceleration

g Peripheral resistance determined mainly by blood viscosity and by arteriole diameter; in general, less blood viscosity, less peripheral resistance, but smaller diameter of arterioles, greater peripheral resistance

h Blood viscosity determined by concentration of blood proteins and of blood cells and directly related to both

i Arteriole diameter regulated mainly by pressoreflexes and chemoreflexes; in general, increase in arterial pressure produces reflex dilatation of arterioles, whereas hypoxia and hypercapnea cause constriction of arterioles in blood reservoir organs but dilatation of them in local structures, notably in skeletal muscles, heart, and brain

j Volume of blood circulating per minute determined by blood pressure gradient and peripheral resistance; according to Poiseuille's law, directly related to pressure gradient and inversely related to peripheral resistance

k Respirations and skeletal muscle contractions tend to increase venous return to heart

BLOOD PRESSURE

1 How arterial blood pressure measured clinically

a Sphygmomanometer

b Systolic pressure normally about 120 mm Hg
and diastolic pressure about 80 mm Hg
2 Relation to arterial and venous bleeding
  a Arterial bleeding in spurts due to difference
in amounts of systolic and diastolic pressures
  b Venous bleeding – slow and steady due to
low, practically constant venous pressure

VELOCITY OF BLOOD

1 Speed with which blood flows
2 Most rapid in arteries and slowest in capillaries

PULSE

1 Definition – alternate expansion and recoil
of artery
2 Cause – intermittent injections of blood from
heart into aorta with each ventricular contraction; pulse can be felt because of elasticity of
arterial walls
3 Pulse wave – pulse starts at beginning of aorta
and proceeds as wave of expansion throughout
arteries
4 Where pulse can be felt – radial, temporal,
common carotid, facial, brachial, femoral,
and popliteal arteries; where near surface and
over firm background, such as bone; pressure
points, points where bleeding can be stopped by
pressure, roughly related to places where pulse
can be felt
5 Venous pulse – in large veins only; due to
changes in venous pressure brought about by
alternate contraction and relaxation of atria

LYMPHATIC SYSTEM

Part of circulatory system – consists of lymph,
interstitial (tissue) fluid, lymphatics, and lymph
nodes

Lymph and interstitial (tissue fluid)

1 Definition
  a Lymph – clear, watery fluid found in lymphatic vessels
  b Interstitial fluid (tissue fluid) – clear liquid
in tissue spaces

Lymphatics

1 Formation and distribution
  a Start as capillaries in tissue spaces
  b Widely distributed throughout body
  c Two or more main lymphatic ducts – thoracic
duct, which drains into left subclavian vein
at junction of internal jugular and subclavian, and one or more right lymphatic
ducts which drain into right subclavian vein
2 Structure
  a Similar to veins except thinner walled
  b Contain more valves and contain lymph
nodes located at intervals
3 Function – return of water and proteins from
interstitial fluid to blood from which they came

Lymph circulation

Water and solutes from capillary blood to interstitial fluid, to lymphatics, to blood at junction of
internal jugular and subclavian veins

Lymph nodes

1 Structure
  a Lymphatic tissue, separated into compartments by fibrous partitions
  b Afferent lymphatics enter each node and
efferent lymphatic leaves each node
2 Location – usually in clusters (see p. 352)
3 Functions
  a Defense functions – filter out injurious substances and phagocytose them
  b Hemopoiesis – formation of lymphocytes
and monocytes

SPLEEN

1 Location – left hypochondrium
2 Structure
  a Similar to lymph nodes, ovoid in shape
  b Size varies
  c Contains numerous venous blood spaces
that serve as blood reservoir
3 Functions
  a Defense – protection by phagocytosis by
reticuloendothelial cells and antibody formation by some lymphocytes
  b Hemopoiesis of nongranular leukocytes
(monocytes and lymphocytes) and of red cells
before birth; spleen also forms plasma cells
  c Red blood cell and platelet destruction –
reticuloendothelial cells phagocytose these
cells
  d Blood reservoir

review questions

1 Compare different kinds of blood cells as to appearance and size; functions; formation, destruction, and life span; number per cubic millimeter of blood.

2 "Graft-rejection cells" is a nickname for which kind of blood cells? Why?

3 What type of cells secrete "circulating" antibodies?

4 What is the function of circulating antibodies? Of cellular antibodies?

5 Since plasma cells synthesize and secrete circulating antibodies, what organelles would you expect to be prominent or abundant in plasma cells? (Review pp. 15 to 16 if you cannot answer this question.)

6 Suppose your doctor has told you that you have a lymphocyte count of 2,000 per cmm. Would you think this was normal or abnormal?

7 Why might a surgeon give antilymphocyte serum to a patient who had had a kidney transplant?

8 Explain what the term hematocrit means. Is it normally less than 50? More than 55?

9 What triggers blood clotting?

10 Write two equations to show the basic chemical reactions that produce a blood clot.

11 Why and how does a vitamin K deficiency affect blood clotting?

12 Explain some principles and methods by which blood clotting may be hastened.

13 Explain what these terms mean: type AB blood; Rh negative blood.

14 Suppose you have type B blood. Which two of the following kinds of blood might be used for transfusing you: Type A? Type B? Type AB? Type O?

15 Which of the following babies would be most likely to have erythroblastosis fetalis: Fourth child of Rh-positive mother and Rh-negative father? First child of Rh-negative mother and Rh-positive father? Second child of Rh-negative mother and Rh-positive father? Explain your reasoning.

16 The cells of what organ synthesize most plasma proteins? What is the normal plasma protein concentration?

17 What are some of the functions served by plasma proteins?

18 Describe the pericardium, differentiating between the fibrous and serous portions.

19 Exactly where is pericardial fluid found? Explain its function.

20 Describe the heart's own blood supply. Explain why occlusion of a large coronary artery branch has serious consequences.

21 Identify, locate, and describe function of each of the following structures: S.A. node; A.V. node; bundle of His.

22 Compare arteries, veins, and capillaries as to structure and functions.

23 Differentiate between systemic, pulmonary, and portal circulation.

24 Explain the differences between fetal and postnatal circulation and the functional reasons for these differences.

25 Explain the heart control mechanism. Include control of both rate and force. Devise a diagram to indicate the different parts of the mechanism.

26 Explain the vasomotor mechanism. Devise a diagram to indicate its various parts.

27 Explain reactive hyperemia and one theory about the mechanism producing it.

28 State in your own words the basic principle of fluid flow.

29 State in your own words Poiseuille's law. Give an example of increased circulation to illustrate application of this law. Give an example of decreased circulation to illustrate application of this law.

30 What mechanisms control arterial blood pressure? Cite an example of the operation of one or more of these mechanisms to increase arterial pressure; to decrease it.

31 What mechanisms control peripheral resistance? Cite an example of the operation of one or more parts of this mechanism to increase resistance; to decrease it.

32 What two factors determine blood viscosity? What does viscosity mean? Give an example of a condition in which blood viscosity decreases. Explain its effect on circulation.

33 What effect, if any, would a respiratory stimulant drug have on circulation. Explain why it would or would not affect circulation.

34 Describe and explain the effects of exercise on circulation.

35 What and where is lymph? Describe its circulation.

36 Compare lymphatics and lymph nodes as to structure and location and function.

37 Describe the location and function of the spleen. What functions is it thought to perform?

38 If cancer cells from breast cancer were to enter the lymphatics of the breast, where do you think they might lodge and start new growths? Explain, using your knowledge of the anatomy of the lymphatic and circulatory systems.

39 Starting with the left ventricle of the heart, list the vessels through which blood would flow in reaching the small intestine; the large intestine; the liver (two ways); the spleen; the stomach; the kidneys; the suprarenal glands; the ovaries or testes; the anterior part of the base of the brain; the little finger of the right hand. List the vessels through which the blood returns from these parts to the right atrium of the heart (see Figs. 10-12, 10-16, and 10-24).

40 Name and explain the action of the heart and its valves.

41 Trace the flow of blood through the heart.

42 Give two reasons why blood is considered a protective agent against infection.

43 Describe one or more mechanisms that probably operate to return circulation to normal a short time after exercise ceases.

44 Give the general location of the following veins: longitudinal sinus; internal jugular vein; innominate vein; portal vein; great saphenous vein; inferior vena cava; basilic vein.

45 Define the following terms briefly: adenitis; anemia; aneurysm; atherosclerosis; basophils; blood pressure; diastole; differential count; embolus; endocardium; eosinophil; erythrocyte; heart block; hemophilia; leukemia; leukocyte; leukopenia; lymphocyte; monocyte; myocardium; neutrophil; pH; phagocytosis; phlebitis; plasma; peripheral resistance; pulse pressure; sphygmomanometer; thrombocyte; thrombus; systole.

11

The respiratory system

Functions and importance

Organs
Nose
  Structure
  Functions
Pharynx
  Structure
  Functions
Larynx
  Location
  Structure
  Function
Trachea
  Structure
  Function
Bronchi
  Structure
  Function
Lungs
  Structure
  Function

Thorax (chest)
Structure
Function

Physiology
Kinds of respirations
Mechanism of respirations
Amount of air exchanged in
  respirations
Types of respirations
Some principles about
  gases
How blood transports gases
Exchange of gases in lungs
Exchange of gases in tissues
Control of respirations

■ Functions and importance

The respiratory system consists of those organs that make it possible for blood to exchange gases with air. They are the nose, pharynx, larynx, trachea, bronchi, and lungs. This group of organs constitutes the lifeline of the body. If anything interferes with the functioning of this anatomical lifeline, death ensues in a very short time.

The exchange of gases between the blood and air is known as respiration, but actually it is only one phase of respiration. The transportation of gases between the lungs and tissues constitutes another phase, and the exchange of gases between the blood and tissues still another. The first phase is a function of the respiratory system, whereas the second and third phases are functions of the circulatory system. The first and second phases are vital only because they make possible the third phase. The all-important requisite is that cells receive oxygen and rid themselves of carbon dioxide—in short, that they "breathe." But most of our billions of cells cannot exchange these gases directly with air. They lie far too distant from it. Hence, they need special structures to make the exchange for them and other structures to transport the gases to and from them. The respiratory and circulatory systems fulfill these needs.

Organs

■ Nose

Structure

The nose consists of an internal and an external portion. The external portion—that is, the part that protrudes from the face—is considerably smaller than the internal portion, which lies over the roof of the mouth. The interior of the nose is hollow and is separated by a partition, the *septum* (Fig. 11-1), into a right and a left cavity. The palatine bones, which form both the floor of the nose and the roof of the mouth, separate the nasal cavities from the mouth cavity. Sometimes the palatine bones fail to unite completely, producing a condition known as *cleft palate*. When this abnormality exists, the mouth is only partially separated from the nasal cavity, and difficulties arise in swallowing.

Each nasal cavity is divided into three passageways (superior, middle, and inferior meati) by the projection of the turbinates (conchae) from the lateral walls of the internal portion of the nose (Figs. 11-2 and 11-3). The superior and middle turbinates are processes of the ethmoid bone, while the inferior turbinates are separate bones.

The external openings into the nasal cavities (nostrils) have the technical name of *anterior nares*. The posterior nares (or choanae) are openings from the internal nose into the nasopharynx.

Ciliated mucous membrane lines the nose and the rest of the respiratory tract down as far as the smaller bronchioles.

Four pairs of sinuses drain into the nose. These paranasal sinuses are the frontal, maxillary, ethmoidal, and sphenoidal. They drain as follows:

1 Into the middle meatus (passageway below middle turbinate)—frontal, maxillary, and anterior ethmoidal sinuses
2 Into the superior meatus—posterior ethmoidal sinuses
3 Into the space above the superior turbinates (sphenoethmoidal recess)—sphenoid sinuses

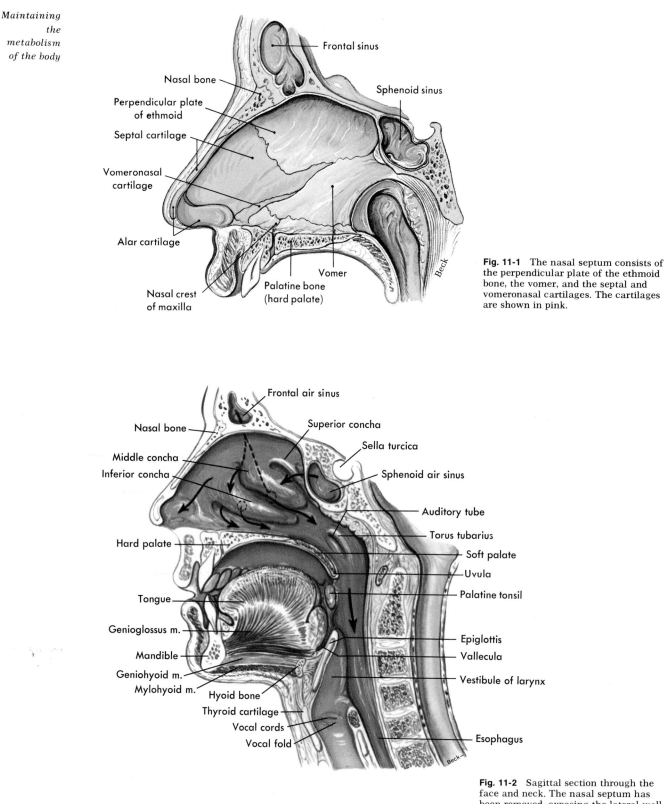

Frontal sinus

Nasal bone

Sphenoid sinus

Perpendicular plate of ethmoid

Septal cartilage

Vomeronasal cartilage

Alar cartilage

Vomer

Palatine bone (hard palate)

Nasal crest of maxilla

Fig. 11-1 The nasal septum consists of the perpendicular plate of the ethmoid bone, the vomer, and the septal and vomeronasal cartilages. The cartilages are shown in pink.

Frontal air sinus

Nasal bone

Superior concha

Sella turcica

Middle concha

Sphenoid air sinus

Inferior concha

Auditory tube

Torus tubarius

Hard palate

Soft palate

Uvula

Palatine tonsil

Tongue

Genioglossus m.

Epiglottis

Mandible

Vallecula

Geniohyoid m.

Vestibule of larynx

Mylohyoid m.

Hyoid bone

Thyroid cartilage

Vocal cords

Vocal fold

Esophagus

Fig. 11-2 Sagittal section through the face and neck. The nasal septum has been removed, exposing the lateral wall of the nasal cavity. Note the position of the conchae (turbinates).

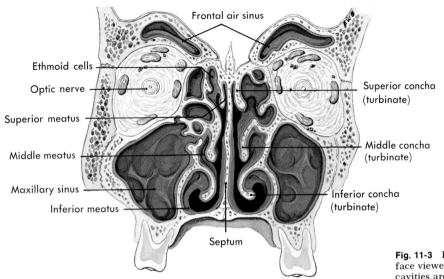

Frontal air sinus

Ethmoid cells

Optic nerve

Superior meatus

Middle meatus

Maxillary sinus

Inferior meatus

Septum

Superior concha
(turbinate)

Middle concha
(turbinate)

Inferior concha
(turbinate)

Fig. 11-3 Frontal section through the face viewed from behind. The sinus cavities are shown in red.

Functions

The nose serves as a passageway for air going to and from the lungs, filtering it of impurities and warming, moistening, and chemically examining it for substances that might prove irritating to the mucous lining of the respiratory tract. It serves as the organ of smell, since olfactory receptors are located in the nasal mucosa, and it aids in phonation.

■Pharynx

Structure

Another name for the pharynx is the throat. It is a tubelike structure about 5 inches long that extends from the base of the skull to the esophagus and lies just anterior to the cervical vertebrae. It is made of muscle, is lined with mucous membrane, and has three parts: one located behind the nose, the *nasopharynx;* one behind the mouth, the *oropharynx;* and another behind the larynx, the *laryngopharynx.*

Seven openings are found in the pharynx (Fig. 11-2):

1 Right and left auditory (eustachian) tubes open into the nasopharynx
2 Two posterior nares into the nasopharynx

3 The opening from the mouth, known as the *fauces,* into the oropharynx
4 The opening into the larynx from the laryngopharynx
5 The opening into the esophagus from the laryngopharynx

The *adenoids* or pharyngeal tonsils are located in the nasopharynx, on its posterior wall opposite the posterior nares. If the adenoids become enlarged, they fill the space behind the posterior nares, making it difficult or impossible for air to travel from the nose into the throat. When this happens, the individual keeps his mouth open to breathe and is described as having an "adenoidy" appearance.

Two pairs of organs are found in the oropharynx: the faucial or *palatine tonsils,* located behind and below the pillars of the fauces, and the *lingual tonsils,* located at the base of the tongue. The palatine tonsils are the ones most commonly removed by a tonsillectomy. Only rarely are the lingual ones also removed.

Functions

The pharynx serves as a hallway for the respiratory and digestive tracts, since both air and food must pass through this structure before reaching the appropriate tubes.

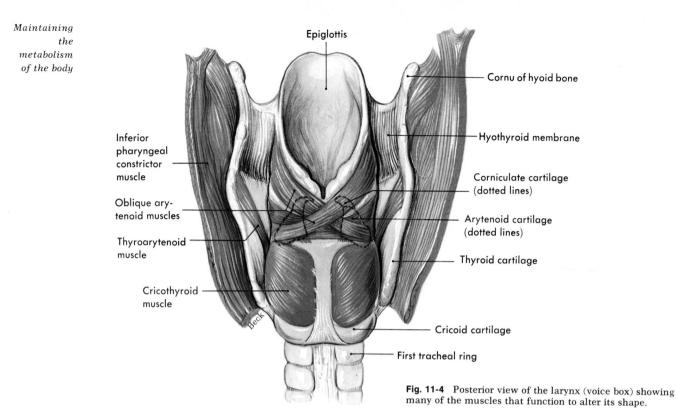

Fig. 11-4 Posterior view of the larynx (voice box) showing many of the muscles that function to alter its shape.

It also plays an important part in phonation. For example, only by the pharynx changing its shape can the different vowel sounds be formed.

■Larynx

Location

The larynx or voice box lies at the upper end of the trachea and just below the pharynx. It might be described as a vestibule opening into the trachea from the pharynx.

Structure

The larynx consists of nine pieces of cartilage so joined that they make a boxlike structure (Fig. 11-4). We shall describe only the three most prominent laryngeal cartilages:

1 The largest and the one that gives the characteristic triangular shape to the anterior wall of the larynx is called the *thyroid cartilage* or, by lay people, the Adam's apple. This cartilage is usually larger in men than in women and has less of a

fat pad lying over it—two reasons why a man's Adam's apple protrudes more than a woman's.

2 A small cartilage, attached along one edge to the thyroid cartilage but free on its other borders, giving it a hingelike action, is named the *epiglottis* or the lid cartilage because during the swallowing act it forms a kind of lid over the opening into the larynx. When the epiglottis fails to close, food or liquids enter the larynx instead of the esophagus, and we say we have "swallowed down our Sunday throat."

3 The *cricoid* or signet ring cartilage, so called because its shape resembles a signet ring (turned so the signet forms part of the posterior wall of the larynx), is the most inferiorly placed of the nine cartilages.

The mucous membrane lining of the larynx forms two horizontal folds known as the *false vocal cords*. The *true vocal cords* are fibrous bands stretched across the hollow interior of the larynx. The space between the true vocal cords (or vocal folds) is called the *glottis*.

Function

The larynx serves a protective function in that the epiglottis closes off the airway during the act of swallowing and thereby prevents particles or liquids from entering the lungs. Its other function is voice production. The length and tension of the vocal cords determines the pitch of the voice. Short tense cords produce high notes and long relaxed cords low notes. Several other structures aid the larynx in voice production by acting as sounding boards or resonating chambers. Thus, the size and shape of the nose, mouth, pharynx, and bony sinuses help to determine the quality of the voice.

■Trachea

Structure

Smooth muscle, in which are embedded C-shaped rings of cartilage at regular in-

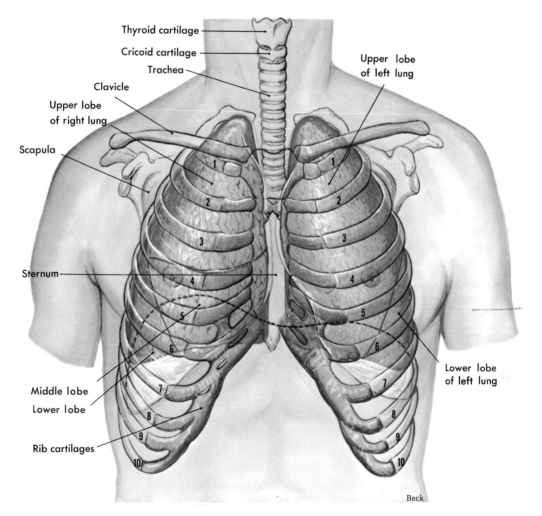

Fig. 11-5 Projection of the lungs and trachea in relation to the rib cage and clavicles. Dotted line indicates location of the dome-shaped diaphragm at the end of expiration and before inspiration. Note that apex of each lung projects above the clavicle.

tervals, fashions the walls of the trachea or windpipe. The cartilaginous rings are incomplete on the posterior surface. They give firmness to the wall, tending to prevent it from collapsing and shutting off the vital airway. Occasionally, certain conditions, such as cervical adenitis, may obstruct the trachea, making necessary emergency measures to open it. Two methods are used. Either an incision is made into the trachea (tracheotomy) and a double metal tube inserted through the opening in the neck, or the tube is inserted by way of the mouth and larynx (intubation).

The trachea is about 4 1/2 inches long and extends from the larynx to the bronchi (Fig. 11-5). It is cylindrical in shape with a diameter of approximately 1 inch.

Function

The trachea performs a simple but vital function—it furnishes part of the open passageway through which air can reach the lungs from the outside. Obstruction of this airway for even a few minutes causes death from asphyxiation.

■ Bronchi

Structure

The trachea divides at its lower end into two *primary bronchi*, of which the right bronchus is slightly larger and more vertical than the left. This anatomical fact explains why aspirated foreign objects frequently lodge in the right bronchus. In structure, the bronchi resemble the trachea. Their walls contain incomplete cartilaginous rings before the bronchi enter the lungs, but they become complete within the lungs. Ciliated mucosa lines the bronchi, as it does the trachea.

Each primary bronchus enters the lung on its respective side and immediately divides into smaller branches called *secondary bronchi*. The secondary bronchi continue to branch, forming small *bron-*

chioles. The trachea and the two primary bronchi and their many branches resemble an inverted tree trunk with its branches and are, therefore, spoken of as the bronchial tree. The bronchioles subdivide into smaller and smaller tubes, eventually terminating in microscopic branches that divide into *alveolar ducts*, which terminate in several alveolar sacs, the walls of which consist of numerous *alveoli* (Figs. 11-6 and 11-7). The structure of an alveolar duct with its branching alveolar sacs can be likened to a bunch of grapes—the stem represents the alveolar duct, each cluster of grapes represents an alveolar sac, and each grape represents an alveolus. Some 300 million alveoli are estimated to be present in our two lungs.

The structure of the secondary bronchi and bronchioles shows some modification of the primary bronchial structure. The cartilaginous rings become irregular and disappear entirely in the smaller bronchioles. By the time the branches of the bronchial tree have dwindled sufficiently to form the alveolar ducts and sacs and the alveoli, only the internal surface layer of cells remains. In other words, the walls of these microscopic structures consist of a single layer of simple, squamous epithelial tissue. As we shall see, this structural fact makes possible the performance of their function.

Function

The tubes composing the bronchial tree perform the same function as the trachea—that of furnishing a passageway by which air can reach the interior of the lung. The alveoli, enveloped as they are by networks of capillaries, provide spaces where gases can diffuse between air and blood. Someone has observed that "the lung passages all serve the alveoli" just as "the circulatory system serves the capillaries." Certain diseases may block the passage of air

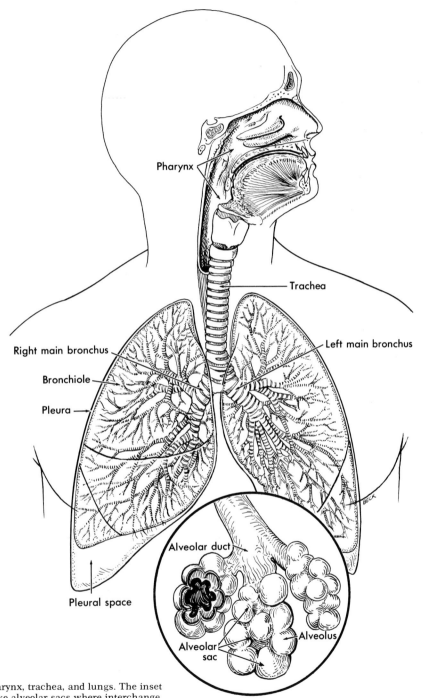

Fig. 11-6 The pharynx, trachea, and lungs. The inset shows the grapelike alveolar sacs where interchange of oxygen and carbon dioxide takes place through the thin walls of the alveoli (see also Fig. 11-7). Capillaries (not shown) surround the alveoli.

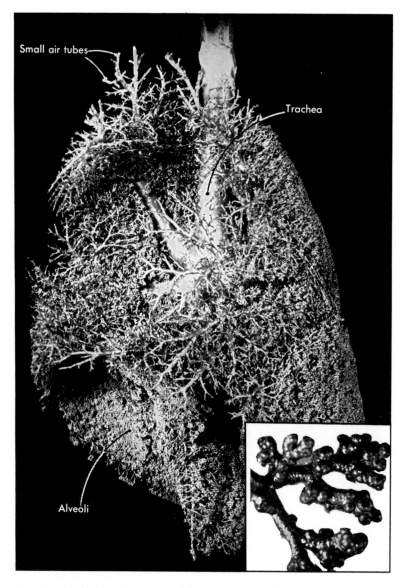

Fig. 11-7 Metal cast of air spaces of the lungs of a dog. The inset shows a cast of clusters of alveoli at the terminations of tiny air tubes. The magnification of the inset is about eleven times the actual size. (From Carlson, A. J., and Johnson, V. E.: The machinery of the body, Chicago, The University of Chicago Press.)

through the bronchioles or through the alveoli. For example, in pneumonia the alveoli become inflamed, and the accompanying wastes plug up these minute air spaces, making the affected part of the lung solid. Whether the victim survives depends largely upon the extent of the solidification.

Lungs

Structure

The lungs are cone-shaped organs, large enough to fill the pleural portion of the thoracic cavity completely. They extend from the diaphragm to a point slightly above the clavicles and lie against the ribs both anteriorly and posteriorly. The medial surface of each lung is roughly concave to allow room for the mediastinal structures and for the heart, but concavity is greater on the left than on the right because of the position of the heart. The primary bronchi and pulmonary blood vessels (bound together by connective tissue to form what is known as the *root* of the lung) enter each lung through a slit on its medial surface called the *hilum*.

The broad inferior surface of the lung, which rests on the diaphragm, constitutes the *base*, whereas the pointed upper margin is the *apex*.

The left lung is partially divided by fissures into two *lobes* (upper and lower) and the right lung into three lobes (superior, middle, and inferior). Internally, each lung consists of millions of microscopic alveoli with their related ducts and bronchioles and bronchi, as described in the discussion of the structure of the bronchi.

Visceral pleura covers the outer surfaces of the lungs and adheres to them much as the skin of an apple adheres to the apple.

Function

The lungs provide a place where an exchange of gases can take place between blood and air. Lung structure makes possible this function. Because an open airway branches into millions of thin-walled alveoli, enveloped by networks of capillaries (with a total surface area of about 90 square meters,* or sixty times that of the whole body), large amounts of oxygen can be quickly loaded into blood and large amounts of carbon dioxide quickly unloaded from it.

Thorax (chest)

Structure

As described on p. 3, the thoracic cavity has three divisions, separated from each other by partitions of pleura. The part of the cavity occupied by the lungs is the pleural division, the space between the lungs occupied by the esophagus, trachea, large blood vessels, etc. is the mediastinum, and the part occupied by the heart and its enveloping sac is the pericardial portion.

The parietal layer of the pleura lines the entire thoracic cavity. It adheres to the internal surface of the ribs and the superior surface of the diaphragm, and it partitions off the mediastinum. A separate pleural sac thus encases each lung. Since the outer surface of each lung is covered by the visceral layer of the pleura, the visceral pleura lies against the parietal pleura, separated only by a potential space (pleural space) which contains just enough pleural fluid for lubrication. Thus, when the lungs inflate with air, the smooth, moist visceral pleura coheres to the smooth, moist parietal pleura. Friction is thereby avoided, and respirations are painless. In pleurisy, on the other hand, the pleura is inflamed and respirations become painful.

*Mountcastle, V. B., editor: Medical physiology, ed. 12, St. Louis, 1968, The C. V. Mosby Co.

Function

The thorax plays a major role in respirations. Because of the elliptical shape of the ribs and the angle of their attachment to the spine, the thorax becomes larger when the chest is raised and smaller when it is lowered. And it is these changes in thorax size which bring about inspiration and expiration (discussed below). Lifting up the chest raises the ribs so they no longer slant downward from the spine, and because of their elliptical shape, this enlarges both depth (from front to back) and width of the thorax. (If this does not sound convincing to you, examine a skeleton to see why it is so.)

Physiology

Kinds of respirations

There are two kinds of respiration: external or "lung breathing" and internal or "cell breathing." Lung breathing is important only because it makes possible cell breathing. It is the necessary preliminary to cell breathing. And it is this—the continual supply of oxygen to cells—that is essential for life.

Mechanism of respirations

Air moves in and out of the lungs for the same basic reason that blood flows in vessels—because of a pressure gradient—a gas pressure gradient in the case of air movement and a blood pressure gradient in the case of blood flow. When atmospheric pressure is greater than pressure within the lung, air flows down this gas pressure gradient. Then air moves from the atmosphere into the lungs. Inspiration occurs, in other words. And when pressure in the lungs is greater than atmospheric pressure, air again moves down a gas pressure gradient. But now, this means that it moves in the opposite direction. This time air moves outward from the lungs into the air. The respiratory mechanism, therefore, must somehow establish these two gas pressure gradients—one in which intrapulmonic pressure (pressure within the lungs) is lower than atmospheric pressure to produce inspiration and one in which it is higher than atmospheric pressure to produce expiration.

These pressure gradients are established by changes in the size of the thoracic cavity, which, in turn, are produced by contraction and relaxation of respiratory muscles. When the diaphragm contracts, for example, it descends, and this enlarges the vertical length of the thorax. Other muscles may also contract at the same time, elevating the sternum and ribs and enlarging the thorax from front to back and from side to side. The increase in size of the thorax causes intrapleural (intrathoracic) pressure to decrease—due to operation of the familiar principle known as Boyle's law. At the end of an expiration and before the beginning of the next inspiration, intrathoracic pressure is about 2.5 mm Hg less than atmospheric pressure, or, in other words, it is −2.5 mm Hg. During quiet inspiration, intrathoracic pressure decreases further to −6 mm Hg; that is, it becomes more negative than it was at the beginning of inspiration. This decrease in intrathoracic pressure and the cohesion of the lungs to the inner wall of the thorax together bring about the expansion of the lungs and a decrease in their inside or intrapulmonic pressure. It decreases from atmospheric level to a subatmospheric level; typically, to −2 mm Hg. The moment intrapulmonic pressure becomes less than atmospheric pressure, a pressure gradient exists between the atmosphere and the interior of the lungs. Air then necessarily moves into the lungs. Figs. 11-8 and 11-9

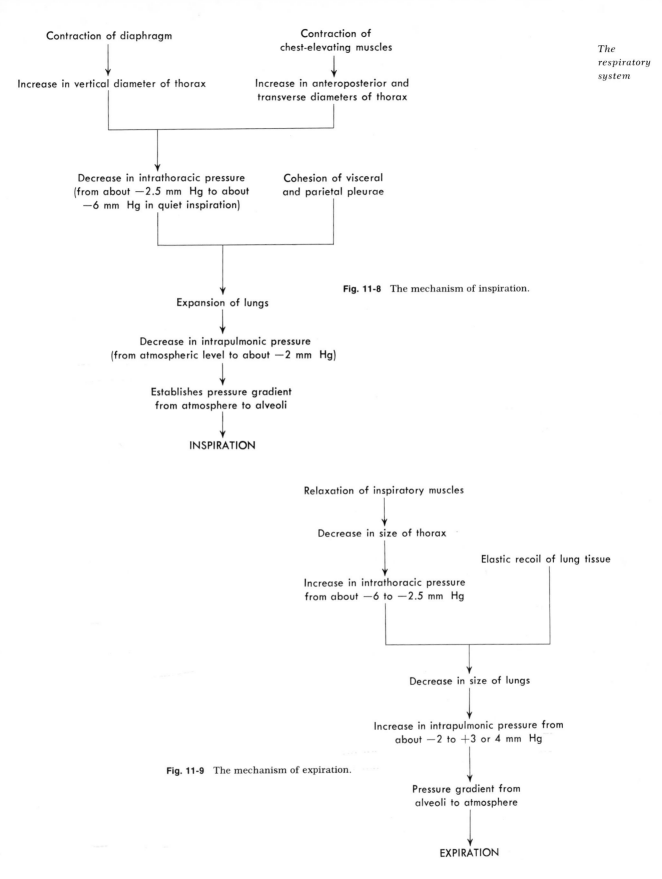

Fig. 11-8 The mechanism of inspiration.

Fig. 11-9 The mechanism of expiration.

show the mechanisms of inspiration and of expiration in diagrammatic form.

To apply some of the information just discussed about the respiratory mechanism, let us suppose that a surgeon makes an incision through the chest wall into the pleural space, as he would in doing one of the dramatic, modern "open-chest" operations. Air would then be present in the thoracic cavity, a condition known as *pneumothorax*. What change, if any, do you think would take place in respirations?

Intrathoracic pressure would, of course, immediately increase from its normal sub-atmospheric level to atmospheric level. More pressure than normal would, therefore, be exerted upon the outer surface of the lung and would cause its collapse. It could even collapse the other lung. This is because the mediastinum is a mobile rather than a rigid partition between the two pleural sacs. This anatomical fact allows the increased pressure in the side of the chest that is open to push the heart and other mediastinal structures over toward the intact side where they exert pressure on the other lung. Pneumothorax results in many respiratory and circulatory changes. They are of great importance in determining medical and nursing care but lie beyond the scope of this book.

Amount of air exchanged in respirations

Poiseuille's law (p. 342) applies to the flow of gases as well as of liquids. This means that the volume of air inspired is directly related to the gas pressure gradient between the atmosphere and the lung alveoli and is inversely related to the resistance opposing air flow. In general, the deeper the inspiration, the lower the intrapulmonic pressure, the greater the pressure gradient from atmosphere to alveoli, and the larger the volume of air inspired. Obstruction of the airway has the opposite effect. For example, if the tongue falls back into the throat—a very real danger in an unconscious patient—it obviously blocks off the airway, increases resistance, and slows air flow. It may even prevent any air at all from moving down into the alveoli.

An apparatus called a *spirometer* is used to measure the amount of air exchanged in breathing. The amount of air exhaled normally after a normal inspiration is termed *tidal air* (Fig. 11-10). The average individual has a tidal air of approximately 500 milliliters (about 1/2 liter). A forcible expiration after a normal inspiration represents the *expiratory reserve volume* (about 1 liter). *Inspiratory reserve volume* is the amount that can be forcibly inspired over and above a normal inspiration. It is measured by having the individual exhale normally after a forced inspiration. The normal inspiratory reserve volume is about 3.3 liters. No matter how forcefully an individual exhales, he cannot squeeze all the air out of his lungs. Some of it remains trapped in the alveoli. This amount of air that cannot be forcibly expired is known as *residual air* and amounts to about 1.2 liters. If, after death, the thoracic cavity is opened and the lungs are collapsed by atmospheric pressure, a small amount of air still remains in the alveoli (*minimal air*). Its presence can be demonstrated by placing a small piece of lung in water. It will float due to the presence of the minimal air, whereas any other soft tissue will sink. Because of this property, the lungs of an animal are known as the "lights" in slaughterhouses. The demonstration of minimal air in the lungs sometimes has legal importance in proving whether or not a child has lived at all or was born dead. If he has once filled his lungs with air, minimal air can be demonstrated at autopsy by the simple method described.

The term *vital capacity* means the approximate volume of the lungs as deter-

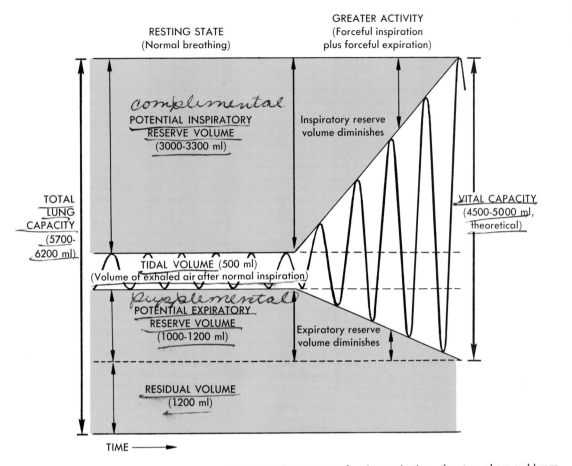

RESTING STATE
(Normal breathing)

GREATER ACTIVITY
(Forceful inspiration
plus forceful expiration)

complemental
POTENTIAL INSPIRATORY
RESERVE VOLUME
(3000-3300 ml)

Inspiratory reserve
volume diminishes

TOTAL
LUNG
CAPACITY
(5700-
6200 ml)

VITAL CAPACITY
(4500-5000 ml,
theoretical)

TIDAL VOLUME (500 ml)
(Volume of exhaled air after normal inspiration)

supplemental
POTENTIAL EXPIRATORY
RESERVE VOLUME
(1000-1200 ml)

Expiratory reserve
volume diminishes

RESIDUAL VOLUME
(1200 ml)

TIME ——→

Fig. 11-10 During normal, quiet respirations, the atmosphere and lungs exchange about 500 ml of air (tidal air). With a forcible inspiration, about 3,300 ml more air can be inhaled (inspiratory reserve volume). After a normal inspiration, approximately 1,000 ml of air can be forcibly expired (expiratory reserve volume). Vital capacity is the amount of air that can be forcibly expired after a maximum inspiration and indicates, therefore, the largest amount of air that can enter and leave the lungs during respiration. Residual air is that which remains trapped in the alveoli.

mined by measuring the largest possible expiration after the largest possible inspiration. Note in Fig. 11-10 that tidal volume plus inspiratory reserve volume plus expiratory reserve volume equals vital capacity. How large a vital capacity one has depends upon many factors—the size of one's thoracic cavity, the size of one's rib cage, one's posture, and various other factors. For example, if the lungs contain more blood than normal, alveolar air space is encroached upon and vital capacity accordingly decreases. This becomes a very important factor in congestive heart disease. It probably explains also the

smaller vital capacity in normal individuals in the supine position. Excess fluid in the pleural or abdominal cavities also decreases vital capacity, as does the disease emphysema. In the latter condition, alveolar walls become stretched—that is, lose their elasticity—and are unable to collapse normally for expiration. This leads to a great increase in the amount of residual air —so much so, in fact, that the chest occupies the inspiratory position even at rest. Excessive muscular effort is necessary, therefore, for inspiration, and because of the loss of elasticity of lung tissue, greater effort is required, too, for expiration.

The term *alveolar ventilation* means the volume of inspired air that actually reaches, "ventilates," the alveoli and takes part in the exchange of gases between air and blood. (Alveolar air exchanges some of its oxygen for some of blood's carbon dioxide.) With every breath you take, part of the entering air necessarily fills your air passageways—nose, pharynx, larynx, trachea, and bronchi. This portion of air does not descend into any alveoli so cannot take part in gas exchange. In this sense, it is "dead air." And appropriately, the larger air passageways it occupies are said to constitute the *anatomic dead space*. One rule of thumb estimates the volume of air in the anatomic dead space as the same number of milliliters as the individual's weight in pounds. Another generalization says that the anatomic dead space approximates 30% of the tidal air volume. Tidal volume minus dead space volume equals alveolar ventilation volume. Suppose you have a normal tidal volume of 500 milliliters and that 30% of this, or 150 milliliters, fills the anatomic dead space. The amount of air reaching your alveoli—your alveolar ventilation volume—is then 350 milliliters per breath. Emphysema and certain other abnormal conditions, in effect, increase the amount of dead space air. Consequently, alveolar ventilation decreases and this, in turn, decreases the amount of oxygen that can enter blood and the amount of carbon dioxide that can leave it. Inadequate air-blood gas exchange, therefore, is the inevitable result of inadequate alveolar ventilation. Or stated differently, the alveoli must be adequately ventilated in order for an adequate gas exchange to take place in the lungs.

Types of respirations

Respirations, when normal, are of either of two types or a combination of both: abdominal or costal. *Abdominal breathing*, sometimes called diaphragmatic or deep breathing, is characterized by an outward movement of the abdominal wall due to the contraction and descent of the diaphragm. *Costal* (shallow or chest) *breathing* is characterized by an upward, outward movement of the chest due to contraction of the external intercostals and other chest-elevating muscles. Normal quiet breathing of either the abdominal or costal type is known as *eupnea*.

There are various types of abnormal respirations, a few of which will be described. *Apnea* is a temporary cessation of respirations. *Dyspnea* is difficult or labored breathing. *Orthopnea* is inability to breathe in the horizontal position. *Cheyne-Stokes* respirations are characterized by a period of dyspnea followed by a period of apnea. The latter type of respiration often precedes death.

Some principles about gases

Before discussing respirations further, we need to understand the following principles.

1 *Dalton's law* (or the law of partial pressures). The term *partial pressure* means the pressure exerted by any one gas in a mixture of gases or in a liquid. The partial pressure of a gas in a mixture of gases is directly related to the concentration of that gas in the mixture and to the total pressure of the mixture. Suppose we apply this principle to compute the partial pressure of oxygen in the atmosphere. The concentration of oxygen in the atmosphere is 20.96% and the total pressure of the atmosphere is 760 mm Hg under standard conditions. Therefore:

Atmospheric $P_{O_2} = 20.96\% \times 760 = 159.2$ mm Hg

The symbol used to designate partial pressure is the capital letter P preceding the chemical symbol for the gas. Examples: alveolar air P_{O_2} is about 100 mm Hg;

**Table 11-1. Oxygen and carbon dioxide
pressure gradients**

	Atmo-sphere	*Alveolar air*	*Arterial blood*	*Venous blood*
P_{O_2}	160*	100	100	37
P_{CO_2}	0.3	40	40	46

*All figures indicate approximate mm Hg pressure under usual conditions.

arterial blood P_{O_2} is also about 100 mm Hg; venous blood P_{O_2} is about 37 mm Hg. The word *tension* is often used as a synonym for the term partial pressure—oxygen tension means the same thing as P_{O_2}.

2 The partial pressure of a gas in a liquid is directly determined by the amount of that gas dissolved in the liquid, which, in turn, is determined by the partial pressure of the gas in the environment of the liquid. Gas molecules diffuse into a liquid from its environment and dissolve in the liquid until the partial pressure of the gas in solution becomes equal to its partial pressure in the environment of the liquid. For example, alveolar air constitutes the environment of blood moving through pulmonary capillaries. Standing between the blood and the air are only the very thin alveolar and capillary membranes, and both of these are highly permeable to oxygen and carbon dioxide. By the time blood leaves the pulmonary capillaries as arterial blood, diffusion and approximate equilibration of oxygen and carbon dioxide across the membranes has occurred. Arterial blood P_{O_2} and P_{CO_2}, therefore, usually equal or very nearly equal alveolar P_{O_2} and P_{CO_2} (Table 11-1).

How blood transports gases

Blood transports oxygen and carbon dioxide as solutes and as parts of molecules of certain chemical compounds. Immediately upon entering the blood, both oxygen and carbon dioxide dissolve in the plasma. But, because fluids can hold only small amounts of gas in solution, most of the

oxygen and carbon dioxide rapidly form a chemical union with some other blood constituent. In this way, comparatively large volumes of the gases can be transported. For example, every 100 milliliters of arterial blood contains about 20 milliliters of oxygen instead of a mere 0.5 of a milliliter, which is all that can stay in solution in that amount of blood. About 19 milliliters of oxygen combines chemically with the hemoglobin present in 100 milliliters of blood to form oxyhemoglobin. Since each gram of hemoglobin can unite with about 1.3 milliliters of oxygen, the exact amount of oxygen in blood depends mainly upon the amount of hemoglobin present—15 grams of hemoglobin per 100 milliliters of blood is a typical normal blood hemoglobin content. With this amount of hemoglobin, 100 milliliters of arterial blood, when 100% saturated with oxygen, contains the following:

19.5 ml oxygen as oxyhemoglobin (15×1.3 ml)
0.5 ml oxygen in solution in plasma
20.0 ml total oxygen content

Perhaps a more common way of expressing blood oxygen content is in terms of volume per cent. Normal arterial blood, for example, contains about 20 vol.% O_2 (meaning 20 milliliters of oxygen in each 100 milliliters of blood).

Blood that contains more hemoglobin can, of course, transport more oxygen and that which contains less hemoglobin can transport less oxygen. Hence, hemoglobin deficiency anemia decreases oxygen transport and may produce marked cellular hypoxia (inadequate oxygen supply).

In order to combine with hemoglobin, oxygen must, of course, diffuse from plasma into the red cells where millions of hemoglobin molecules are located. Several factors influence the rate at which hemoglobin combines with oxygen in lung capillaries. For instance, as the following equation and Fig. 11-11 show, an increasing blood P_{O_2}

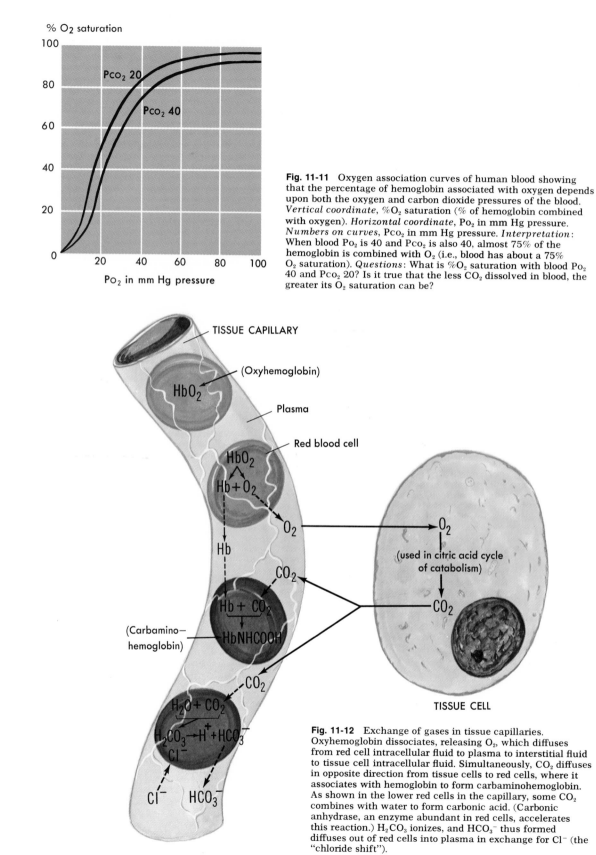

Fig. 11-11 Oxygen association curves of human blood showing that the percentage of hemoglobin associated with oxygen depends upon both the oxygen and carbon dioxide pressures of the blood. *Vertical coordinate,* %O_2 saturation (% of hemoglobin combined with oxygen). *Horizontal coordinate,* P_{O_2} in mm Hg pressure. *Numbers on curves,* P_{CO_2} in mm Hg pressure. *Interpretation:* When blood P_{O_2} is 40 and P_{CO_2} is also 40, almost 75% of the hemoglobin is combined with O_2 (i.e., blood has about a 75% O_2 saturation). *Questions:* What is %O_2 saturation with blood P_{O_2} 40 and P_{CO_2} 20? Is it true that the less CO_2 dissolved in blood, the greater its O_2 saturation can be?

Fig. 11-12 Exchange of gases in tissue capillaries. Oxyhemoglobin dissociates, releasing O_2, which diffuses from red cell intracellular fluid to plasma to interstitial fluid to tissue cell intracellular fluid. Simultaneously, CO_2 diffuses in opposite direction from tissue cells to red cells, where it associates with hemoglobin to form carbaminohemoglobin. As shown in the lower red cells in the capillary, some CO_2 combines with water to form carbonic acid. (Carbonic anhydrase, an enzyme abundant in red cells, accelerates this reaction.) H_2CO_3 ionizes, and HCO_3^- thus formed diffuses out of red cells into plasma in exchange for Cl^- (the "chloride shift").

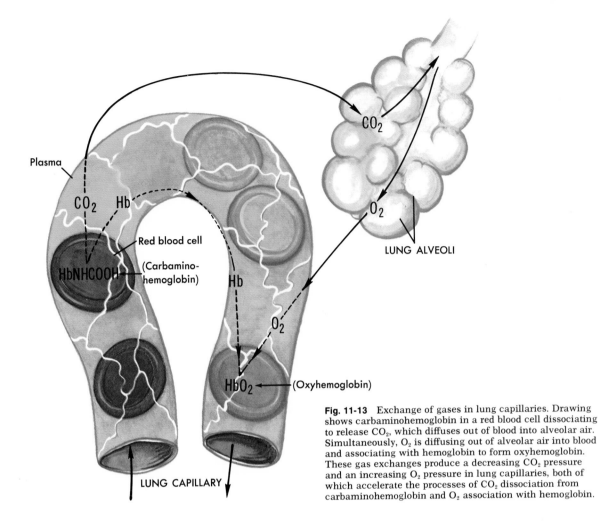

Fig. 11-13 Exchange of gases in lung capillaries. Drawing shows carbaminohemoglobin in a red blood cell dissociating to release CO_2, which diffuses out of blood into alveolar air. Simultaneously, O_2 is diffusing out of alveolar air into blood and associating with hemoglobin to form oxyhemoglobin. These gas exchanges produce a decreasing CO_2 pressure and an increasing O_2 pressure in lung capillaries, both of which accelerate the processes of CO_2 dissociation from carbaminohemoglobin and O_2 association with hemoglobin.

and a decreasing P_{CO_2} both accelerate hemoglobin association with oxygen.

$$Hb + O_2 \xrightarrow[\text{(Decreasing } P_{CO_2})]{\text{(Increasing } P_{O_2})} HbO_2$$

Decreasing P_{O_2} and increasing P_{CO_2}, on the other hand, accelerate oxygen dissociation from oxyhemoglobin—that is, the reverse of the preceding equation. Oxygen associates with hemoglobin rapidly—so rapidly, in fact, that about 97% of the blood's hemoglobin has united with oxygen by the time the blood leaves the lung capillaries to return to the heart. In other words, the average *oxygen saturation* of arterial blood is about 97%.

Carbon dioxide is carried in the blood in several ways, the most important of which are described briefly as follows.

1 A small amount of carbon dioxide dissolves in plasma and is transported as a solute (dissolved carbon dioxide produces the P_{CO_2} of blood).

2 More than one-half of the carbon dioxide is carried in the plasma as bicarbonate ions (Fig. 11-12).

3 Somewhat less than one-third of blood carbon dioxide unites with the NH_2 group of hemoglobin and certain other proteins to form carbamino compounds. Most of these are formed and transported in the red cells, since hemoglobin is the main protein to combine with carbon dioxide. The compound formed has a tongue-twisting name—carbaminohemoglobin. Carbon dioxide association with hemoglobin is accelerated by an increasing P_{CO_2} and a decreasing P_{O_2}

and is slowed by the opposite conditions. How do the conditions that accelerate carbon dioxide association affect the rate of oxygen association with hemoglobin?

Exchange of gases in lungs

The exchange of gases in the lungs takes place between alveolar air and venous blood flowing through lung capillaries. Gases move in both directions through the alveolar-capillary membrane (Fig. 11-13). Oxygen enters blood from the alveolar air because the Po_2 of alveolar air is greater than the Po_2 of venous blood. Another way of saying this is that oxygen diffuses "down" its pressure gradient. Simultaneously, carbon dioxide molecules exit from the blood by diffusing down the carbon dioxide pressure gradient out into the alveolar air. The Pco_2 of venous blood is much higher than the Pco_2 of alveolar air. This two-way exchange of gases between alveolar air and venous blood converts venous blood to arterial blood (examine Fig. 11-14).

The amount of oxygen that diffuses into blood each minute depends upon several factors, notably upon these four:

1 The oxygen pressure gradient between alveolar air and venous blood (alveolar Po_2 – venous blood Po_2)

2 The total functional surface area of the alveolar-capillary membrane

3 The respiratory minute volume (respiratory rate per minute times volume of air inspired per respiration)

4 Alveolar ventilation (discussed on p. 376)

All four of these factors bear a direct relation to oxygen diffusion. Anything that decreases alveolar Po_2, for instance, tends to decrease the alveolar-venous oxygen pressure gradient and therefore tends to decrease the amount of oxygen entering the blood. Application: alveolar air Po_2 decreases as altitude increases, and therefore less oxygen enters the blood at high altitudes. At a certain high altitude, alveolar air Po_2 equals venous blood Po_2. How would this affect oxygen diffusion into blood?

Anything that decreases the total functional surface area of the alveolar-capillary membrane also tends to decrease oxygen diffusion into the blood (by functional surface area is meant that which is freely permeable to oxygen). Application: in emphysema, this total functional area decreases and is one of the factors responsible for poor blood oxygenation in the condition.

Anything that decreases the respiratory minute volume also tends to decrease blood oxygenation. Application: morphine slows respirations and therefore decreases the respiratory minute volume (volume of air inspired per minute) and tends to lessen the amount of oxygen entering the blood. The main factors influencing blood oxygenation are shown in Fig. 11-15.

Several times we have stated the principle that structure determines functions. You may find it interesting to note the application of this principle to gas exchange in the lungs. Several structural facts facilitate oxygen diffusion from the alveolar air into the blood in lung capillaries:

1 The fact that the walls of the alveoli and of the capillaries together form a very thin barrier for the gases to cross (estimated at not more than 0.004 mm thick)

2 The fact that both alveolar and capillary surfaces are extremely large

3 The fact that the lung capillaries accommodate a large amount of blood at one time (about 900 ml)

4 The fact that the blood is distributed through the capillaries in a layer so thin (equal only to the diameter of one red corpuscle) that each corpuscle comes in close proximity to alveolar air

The
respiratory
system

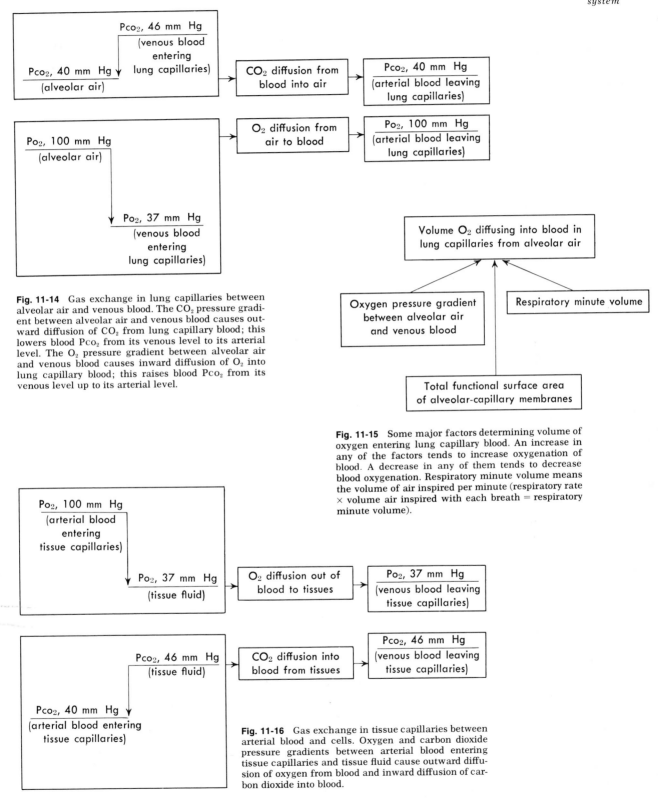

Fig. 11-14 Gas exchange in lung capillaries between alveolar air and venous blood. The CO_2 pressure gradient between alveolar air and venous blood causes outward diffusion of CO_2 from lung capillary blood; this lowers blood P_{CO_2} from its venous level to its arterial level. The O_2 pressure gradient between alveolar air and venous blood causes inward diffusion of O_2 into lung capillary blood; this raises blood P_{CO_2} from its venous level up to its arterial level.

Fig. 11-15 Some major factors determining volume of oxygen entering lung capillary blood. An increase in any of the factors tends to increase oxygenation of blood. A decrease in any of them tends to decrease blood oxygenation. Respiratory minute volume means the volume of air inspired per minute (respiratory rate × volume air inspired with each breath = respiratory minute volume).

Fig. 11-16 Gas exchange in tissue capillaries between arterial blood and cells. Oxygen and carbon dioxide pressure gradients between arterial blood entering tissue capillaries and tissue fluid cause outward diffusion of oxygen from blood and inward diffusion of carbon dioxide into blood.

Exchange of gases in tissues

The exchange of gases in tissues takes place between arterial blood flowing through tissue capillaries and cells (Fig. 11-12). It occurs because of the principle already noted that gases move down a gas pressure gradient. More specifically, in the tissue capillaries oxygen diffuses out of arterial blood because the oxygen pressure gradient favors its outward diffusion (see Fig. 11-16). Arterial blood Po_2 is about 100 mm Hg, interstitial fluid Po_2 is considerably lower, and intracellular fluid Po_2 is still lower. Although interstitial fluid and intracellular fluid Po_2 are not definitely established, they are thought to vary considerably—perhaps from around 60 mm Hg down to about 1 mm Hg. As activity increases in any structure, its cells necessarily utilize oxygen more rapidly. This decreases intracellular and interstitial Po_2 which, in turn, tends to increase the oxygen pressure gradient between blood and tissues and to accelerate oxygen diffusion out of the tissue capillaries. In this way, the rate of oxygen utilization by cells automatically tends to regulate the rate of oxygen delivery to cells. As dissolved oxygen diffuses out of arterial blood, blood Po_2 decreases, and this accelerates oxyhemoglobin dissociation to release more oxygen into the plasma for diffusion out to cells, as indicated in the following equation.

$$\text{Hb} + \text{O}_2 \xrightleftharpoons[\text{(Increasing Pco}_2)]{\text{(Decreasing Po}_2)} \text{HbO}_2$$

Because of oxygen release to tissues from tissue capillary blood, Po_2, oxygen saturation, and total oxygen content are all less in venous blood than in arterial blood, as shown in Table 11-2.

Carbon dioxide exchange between tissues and blood takes place in the opposite direction from oxygen exchange. Catabolism produces large amounts of carbon dioxide inside cells. So, intracellular and interstitial Pco_2 are higher than arterial blood Pco_2. This means that the carbon dioxide pressure gradient causes diffusion of carbon dioxide from the tissues into the blood flowing along through tissue capillaries (Fig. 11-16). Consequently, the Pco_2 of blood increases in tissue capillaries from its arterial level of about 40 mm Hg to its venous level of about 46 mm Hg. This increasing Pco_2 and decreasing Po_2 together produce two effects—they favor both oxygen dissociation from oxyhemoglobin and carbon dioxide association with hemoglobin to form carbaminohemoglobin (reexamine Fig. 11-11).

Control of respirations

The mechanism for the control of respirations has many parts (Fig. 11-17). A brief description of its main features follows.

1 The Pco_2, Po_2, and pH of *arterial blood* all influence respirations. The Pco_2 influences respiratory centers in the medulla via its action on chemoreceptors. Chemo-

Table 11-2. Blood oxygen

	Venous blood	Arterial blood
Po_2	37 mm Hg	100 mm Hg
Oxygen saturation	75%	97%
Oxygen content	15 ml O_2 per 100 ml blood	20 ml O_2 per 100 ml blood*
*Oxygen utilization by tissues = Difference between oxygen contents of arterial and venous blood (20-5) = 5 ml O_2 per 100 ml blood circulated per minute.		

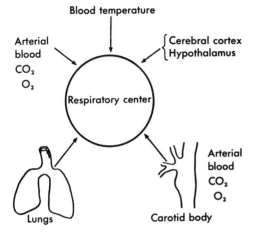

Blood temperature

Arterial
blood
CO_2
O_2

Cerebral cortex
Hypothalamus

Respiratory center

Lungs

Arterial
blood
CO_2
O_2

Carotid body

Fig. 11-17 Respiratory control
mechanism. Scheme to show the
main factors that influence the
respiratory center and thereby
control respirations. (See text for
discussion.)

receptors in this case are cells sensitive to
changes in arterial blood's oxygen, carbon
dioxide, and hydrogen ion concentrations.
The normal range for arterial P_{CO_2} is about
38 to 40 mm Hg. When it increases even
slightly above this, it has a stimulating
effect mainly on central chemoreceptors
(postulated to be present on the medulla).
To a lesser degree, increased arterial P_{CO_2}
also stimulates peripheral chemoreceptors
that are known to be present in the carotid
bodies and aorta. Stimulation of chemoreceptors by increased arterial P_{CO_2} results
in faster breathing with a greater volume
of air moving in and out of the lungs per
minute. Decreased arterial P_{CO_2} produces
opposite effects—it inhibits central and
peripheral chemoreceptors, which leads
to inhibition of medullary respiratory centers and slower respirations. In fact, breathing stops entirely for a few moments (apnea) when arterial P_{CO_2} drops moderately
—to about 35 mm Hg, for example.

The role of *arterial blood* P_{O_2} in controlling respirations is not entirely clear. Presumably, it has little influence as long as

it stays above a certain level. But, neurons
of the respiratory centers, like all body
cells, require adequate amounts of oxygen
in order to function optimally. Consequently, if they become hypoxic, they become depressed and send fewer impulses
to respiratory muscles. Respirations then
decrease or fail entirely. This principle has
important clinical significance. For example, the respiratory centers cannot respond
to stimulation by an increasing blood CO_2
if, at the same time, blood P_{O_2} falls below a
critical level—a fact that may become of
life and death importance during anesthesia.

However, a decrease in arterial blood P_{O_2}
below 70 mm Hg but not so low as the critical level stimulates chemoreceptors in the
carotid and aortic bodies and causes reflex
stimulation of the inspiratory center. This
constitutes an emergency respiratory control mechanism. It does not help regulate
respirations under usual conditions when
arterial blood P_{O_2} remains considerably
higher than 70 mm Hg—the level necessary to stimulate the chemoreceptors.

A decrease in *arterial blood* pH (increase
in acid), within certain limits, has a stimulating effect on carotid and aortic chemoreceptors and also on respiratory centers and
therefore tends to increase respirations.

2 *Arterial blood pressure* helps control
respirations through the respiratory pressoreflex mechanism. A sudden rise in arterial pressure, by acting on aortic and
carotid baroreceptors, results in reflex
slowing of respirations. And a sudden drop
in arterial pressure brings about a reflex
increase in rate and depth of respirations.
The pressoreflex mechanism is probably
not of great importance in the control of
respirations. It is, however, as you will
recall, of major importance in the control
of circulation.

3 The *Hering-Breuer reflexes* help control respirations—particularly their depth

and rhythmicity. They are believed to regulate the normal depth of respirations (extent of lung expansion) and, therefore, the volume of tidal air in the following way. Presumably, when the tidal volume of air has been inspired, the lungs are expanded enough to stimulate baroreceptors located within them. The baroreceptors then send inhibitory impulses to the inspiratory center, relaxation of inspiratory muscles occurs, and expiration follows — the Hering-Breuer expiratory reflex. Then, when the tidal volume of air has been expired, the lungs are sufficiently deflated to inhibit the lung baroreceptors and allow inspiration to start again — the Hering-Breuer inspiratory reflex.

4 The *pneumotaxic center* in the upper part of the pons is postulated to function mainly to maintain rhythmicity of respirations. Whenever the inspiratory center is stimulated, it sends impulses to the pneumotaxic center as well as to inspiratory muscles. The pneumotaxic center, after a moment's delay, stimulates the expiratory center, which then feeds back inhibitory impulses to the inspiratory center. Inspiration, therefore, ends and expiration starts. Lung deflation soon initiates the Hering-Breuer inspiratory reflex, and inspiration starts again. In short, the pneumotaxic center and Hering-Breuer reflexes together constitute an automatic device for producing rhythmic respirations.

5 The *cerebral cortex* helps control respirations. Impulses to the respiratory center from the motor area of the cerebrum may either increase or decrease the rate and strength of respirations. In other words, an individual may voluntarily speed up or slow down his breathing rate. This voluntary control of respirations, however, has certain limitations. For example, one may will to stop breathing and do so for a few minutes. But holding the breath results in an increase in the carbon dioxide content of the blood since it is not being removed by respirations. Carbon dioxide is a powerful respiratory stimulant. So when arterial blood P_{CO_2} increases to a certain level, it stimulates the inspiratory center both directly and reflexly to send motor impulses to the respiratory muscles, and breathing is resumed even though the individual may still will contrarily. This knowledge that the carbon dioxide content of the blood is a more powerful regulator of respirations than cerebral impulses is of practical value when dealing with a child who holds his breath to force the granting of his wishes. The best treatment is to ignore such behavior, knowing that respirations will start again as soon as the amount of carbon dioxide in arterial blood increases to a certain level.

6 Miscellaneous factors also influence respirations. Among these are blood temperature and sensory impulses from skin thermal receptors and from superficial or deep pain receptors.

a *Sudden painful stimulation* produces a reflex apnea, but continued painful stimuli cause faster and deeper respirations.

b *Sudden cold stimuli* applied to the skin cause reflex apnea.

c *Afferent impulses* initiated by stretching the anal sphincter produce reflex acceleration and deepening of respirations. Use has sometimes been made of this mechanism as an emergency measure to stimulate respirations during surgery.

d *Stimulation of the pharynx or larynx* by irritating chemicals or by touch causes a temporary apnea. This is the choking reflex, a valuable protective device. It operates, for example, to prevent aspiration of food or liquids during swallowing.

Control of respirations during exercise. Respirations increase greatly during strenuous exercise. The mechanism that accomplishes this, however, is not identical to the one that produces moderate increases in breathing. A number of studies have shown that arterial blood P_{CO_2}, P_{O_2}, and pH do not change enough during exercise to produce the degree of hyperpnea (faster, deeper respirations) observed. Venous blood P_{CO_2}, however, is known to increase with strenuous muscle exertion. Krahl and Armstrong reported finding chemoreceptors in the walls of the pulmonary artery—a strategic location for them to be acted on by changes in venous blood.* It may be, they suggest, that the higher venous P_{CO_2} present during exercise acts as a stimulant to these receptors and helps bring about the characteristic faster, deeper breathing.

*Physiology—Breathing regulated by blood composition, Sci. Newsletter **82:**56 (July 28), 1962.

outline summary

FUNCTIONS AND IMPORTANCE
Exchange of gases between blood and air; of vital importance

ORGANS

Nose
1 Structure
  a Portions—internal, in skull, above roof at mouth; external, protruding from face
  b Cavities
    1 Divisions—right and left
    2 Meati—superior, middle, and lower; named for turbinate located above each meatus
    3 Openings—to exterior, anterior nares; to nasopharynx, posterior nares
    4 Turbinates (conchae)—superior and middle, processes of ethmoid bone; inferior turbinates separate bones; divide internal nasal cavities into three passageways or meati
    5 Floor—formed by palatine bones that also act as roof of mouth
  c Lining—ciliated mucous membrane
  d Sinuses draining into nose (or paranasal sinuses)—frontal, maxillary (or antrum of Highmore), sphenoidal, and ethmoidal
2 Functions
  a Serves as passageway for incoming and outgoing air, filtering, warming, moistening, and chemically examining it

  b Organ of smell because olfactory receptors located in nasal mucosa
  c Aids in phonation

Pharynx
1 Structure—made of muscle with mucous lining
  a Divisions—nasopharynx, behind nose; oropharynx, behind mouth; laryngopharynx, behind larynx
  b Openings—four in nasopharynx: two auditory tubes and two posterior nares; one in oropharynx: fauces from mouth; and two in laryngopharynx: into esophagus and into larynx
  c Organs in pharynx—adenoids or pharyngeal tonsils in nasopharynx; palatine and lingual tonsils in oropharynx
2 Functions—serves both respiratory and digestive tracts as passageway for air, food, and liquids; aids in phonation

Larynx
1 Location—at upper end of trachea, just below pharynx
2 Structure
  a Cartilages—nine pieces arranged in boxlike formation; thyroid largest, known as "Adam's apple"; epiglottis, "lid" cartilage; cricoid, "signet ring" cartilage

b Vocal cords—false cords, folds of mucous lining; true cords, fibroelastic bands stretched across hollow interior of larynx; glottis, opening between true vocal cords

c Lining—ciliated mucous membrane

d Sexual differences—male larynx larger, covered with less fat, and therefore more prominent than female larynx

3 Function—expired air causes true vocal cords to vibrate, producing voice; pitch determined by length and tension of cords

Trachea

1 Structure

a Walls—smooth muscle; contain C-shaped rings of cartilage at intervals, which keeps the tube open at all times; lining—ciliated mucous membrane

b Extent—from larynx to bronchi; about 4 1/2 inches long

2 Function—furnishes open passageway for air going to and from lungs

Bronchi

1 Structure—formed by division of trachea into two tubes; right bronchus slightly larger and more vertical than left; same structure as trachea; each primary bronchus branches as soon as enters lung into secondary bronchi, which branch into bronchioles, which branch into microscopic alveolar ducts, which terminate in cluster of blind sacs called alveoli; trachea and two primary bronchi and all their branches compose "bronchial tree"; alveolar walls composed of single layer of cells

2 Function—bronchi and their many branching tubes furnish passageway for air going to and from lungs; alveoli provide large, thin-walled surface area where blood and air can exchange gases

Lungs

1 Structure

a Size, shape, location—large enough to fill pleural divisions of thoracic cavity; cone-shaped; extend from base, on diaphragm, to apex, located slightly above clavicle

b Divisions—three lobes in right lung, two in left; root of lung consists of primary bronchus and pulmonary artery and veins, bound together by connective tissue; hilum is vertical slit on mesial surface of lung through which root structures enter lung; base is broad, inferior surface of lung; apex is pointed upper margin

c Covering—visceral layer of pleura

2 Function—furnish place where large amounts of air and blood can come in close enough contact for rapid exchange of gases to occur

THORAX (CHEST)

1 Structure

a Three divisions

1 Pleural portion—contains lungs

2 Mediastinum—area between two lungs; contains esophagus, trachea, great blood vessels, etc.

3 Pericardial portion—space occupied by heart and pericardial sac

b Lining

1 Parietal layer of pleura lines entire chest cavity and covers superior surface of diaphragm

2 Forms separate sac encasing each lung

3 Separated from visceral pleura, covering lungs, only by potential space—pleural space—which contains few drops of pleural fluid

c Shape of ribs and angle of their attachment to spine such that elevation of rib cage enlarges two dimensions of thorax, its width and depth from front

2 Function

a Increase in size of thorax leads to inspiration

b Decrease in size of thorax leads to expiration

PHYSIOLOGY

1 Kinds of respirations

a External or lung breathing

b Internal or cell breathing

2 Mechanism of respirations

a Contraction of diaphragm and chest elevating muscles enlarges thorax, thereby decreases intrathoracic pressure, which causes expansion of lungs, which decreases intrapulmonic pressure to subatmospheric level, which establishes gas pressure gradient, which causes air to move into lungs

b Relaxation of inspiratory muscles produces opposite effects; see Fig. 11-9

3 Amount of air exchanged in respirations

a Directly related to gas pressure gradient between atmosphere and lung alveoli and inversely related to resistance opposing air flow

1 Measured by apparatus called spirometer

2 Tidal air—average amount expired after normal inspiration; approximately 500 ml or 1 pt

3 Supplemental air—amount that can be forcibly expired after normal inspiration

4 Complemental air—amount that can be forcibly inspired after normal inspiration; measured by having individual expire normally after forced inspiration

5 Residual air—that which cannot be forcibly expired from lungs

6 Minimal air—that which can never be removed from alveoli, even when lungs subjected to atmospheric pressure

7 Vital capacity—approximate capacity of lungs (limited by size of thoracic cavity and various other factors, such as amount of blood in lungs and condition of alveoli); air that can be forcibly expired after forcible inspiration represents vital capacity

8 Alveolar ventilation—volume of inspired air that actually reaches alveoli; computed by subtracting dead space air volume from tidal air volume

4 Types of respirations
  a Normal (eupnea)
    1 Abdominal (also called deep or diaphragmatic breathing)—characterized by outward movement of abdominal wall due to contraction and descent of diaphragm
    2 Costal (also called shallow or chest breathing)—characterized by upward, outward movement of chest due to contraction of chest-elevating muscles
  b Abnormal
    1 Apnea—temporary cessation of respirations
    2 Dyspnea—difficult or painful respirations
    3 Orthopnea—inability to breathe in horizontal position
    4 Cheyne-Stokes—alternate periods of dyspnea and apnea

5 Some principles about gases
  a Dalton's law—partial pressure of gas in mixture of gases directly related to concentration of that gas in mixture and to total pressure of mixture
  b Partial pressure of gas in liquid directly related to amount of gas dissolved in liquid; becomes equal to partial pressure of that gas in environment of liquid

6 How blood transports gases
  a Oxygen
    1 About 0.5 ml transported as *solute*—i.e., dissolved in 100 ml blood
    2 About 19.5 ml O_2 per 100 ml blood transported as *oxyhemoglobin* in red blood cells
    3 About 20 ml = *Total O_2 content* per 100 ml blood (100% saturation of 15 gm hemoglobin)
  b Carbon dioxide
    1 Small amount dissolves in plasma and transported as true *solute*
    2 More than half of CO_2 transported as *bicarbonate ion* in plasma
    3 Somewhat less than one-third of CO_2 transported in red blood cells as *carbaminohemoglobin*

7 Exchange of gases in lungs (between alveolar air and venous blood)
  a Where it occurs—in lung capillaries; across alveolar-capillary membrane

  b What exchange consists of—oxygen diffuses out of alveolar air into venous blood; carbon dioxide diffuses in opposite direction
  c Why it occurs—oxygen pressure gradient causes inward diffusion of oxygen; carbon dioxide pressure gradient causes outward diffusion of carbon dioxide

	Alveolar air	*Venous blood*
P_{O_2}	100 mm Hg	37 mm Hg
P_{CO_2}	40 mm Hg	46 mm Hg

  d Results of gas exchange
    1 P_{O_2} of blood increases to arterial blood level as blood moves through lung capillaries
    2 P_{CO_2} of blood decreases to arterial blood level as blood moves through lung capillaries
    3 Oxygen association with hemoglobin to form oxyhemoglobin and carbon dioxide dissociation from carbaminohemoglobin both accelerated by increasing P_{O_2} and decreasing P_{CO_2}

8 Exchange of gases in tissues (between arterial blood and cells)
  a Where it occurs—tissue capillaries
  b What exchange consists of—oxygen diffuses out of arterial blood into interstitial fluid and on into cells, whereas carbon dioxide diffuses in opposite direction
  c Why it occurs—O_2 pressure gradient causes outward diffusion of O_2; CO_2 pressure gradient causes inward diffusion of CO_2

	Arterial blood	*Interstitial fluid*
P_{O_2}	100 mm Hg	60(?) mm Hg down to 1(?) mm Hg
P_{CO_2}	46 mm Hg	50(?) mm Hg

  d Results of oxygen diffusion out of blood and carbon dioxide diffusion into blood
    1 P_{O_2} blood decreases as blood moves through tissue capillaries; arterial P_{O_2}, 100 mm Hg, becomes venous P_{O_2}, 40 mm Hg (figures vary)
    2 P_{CO_2} blood increases; arterial P_{CO_2}, 40 mm Hg, beomes venous P_{CO_2}, 46 mm Hg (figures vary)
    3 Oxygen dissociation from hemoglobin and carbon dioxide association with hemoglobin to form carbaminohemoglobin both accelerated by decreasing P_{O_2} and increasing P_{CO_2}

9 Control of respirations (see Fig. 11-17)

    a Respiratory centers—inspiratory and expiratory centers in medulla; pneumotaxic center in pons

    b Control of respiratory centers

        1 Carbon dioxide major regulator of respirations; increased blood carbon dioxide content, up to certain level, stimulates respiration and above this level depresses respirations; decreased blood carbon dioxide decreases respirations

        2 Oxygen content of blood influences respiratory center—decreased blood O_2, down to a certain level, stimulates respirations and below this critical level depresses them; O_2 control of respirations nonoperative under usual conditions

        3 Hering-Breuer mechanism helps control rhythmicity of respirations; increased alveolar pressure inhibits inspiration and starts expiration; decreased alveolar pressure stimulates inspiration and ends expiration

        4 Miscellaneous factors influence respiratory center—e.g., body temperature, pain, emotions, etc.

    c Control of respirations during exercise—not yet established

  1 What anatomical feature favors the spread of the common cold through the respiratory passages and into the middle ear and mastoid sinus?

  2 How are the turbinates arranged in the nose? What are they?

  3 What organs are found in the nasopharynx?

  4 What tubes open into the nasopharynx?

  5 Make a diagram showing the termination of a bronchiole in an alveolar duct with alveoli.

  6 What kind of membrane lines the respiratory system?

  7 What is the serous covering of the lungs called? Where else, besides covering the lungs, is this same membrane found?

  8 The pharynx is common to what two systems?

  9 Are the lungs active or passive organs during breathing? Explain.

10 What is the main inspiratory muscle?

11 How is inspiration accomplished? Expiration?

12 If an opening is made into the pleural cavity from the exterior, what happens? Why?

13 What is the pleural space? What does it contain?

14 What substance found in blood is the natural chemical stimulant for the respiratory center?

15 Compare mechanisms that achieve internal and external respiration.

16 Respirations increase during exercise. Explain the mechanisms involved.

17 What is the voice box? Of what is it composed? What is the Adam's apple?

18 What is the epiglottis? What is its function?

19 What are the true vocal cords? Where are they? What name is given to the opening between the cords?

20 Name the three divisions of the thorax and their contents.

21 One gram of hemoglobin combines with how many milliliters of oxygen?

22 Suppose your blood has a hemoglobin content of 15 gm per 100 ml and an oxygen saturation of 97%. How many milliliters of oxygen would 100 ml of your arterial blood contain?

23 Compare the mechanisms that accelerate respiration with those that accelerate circulation during exercise.

24 Make a generalization about the effect of a moderate increase in the amount of blood CO_2 on circulation and respiration. What advantage can you see in this effect?

25 Make a generalization about the effect of a moderate decrease in blood O_2 on circulation and respiration.

26 Define the following terms briefly: alveolus; apnea; asphyxia; complemental air; cyanosis; dyspnea; minimal air; orthopnea; Pco_2; Po_2; pleurisy; residual air; respiration; spirometer; supplemental air; thorax; tidal air; vital capacity.

Functions and importance

Organs
Walls of organs
  Coats
  Modifications of coats
Mouth (buccal cavity)
Salivary glands
Teeth
Pharynx
Esophagus
Stomach
  Size, shape, and position
  Divisions
  Curves
  Sphincter muscles
  Coats
  Glands
  Functions
Small intestine
  Size and position
  Divisions
  Coats
  Functions
Large intestine (colon)
  Size
  Divisions
  Coats
  Functions
Liver
  Location and size
  Lobes
  Ducts
  Functions
Gallbladder
  Size, shape, and location
  Structure
  Functions
  Correlations
Pancreas
  Size, shape, and location
  Structure
  Functions
Vermiform appendix
  Size, shape, and location
  Structure

Digestion
Definition
Purpose
Kinds
Control of digestive gland
  secretion

Absorption
Definition
How accomplished

Metabolism
Meaning
Ways in which foods
  metabolized
Carbohydrate metabolism
  Glucose transport through
    cell membranes and
    phosphorylation
  Glycogenesis and
    glycogenolysis
  Gluconeogenesis
  Control of glucose
    metabolism
Fat metabolism
  Control
Protein metabolism
  Control
Metabolism of vitamins,
  mineral salts, and water
Metabolic rates
  Meaning
  Ways of expressing
  Basal metabolic rate
    Factors influencing
    How determined
    Total metabolic rate
  Energy balance and its
    relationship to body
    weight
**Mechanisms for regulating
food intake**

**Homeostasis of body
temperature**
Heat production
Heat loss
  Evaporation
  Radiation
  Conduction
  Convection
Thermostatic control of heat
  production and loss
  Heat-dissipating
    mechanism
  Heat-gaining mechanism
Skin thermal receptors
Correlations

12

The digestive system

▪ Functions and importance

The organs of the digestive system together perform a vital function—that of preparing food for absorption and for use by the millions of body cells. Most food when eaten is in a form that cannot reach the cells (because it cannot pass through the intestinal mucosa into the bloodstream) nor could it be used by the cells even if it could reach them. It must, therefore, be modified as to both chemical composition and physical state. This process of altering the chemical and physical composition of food so that it can be absorbed and utilized by body cells is known as digestion and is the function of the digestive system. Part of the digestive system, the large intestine, serves also as an organ of elimination, ridding the body of the wastes resulting from the digestive process.

Organs

The main organs of the digestive system form a tube all the way through the ventral cavities of the body. It is open at both

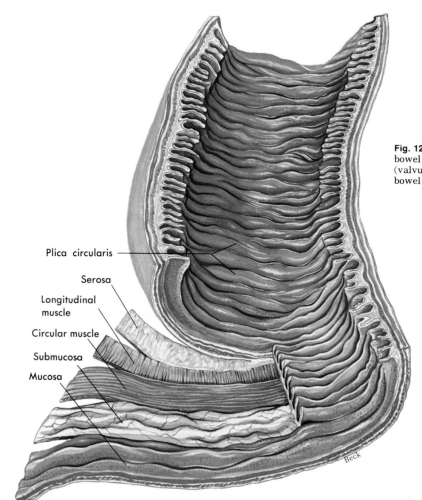

Fig. 12-1 Section of the small bowel showing the circular folds (valvulae conniventes) and layers of the bowel wall.

Plica circularis

Serosa

Longitudinal muscle

Circular muscle

Submucosa

Mucosa

Beck

ends. This tube is usually referred to as the *alimentary canal* (or tract) or the *gastrointestinal* or GI tract. The following organs from the gastrointestinal tract: mouth, pharynx, esophagus, stomach, and intestines. Several accessory organs are located in the main digestive organs or open into them. They are the salivary glands, teeth, liver, gallbladder, pancreas, and vermiform appendix.

■ Walls of organs
Coats

The alimentary canal is essentially a tube whose walls are fashioned of four

layers of tissues: a mucous lining, a submucous coat of connective tissue in which are embedded the main blood vessels of the tract, a muscular coat, and a fibroserous coat.

Modifications of coats

Although the same four tissue coats form the various organs of the alimentary tract, their structure varies in different organs. Some of these modifications are listed in Table 12-1.

The parietal peritoneum, which lines the posterior wall of the abdominal cavity, projects from the lumbar region into the

Table 12-1. Modifications of coats of digestive tract

Organ	Mucous coat	Muscle coat	Fibroserous coat
Esophagus		Two layers—inner one of circular fibers and outer one of longitudinal fibers; striated muscle in upper part and smooth in lower part of esophagus and in rest of tract	Outer coat fibrous; serous around part of esophagus in thoracic cavity
Stomach	Arranged in temporary longitudinal folds called *rugae;* allow for distention (Fig. 12-10) Contains microscopic gastric and hydrochloric acid glands	Has three layers instead of usual two—circular, longitudinal, and oblique fibers; two sphincters—cardiac at entrance of stomach and pyloric at its exit formed by circular fibers	Outer coat visceral peritoneum; hangs in double fold from lower edge of stomach over intestines, forming apronlike structure or "lace apron," *greater omentum* (Fig. 12-3)
Small intestine	Contains permanent circular folds, *valvulae conniventes* (or plica circularis) (Fig. 12-1) Microscopic fingerlike projections, *villi* (Fig. 12-2) Microscopic intestinal glands (of Lieberkühn) Microscopic duodenal (Brunner's) glands Clusters of lymph nodes, *Peyer's patches* Numerous single lymph nodes called solitary nodes	Two layers—inner one of circular fibers and outer one of longitudinal fibers	Outer coat visceral peritoneum
Large intestine	Solitary nodes Intestinal glands	Incomplete outer longitudinal coat; present only in three tapelike strips (taenia libera); small sacs (haustra) give rest of wall of large intestine puckered appearance (Fig. 12-11); internal anal sphincter formed by circular smooth fibers and external anal sphincter by striated fibers	Outer coat visceral peritoneum

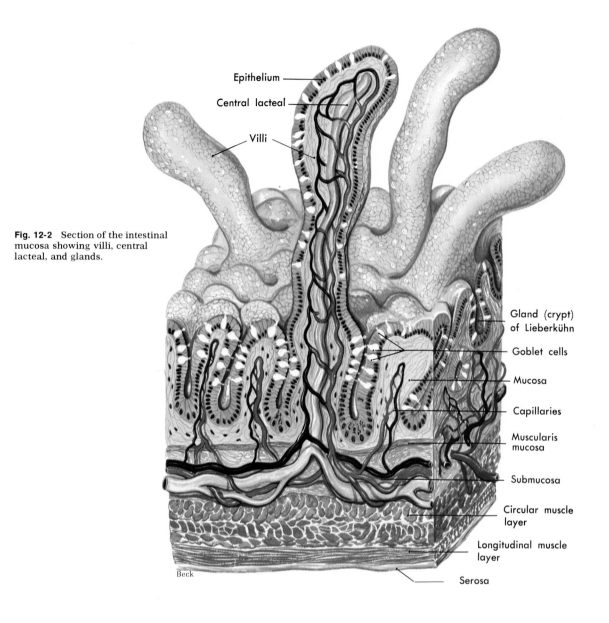

Epithelium

Central lacteal

Villi

Fig. 12-2 Section of the intestinal mucosa showing villi, central lacteal, and glands.

Gland (crypt) of Lieberkühn

Goblet cells

Mucosa

Capillaries

Muscularis mucosa

Submucosa

Circular muscle layer

Longitudinal muscle layer

Beck

Serosa

abdominal cavity in a double fold, shaped like a plaited fan. It is named the *mesentery* (Fig. 12-4). The loose outer edge of this great fan measures approximately 20 feet, whereas its attached posterior border has a length of from only 6 to 8 inches. Most of the small intestine is attached to its outer edge. The mesentery, then, may be defined as a fan-shaped double fold of parietal peritoneum by which the small intestine is anchored to the posterior abdominal wall. It should not be confused

*viseval
peritre*

with the *greater omentum,** an apron-shaped double fold of peritoneum that is attached at its upper border to the first part of the duodenum, the lower edge of the stomach, and the transverse colon and hangs down loosely over the intestines. In case of a localized abdominal inflammation, such as appendicitis, the omentum

*The *lesser omentum* is a fold of peritoneum that attaches the liver to the lesser curvature of the stomach and beginning of the duodenum.

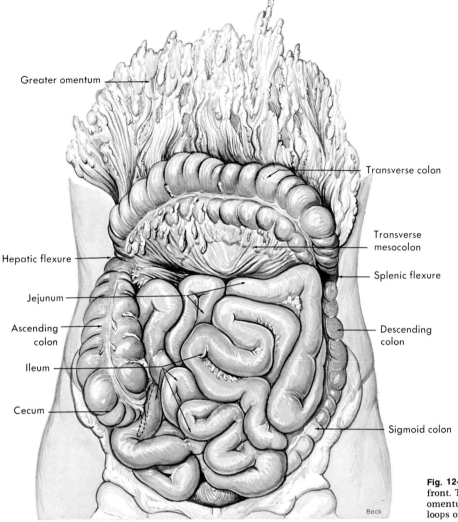

Greater omentum

Hepatic flexure

Jejunum

Ascending colon

Ileum

Cecum

Transverse colon

Transverse mesocolon

Splenic flexure

Descending colon

Sigmoid colon

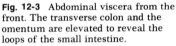

Fig. 12-3 Abdominal viscera from the front. The transverse colon and the omentum are elevated to reveal the loops of the small intestine.

envelops the inflamed area, walling it off from the rest of the abdomen. Spotty deposits of fat accumulate in the omentum, giving it a lacy appearance (Fig. 12-3).

■ Mouth (buccal cavity)

The following structures form the buccal cavity: the cheeks (sidewalls), the tongue and its muscles (floor), and the hard and soft palates (roof). Of these, only the palates and the tongue will be included in this discussion.

The *hard palate* consists of the two palatine bones and parts of the two superior maxillary bones. The *soft palate*, which forms a partition between the mouth and nasopharynx, is fashioned of muscle arranged in the shape of an arch. The opening in the arch leads from the mouth into the oropharynx and is named the *fauces,* whereas the two vertical side portions of the arch are appropriately termed the pillars of the fauces. Suspended from the midpoint of the posterior border of the arch is a small cone-shaped process, the *uvula.*

The entire buccal cavity, like the rest of the digestive tract, is lined with mucous membrane.

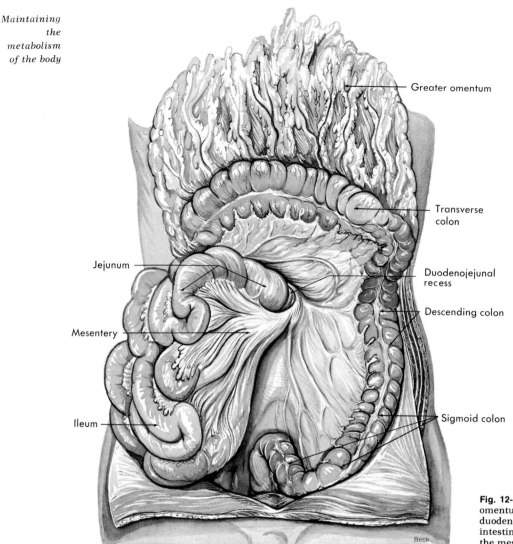

Greater omentum

Transverse
colon

Jejunum

Duodenojejunal
recess

Mesentery

Descending colon

Ileum

Sigmoid colon

Beck

Fig. 12-4 The transverse colon and
omentum are raised to demonstrate the
duodenojejunal recess. The small
intestine is pulled to the side to show
the mesentery. (Also see Fig. 12-3.)

The salivary glands and teeth are accessory organs of the mouth.

Skeletal muscle covered with mucous
membrane composes the *tongue.* Several
muscles that originate on skull bones insert into the tongue. The rough elevations
on the tongue's surface are called *papillae.*
They contain the taste buds. Three types,
filiform, fungiform, and circumvallate, can
be observed (Fig. 12-5).

The filiform papillae are numerous
threadlike structures distributed over the
anterior two-thirds of the tongue. Fungiform papillae are knoblike elevations most
numerous near the edges of the tongue.

Circumvallate papillae form an inverted V
at the posterior part of the tongue.

The *frenum* (or frenulum) is a fold of
mucous membrane in the midline of the
undersurface of the tongue that helps to
anchor the tongue to the floor of the
mouth. If the frenum is too short for free-
dom of tongue movements, the individual
is said to be tongue-tied, and his speech
is faulty.

■Salivary glands

See Table 12-2 for the names, location,
and duct openings of the salivary glands.

Mumps is an acute infection of the par-

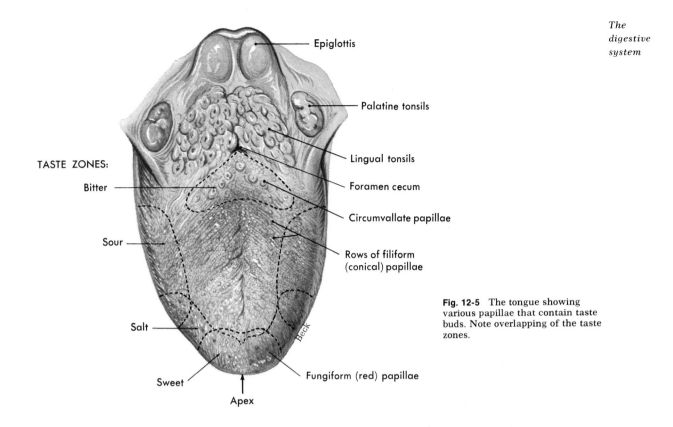

Fig. 12-5 The tongue showing various papillae that contain taste buds. Note overlapping of the taste zones.

Table 12-2. Salivary glands

Name of gland	Location	Duct openings
Parotid	Below and in front of ear	On inside of cheek, opposite upper second molar tooth; known as Stensen's duct
Submaxillary	Posterior part of floor of mouth	Floor of mouth, at sides of frenum; known as Wharton's duct
Sublingual	Anterior part of floor of mouth, under tongue	Several ducts open into floor of mouth

otid glands characterized by swelling of the glands. The act of opening the mouth causes pain because it squeezes that part of the gland that projects between the temporomandibular joint and the mastoid process.

Teeth

The so-called baby teeth or the set that appears first and is later shed are technically known as the *deciduous teeth* (Fig. 12-6), whereas those that replace these are the *permanent teeth* (Fig. 12-7). The names and numbers of teeth present in both sets are given in Table 12-3.

The first deciduous tooth erupts usually at the age of about 6 months. The rest follow at the rate of one or more a month until all twenty have appeared. There is, however, great individual variation in the age at which teeth erupt. Deciduous teeth are shed generally between the ages of 6 and

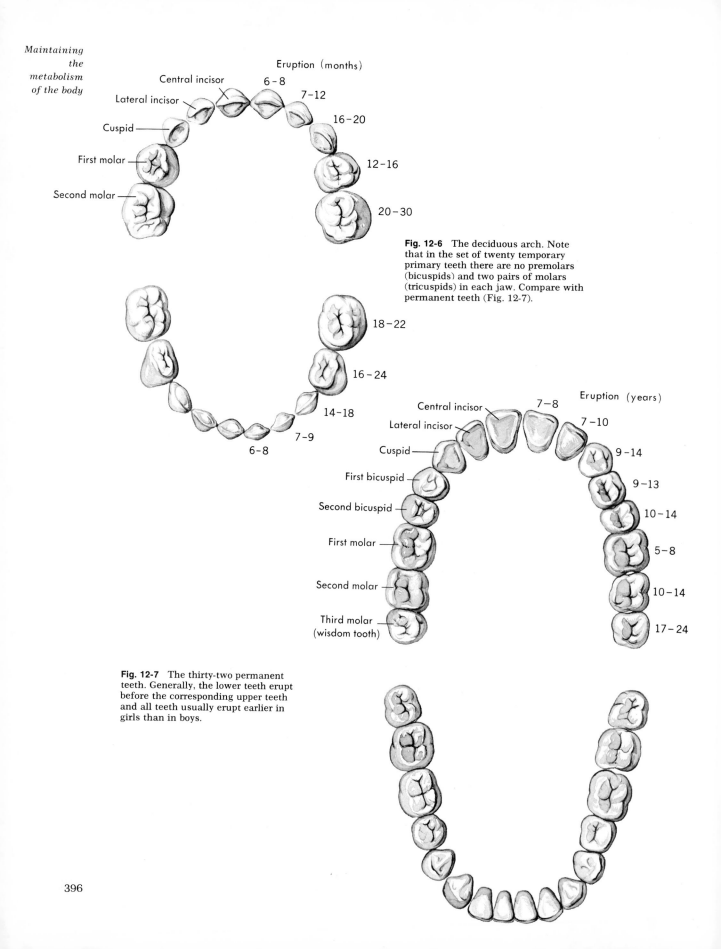

Eruption (months)

Central incisor — 6–8

Lateral incisor — 7–12

Cuspid — 16–20

First molar — 12–16

Second molar — 20–30

Fig. 12-6 The deciduous arch. Note that in the set of twenty temporary primary teeth there are no premolars (bicuspids) and two pairs of molars (tricuspids) in each jaw. Compare with permanent teeth (Fig. 12-7).

18–22

16–24

14–18

7–9

6–8

Eruption (years)

Central incisor — 7–8

Lateral incisor — 7–10

Cuspid — 9–14

First bicuspid — 9–13

Second bicuspid — 10–14

First molar — 5–8

Second molar — 10–14

Third molar (wisdom tooth) — 17–24

Fig. 12-7 The thirty-two permanent teeth. Generally, the lower teeth erupt before the corresponding upper teeth and all teeth usually erupt earlier in girls than in boys.

Table 12-3. Dentition

Name of tooth	Number per jaw	
	Deciduous set	Permanent set
Central incisors	2	2
Lateral incisors	2	2
Cuspids (canines)	2	2
Premolars (bicuspids)	0	4
Molars (tricuspids)	4	6
Total per jaw	10	16
Total per set	20	32

13 years. The third molars (wisdom teeth) are the last to appear, erupting usually sometime after 17 years of age.

Intact enamel resists bacterial attack, but once it is broken, the softer dentine decays. Pyorrhea is an inflammation of the gums (gingivae) and periodontal membrane (Fig. 12-8).

■ Pharynx

For a discussion of the pharynx, see pp. 365 to 366.

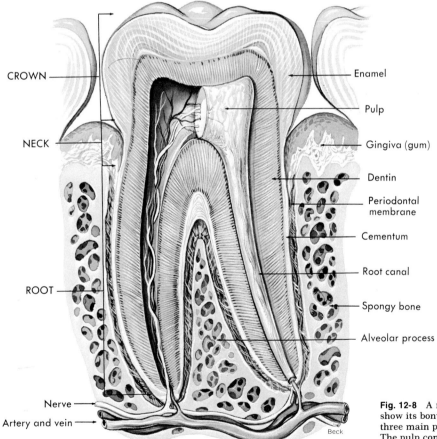

CROWN — Enamel

— Pulp

NECK — Gingiva (gum)

— Dentin

— Periodontal membrane

— Cementum

— Root canal

ROOT — Spongy bone

— Alveolar process

Nerve —

Artery and vein —

Beck

Fig. 12-8 A molar tooth sectioned to show its bony socket and details of its three main parts: crown, neck, and root. The pulp contains nerves and blood vessels.

397

■Esophagus

The esophagus, a collapsible tube about 10 inches long, extends from the pharynx to the stomach, piercing the diaphragm in its descent from the thoracic to the abdominal cavity. It lies posterior to the trachea and heart.

Unlike the trachea, the esophagus is a collapsible tube, its muscle walls lacking the cartilaginous rings found in the trachea.

■Stomach

Size, shape, and position

Just below the diaphragm, the alimentary tube dilates into an elongated pouch-like structure, the stomach (Fig. 12-9), the size of which varies according to several factors, notably sex and the amount of distention. In general, the female stomach is usually more slender and smaller than the male stomach. For some time after a meal, the stomach is enlarged due to distention

Bolus — ball of semi-digested food going down esophagus

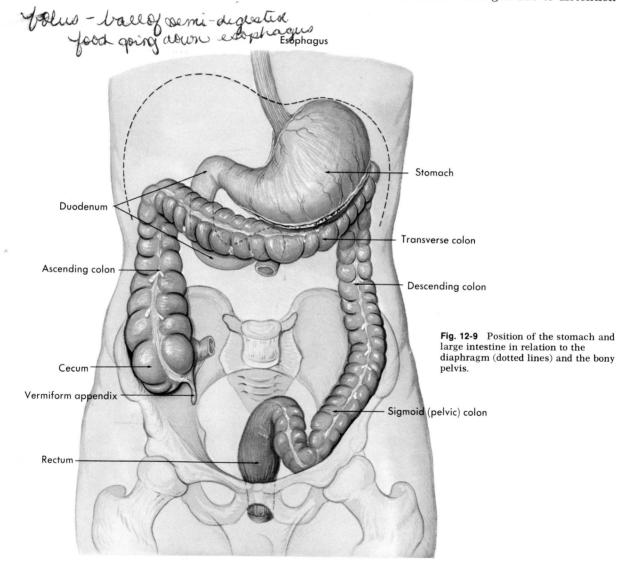

Esophagus

Stomach

Duodenum

Transverse colon

Ascending colon

Descending colon

Cecum

Vermiform appendix

Rectum

Sigmoid (pelvic) colon

Fig. 12-9 Position of the stomach and large intestine in relation to the diaphragm (dotted lines) and the bony pelvis.

of its walls, but as food leaves, the walls partially collapse, leaving the organ about the size of a large sausage.

The stomach lies in the upper part of the abdominal cavity under the liver and diaphragm, with approximately five-sixths of its mass to the left of the median line. In other words, it is described as lying in the epigastrium and left hypochondrium (Fig. 1-3, p. 7). Its position, however, alters frequently. For example, it is pushed downward with each inspiration and upward with each expiration. When it is greatly distended from an unusually large meal, its size interferes with the descent of the diaphragm on inspiration, producing the familiar feeling of dyspnea that accompanies overeating. In this state, the stomach also pushes upward against the heart, giving rise to the sensation that the heart is being crowded.

Divisions

The *fundus,* the *body,* and the *pylorus* are the three divisions of the stomach. The fundus is the enlarged portion to the left and above the opening of the esophagus into the stomach. The body is the central part of the stomach, and the pylorus is its lower portion (Fig. 12-10).

Fig. 12-10 Muscle layers and interior of the stomach.

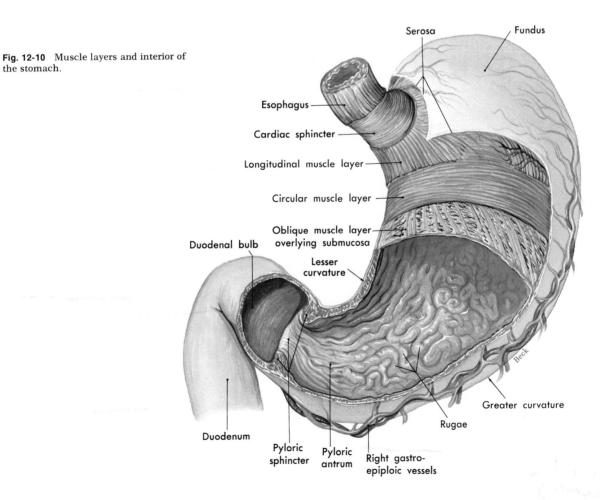

Serosa
Fundus
Esophagus
Cardiac sphincter
Longitudinal muscle layer
Circular muscle layer
Oblique muscle layer overlying submucosa
Duodenal bulb
Lesser curvature
Greater curvature
Duodenum
Pyloric sphincter
Pyloric antrum
Right gastro-epiploic vessels
Rugae

Curves

The upper right border of the stomach presents what is known as the *lesser curvature* and the lower left border the *greater curvature*.

Sphincter muscles

Sphincter muscles guard both stomach openings. A sphincter muscle consists of circular fibers so arranged that there is an opening in the center of them (like the hole in a doughnut) when they are relaxed and no opening when they are contracted.

The *cardiac sphincter* guards the opening of the esophagus into the stomach, and the *pyloric sphincter* guards the opening from the pyloric portion of the stomach into the first part of the small intestine (duodenum). This latter muscle is of clinical importance because *pylorospasm* is a fairly common condition in babies. The pyloric fibers do not relax normally to allow food to leave the stomach, and the baby vomits his food instead of digesting and absorbing it. The condition is relieved by the administration of a drug that relaxes smooth muscle. Another abnormality of the pyloric sphincter is pyloric stenosis, an obstructive narrowing of its opening.

Coats

See Table 12-1 and Fig. 12-10 for information on coats of the stomach.

Glands

Numerous microscopic tubular glands are embedded in the gastric mucosa. Those in the mucosa that lines the fundus and body of the stomach secrete most of the gastric juice, a fluid composed of mucus, enzymes, and hydrochloric acid. *Epithelial cells* that form the surface of the gastric mucosa (next to the lumen of the stomach) secrete mucus. *Parietal cells* secrete hydrochloric acid, and *chief cells* (or zymogen cells) secrete the enzymes of the gastric

juice. In pernicious anemia, the gastric mucosa atrophies, with the result that hydrochloric acid and a mysterious substance known as the intrinsic factor are not produced. Achlorhydria, therefore, is a characteristic finding in pernicious anemia. Absence of the intrinsic factor produces anemia. Without it, vitamin B_{12} cannot be absorbed, and without vitamin B_{12} the red bone marrow is not sufficiently stimulated to produce enough normal red blood cells.

Functions

The stomach carries on the following functions:

1 It serves as a reservoir, storing food until it can be partially digested and moved farther along the gastrointestinal tract.

2 It secretes gastric juice, one of the juices whose enzymes digest food.

3 Through contractions of its muscular coat, it churns the food, breaking it into small particles and mixing them well with the gastric juice. In due time, it moves the gastric contents on into the duodenum.

4 It secretes the intrinsic factor just mentioned.

5 It carries on a limited amount of absorption—of some water, alcohol, and certain drugs.

Small intestine
Size and position

The small intestine is a tube measuring approximately 1 inch in diameter and 20 feet in length. Its coiled loops fill most of the abdominal cavity.

Divisions

The small intestine consists of three divisions: the duodenum, the jejunum, and the ileum. The *duodenum** is the uppermost

---

*Derivation of the word duodenum may interest you. It comes from words which means 12 fingerbreadths, a distance of about 11 inches, the approximate length of the duodenum.

division and is the part to which the pyloric end of the stomach attaches. It is about 10 inches long and is shaped roughly like the letter C. The duodenum becomes *jejunum* at the point where the tube turns abruptly forward and downward. The jejunal portion continues for approximately the next 8 feet, where it becomes *ileum*, but without any clear line of demarcation between the two divisions. The ileum is about 12 feet long.

Coats

See Table 12-1 and Fig. 12-1 for information on the coats of the small intestine.

Functions

The small intestine carries on three main functions.

1 It completes the digestion of foods. The digestive intestinal juice contains mucus and many digestive enzymes. The glands of Lieberkühn secrete the digestive enzymes, whereas the glands of Brunner and innumerable goblet cells secrete the mucus.

2 It absorbs the end products of digestion into blood and lymph.

3 It secretes hormones—for example, some which help control the secretion of pancreatic juice, bile, and intestinal juice.

[handwritten notes: Peyers Patches, spot of Lymphoid tissue, secretes mucus]

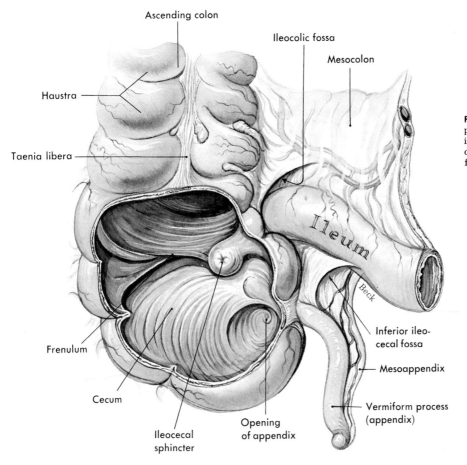

Ascending colon

Ileocolic fossa

Mesocolon

Haustra

Taenia libera

Frenulum

Cecum

Ileocecal sphincter

Opening of appendix

Ileum

Beck

Inferior ileocecal fossa

Mesoappendix

Vermiform process (appendix)

Fig. 12-11 The vermiform process (appendix) and the ileocecal region. The cecum is opened to reveal the papillary form of ileocecal sphincter.

■ Large intestine (colon)

Size

The lower part of the alimentary canal bears the name *large intestine* because its diameter is noticeably larger than that of the small intestine. Its length, however, is much less, being about 5 or 6 feet. Its average diameter is approximately 2 1/2 inches but decreases toward the lower end of the tube.

Divisions

The large intestine is divided into the cecum, colon, and rectum.

Cecum. The first two or three inches of the large intestine are named the cecum. It is located in the lower right quadrant of the abdomen (Fig. 12-9).

Colon. The colon is divided into the following portions: ascending, transverse, descending, and sigmoid.

1 The *ascending colon* lies in the vertical position, on the right side of the abdomen, extending up to the lower border of the liver. The ileum joins the large intestine at the junction of the cecum and ascending colon, the place of attachment resembling the letter T in formation (Fig. 12-11). The ileocecal valve guards the opening of the ileum into the large intestine, permitting material to pass from the former into the latter but not in the reverse direction.

2 The *transverse colon* passes horizontally across the abdomen, below the liver and stomach and above the small intestine.

3 The *descending colon* lies in the vertical position, on the left side of the abdomen, extending from a point below the stomach to the level of the iliac crest.

4 The *sigmoid colon* is that portion of the large intestine that courses downward below the iliac crest. It describes an S-shaped curve. The lower part of the curve, which joins the rectum, bends toward the left, the anatomical reason for placing a patient on the left side when giving an enema. In this position, gravity aids the flow of the water from the rectum into the sigmoid flexure.

Rectum. The last seven or eight inches of the intestinal tube is called the rectum. The terminal inch of the rectum is called the *anal canal*. Its mucous lining is arranged in numerous vertical folds known as *rectal columns*, each of which contains an artery and a vein. *Hemorrhoids* (or piles) are enlargements of the veins in the anal canal. The opening of the canal to the exterior is guarded by two sphincter muscles—an internal one of smooth muscle and an external one of striated muscle. The opening itself is called the *anus*. The general direction of the rectum is up, in, and back.

Coats

See Table 12-1 for information on the coats of the large intestine.

Functions

The main functions of the large intestine are absorption of water and elimination of the wastes of digestion.

■ Liver

Location and size

The liver is the largest gland in the body. It weighs between 3 and 4 pounds, lies immediately under the diaphragm, and occupies most of the right hypochondrium and part of the epigastrium (Fig. 12-12).

Lobes

The falciform ligament divides the liver into two main lobes, right and left, with the right lobe having three parts designated as the right lobe proper, the caudate lobe (a small four-sided area on the posterior surface), and the quadrate lobe (an approximately oblong section on the undersurface). Each lobe is divided into numerous lobules by small blood vessels and by

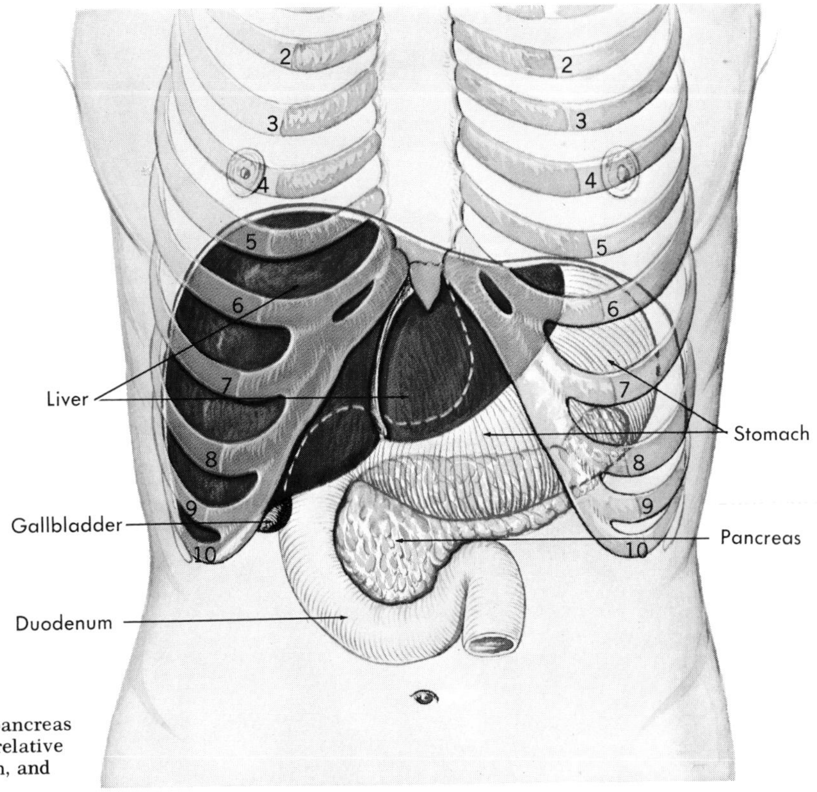

Fig. 12-12 The liver and pancreas
in their normal positions relative
to the rib cage, diaphragm, and
stomach.

fibrous strands that form a supporting
framework (the capsule of Glisson) for
them. The capsule of Glisson is an extension of the heavy connective tissue capsule
that envelops the entire liver. The hepatic
lobules, the anatomical units of the liver,
are tiny hexagonal or pentagonal cylinders
about 2 millimeters high and 1 millimeter
in diameter. A small branch of the hepatic
vein extends through the center of each
lobule. Around this central (intralobular)
vein, in columns radiating outward, are
arranged the hepatic cells. Three separate
sets of tiny tubes—branches of the hepatic
artery, of the portal vein (interlobular
veins), and of the hepatic duct (interlobular bile ducts)—are arranged around each
lobule. From these, irregular branches (sinusoids) of the interlobular veins extend
between the radiating columns of hepatic
cells to join the central vein. Branches of
the interlobular bile ducts also run between each two rows of hepatic cells.

Ducts

The small bile ducts within the liver join
to form two larger ducts which emerge
from the undersurface of the organ as
the right and left hepatic ducts but which
immediately join to form one *hepatic duct.*
The hepatic duct merges with the *cystic
duct* from the gallbladder, forming the
common bile duct (Fig. 12-13), which
opens into the duodenum in a small raised
area, called the *ampulla* or *papilla of
Vater* or duodenal papilla. This papilla
is located 3 to 4 inches below the pyloric
opening from the stomach.

Functions

The liver is one of the most vital organs
of the body. Although its cells are merely
microscopic dots in size, they do so many
things at once as to seem incredible—or
perhaps miraculous would be a better
word. Because of the diversity of its activities, a single liver cell might be likened

403

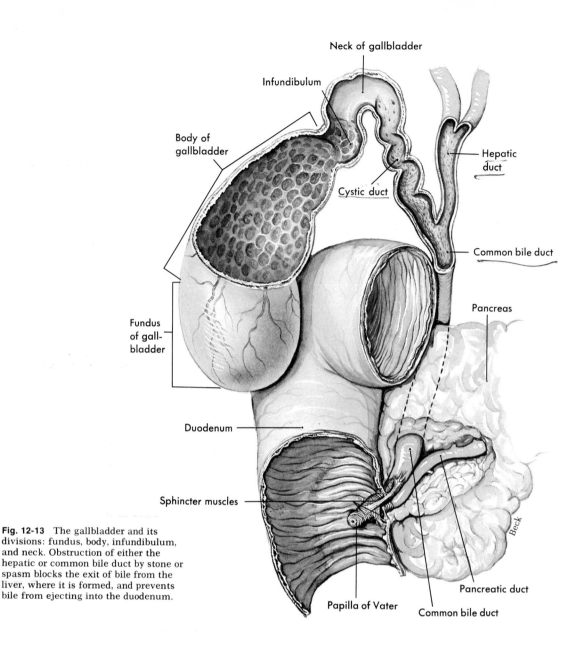

Neck of gallbladder

Infundibulum

Body of
gallbladder

Hepatic
duct

Cystic duct

Common bile duct

Pancreas

Fundus
of gall-
bladder

Duodenum

Sphincter muscles

Fig. 12-13 The gallbladder and its
divisions: fundus, body, infundibulum,
and neck. Obstruction of either the
hepatic or common bile duct by stone or
spasm blocks the exit of bile from the
liver, where it is formed, and prevents
bile from ejecting into the duodenum.

Papilla of Vater

Pancreatic duct

Common bile duct

to a *factory* (it makes many chemical
compounds), a *warehouse* (it stores such
valuables as glycogen, iron, and certain
vitamins), a *waste disposal plant* (it excretes bile pigments, urea, and various de-
toxication products), and a *power plant* (its
catabolism produces considerable heat).

Here, in brief, are the liver's main functions.

1 It secretes about a pint of bile a day.
Bile contains bile salts, useful substances
that facilitate fat digestion and absorption.

However, bile also contains various waste
products.

2 The liver plays an essential role in
the metabolism of all three kinds of foods.

a For its special part in carbohydrate
metabolism, the liver carries on three processes: glycogenesis, glycogenolysis, and
gluconeogenesis (pp. 417 to 418). Briefly,
by these processes, the liver plays an important part in maintaining homeostasis of
blood sugar—helped, as we shall see, by
various other processes.

b Liver cells, more than any others, carry out the first step in both protein and fat catabolism (see pp. 425 and 426).

c Liver cells perform an essential part of protein anabolism by synthesizing various blood proteins such as prothrombin and fibrinogen, albumins, and many globulins. This function alone stamps the liver as one of the highly essential workers for bodily welfare and survival. Consider these three vital functions of the blood proteins synthesized by the liver: (1) prothrombin and fibrinogen, as you already know, play essential parts in blood clotting, (2) all blood proteins contribute to blood osmotic pressure and therefore are important for maintaining water balance, and (3) all blood proteins contribute to blood viscosity and therefore are essential for maintenance of normal blood pressure and circulation.

Gallbladder

Size, shape, and location

The gallbladder is a pear-shaped sac from 3 to 4 inches long and an inch or more wide (Fig. 12-13). It lies on the undersurface of the liver and is attached to this organ by areolar tissue.

Structure

Serous, muscular, and mucous coats compose the wall of the gallbladder. The mucosal lining is arranged in rugae, similar in structure and function to those of the stomach.

Functions

The gallbladder concentrates and stores the bile which enters it by way of the hepatic and cystic ducts. Then later, when digestion is going on in the stomach and intestines, the gallbladder contracts, ejecting the concentrated bile into the duodenum.

Correlations

Inflammation of the lining of the gallbladder is called *cholecystitis*. *Cholecystectomy* is the surgical removal of the gallbladder. *Jaundice,* a yellow discoloration of the skin and mucosa, results whenever obstruction of the hepatic or common bile duct occurs. Bile is thereby denied its normal exit from the body in the feces. Instead, it is absorbed into the blood. As more bile pigments than normal accumulate in blood, the latter takes on a yellow hue. Without the normal amount of bile pigments, the feces become a grayish, so-called clay color.

Pancreas — *exocrine & endocrine*

Size, shape, and location

The pancreas is roughly fish shaped. It lies behind the stomach, with its head and neck in the C-shaped curve of the duodenum, its body extending horizontally across the posterior abdominal wall, and its tail touching the spleen (Fig. 12-12). According to an old anatomical witticism, the "romance of the abdomen" is the pancreas lying "in the arms of the duodenum."

This gland varies in size according to sex and individuals, being larger in men than in women. Usually its length is 6 to 9 inches, its width 1 to 1 1/2 inches, and its thickness 1/2 to 1 inch. It weighs about 3 ounces.

Structure

The pancreas is classified as a compound tubuloacinar gland. The word compound tells us that the gland has a branching duct. The term tubuloacinar, on the other hand, tells us that some of the secreting units of the pancreas resemble tiny tubes and some tiny grapes in shape. Secreting cells constitute the walls of these tubular and acinar units. These are exocrine glands since they release their secretion into the microscopic duct within each unit

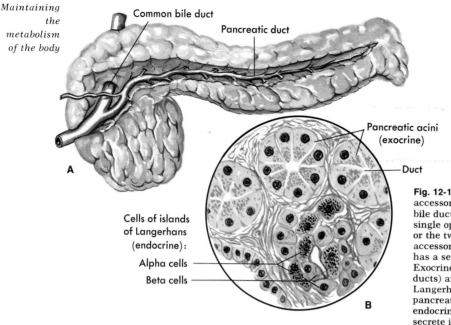

Common bile duct

Pancreatic duct

Pancreatic acini (exocrine)

Duct

A

Cells of islands of Langerhans (endocrine):

Alpha cells

Beta cells

B

Fig. 12-14 A, Pancreas dissected to show main and accessory ducts. The main duct may join the common bile duct, as shown here, to enter the duodenum by a single opening at the papilla of Vater (see Fig. 12-13), or the two ducts may have separate openings. The accessory pancreatic duct is usually present and has a separate opening into the duodenum. **B,** Exocrine glandular cells (around small pancreatic ducts) and endocrine glandular cells of islands of Langerhans (adjacent to blood capillaries). Exocrine pancreatic cells secrete pancreatic juice, alpha endocrine cells secrete glucagon, and beta cells secrete insulin.

(Fig. 12-14). These tiny ducts unite to form larger ducts that eventually join the main pancreatic duct, the *duct of Wirsung,* which extends throughout the length of the gland from its tail to its head. It empties into the duodenum at the same point as the common bile duct—that is, at the ampulla of Vater. (An accessory duct, the *duct of Santorini,* is frequently found extending from the head of the pancreas into the duodenum, about an inch above the duodenal papilla.)

In between the tubuloacinar units of the pancreas, like so many little islands isolated from one another and from the ducts of the pancreas, lie clusters of cells called *islets* or *islands of Langerhans* (Fig. 12-14). Special staining techniques have revealed that two kinds of cells—alpha cells and beta cells—chiefly compose the islands of Langerhans. They are secreting cells, but their secretion passes into blood capillaries rather than into ducts. Thus the pancreas is a dual gland—an exocrine or duct gland because of the tubuloacinar units and an endocrine or

ductless gland because of the islands of Langerhans.

Functions

1 The tubuloacinar units of the pancreas secrete the digestive enzymes found in pancreatic juice. Hence the pancreas plays an important part in digestion (p. 409).

2 Beta cells of the pancreas secrete *insulin,* a hormone that exerts a major control over carbohydrate metabolism (p. 418 and Fig. 12-22).

3 Alpha cells, as indicated by considerable evidence, secrete *glucagon,* another hormone involved in carbohydrate metabolism (p. 419).

■Vermiform appendix
Size, shape, and location

The appendix is a blind tube branching from the lower portion of the cecum (Fig. 12-11). As its name suggests, it resembles a large angle worm in size and shape, although its size varies greatly in different individuals.

Structure

The structure of the appendix is similar to that of the rest of the intestine. Its mucous lining frequently becomes inflamed, a condition well known as *appendicitis*.

Digestion

Definition

Digestion is the sum of all the changes food undergoes in the alimentary canal.

Purpose

The purpose of digestion as suggested in the dictionary definition is "the conversion of food into assimilable matter." Digestion is necessary because foods, as eaten, are too complex in physical and chemical composition to pass through the intestinal mucosa into the blood or for cells to utilize them for energy and tissue building. In other words, digestion is the necessary preliminary to both absorption and metabolism of foods.

Kinds

Since both the physical and chemical composition of ingested food makes its absorption impossible, two kinds of digestive changes are necessary, *mechanical* and *chemical.*

Mechanical digestion. Mechanical digestion consists of all those movements of the alimentary tract that bring about the following:

1 Change the physical state of ingested food from comparatively large, solid pieces into minute dissolved particles, thereby facilitating chemical digestion
2 Propel the food forward along the alimentary tract, finally eliminating the digestive wastes from the body
3 Churn the intestinal contents in such

a way that it becomes well mixed with the digestive juices and that all parts of it come in contact with the surface of the intestinal mucosa, thereby facilitating absorption

A list of definitions of the different processes involved in mechanical digestion, together with the organs that accomplish them, are given in Table 12-4. Swallowing, the movement of food out of the stomach, and defecation are considered in more detail below.

Swallowing consists of three steps: movement of the food through the mouth into the pharynx, through the pharynx into the esophagus, and through the esophagus into the stomach. Only the first act is voluntary—that is, can be controlled by the will. The other two actions are reflexes that occur automatically after food enters the pharynx. Stimulation of the mucosa of the back of the mouth, pharynx, or laryngeal region initiates the second step —reflex contractions of the muscular pharyngeal walls. Consequently, paralysis of the sensory nerves to the mucosa of the back of the mouth, pharynx, or laryngeal region by a drug such as Novocain makes swallowing impossible. Since the pharynx serves both the digestive and respiratory systems, it has connections with the parts of the digestive tract above and below it (mouth and esophagus) and with the parts of the respiratory tract above and below (nasal cavity and larynx). When food enters the pharynx and its walls contract, theoretically its contents could be squeezed in any of these four directions— that is, back up into the nose or mouth or downward into the larynx or esophagus. That it takes only one of these routes (into the esophagus) is due to the fact that the openings into the other organs become blocked off during swallowing. Elevation of the tongue (so that it presses against the roof of the mouth) bars entry into the mouth, elevation of the soft palate obstructs the passageway into the nose, and elevation of the larynx closes over the opening into this organ. As the larynx moves upward, the base of the tongue pushes the epiglottis downward over the laryngeal opening like a lid. If solids or liquids enter the larynx and lower respiratory tract, we say we have "swallowed down our Sunday throats." Infection or even fatal asphyxiation can result. But an additional mechanism normally prevents such an occurrence. Stimulation of the mucosa of the back of the mouth, pharynx, and laryngeal region produces not only the second phase of the swallowing act, but also momentary inhibition of respirations. Obviously, prevention of inspiration during swallowing greatly lessens the danger of foods entering the respiratory tract.

Stimulation of the esophageal mucosa by material entering it from the pharynx initiates the reflex of esophageal peristalsis. Liquids and well-chewed foods are precipitated down the esophagus to the cardiac sphincter by the force of pharyn-

Table 12-4. Processes in mechanics of digestion

Organ	Mechanical process	Nature of process
Mouth (teeth and tongue)	Mastication	Chewing movements—reduce size of food particles and mix them with saliva
	Deglutition	Swallowing—movement of food from mouth to stomach
Pharynx	Deglutition	
Esophagus	Deglutition Peristalsis	Wormlike movements that squeeze food downward in tract; constricted ring forms first in one section, the next, etc., causing waves of contraction to spread throughout canal
Stomach	Churning	Forward and backward movement of gastric contents; peristalsis propels it forward; closed pyloric sphincter deflects it backward
	Peristalsis	Moves material through stomach and at intervals into duodenum
Small intestine	Churning (rhythmic segmentation) Peristalsis	Forward and backward movement within segment of intestine; purpose, to mix food and digestive juices thoroughly and to bring all digested food in contact with intestinal mucosa to facilitate absorption; purpose of peristalsis, on the other hand, to propel intestinal contents along digestive tract
Large intestine Colon	Haustral churning Peristalsis	Churning movements within haustral sacs
Descending colon	Mass peristalsis	Entire contents moved into sigmoid colon and rectum; usually occurs after a meal
Rectum	Defecation	Emptying of rectum, so-called bowel movement

geal contraction and by gravity. They are then moved through the sphincter into the stomach by means of esophageal peristalsis. The latter is responsible for propelling large solid particles throughout the length of the esophagus.

Emptying of the stomach after a meal requires a considerable period of time (about one to four hours for the average meal). Many theories have been proposed about the mechanism regulating gastric emptying. It is now known that as small amounts of gastric contents become liquefied, they are ejected bit by bit every twenty seconds or so into the duodenum until a certain amount has accumulated there. From then on, the emptying process slows down due to operation of a mechanism known as the *enterogastric reflex*. Both nerve impulses and a hormone initiate this

Table 12-5. Chemical digestion

Digestive juices and enzymes	Food enzyme digests (or hydrolyzes)	Resulting product*
Saliva 　Amylase (ptyalin)	Starch (polysaccharide or complex sugar)	Maltose (a disaccharide or double sugar)
Gastric juice 　Protease (pepsin) plus hydrochloric acid	Proteins, including casein	Proteoses and peptones (partially digested proteins)
Lipase (of little importance)	Emulsified fats (butter, cream, etc.)	*Fatty acids and glycerol*
Bile (contains no enzymes)	Large fat droplets (unemulsified fats)	Small fat droplets or emulsified fats
Pancreatic juice 　Protease (trypsin)†	Proteins (either intact or partially digested)	Proteoses, peptides, and *amino acids*
Lipase (steapsin)	Bile-emulsified fats	*Fatty acids and glycerol*
Amylase (amylopsin)	Starch	Maltose
Intestinal juice (succus entericus) 　Peptidases	Peptides	*Amino acids*
Sucrase	Sucrose (cane sugar)	*Glucose and fructose*‡ (simple sugars or monosaccharides)
Lactase	Lactose (milk sugar)	*Glucose and galactose* (simple sugars)
Maltase	Maltose (malt sugar)	*Glucose* (grape sugar)

*Substances in italics are end products of digestion or, in other words, completely digested foods ready for absorption.
†Secreted in inactive form (trypsinogen); activated by enterokinase, an enzyme in the intestinal juice.
‡Glucose is also called dextrose; fructose is called levulose.

reflex. Fats and sugars present in the small intestine stimulate the intestinal mucosa to release a hormone (called enterogastrone) into the bloodstream. When it reaches the stomach wall via the circulation, it has an inhibitory effect on gastric muscle, decreasing its peristalsis (the enterogastric reflex) and thereby slowing down stomach emptying. Proteins and acid also initiate the enterogastric reflex but not by means of a hormone. They stimulate vagal nerve receptors in the intestinal mucosa.

Defecation is a reflex brought about by stimulation of receptors in the rectal mucosa. Recent distention of the rectum constitutes the usual stimulus. Normally the rectum is empty until mass peristalsis moves fecal matter out of the colon into the rectum. This distends the rectum and produces the desire to defecate. Also, it stimulates colonic peristalsis and initiates reflex relaxation of the internal sphincter of the anus. Voluntary straining efforts and relaxation of the external anal sphincter may then follow as a result of the desire to defecate. And together these several responses bring about defecation. Note that this is a reflex partly under voluntary control. If one voluntarily inhibits it, the rectal receptors soon become depressed and the urge to defecate usually does nor recur until about twenty-four hours later, when mass peristalsis again takes place. During the interim, water is absorbed from the fecal mass, producing a hardened or constipated stool.

Chemical digestion. Chemical digestion consists of all the changes in chemical composition that foods undergo in their travel through the alimentary canal. These changes result from the hydrolysis of foods. (*Hydrolysis* is a chemical process in which a compound unites with water and then splits into simpler compounds.)

Numerous enzymes* present in the various digestive juices catalyze the hydrolysis of foods (Table 12-5).

Although we eat six kinds of chemical substances (carbohydrates, proteins, fats, vitamins, mineral salts, and water), only the first three named have to be chemically digested in order to be absorbed.

Different enzymes require different hydrogen ion concentrations in their environment for optimal functioning. Ptyalin, the main enzyme in saliva, functions best in the neutral to slightly acid pH characteristic of saliva. It is gradually inactivated by the marked acidity of gastric juice. In contrast, pepsin, an enzyme in gastric juice, is inactive unless sufficient hydrochloric acid is present. Therefore, in dis-

---

*The term enzyme means literally "in yeast." It was derived from the fact that these substances were first discovered in yeast cells. Enzymes are usually defined simply as "organic catalysts"—that is, they are organic compounds, and they accelerate chemical reactions without appearing in the final products of the reaction. Enzymes are vital substances. Without them, the chemical reactions necessary for life could not take place. So important are they that someone has even defined life as the "orderly functioning of hundreds of enzymes."

Chemical structure: Enzymes are proteins. Frequently their molecules also contain a nonprotein part called the *prosthetic group* of the enzyme molecule (if this group readily detaches from the rest of the molecule, it is spoken of as the *coenzyme*). Some prosthetic groups contain inorganic ions (Ca^{++}, Mg^{++}, Mn^{++}, etc.). Many of them contain vitamins. In fact, every vitamin of known function constitutes part of a prosthetic group of some enzyme. Nicotinic acid, thiamine, riboflavin, and other B complex vitamins, for example, function in this way.

Classification and naming: Two of the systems used for naming enzymes are as follows: suffix *ase* is used either with the root name of the substance whose chemical reaction is catalyzed (the substrate chemical, that is) or with the word that describes the kind of chemical reaction catalyzed. Thus, according to the first method, sucrase is an enzyme that catalyzes a chemical reaction in which sucrose takes part. According to the second method, sucrase might also be called hydrolase because it hydrolyzes sucrose. Enzymes investigated before these methods of nomen-

eases characterized by gastric hypoacidity (pernicious anemia, for example), hydrochloric acid is given orally before meals.

Carbohydrate digestion (Fig. 12-15). Carbohydrates are saccharide compounds. This means that their molecules contain one or more saccharide groups ($C_6H_{10}O_5$). Polysaccharides, notably starches, contain many of these groups, disaccharides (sucrose, lactose, and maltose) contain two of them, and monosaccharides (glucose, fructose, and galactose) contain only one. Polysaccharides are hydrolyzed to disaccharides by enzymes known as *amylases* found in saliva and pancreatic juice (salivary amylase is also called ptyalin and pancreatic amylase is also known as amylopsin). Disaccharides are hydrolyzed to monosaccharides by the intestinal juice enzymes sucrase, lactase, and maltase.

Protein digestion (Fig. 12-16). Protein compounds have very large molecules made up of amino acids. Enzymes called *proteases* catalyze the hydrolysis of proteins into intermediate compounds—for example, proteoses and peptides—and finally into amino acids. The main proteases are pepsin in gastric juice, trypsin in pancreatic juice, and peptidases in intestinal juice.

Fat digestion (Fig. 12-17). Because fats are insoluble in water, they must be emulsified—that is, dispersed as very small droplets—before they can be digested. Bile emulsifies fats in the small intestine. This facilitates fat digestion by providing a

---

clature were adopted still are called by older names, such as ptyalin, pepsin, trypsin, etc.

Classified according to the kind of chemical reactions catalyzed, enzymes fall into several groups:

1 *Oxidation-reduction enzymes.* These are known as oxidases, hydrogenases, and dehydrogenases. Energy release for muscular contraction and all physiological work depends upon these enzymes.
2 *Hydrolyzing enzymes* or hydrolases. Digestive enzymes belong to this group. These are generally named after the substrate acted upon: for example, lipase, sucrase, maltase, etc.
3 *Phosphorylating enzymes.* These add or remove phosphate groups and are known as phosphorylases or phosphatases.
4 *Enzymes that add or remove carbon dioxide.* These are known as carboxylases or decarboxylases.
5 *Enzymes that rearrange atoms within a molecule.* These are known as mutases or isomerases.
6 *Hydrases.* These add water to a molecule without splitting it, as hydrolases do.

Enzymes are also classified as intracellular or extracellular, depending upon whether they act within cells or outside of them in the surrounding medium. Most enzymes act intracellularly in the body, an important exception being the digestive enzymes.

Properties: In general, the properties of enzymes are the same as those of proteins, since enzymes are proteins. For example, they form colloidal solutes in water and are precipitated or coagulated by various agents, such as high temperatures and salts of heavy metals. Hence, these agents inactivate enzymes. Other important enzyme properties are as follows.

1 Most enzymes are *specific in their action*—that is, act only on a specific substrate. This is attributed to a "key-in-a-lock" kind of action, the configuration of the enzyme molecule fitting the configuration of some part of the substrate molecule.
2 Enzymes *function optimally at a specific pH* and become inactive if this deviates beyond narrow limits.
3 A *variety of physical and chemical agents inactivate or inhibit enzyme action*—for example, x-rays and radium rays (this presumably accounts for some of the ill effects of excessive radiation), certain antibiotic drugs, unfavorable pH, etc.
4 *Most enzymes catalyze a chemical reaction in both directions,* the direction and rate of the reaction being governed by the law of mass action. An accumulation of a product slows the reaction and tends to reverse it. A practical application of this fact is the slowing of digestion when absorption is interfered with and the products of digestion accumulate.
5 *Enzymes are continually being destroyed in the body* and therefore have to be continually synthesized, even though they are not used up in the reactions they catalyze.
6 *Many enzymes are synthesized in an inactive form* that must be activated by some other substance before the enzyme can function. The kinases are such enzyme activators. For example, thrombokinase (thromboplastin) converts inactive prothrombin into active thrombin and enterokinase changes inactive trypsinogen into active trypsin, etc.

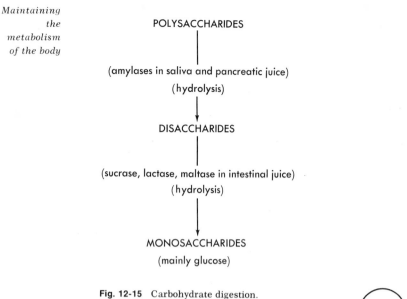

Fig. 12-15 Carbohydrate digestion.

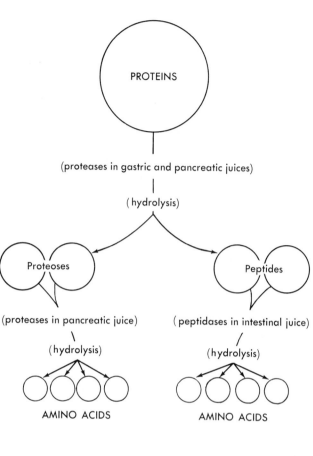

Fig. 12-16 Protein digestion. Gastric juice protease (pepsin) and pancreatic juice protease (trypsin) hydrolyze proteins to proteoses and peptides. Protein digestion is then completed by pancreatic proteases, which hydrolyze proteoses to amino acids, and intestinal juice peptidases, which hydrolyze peptides to amino acids.

greater contact area between fat molecules and pancreatic lipase, the main fat-digesting enzyme (pancreatic lipase was formerly known as steapsin). For a summary of the actions of each digestive juice, see Table 12-5.

Residues of digestion. Certain components of food resist digestion and are eliminated from the intestines in the feces. These *residues of digestion* are cellulose from carbohydrates, undigested connective tissue and toxins from meat proteins, and undigested fats. In addition to these wastes, feces consists of bacteria, pigments, water, and mucus.

Control of digestive gland secretion

Digestive glands secrete when food is present in the alimentary tract or when it is seen, smelled, or imagined. Complicated

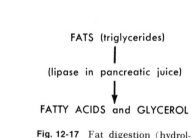

Fig. 12-17 Fat digestion (hydrolysis) by lipase, facilitated first by emulsion of fats by bile.

reflex and chemical (hormonal) mechanisms control the flow of digestive juices in such a way that they appear in proper amounts when and for as long as needed.

Saliva. As far as is known, only reflex mechanisms control the secretion of saliva. Chemical, mechanical, olfactory, and visual stimuli initiate afferent impulses to centers in the brainstem that send out efferent impulses to salivary glands, stimulating them. Chemical and mechanical stimuli come from the presence of food in the mouth, and olfactory and visual stimuli come from the smell and sight of food.

Gastric secretion. Stimulation of gastric juice secretion occurs in three phases controlled by reflex and chemical mechanisms. Because stimuli that activate these mechanisms arise in the head, stomach, and intestines, the three phases are known as the cephalic, gastric, and intestinal phases, respectively.

The *cephalic phase* is also spoken of as the reflex phase and as the psychic phase because a reflex mechanism controls gastric juice secretion at this time and psychic factors activate the mechanism. For example, the sight or smell or taste of food that is pleasing to an individual stimulates various head receptors and thereby initiates reflex stimulation of the gastric glands. Parasympathetic fibers in branches of the vagus nerve conduct the stimulating efferent impulses to the glands.

During the *gastric phase* of gastric juice secretion, the following chemical control

mechanism dominates. Substances (such as meat extractives and products of protein digestion) in foods that have reached the pyloric portion of the stomach stimulate its mucosa to release a hormone called *gastrin* into the blood in stomach capillaries. When it circulates to the gastric glands, it greatly accelerates their secretion of gastric juice which has a high pepsin and hydrochloric acid content (Table 12-6). Hence, this seems to be a device for ensuring that when food is in the stomach there will be enough enzymes there to digest it.

The *intestinal phase* of gastric juice secretion is less clearly understood than the other two. A chemical control mechanism, however, is believed to operate. And it is known, too, that the hormone *enterogastrone* (released by intestinal mucosa when fat is in the intestine) causes a lessening of both gastric secretion and motility.

Pancreatic secretion. Chemical control of pancreatic secretion has been established. Also, reflex control is postulated. Chemical control ensures continued pancreatic secretion while food is in the duo-

Table 12-6. Actions of some digestive hormones summarized

Hormone	Source	Action
Gastrin	Formed by gastric mucosa in presence of partially digested proteins	Stimulates secretion of gastric juice rich in pepsin and HCl
Secretin	Formed by action of hydrochloric acid on prosecretin (chemical normally present in intestinal mucosa)	Stimulates secretion of pancreatic juice low in enzymes Stimulates secretion of bile by liver May stimulate secretion of intestinal juice
Pancreozymin	Formed by intestinal mucosa in presence of hydrochloric acid and partially digested proteins	Stimulates pancreatic cells to produce enzymes
Cholecystokinin	Formed by intestinal mucosa in presence of fats	Stimulates ejection of bile from gallbladder

denum. The presence in the small intestine of hydrochloric acid, protein, and fat digestion products causes the intestinal mucosa to release a hormone, _secretin,_ into the blood. When secretin circulates to the pancreas, it stimulates pancreatic cells to secrete at a faster rate. But the juice they produce has a low enzyme content. Another hormone, pancreozymin, stimulates them to produce enzymes. The presence of hydrochloric acid and partially digested proteins in the intestine serves as the stimulus for pancreozymin release by the intestinal mucosa.

Secretin has an interesting claim to fame. Not only was it the first hormone to be discovered, but its discovery gave rise to the broad concept of hormonal control of body activities.

Secretion of bile. Chemical mechanisms dominate the control of bile. Secretin, the same hormone that stimulates pancreatic activity, stimulates the liver to secrete bile. The ejection of bile, however, from the gallbladder into the duodenum is largely controlled by another hormone, _cholecysto-kinin_, which is formed by the intestinal mucosa when fats are present in the duodenum.

Intestinal secretion. Knowledge concerning the regulation of the secretion of _intestinal juice_ is still somewhat obscure. It is thought that the intestinal mucosa, stimulated by hydrochloric acid and food products, releases into the blood a hormone, _enterocrinin_, which brings about increased intestinal juice secretion. Presumably, neural mechanisms also help control the secretion of this digestive juice.

Absorption

Definition

Absorption is the passage of substances (notably digested foods, water, salts, and vitamins) through the intestinal mucosa into the blood or lymph.

How accomplished

Absorption is not entirely a passive process explainable on the basis of the physical laws of diffusion, filtration, and osmosis alone. Investigation has shown that these phenomena play a part in absorption but that they are aided by active transport mechanisms. These are not clearly understood processes carried on by epithelial cells of the intestinal mucosa. They move substances in the opposite direction from that expected according to the laws of osmosis and diffusion.

The mechanism postulated for glucose absorption (Fig. 12-18) consists of these steps:

1 Glucose moves into cells of the intestinal mucosa presumably by diffusion and active transport.

2 Inside the mucosal cells, phosphorylation of the glucose molecule occurs immediately, presumably by the following reaction catalyzed by the enzyme glucokinase:

$$\text{Glucose} + \text{ATP} \xrightarrow{\text{(Glucokinase)}} \text{Glucose-6-phosphate} + \text{ADP}$$

3 Glucose-6-phosphate diffuses across the mucosal cell, and near its surface adjacent to a blood capillary, the preceding equation is reversed, but by another enzyme, phosphatase. Glucose then moves out of the cell into the blood.

That an active transport mechanism operates for the absorption of amino acids is a conclusion supported by considerable evidence. But a definite description of the mechanism is still lacking. And the mechanism for fat absorption is also not clear, although it, too, is thought to involve some kind of active transport. It is known, how-

Table 12-7. Food absorption

Form absorbed	Structures into which absorbed	Circulation
Protein – as amino acids Perhaps minute quantities of some whole proteins absorbed – e.g., in allergic reactions	Blood in intestinal capillaries	Portal vein, liver, hepatic vein, inferior vena cava to heart, etc.
Carbohydrates – as simple sugars	Same as amino acids	Same as amino acids

Fats
  Glycerol
 	
  Fatty acids combine with bile salts to form water-soluble substance
 	
  Some finely emulsified undigested fats absorbed
	Lymph in intestinal lacteals
	
	Lymph in intestinal lacteals
	
	Small fraction enters intestinal blood capillaries
		During absorption, i.e., while in epithelial cells of intestinal mucosa, glycerol and fatty acids recombine to form microscopic particles of fats (chylomicrons); lymphatics carry them by way of thoracic duct to left subclavian vein, superior vena cava, heart, etc.
		
		Some fats transported by blood in form of phospholipids or cholesterol esters

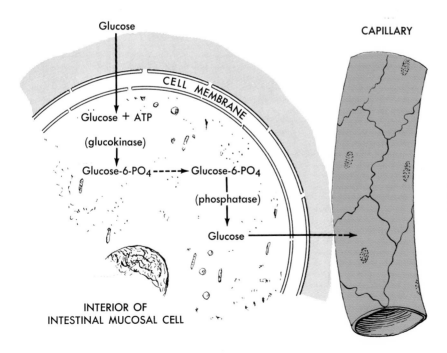

Fig. 12-18 Glucose enters mucosal cells of intestinal villi (presumably by diffusion and active transport) from the lumen of the intestine and exits from the opposite surfaces of these cells into blood capillaries.

ever, that some fatty acids and glycerol enter intestinal mucosal cells and there recombine to form neutral fats which then move out of the cells into lymphatic capillaries (lacteals) in the intestinal villi. In addition, some fatty acids are absorbed as such into blood capillaries. Also, some neutral fats are absorbed without first being hydrolyzed to fatty acids and glycerol.

Active transport mechanisms seem to be available also for water and salt absorption. And apparently some sort of mechanism operates to prevent absorption of certain substances. Magnesium sulfate (Epsom salts), for example, is not absorbed from the intestine even though its molecules are smaller than glucose molecules which can diffuse through the intestinal mucosa. In fact, it is this nonabsorbability of magnesium sulfate that makes it an effective cathartic. (Since magnesium sulfates do not diffuse freely through the intestinal mucosa, do you think their presence in intestinal fluid would create an osmotic pressure gradient between the intestinal fluid and blood? If so, what effect would this have on water movement between these two fluids? Reread pp. 23 to 27 if you need help in answering these questions.)

Note that after absorption food does not pass directly into the general circulation. Instead, it first travels by way of the portal system to the liver. During intestinal absorption, blood entering the liver via the portal vein contains greater concentrations of glucose, amino acids, and fats than does blood leaving the liver via the hepatic vein for the systemic circulation. Clearly, the excess of these food substances, over and above the normal blood levels, has remained behind in the liver.

What the liver does with them is part of the story of metabolism, our next topic for discussion.

Metabolism
■Meaning

Foods are first digested, then absorbed, and, finally, metabolized. Metabolism, as the word is usually defined, means the chemical changes absorbed foods undergo within the body cells. More simply, metabolism is the body's utilization of foods. Digestion and absorption are merely preliminary steps or preparations for metabolism.

■Ways in which foods metabolized

The body utilizes — that is, metabolizes — carbohydrates, proteins, fats, vitamins, inorganic salts (or minerals), and water in two general ways. It catabolizes foods to make their stored energy available for cellular work and anabolizes them to build protoplasm, enzymes, hormones, and other complex compounds.

Research has unlocked some of the secrets of anabolism. Brilliant work by many investigators has disclosed some of the actual events that take place inside cells as they build the small, simple food molecules up into the large, complex molecules of enzymes and hosts of other compounds. A full description of this does not seem appropriate in a book of this kind. But if you enjoy reading exciting though not easy material, by all means read the articles cited in the footnote.* And for a brief description of protein anabolism, see p. 426. Anabolism is one of the most important kinds of work that cells do. Such vital functions as cell growth and repair and reproduction, for

*Rich, A.: Polyribosomes, Sci. Amer. **209**:44-53 (Dec.), 1963; Nossal, G. J. V.: How cells make antibodies, Sci. Amer. **211**:106-115 (Dec.), 1964; Hurwitz, J., and Furth, J. J.: Messenger RNA, Sci. Amer. **206**:41-49 (Feb.), 1962.

example, are achieved through the process of anabolism.

The relationship between anabolism and catabolism is worth noting. Anabolism depends upon catabolism. Catabolism must occur in order that anabolism may occur. Why? Because catabolism provides the energy for the work of anabolism. Energy is stored in ATP molecules during catabolism and is released from them during anabolism.

■Carbohydrate metabolism

A physiological principle worth noting is that carbohydrates are the body's "preferred fuel." Normally, body cells "burn"—that is, catabolize—carbohydrates in order to carry on their many kinds of work and survive. They use fats and proteins as secondary or supplementary fuels if their glucose supply becomes inadequate for one reason or another. Physiologists sometimes express this idea by saying that cells "shift" from carbohydrate to fat or protein utilization when the amount of glucose entering them proves too meager for their energy needs. This most commonly occurs when a person goes without food for many hours or when he has untreated diabetes mellitus. How the shift from carbohydrate to fat or protein utilization is brought about is discussed on pp. 419 and 426. Now we shall consider the processes involved in carbohydrate metabolism and then turn to the mechanisms that control them.

Glucose transport through cell membranes and phosphorylation

Carbohydrate metabolism starts with the active transport of glucose through cell membranes. Immediately upon reaching the interior of the cell, glucose reacts with ATP to produce glucose-6-phosphate. The reaction is referred to as glucose phosphorylation and is catalyzed by the enzyme glucokinase (or glucose hexokinase). Next,

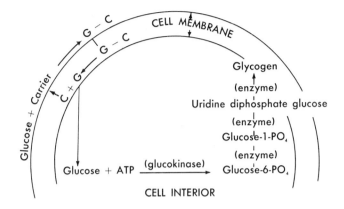

Fig. 12-19 Glycogenesis.

glucose-6-phosphate is either anabolized or catabolized.

Glycogenesis and glycogenolysis

Glycogenesis is a process of glucose anabolism. Simply defined, glycogenesis is the formation of glycogen from glucose. Note in Fig. 12-19 that glycogenesis consists of a series of chemical reactions, each one catalyzed by a different enzyme. Glycogen molecules do not remain permanently in the cell but from time to time are broken down or hydrolyzed back to glucose-6-phosphate or, in certain cells, all the way back to glucose. This reversal of the process of glycogenesis, this hydrolysis of glycogen, is called *glycogenolysis*. Now look at Fig. 12-20. Phosphorylase, the enzyme that catalyzes the first reaction of glycogenolysis, is presumably present in all cells. But phosphatase (glucose phosphatase), the enzyme that catalyzes the final step of glycogenolysis—the changing of glucose-6-phosphate back to glucose—is lacking in most cells. Only liver cells, intestinal mucosal cells, and kidney tubule cells contain glucose phosphatase. Therefore, the term glycogenolysis means different things in different cells. *Muscle cell glycogenolysis*, for example, means the

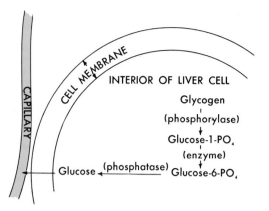

Fig. 12-20 Glycogenolysis in a liver cell.

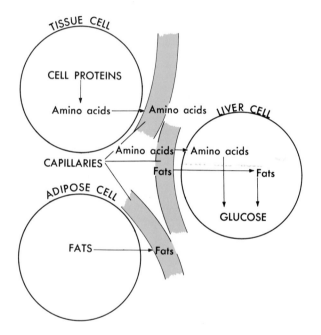

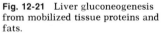

Fig. 12-21 Liver gluconeogenesis from mobilized tissue proteins and fats.

conversion of glycogen to glucose-6-phosphate, but *liver glycogenolysis* means the changing of glycogen to glucose and the addition of this glucose to the bloodstream. Carbohydrate catabolism was described on pp. 29 to 34. Reread these pages carefully if you are not sure of the answers to the following questions:

1 What two processes make up the larger process of glucose catabolism?

2 What overall chemical change occurs in glycolysis? In the citric acid cycle?

3 What energy change occurs in glycolysis? In the citric acid cycle?

4 Where do glycolysis and the citric acid cycle occur?

Gluconeogenesis

Gluconeogenesis means literally the formation of "new" glucose—"new" in the sense that it is made from proteins or fats, not from carbohydrates. The process, which occurs in the liver, consists of many complex chemical reactions. The new glucose produced from fats or proteins by gluconeogenesis diffuses out of liver cells into the blood (Fig. 12-21). Gluconeogenesis, therefore, can add glucose to the blood when needed, as can the process of liver glycogenolysis. Obviously, then, the liver is a most important organ for maintaining blood glucose homeostasis.

Control of glucose metabolism

The complex mechanism that normally maintains homeostasis of blood glucose concentration consists of hormonal and neural devices. At least five endocrine glands—islands of Langerhans, anterior pituitary gland, adrenal cortex, adrenal medulla, and thyroid gland—and at least eight hormones secreted by those glands function as key parts of the glucose homeostatic mechanism.

Beta cells of the islands of Langerhans in the pancreas secrete the most famous sugar-regulating hormone of them all, *insulin*. Although no one yet knows exactly how insulin acts, several of its effects are well known. For instance, insulin is known to act in some way to accelerate glucose transport through cell membranes and to increase the activity of the enzyme glucokinase. As shown in Fig. 12-19, glucokinase catalyzes glucose phosphorylation, the reaction which must occur before either glycogenesis or glucose catabolism can take place. By applying these facts, you can deduce for yourself some of the prominent metabolic defects resulting from insulin deficiency such as occurs in diabetes mellitus. Slow glycogenesis with resulting low glycogen storage, decreased glucose catabolism, and increased blood glucose all result from the slower glucose transport into cells and the decreased glucokinase activity produced by insulin deficiency.

The islands of Langerhans secrete two sugar-regulating hormones—insulin from the beta cells and glucagon from the alpha cells. Whereas insulin tends to decrease blood glucose, glucagon tends to increase it. *Glucagon* increases the activity of the enzyme phosphorylase (Fig. 12-20), and this necessarily accelerates liver glycogenolysis and releases more of its product, glucose, into the blood.

The anterior pituitary gland (adenohypophysis) also secretes three hormones that help regulate carbohydrate metabolism particularly in times of stress. Growth hormone, adrenocorticotropic hormone (ACTH), and thyroid-stimulating hormone are their names.

Growth hormone tends to bring about a shift from carbohydrate utilization to fat utilization. Less glucose, therefore, leaves the blood to enter cells or, stated the other way around, more glucose remains in the blood. In other words, blood glucose tends to increase. Growth hormone presumably brings about this shift to fat utilization by influencing several processes. It appears to act in some way to decrease the deposition of fats in adipose tissue, to increase the mobilization of fats from adipose tissue, and to decrease carbohydrate utilization by all tissues.

ACTH also tends to increase blood glucose concentration, but it does it in a different and an even more indirect way than does growth hormone. ACTH stimulates the adrenal cortex to increase its secretion of glucocorticoids. Glucocorticoids accelerate the mobilization of proteins—that is, the breakdown or hydrolysis of tissue proteins to amino acids. More amino acids enter the circulation and are carried to the liver. Liver cells step up their production of "new" glucose from the mobilized amino acids—in technical language, gluconeogenesis accelerates. More glucose streams out of liver cells into the blood and adds to the blood glucose level. In short, growth hormone and glucocorticoids are hyperglycemic hormones. They both tend to increase blood glucose concentration, in other words. And ACTH also is hyperglycemic, but only indirectly through its effects on glucocorticoid secretion.

The anterior pituitary gland's thyroid-stimulating hormone (thyrotropin) stimulates the thyroid gland to increase its secretion of thyroid hormone. Thyroid hormone accelerates catabolism—usually glucose catabolism since glucose is the body's "preferred fuel."

Epinephrine is a hormone secreted in large amounts by the adrenal medulla in times of emotional or physical stress. Like the hormone glucagon secreted by the alpha islet cells, epinephrine increases phosphorylase activity (Fig. 12-20). This makes liver glycogenolysis go on at a faster rate and causes more of its product, glucose, to enter the blood. Epinephrine accelerates both liver and muscle glycogenolysis, whereas glucagon accelerates only liver glycogenolysis.

In summary, glucose metabolism is regulated mainly by the following hormones:

1 *Insulin,* a hypoglycemic hormone—that is, it tends to decrease blood glucose
2 Glucagon, epinephrine, growth hormone, ACTH, and glucocorticoids

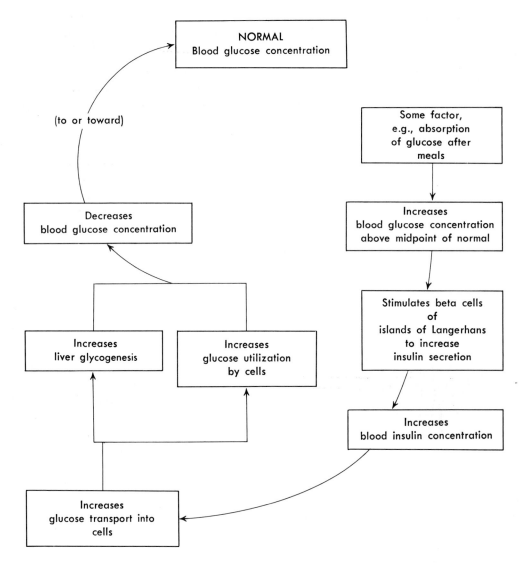

Fig. 12-22 Homeostatic mechanism which, under usual conditions, prevents blood glucose concentration from increasing above the upper limit of normal. This insulin mechanism and the glucagon mechanism shown in Fig. 12-24 operate together to maintain homeostasis of blood glucose under usual circumstances in the normal body.

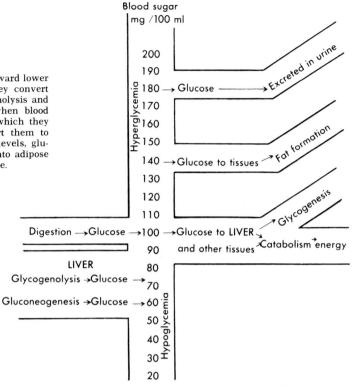

Fig. 12-23 When blood glucose starts to decrease toward lower normal, liver cells increase the rate at which they convert glycogen, amino acids, and fats to glucose (glycogenolysis and gluconeogenesis) and release it into blood. But when blood glucose increases, liver cells increase the rate at which they remove glucose molecules from blood and convert them to glycogen for storage (glycogenesis). At still higher levels, glucose leaves blood for tissue cells to be anabolized into adipose tissue and at still higher levels is excreted in the urine.

are hyperglycemic hormones—that is, they tend to increase blood glucose

3 Thyroid-stimulating hormone and thyroid hormone—ordinarily tend to decrease blood glucose

• • •

Some important principles about normal carbohydrate metabolism may be summarized as follows.

1 *Principle of "preferred energy fuel."* All body cells catabolize glucose for their needed energy supply when the amount of glucose and the amount of insulin in the blood are adequate. Glucose is the "preferred energy fuel" of human cells—preferred to fat and protein, that is. Therefore, cells get their energy first from glucose—for as long as enough of it continues to enter them—then next from fats, and last from proteins.

2 *Principle of glycogenesis.* The process of glycogenesis is part of a homeostatic mechanism that operates when blood glucose concentration increases above the

midpoint of its normal range. When blood glucose concentration increases, glycogenesis also increases. This soon decreases blood glucose, ordinarily enough to maintain its normal level. Example: soon after meals, while glucose is being absorbed rapidly, a great many glucose molecules leave the blood for storage as glycogen—mainly in liver cells but also in muscle and various other cells (Figs. 12-22 and 12-23).

3 *Principle of glycogenolysis.* The process of liver glycogenolysis is part of a homeostatic mechanism that operates when blood glucose concentration decreases below the midpoint of its normal range. When blood glucose concentration decreases, liver glycogenolysis increases (Figs. 12-23 and 12-24). Example: a few hours after meals, when blood glucose decreases below the midpoint of normal, liver glycogenolysis accelerates. This, of course, adds glucose molecules to blood and tends to raise glucose concentration back up to the midpoint of normal. However, glycogenolysis alone can probably maintain homeostasis

421

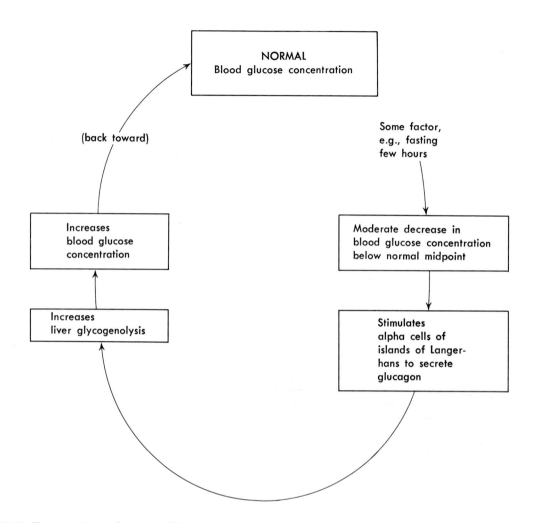

Fig. 12-24 Homeostatic mechanism which, under usual conditions, is chiefly responsible for preventing blood glucose from falling below the lower limit of normal. This glucagon mechanism and the insulin mechanism shown in Fig. 12-22 work together to maintain homeostasis of blood glucose in the normal body under usual circumstances. See Fig. 12-25 for mechanism that controls carbohydrate metabolism under stress conditions.

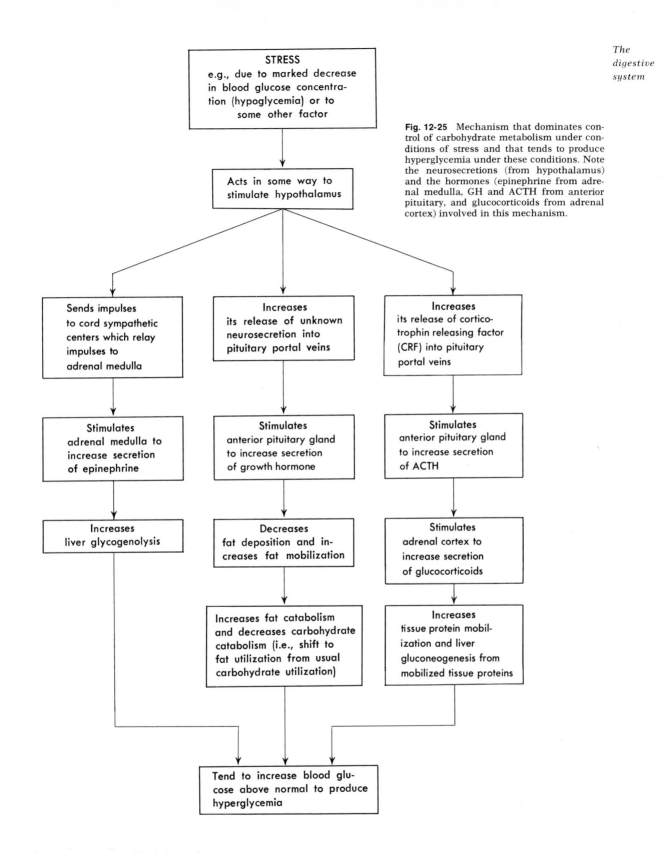

Fig. 12-25 Mechanism that dominates control of carbohydrate metabolism under conditions of stress and that tends to produce hyperglycemia under these conditions. Note the neurosecretions (from hypothalamus) and the hormones (epinephrine from adrenal medulla, GH and ACTH from anterior pituitary, and glucocorticoids from adrenal cortex) involved in this mechanism.

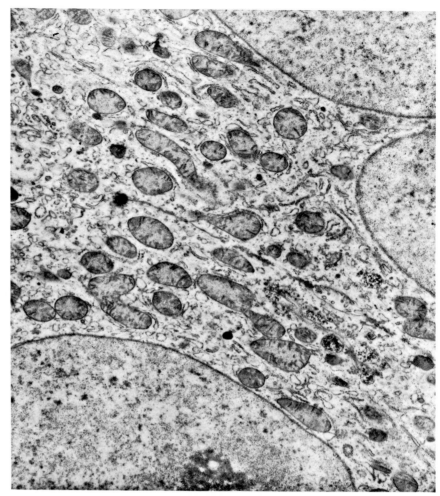

Fig. 12-26 Electron photomicrograph of mouse liver. Numerous mitochondria are visible in the midportion of the photograph. (×25,000; Courtesy Department of Pathology, Case Western Reserve University, Cleveland, Ohio.)

of blood glucose concentration only a few hours since the body can store only modest amounts of glycogen.

4 *Principle of gluconeogenesis.* Liver gluconeogenesis is another part of the homeostatic mechanism that operates when blood glucose decreases below the midnormal point (Figs. 12-21 and 12-23). When this happens or when the amount of glucose entering cells is inadequate (for example, in diabetes mellitus), liver gluconeogenesis accelerates and tends to raise blood glucose concentration back to or above normal.

5 *Principle of glucose storage as fat.* If blood glucose still remains higher than normal after glycogenesis and catabolism and if the blood insulin content is adequate,

the excess glucose is converted to fat (mainly by liver cells) and is stored as such in fat depots—in adipose tissue, in other words. Unfortunately from the standpoint of health and beauty, the amount of glucose the body can store as fat is virtually unlimited.

6 *Principle of hyperglycemia in stress.* Under conditions of stress, blood glucose concentration tends to increase above normal (Fig. 12-25).

■Fat metabolism

Body cells both catabolize and anabolize fats. Fats constitute a more concentrated energy food than carbohydrates. Catabolism of 1 gram of fat yields 9 kilocalories of heat; catabolism of 1 gram of carbohy-

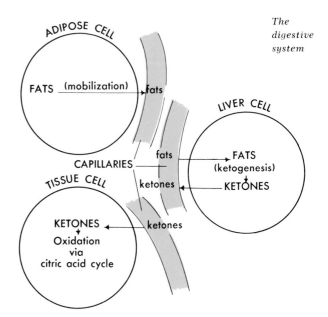

Fig. 12-28 Fat mobilization from adipose cell followed by catabolism (ketogenesis and citric acid cycle).

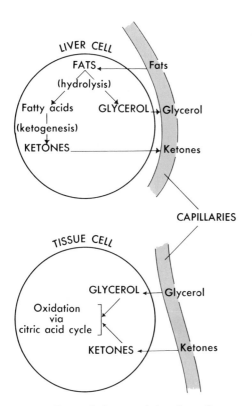

Fig. 12-27 Fat catabolism consisting, first, of ketogenesis by liver cells and, second, of oxidation by tissue and liver cells.

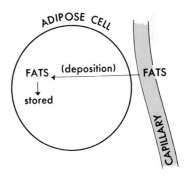

Fig. 12-29 Fat deposition (lipogenesis or fat anabolism).

drates yields only 4.1 kilocalories. *Fat catabolism*, like carbohydrate catabolism, consists of two main processes, each of which, in turn, consists of a series of chemical reactions. The first process in fat catabolism is called *ketogenesis*, which means the formation of ketone bodies. Liver cells carry on most of the ketogenesis for the body. It consists of the following steps: first, the hydrolysis of fat to form fatty acids and glycerol; next, a series of chemical reactions (known as beta oxidations) by which fatty acids are converted to coenzyme A; and, last, the conversion of acetyl coenzyme A to acetoacetic acid (by a process called condensation). Acetoacetic acid is a ketone body. Some of it undergoes a chemical

change to become two other ketone bodies, acetone and β-hydroxybutyric acid. Ketone bodies are oxidized via the citric acid cycle. In other words, the final step of fat catabolism and of glucose catabolism are the same. Liver cells oxidize a small portion of the ketone bodies for their own energy needs, but most of them are transported by the blood to other tissue cells for the final step of catabolism. Glycerol is metabolized similarly to glucose.

Fat anabolism (fat deposition or lipogenesis) consists of the storage of fats mainly in adipose tissue. It also includes the use

425

of fats in the synthesis of protoplasm and various complex compounds. Fats stored in the fat depots constitute the body's largest reserve energy source—too large too often, unfortunately.

The main facts about fat metabolism are summarized in Figs. 12-27 to 12-29.

Control

Fat metabolism is controlled mainly by the following hormones: insulin, growth hormone, ACTH, and glucocorticoids. You probably recall from our discussion of these hormones in connection with carbohydrate metabolism that they regulate fat metabolism in such a way that the rate of fat utilization is inversely related to the rate of carbohydrate utilization. If some condition such as diabetes mellitus causes carbohydrate utilization to decrease below energy needs, increased secretion of growth hormone, ACTH, and glucocorticoids soon follows (Fig. 12-25). And these hormones, in turn, bring about an increase in fat catabolism. (More details are given on pp. 419 and 428.) But when carbohydrate catabolism equals energy needs, fats are not mobilized out of storage and catabolized. Instead, they are spared and stored in adipose tissue. "Carbohydrates have a 'fat-sparing' effect," so says an old physiological maxim. Or, stating this truth more descriptively, as anyone who indulges in too many sweets would agree, "carbohydrates have a 'fat-storing' effect."

■Protein metabolism

In protein metabolism, anabolism is primary and catabolism is secondary. In carbohydrate and fat metabolism, the opposite is true—catabolism is primary and anabolism is secondary. Proteins are primarily tissue-building foods. Carbohydrates and fats are primarily energy-supplying foods.

The cellular process called protein synthesis (or protein anabolism) produces many substances—enzymes, antibodies, and secretions, to name some noteworthy ones. Protein anabolism plays a major role in the growth and reproduction both of cells and of the body as a whole. And protein anabolism is also the chief process of repair. It accomplishes the healing of wounds, the formation of scar tissue, and the replacement of cells destroyed by daily wear and tear. Protein anabolism is truly "big business" in the body. Red blood cell replacement alone, for instance, runs into millions of cells per second.

A brief synopsis of the process of protein anabolism as now visualized is as follows. First, a gene (a segment of a DNA molecule) directs the formation of a molecule of "messenger RNA," thereby transferring to it the gene's instructions for synthesizing a specific protein. As soon as it is formed, the messenger RNA diffuses out of the nucleus and goes to one of the many ribosomes located in the cytoplasm of the cell. (Ribosomes, you will recall, are the cell's tiny protein factories. They consist of protein and RNA, the latter identified as ribosomal or template RNA.) Next, a molecule of "transfer RNA," present in cytoplasm, attaches itself to a molecule of a specific amino acid and transfers it to a ribosome, where it fits it into its proper position as indicated by the messenger RNA. More transfer RNA molecules, one after the other in rapid sequence, bring more amino acids to the ribosome and fit them into their proper positions. Result? A chain of amino acids joined to each other in a definite sequence—a protein, in other words. Thus is protein anabolism achieved through the combined work of messenger RNA, transfer RNA, and template or ribosomal RNA. Protein anabolism is a major kind of cellular work. One human cell, for example, is estimated to synthesize perhaps two thousand different enzymes. And, in addition, many cells produce special proteins such as antibodies by plasma cells, fibrinogen by liver cells, and various protein hormones by endocrine gland cells.*

Protein catabolism, like the catabolism of fats, consists of two processes. The first takes place mainly in liver cells and the second is the citric acid cycle and occurs in all cells. The first step in protein catabolism is known as *deamination*, a reaction in which an amino (NH_2) group is split off from an amino acid molecule to form a molecule of ammonia and one of keto acid. Most of the ammonia is converted to urea and is excreted via the urine. The keto acid may be oxidized via the tricarboxylic acid cycle or may be converted to glucose (gluconeogenesis) or to fat (lipogenesis). Formerly, protein catabolism was believed to occur only when the amount of protein ingested exceeded that needed for anabolism. But now it is known that both protein catabolism and anabolism go on continually. Only their rates differ from time to time. With a protein-deficient diet, for example, protein catabolism exceeds protein anabolism. And various hormones, as we shall see, also influence the rates of protein catabolism and anabolism.

Usually a state of *protein balance* exists in the normal healthy adult body—that is, the rate of protein anabolism equals or balances the rate of protein catabolism. And when the body is in protein balance, it is also in a state of *nitrogen balance*. For then, the amount of nitrogen taken into the body (in protein foods) equals the amount of nitrogen in protein catabolic waste products excreted in the urine, feces, and sweat. Two kinds of protein or nitrogen imbalance exist. When protein catabolism exceeds protein anabolism, the amount of nitrogen in the urine exceeds the amount of nitrogen in the protein foods ingested. The individual is then said to be in a state of *negative nitrogen balance*. Or, in a

*For more details about protein synthesis, see supplementary readings for Chapter 12, reference 5.

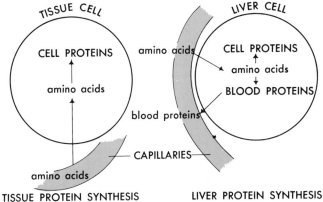

Fig. 12-30 Protein synthesis (anabolism). Growth hormone and testosterone tend to accelerate the processes shown so are called anabolic hormones.

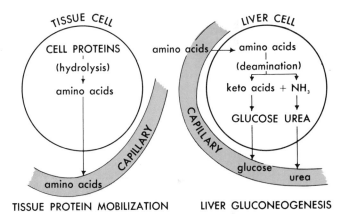

Fig. 12-31 Protein mobilization and catabolism. Glucocorticoids tend to accelerate these processes so are classed as protein catabolic hormones. (Also see Figs. 12-25 and 12-32.)

state of "tissue wasting"—because more of his tissue proteins are being catabolized than are being replaced by protein synthesis. Protein-poor diets, starvation, and wasting illnesses, for example, produce a negative nitrogen balance.

A *positive nitrogen balance* (nitrogen intake in foods greater than nitrogen output in urine) indicates that protein anabolism is going on at a faster rate than protein catabolism. A state of positive nitrogen

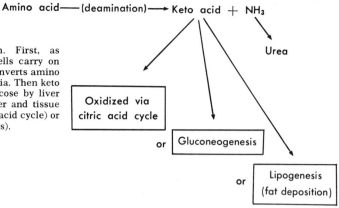

Fig. 12-32 Protein catabolism. First, as shown in Fig. 12-31, liver cells carry on deamination, a process that converts amino acids to keto acids and ammonia. Then keto acids may be changed to glucose by liver cells (gluconeogenesis), or liver and tissue cells may oxidize them (citric acid cycle) or convert them to fat (lipogenesis).

balance, therefore, characterizes any condition in which large amounts of tissue are being synthesized, as during growth, pregnancy, and convalescence from an emaciating illness.

The main facts about protein metabolism are summarized in Figs. 12-30 to 12-32. Compare them with Figs. 12-23 and 12-27 to 12-29. Note the important part played by the liver in the metabolism of all three kinds of foods.

Control

Protein metabolism, like that of carbohydrates and fats, is controlled largely by hormones rather than by the nervous system. Growth hormone and the male hormone testosterone both have a stimulating effect on protein synthesis or anabolism. For this reason, they are referred to as anabolic hormones. Protein catabolic hormones of greatest consequence are glucocorticoids. They are thought to act in some way, still unknown, to speed up tissue protein mobilization—that is, the hydrolysis of cell proteins to amino acids, their entry into the blood, and their subsequent catabolism (Fig. 12-31). ACTH functions indirectly as a protein catabolic hormone because of its stimulating effect on glucocorticoid secretion.

Thyroid hormone is necessary for and tends to promote protein anabolism and, therefore, growth when plenty of carbohydrates and fats are available for energy production. On the other hand, under different conditions, for example, when the amount of thyroid hormone is excessive or when the energy foods are deficient, this hormone may then promote protein mobilization and catabolism.

• • •

Some of the facts about metabolism set forth in the preceding paragraphs are summarized in Table 12-8.

Metabolism of vitamins, mineral salts, and water

Vitamins are substances that were recognized as necessary for life and health before their chemical composition was known. They constitute part of the chemical mechanisms for controlling body activities. All those of known function are components of enzyme molecules. Therefore, they are necessary for numerous physiological reactions. Without adequate amounts of the various vitamins, normal energy production, growth and development, reproduction, resistance to infection, and health in general are not possible. If the diet is too deficient in vitamins, life itself cannot continue. As their name suggests, vitamins are vital.

Mineral salts are also essential for both health and life. Not only do they function as part of the body's chemical control mechanism, but they also constitute structural components of body tissues. For example, calcium and phosphorus are important ingredients of bones and teeth.

Table 12-8. Metabolism

Food	Anabolism	Catabolism
1) Carbohydrates *mono di poly*	Temporary excess changed into glycogen by liver cells in presence of insulin; stored in liver and skeletal muscles until needed and then changed back to glucose (Figs. 12-19 and 12-20) True excess beyond body's energy requirements converted into adipose tissue; stored in various fat depots of body	Oxidized, in presence of insulin, to yield energy (4.1 kilocalories per gram) and wastes (carbon dioxide and water) $C_6H_{12}O_6 + 6O_2 \longrightarrow energy + 6CO_2 + 6H_2O$
2) Fats	Built into adipose tissue; stored in fat depots of body	Fatty acids $\downarrow$ (liver ketogenesis) ketone bodies $\downarrow$ (tissues; citric acid cycle) energy (9.3 kilocalories per gram) $+ CO_2 + H_2O$ Glycerol $\downarrow$ (liver gluconeogenesis) glucose
3) Proteins	Temporary excess stored in liver and skeletal muscles Synthesized into tissue proteins, blood proteins, enzymes, hormones, etc.	Deaminated by liver forming ammonia (which is converted to urea) and keto acids (which are either oxidized or changed to glucose or fat)

Examples of minerals functioning as control mechanisms are the following. Sodium and potassium play critical parts in the maintenance of the acid-base balance of the body and in the maintenance of water balance. Potassium is especially important for normal functioning of muscles (including the heart) and nerves. Calcium salts help sustain rhythmic heart and intestinal contractions and are necessary for normal functioning of skeletal muscles and for normal growth. A number of enzymes contain minerals.

Water serves many functions in the body. Some of the chief ones are as follows:

1 Water is an important constituent of every cell in the body. It composes well over one-half of the cells' substance. When dehydrated slightly, cells lose their power to resist infection and die if they are more completely dehydrated.

2 Water plays an essential part in all body functions—for example, in glandular secretion, digestion, absorption, anabolism, elimination of wastes, heat regulation, respiration, circulation, resistance to disease, etc. All cellular activity, and therefore life itself, depends upon the presence of adequate amounts of water. Materials must be dissolved in order to cross the cell membrane and in order to facilitate chemical reactions between them.

3 Water is important for the dilution of toxic wastes, thereby preventing damage of the kidney cells when they eliminate these toxins.

How important water is to life can be estimated by the fact that life can be maintained for a much longer time without food than without water. Men adrift in the ocean, for example, will live much longer with fresh water and no food than they will with food and no water. For a summary of some of the mechanisms involved in maintaining the water balance of the body, see Chapter 15.

■Metabolic rates
Meaning

The term *metabolic rate* means the amount of energy (heat) released in the body in a given time by catabolism. It represents energy expended or used for accomplishing various kinds of work. In short, metabolic rate actually means catabolic rate or rate of energy release. (It is impossible to measure the rate at which foods are anabolized but fairly easy to determine the rate at which they are catabolized—see pp. 431 to 432.)

Heat energy is measured in units named calories. One *kilocalorie* (a so-called "large" Calorie) is the amount of heat used to raise the temperature of 1 kilogram (liter) of water 1° Centigrade.

Ways of expressing

Metabolic rates are expressed in either of two ways: (1) in terms of the number of kilocalories of heat energy expended per hour or per day or (2) as normal or as a definite percentage above or below normal.

Basal metabolic rate

The basal metabolic rate (BMR) is the body's rate of energy expenditure under "basal conditions"—namely, when the individual:

1 Is awake but resting—that is, lying down and, so far as possible, not moving a muscle
2 Is in the postabsorptive state (twelve to eighteen hours after the last meal)
3 Is in a comfortably warm environment

Note that the basal metabolic rate is not the minimum metabolic rate. It does not indicate the smallest amount of energy that must be expended to sustain life. It does, however, indicate the smallest amount of energy expenditure that can sustain life and also maintain the waking state and a normal body temperature in a comfortably warm environment.

Factors influencing

The basal metabolic rate is not identical for all individuals because of the influence of various factors (Fig. 12-33), some of which are described in the following paragraphs.

Size. In computing the basal metabolic rate, size is usually considered the amount of the body's surface area. It is computed from the individual's height and weight. Per square meter of body surface, if other conditions are equal, a large individual has the same basal metabolic rate as a small one, but because a large individual has more square meters of surface area, his basal metabolism is greater than that of a small individual. For example, the BMR for a man in his twenties is about 40 kilocalories per square meter of body surface per hour (Table 12-9). A large man with a body surface area of 1.9 square meters would, therefore, have a basal metabolism of 76 kilocalories per hour, whereas a smaller man with a surface area of perhaps 1.6 square meters would have a basal metabolism of only 64 kilocalories per hour. The average surface area for American adults is 1.6 square meters for women and 1.8 square meters for men.

Sex. Men oxidize their food approximately 5% to 7% faster than women. Therefore, their basal metabolic rates are about 5% higher for a given size and age. A man 5 feet, 6 inches tall, weighing 140 pounds, for example, has a 5% higher basal metabolic rate than a woman of the same height, weight, and age.

Age. That the fires of youth burn more brightly than those of age is a physiological as well as a psychological fact. In general, the younger the individual, the higher his basal metabolic rate for a given size and sex. Exception: the BMR is slightly lower at birth than a few years later. That is to say, the rate increases slightly during the first three to six years and than starts to

Table 12-9. Basal metabolism (Aub-DuBois)

Age (yr)	Kilocalories per hour per square meter body surface	
	Male	Female
10–12	51.5	50.0
12–14	50.0	46.5
14–16	46.0	43.0
16–18	43.0	40.0
18–20	41.0	38.0
20–30	39.5	37.0
30–40	39.5	36.5
40–50	38.5	36.0
50–60	37.5	35.0
60–70	36.5	34.0

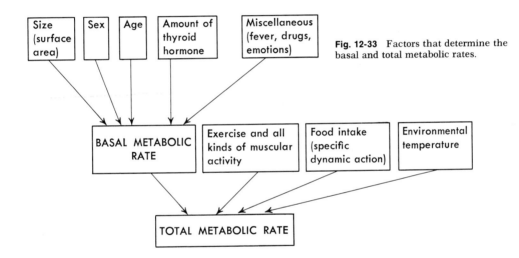

Fig. 12-33 Factors that determine the basal and total metabolic rates.

decrease and continues to do so throughout life. The basal metabolism per hour per square meter of surface area for different age groups is given in Table 12-9.

Thyroid hormone. Thyroid hormone stimulates basal metabolism. Without a normal amount of this hormone in the blood, a normal basal metabolic rate cannot be maintained. When an excess of thyroid hormone is secreted, foods are catabolized faster, much as coal is burned faster when a furnace draft is open. Deficient thyroid secretion, on the other hand, slows the rate of metabolism.

Fever. Fever increases the basal metabolic rate. According to DuBois, metab-olism increases about 13% per degree Centigrade rise in body temperature.*

Drugs. Certain drugs, such as caffeine, Benzedrine, and dinitrophenol, increase the basal metabolic rate.

Other factors. Other factors, such as *emotions* and *pregnancy*, also influence basal metabolism. Both of these factors increase the basal rate.

How determined

Originally, basal metabolic rates were determined by a method known as direct

*Mountcastle, V. B., editor: Medical physiology, ed. 12, St. Louis, 1968, The C. V. Mosby Co.

calorimetry, a method too time consuming and costly for use on large numbers of people. Now, a rapid, inexpensive method, *indirect calorimetry*, is used in practically all hospitals, as well as in many doctors' offices and nutrition laboratories. This method of determining metabolism consists simply of an oxygen tank into which the patient breathes by means of a rubber tube leading to his mouth, his nose being clamped off. He receives oxygen from the tank and expires carbon dioxide into it, the latter being removed by soda lime contained in a tank within the oxygen tank. The amount of oxygen consumed in a given length of time is measured. Research has shown that for every liter of oxygen consumed, an average of 4.825 kilocalories of heat are produced under basal conditions. Multiplying the amount of oxygen consumed by 4.825 therefore gives the number of kilocalories produced in the given time. From that figure the number produced in twenty-four hours can, of course, be readily computed. This number represents the patient's basal metabolic rate per day expressed in kilocalories. Usually the basal metabolic rate is expressed as normal or as a definite percent above or below normal. The percentage is computed by comparing the actual basal metabolic rate (in kilocalories) with what is known to be an average basal metabolic rate (in kilocalories) for normal individuals of the given size, sex, and age. Statistical tables, based on research, give these normal rates for different sizes, sexes, and ages. For instance, you can compute the average BMR for a person of your own size, sex, and age in this way:

1 Start with your weight in kilograms and your height in centimeters. (Convert pounds to kilograms by dividing pounds by 2.2. Convert inches to approximate centimeters by multiplying inches by 2.5.) For example, 110 lb = 50 kg; 5 ft, 3 in = 158 cm.

2 Convert your weight and height to square meters, using Fig. 12-34. For example, weight 50 kg and height 158 cm = about 1.5 sq m surface area of body.

3 Find your age and sex in Table 12-9 and then multiply the number of kilocalories per square meter per hour given there by your square meters of surface area and then by 24. For example, average BMR per day for a 25-year-old female, weight 110 lb and height 5 ft, 3 in = 1,332 kilocalories (37 × 1.5 × 24).

A quick rule of thumb for estimating a young woman's BMR is to multiply her weight in pounds by twelve.

Total metabolic rate

The <u>total metabolic rate</u> is the amount of energy used or expended by the body in a given time. It is expressed in kilocalories per hour or per day. Most of the many factors that together determine the total metabolic rate are shown in Fig. 12-33. Of these, the main direct determinants are the following:

1 The basal metabolic rate—that is, the energy used to do the work of maintaining life under the basal conditions previously described, plus

2 The energy used to do all kinds of skeletal muscle work—from the simplest activities such as feeding oneself or sitting up in bed to the most strenuous kind of physical labor or exercise, plus

3 The energy expended by the specific dynamic action (SDA) of foods

The SDA of a food is a mysterious action which increases the metabolic rate, which speeds up the rate of energy expenditure. A food starts performing its specific dynamic action soon after ingestion and continues for several hours. Proteins have an SDA of about 30% of their caloric value. More specifically, ingestion of 100 kilocalories of protein increases the total metabolic

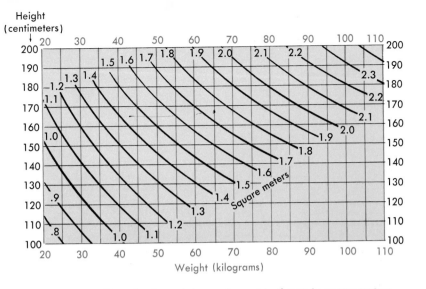

Fig. 12-34 Chart for determining surface area of man in square meters from weight in kilograms and height in centimeters according to the following formula: area (sq m) = $wt^{0.425} \times ht^{0.725} \times 7184$. (Redrawn after DuBois, D., and DuBois, E. F.: Clinical calorimetry, Arch. Intern. Med. [Chicago] **17**:863-871, 1916.)

rate by 30 kilocalories. Carbohydrates and fats have a much smaller SDA—reportedly about 5% of their caloric value in contrast to protein's 30%. Reducing diets, as you probably know, are often high-protein diets for this reason—the high SDA of protein increases the total metabolic rate. It increases the amount of energy used or expended by the body, in other words.

Energy balance and its relationship to body weight

When we say that the body maintains a state of energy balance, we mean that its energy input equals its energy output. Energy input per day equals the total calories (kilocalories) in the food ingested per day. And energy output equals the total metabolic rate expressed in kilocalories. But you may be wondering what energy intake, output, and balance have to do with body weight. "Everything" would be a fairly good one-word answer. Or, to be somewhat more explicit, the following basic principles describe the relationships between these factors:

1 Body weight remains constant (except for possible variations in water con-

tent) when the body maintains energy balance—when the total calories in the food ingested equals the total metabolic rate, that is. Example: if you have a total metabolic rate of 2,000 kilocalories per day and if the food you eat per day yields 2,000 kilocalories, your body will be maintaining energy balance and your weight will stay constant.

2 Body weight increases when energy input exceeds energy output—when the total calories of food intake per day is greater than the total calories of the metabolic rate. A small amount of the excess energy input is used to synthesize glycogen for storage in the liver and muscles. But the rest of it is used for synthesizing fat and storing it in adipose tissue. If you were to eat 3,000 kilocalories each day for a week and if your total metabolic rate were 2,000 kilocalories per day, you would gain weight. How much you would gain, you can discover by doing a little simple arithmetic:

Total energy input for week = 21,000 kilocalories
Total energy output for week = 14,000 kilocalories
Excess energy input for week = 7,000 kilocalories

Approximately 3,500 kilocalories are used to synthesize 1 pound of adipose tissue. Hence, at the end of this one week of "overeating"—of eating 7,000 kilocalories over and above your total metabolic rate—you would have gained about two pounds.

3 Body weight decreases when energy input is less than energy output—when the total number of calories in the food eaten is less than the total metabolic rate. Suppose you were to eat only 1,000 kilocalories a day for a week and that you have a total metabolic rate of 2,000 kilocalories per day. By the end of the week your body would have used a total of 14,000 kilocalories of energy for maintaining life and doing its many kinds of work. All 14,000 kilocalories of this actual energy expenditure had to come from catabolism of foods since this is the body's only source of energy. Catabolism of ingested food supplied 7,000 kilocalories and catabolism of stored food supplied the remaining 7,000 kilocalories. That week your body would not have maintained energy balance. Nor would it have maintained weight balance. It would have incurred an energy deficit paid out of the energy stored in approximately 2 pounds of body fat. In short, you would have lost about 2 pounds.

· · ·

Anyone who wants to reduce should remember this cardinal principle: eat fewer calories than your total metabolic rate. Obey this law and you will lose weight. Ignore it and you will not lose weight. Unless caloric intake is less than caloric output (total metabolic rate), weight loss is impossible. But we ought to note one other point about this principle. There are two ways to make your caloric intake less than your total metabolic rate. The approach commonly advised to would-be reducers is simply "cut down on your calories"—a principle that proves easy to understand but difficult to apply. The other approach to weight reduction seems to be thought of or emphasized less often. It is this: increase your total metabolic rate (your caloric output, that is) and also decrease your caloric intake. Simple arithmetic shows which method produces a faster weight loss. Suppose you have a total metabolic rate of 2,000 kilocalories a day and that you ingest 1,500 kilocalories a day for a week. By the end of the week you will have taken in 3,500 kilocalories less than your total metabolic rate. That deficit will have been supplied by the catabolism of about one pound of body fat. Now suppose that the next week you still eat 1,500 kilocalories a day but you also exercise briskly for a half hour each day. Suppose you swim sidestroke at the rate of 1.6 miles per hour for a half hour. This will increase your total metabolic rate from 2,000 to 2,600 kilocalories.* Your calorie deficit will then be 2,600 minus 1,500, or 1,100 kilocalories a day, or 7,700 kilocalories for the week. Catabolism of more than 2 pounds of body fat will have supplied those 7,700 kilocalories. In short, you will have lost more than twice as much the second week as the first even though you ate the same amount both weeks.

Foods are stored as glycogen, fats, and tissue proteins. And as you will recall, cells catabolize them preferentially in this same order: carbohydrates, fats, and proteins. If there is no food intake, almost all of the glycogen is estimated to be used up in a matter of one or two days. Then, with no more carbohydrate to act as a fat sparer, fat is catabolized. How long it takes to deplete all of this reserve food depends, of course, upon how much adipose tissue the individual has when he starts his starva-

---

*See supplementary readings for Chapter 12, reference 4.

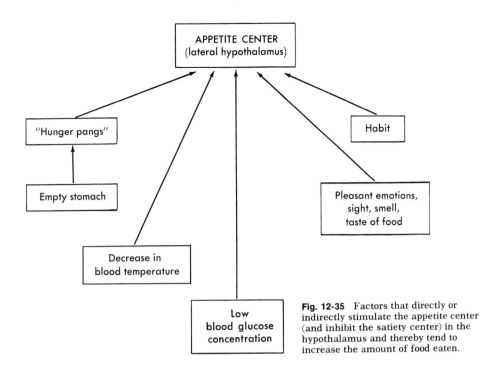

Fig. 12-35 Factors that directly or indirectly stimulate the appetite center (and inhibit the satiety center) in the hypothalamus and thereby tend to increase the amount of food eaten.

tion diet. Finally, with no more fat available as a protein sparer, tissue proteins are catabolized rapidly, and death soon ensues.

Mechanisms for regulating food intake

The hypothalamus contains the centers that regulate the amount of food we eat. A cluster of cells in the lateral hypothalamus serves as the *appetite center.* And another group of cells, located in the ventral medial nucleus of the hypothalamus, functions as a *satiety center.* What acts directly on these centers to stimulate or depress them is still a matter of theory rather than fact. One theory (the "thermostat theory") holds that it is the temperature of the blood circulating to the hypothalamus that influences the centers. A moderate decrease in blood temperature stimulates the appetite center (and inhibits the satiety center). Result: the individual has an appetite. He wants to eat. And probably does. An increase in blood temperature produces the opposite effect, a depressed appetite (anorexia).

One well-known instance of this is the loss of appetite in persons who have a fever.

Another theory (the "glucostat theory") says that it is the blood glucose concentration and rate of glucose utilization that influences the hypothalamic feeding centers. A low blood glucose concentration or low glucose utilization stimulates the appetite center, whereas a high blood glucose concentration inhibits it. Unquestionably, a great many factors operate together as a complex mechanism for regulating food intake. Some of these factors are indicated in Fig. 12-35.

Homeostasis of body temperature

Warm-blooded animals, such as man, maintain a remarkably constant temperature despite sizable variations in environmental temperatures.

Normally in most people, body temperature moves up and down very little in the course of a day. It hovers close to a midpoint of about 37° C., increasing perhaps to 37.6° C. by late afternoon and decreasing to around 36.2° C. by early morning. This homeostasis of body temperature is of the utmost importance. Why? Because healthy survival depends upon biochemical reactions taking place at certain rates. And these rates, in turn, depend upon normal enzyme functioning, which depends upon body temperature staying within the narrow range of normal.

In order to maintain an even temperature, the body must, of course, balance the amount of heat it produces with the amount it loses. This means that if extra heat is produced in the body, this same amount of heat must then be lost from it. Obviously if this does not occur, if increased heat loss does not follow close upon increased heat production, body temperature will climb steadily upward.

■Heat production

Heat is produced by one means—catabolism of foods. Because the muscles and glands (liver, especially) are the most active tissues, they carry on more catabolism and therefore produce more heat than any of the other tissues. So the chief determinant of how much heat the body produces is the amount of muscular work it does. During exercise and shivering, for example, catabolism and heat production increase greatly. But during sleep, when very little muscular work is being done, catabolism and heat production decrease.

■Heat loss

Heat is lost from the body by the physical processes of evaporation, radiation, conduction, and convection. Some 80% or more of this heat transfer occurs through the skin. The rest takes place through the mucous membranes of the respiratory, digestive, and urinary tracts.

Evaporation

Heat energy must be expended to evaporate any fluid. Evaporation of water, there-fore, constitutes one method by which heat is lost from the body, especially from the skin. At moderate temperatures it accounts for about half as much heat loss as does radiation. But at high environmental temperatures, evaporation constitutes the only method by which heat can be lost from the skin. A humid atmosphere necessarily retards evaporation and therefore lessens the cooling effect derived from it—the explanation for the fact that the same degree of temperature seems hotter in humid climates than in dry ones.

Radiation

Radiation is the transfer of heat from the surface of one object to that of another without actual contact between the two. Heat radiates from the body surface to nearby objects that are cooler than the skin and radiates to the skin from those that are warmer than the skin. This is, of course, the principle of heating and cooling systems. From surfaces that have been heated to temperatures warmer than the skin, heat radiates to the skin, thereby warming it, whereas with cooled surfaces, heat radiates from the skin to them, thereby cooling the skin. The amount of heat lost by radiation from the skin is made to vary as needed by dilatation of surface blood vessels when more heat needs to be lost and by vasoconstriction when heat loss needs to be decreased. In cool environmental temperatures, radiation accounts for a greater percentage of heat loss from the skin than both conduction and evaporation combined. In hot environments, on the other hand, no heat is lost by radiation but instead may even be gained by radiation from warmer surfaces to the skin.

Conduction

Conduction means the transfer of heat to any substance actually in contact with the body—to clothing or jewelry, for example, or even to cold foods or liquids ingested. This process accounts for a relatively small amount of heat loss compared to the amount lost by evaporation and radiation.

Convection

Convection is the transfer of heat away from a surface by movement of heated air or fluid particles. Usually, convection causes very little heat loss from the body's surface. But it can account for considerable heat loss—as you know from experience if you have ever stepped from your bath into even slightly moving air from an open window.

■Thermostatic control of heat production and loss

The control mechanism that normally maintains homeostasis of body temperature consists of two parts:

1 A *heat-dissipating mechanism* which acts to increase heat loss when blood temperature increases above a certain point (Fig. 12-36). This mechanism, therefore, prevents body temperature from rising above normal under usual circumstances.

2 A *heat-gaining mechanism* which acts to accelerate catabolism and thereby to increase heat production when blood temperature decreases below a certain point. Under ordinary conditions, this mechanism prevents body temperature from falling below normal.

Heat-dissipating mechanism

In the anterior part of the hypothalamus, behind the sphenoid sinuses, lies a group of cells referred to collectively as the "human thermostat." These neurons are thermal receptors—that is, they are stimulated by a very slight increase in the temperature

of the blood above the point at which the human thermostat is set—normally about 37° C. In a sense, one might say that these cells of the hypothalamus take the temperature of the blood circulating to them. Whenever it increases by as little as 0.01° above 37° C* (or some other set point), these neurons send out impulses that eventually reach sweat glands and blood vessels of the skin. They stimulate the body's two million or more sweat glands to increase their rate of secretion, and they also cause dilatation of surface blood vessels. Evaporation of the larger amount of sweat causes a greater heat loss from the skin. And also more heat is lost by radiation from the larger quantity of blood circulating near the surface in the dilated skin vessels.

Heat-gaining mechanism

In a cold environment, the mechanism that tries to maintain homeostasis of body temperature includes two kinds of responses—those that decrease heat loss and those that increase heat production. Together, they almost always succeed in preventing a decrease in blood temperature below the lower limit of normal. Skin blood vessel constriction decreases the volume of blood circulating near the surface so decreases heat loss by radiation. In addition, shivering and voluntary muscle contractions occur, thereby accelerating catabolism and heat production. Further details about the mechanism for preventing body temperature from falling below normal are not yet established.

■Skin thermal receptors

In addition to the thermal receptors in the hypothalamus, there are many heat and cold receptors located in the skin. Impulses initiated in skin thermal receptors

*See supplementary readings for Chapter 12, reference 1.

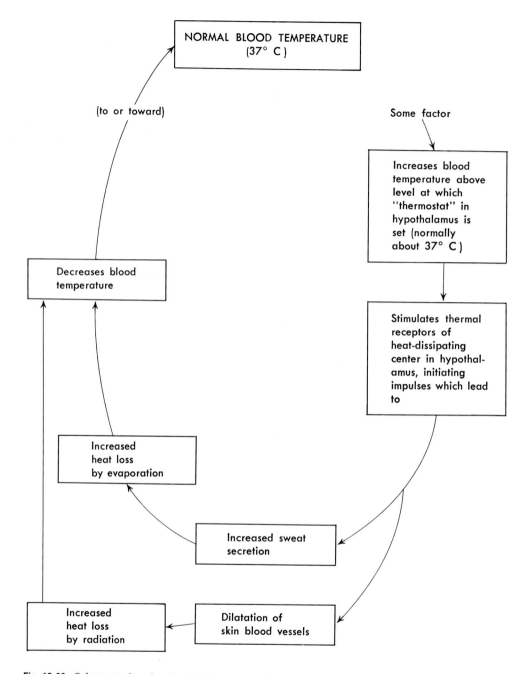

Fig. 12-36 Scheme to show how heat-dissipating mechanism operates to maintain normal body temperature. In principle, it cancels out any heat gain by bringing about an equal heat loss. Under usual circumstances, this mechanism succeeds in preventing body temperature from rising above the upper limit of normal. When it fails, fever develops. According to one theory, certain factors can increase the threshold of stimulation of hypothalamic thermal receptors – in more picturesque language, they reset the hypothalamic thermostat at a level higher than normal. Then, blood temperature rises above normal before it activates the heat-dissipating mechanism shown in this diagram.

travel to the cerebral cortex sensory area. Here, they give rise to sensations of skin temperature and are relayed out over voluntary motor paths to produce skeletal muscle movements that affect skin temperature. For example, on a hot day you "feel hot" because of stimulation of your skin heat receptors. Often you make some sort of movements to "cool yourself off." You may start fanning yourself, or perhaps you turn on an air conditioner or go swimming. And as a result, your skin temperature decreases back to a more comfortable level. Thus, the thermal receptors in the skin will have taken part in a conscious mechanism that helps regulate skin temperature. Convincing evidence supports the view that this is their only function and that impulses from them do not travel to the heat-regulating centers in the hypothalamus and therefore do not take part in the automatic regulation of internal body temperature.

■Correlations

When a fever or higher than normal temperature exists, it is thought to be due primarily to inability of the heat-dissipating mechanisms to keep pace with heat production. Some factor—perhaps chemicals from microorganisms or injured tissue cells—stimulates catabolism, thereby producing more heat in the body in a given time. Heat-dissipating mechanisms (Fig. 12-36) operate in an effort to compensate for the heat gain. But presumably they cannot increase heat loss as much as heat production has increased. So body temperature necessarily increases. The patient "has a fever," in other words.

The heat-regulating centers are present at birth but do not function well for a short time after birth. Therefore, newborn babies need to be kept somewhat warmer than adults. If the baby is born prematurely, the heat-regulation centers do not function for a longer time, perhaps several weeks.

outline summary

FUNCTIONS AND IMPORTANCE
1 Prepare food for absorption and metabolism
2 Absorption
3 Elimination of wastes
4 Vital importance

ORGANS
1 Main organs
  a Compose alimentary canal—mouth, pharynx, esophagus, stomach, and intestines
2 Accessory organs
  a Salivary glands, teeth, liver, gallbladder, pancreas, and vermiform appendix

Walls of organs
1 Coats
  a Mucous lining
  b Submucous coat of connective tissue—main blood vessels here
  c Muscular coat
  d Fibroserous coat

2 Modifications of coats
  a Mucous lining
    1 Rugae and microscopic gastric and hydrochloric acid glands in stomach
    2 Circular folds, villi, intestinal glands, Peyer's patches, and solitary lymph nodes in small intestine
    3 Solitary nodes and intestinal glands in large intestine
  b Muscle coat
    1 Three layers (circular, longitudinal, oblique) in stomach instead of only two layers as in rest of tract
    2 Three tapelike strips make up outer, longitudinal layer and pucker large intestine into small sacs called haustra
  c Fibroserous coat
    1 Peritoneum covers stomach and intestines
    2 Greater omentum or lace apron—double fold of peritoneum that hangs from lower edge of stomach like an apron over intestines; should not be confused with mesentery, which also is double fold of peritoneum but fan shaped and attached at short side to posterior wall of abdominal cavity; small intestines anchored to posterior abdominal wall by means of mesentery

Mouth (buccal cavity)

1 Formed by cheeks, hard and soft palates, tongue, and muscles
2 Hard palate—formed by two palatine bones and parts of two maxillary bones
3 Soft palate—formed of muscle in shape of arch; forms partition between mouth and nasopharynx; fauces is archway or opening from mouth to oropharynx; uvula is conical-shaped process suspended from midpoint of arch
4 Tongue
  a Many rough elevations on tongue's surface called papillae; contain taste buds
  b Frenum—a fold of mucous membrane that helps anchor tongue to mouth floor

Salivary glands

1 Parotid—below and in front of ear; duct opens on inside of cheek, opposite upper second molar tooth
2 Submaxillary—posterior part of mouth floor
3 Sublingual—anterior part of mouth floor

Teeth

1 Deciduous or baby teeth—ten per jaw or twenty in set
2 Permanent—sixteen per jaw or thirty-two per set
3 Structure of typical tooth—see Fig. 12-8

Pharynx

See pp. 365 to 366

Esophagus

1 Position and extent
  a Posterior to trachea and heart; pierces diaphragm
  b Extends from pharynx to stomach, distance of approximately 10 inches
2 Structure—collapsible, muscle tube

Stomach

1 Size, shape, and position
  a Size varies in different individuals; also according to whether distended or not
  b Elongated pouch
  c Lies in epigastric and left hypochondriac portions of abdominal cavity
2 Divisions
  a Fundus—portion above esophageal opening
  b Body—central portion
  c Pylorus—constricted, lower portion
3 Curves
  a Lesser—upper, right border
  b Greater—lower, left border
4 Sphincter muscles
  a Cardiac—guarding opening of esophagus into stomach
  b Pyloric—guarding opening of pylorus into duodenum

5 Coats—see Table 12-1
6 Glands
  a Epithelial cells of gastric mucosa secrete mucus
  b Parietal cells secrete hydrochloric acid
  c Chief cells (zymogen cells) secrete enzymes of gastric juice
7 Functions
  a Serves as food reservior
  b Secretes gastric juice
  c Contractions break food into small particles, mix them well with gastric juice, and move contents on into duodenum
  d Secretes the antianemic intrinsic factor
  e Carries on a limited amount of absorption— some water, alcohol, and certain other drugs

Small Intestine

1 Size and position
  a Approximately 1 inch in diameter and 20 feet in length
  b Its coiled loops fill most of abdominal cavity
2 Divisions
  a Duodenum
  b Jejunum
  c Ileum
3 Coats—see Table 12-1
4 Functions
  a Completes digestion of foods
  b Absorbs end products of digestion
  c Secretes hormones that help control secretion of pancreatic juice, bile, and intestinal juice

Large intestine (colon)

1 Size—approximately 2 1/2 inches in diameter and 5 or 6 feet in length
2 Divisions
  a Cecum
  b Colon
    1 Ascending
    2 Transverse
    3 Descending
    4 Sigmoid
  c Rectum
3 Coats—see Table 12-1
4 Functions
  a Absorption of water
  b Elimination of digestive wastes

Liver

1 Location and size
  a Occupies most of right hypochondrium and part of epigastrium
  b Largest gland in body
2 Lobes
  a Right lobe, subdivided into three smaller lobes —right lobe proper, caudate, and quadrate
  b Left lobe
  c Lobes divided into lobules by blood vessels and fibrous partitions

3 Ducts
  a Hepatic duct from liver
  b Cystic duct from gallbladder
  c Common bile duct formed by union of hepatic and cystic ducts and opens into duodenum at ampulla of Vater (duodenal papilla)
4 Functions
  a Secretes bile
  b Plays essential role in metabolism of carbohydrates, proteins, and fats—e.g., carries on glycogenesis, glycogenolysis, and gluconeogenesis; also deamination and ketogenesis; synthesizes various blood proteins

Gallbladder

1 Size, shape, and location
  a Approximately size and shape of small pear
  b Lies on undersurface of liver
2 Structure—sac of smooth muscle with mucous lining arranged in rugae
3 Functions
  a Concentrates and stores bile
  b During digestion, ejects bile into duodenum

Pancreas

1 Size, shape, and location
  a Larger in men than in women but varies in different individuals
  b Shaped something like a fish with head, body, and tail
  c Lies in C-shaped curve of duodenum
2 Structure—similar to salivary glands
  a Divided into lobes and lobules
  b Pancreatic cells pour their secretion into duct that runs length of gland and empties into duodenum at ampulla of Vater
  c Clusters of cells, not connected with any ducts, lie between pancreatic cells—called islets or islands of Langerhans—composed of alpha and beta type cells—latter thought to secrete insulin, former to secrete glucagon
3 Functions
  a Secretes pancreatic juice
  b Beta cells of islands of Langerhans secrete insulin
  c Alpha cells of islands of Langerhans secrete glucagon

Vermiform appendix

1 Size, shape, and location
  a About size and shape of large angle worm
  b Blind-end tube off cecum
2 Structure—similar to rest of intestine

DIGESTION

1 Definition—all changes food undergoes in alimentary canal
2 Purpose—conversion of foods into chemical and physical forms that can be absorbed and metabolized
3 Kinds
  a Mechanical—movements that change physical state of foods, propel them forward in alimentary tract, eliminate digestive wastes from tract and facilitate absorption (see Table 12-4 for description of processes involved in mechanical digestion)
    1 Mastication (chewing)
    2 Swallowing (deglutition)
      a Movement of food through mouth into pharynx—voluntary act
      b Movement of food through pharynx into esophagus—involuntary or reflex act initiated by stimulation of mucosa of back of mouth, pharynx, or laryngeal region; paralysis of receptors here, e.g., by Novocain, makes swallowing impossible
      c Movement of food through esophagus into stomach; accomplished by esophageal peristalsis—reflex initiated by stimulation of esophageal mucosa
    3 Peristalsis—wormlike movements that squeeze food downward in tract; emptying of stomach or opening of pyloric sphincter regulated by enterogastric reflex; fats and sugars in intestine stimulate mucosa to release enterogastrone into blood which, in turn, inhibits gastric peristalsis and slows stomach emptying; proteins and acid stimulate vagal nerve receptors in intestinal mucosa and thereby initiate reflex emptying of stomach
    4 Churning
    5 Mass peristalsis
    6 Defecation—reflex initiated by stimulation of rectal mucosa
  b Chemical—series of hydrolytic processes dependent upon specific enzymes (see Table 12-5 for description of chemical changes)
4 Control of digestive gland secretion—see Table 12-6
  a Saliva—secretion is reflex initiated by stimulation of taste buds, other receptors in mouth and esophagus, olfactory receptors, and visual receptors
  b Gastric secretion—controlled reflexly by same stimuli that initiate salivary secretion; also controlled chemically by hormone, gastrin, released by gastric mucosa in presence of partially digested proteins; enterogastrone (hormone just mentioned as slowing stomach emptying) also has inhibitory effect on gastric secretion

c Pancreatic secretion – controlled chemically by hormones secretin and pancreozymin formed by intestinal mucosa when hydrochloric acid enters duodenum

d Bile

   1 Secretion of – controlled chemically by same hormone (secretin) that regulates pancreatic secretion

   2 Ejection of into duodenum – controlled chemically by hormone cholecystokinin formed by intestinal mucosa when fats present in duodenum

e Intestinal secretion – control still obscure, although believed to be both reflex and chemical

ABSORPTION

1 Definition – passage of substances through intestinal mucosa into blood or lymph
2 How accomplished – probably mainly by active transport mechanisms

METABOLISM

Meaning

Chemical changes foods undergo inside cells or utilization of foods by body cells

Ways in which foods metabolized

1 Catabolism – breaks down food molecules to simpler compounds (carbon dioxide, water, and nitrogenous wastes), transferring some of their energy to phosphate compounds, notably, ATP, and releasing some of it as heat
2 Anabolism – building up food molecules into more complex compounds – notably, glycogen, enzymes and other cell proteins, hormones, etc.

Carbohydrate metabolism

1 Glucose transport through cell membranes and phosphorylation

  a Insulin promotes this transport through cell membranes

  b *Glucose phosphorylation* – conversion of glucose to glucose-6-phosphate, catalyzed by enzyme glucokinase; *insulin* increases activity of glucokinase so promotes glucose phosphorylation

2 Glycogenesis – conversion of glucose to glycogen for storage; occurs mainly in liver and muscle cells

3 Glycogenolysis

  a In muscle cells – glycogen changed back to glucose-6-phosphate, preliminary to catabolism

  b In liver cells – glycogen changed back to glucose; enzyme, glucose phosphatase, present in liver cells catalyzes final step of glycogenolysis, changing of glucose-6-phosphate to glucose; this enzyme lacking in most other cells; *glucagon* increases activity of phos-

phorylase so accelerates liver glycogenolysis; *epinephrine* accelerates liver and muscle glycogenolysis

4 Glucose catabolism

  a *Glycolysis* – series of anaerobic reactions that break 1 glucose molecule down into 2 pyruvic acid molecules with conversion of small amount of energy stored in glucose to heat and to ATP

  b *Krebs' citric acid cycle* – series of aerobic chemical reactions by which 2 pyruvic acid molecules (from 1 glucose molecule) broken down to 6 carbon dioxide and 6 water molecules with release of energy as heat and ATP

5 Gluconeogenesis – sequence of chemical reactions carried on in liver cells; converts protein or fat compounds into glucose; growth hormone, ACTH, and glucocorticoids have stimulating effect on rate of gluconeogenesis

6 Control of glucose metabolism

  a By hormones secreted by islands of Langerhans in pancreas

    1 Insulin (from beta cells) – tends to accelerate glucose utilization by cells because it accelerates glucose transport through cell membranes and glucose phosphorylation; hence, insulin tends to decrease blood glucose concentration – i.e., has hypoglycemic effect

    2 Glucagon (from alpha cells) – increases activity of enzyme phosphorylase, thereby accelerating liver glycogenolysis with release of glucose into blood; hence, glucagon tends to increase blood glucose – i.e., has hyperglycemic effect

  b By hormones secreted by anterior pituitary gland, adrenal cortex, and thyroid gland

    1 Growth hormone – decreases fat deposition, increases fat mobilization and catabolism; hence, tends to bring about shift to fat utilization from "preferred" glucose utilization

    2 ACTH and glucocorticoids – ACTH stimulates adrenal cortex to increase secretion of glucocorticoids, which accelerate tissue protein mobilization and subsequent liver gluconeogenesis from mobilized proteins; therefore, ACTH and glucocorticoids tend to increase blood glucose – i.e., have hyperglycemic effect

    3 Thyrotropin – stimulates thyroid gland to increase secretion of thyroid hormone which accelerates catabolism, usually glucose catabolism since glucose "preferred fuel"

  c By hormone secreted by adrenal medulla – epinephrine increases phosphorylase activity so accelerates muscle and liver glycogenolysis; hence, tends to increase blood glucose – i.e., has hyperglycemic effect

7 Principles about normal carbohydrate metabolism
  a Principle of "preferred energy fuel" – cells catabolize first glucose, sparing fats and proteins; when their glucose supply becomes inadequate, next catabolize fats, sparing proteins, and last catabolize proteins
  b Principle of glycogenesis – glucose in excess of about 120 to 140 mg/100 ml blood brought to liver by portal veins enters liver cells, where it undergoes glycogenesis
  c Principle of glycogenolysis – when blood glucose decreases below midpoint of normal, liver glycogenolysis accelerates and tends to raise blood glucose concentration back toward midpoint of normal
  d Principle of gluconeogenesis – when blood glucose decreases below normal or when amount of glucose entering cells inadequate, liver gluconeogenesis accelerates and tends to raise blood glucose concentration
  e Principle of glucose storage as fat – when blood sugar glucose higher than normal and blood insulin content adequate, excess glucose converted to fat, mainly by liver cells, and stored as such in fat depots
  f Principle of hyperglycemia in stress – under conditions of stress, blood glucose concentration tends to increase above normal

Fat metabolism

1 Catabolism
  a Hydrolysis of fats to fatty acids and glycerol, primarily in liver cells
  b Glycerol oxidized same as carbohydrates
  c Fatty acids converted to ketone bodies (ketogenesis); occurs mainly in liver; largest proportion of ketones enters blood from liver cells to be transported to tissues for oxidation to carbon dioxide and water via tricarboxylic acid cycle
2 Anabolism – for tissue synthesis and for building various compounds; fats deposited in connective tissue converts it to adipose tissue
3 Fat mobilization – release of fats from adipose tissue cells, followed by their catabolism; occurs when blood contains less glucose than normal or when it contains less insulin than normal; if excessive, leads to ketosis
4 Control – by following major factors
  a Rate of glucose catabolism one of main regulators of fat metabolism; in general, normal or high rates of glucose catabolism accompanied by low rates of fat mobilization and catabolism and high rates of fat deposition; converse also true
  b Insulin helps control fat metabolism by its effects on glucose metabolism; in general, normal amounts of insulin and blood glucose tend to decrease fat mobilization and catabolism and to increase fat deposition; insulin deficiency increases fat mobilization and catabolism; converse also true
  c Growth hormone – decreases fat deposition and increases fat mobilization and utilization – i.e., growth hormone tends to bring about shift from glucose to fat utilization
  d Glucocorticoids help control fat metabolism; in general, when blood glucose lower than normal and in various stress situations, more glucocorticoids secreted and accelerate fat mobilization and gluconeogenesis from them; when blood glucose higher than normal but rate of glucose catabolism low (as in diabetes mellitus), glucocorticoids also increase fat mobilization, but it is followed by ketogenesis from them; when blood glucose higher than normal, and provided its insulin content adequate, glucocorticoids accelerate fat deposition

Protein metabolism

1 Anabolism of proteins of primary importance; their catabolism, secondary, amino acids used to synthesize all kinds of tissue (e.g., for growth and repair); also used to synthesize many other substances such as enzymes, hormones, antibodies, and blood proteins
2 Catabolism
  a Deamination of amino acid molecule to form ammonia and keto acid; mainly in liver cells
  b Ammonia converted to urea (mainly in liver) and excreted via urine
  c Keto acids may be converted to glucose in liver, or to fat, or oxidized in liver or tissue cells via tricarboxylic acid cycle
3 Control
  a STH and testosterone both have stimulating effect on protein synthesis or anabolism
  b ACTH and glucocorticoids – protein catabolic hormones; accelerate tissue protein mobilization – i.e., hydrolysis of tissue proteins to amino acids and their release into blood; liver converts amino acids to glucose (gluconeogenesis) or deaminates them
  c Thyroid hormone promotes protein anabolism when nutrition adequate and amount of hormone normal; therefore, adequate amounts necessary for normal growth

Metabolism of vitamins, mineral salts, and water

1 Vitamins essential for many physiological processes such as normal metabolism, growth, reproduction, resistance to infection, etc.
2 Mineral salts essential for normal metabolism and for maintenance of favorable internal environment
3 Water essential for all physiological reactions

Metabolic rates

Meaning

Amount of heat energy expended in given time

Ways of expressing

In kilocalories or as "normal" or as a definite percentage above or below normal—e.g., +10% or −10%

Basal metabolic rate

Amount of heat produced (energy expended) in waking state when body at complete rest, twelve to eighteen hours after last meal, in comfortably warm environment

1 Factors influencing
  a Size—greater surface area, higher BMR (surface area computed from height and weight)
  b Sex—approximately 5% higher in males
  c Age—higher in youth than in age
  d Abnormal functioning of certain endocrines, particularly thyroid gland
  e Fever—each Centigrade degree rise in temperature increases BMR approximately 13%
  f Certain drugs (e.g., dinitrophenol) increase BMR
  g Other factors (e.g., pregnancy and emotions) increase BMR
2 How determined
  a Direct calorimetry—too expensive and too time consuming for wide use
  b Indirect calorimetry—measures amount of oxygen inspired in given time; 4.825 kilocalories of heat produced for each liter of oxygen consumed
3 Total metabolic rate—amount of heat produced by body in average twenty-four hours; equal to basal rate plus number of kilocalories produced by muscular work, eating and digesting food, and adjusting to cool temperatures; expressed in kilocalories per twenty-four hours

Energy balance and its relationship to body weight

1 Energy balance means that energy input (total kilocalories in food ingested) equals energy output (i.e., total metabolic rate expressed in kilocalories)
2 In order for body weight to remain constant (except for variations in water content), energy balance must be maintained—total kilocalories ingested must equal total metabolic rate
3 Body weight increases when energy input exceeds energy output—when total kilocalories ingested greater than total metabolic rate
4 Body weight decreases when energy input less than energy output—when total kilocalories ingested less than total metabolic rate; no diet will reduce weight unless it contains fewer kilocalories than the total metabolic rate of the individual eating diet

MECHANISMS FOR REGULATING FOOD INTAKE

1 Thermostat theory holds that moderate decrease in blood temperature acts as stimulant to appetite center so increases appetite; and increase in blood temperature (fever) produces opposite effect
2 Glucostat theory postulates that low blood glucose concentration or low rate of glucose utilization (e.g., diabetes mellitus) acts as stimulant to appetite center so increases appetite; and high blood glucose level produces opposite effect

HOMEOSTASIS OF BODY TEMPERATURE

In order to maintain homeostasis of body temperature heat production must equal heat loss

1 Heat production—by catabolism of foods in skeletal muscles and liver especially
2 Heat loss
  a By physical processes of evaporation, radiation, conduction, and convection
  b About 80% of heat loss occurs through skin; rest takes place through mucosa of respiratory, digestive, and urinary tracts
3 Thermostatic control of heat production and loss
  a Heat-dissipating mechanism—see Fig. 12-36
  b Heat-gaining mechanism
    1 Details not established but mechanism activated by decrease in blood temperature
    2 Responses—skin blood vessel constriction, shivering, and voluntary muscle contractions
4 Skin thermal receptors—stimulation of skin thermal receptors gives rise to sensations of heat or cold; also initiates voluntary movements to reduce these sensations—e.g., fanning oneself to cool off or exercising to warm up

review questions

1 Name and describe the coats that compose the walls of the esophagus, stomach, and intestines.
2 Differentiate between the peritoneum, the mesentery, and the omentum.
3 Explain what the tonsils and adenoids are and their locations.
4 Give the names and number of deciduous teeth and of permanent teeth.
5 Discuss the functions of gastric juice.
6 What juices digest proteins? Carbohydrates? Fats?
7 Differentiate between the end products of digestion and the end products of metabolism.
8 Contrast the function or purpose of digestion with the function or purpose of metabolism.
9 Compare anabolism and catabolism as to type of chemical reactions involved and end products.
10 Compare digestion and anabolism as to type of chemical reactions.
11 Discuss the functions of the pancreas in digestion and metabolism.
12 Compare proteins, carbohydrates, and fats as to functions they serve in the body. Mention both similarities and differences.
13 Compare proteins, carbohydrates, and fats as to their metabolic wastes.
14 Differentiate between digestive and metabolic wastes.
15 Differentiate between basal and total metabolic rates.
16 Compare absorption of the three kinds of foods.

Situation: A patient has obstructive jaundice.

17 Which duct, or ducts, might be obstructed to produce the symptom jaundice? Explain.
18 Would this affect this patient's digestion, absorption, or metabolism? Explain.
19 Explain why vitamin K might be given to this patient.

Situation: A patient has advanced liver disease.

20 Discuss possible effects on digestion, absorption, or metabolism.
21 What other functions might be affected?

Situation: Mrs. A., who weighs 160 lb, is 5 ft, 5 in tall, and is 50 years of age, has been given a reducing diet by her doctor. She complains to you that she "knows it won't work" because "it isn't what she eats that makes her fat." As proof, she says that both her husband and son "eat twice as much" as she and so does her younger sister.

22 How would you answer Mrs. A.?
23 Mrs. A. also tells you that she does not like so much meat and so many eggs and that she would rather do without some of these and have some pie or cake each day. How would you answer her?

13

The urinary
system

Organs
Kidneys
　Gross anatomy
　　Size, shape, and
　　　location
　　External structure
　　Internal structure
　Microscopic anatomy
　Physiology
　　Functions
　　How kidneys excrete
　　　urine
　　Mechanisms that control
　　　volume of urine
　　　excreted
　　Influence of kidneys on
　　　blood pressure
Ureters
　Location and structure
　Function
　Correlations
Bladder
　Structure and location
　Functions
Urethra
　Structure and location
　Functions

Urine
Physical characteristics
Chemical composition
Definitions

The urinary system consists of those organs that produce urine and eliminate it from the body. They are two kidneys, two ureters, one bladder, and one urethra (Fig. 13-1). The excretion of urine and its elimination from the body are vital functions since together they constitute one of the most important mechanisms for maintaining homeostasis. As one individual has phrased it, "The composition of the blood (and internal environment) is determined not by what the mouth ingests but by what the kidney keeps."[*]

The substances excreted from the kidneys and other excretory organs are listed in Table 13-1.

Organs
■Kidneys
Gross anatomy
Size, shape, and location

The kidneys resemble lima beans in shape. An average-sized kidney measures approximately 4 1/2 inches in length, 2 to 3 inches in width, and 1 inch in thickness. Usually the left kidney is slightly larger than the right.

The kidneys lie behind the parietal peritoneum, against the posterior abdominal wall, at the level of the last thoracic and first three lumbar vertebrae (or, as you

[*]From Smith, H. W.: Lectures on the kidney, Lawrence, Kansas, 1943, University of Kansas, p. 3.

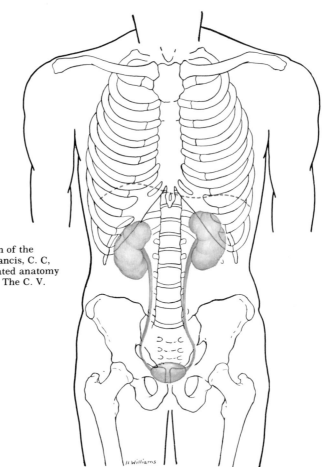

Fig. 13-1 Normal position of the urinary organs. (From Francis, C. C, and Farrell, G. L.: Integrated anatomy and physiology, St. Louis, The C. V. Mosby Co.)

Table 13-1. Excretory organs of the body

Excretory organ	*Substance excreted*
Kidneys	Nitrogenous wastes (from protein catabolism) Toxins (e.g., from bacteria) Water (from ingestion and from catabolism) Mineral salts
Skin (sweat glands)	Water Mineral salts Small amounts of nitrogenous wastes
Lungs	Carbon dioxide (from catabolism) Water
Intestine	Wastes from digestion (cellulose, connective tissue, etc.) Some metabolic wastes (e.g., bile pigments; also salts of calcium and other heavy metals)

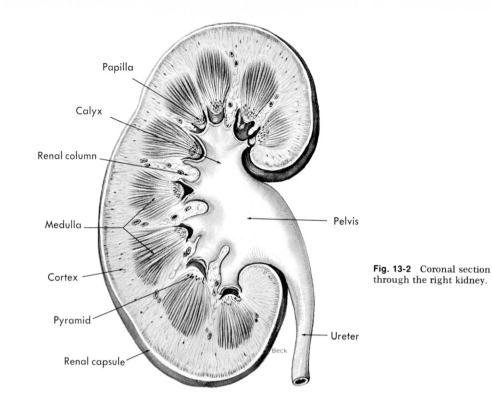

Papilla

Calyx

Renal column

Medulla

Cortex

Pyramid

Renal capsule

Pelvis

Ureter

Beck

Fig. 13-2 Coronal section through the right kidney.

can see in Fig. 13-1, just above the waist-line). The liver pushes the right kidney down to a level somewhat lower than the left. A heavy cushion of fat normally keeps the kidneys up in position. Very thin individuals may suffer from ptosis (dropping) of one or both of these organs. Connective tissue (renal fasciae) anchors the kidneys to surrounding structures and helps to maintain their normal position.

External structure

The mesial surface of each kidney presents a concave notch called the *hilum*. Structures enter the kidneys through this notch just as they enter the lung through its hilum. A tough white fibrous capsule encases each kidney.

Internal structure

When a coronal section is made through the kidney, two kinds of substances are seen composing its interior: an outer layer, the *cortex*, and an inner portion, the *medulla* (Fig. 13-2). The latter is divided into

a dozen or more triangular wedges, the *renal pyramids*. The bases of the pyramids face the cortex, and their apices or *renal papillae* face the center of the kidney. The pyramids have a striated appearance as contrasted with the smooth texture of the cortical substance. The cortex extends inward between each two pyramids, forming the *renal columns*.

Microscopic anatomy

Microscopic examination reveals the kidney to be composed of peculiarly shaped structures resembling tiny funnels with proportionately long convoluted stems. As you read about these anatomical funnels, identify each of their parts in Fig. 13-3. Note that *Bowman's capsule* consists of two layers of flat epithelial cells with a space between the layers. Observe also that Bowman's capsule has invaginated in it a cluster of capillaries designated as a *glomerulus*. A Bowman's capsule and its partially encased glomerulus is named a *renal corpuscle* (or *malpighian corpuscle*).

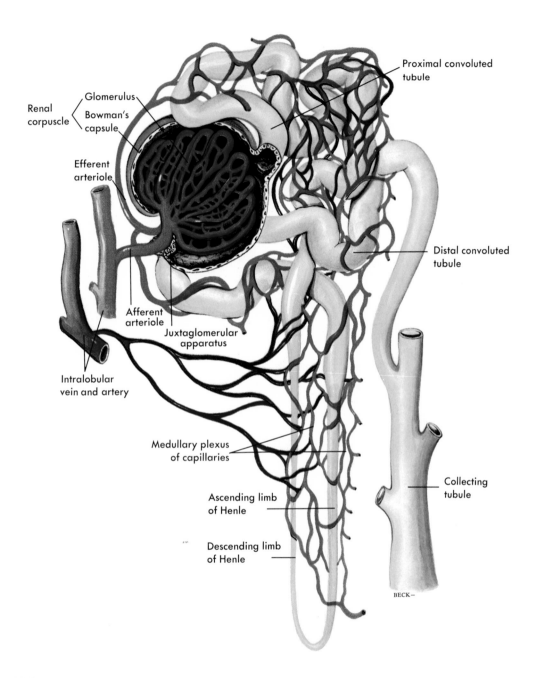

Renal corpuscle { Glomerulus / Bowman's capsule

Efferent arteriole

Afferent arteriole

Juxtaglomerular apparatus

Intralobular vein and artery

Proximal convoluted tubule

Distal convoluted tubule

Medullary plexus of capillaries

Ascending limb of Henle

Descending limb of Henle

Collecting tubule

BECK—

Fig. 13-3 The nephron unit with its blood vessels. Blood flows through nephron vessels as follows: intralobular artery → afferent arteriole → glomerulus → efferent arteriole → peritubular capillaries (around the tubules) → venules → intralobular vein.

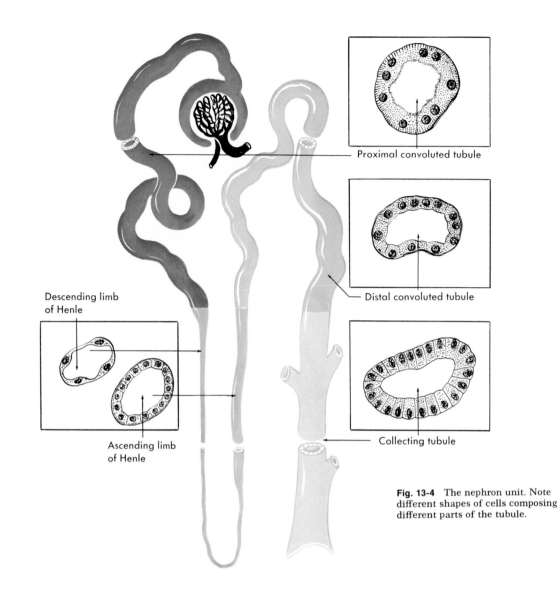

Proximal convoluted tubule

Distal convoluted tubule

Descending limb
of Henle

Ascending limb
of Henle

Collecting tubule

Fig. 13-4 The nephron unit. Note
different shapes of cells composing
different parts of the tubule.

Blood flows into each glomerulus by way
of an afferent arteriole and out of it by way
of an efferent arteriole – a unique arrangement. Blood usually flows out of capillaries
into what kind of vessels? Extending from
each Bowman's capsule is a renal tubule.
It consists of several sections. The first
is known as the *proximal convoluted
tubule;* proximal because it is the segment
nearest the tubule's origin from the Bowman's capsule and convoluted because it
pursues a tortuous rather than a straight
course. The proximal tubule becomes the
descending limb of the *loop of Henle,*

which becomes the *ascending limb,* which
becomes the *distal convoluted tubule,*
which terminates in a straight or *collecting tubule.* Now look at Fig. 13-4. The
fact that the cells composing different
parts of the renal tubule vary in shape
suggests that they perform different functions. Whether or not they do you will discover in later paragraphs.

Collecting tubules join larger tubules,
and all the larger collecting tubules of
one renal pyramid converge to form one
tube that opens at a renal papilla into
one of the small calyces. Bowman's cap-

sules and both convoluted tubules lie in the cortex of the kidney, whereas the loops of Henle and collecting tubules lie in its medulla.

A glomerulus, Bowman's capsule, and its tubule – proximal, convoluted, loop of Henle, and distal convoluted portions – together constitute a *nephron,* the structural and functional unit of the kidney. *Gray's Anatomy* says that there are about a million and a quarter nephrons in each kidney.

Physiology
Functions

The function of the kidneys is to excrete urine, a life-preserving function because homeostasis depends upon it. More than any other organ in the body, the kidneys can adjust the amounts of water and electrolytes leaving the body so that they equal the amounts of these substances entering the body. The vital conditions of fluid and electrolyte balance and acid-base balance, therefore, depend most of all upon adequate kidney functioning. Here are just a few of the blood constituents that cannot be held to their normal concentration ranges if the kidneys fail: sodium, potassium, chloride, and nitrogenous wastes from protein metabolism such as urea. In short, kidney failure means homeostasis failure and, if not relieved, inevitable death.

In addition to excreting urine, the kidneys are now known to influence blood pressure (Fig. 9-13, p. 282).

How kidneys excrete urine

Three processes – glomerular filtration, tubular reabsorption, and tubular secretion – together accomplish the kidneys' function of urine excretion (Figs. 13-5 and 13-6).

Glomerular filtration. Filtration, the first step in the formation of urine, takes place from the blood in the glomeruli out into the Bowman's capsules. Water and solutes filter out of the glomeruli even faster than out of ordinary capillaries. At least two structural features make renal corpuscles especially effective filtration membranes. For one thing, glomeruli have many more pores than other capillaries. And the fact that the efferent arteriole is smaller in diameter than the afferent arteriole makes for a higher resistance to blood flow out of glomeruli than out of other capillaries and, therefore, for a higher blood pressure (hydrostatic pressure, that is) in glomeruli than in ordinary capillaries. Glomerular hydrostatic pressure, for instance, averages about 70 mm Hg, whereas capillary hydrostatic pressure averages only about 30 mm Hg.

Fluid moves out of the glomeruli into the Bowman's capsules for the same reason that it moves out of other capillaries into interstitial fluid or moves from any area to another – because a pressure gradient exists between the two areas. Normally, glomerular hydrostatic pressure, blood colloidal osmotic pressure, and capsular hydrostatic pressure together determine the pressure gradient (commonly spoken of as the *effective filtration pressure* or EFP) between glomeruli and capsules. And of these three pressures, the main driving force, the main determinant of the effective filtration pressure, is glomerular hydrostatic pressure. It tends to move fluid out of the glomeruli. In contrast, the *capsular* hydrostatic pressure and the blood colloidal osmotic pressure both exert force in the opposite direction. (If you do not recall why osmotic pressure is a "water-pulling" rather than a "water-pushing" force, reread p. 24.) Suppose that there is a glomerular hydrostatic pressure of 70 mm Hg and that it is opposed by a capsular hydrostatic pressure of 20 and a blood colloidal osmotic pressure of 30. There would then be a net or effective filtration pressure of 20 mm Hg (70 – 20 – 30). In other words, there would be a

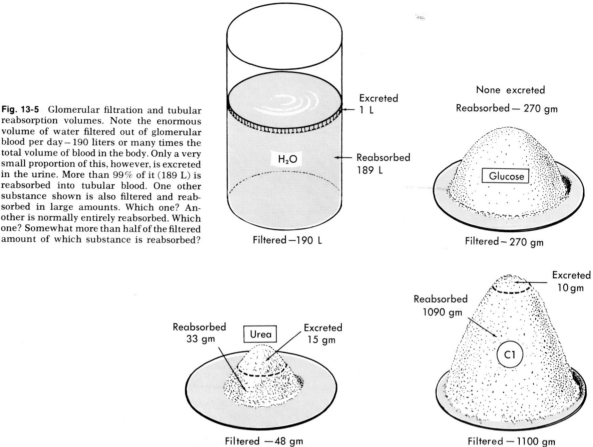

Fig. 13-5 Glomerular filtration and tubular reabsorption volumes. Note the enormous volume of water filtered out of glomerular blood per day—190 liters or many times the total volume of blood in the body. Only a very small proportion of this, however, is excreted in the urine. More than 99% of it (189 L) is reabsorbed into tubular blood. One other substance shown is also filtered and reabsorbed in large amounts. Which one? Another is normally entirely reabsorbed. Which one? Somewhat more than half of the filtered amount of which substance is reabsorbed?

pressure gradient of 20 mm Hg, causing fluid to filter out of the glomeruli into the capsules.

Another factor, *capsular* colloidal osmotic pressure, operates when disease has increased glomerular permeability enough to allow blood protein molecules to diffuse out of the blood into Bowman's capsule. Under these circumstances the capsular filtrate exerts an osmotic pressure in opposition to blood osmotic pressure. Capsular osmotic pressure tends to draw water out of the blood and so constitutes another force to be added to glomerular hydrostatic pressure in determining the effective filtration pressure. By using the following formula and figures given in Table 13-2, you will be able to figure out for yourself how changes in the various pressures cause

changes in the glomerular filtration rate.

$$
\begin{aligned}
\text{Effective} \\
\text{filtration} \\
\text{pressure}
\end{aligned}
=
\left(
\begin{aligned}
&\text{Glomerular} &&\text{Capsular} \\
&\text{hydrostatic} &+\; &\text{osmotic} \\
&\text{pressure} &&\text{pressure}
\end{aligned}
\right)
$$

$$
\text{minus}
\left(
\begin{aligned}
&\text{Glomerular} &&\text{Capsular} \\
&\text{osmotic} &+\; &\text{hydrostatic} \\
&\text{pressure} &&\text{pressure}
\end{aligned}
\right)
$$

For example, glomerular hydrostatic pressure may decrease sharply under stress conditions such as following severe hemorrhage. Table 13-2 gives pressures which might exist under these circumstances. Using these figures, what do you compute the effective glomerular filtration pressure to be? Do you think it is true, after doing this computation, that glomerular filtration ceases entirely when glomerular hydrostatic pressure falls below a certain critical level?

Table 13-2. Normal and abnormal glomerulocapsular pressures

	Hydrostatic pressure	*Colloidal osmotic pressure*
Normal		
Glomerular blood	70 mm Hg	30 mm Hg
Capsular filtrate	20 mm Hg	———
Abnormal		
Glomerular blood	44 mm Hg	28 mm Hg
Capsular filtrate	20 mm Hg	4 mm Hg

Kidney disease, as previously noted, sometimes causes a loss of blood proteins in the urine. Blood protein concentration then decreases and causes blood colloidal osmotic pressure to decrease. Suppose that blood colloidal osmotic pressure were to decrease to 20 mm Hg while the other normal figures given in Table 13-2 remained the same. Use these figures to compute the effective filtration pressure. If, based on this example, you were to formulate a principle, would you say that a decrease in blood protein concentration tends to increase or decrease the glomerular concentration rate? (Check your answer in the paragraph after the next one.)

Glomerular hydrostatic pressure is regulated by mechanisms that change the size of the afferent and efferent arterioles and is also influenced by changes in systemic blood pressure. For example, sympathetic impulses cause constriction of both afferent and efferent arterioles. But with intense sympathetic stimulation, the afferent arteriole becomes much more constricted than the efferent. Consequently, glomerular hydrostatic pressure falls. Sometimes under severe stress conditions, it drops to a level too low to maintain filtration, and the kidneys "shut down" completely. In more technical language, renal suppression occurs (for example, see Table 13-2). Glomerular hydrostatic pressure and filtration are directly related to systemic blood pressure. By this, we mean that an increase in blood pressure tends to produce an increase in glomerular pressure and in the filtration rate. The converse is also true.

Glomerular filtration is inversely related to blood colloidal osmotic pressure and blood protein concentration. Specifically, a decrease in blood protein concentration causes a decrease in blood osmotic pressure which, in turn, causes an increase in glomerular filtration. Normally the glomerular filtration rate averages about 125 milliliters per minute in men and somewhat less in women.

Tubular reabsorption (Fig. 13-6). The second step in urine formation is reabsorption of substances needed by the body — that is, of most (usually from 97% to 99%) of the water and part of the solutes from the glomerular filtrate back into the blood. Reabsorption is the function of the cells composing the walls of the convoluted tubules, loop of Henle, and collecting tubules. In the execution of this function these cells display astonishing powers of selection and discrimination. For example, they absorb most efficiently the substances that the body needs most vitally, such as Na^+, Cl^-, HCO_3^-, H_2O, and glucose.

Reabsorption is not merely a physical matter of diffusion and osmosis but consists also of somewhat obscure active transport mechanisms that require energy release by tubule cells. Because reabsorption is achieved partly by active transport mechanisms, it is one of the first powers diminished in kidney disease. Practical use is made of this fact in tests that measure the concentration of urine at different times of the day and thereby evaluate the active transport powers of the tubules.

Let us consider first the mechanism for reabsorption of solutes. Active transport mechanisms absorb glucose, amino acids, and other nutrient substances and at least some electrolytes (salts). Proximal tubule

453

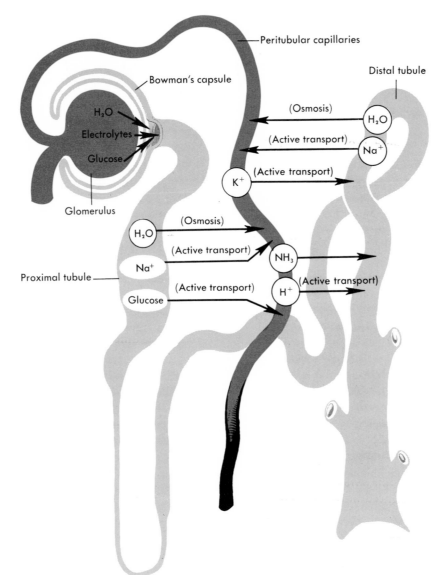

Fig. 13-6 Diagram showing glomerular filtration, tubular reabsorption, and tubular secretion—the three processes by which the kidneys excrete urine. In the proximal tubule, note that water is reabsorbed from the tubular filtrate into blood by osmosis but sodium and glucose are reabsorbed mainly by active transport mechanisms. Note, too, that water and sodium are also reabsorbed from the distal tubule. But potassium and hydrogen ions and ammonia are secreted into the distal tubule from the blood.

cells transport glucose and other nutrients. So these vitally important materials, after filtering out of the glomeruli, return to the blood from the first part of the renal tubule. The transport mechanisms function so effectively that normally no glucose at all is lost in the urine (Fig. 13-5). If, however, blood glucose exceeds a certain threshold amount (often around 150 milligrams per 100 milliliters of blood), the glucose transport mechanism cannot reabsorb all of it, and the excess remains in the urine. In other words, this mechanism has a maximum capacity for moving glucose molecules back into the blood. Occasionally this capacity is greatly reduced, in which case glucose appears in the urine (glycosuria or glucosuria), even though the blood sugar may be normal. This condition is known as renal diabetes or renal glycosuria.

Electrolytes are reabsorbed partly by active transport mechanisms and partly by diffusion. Sodium ions, for example, are known to be actively transported from all parts of the tubule. And this has important effects on the reabsorption of certain other ions and on water reabsorption. Sodium ions are positive ions (cations). Therefore, when they diffuse from the tubules into the peritubular blood, the latter momentarily becomes electropositive to the tubular filtrate. This attracts negative ions (notably chloride) and causes an equal number of them to diffuse out of the tubule. Sodium movement out of the tubule also produces a momentary disequilibrium between the tubular filtrate and peritubular blood osmotic pressures. It subtracts sodium ions from the filtrate but adds them to the blood — hence, the osmotic pressure of peritubular blood increases above the level of the filtrate osmotic pressure. Water, obeying the law of osmosis, follows rapidly along to reestablish osmotic equilibrium between the filtrate and blood. In short, the active transport of sodium out of the tubule is the main factor causing osmosis of water out of it. And, therefore, the water reabsorbtion from the proximal tubule by this mechanism has been described as "obligatory water reabsorption" — obligatory because demanded by the law of osmosis. The distal and collecting tubules, in contrast, have the ability or faculty to vary the volume of water they reabsorb. Hence the term "facultative water reabsorption" indicates a function of distal and collecting tubules.

How much water finally moves back into the blood depends upon the presence of a hormone released by the neurohypophysis (posterior pituitary gland). Its name, antidiuretic hormone (ADH), is appropriate because it works against diuresis (large volume of urine). In other words, ADH decreases the amount of urine produced. The dynamics of the ADH mechanism are extremely complex, but the results are clear — increased amounts of water are reabsorbed from both the distal and collecting tubules by osmosis (Fig. 13-7). Therefore, urine volume decreases while its concentration increases. Urine becomes hypertonic to blood under the ADH influence. How important the ADH-tubule mechanism is can be appreciated from the fact that water balance depends upon it and cannot be maintained if it fails (Fig. 15-9, p. 509).

Certain adrenal cortex hormones, classified as mineralocorticoids (M-C's), regulate the reabsorption of electrolytes — of ions, that is. Of the natural mineralocorticoids, aldosterone is the most potent. See p. 280 for a discussion of its action.

Tubular secretion. In addition to reabsorption, tubule cells also secrete certain substances. Tubular secretion (or tubular excretion, as it is also called) means the movement of substances out of the blood into the filtrate in the kidney tubules. Tubular reabsorption, you recall, means

Table 13-3. Reabsorption and secretion mechanisms (for moving substances out of and into renal tubules)

Substance	Moved out of	By mechanism of	Into
Electrolytes, e.g.: Sodium ions	Blood in glomeruli	*Filtration*	Bowman's capsules
	Filtrate in tubules (all parts)	*Active transport* (reabsorption)	Blood in peritubular capillaries
Potassium ions	Blood in glomeruli Blood in peritubular capillaries	*Filtration* *Active transport* (secretion)	Bowman's capsules Filtrate in tubules
Chloride ions	Blood in glomeruli	*Filtration*	Bowman's capsules
	Filtrate in tubules	*Diffusion* (secondary to Na^+-transport)	Blood in peritubular capillaries
Nutrients, e.g.: Glucose	Blood in glomeruli	*Filtration*	Bowman's capsules
	Filtrate in proximal tubules	*Active transport* (reabsorption)	Blood in peritubular capillaries
Wastes, e.g.: Urea (most abundant solute in urine)	Blood in glomeruli	*Filtration*	Bowman's capsules
	Filtrate in tubules	*Diffusion*	Blood in peritubular capillaries
Water	Blood in glomeruli	*Filtration*	Bowman's capsules
	Filtrate in proximal tubules	*Osmosis* (obligatory water reabsorption)	Blood in peritubular capillaries
	Filtrate in distal and collecting tubules	*Osmosis* (facultative water reabsorption; ADH-controlled)	Blood in peritubular capillaries
Hydrogen ions	Blood in peritubular capillaries	*Active transport* (secretion)	Filtrate in distal tubules
Ammonia	Distal tubule cells	*Diffusion*	Filtrate in distal tubules

the movement of substances in the opposite direction—that is, out of the tubule filtrate into the blood. In both reabsorption and secretion, some substances are moved by active transport and some by passive mechanisms (diffusion and osmosis). For example, sodium ions and glucose molecules are reabsorbed and potassium and hydrogen ions are secreted by active transport mechanisms. But water is reabsorbed by osmosis and ammonia is secreted by diffusion, both passive processes. Tubule cells also secrete certain drugs—for example, penicillin and para-aminohippuric

acid (PAH), two clinically important substances. For example, by administering a known amount of PAH and then measuring the amount excreted in the urine, tubular secretion can be evaluated. If the tubules are impaired, they, of course, cannot move as much PAH out of the blood, and less appears in the urine within a given time.

The reabsorption and secretion mechanisms are summarized in Table 13-3. Table 13-4 and Fig. 13-6 summarize the functions of the different parts of the nephron in forming urine.

Table 13-4. Functions of different parts of nephron in urine formation

Part of nephron	*Function*	*Substance moved*
Glomerulus	Filtration	Water All solutes except colloids such as blood proteins
Proximal tubule and loop of Henle	Reabsorption by active transport	Na^+ and probably some other ions; nutrients—glucose and amino acids
	Reabsorption by diffusion (secondary to active transport)	Cl^-, HCO_3^-, and probably some other ions
	Obligatory water reabsorption by osmosis	Water
Distal and collecting tubules	Reabsorption by active transport	Na^+ and probably some other ions
	Facultative water reabsorption by osmosis (ADH-controlled)	Water
	Secretion by diffusion	Ammonia
	Secretion by active transport	K^+, H^+, and some drugs

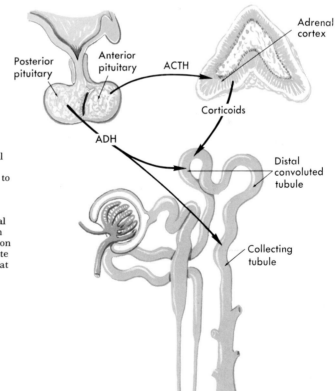

Fig. 13-7 **ADH** and corticoid control of distal renal tubule reabsorption. **ADH,** a posterior pituitary hormone, stimulates distal and collecting tubules to increase water reabsorption into blood. **ACTH,** an anterior pituitary hormone, stimulates the adrenal cortex to secrete chiefly glucocorticoids. Glucocorticoids, in turn, mildly stimulate the distal tubules to reabsorb sodium and excrete potassium more rapidly. They also have a sensitizing action on the tubules which permits them to adjust their rate of water reabsorption to fluid intake volume so that diuresis occurs following a water load.

Mechanisms that control volume of urine excreted

The amount of urine excreted is regulated chiefly by ADH and aldosterone, hormones that influence the amount of water reabsorbed by the distal tubule cells (pp. 272 and 280). It is thought that only under pathological conditions does the rate of glomerular filtration change enough to alter the volume of urine produced.

The total volume of extracellular fluid is known to influence the volume of urine secreted, presumably by increasing or decreasing ADH secretion. The exact mechanism, however, by which it accomplishes this is still unproved. It may operate through its effect on ECF sodium concentration. An increase in ECF volume tends to decrease blood sodium concentration, and a decrease in ECF volume tends to increase blood sodium concentration. For the effect of the latter on ADH secretion, see Fig. 15-9 (p. 509). The common experience of increased output following rapid ingestion of a large amount of fluid attests to the fact that an increased ECF volume is soon followed by increased urinary output. Evidence of the second principle is the oliguria or even anuria in dehydrated patients.

Another factor that helps regulate the amount of urine secreted is the total amount of solutes excreted through the kidneys. The more solutes to be excreted, the greater the volume of urine. In diabetes, for example, more solids are excreted than normally due to the excess glucose that "spills over" into the urine. And therefore the volume of urine secreted daily is greater than in the normal individual.

Influence of kidneys on blood pressure

Clinical observation and animal experiments have established the fact that destruction of a large proportion of total kidney tissue usually results in the development of hypertension. This happens frequently, for example, in patients who have severe renal arteriosclerosis or glomerulonephritis. Many experiments have been performed and various theories devised to explain the mechanism responsible for "renal hypertension." Ischemic kidneys are known to produce a proteolytic enzyme, *renin*, which hydrolyzes one of the blood proteins (a globulin) to produce *angiotonin* (hypertensin), which causes arteriolar constriction and a rise in blood pressure (Fig. 9-13, p. 282).

Ureters

Location and structure

The two ureters are tubes from 10 to 12 inches long. At their widest point they measure less than 1/2 inch in diameter. They lie behind the parietal peritoneum and extend from the kidneys to the posterior surface of the bladder. As the upper end of each ureter enters the kidney, it enlarges into a funnel-shaped basin named the *renal pelvis*. The pelvis expands into several branches called *calyces*, Each calyx contains a renal papilla. As urine is secreted, it drops out of the collecting tubules, whose openings are in the papillae, into the calyces, thence into the pelvis, and down the ureters into the bladder.

The walls of the ureters are composed of three coats: a lining coat of mucous membrane, a middle coat of two layers of smooth muscle, and an outer fibrous coat.

Where the ureters empty into the bladder, there is a fold of mucous membrane that serves as a valve preventing the backflow of urine into the ureter when the bladder contracts.

Function

The ureters, together with their expanded upper portions, the pelves and calyces, collect the urine as it forms and drain it

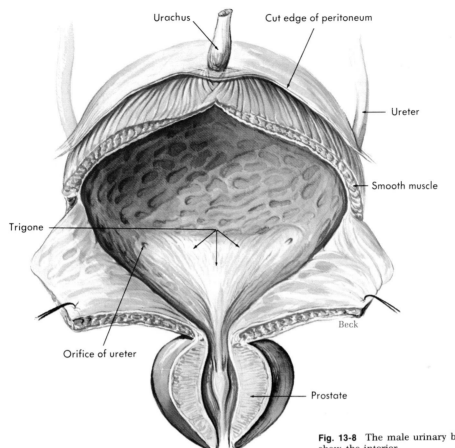

Urachus

Cut edge of peritoneum

Ureter

Smooth muscle

Trigone

Beck

Orifice of ureter

Prostate

Fig. 13-8 The male urinary bladder cut to show the interior.

into the bladder. Peristaltic waves (about one to five per minute) force the urine down the ureters and into the bladder.

Correlations

Stones known as *renal calculi* sometimes develop within the kidney. Urine may wash them into the ureter, where they cause extreme pain if they are large enough to distend its walls.

■Bladder
Structure and location

The bladder is a collapsible bag located directly behind the symphysis pubis. It lies below the parietal peritoneum, which covers only its superior surface. Three layers of smooth muscle (known collectively as the detrusor muscle) fashion its walls,

while mucous membrane, arranged in rugae, forms its lining. Because of the rugae and the elasticity of its walls, the bladder is capable of considerable distention, although its capacity varies greatly with individuals. There are three openings in the floor of the bladder—two from the ureters and one into the urethra. The ureter openings lie at the posterior corners of the triangular-shaped floor (the trigone) and the urethral opening at the anterior and lower corner (Fig. 13-8).

Functions

The bladder performs two functions.

1 It serves as a reservoir for urine before it leaves the body.

2 Aided by the urethra, it expels urine from the body. Distention of the bladder

459

with urine stimulates stretch receptors in the bladder wall. This initiates reflex contraction of the bladder wall but simultaneous relaxation of the internal sphincter, followed rapidly by relaxation of the external sphincter and emptying of the bladder. Parasympathetic fibers transmit the impulses that cause contractions of the bladder and relaxation of the internal sphincter. Voluntary contraction of the external sphincter, however, to prevent or terminate micturition is learned. Voluntary control of micturition is possible only if the nerves supplying the bladder and urethra, the projection tracts of the cord and brain, and the motor area of the cerebrum are all intact. Injury to any of these parts of the nervous system, by a cerebral hemorrhage or cord injury, for example, results in involuntary emptying of the bladder at intervals. Involuntary micturition is called *incontinence.* In the average bladder, 250 milliliters of urine will cause a moderately distended sensation and therefore the desire to void.

Occasionally an individual is unable to void even though the bladder contains an excessive amount of urine. This condition is known as *retention.* It often follows pelvic operations and childbirth. Catheterization (introduction of a rubber tube through the urethra into the bladder to remove urine) is used to relieve the discomfort accompanying retention. A more serious complication, which is also characterized by the inability to void, is called *suppression.* In this condition, the patient cannot void because the kidneys are not secreting any urine, and therefore the bladder is empty. Catheterization, of course, gives no relief for this condition.

■Urethra

Structure and location

The urethra is a small tube leading from the floor of the bladder to the exterior. In the female it lies directly behind the symphysis pubis and anterior to the vagina. It extends up, in, and back for a distance of about 1 to 1 1/2 inches. The male urethra follows a tortuous course for a distance of approximately 8 inches. Immediately below the bladder, it passes through the center of the prostate gland, then between two sheets of white fibrous tissue connecting the pubic bones, and then through the penis, the external male reproductive organ. These three parts of the urethra are known, respectively, as the prostatic portion, the membranous portion, and the cavernous portion.

The opening of the urethra to the exterior is named the *urinary meatus.*

Mucous membrane lines the urethra as well as the rest of the urinary tract.

Functions

As the terminal portion of the urinary tract, the urethra serves as the passageway for eliminating urine from the body. In addition, the male urethra is the terminal portion of the reproductive tract and serves as the passageway for eliminating the reproductive fluid (semen) from the body. The female urethra serves only the urinary tract.

Urine

Physical characteristics

The physical characteristics of normal urine are listed in Table 13-5.

Chemical composition

Urine is approximately 95% water, in which are dissolved several kinds of substances, the most important of which are listed as follows:

1 *nitrogenous wastes* from protein metabolism—such as urea (most abun-

Table 13-5. Physical characteristics of normal urine

Amount (24 hours)	Three pints (1,500 ml) but varies greatly according to fluid intake, amount of perspiration, and several other factors
Clearness	Transparent or clear; upon standing, becomes cloudy
Color	Amber or straw-colored; varies according to amount voided—less voided, darker the color, usually; diet also may change color—e.g., reddish color from beets
Odor	"Characteristic"; upon standing, develops ammonia odor due to formation of ammonium carbonate
Specific gravity	1.015 to 1.020; highest in morning specimen
Reaction	Acid but may become alkaline if diet is largely vegetables; high-protein diet increases acidity; stale urine has alkaline reaction due to decomposition of urea forming ammonium carbonate; normal range for urine pH 4.8-7.5; average, about 6; rarely becomes more acid than 4.5 or more alkaline than 8

dant solute in urine), uric acid, ammonia and creatinine

2 *electrolytes*—mainly the following ions: sodium, potassium, ammonium, chloride, bicarbonate, phosphate, and sulfate; amounts and kinds of minerals vary with diet and other factors

3 *toxins*—during disease, bacterial poisons leave the body in the urine—an important reason for "forcing fluids" on patients suffering with infectious diseases, so as to dilute the toxins that might damage the kidney cells if they were eliminated in a concentrated form

4 *pigments*

5 *hormones*

6 various *abnormal constituents* some-times found in urine—such as glucose, albumin, blood, casts, calculi, etc.

Definitions

1 *glycosuria*, or *glucosuria*—sugar (glucose) in the urine

2 *hematuria*—blood in the urine

3 *pyuria*—pus in the urine

4 *cast*—substances, such as mucus, that harden and form molds inside the uriniferous tubules and then are washed out into the urine; microscopic in size

5 *dysuria*—painful urination

6 *polyuria*—unusually large amounts of urine

7 *oliguria*—scanty urine

8 *anuria*—absence of urine

outline summary

ORGANS
Kidneys
  Gross anatomy
1 Size, shape, and location
  a 4 1/2 in × 2 in × 3 in × 1 in
  b Shaped like lima beans
  c Lie against posterior abdominal wall, behind peritoneum, at level of last thoracic and first three lumbar vertebrae; right kidney slightly lower than left

2 External structure
  a Hilum, concave notch on mesial surface
  b Enveloping capsule of white fibrous tissue
3 Internal structure
  a Outer layer called cortex
  b Inner portion called medulla
  c Renal pyramids triangular wedges of medullary substance, apices of which called papillae
  d Renal columns inward extensions of cortex between pyramids

Microscopic anatomy

1 Cluster of capillaries invaginated in Bowman's capsule called *glomerulus*
2 Bowman's capsule together with glomerulus constitute renal (malpighian) corpuscle
3 Physiological unit of kidney called nephron — consists of renal corpuscle, convoluted tubules, loop of Henle, and straight tubule

Physiology

1 Functions
  a Excrete urine, by which various toxins and metabolic wastes excreted and composition and volume of blood regulated
  b Influence blood pressure
2 How kidneys excrete urine (see Table 13-3 and Fig. 13-7)
  a Filtration of substances from blood in glomeruli into Bowman's capsules
  b Reabsorption of most of water and part of solutes from tubular filtrate back into blood
  c Secretion of K^+, H^+, NH_3, and some other substances into tubular filtrate from blood
3 Mechanisms that control volume of urine excreted are factors that change
  a Amount of water reabsorbed from tubular filtrate into blood — most important determinant of amount of urine formed; posterior pituitary ADH increases water reabsorption from distal and collecting tubules; corticoids (especially, aldosterone) increase sodium reabsorption and therefore also increase water reabsorption
  b Rate of filtration from glomeruli — normally quite constant at about 125 ml per minute; varies directly with changes in glomerular blood pressure — i.e., increased glomerular blood pressure tends to increase glomerular filtration rate and therefore urine volume and vice versa but glomerular blood pressure normally quite constant; decreased blood colloidal osmotic pressure tends to increase glomerular filtration and urine volume and vice versa
  c Total extracellular fluid volume; urine output increases following increase in total ECF and decreases following decrease in ECF
  d Amount of solutes excreted in urine; urine output increases when solutes increase

Ureters

1 Location and structure
  a Lie retroperitoneally
  b Extend from kidneys to posterior part of bladder floor
  c Ureter expands as it enters kidney, becoming renal pelvis, which is subdivided into calyces, each of which contains renal papilla
  d Walls of smooth muscle with mucous lining and fibrous outer coat
2 Function — collect urine and drain it into bladder

Bladder

1 Structure and location
  a Collapsible bag of smooth muscle lined with mucosa
  b Lies behind symphysis pubis, below parietal peritoneum
  c Three openings — one into urethra and two into ureters
2 Functions
  a Reservoir for urine
  b Expels urine from body by way of urethra, called micturition, urination, or voiding; retention, inability to expel urine from bladder; suppression, failure of kidneys to form urine

Urethra

1 Structure and location
  a Musculomembranous tube lined with mucosa
  b Lies behind symphysis, in front of vagina in female
  c Extends through prostate gland, fibrous sheet, and penis in male
  d Opening to exterior called urinary meatus
2 Functions
  a Female — passageway for expulsion of urine from body
  b Male — passageway for expulsion of urine and of reproductive fluid (semen)

URINE

1 Physical characteristics — see Table 13-5
2 Chemical composition — consists of approximately 95% water in which are dissolved:
  a Wastes from protein metabolism (urea, uric acid, creatinine, etc.)
  b Mineral salts (sodium chloride main one but various others according to diet)
  c Toxins — e.g., from bacteria
  d Pigments
  e Sex hormones
  f Abnormal constituents — e.g., glucose in diabetes and numerous others such as albumin, blood, casts, and calculi
3 Definitions
  a Glycosuria — sugar in urine
  b Hematuria — blood in urine
  c Pyuria — pus in urine
  d Casts — microscopic bits of substances that harden and form molds inside tubules
  e Dysuria — painful urination
  f Polyuria — excessive amounts of urine
  g Oliguria — scanty urine
  h Anuria — absence of urine

review questions

1 What four organs are excretory organs?
2 Which organs eliminate wastes of protein metabolism? Of digestion? Of carbohydrate and fat metabolism?
3 Name, locate, and give main function(s) of each organ of the urinary system.
4 How far and in which direction must a catheter be inserted to reach the bladder in the female? In the male?
5 Describe the microscopic structure of the kidney.
6 Describe the mechanism of urine formation, relating each step to part of the nephron that performs it.

Situation: An artificial kidney consists of a device in which blood flows directly from a patient's body through cellophane tubing immersed in a dialyzing fluid that contains prescribed amounts of various electrolytes and other substances. The two columns below show the composition of one patient's blood and of the dialyzing fluid used for his treatment with the artificial kidney.

	Blood plasma (in coiled tube) mEq/L	Dialyzing fluid (around coiled tube) mEq/L
Na^+	142	126
K^+	5	5
Ca^{++}	5	0
Mg^{++}	3	0
Cl^-	103	110
HCO_3^-	27	25
$HPO_4^=$	2	0
$SO_4^=$	1	0

	mg/100 ml	mg/100 ml
Glucose	100	1,750
Urea	26	0
Uric acid	4	0
Creatinine	1	0

7 What body structures do you think the cellophane tube substitutes for?
8 Which, if any, of the above substances diffuse out of the blood into the dialyzing fluid? Give your reasons.
9 Which, if any, of the above substances diffuse into the blood from the dialyzing fluid? Give your reasons.
10 Which, if any, of the above substances do not pass through the cellophane membrane in either direction? Give your reasons.
11 What reasons can you see for having the dialyzing fluid contain the concentration of each substance given above? What does it accomplish?

Situation: Results of a blood urea clearance test indicate that the glomerular filtration rate of a patient who has had a severe hemorrhage is less than 50% of normal.

12 Explain the mechanism responsible for the drop in the glomerular filtration rate.
13 Do you consider this a homeostatic mechanism? Does it serve a useful purpose? If so, what purpose?
14 What is the normal glomerular filtration rate?
15 What would you expect to be true of the volume of urine this patient would excrete — normal, polyuria, oliguria?
16 Define the following terms briefly: Bowman's capsule; calculi; calyces; casts; cystitis; dysuria; glomerulus; glycosuria; hematuria; incontinence; nephritis; oliguria; polyuria; ptosis; pyelitis; renal capsule; renal cortex; renal hilum; renal medulla; renal papilla; renal pelvis; renal pyramids; retention; suppression.

Reproduction of the human being

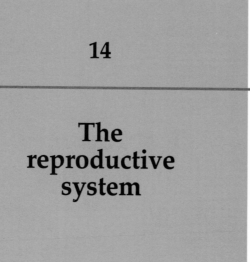

14

The reproductive system

Meaning and function

Male reproductive organs
Testes
    Structure and location
    Functions
    Structure of
      spermatozoa
Ducts of testes
    Epididymis
      Structure and location
      Functions
    Seminal duct (vas
      deferens; ductus
      deferens)
      Structure and location
      Function
      Correlation
    Ejaculatory duct
    Urethra
Accessory reproductive
  glands
    Seminal vesicles
      Structure and location
      Function
    Prostate gland
      Structure and location
      Function

Bulbourethral glands
    Structure and location
    Function
Supporting structures
  Scrotum (external)
  Penis (external)
    Structure
    Functions
  Spermatic cords (internal)
Composition and course of
  seminal fluid
Male fertility

Female reproductive organs
Uterus
  Structure
  Location
  Position
  Functions
Uterine tubes—fallopian
  tubes or oviducts
  Location
  Structure
  Function

Ovaries—female gonads
  Location and size
  Microscopic structure
  Functions
Vagina
  Location
  Structure
  Functions
Vulva
Perineum
Breasts
  Location and size
  Structure
  Function
  Mechanism controlling
    lactation

Female sexual cycles
Recurring cycles
  Ovarian cycles
  Endometrial or menstrual
    cycle
  Myometrial cycle
  Gonadotropic cycles
Control
  Cyclical changes in
    ovaries
  Cyclical changes in
    uterus
  Cyclical changes in
    amounts of FSH and
    estrogens secreted

Functions served by cycles
Menarche and
  menopause

Embryology
Meaning and scope
Value of knowlege of
  embryology
Steps in development of new
  individual
  Preliminary processes
  Developmental processes

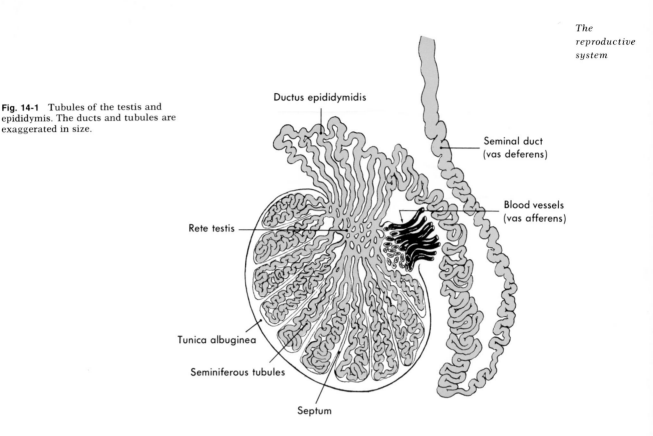

Fig. 14-1 Tubules of the testis and epididymis. The ducts and tubules are exaggerated in size.

Ductus epididymidis

Seminal duct (vas deferens)

Blood vessels (vas afferens)

Rete testis

Tunica albuginea

Seminiferous tubules

Septum

■Meaning and function

The reproductive system consists of those organs whose function is to produce a new individual—that is, to accomplish reproduction. And some reproductive organs produce hormones. Different forms of life reproduce in different ways, but all living organisms, no matter how simple or how complex, are able to perform this miracle.

Reproduction of cells, of course, must occur in order that reproduction of a multicellular organism may occur. But before pursuing this idea any further, we shall describe the various reproductive organs.

Male reproductive organs

Glands, ducts, and supporting structures compose the male reproductive system. The glands are the testes (paired), sem-inal vesicles (paired), prostate, and bulbo-urethral glands (paired). The ducts are the epididymis (paired), seminal ducts (vas deferens; ductus deferens) (paired), ejaculatory ducts (paired), and urethra. And the supporting structures are the scrotum, penis, and spermatic cords (paired). (See Plate XIV of the color insert, p. 60.)

■Testes
Structure and location

The testes are small ovoid glands that lie in a pouchlike, skin-covered structure called the *scrotum*. A white fibrous capsule encases each testis and sends partitions through its interior, dividing it into lobules. Each lobule contains a tiny, coiled *semi-niferous tubule* (Fig. 14-1) and numerous interstitial cells (of Leydig). The seminiferous tubules come together to form a plexus from which a few ducts emerge and enter the head of the epididymis.

Functions

The testes perform two primary functions: spermatogenesis and secretion of hormones.

1 Spermatogenesis, the production of spermatozoa (sperm), the male gametes or reproductive cells. The seminiferous tubules produce the sperm.

2 Secretion of hormones, chiefly testosterone (androgen or masculinizing hormone) by interstitial cells (Leydig cells). Testosterone serves the following general functions:

a It promotes "maleness"—that is, the development and maintenance of male secondary sex characteristics, of male accessory organs such as the prostate, seminal vesicles, etc., and of adult male sexual behavior.

b It helps regulate metabolism and is sometimes referred to as "the anabolic hormone" because of its marked stimulating effect on protein anabolism. And by stimulating protein anabolism, testosterone promotes growth of skeletal muscles (responsible for greater muscular development and strength of male) and promotes growth of bone. However, testosterone also promotes closure of the epiphyses. Early sexual maturation, therefore, generally leads to early epiphyseal closure and shortness. The converse also holds true: late sexual maturation, delayed epiphyseal closure, and tallness tend to go together.

c It plays a part in fluid and electrolyte metabolism. Testosterone has a mild stimulating effect on kidney tubule reabsorption of sodium and therefore water; it also promotes kidney tubule excretion of potassium.

d It inhibits anterior pituitary secretion of gonadotropins—namely, FSH and ICSH (interstitial cell–stimulating hormore; called LH or luteinizing hormone in the female).

The anterior pituitary gland controls the testes by means of its gonadotropic hormones—specifically, follicle-stimulating hormone (FSH) and ICSH just mentioned. FSH stimulates the seminiferous tubules to produce sperm more rapidly. ICSH stimulates interstitial cells to increase their secretion of testosterone. Soon the blood concentration of testosterone reaches a high level which inhibits anterior pituitary secretion of FSH and ICSH. Thus, a negative feedback mechanism operates between the anterior pituitary gland and the testes. A high blood concentration of gonadotropins stimulates testosterone secretion. But a high blood concentration of testosterone inhibits (has a negative effect on) gonadotropin secretion (Fig. 14-2).

Structure of spermatozoa

Fig. 14-3 shows the characteristic parts of a spermatozoon: head, neck, and elongated, lashlike tail.

Ducts of testes

Epididymis

Structure and location

Each epididymis consists of a single tightly coiled tube enclosed in a fibrous casing. The tube has a very small diameter (just barely macroscopic) but measures approximately 20 feet in length. It lies along the top and side of the testis. As shown in Fig. 14-1, several small ducts connect the seminiferous tubules of the testis with the epididymis. (Also see Fig. 14-6.)

Functions

The epididymis serves the following functions:

1 It serves as one of the ducts through which sperm pass in their journey from the testis to the exterior.

2 It stores a small quantity of sperm prior to ejaculation.

3 It secretes a small part of the seminal fluid (semen).

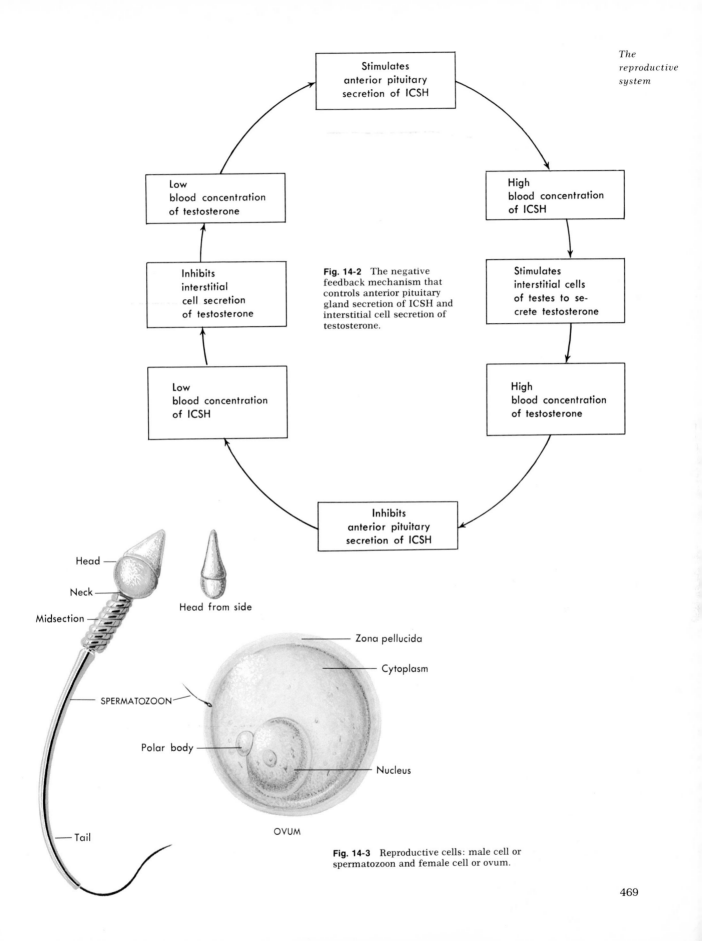

Stimulates anterior pituitary secretion of ICSH

High blood concentration of ICSH

Low blood concentration of testosterone

Stimulates interstitial cells of testes to secrete testosterone

Inhibits interstitial cell secretion of testosterone

High blood concentration of testosterone

Low blood concentration of ICSH

Inhibits anterior pituitary secretion of ICSH

Fig. 14-2 The negative feedback mechanism that controls anterior pituitary gland secretion of ICSH and interstitial cell secretion of testosterone.

Head

Neck

Midsection

SPERMATOZOON

Tail

Head from side

Zona pellucida

Cytoplasm

Polar body

Nucleus

OVUM

Fig. 14-3 Reproductive cells: male cell or spermatozoon and female cell or ovum.

469

Seminal duct (vas deferens; ductus deferens)
Structure and location

The seminal duct, like the epididymis, is a tube. In fact, it is an extension of the epididymis. It passes through the inguinal canal—where it is enclosed in a fibrous cylinder, the spermatic cord—into the abdominal cavity. Here, it extends over the top and down the posterior surface of the bladder, where it joins the duct from the seminal vesicle to form the ejaculatory duct (Figs. 14-4 and 14-5).

Function

The seminal duct serves as one of the ducts for the testis, connecting the epididymis with the ejaculatory duct.

Correlation

Severing of the seminal ducts (usually through incision in the groin) makes a man sterile. Why? Because it interrupts the route to the exterior from the epididymis. To leave the body, sperm must journey in succession through the epididymis, seminal duct, ejaculatory duct, and urethra.

Ejaculatory duct

The two ejaculatory ducts are short tubes that pass through the prostate gland to terminate in the urethra. As Fig. 14-4 shows, they are formed by the union of the seminal ducts with the ducts from the seminal vesicles.

Urethra

For a discussion of the urethra, see p. 460.

■ Accessory reproductive glands
Seminal vesicles
Structure and location

The seminal vesicles are convoluted pouches that lie along the lower part of the posterior surface of the bladder, directly in front of the rectum.

Function

The seminal vesicles secrete the viscous liquid portion of the semen, which contains nutrients that support sperm metabolism. The seminal vesicles, it is now known, do not store sperm.

Prostate gland
Structure and location

The prostate is a compound tubuloalveolar gland that lies just below the bladder and is shaped like a doughnut. The fact that the urethra passes through the small hole in the center of the prostate is a matter of considerable clinical significance. Many older men suffer from enlargement of this gland. As it enlarges, it squeezes the urethra, frequently closing it so completely that urination becomes impossible. Urinary retention results. Surgical removal of the gland (prostatectomy) is resorted to as a cure for this condition when other less radical methods of treatment fail.

Function

The prostate secretes a thin alkaline substance that constitutes the largest part of the seminal fluid. Its alkalinity helps protect the sperm from acid present in the male urethra and female vagina and thereby increases sperm motility. (Acid depresses or, if strong enough, kills sperm. Sperm motility is greatest in neutral or slightly alkaline media.)

Bulbourethral glands
Structure and location

The two bulbourethral or Cowper's glands resemble peas in both size and shape. You can see the location of these compound tubuloalveolar glands below the prostate glands in Fig. 14-5. A duct approximately 1 inch long connects them with the membranous portion of the urethra.

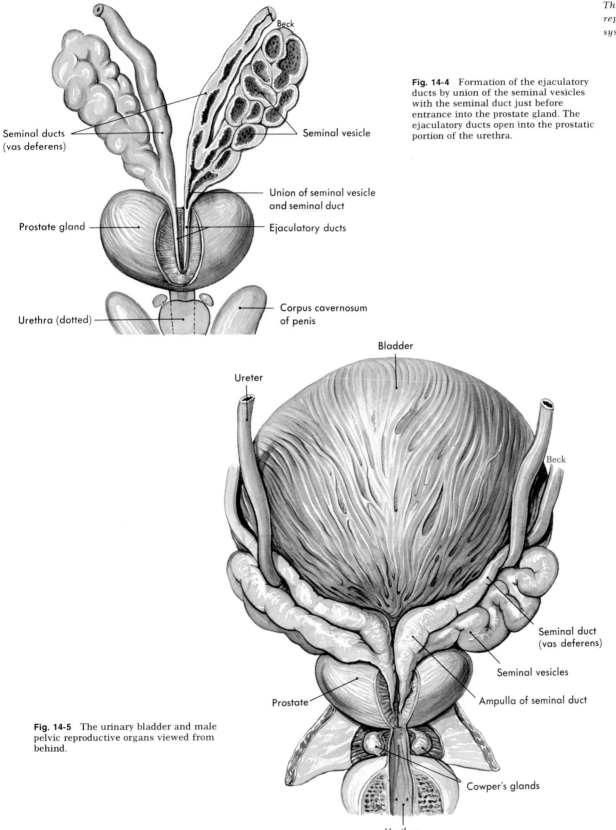

Beck

Seminal ducts
(vas deferens)

Seminal vesicle

Union of seminal vesicle
and seminal duct

Prostate gland

Ejaculatory ducts

Urethra (dotted)

Corpus cavernosum
of penis

Fig. 14-4 Formation of the ejaculatory ducts by union of the seminal vesicles with the seminal duct just before entrance into the prostate gland. The ejaculatory ducts open into the prostatic portion of the urethra.

Bladder

Ureter

Beck

Seminal duct
(vas deferens)

Seminal vesicles

Ampulla of seminal duct

Prostate

Cowper's glands

Urethra

Fig. 14-5 The urinary bladder and male pelvic reproductive organs viewed from behind.

471

Function

Like the prostate, the bulbourethral glands secrete an alkaline fluid which is important for counteracting the acid present in both the male urethra and the female vagina.

■Supporting structures

Scrotum (external)

The scrotum is a skin-covered pouch suspended from the perineal region. Internally, it is divided into two sacs by a septum, each sac containing a testis, epididymis, and lower part of the spermatic cord.

Penis (external)

Structure

Three cylindrical masses of erectile or cavernous tissue, enclosed in separate fibrous coverings and held together by a covering of skin, compose the penis. The two larger and uppermost of these cylinders are named the *corpora cavernosa penis*, whereas the smaller, lower one, which contains the urethra, is called the *corpus cavernosum urethrae*.

Erectile tissue that composes these structures resembles a rubber sponge in structure, consisting, as it does, of many irregular cavernous spaces, the venous sinuses. Under the influence of the sexual emotion, the arteries and arterioles of the penis dilate, flooding and distending the cavernous spaces with blood and thereby causing the organ to become enlarged, rigid, and erect.

At the distal end of the penis, there is a slightly bulging structure, the *glans penis*, over which the skin is folded doubly to form a more or less loose-fitting, retractable casing known as the *prepuce* or foreskin. If the foreskin fits too tightly about the glans, a circumcision is usually performed to prevent irritation.

Functions

The penis contains the urethra, the terminal duct for both urinary and reproductive tracts, and it is the copulatory organ by means of which spermatozoa are introduced into the female vagina. The scrotum and penis together constitute the *external genitals* of the male.

Spermatic cords (internal)

The _spermatic cords_ are cylindrical casings of white fibrous tissue located in the inguinal canals between the scrotum and the abdominal cavity. They enclose the seminal ducts, blood vessels, lymphatics, and nerves (Fig. 14-6).

■Composition and course of seminal fluid

The following structures secrete the substances which, together, make up the seminal fluid or semen:

1 Testes and epididymides – their secretions, according to one estimate, constitute less than 5% of the seminal fluid volume.

2 Seminal vesicles – their secretions are reported to contribute about 30% of the seminal fluid volume.

3 Prostate gland – its secretions constitute the bulk of the seminal fluid volume, reportedly about 60%.

4 Bulbourethral glands – their secretions are said to constitute less than 5% of the seminal fluid volume.

Besides contributing slightly to the fluid part of semen, the testes also add hundreds of millions of sperm. In traversing the distance from their place of origin to the exterior, the sperm must pass from the testis through the epididymis, seminal duct, ejaculatory duct, and urethra. Note that the male gametes originate in an organ located outside the body (that is, not within a body cavity), travel inside, and finally are expelled outside.

Early in fetal life the testes are located

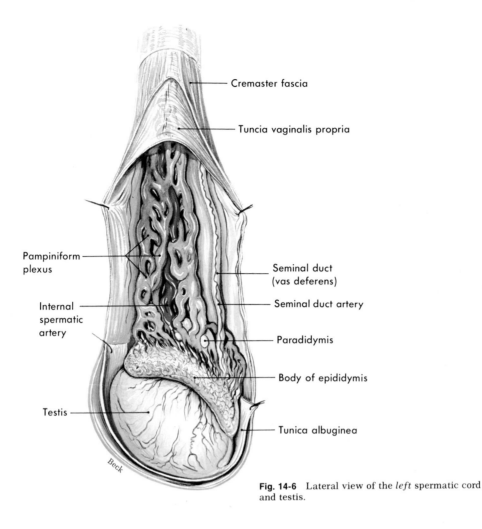

Cremaster fascia

Tuncia vaginalis propria

Pampiniform plexus

Seminal duct (vas deferens)

Internal spermatic artery

Seminal duct artery

Paradidymis

Body of epididymis

Testis

Tunica albuginea

Beck

Fig. 14-6 Lateral view of the *left* spermatic cord and testis.

in the abdominal cavity but normally descend in the spermatic cord (located in the inguinal canal) into the scrotum some time before birth. Occasionally a baby is born with undescended testes, a condition readily observed by palpation of the scrotum. Because the higher temperature inside the abdominal cavity makes sperm infertile, measures are taken to bring the testes down into the scrotum in order to prevent sterility.

Ejaculation of the seminal fluid occurs at irregular intervals due to arousal of the sexual emotion.

■Male fertility

Male fertility relates to many factors—most of all to the number of sperm ejac-ulated but also to their size and shape. Fertile sperm have a uniform size and shape. They are highly motile. Although only one sperm fertilizes an ovum, millions of sperm seem to be necessary for fertilization to occur. According to one estimate, when the sperm count falls below about 50 million per milliliter of semen, sterility results.

One hypothesis suggested to explain this puzzling fact is this: semen that contains an adequate number of sperm also contains enough hyaluronidase to liquefy the intercellular cement between the cells that encase each ovum. Without this, a single sperm cannot penetrate the layers of cells (corona radiata) around the ovum and hence cannot fertilize it.

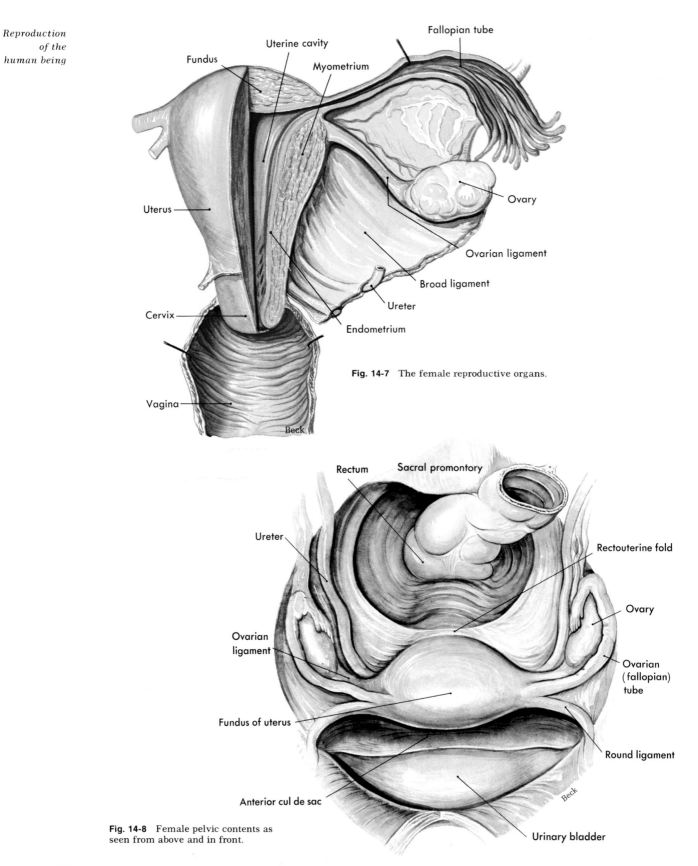

Fundus

Uterine cavity

Myometrium

Fallopian tube

Uterus

Ovary

Ovarian ligament

Broad ligament

Ureter

Cervix

Endometrium

Vagina

Beck

Fig. 14-7   The female reproductive organs.

Rectum

Sacral promontory

Ureter

Rectouterine fold

Ovary

Ovarian
ligament

Ovarian
(fallopian)
tube

Fundus of uterus

Round ligament

Anterior cul de sac

Beck

Urinary bladder

Fig. 14-8   Female pelvic contents as
seen from above and in front.

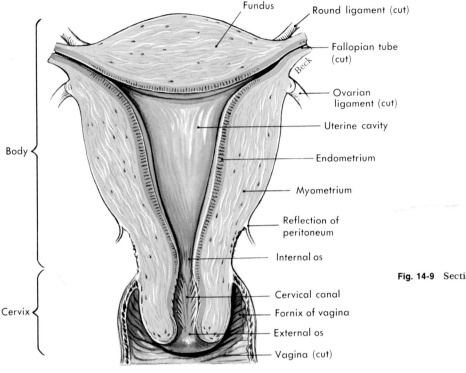

Fundus
Round ligament (cut)
Fallopian tube (cut)
Beck
Ovarian ligament (cut)
Uterine cavity
Endometrium
Myometrium
Reflection of peritoneum
Internal os
Cervical canal
Fornix of vagina
External os
Vagina (cut)
Body
Cervix

Fig. 14-9 Sectioned view of the uterus.

Female reproductive organs

Examine Figs. 14-7 and 14-8 and Plate XIV of the color insert (p. 60) to identify the following organs of the female reproductive system:

1 *primary sex organs* — the two ovaries (female gonads)
2 *secondary sex organs* — two uterine tubes (fallopian tubes or oviducts), one uterus, one vagina, one vulva (pudendum or external genitalia), and two breasts or mammary glands

■ Uterus

Structure

Size, shape, and divisions. The uterus is pear-shaped in its virgin state and measures approximately 3 inches in length, 2 inches in width at its widest part, and 1 inch in thickness. Note in Fig. 14-9 that the uterus has two main parts: an upper portion, the *body*, and a lower, narrow sec-

tion, the *cervix*. Did you notice that the body of the uterus rounds into a bulging prominence above the level at which the uterine tubes enter? This bulging upper surface is the *fundus.*

Wall. Three coats compose the walls of the uterus: endometrium, myometrium, and parietal peritoneum.

1 A lining of mucous membrane, called the *endometrium*, composed of three layers of tissues: a compact surface layer of columnar epithelium, a spongy middle layer of loose connective tissue, and a basal layer of dense connective tissue that attaches the endometrium to the underlying myometrium. During menstruation and following delivery of a baby, the compact and spongy layers slough off.

2 A thick, middle coat (the *myometrium*) consists of three layers of smooth muscle fibers that extend in all directions, longitudinally, transversely, obliquely, and give the uterus great strength. The myometrium is thickest in the fundus and thinnest

370,000 oogonia in ovary at birth.
190,000 "   "   "   "   at puberty.

475

in the cervix – a good example of the principle of structural adaptation to function. In order to expel a fetus – that is, move it down and out of the uterus – the fundus must contract more forcibly than the lower part of the uterine wall and the cervix must be stretched or dilated.

3 An external coat of serous membrane, the *parietal peritoneum*, is incomplete since it covers none of the cervix and only part of the body (all except the lower one-fourth of its anterior surface). The fact that the entire uterus is not covered with peritoneum has clinical value because it makes it possible to perform operations on this organ without the risk of infection that attends cutting into the peritoneum.

Cavities. The cavities of the uterus are small due to the thickness of its walls (see Fig. 14-9). The body cavity is flat and triangular. Its apex is directed downward and constitutes the *internal os*, which opens into the *cervical canal*. The cervical canal is constricted on its lower end also, forming the *external os*, which opens into the vagina. The uterine tubes open into the body cavity at its upper, outer angles.

Blood supply. The uterus receives a generous supply of blood from uterine arteries, branches of the internal iliac arteries.

Location

Fig. 14-8 shows the location of the uterus in the pelvic cavity between the bladder and the rectum.

Position

1 *Normally* the uterus is flexed between the body and cervix, with the body lying over the superior surface of the bladder, pointing forward and slightly upward (Fig. 14-10). The cervix points downward and backward from the point of flexion, joining the vagina at approximately a right angle. Several ligaments hold the uterus in

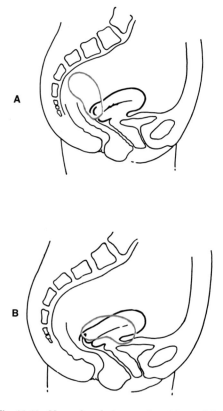

Fig. 14-10 Normal and abnormal positions of the uterus. Red lines show the abnormal positions. **A,** Retroflexion. **B,** Anteflexion.

place but allow its body considerable movement, a characteristic that often leads to malpositions of the organ.

2 The uterus may lie in any one of several *abnormal positions.* A common one is retroversion or backward tilting of the entire organ.

3 *Eight ligaments* (three pairs, two single ones) anchor the uterus in the pelvic cavity: broad (paired), uterosacral (paired), posterior (single), anterior (single), and round (paired). Six of these so-called ligaments are actually extensions of the parietal peritoneum in different directions. The round ligaments are fibromuscular cords.

a The *two broad ligaments* (Fig. 14-7) are double folds of parietal peritoneum that form a kind of partition across the pelvic cavity. The uterus is suspended between these two folds.

b The *two uterosacral ligaments* are foldlike extensions of the peritoneum from the posterior surface of the uterus to the sacrum, one on each side of the rectum.

c The *posterior ligament* is a fold of peritoneum extending from the posterior surface of the uterus to the rectum. This ligament forms a deep pouch known as the *cul-de-sac of Douglas* (or rectouterine pouch) between the uterus and rectum. Since this is the lowest point in the pelvic cavity, pus collects here in pelvic inflammations. To secure drainage, an incision may be made at the top of the posterior wall of the vagina (posterior colpotomy).

d The *anterior ligament* is the fold of peritoneum formed by the extension of the peritoneum on the anterior surface of the uterus to the posterior surface of the bladder. This fold also forms a cul-de-sac but one that is less deep than the posterior pouch.

e The *two round ligaments* (Fig. 14-8) are fibromuscular cords extending from the upper, outer angles of the uterus through the inguinal canals and disappearing in the labia majora.

Functions

The uterus or womb plays a role in the accomplishment of three highly important, though nonvital, functions: menstruation, pregnancy, and labor.

1 *Menstruation* is a sloughing away of the compact and spongy layers of the endometrium, attended by bleeding from the torn vessels.

2 In *pregnancy,* the embryo implants itself in the endometrium and there lives as a parasite throughout the fetal period.

3 *Labor* consists of powerful, rhythmic contractions of the muscular uterine wall that result in expulsion of the fetus or birth.

■ Uterine tubes—fallopian tubes or oviducts
Location

The uterine tubes are attached to the uterus at its upper outer angles (see Figs. 14-7 and 14-8). They lie between the folds of the broad ligaments and extend upward and outward toward the sides of the pelvis and then curve downward and backward.

Structure

The same three coats (mucous, smooth muscle, serous) of the uterus compose the tubes. The mucosa of the tubes, however, is ciliated. At the distal end, each tube expands into a funnel-like portion called the *infundibulum.* The open outer margin of the infundibulum resembles a fringe in its irregular outline. The fringelike projections are known as *fimbriae.* Here, the mucous lining of the tubes is directly continuous with the peritoneum—a fact of great clinical significance because the tubal mucosa is continuous with that of the uterus and vagina and, therefore, often becomes infected by gonococci or other organisms introduced into the vagina. And inflammation of the tubes (salpingitis) may readily spread to become inflammation of the peritoneum (peritonitis), a serious condition. In the male, there is no such direct route by which microorganisms can reach the peritoneum from the exterior.

Each uterine tube is approximately 4 inches long.

Function

The tubes serve as ducts for the ovaries, although they are not attached to them. Ovaries are the organs that produce ova (the female gametes). Fertilization, the union of a spermatozoon with an ovum, normally occurs in the tubes.

Fig. 14-11 Mammalian ovary showing successive stages of ovarian follicle and ovum development. Begin with the first stage (egg nest near arrow at lower left) and follow the arrow counterclockwise to the final stage (corpus albicans).

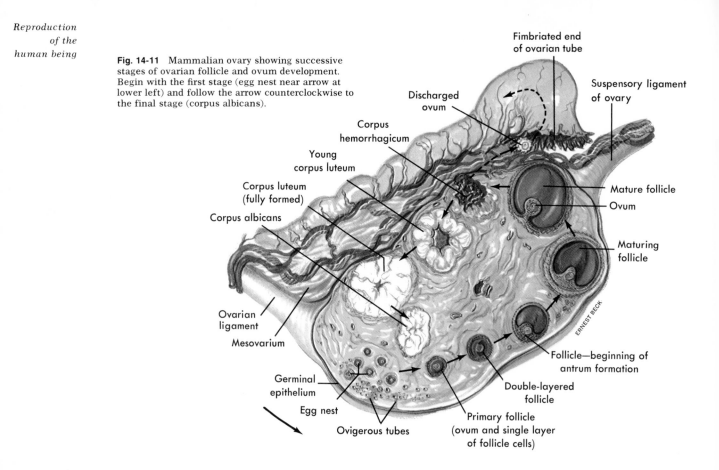

■Ovaries—female gonads
Location and size

The ovaries are glands that resemble large almonds in size and shape and are located one on either side of the uterus, below and behind the uterine tubes. Each ovary attaches to the posterior surface of the broad ligament by the mesovarian ligament. The ovarian ligament anchors it to the uterus. The distal portion of the tube curves about the ovary in such a way that the fimbriae cup over the ovary but do not actually attach to it. Here, then, is a gland whose duct is detached from it, a fact that makes possible pregnancy in the pelvic cavity instead of in the uterus as is normal.

Microscopic structure

The surface of the ovary consists of a single layer of germinal epithelial cells, whereas its interior is made up of connective tissue in which are embedded thousands of microscopic structures known as *graafian follicles*. After puberty, the follicles are present in varying stages of development (Fig. 14-11). The primordial follicles consist of an *ovum* encased in a nest of epithelial cells. Before puberty, all the follicles are in this stage. Development of the follicles after puberty is discussed on p. 483.

Functions

The ovaries perform two functions: ovulation and secretion. Ova develop and mature in the ovaries and are discharged from them into the pelvic cavity between the folds of the broad ligament. The ovaries also secrete the female hormones—estrogens (chiefly estradiol and estrone) and progesterone. More details about their

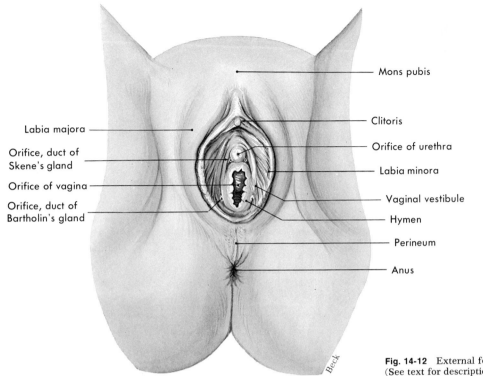

Labia majora

Orifice, duct of
Skene's gland

Orifice of vagina

Orifice, duct of
Bartholin's gland

Mons pubis

Clitoris

Orifice of urethra

Labia minora

Vaginal vestibule

Hymen

Perineum

Anus

Beck

Fig. 14-12 External female genitalia.
(See text for description.)

secretion and their functions appear on
pp. 483, 485, and 486.

Vagina

Location

The vagina is situated between the rectum, which lies posterior to it, and the urethra and bladder, which lie anterior to it. It extends upward and backward from its external orifice.

Structure

The vagina is a collapsible tube, capable of great distention, is composed mainly of smooth muscle, and is lined with mucous membrane arranged in rugae. Its anterior wall, which measures from 2 1/2 to 3 inches in length, is about 1 inch shorter than the posterior wall because the cervix protrudes into the uppermost portion of the anterior wall. In the virginal state, a fold of mucous membrane, the *hymen* ("maidenhead"), forms a border around the external opening of the vagina, par-

tially closing the orifice. Occasionally, this structure completely covers the vaginal outlet, a condition referred to as *imperforate hymen*. Perforation has to be done before the menstrual flow can escape.

Functions

The vagina constitutes an essential part of the reproductive tract because of the following:

1 It is the organ that receives the seminal fluid from the male.

2 It serves as the lower part of the birth canal.

3 It acts as the excretory duct for uterine secretions and the menstrual flow.

Vulva

Fig. 14-12 shows the structures which, together, constitute the female external genitals (reproductive organs) or vulva: mons veneris, labia majora, labia minora, clitoris, urinary meatus (urethral orifice),

479

vaginal orifice, and Bartholin's or the greater vestibular glands.

1 The *mons veneris* or mons pubis is a skin-covered pad of fat over the symphysis pubis. Coarse hairs appear on this structure at puberty and persist throughout life.

2 The *labia majora* or "large lips" are covered with pigmented skin and hair on the outer surface and are smooth and free from hair on the inner surface. They are composed mainly of fat and numerous glands.

3 The *labia minora* or "small lips" are located within the labia majora and are covered with modified skin. These two lips come together anteriorly in the midline. The area between the labia minora is the *vestibule*.

4 The *clitoris* is a small organ composed of erectile tissue, located just behind the junction of the labia minora, and homologous to the corpora cavernosa and glans of the penis. The *prepuce* or foreskin covers the clitoris, as it does the glans penis in the male.

5 The *urinary meatus* (urethral orifice) is the small opening of the urethra, situated between the clitoris and the vaginal orifice.

6 The *vaginal orifice* is an opening which, in the virginal state, is usually only slightly larger than the urinary meatus because of the constricting border formed by the hymen. In the marital state, the vaginal orifice is noticeably larger than the urinary meatus. It is located posterior to the meatus.

7 *Bartholin's* or the *greater vestibular glands* are two bean-shaped glands, one on either side of the vaginal orifice. Each gland opens by means of a single, long duct into the space between the hymen and the labium minus. These glands are of clinical importance because they are frequently infected (bartholinitis or Bartholin's abscess), particularly by the gonococcus. They are homologous to the bul-bourethral glands in the male, and they secrete a lubricating fluid. Opening into the vestibule near the urinary meatus by way of two small ducts is a group of tiny mucous glands, the *lesser vestibular* or *Skene's glands*. These have clinical interest because gonococci that lodge there are difficult to eradicate.

Perineum

The perineum is the skin-covered muscular region between the vaginal orifice and the anus. This area has great clinical importance because of the danger of its being torn during childbirth. If the tear is deep, it may extend all the way through the perineum and even through the anal sphincter, resulting in involuntary seepage from the rectum until the laceration is repaired. To avoid this possibility, an incision known as an *episiotomy* is usually made in the perineum, particularly at the birth of a first baby.

Breasts

Location and size

The breasts lie over the pectoral muscles and are attached to them by a layer of connective tissue (fascia). Estrogens and progesterone, two ovarian hormones, control their development during puberty. Estrogens stimulate growth of the ducts of the mammary glands, whereas progesterone stimulates development of the alveoli, the actual secreting cells. Breast size is determined more by the amount of fat around the glandular tissue than by the amount of glandular tissue itself. Hence, the size of the breast does not relate to its functional ability.

Structure

Each breast consists of several lobes separated by septa of connective tissue. Each lobe consists of several lobules, which, in turn, are composed of connective tissue in

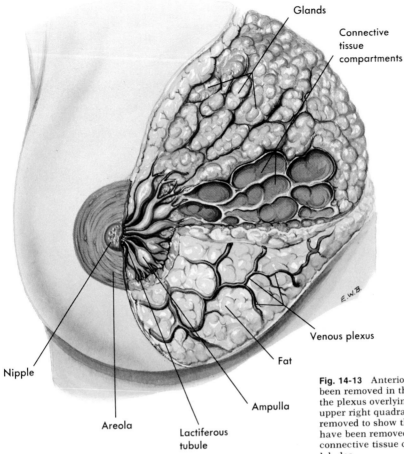

Glands

Connective
tissue
compartments

E.W.B.

Venous plexus

Fat

Nipple

Ampulla

Areola

Lactiferous
tubule

Fig. 14-13 Anterior view of the breast. The skin has been removed in the lower right quadrant to reveal the plexus overlying the adipose (fatty) tissue. In the upper right quadrant the adipose tissue has been removed to show the alveoli of the glands. The glands have been removed in a small area to reveal the connective tissue compartments that separate the lobules.

which are embedded the secreting cells (alveoli) of the gland, arranged in grapelike clusters around minute ducts. The ducts from the various lobules unite, forming a single excretory duct for each lobe, or between fifteen and twenty in each breast. These main ducts converge toward the nipple, like the spokes of a wheel. They enlarge slightly before reaching the nipple into ampullae or small "reservoirs" (Fig. 14-13). Each of these main ducts terminates in a tiny opening on the surface of the nipple. Adipose tissue is deposited around the surface of the gland, just under the skin, and between the lobes. The nipples are bordered by a circular pigmented area, the *areola*. It contains numerous sebaceous glands that appear as small nodules under the skin. In Caucasians (other than those with very dark complexions),

the areola and nipple change color from delicate pink to brown early in pregnancy – a fact of value in diagnosing a first pregnancy. The color decreases after lactation has ceased but never entirely returns to the virginal hue. In darker skinned women, no noticeable color change in the areola or nipple heralds the first pregnancy.

Function

The function of the mammary glands is lactation – that is, the secretion of milk for the nourishment of newborn infants.

Mechanism controlling lactation

Very briefly, lactation is controlled as follows:

1 The ovarian hormones, estrogens and progesterone, act on the breasts to make them structurally ready to secrete milk.

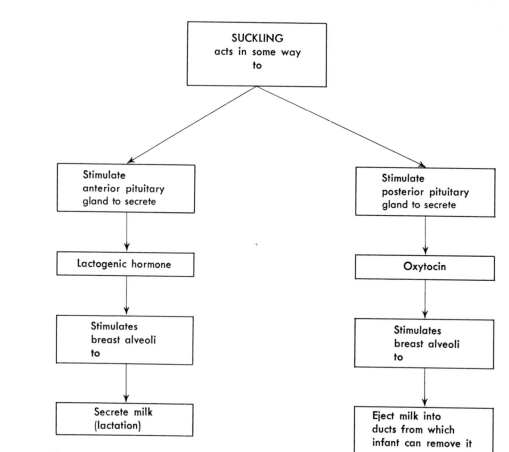

Fig. 14-14 Mechanism for controlling lactation and milk ejection.

Estrogens promote development of the ducts of the breasts. Progesterone acts on the estrogen-primed breasts, to promote completion of the development of the ducts and development of the alveoli, the secreting cells of the breasts. A high blood concentration of estrogens (for example, during pregnancy) inhibits anterior pituitary secretion of lactogenic hormone.

2 Shedding of the placenta following delivery of the baby cuts off a major source of estrogens. The resulting rapid drop in the blood concentration of estrogens stimulates anterior pituitary secretion of lactogenic hormone. Also, the suckling movements of a nursing baby act in some way to stimulate both anterior pituitary secretion of lactogenic hormone and posterior pituitary secretion of oxytocin (Fig. 14-14).

3 Lactogenic hormone stimulates lactation—that is, stimulates alveoli of the mammary glands to secrete milk. Milk secretion starts about the third or fourth day after delivery of a baby, supplanting a thin, yellowish secretion called *colostrum*. With repeated stimulation by the suckling infant, plus various favorable mental and physical conditions, lactation may continue almost indefinitely.

4 Oxytocin stimulates the alveoli of the breasts to eject milk into the ducts, thereby making it accessible for the infant to remove by suckling.

Female sexual cycles
■Recurring cycles

Many changes recur periodically in the female during the years between the onset of the menses (menarche) and their cessation (menopause or climacteric). Most obvious, of course, is menstruation – the outward sign of changes in the endometrium. Most women also note periodic changes in the breasts. But these are only two of many changes that occur over and over again at fairly uniform intervals during the thirty some years of female reproductive maturity. Rhythmic changes also take place in the ovaries, the myometrium, the vagina, hormone secretion, body temperature, and even in mood or "emotional tone." We shall investigate some of the details known or postulated about changes in the ovaries, endometrium, myometrium, and hormones and about the relations between these changes.

Ovarian cycles

Once each month, on about the first day of menstruation, several primitive graafian follicles and their enclosed ova begin to grow and develop. The follicular cells proliferate and start to secrete estrogens (one kind of female hormone) in increasing amounts for about two weeks. Usually, only one follicle matures and migrates to the surface of the ovary. The surface of the follicle degenerates, causing expulsion of the mature ovum into the pelvic cavity (ovulation).

When does ovulation occur? This is a question of great practical importance and one that has been given many answers. Present-day physiologists generally agree that ovulation usually occurs fourteen days before the next menstrual period begins. But, they quickly add, there are exceptions to this general rule. Ovulation may even take place nineteen or more days before the onset of the next menses (see item 3, p. 484). A few women experience pain within a few hours after ovulation. This is referred to as *mittelschmerz* – German for middle pain. It has been ascribed to irritation of the peritoneum by hemorrhage from the ruptured follicle.

Shortly before ovulation, the ovum undergoes a special type of mitosis (called meiosis) in which its number of chromosomes is reduced by half. Immediately after ovulation, cells of the ruptured follicle enlarge and, due to the appearance of lipoid substances in them, become transformed into a golden-colored body, the *corpus luteum*. The corpus luteum grows for seven or eight days. During this time, it secretes both progesterone and estrogens in increasing amounts. Then, provided fertilization of the ovum did not occur, the size of the corpus luteum and the amount of its secretions gradually diminish. By the twenty-sixth or twenty-seventh day following onset of the menses, progesterone secretion has ceased entirely and estrogen production has reached a minimum. About two days later, another menses starts presumably because of the low blood levels of progesterone and estrogens.

Endometrial or menstrual cycle

During menstruation, necrotic bits of the compact and spongy layers of the endometrium slough off, leaving denuded bleeding areas. Following menstruation, the cells of these layers proliferate, causing the endometrium to reach a thickness of 2 or 3 millimeters by the time of ovulation. During this period, endometrial glands and arterioles grow longer and more coiled – two factors that also contribute to the thickening of the endometrium. After ovulation, the endometrium grows still thicker (reaching a maximum of about 4 to 6 millimeters). Most of this increase, however, is believed due to swelling produced by fluid retention rather than to further proliferation of endometrial cells. The increasingly tortuous endometrial glands start to secrete during the time between ovulation and the next menses. Then, the day before menstruation starts again, the tightly coiled arterioles constrict, producing endometrial ischemia. This leads to necrosis, sloughing, and, once again, menstrual bleeding.

The menstrual cycle is customarily divided into phases, named for major events occurring in them: menses, postmenstrual or preovulatory phase, ovulation, and premenstrual phase.

1 The *menses* or *menstrual period* occurs on cycle days 1 to 5. There is some individual variation, however.

2 The *postmenstrual* or *preovulatory phase* occurs between the end of the menses and ovulation. It usually includes cycle days 6 to 13 or 14 in a 28-day cycle. But the length of this phase varies more than do the others. It lasts longer in long cycles and ends sooner in short ones. This phase is also called the *estrogenic* or *follicular phase* because of the high blood estrogen content due to secretion by the developing follicle. And *proliferative phase* is still another name for it because proliferation of endometrial cells occurs at this time.

3 *Ovulation* (that is, rupture of the mature follicle with expulsion of its ovum into the pelvic cavity) occurs frequently on cycle day 15 in a 28-day cycle. However, it occurs on different days in different length cycles depending on the length of the preovulatory phase. For example, in a 32-day cycle, the preovulatory phase would probably last until cycle day 18 and ovulation would then occur on cycle day 19 instead of 15. In short, the day of ovulation cannot be predicted with certainty. One cannot know ahead of time precisely how many days the preovulatory phase will last. And this physiological fact probably accounts for most of the unreliability of the rhythm method of contraception.

4 The *premenstrual phase* occurs between ovulation and the onset of the menses. This phase is also called the *luteal phase* because the corpus luteum secretes

during this time and *progesterone phase* because this hormone is secreted only during this phase. The length of this premenstrual phase is pretty constant, lasting usually 14 days—that is, cycle days 15 to 28. Differences in length of total menstrual cycle, therefore, exist mainly because of differences in duration of the preovulatory rather than of the premenstrual phase.

Myometrial cycle

The myometrium contracts mildly but with increasing frequency during the two weeks preceding ovulation. Contractions decrease or disappear between ovulation and the next menses, thereby lessening the probability of expulsion of an implanted ovum.

Gonadotropic cycles

The adenohypophysis (anterior pituitary gland) secretes two hormones called gonadotropins that influence female reproductive cycles. Their names are follicle-stimulating hormone (FSH) and luteinizing hormone (LH). The amount of each gonadotropin secreted varies with a rhythmic regularity that can be related, as we shall see, to the rhythmic ovarian and uterine changes just described.

■ Control

Physiologists agree that hormones play the major role in producing the cyclic changes characteristic in the female during the years of reproductive maturity. By correlating the changing blood concentrations of the pituitary gonadotropins with the monthly ovarian and uterine changes, investigators have arrived at a working hypothesis about the main features of the control mechanisms.

A brief description follows of the mechanisms that produce cyclical changes in the

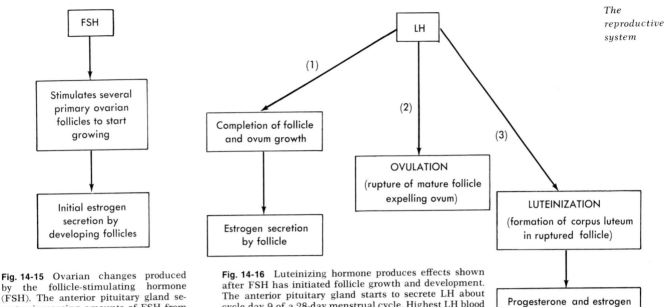

Fig. 14-15 Ovarian changes produced by the follicle-stimulating hormone (FSH). The anterior pituitary gland secretes increasing amounts of FSH from the first to about the seventh day of the menstrual cycle and then decreasing amounts. The lowest FSH blood level is reached a day or so after ovulation and remains low for the rest of the cycle.

Fig. 14-16 Luteinizing hormone produces effects shown after FSH has initiated follicle growth and development. The anterior pituitary gland starts to secrete LH about cycle day 9 of a 28-day menstrual cycle. Highest LH blood level is reached about the time FSH reaches its lowest level—that is, a day or so after ovulation. From then on (during the premenstrual phase), LH concentration decreases gradually and reaches its lowest level just before menstruation.

ovaries and uterus and in the amounts of gonadotropins secreted.

Cyclical changes in ovaries

Cyclical changes in the ovaries result from cyclical changes in the amounts of gonadotropins secreted by the anterior pituitary gland. An increasing blood concentration of FSH has two effects: it stimulates one or more primitive graafian follicles and ova to start growing, and it stimulates the follicles to secrete estrogens. The anterior pituitary gland secretes increasing amounts of FSH from the first to about the seventh day of the menstrual cycle and decreasing amounts thereafter. By a day or so after ovulation, blood FSH concentration has reached its lowest point. It remains low for the rest of the cycle—that is, throughout the premenstrual phase. (Fig. 14-15 shows ovarian change produced by FSH.)

Several days before ovulation, the anterior pituitary gland starts to release increasing amounts of LH into the blood.

As Fig. 14-16 shows, LH brings about the following changes:

1 Completion of growth of the follicle and ovum with increasing secretion of estrogens by the follicle during the preovulatory phase in the menstrual cycle.

2 Rupturing of the mature follicle with expulsion of its ripe ovum (process known as *ovulation*). Because of this function, LH is sometimes called the ovulating hormone.

3 Formation of a golden body, the corpus luteum, in the ruptured follicle (process called *luteinization*). The name luteinizing hormone refers, obviously, to this LH function. Blood LH concentration increases from about cycle day 9 until a day or so after ovulation. Hence blood's LH concentration is highest when its FSH concentration is lowest. LH concentration decreases gradually during the premenstrual phase and reaches its lowest level just before menstruation.

If pregnancy does not occur, the corpus luteum reaches its maximum development in about 8 days and then starts to regress. Gradually, fibrous tissue (corpus albicans) replaces it. On or about cycle day 25 the corpus luteum stops secreting both progesterone and estrogens. Their blood concentrations drop precipitously. Without stimulation by these hormones, the endometrium degenerates rapidly and starts to bleed. Menstruation, in other words, occurs, and another menstrual cycle has begun.

Cyclical changes in uterus

Cyclical changes in the uterus are brought about by changing blood concentrations of estrogens and progesterone. As blood estrogens increase during the preovulatory phase of the menstrual cycle, they produce the following main changes in the uterus:

1 Proliferation of endometrial cells, producing a thickening of the endometrium
2 Growth of endometrial glands
3 Increase in the water content of the endometrium
4 Increased myometrial contractions

Increasing blood progesterone concentration during the premenstrual phase of the menstrual cycle produces progestational changes in the uterus—that is, changes which are favorable for pregnancy, specifically:

1 Secretion by endometrial glands, thereby preparing the endometrium for implantation of a fertilized ovum
2 Increase in the water content of the endometrium
3 Decreased myometrial contractions

Cyclical changes in amounts of FSH and estrogens secreted

Examine Fig. 14-17 carefully. Note particularly the effects of a high blood concentration of estrogens on anterior pituitary gland secretion and the effect of a low blood concentration of FSH on the development of a graafian follicle and ovum. Establishment of these two facts led eventually to the development of "the pill" for preventing pregnancy. Contraceptive pills contain estrogen-like compounds, or progesterone, or both. By building up a high blood concentration of the estrogen-like substance, they prevent the development of a follicle and its ovum that month. With no mature ovum to be expelled, ovulation does not occur, and therefore pregnancy cannot occur. The next menses, however, does take place—because the progesterone and estrogen dosage is stopped in time to allow their blood levels to decrease as they normally do near the end of the cycle to bring on menstruation. Actually the effects of contraceptive pills are much more complex than our explanation indicates. They still are not completely understood.

■Functions served by cycles

The major function seems to be to prepare the endometrium each month for a pregnancy. If it does not occur, the thick vascular lining, no longer needed, is shed.

If fertilization of the ovum (pregnancy) occurs, the menstrual cycle is modified as follows:

1 The corpus luteum does not disappear but persists and continues to secrete progesterone and estrogens for six months or more of pregnancy. If it is removed by any means during the early months of pregnancy, spontaneous abortion results.
2 The fertilized ovum, which immediately starts developing into an embryo, travels down the tube and implants itself in the endometrium, so carefully prepared for this event.

■Menarche and menopause

The menstrual flow first occurs (menarche) at puberty, at about the age of 13

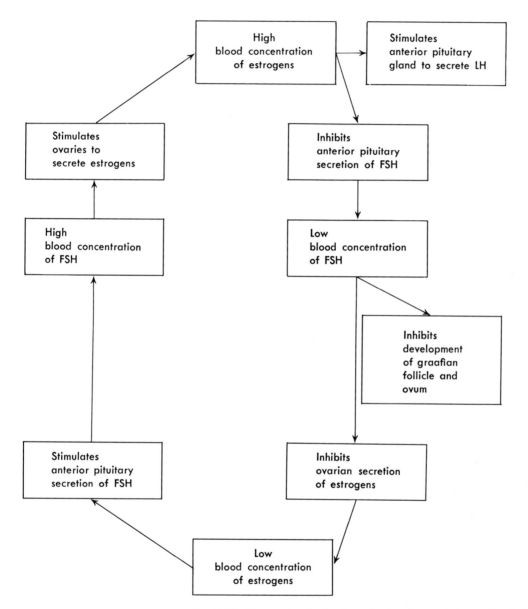

Fig. 14-17 The negative feedback mechanism that controls anterior pituitary
secretion of follicle-stimulating hormone (FSH) and ovarian secretion of estrogens.
A high blood level of FSH stimulates estrogen secretion, whereas the resulting high
estrogen level inhibits FSH secretion. How does this compare with interstitial
cell–stimulating hormone (ICSH)–testosterone feedback mechanism? (See Fig. 14-2
if you want to check your answer.) According to the above diagram, what effect
does a high blood concentration of estrogens have on anterior pituitary secretion
of FSH? of LH?

years, although there is wide individual variation according to race, nutrition, health, heredity, etc. Normally, it recurs about every twenty-eight days for some thirty years or so, except during pregnancy, and then ceases (menopause or climacteric) at about the age of 45 years.

Embryology
■Meaning and scope

Embryology is the science of the development of the individual before birth. It is a story of miracles, describing the means by which a new human life is started and the steps by which a single microscopic cell is transformed into a complex human being. In a work of this kind, it seems feasible to include only a few of the main points in the development of the new individual since the facts amassed by the science of embryology are so many and so intricate that to tell them requires volumes, not paragraphs.

■Value of knowledge of embryology

The main value to the medical profession of knowing the steps by which a new individual is evolved lies in their explanation of various congenital deformities. For example, one of the most common malformations is harelip, a condition which results from imperfect fusion of the frontal and maxillary processes during embryonic development.

■Steps in development of new individual
Preliminary processes

Production of a new human being starts with the union of a spermatozoon and an ovum to form a single cell. Two preliminary steps, however, are necessary before such a union can take place: maturation of the sex cells (meiosis) and ovulation and insemination.

Maturation of sex cells (meiosis*)—reduction in number of chromosomes to one-half original number. Only when maturation has occurred are the ovum and spermatozoon mature and ready to unite with each other. The necessity for chromosome reduction as a preliminary to union of the sex cells is explained by the fact that the cells of each species of living organisms contain a specific number of chromosomes. Human cells, for example, contain 46 chromosomes. If the male and female cells united without first halving their respective chromosomes, the resulting cell would contain twice as many chromosomes as is specific for human beings. Mature ova and sperm, therefore, contain only 23 chromosomes or one-half as many as other human cells. Of these, one is the sex chromosome and may be either one of two types, known as X or Y. All ova contain an X chromosome. Sperm, on the other hand, have either an X or a Y chromosome. A female child results from the union of an X chromosome–bearing sperm with an ovum and a male child from the union of a Y chromosome–bearing sperm with an ovum.

Ovulation and insemination. The second preliminary step necessary for conception of a new individual consists in bringing the sperm and ovum into proximity with each other so that the union of the two can take place. Two processes are involved in the accomplishment of this step.

1 Ovulation or expulsion of the mature ovum from the graafian follicle into the pelvic cavity, from which it enters one of the uterine tubes.

2 Insemination or expulsion of the seminal fluid from the male urethra into the female vagina. Several million sperm en-

---

*For details of the mechanism of meiosis, see supplementary readings for Chapter 14, reference 3.

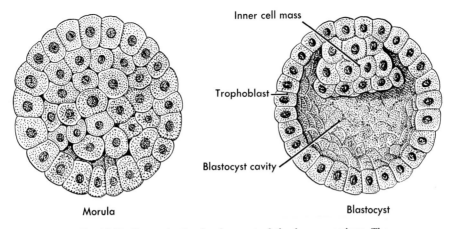

Fig. 14-18 Stages in the development of the human embryo. The morula consists of an almost solid spherical mass of cells. The embryo reaches this stage about three days after fertilization. The blastocyst (hollowing) stage develops later, after implantation in the uterine lining.

ter the female reproductive tract with each ejaculation of semen. By lashing movements of their flagella-like tails, assisted somewhat by muscular contractions of surrounding structures, the sperm make their way into the external os of the cervix, through the cervical canal and uterine cavity, and into the tubes.

Developmental processes

Fertilization or union of male and female gametes to produce one-celled individual called zygote. The sperm "swim" up the tube toward the ovum. Although numerous sperm surround the ovum, only one penetrates it. As soon as the head and neck of one spermatozoon enter the ovum (the tail drops off), the remaining sperm seem to be repulsed. The sperm head then forms itself into a nucleus that approaches and eventually fuses with the nucleus of the ovum, producing, at that moment, a new single-celled individual or *zygote*. One-half of the 46 chromosomes in the zygote nucleus have come from the sperm and one-half from the ovum. Since chromosomes are composed of *genes* or inheritance determinants, the new being inherits one-half of its characteristics from its father and one-half from its mother.

Normally, fertilization occurs in the uterine tube. Occasionally, however, it takes place in the pelvic cavity, as evidenced by pregnancies that start to develop in the pelvic cavity instead of in the uterus.

Inasmuch as the ovum lives only a short time (probably less than forty-eight hours) after leaving the graafian follicle, fertilization can occur only around the time of ovulation (p. 484). Sperm also have short-lived fertility, probably only about twenty-four hours after entering the female tract.

Cleavage or segmentation. Cleavage consists of repeated mitotic divisions, first of the zygote to form two cells, then of those two cells to form four cells, etc., resulting, in about three days' time, in the formation of a solid, spherical mass of cells known as a *morula* (Fig. 14-18). About this time, the embryo reaches the uterus, where it starts to implant itself in the endometrium. Occasionally, implantation occurs in the tube or pelvic cavity instead of in the uterus. The condition is known as an *ectopic pregnancy*.

As the cells of the morula continue to divide, a hollow ball of cells or *blastocyst*, consisting of an outer layer of cells and an inner cell mass, is formed. Implantation in the uterine lining is now complete (Fig.

14-18). About ten days have elapsed since fertilization. The cells that compose the outer wall of the blastocyst are known as trophoblasts. They eventually become part of the placenta, the structure that anchors the fetus to the uterus.

Differentiation. As the cells composing the inner mass of the blastocyst continue to divide, they arrange themselves into a structure shaped like a figure eight, containing two cavities separated by a double-layered plate of cells known as the *embryonic disc*. The youngest human embryos examined have been at this stage or about two weeks old dated from the time of fertilization. The cells that form the cavity above the embryonic disc eventually become a fluid-filled, shock-absorbing sac (the amnion) in which the fetus floats. The cells of the lower cavity form the yolk sac, a small vesicle attached to the belly of the embryo until about the middle of the second month, when it breaks away. Only the double layer of cells that compose the embryonic disc is destined to form the new individual. The upper layer of cells is called the *ectoderm* and the lower layer the *entoderm*. A third layer of cells, known as the *mesoderm*, develops between the ectoderm and entoderm. Up to this time, all the cells have appeared alike, but now they are dif-

ferentiated into three distinct types, ecto-dermal, mesodermal, and entodermal, known as the *primary germ layers*, each of which will give rise to definite structures. For example, the ectoderm cells will form the skin and its appendages and the nervous system; the mesoderm, the muscles, bones, and various other connective tissues; the entoderm, the epithelium of the digestive and respiratory tracts, etc.

Histogenesis and organogenesis. The story of how the primary germ layers develop into many different kinds of tissues (histogenesis) and how those tissues arrange themselves into organs (organogenesis) is long and complicated. Its telling belongs to the science of embryology. But for the beginning student of anatomy it seems sufficient to appreciate that life begins when two sex cells unite to form a single cell, that the new human body evolves by a series of processes consisting of cell multiplication, cell growth, cell differentiation, and cell rearrangements, all of which take place in definite, orderly sequence. By the end of the second month, a recognizable human figure has been formed. At the end of another month, the sex is clearly distinguishable, and from then until birth, development is mainly a matter of growth.

outline summary

MEANING AND FUNCTION
Consists of organs which, together, produce new individual

MALE REPRODUCTIVE ORGANS
1 Glands
  a Testes, gonads (paired)
  b Accessory glands
    1 Seminal vesicles (paired)
    2 Prostate gland
    3 Bulbourethral (Cowper's) glands (paired)
2 Ducts of testes
  a Epididymis (paired)
  b Seminal ducts (vas deferens; ductus deferens) (paired)

  c Ejaculatory ducts (paired)
  d Urethra
3 Supporting structures
  a External—scrotum and penis
  b Internal—spermatic cords (paired)

Testes
1 Structure and location
  a Several lobules composed of seminiferous tubules and interstitial cells (of Leydig), separated by septa, encased in fibrous capsule
  b Few ducts emerge from top of organ and enter head of epididymis
  c Located in scrotum, one testis in each of two scrotal compartments

2 Functions
  a Spermatogenesis – formation of mature male gametes (spermatozoa) by seminiferous tubules
  b Secretion of hormone (testosterone) by interstitial cells
3 Structure of spermatozoa – consists of head, neck (middle piece), and whiplike tail

Ducts of testes
Epididymis
1 Structure and location
  a Single tightly coiled tube enclosed in fibrous casing
  b Lies along top and side of each testis
2 Functions
  a Duct for seminal fluid
  b Also secretes part of seminal fluid
  c Sperm become capable of motility while they are stored in epididymis

Seminal duct (vas deferens; ductus deferens)
1 Structure and location
  a Tube, extension of epididymis
  b Extends through inguinal canal, into abdominal cavity, over top and down posterior surface of bladder to join duct from seminal vesicle
2 Function
  a One of excretory ducts for seminal fluid
  b Connects epididymis with ejaculatory duct

Ejaculatory duct
1 Formed by union of seminal duct with duct from seminal vesicle
2 Passes through prostate gland, terminating in urethra

Urethra
See p. 460

Accessory reproductive glands
Seminal vesicles
1 Structure and location – convoluted pouches on posterior surface of bladder
2 Function – secrete nutrient-rich part of seminal fluid

Prostate gland
1 Structure and location
  a Doughnut-shaped
  b Encircles urethra just below bladder
2 Function – adds alkaline secretion to seminal fluid

Bulbourethral glands
1 Structure and location
  a Small, pea-shaped structures with 1-in long ducts leading into urethra
  b Lie below prostate gland
2 Function – secrete alkaline fluid that is part of semen

Supporting structures
External
1 Scrotum
  a Skin-covered pouch suspended from perineal region
  b Divided into two compartments
  c Contains testis, epididymis, and first part of seminal duct
2 Penis – composed of three cylindrical masses of erectile tissue, one of which contains urethra

Internal
1 Spermatic cords
  a Fibrous cylinders located in inguinal canals
  b Enclose seminal ducts, blood vessels, lymphatics and nerves

Composition and course of seminal fluid
1 Consists of secretions from testes, epididymides, seminal vesicles, prostate, and bulbourethral glands
2 Each drop contains millions of sperm
3 Passes from testes through epididymis, seminal duct, ejaculatory duct, and urethra

Male fertility
Depends chiefly on large numbers of normal-sized and normal-shaped sperm being ejaculated

FEMALE REPRODUCTIVE ORGANS
1 Glands
  a Ovaries (gonads)
  b Accessory glands
    1 Bartholin's (greater vestibular)
    2 Skene's (lesser vestibular)
    3 Mammary (breasts)
2 Ducts
  a Uterine (fallopian) tubes
  b Uterus
  c Vagina
3 External genitalia (vulva)

Uterus
1 Structure
  a Size, shape, and divisions
    1 In virginal state, 3 in × 2 in × 1 in
    2 Pear-shaped
    3 Consists of body and cervix; fundus bulging upper surface of body
  b Wall – lining of mucosa called endometrium; thick, middle coat of muscle called myometrium; partial external coat of peritoneum
  c Cavities – body cavity small and triangular in shape with three openings – two from tubes and one (internal os) into cervical canal; external os opening of cervical canal into vagina
  d Blood supply – generous, from uterine arteries

2 Location – in pelvic cavity between bladder and rectum

3 Position

   a Flexed between body and cervix, with body lying over bladder pointing forward and slightly upward

   b Cervix joins vagina at right angles

   c Capable of considerable mobility; therefore, often in abnormal positions, such as retroverted

   d Eight ligaments anchor it – two broad, two uterosacral, one posterior, one anterior, and two round

4 Functions

   a Menstruation

   b Pregnancy

   c Labor and expulsion of fetus

Uterine tubes – fallopian tubes or oviducts

1 Location – attached to uterus at upper, outer angles

2 Structure

   a Same three coats as uterus

   b Distal ends open with fimbriated margins

   c Mucosa and peritoneum in direct contact here

3 Function

   a Serve as ducts for ovaries, providing passageway by which ova can reach uterus

   b Fertilization occurs here normally

Ovaries – female gonads

1 Location and size

   a Size and shape of large almonds

   b Lie behind and below uterine tubes

   c Anchored to uterus and broad ligament

2 Microscopic structure

   a Consists of several thousand graafian follicles embedded in connective tissue base

   b Follicles in all stages of development; usually each month one matures, ruptures, and expels its ovum into abdominal cavity

3 Functions

   a Oogenesis – formation of mature female gametes (ova)

   b Secretion of hormones – estrogens and progesterone

Vagina

1 Location – between rectum and urethra

2 Structure

   a Collapsible, musculomembranous tube, capable of great distention

   b External outlet protected by fold of mucous membrane, hymen

3 Functions

   a Receive seminal fluid

   b Is lower part of birth canal

   c Is excretory duct for uterine secretions and menstrual flow

Vulva

1 Consists of numerous structures which, together, constitute external genitalia

2 Main structures

   a Mons veneris – skin-covered pad of fat over symphysis pubis

   b Labia majora – hairy, skin-covered lips

   c Labia minora – small lips covered with modified skin

   d Clitoris – small mound of erectile tissue just below junction of two labia minora

   e Urinary meatus – just below clitoris; opens into urethra

   f Vaginal orifice – below urethra

   g Bartholin's or greater vestibular glands

      1 Comparable to bulbourethral glands of male

      2 Open by means of long duct in space between hymen and labia minora

      3 Ducts from lesser vestibular or Skene's glands open near urinary meatus

Perineum

1 Region between vaginal orifice and anus

2 Frequently torn at childbirth

Breasts

1 Location and size

   a Just under skin, over pectoral muscles

   b Size depends on deposits of adipose tissue

2 Structure

   a Divided into lobes and lobules; latter composed of glands

   b Single excretory duct per lobe opens in nipple

   c Circular, pigmented area called areola borders nipple

3 Function – secrete milk for infant

4 Mechanism controlling lactation – see Fig. 14-14

FEMALE SEXUAL CYCLES

Recurring cycles

1 Ovarian cycles

   a Each month, follicle and ovum develop

   b Follicle secretes estrogens

   c Ovum matures and follicle ruptures (ovulation)

   d Corpus luteum forms and secretes progesterone and estrogens

   e If no pregnancy, corpus luteum gradually degenerates and is replaced by fibrous tissue; remains if there is pregnancy

2 Endometrial or menstrual cycle

   a Surface of endometrium sloughs off during menses, with bleeding from denuded area

   b Regeneration and proliferation of lining occurs

   c Endometrial glands and arterioles become longer and more tortuous

   d Glands secrete viscous mucus and cycle repeats

3 Myometrial cycle – contractility increases before ovulation and subsides following
4 Gonadotropic cycles
  a FSH – a high rate of secretion for about ten days after menses
  b LH secretion increasing from ninth cycle day to day of ovulation (Fig. 14-16)

Control

1 Cyclical changes in ovaries controlled as follows:
  a Increasing blood concentration of FSH (for about ten days after menses) stimulates one or more primitive graafian follicles to start growing and to start secreting estrogens (Fig. 14-15)
  b Increasing blood concentration of LH (from cycle day 9 to day of ovulation) causes completion of growth of follicle and ovum, stimulates follicle to secrete estrogens, causes ovulation, and causes luteinization (Fig. 14-16)
2 Cyclical changes in uterus controlled as follows:
  a Increasing blood concentration of estrogens during preovulatory phase causes proliferation of endometrial cells with resultant thickening of endometrium, growth of endometrial glands, increased water content of endometrium, and increased contractions by myometrium
  b Increasing blood concentration of progesterone after ovulation – i.e., during premenstrual phase, causes secretion by endometrial glands and preparation of endometrium for implantation, increased water content of endometrium, and decreased contractions by myometrium
3 Cyclical changes in FSH and estrogen secretion – see Fig. 14-17

Functions served by cycles

1 Preparation of endometrium for pregnancy during preovulatory and premenstrual periods
2 Shedding of progestational endometrium if no pregnancy occurs

Menarche and menopause

1 Menarche – onset of menses; about 13 years of age
2 Menopause (climacteric) – cessation of menses; about 45 years of age

EMBRYOLOGY
Meaning and scope

1 Science of development of individual before birth
2 Long and complex study

Value of knowledge of embryology

Interprets many abnormal formations

Steps in development of new individual

1 Preliminary processes
  a Maturation of ovum and sperm or reduction of their chromosomes to one-half original number – i.e., one-half of 46
  b Ovulation and insemination
    1 Ovulation once every 28 days, usually about 14 days before beginning of next menstrual period
    2 Insemination, indefinite occurrence; sperm introduced into vagina, swim up to meet ovum in tube
2 Developmental processes
  a Fertilization or union of ovum and sperm
    1 Normally occurs in tube
    2 Only one sperm penetrates ovum
    3 One-celled new individual called zygote formed by union
  b Cleavage or segmentation
    1 Multiplication of cells from zygote by repeated mitosis
    2 Morula or solid ball of cells formed; becomes hollow with a cluster of cells attached at one point on inner surface of sphere; called blastocyst at this stage; now implanted in endometrium
  c Differentiation of cells into three primary germ layers – ectoderm, mesoderm, and entoderm
  d Histogenesis or formation of various tissues from primary germ layers; organogenesis or formation of various organs by rearrangements of tissues, such as fusions, shiftings, foldings, etc.

review questions

1 Name the male sex glands. Where are they located?
2 Of what is the seminal fluid composed? Trace its course from its formation in the gonads to the exterior.
3 What and where are the prostate glands?
4 What and where are Cowper's glands?
5 What is the spermatic cord? From what to what does it extend, and what does it contain?
6 Removal of the testes (orchiectomy or castration) results in both sterility and various changes in the secondary sex characteristics. Why?
7 Name the female sex glands. Name all the internal female reproductive organs.
8 What is the perineum? Of what clinical importance is it in the female?
9 Name the three openings to the exterior from the female pelvis. How many openings to the exterior are there in the male?
10 What is a graafian follicle? What does it contain?
11 How many ova mature in a month, usually?
12 When, in the menstrual cycle, is ovulation thought to occur?
13 Discuss the mechanism thought to control menstruation.
14 Name the periods in the menstrual cycle with the approximate length in days of each and the main events.
15 What two organs are necessary for menstruation to occur?
16 Name the divisions of the uterus.
17 What and where are the fallopian tubes? Approximately how long are they? With what are they lined? Their lining is continuous on their distal ends with what? On their proximal ends? Why is an infection of the lower part of the female reproductive tract likely to develop into a very serious condition?
18 What is the cul-de-sac of Douglas?
19 Describe briefly the hormonal control of breast development and lactation.
20 Explain how contraceptive pills act to prevent pregnancy.
21 Where does fertilization normally take place?
22 Which of the following patients will no longer menstruate? Why? (1) One who has had a bilateral salpingectomy (removal of the tubes)? (2) One who has had a panhysterectomy (removal of the entire uterus)? (3) One who has had a bilateral oophorectomy (removal of both ovaries)? (4) One who has had a cervical hysterectomy (body and fundus removed)? (5) One who has had a unilateral oophorectomy?
23 Which of the patients described in question 22 could no longer become pregnant? Explain.
24 Following menopause, which of the following hormones, estrogens, FSH, progesterone, would you expect to have a high blood concentration? Which a low blood concentration? Explain your reasoning.
25 Define or make an identifying statement about each of the following: adolescence; climacteric; colostrum; corpus luteum; ectopic pregnancy; embryology; endometrium; fertilization; fimbriae; gamete; genes; gonad; hysterectomy; maturation; meiosis; menarche; mitosis; morula; oophorectomy; ovum; ovulation; puberty; salpingitis; semen; spermatozoon; vulva; zygote.

Fluid, electrolyte, and acid-base balance

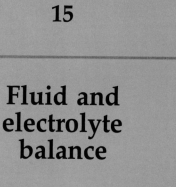

15

Fluid and electrolyte balance

Introduction

Some general principles
   about fluid balance

Avenues by which water
   enters and leaves body

Mechanisms that maintain
   homeostasis of total
   fluid volume
Regulation of urine volume
Factors that alter fluid
   loss under abnormal
   conditions
Regulation of fluid intake

Mechanisms that maintain
   homeostasis of fluid
   distribution
Comparison of plasma,
   interstitial fluid, and
   intracellular fluid
Control of water movement
   between plasma and
   interstitial fluid
Control of water movement
   through cell membranes
   between interstitial
   and intracellular fluids

▇Introduction

The term fluid balance means several things. It, of course, means the same thing as homeostasis of fluids. To say that the body is in a state of fluid balance is to say that the total amount of water in the body is normal and that it remains relatively constant. But fluid balance also means something more. It also means relative constancy of the distribution of water in the body's three fluid compartments. The volume of water inside the cells, in the interstitial spaces, and in the blood vessels all remains relatively constant when a condition of fluid balance exists. Fluid imbalance, then, means that both the total volume of water in the body and the amount in one or more of its fluid compartments have increased or decreased beyond normal limits.

Fluid balance and electrolyte balance are interdependent. If one deviates from normal, so does the other. A discussion of one, therefore, necessitates a discussion of the other.

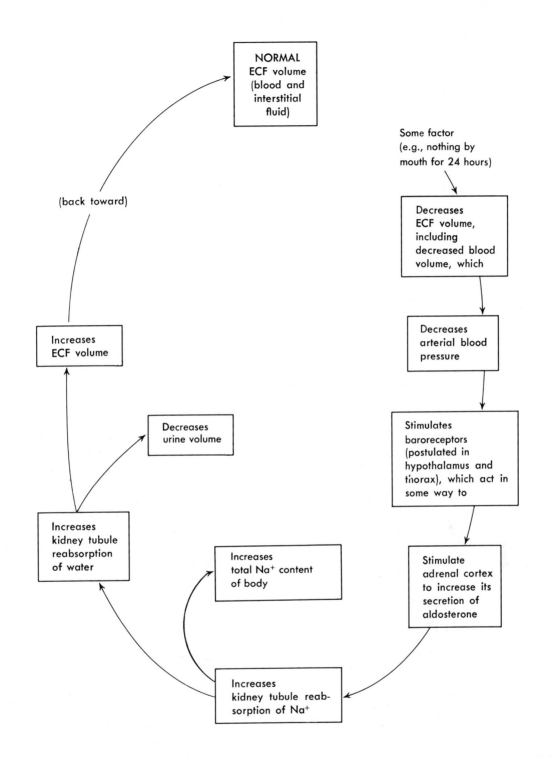

Fig. 15-1 Aldosterone mechanism that tends to restore normal extracellular fluid (ECF) volume when it decreases below normal. Excess aldosterone, however, leads to excess extracellular fluid volume—i.e., excess blood volume (hypervolemia) and excess interstitial fluid volume (edema)—and also to an excess of the total NA⁺ content of the body.

Table 15-1. Typical normal values for each portal of water entry and exit

Intake		*Output*	
Ingested liquids	1,500 ml	Kidneys (urine)	1,400 ml
Water in foods	700 ml	Lungs (water in expired air)	350 ml
Water formed by catabolism	200 ml	Skin	
		By diffusion	350 ml
		By sweat	100 ml
		Intestines (in feces)	200 ml
Totals	2,400 ml		2,400 ml

Modern medicine attaches great importance to fluid and electrolyte balance. Today, a large proportion of hospital patients receive some kind of fluid and electrolyte therapy. To help you understand the rationale underlying such treatment, this chapter is included. We shall consider successively some general principles about fluid balance, the avenues by which water enters and leaves the body, the mechanisms that maintain homeostasis of total fluid volume, and the mechanisms that maintain homeostasis of fluid distribution.

Some general principles about fluid balance

1 The cardinal principle about fluid balance is this: fluid balance can be maintained only if intake equals output. Obviously, if more water or less leaves the body than enters it, imbalance will result. Total fluid volume will increase or decrease, but cannot remain constant under those conditions.

2 Devices for varying output so that it equals intake constitute the most crucial mechanism for maintaining fluid balance, but mechanisms for adjusting intake to output also operate. Fig. 15-1 illustrates the aldosterone mechanism for decreasing fluid output (urine volume) to compensate for decreased intake. Fig. 15-2 diagrams a postulated mechanism for adjusting intake to compensate for excess output.

3 Mechanisms for controlling water movement between the fluid compartments of the body constitute the most rapid-acting fluid balance devices. They serve first of all to maintain normal blood volume at the expense of interstitial fluid volume.

Avenues by which water enters and leaves body

Water enters the body, as everyone knows, from the digestive tract—in the liquids one drinks and in the foods one eats. But, in addition, and less universally known, water enters the body—that is, is added to its total fluid volume—from its billions of cells. Each cell produces water by catabolizing foods, and this water enters the bloodstream. Water normally leaves the body by four exits: kidneys (urine), lungs (water in expired air), skin (by diffusion and by sweat), and intestines (feces). In accord with the cardinal principle of fluid balance, the total volume of water entering the body normally equals the total volume leaving. In short, fluid intake normally equals fluid output. Fig. 15-3 illustrates the portals of water entry and exit, and Table 15-1 gives their normal volumes. These, however, can vary considerably and still be considered normal.

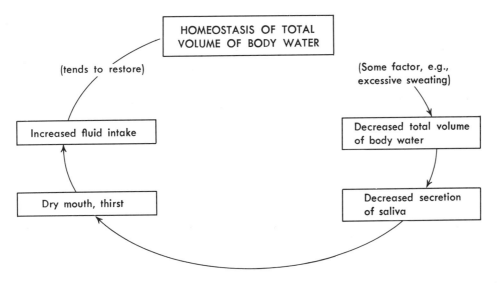

Fig. 15-2 The basic principle of a postulated homeostatic mechanism for adjusting intake to compensate for excess output.

■ Mechanisms that maintain homeostasis of total fluid volume

Under normal conditions, homeostasis of the total volume of water in the body is maintained or restored primarily by devices that adjust output (urine volume) to intake and secondarily by mechanisms that adjust fluid intake.

Regulation of urine volume

Two factors together determine urine volume: the glomerular filtration rate and the rate of water reabsorption by the renal tubules. The glomerular filtration rate, except under abnormal conditions, remains fairly constant—hence does not normally cause urine volume to fluctuate. The rate of tubular reabsorption of water, on the other hand, fluctuates considerably. The rate of tubular reabsorption, therefore, rather than the glomerular filtration rate, normally adjusts urine volume to fluid intake. And the amount of ADH and of aldosterone secreted regulates the amount of water reabsorbed by the kidney tubules (discussed on p. 458; also see Fig. 15-1). In other words, urine volume is regulated chiefly by hormones secreted by the posterior lobe of the pituitary gland (ADH) and by the adrenal cortex (aldosterone).

Although changes in the volume of fluid loss via the skin, the lungs, and the intestines also affect the fluid intake-output ratio, these volumes are not automatically adjusted to intake volume, as is the volume of urine.

Factors that alter fluid loss under abnormal conditions

Both the rate of respiration and the volume of sweat secreted may greatly alter fluid output under certain abnormal conditions. For example, a patient who hyperventilates for an extended time loses an excessive amount of water via the expired air. If, as frequently happens, he also takes in less water by mouth than normal, his fluid output then exceeds his intake and he develops a fluid imbalance—namely, dehydration (that is, a decrease in total body water). Other abnormal conditions such as vomiting, diarrhea, intestinal drainage, etc. also cause fluid and electrolyte output to exceed intake so produce fluid and electrolyte imbalances.

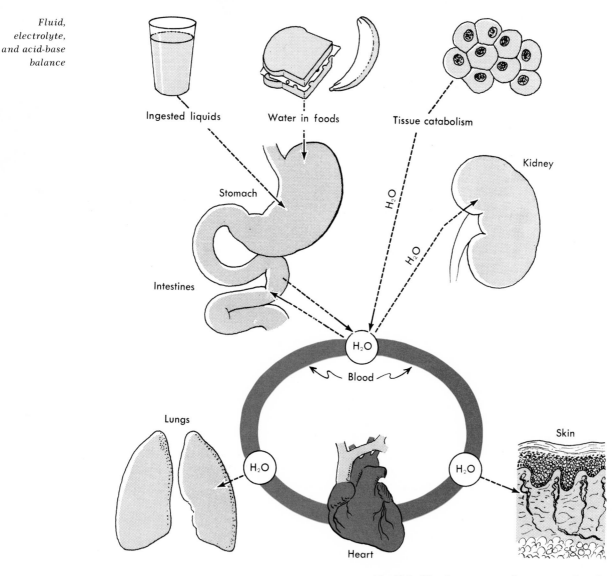

Fig. 15-3 The three sources of water entry into the body and its four avenues of exit.

Regulation of fluid Intake

Physiologists disagree about the details of the mechanism for controlling intake so that it increases when output increases and decreases when output decreases. In general, it operates in this way: when dehydration starts to develop, salivary secretion decreases, producing a "dry mouth feeling" and the sensation of thirst. The individual then drinks water, thereby increasing his fluid intake to offset his increased output, and this tends to restore fluid balance (Fig. 15-2). If, however, an individual takes nothing by mouth for several days, fluid balance cannot be maintained despite every effort of homeostatic mechanisms to compensate for the zero intake. Obviously, under this condition, the only way balance could be maintained would be for fluid output to also decrease to zero. But this cannot occur. Some output is obligatory. Why? Because as long as respirations continue, some water leaves the body by way of the expired air. Also, as long as life continues, an irreducible minimum of water diffuses through the skin.

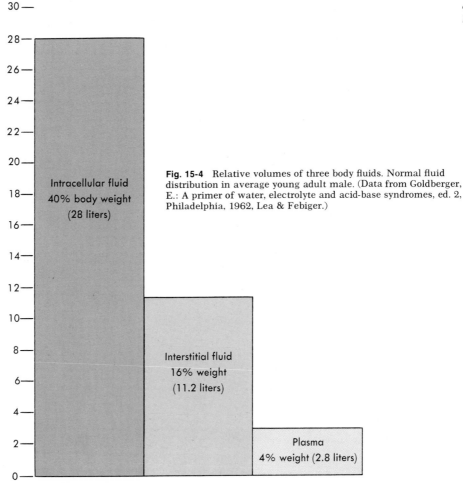

Fig. 15-4 Relative volumes of three body fluids. Normal fluid distribution in average young adult male. (Data from Goldberger, E.: A primer of water, electrolyte and acid-base syndromes, ed. 2, Philadelphia, 1962, Lea & Febiger.)

■ Mechanisms that maintain homeostasis of fluid distribution

Comparison of plasma, interstitial fluid, and intracellular fluid

Structurally speaking, body fluids occupy three fluid compartments: blood vessels, tissue spaces, and cells. Thus, we speak of the plasma, interstitial fluid, and intracellular fluid. Functionally, however, these three fluids may be considered as only two fluids: that which lies outside the cells and that which lies within them. Functionally, then, we speak of the extracellular fluid (ECF), meaning both the plasma and interstitial fluid, and the intracellular fluid (ICF), meaning the water inside all the cells. Extracellular fluid, as you already know, constitutes the internal environment of the body. It, therefore, serves the dual vital functions of providing a relatively constant environment for cells and of transporting substances to and from them. Intracellular fluid, on the other hand, because it is a solvent, functions to facilitate intracellular chemical reactions that maintain life. Compared as to volume, intracellular fluid is the largest, plasma the smallest, and interstitial fluid in between. Fig. 15-4 gives typical normal fluid volumes in a young adult male. Note that intracellular fluid constitutes 40% of body weight, interstitial fluid 16%, and blood plasma 4%. Or, expressed differently, for every kilogram of its weight, the body contains about 400 milliliters of intracellular fluid, 160 milliliters of inter-

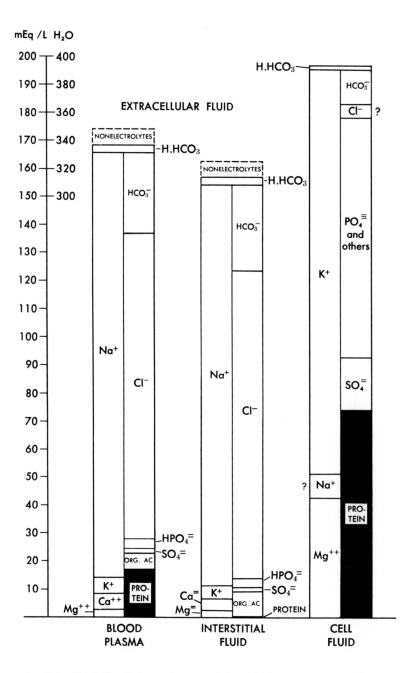

Fig. 15-5 Chief chemical constituents of three fluid compartments. The column of figures on the left (200, 190, 180, etc.) indicates amounts of cations or of anions, while the figures on the right (400, 380, 360, etc.) indicate the sum of cations and anions. Note that chloride and sodium values in cell fluid are questioned. It is probable that at least muscle intracellular fluid contains some sodium but no chloride. (Slightly modified from Mountcastle, V. B., editor: Medical physiology, St. Louis, The C. V. Mosby Co.; after Gamble, J. L.: Physiological information gained from studies on the life raft ration, Harvey Lect. **42**:247-273, 1946-1947.)

Table 15-2. Electrolyte composition of blood plasma*

Cations	Anions
142 mEq Na$^+$	102 mEq Cl$^-$
4 mEq K$^+$	26 HCO$_3^-$
5 mEq Ca^{++}	17 protein$^-$
2 mEq Mg^{++}	6 other
	2 HPO$_4^=$
153 mEq/L plasma	153 mEq/L plasma

*Data from Ruch, T. C., and Patton, H. D.: Physiology and biophysics, ed. 19, Philadelphia, 1965, W. B. Saunders Co.

stitial fluid, and 40 milliliters of plasma. Normal fluid volumes, however, vary considerably—mainly according to the fat content of the body. Fat people have a lower water content than slender ones. And women have a lower water content than men—but chiefly because a woman's body contains a higher percentage of fat. Total fluid volume and fluid distribution also vary with age. As a person grows older, the total amount of water in his body decreases. Fluid makes up a smaller percent of his body weight. Extracellular fluid volume is proportionately larger in infants and children than in adults.

Compared chemically, plasma and interstitial fluid (the two extracellular fluids) are almost identical. Intracellular fluid, on the other hand, shows striking differences from either of the two extracellular fluids. Let us examine first the chemical structure of plasma and interstitial fluid as shown in Fig. 15-5 and Table 15-2.

Perhaps the first difference between the two extracellular fluids to catch your eye (in Fig. 15-5) is that blood contains a slightly larger total of electrolytes (ions) than does interstitial fluids. If you compare the two fluids, ion for ion, you will discover the most important difference between blood plasma and interstitial fluid. Look at the anions (negative ions) in these two extracellular fluids. Note that blood contains an appreciable amount of protein anions. Interstitial fluid, in contrast, contains hardly any protein anions. This is the only functionally important difference between blood and interstitial fluid. It exists because the normal capillary membrane is practically impermeable to proteins. Hence, almost all protein anions remain behind in the blood instead of filtering out into the interstitial fluid. And because proteins remain in the blood, certain other differences also exist between blood and interstitial fluid: notably, blood contains more sodium ions and fewer chloride ions than does interstitial fluid.*

Extracellular fluids and intracellular fluid are more unlike than alike chemically. Chemical difference predominates between the extracellular and intracellular fluids. Chemical similarity predominates between the two extracellular fluids. Study Fig. 15-5 and make some generalizations about the main chemical differences between the extracellular and intracellular fluids. For example: What is the most abundant cation in the extracellular fluids? In the intracellular fluid? What is the most abundant anion in the extracellular fluids? In the intracellular fluid? What about the relative concentrations of protein anions in extracellular fluid and intracellular fluid?

The only reason we have called attention to the chemical structure of the three body fluids is that here, as elsewhere, structure determines function. In this instance, the chemical structure of the three fluids helps control water and electrolyte movement between them. Or, phrased differently, the

*According to the Donnan equilibrium principle, when nondiffusible anions (negative ions) are present on one side of a membrane, there are on that side of the membrane fewer diffusible anions and more diffusible cations (positive ions) than on the other side. Applying this principle to the blood and interstitial fluid: because blood contains nondiffusible protein anions, it contains fewer chloride ions (diffusible anions) and more sodium ions (diffusible cations) than does interstitial fluid.

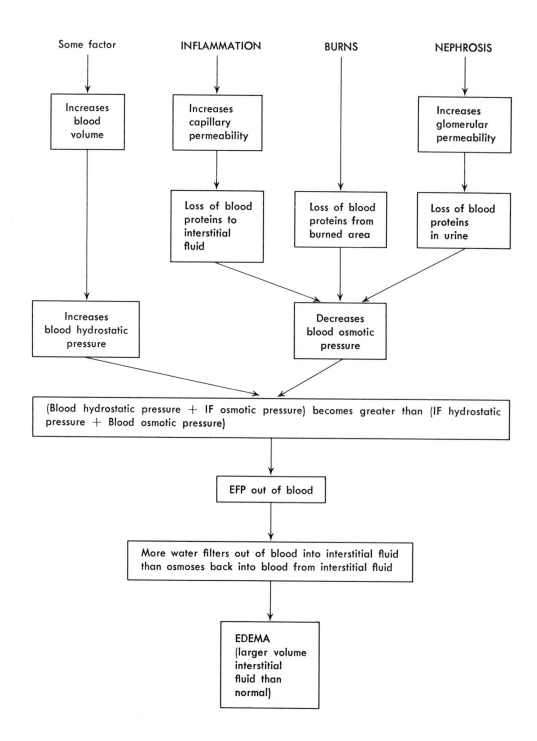

Fig. 15-6 Mechanisms of edema formation in some common conditions (IF, interstitial fluid; EFP, effective filtration pressure). (Also see Fig. 15-7.)

chemical structure of body fluids, if normal, helps maintain homeostasis of fluid distribution and, if abnormal, results in fluid imbalance. Hypervolemia (excess blood volume) is a case in point. Edema, too, frequently stems from changes in the chemical structure of body fluids (Fig. 15-6).

We are ready now to try to answer the following question: How does the chemical structure of body fluids control water movement between them and thereby control fluid distribution in the body?

Control of water movement between plasma and interstitial fluid

Over sixty years ago, Starling advanced a hypothesis about the nature of the mechanism that controls water movement between plasma and interstitial fluid—that is, across the capillary membrane. This hypothesis has since become one of the major premises of physiology and is often spoken of as Starling's "law of the capillaries." According to this law, the control mechanism for water exchange between plasma and interstitial fluid consists of four pressures: blood hydrostatic and colloid osmotic pressures* on one side of the capillary membrane and interstitial fluid hydrostatic and colloid† osmotic pressures on the other side.

According to the physical laws governing filtration and osmosis (p. 24), blood hydrostatic pressure (HP) tends to force fluid out of capillaries into interstitial fluid (IF), but

blood colloid osmotic pressure (OP) tends to draw it back into them. Interstitial fluid hydrostatic pressure, in contrast, tends to force fluid out of the interstitial fluid into the capillaries, and interstitial fluid colloid osmotic pressure tends to draw it back out of capillaries. In short, two of these pressures constitute vectors in one direction and two in the opposite direction. Does this remind you of another mechanism studied earlier? (To check your answer, see p. 451.)

The difference between the two sets of opposing forces obviously represents the net or effective filtration pressure—in other words, the effective force tending to produce the net fluid movement between blood and interstitial fluid. In general terms, therefore, we may state Starling's law of the capillaries this way: the rate and direction of fluid exchange between capillaries and interstitial fluid is determined by the hydrostatic and colloid osmotic pressures of the two fluids. Or, we may state it more specifically as a formula:

$$(BHP + IFOP) - (IFHP + BOP) = EFP^*$$

Note that the factors enclosed in the first set of parentheses tend to move fluid out of capillaries and that those in the second set oppose this movement—they tend to move fluid into the capillaries.

To illustrate operation of Starling's law, let us consider how it controls water exchange at the arterial ends of tissue capillaries. Table 15-3 gives typical normal pressures. Using these figures in Starling's law of the capillaries:

$(35 + 0) - (2 + 25) = 8$ mm Hg net pressure (EFP), causing water to filter out of blood at arterial ends of capillaries into interstitial fluid

---

*Osmotic pressure due to concentrations of protein in blood and interstitial fluid. Since the capillary membrane is permeable to other plasma solutes, they quickly diffuse through the membrane so cause no osmotic pressure to develop against it. Only the proteins, to which the capillary membrane is practically impermeable, cause an actual osmotic pressure against the capillary membrane (explained on pp. 23 and 27).
†A small amount of blood protein passes through the capillary membrane and tends to concentrate around the venous end of capillaries, hence this pressure.

---

*BHP, Blood hydrostatic pressure. IFOP, Interstitial fluid osmotic pressure. IFHP, Interstitial fluid hydrostatic pressure. BOP, Blood osmotic pressure. EFP, Effective filtration pressure between blood and interstitial fluid.

Table 15-3. Pressures at arterial end of tissue capillaries

Arterial end of capillary	Hydrostatic pressure	Colloid osmotic pressure
Blood	35 mm Hg	25 mm Hg
Interstitial fluid	2 mm Hg	0 mm Hg

Table 15-4. Pressures at venous end of tissue capillaries

Venous end of capillary	Hydrostatic pressure	Colloid osmotic pressure
Blood	15 mm Hg	25 mm Hg
Interstitial fluid	1 mm Hg	3 mm Hg*

*A small amount of blood proteins passes through the capillary membrane and tends to concentrate around the venous ends of capillaries—hence this pressure.

The same law operates at the venous end of capillaries (Table 15-4). Again apply Starling's law of the capillaries. What is the net effective pressure at the venous ends of capillaries? In which direction does it cause water to move? Assuming that the figures given in Table 15-4 are normal, do you agree that "about the same amount of water returns to the blood at the venous ends of capillaries as left it from the arterial ends"?

On the basis of our discussion thus far, we can formulate some principles about the transfer of water between blood and interstitial fluid:

1 No net transfer of water occurs between blood and interstitial fluid so long as the effective filtration pressure (EFP) equals 0—that is, when:

$$(Blood\ HP + IFOP) = (IFHP + Blood\ OP)$$

2 A net transfer of water, a "fluid shift," occurs between blood and interstitial fluid whenever the EFP does not equal 0—that is, when:

$$(Blood\ HP + IFOP)$$
$$does\ not\ equal$$
$$(IFHP + Blood\ OP)$$

3 Since (Blood HP + IFOP) is a force that tends to move water out of capillary blood, fluid shifts out of blood into interstitial fluid whenever:

$$(Blood\ HP + IFOP)$$
$$is\ greater\ than$$
$$(IFHP + Blood\ OP)$$

4 Since (IFHP + Blood OP) is a force that tends to move water out of interstitial fluid into capillary blood, fluid shifts out of interstitial fluid into blood whenever:

$$(IFHP + Blood\ OP)$$
$$is\ greater\ than$$
$$(Blood\ HP + IFOP)$$

Or, stated the other way around, whenever:

$$(Blood\ HP + IFOP)$$
$$is\ less\ than$$
$$(IFHP + Blood\ OP)$$

To apply these principles, answer questions 8 to 11 on p. 511 and examine Figs. 15-6 and 15-7.

Control of water movement through cell membranes between interstitial and intracellular fluids

The mechanism that regulates water movement through cell membranes is similar to the one that regulates water movement through capillary membranes. In other words, interstitial fluid and intracellular fluid hydrostatic and osmotic pressures regulate water transfer between these two fluids. But because their osmotic pressures vary more than do their hydrostatic pressures, interstitial fluid and intracellular fluid osmotic pressures serve as the chief regulators of water transfer across cell membranes. Their osmotic pressures, in turn, are directly related to the electrolyte concentration gradients—notably sodium and potassium—maintained

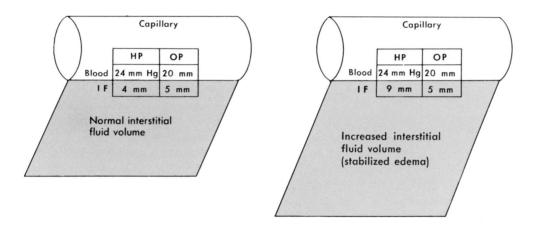

Fig. 15-7 The mechanism of edema formation initiated by a decrease in blood protein concentration and, therefore, in blood osmotic pressure. Left diagram, blood osmotic pressure has just decreased to 20 from the normal 25 mm Hg. This increases the effective filtration pressure (EFP) to 5 mm Hg from a normal of 0 (see Starling's formula, p. 505). The EFP of 5 mm Hg causes fluid to shift out of blood into interstitial fluid (IF) until the EFP again equals 0 – in this case, when the interstitial fluid volume has increased enough the raise interstitial fluid hydrostatic pressure to 9 mm Hg as shown in the diagram on the right. At this point, a new equilibrium is established and equal amounts of water once more are exchanged between the blood and interstitial fluid. Thus, the increased interstitial fluid volume – that is, the edema – becomes stabilized.

across cell membranes. As Fig. 15-5 shows, the chief electrolyte by far in interstitial fluid is sodium salt. And intracellular fluid's main electrolyte is potassium salt. Therefore, a change in the sodium or the potassium concentrations of either of these fluids causes the exchange of fluid between them to become unbalanced. For example, a decrease in interstitial fluid sodium concentration immediately decreases interstitial fluid osmotic pressure, making it hypotonic to intracellular fluid osmotic pressure. In other words, a decrease in interstitial fluid sodium concentration establishes an osmotic pressure gradient between interstitial and intracellular fluids. And this causes net osmosis to occur out of interstitial fluid into cells (a discussion of osmosis appears on pp. 23 to 27). In short, interstitial fluid and intracellular fluid electrolyte concentrations are the main determinants of their osmotic

pressures; their osmotic pressures regulate the amount and direction of water transfer between the two fluids, and this regulates their volumes. Hence, fluid balance depends upon electrolyte balance. And conversely, electrolyte balance depends upon fluid balance. An imbalance in one produces an imbalance in the other (Fig. 15-8).

Normal sodium concentration in interstitial fluid and potassium concentration in intracellular fluid depend upon many factors but especially upon the amount of ADH and aldosterone secreted. As shown in Fig. 15-9, ADH regulates extracellular fluid electrolyte concentration and osmotic pressure by regulating the amount of water reabsorbed into blood by renal tubules. Aldosterone, on the other hand, regulates extracellular fluid volume by regulating the amount of sodium reabsorbed into blood by renal tubules (Fig. 15-1).

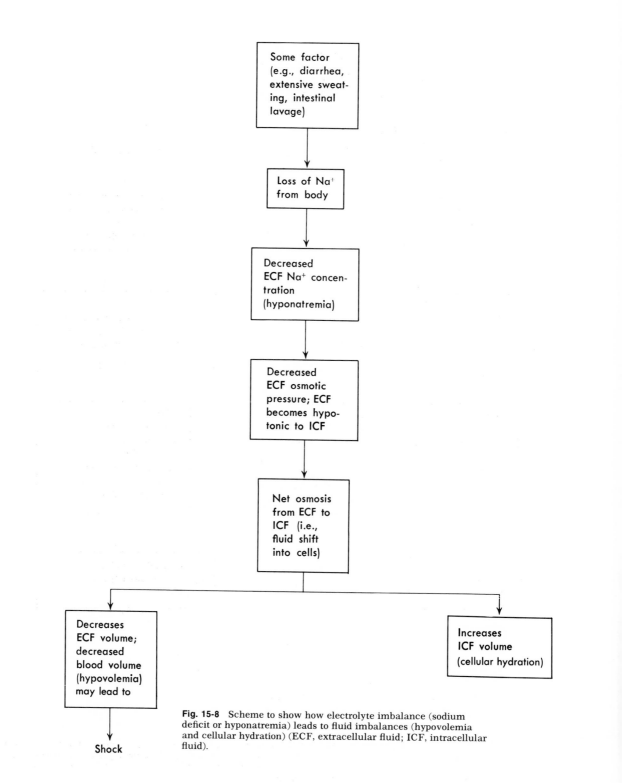

Fig. 15-8 Scheme to show how electrolyte imbalance (sodium deficit or hyponatremia) leads to fluid imbalances (hypovolemia and cellular hydration) (ECF, extracellular fluid; ICF, intracellular fluid).

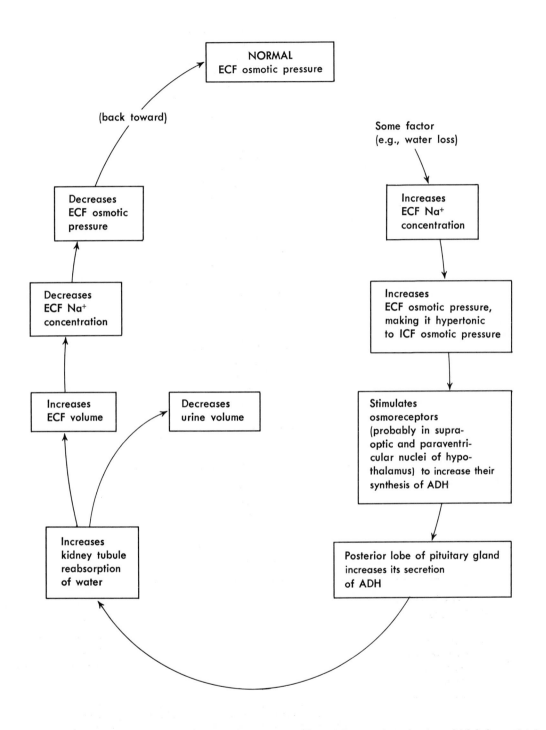

Fig. 15-9 ADH (antidiuretic hormone) mechanism which helps maintain homeostasis of extracellular fluid (ECF) osmotic pressure by regulating its volume and thereby its electrolyte concentration—that is, mainly ECF Na$^+$ concentration.

outline summary

Introduction

1 Meaning of fluid balance
   a Same as homeostasis of fluids—i.e., total volume of water in body normal and remains relatively constant
   b Volume of blood plasma, interstitial fluid, and intracellular fluid all remain relatively constant—i.e., homeostasis of distribution of water as well as of total volume
2 Fluid balance and electrolyte balance interdependent—see Figs. 15-1 and 15-9

Some general principles about fluid balance

1 Cardinal principle—intake must equal output
2 Fluid and electrolyte balance maintained primarily by mechanisms that adjust output to intake; secondarily by mechanisms that adjust intake to output
3 Fluid balance also maintained by mechanisms that control movement of water between fluid compartments

Avenues by which water enters and leaves body

1 Water enters body through digestive tract in
   a Liquids
   b Foods
2 Water formed in body by metabolism of foods
3 Water leaves body via kidneys, lungs, skin, and intestines

Mechanisms that maintain homeostasis of total fluid volume

1 Regulation of urine volume—under normal conditions, by factors that control reabsorption of water by distal and collecting tubules
   a ECF electrolyte concentration (crystalloid osmotic pressure) controls ADH secretion which controls tubule H_2O reabsorption
   b ECF volume controls aldosterone secretion which controls tubule Na^+ reabsorption and therefore water reabsorption
2 Factors that alter fluid loss under abnormal conditions—hyperventilation, hypoventilation, vomiting, diarrhea, circulatory failure, etc.
3 Regulation of fluid intake—see Fig. 15-2
   a Mechanism by which intake adjusted to output not completely known
   b One controlling factor seems to be degree of moistness of mucosa of mouth—if output exceeds intake, mouth feels dry, sensation of thirst occurs, and individual ingests liquids

Mechanisms that maintain homeostasis of fluid distribution

1 Comparison of plasma, interstitial fluid, and intracellular fluid
   a Plasma and interstitial fluid constitute extracellular fluid (ECF), internal environment of body or, in other words, environment of cells
   b Intracellular fluid (ICF) volume largest, plasma volume smallest; ICF about 40% of body weight, IF about 16%, and plasma about 4%, or ICF volume about ten times that of plasma and IF volume about four times plasma volume
   c Chemically, plasma and IF almost identical except that plasma contains slightly more electrolytes and considerably more proteins than IF; also blood contains somewhat more sodium and fewer chloride ions
   d Chemically, ECF and ICF strikingly different; sodium main cation of ECF, potassium main cation of ICF; chloride main anion of ECF; phosphate main anion of ICF; protein concentration much higher in ICF than in IF
2 Control of water movement between plasma and interstitial fluid
   a By four pressures—blood hydrostatic and colloid osmotic pressures and interstitial fluid hydrostatic and colloid osmotic pressures
   b Effect of these pressures on water movement between plasma and interstitial fluid expressed in Starling's "law of the capillaries"; only when (blood hydrostatic pressure + IF colloid osmotic pressure) − (blood colloid osmotic pressure + IF hydrostatic pressure) = 0, do equal amounts of water filter out of blood into IF and osmose back into blood from IF; in other words, water balance exists between these two fluids under these conditions
3 Control of water movement through cell membranes between interstitial and intracellular fluids—primarily by relative crystalloid osmotic pressures of ECF and ICF, which depend mainly upon sodium concentration of ECF and potassium concentration of ICF which, in turn, depend upon intake and output of sodium and potassium and upon sodium-potassium transport mechanisms

review questions

1 Explain in your own words the meaning of the term fluid balance.
2 How is total volume of body fluids kept relatively constant—i.e., what other factors must be controlled in order to keep the total volume of water in the body relatively constant?
3 What, if any, are functionally important differences between the chemical composition of plasma and interstitial fluid?
4 What, if any, are functionally important differences between the chemical composition of extracellular and intracellular fluids?
5 Are plasma and interstitial fluid more accurately described as "similar" or "different" as to chemical composition? Volume?
6 Support your answer to question 5 with some specific facts.
7 Are interstitial fluid and intracellular fluid more accurately described as "similar" or "different" as to chemical composition? Volume?
8 Explain Starling's law of the capillaries in your own words. Be as brief and clear as you can. This law describes the mechanism for controlling what?
9 Suppose that in one individual the normal average or mean pressures (in mm Hg) are capillary blood hydrostatic pressure 24 and osmotic pressure 25; interstitial fluid hydrostatic pressure 4 and osmotic pressure 5. Following a hemorrhage, this patient's blood hydrostatic pressure falls to 18. (a) Assuming that the other pressures momentarily stay the same, what is the EFP now? (b) The new EFP causes a fluid shift in which direction? (c) When will the fluid shift stop and an even exchange of water again go on between blood and interstitial fluids?
10 Formulate a principle by filling in the blanks in the following sentences: Anything that decreases blood hydrostatic pressure in the capillaries tends to cause a fluid shift from _____ into _____. Conversely, anything that increases blood hydrostatic pressure in the capillaries tends to cause a fluid shift out of _____ into _____.
11 Formulate a principle by completing the blanks in the following sentence: A decrease in blood protein concentration tends to decrease blood _____ pressure and therefore tends to cause a fluid shift from _____ to _____ _____.

Situation: A patient has had marked diarrhea for several days.

12 Explain the homeostatic mechanisms that would tend to compensate for this excessive fluid loss.
13 Do you think they could succeed in maintaining fluid balance or would fluid therapy probably be necessary?
14 What, if any, abnormality of fluid distribution do you think would occur in this patient without fluid therapy? Explain your reasoning.
15 Why does dehydration necessarily develop if an individual takes nothing by mouth for several days and receives no fluid therapy? Why cannot homeostatic mechanisms prevent this?

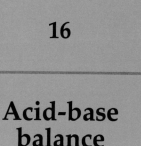

16

Acid-base balance

Mechanisms that control pH of body fluids
Meaning of term pH
Types of pH control mechanisms
Effectiveness of pH control mechanisms; range of pH

Buffer mechanism for controlling pH of body fluids
Buffers defined
Buffer pairs present in body fluids
Action of buffers to prevent marked changes in pH of body fluids
Evaluation of role of buffers in pH control

Respiratory mechanism of pH control
Explanation of mechanism
Adjustment of respirations to pH of arterial blood
Some principles relating respirations and pH of body fluids

Urinary mechanism of pH control
General principles about mechanism
Mechanisms that control urine pH

Acid-base imbalances
Acidosis
Alkalosis

Acid-base balance is vitally important. Acid-base balance means maintenance of homeostasis of the hydrogen ion concentration of body fluids. Even a slight deviation from normal causes pronounced changes in the rate of cellular chemical reactions. This, in turn, threatens survival.

Mechanisms that control pH of body fluids

Meaning of term pH

The term pH is a symbol used to mean the hydrogen ion concentration of a solution. Actually, pH stands for the negative logarithm of the hydrogen ion concentration.* pH indicates the degree of acidity and alkalinity of a solution—the latter because as hydrogen ion concentration increases, OH ion concentration (alkalinity) necessarily decreases. A pH of 7 indicates neutrality (equal amounts of H^+ and OH^-),

---

*A pH of 7, for example, means that a solution contains 10^{-7} grams hydrogen ions per liter. Or, translating this logarithm into a number, a pH of 7 means that a solution contains 0.0000001 (that is, 1/10,000,000) of a gram of hydrogen ions per liter. A solution of pH 6 contains 0.000001 (1/1,000,000) of a gram of hydrogen ions per liter and one of pH 8 contains 0.00000001 (1/100,000,000) of a gram of hydrogen ions per liter. Note that a solution with pH 7 contains ten times as many hydrogen ions as a solution with pH 8 and that pH decreases as hydrogen ion concentration increases.

a pH of less than 7 indicates acidity (more H^+ than OH^-), and one greater than 7 indicates alkalinity (more OH^- than H^+).

Types of pH control mechanisms

Since various acids and bases continually enter the blood from absorbed foods and from the catabolism of foods, some kind of mechanism for neutralizing or eliminating these substances is necessary if blood pH is to remain constant. Actually, three different devices operate together to maintain constancy of pH. Collectively, these devices—buffers, respirations, and kidney excretion acids and bases—might be said to constitute the pH homeostatic mechanism.

Effectiveness of pH control mechanisms; range of pH

The most eloquent evidence of the effectiveness of the pH control mechanism is the extremely narrow range of pH, normally 7.35 to 7.45. In terms of hydrogen ion concentration, this means that normally there is a little more than 1/100,000,000 of a gram of hydrogen ions (pH 8) in a liter of blood but a little less than 1/10,000,000 of a gram (pH 7). Also, the greatest normal amount of hydrogen ions is only about 1/100,000,000 of a gram more than the smallest normal amount. What incredible constancy! Acids continually stream into capillary blood from cell metabolism and yet a liter of venous blood (pH 7.35) contains only about 1/100,000,000 of a gram more hydrogen ions than does a liter of arterial blood (pH 7.45)! The pH homeostatic mechanism does indeed control effectively—astonishingly so.

■ Buffer mechanism for controlling pH of body fluids

Buffers defined

In terms of action, a buffer is a substance that prevents marked changes in the pH of a solution when an acid or a base is added to it. Let us suppose that a small amount of the strong acid HCl is added to a solution that contains a buffer (to blood, for example) and that its pH decreases from 7.41 to 7.27. But if the same amount of HCl were added to pure water containing no buffers, its pH would decrease much more markedly, from 7 to perhaps 3.4. In both instances, pH decreased upon addition of the acid, but much less so with buffers present than without them.

In terms of chemical composition, buffers consist of two kinds of substances and are, therefore, often referred to as "buffer

513

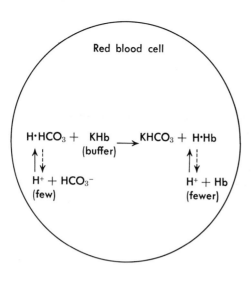

Fig. 16-1 Buffering of carbonic acid (H · HCO$_3$ or H$_2$CO$_3$) inside red blood cell by potassium salt of hemoglobin. Note that each molecule of carbonic acid is replaced by a molecule of the acid hemoglobin. Since hemoglobin is a weaker acid than carbonic acid, fewer of these hemoglobin molecules dissociate to form hydrogen ions. Hence, fewer hydrogen ions are added to red blood cell intracellular fluid than would have been added by unbuffered carbonic acid. Also, since some of the carbonic acid in the red blood cells has come from plasma, fewer hydrogen ions remain in blood than would have if there were no buffering of carbonic acid.

pairs." Most of the body fluid buffer pairs consist of a weak acid and a salt of that acid, as shown below.

Buffer pairs present in body fluids

The main buffer pairs present in body fluids are as follows:

Bicarbonate pairs $\dfrac{\text{NaHCO}_3}{\text{H}_2\text{CO}_3}, \dfrac{\text{KHCO}_3}{\text{H}_2\text{CO}_3},$ etc.

Plasma protein pair $\dfrac{\text{Na} \cdot \text{proteinate}}{\text{Proteins (weak acids)}}$

Hemoglobin pairs $\dfrac{\text{K} \cdot \text{Hb}}{\text{Hb}}$ and $\dfrac{\text{K} \cdot \text{HbO}_2}{\text{HbO}_2}$

(Hb and HbO$_2$ are weak acids)

Phosphate buffer pair $\dfrac{\text{Na}_2\text{HPO}_4 \text{ (basic phosphate)}}{\text{NaH}_2\text{PO}_4 \text{ (acid phosphate)}}$

Action of buffers to prevent marked changes in pH of body fluids

Buffers react with a relatively strong acid (or base) to replace it by a relatively weak acid (or base). That is to say, an acid which highly dissociates to yield many H ions is replaced by one which dissociates less highly to yield fewer H ions. Thus, by the buffer reaction, instead of the strong acid remaining in the solution and contributing many H ions to drastically lower the pH of the solution, a weaker acid takes its place, contributes fewer additional H ions to the solution, and thereby lowers its pH only slightly. Therefore, because blood contains buffer pairs, its pH fluctuates much less widely than it would without

them. In other words, blood buffers constitute one of the devices for preventing marked changes in blood pH. Let us consider some specific examples of buffer action. More carbonic acid is formed in the body than any other acid. It continuously enters tissue capillaries, where it is buffered primarily by the potassium salt of hemoglobin inside the red blood cells, as shown in Fig. 16-1.

Note how this reaction between carbonic acid and the potassium salt of hemoglobin (a buffer) applies the principle of buffering. As a result of the buffering action of KHb, the weaker acid, H · Hb, replaces the stronger acid, H$_2$CO$_3$, and, therefore, the hydrogen ion concentration of blood increases much less than it would have if carbonic acid were not buffered.

The potassium salt of oxyhemoglobin, sodium proteinate (sodium salts of blood proteins), and basic sodium phosphate also buffer carbonic acid.

Nonvolatile, or fixed acids, such as lactic acid and ketone bodies, are buffered mainly by the basic member of the bicarbonate buffer pair, mainly by sodium bicarbonate — ordinary baking soda (Fig. 16-2).

Carbonic acid buffers strong bases as is illustrated in Fig. 16-3.

When blood pH is normal and a state

$$H \cdot lactate + NaHCO_3 \longrightarrow Na \cdot lactate + H \cdot HCO_3$$

$$H^+ + lactate^-$$
(few)

$$H^+ + HCO_3^-$$
(fewer)

Fig. 16-2 Lactic acid (H · lactate) and other nonvolatile or "fixed" acids are buffered by sodium bicarbonate, the most abundant base bicarbonate in the blood. Carbonic acid (H · HCO₃ or H₂CO₃, a weaker acid than lactic acid) replaces lactic acid. Result: fewer hydrogen ions are added to blood than would be if lactic acid were not buffered. Ketone bodies (e.g., acetoacetic acid) from fat catabolism are also buffered by base bicabonate.

$$NaOH + H \cdot HCO_3 \longrightarrow NaHCO_3 + HOH$$

$$Na^+ + OH^-$$
(many)

$$H^+ + OH^-$$
(very few)

Fig. 16-3 Buffering of base NaOH by carbonic acid.

of acid-base balance exists, components of the bicarbonate buffer pair are present in the extracellular fluid in a ratio of 20 parts of base bicarbonate (primarily $NaHCO_3$) to 1 part of carbonic acid. Actually, a liter of plasma normally contains about 27 milliequivalents of base bicarbonate ($B \cdot HCO_3$) and 1.3 milliequivalents of carbonic acid:

$$\frac{27 \text{ mEq} \cdot B \cdot HCO_3}{1.3 \text{ mEq} \cdot H_2CO_3} = \frac{20}{1} = pH\ 7.4$$

An increase in this ratio causes pH to increase (uncompensated alkalosis), and a decrease in it causes pH to decrease (uncompensated acidosis).

Both carbonic acid and nonvolatile acids enter the blood in tissue capillaries and are immediately buffered. If you examine Fig. 16-2, you can discover how the buffering of nonvolatile acids changes blood's bicarbonate–carbonic acid ratio. Note what this buffering action does. It replaces some of the sodium bicarbonate in blood with carbonic acid. So blood leaving the capillaries—venous blood, that is—contains less sodium bicarbonate and more carbonic acid than does arterial blood. And venous blood's base bicarbonate–carbonic acid ratio, therefore, is somewhat lower than arterial blood's. And this necessarily means that venous blood's pH is also lower than arterial blood's pH.

Evaluation of role of buffers in pH control

Buffering alone cannot maintain homeostasis of pH. As we have seen, hydrogen ions continually are added to capillary blood despite buffering. If even a few more hydrogen ions were added every time blood circulated and no way were provided for eliminating them, blood hydrogen ion concentration would necessarily increase and thereby decrease blood pH. "Acid blood," in other words, would soon develop. Respiratory and urinary devices must, therefore, function concurrently with buffers in order to remove from the blood and from the body the hydrogen ions continually being added to blood. Only then can the body maintain constancy of pH.

■Respiratory mechanism of pH control

Explanation of mechanism

Respirations play a vital part in controlling pH. With every expiration, carbon dioxide and water leave the body in the expired air. The carbon dioxide has come from the venous blood—has diffused out of it as it moves through the lung capillaries. Less carbon dioxide remains, therefore, in the arterial blood leaving the lung capillaries and fewer hydrogen ions can be formed in it by the following reactions:

$$CO_2 + H_2O \xrightarrow{\text{(carbonic anhydrase)}} H_2CO_3$$
$$H_2CO_3 \longrightarrow H^+ + HCO_3^-$$

So arterial blood has a lower hydrogen ion concentration and a higher pH than venous blood. A typical average pH for venous blood is 7.36, and 7.41 is a typical average pH for arterial blood.

Adjustment of respirations to pH of arterial blood

Obviously, in order for respirations to serve as a mechanism of pH control, there must be some arrangement so respirations can vary as needed to maintain or restore normal pH. Suppose that blood pH has decreased (that is, hydrogen ion concentration has increased). Respirations then need to increase in order to eliminate more carbon dioxide from the body, and thereby leave less carbonic acid and fewer hydrogen ions in the blood.

The mechanism for adjusting respirations to arterial blood carbon dioxide content or pH operates in this way. Neurons of the respiratory center are sensitive to changes in arterial blood carbon dioxide content and to changes in its pH. If the amount of carbon dioxide in arterial blood increases beyond a certain level, or if arterial blood pH decreases below about 7.38, the respiratory center is stimulated and respirations accordingly increase in rate and depth. This, in turn, eliminates more carbon dioxide, reduces carbonic acid and hydrogen ions, and increases pH back toward the normal level (Fig. 16-4). The carotid chemoreflexes (p. 383) are also devices by which respirations adjust to blood pH and, in turn, adjust pH.

Some principles relating respirations and pH of body fluids

1 A decrease in blood pH below normal (that is, acidosis) tends to cause increased respirations (hyperventilation) which tends to increase pH back toward

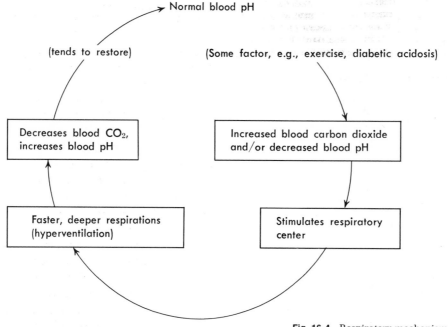

Fig. 16-4 Respiratory mechanism of pH control. A rise in arterial blood CO_2 content or a drop in its pH (below about 7.38) stimulates respiratory center neurons. Hyperventilation results. Less CO_2 and therefore less carbonic acid and fewer hydrogen ions remain in the blood so that blood pH increases, often reaching the normal level.

normal. In other words, acidosis causes hyperventilation which, in turn, acts as a compensating mechanism for the acidosis.

2 Prolonged hyperventilation may increase blood pH enough to produce alkalosis.

3 An increase in blood pH above normal (or alkalosis) causes hypoventilation, which serves as a compensating mechanism for the alkalosis by decreasing blood pH back toward normal.

4 Prolonged hypoventilation may decrease blood pH enough to produce acidosis.

■Urinary mechanism of pH control

General principles about mechanism

Because the kidneys can excrete varying amounts of acid and base, they, like the lungs, play a vital role in pH control. Kidney tubules, by excreting many or few hydrogen ions, in exchange for reabsorbing many or few sodium ions, control urine pH and thereby help control blood pH. If, for example, blood pH decreases below normal, kidney tubules remove more hydrogen ions from the blood to the urine and reabsorb more sodium ions from the urine back into the blood. This, of course, decreases urine pH. But simultaneously— and of far more importance—it increases blood pH back toward normal. This urinary mechanism of pH control is a device for excreting varying amounts of hydrogen ions from the body to match the amounts entering the blood. It constitutes a much more effective device for adjusting hydrogen output to hydrogen input than does the body's only other mechanism for expelling hydrogen ions—namely, the respiratory mechanism previously described. But abnormalities of any one of the three pH control mechanisms soon throws the body into a state of acid-base imbalance. Only when all three parts of this complex mechanism—buffering, respirations, and

urine secretion—function adequately can acid-base balance be maintained.

Let us turn our attention now to the mechanisms that adjust urine pH to counteract changes in blood pH.

Mechanisms that control urine pH

A decrease in blood pH accelerates the renal tubule ion-exchange mechanisms that tend to increase the blood pH back to normal by acidifying urine. The following paragraphs describe these mechanisms.

1 Distal and collecting tubules secrete hydrogen ions into the urine in exchange for basic ions, which they reabsorb. Refer to Fig. 16-5 as you read the rest of this paragraph. Note that carbon dioxide diffuses from tubule capillaries into distal tubule cells, where the enzyme carbonic anhydrase accelerates the combining of carbon dioxide with water to form carbonic acid. The latter dissociates into hydrogen ions and bicarbonate ions. The hydrogen ions then diffuse into the tubular urine, where they displace basic ions (most often sodium) from a basic salt of a weak acid and thereby change the basic salt to an acid salt or to a weak acid that is eliminated in the urine. While this is happening, the displaced sodium or other basic ion diffuses into a tubule cell. Here, it combines with the bicarbonate ion left over from the carbonic acid dissociation to form sodium bicarbonate. The sodium bicarbonate then diffuses—is reabsorbed, that is—into the blood. Consider the various results of this mechanism. Sodium bicarbonate (or other base bicarbonate) is conserved for the body. Instead of all the basic salts that filter out of glomerular blood leaving the body in the urine, considerable amounts are recovered into peritubular capillary blood. In addition, extra hydrogen ions are added to the urine and thereby eliminated from the body. Both the reabsorption of base bicarbonate into blood and the ex-

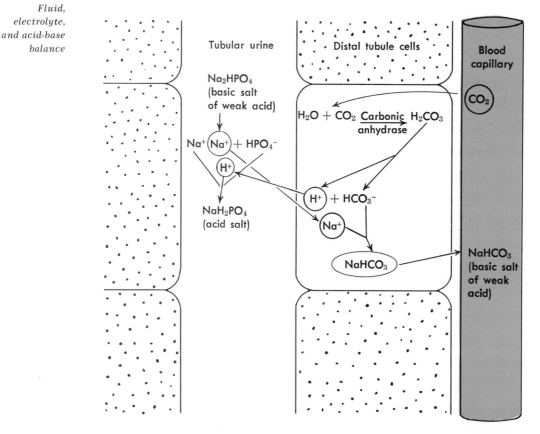

Fig. 16-5 Acidification of urine and conservation of base by distal renal tubule excretion of H ions (discussed on p. 517).

cretion of hydrogen ions into urine tend to increase the ratio of the bicarbonate buffer pair

$$\frac{B \cdot HCO_3}{H \cdot HCO_3}$$

present in blood. This automatically increases blood pH. In short, kidney tubule base bicarbonate reabsorption and hydrogen-ion excretion both tend to alkalinize blood by acidifying urine.

Renal tubules can excrete either hydrogen or potassium in exchange for the sodium they reabsorb. Therefore, in general, the more hydrogen ions they excrete, the fewer the potassium ions they can excrete. For example, in acidosis, tubule excretion of hydrogen ions increases markedly and potassium ion excretion decreases—an important fact because it may lead to *hy-*

perkalemia (excessive blood potassium), a dangerous condition because it can cause heart block and death.

2 Distal and collecting tubule cells excrete ammonia into the tubular urine. As Fig. 16-6 shows, the ammonia combines with hydrogen to form an ammonium ion. The ammonium ion displaces sodium or some other basic ion from a salt of a fixed (nonvolatile) acid to form an ammonium salt. The basic ion then diffuses back into a tubule cell and combines with bicarbonate ion to form a basic salt which, in turn, diffuses into tubular blood. Thus, like the renal tubules' excretion of hydrogen ions, their excretion of ammonia and its combining with hydrogen to form ammonium ions also tends to increase the blood bicarbonate buffer pair ratio and, therefore, tends to

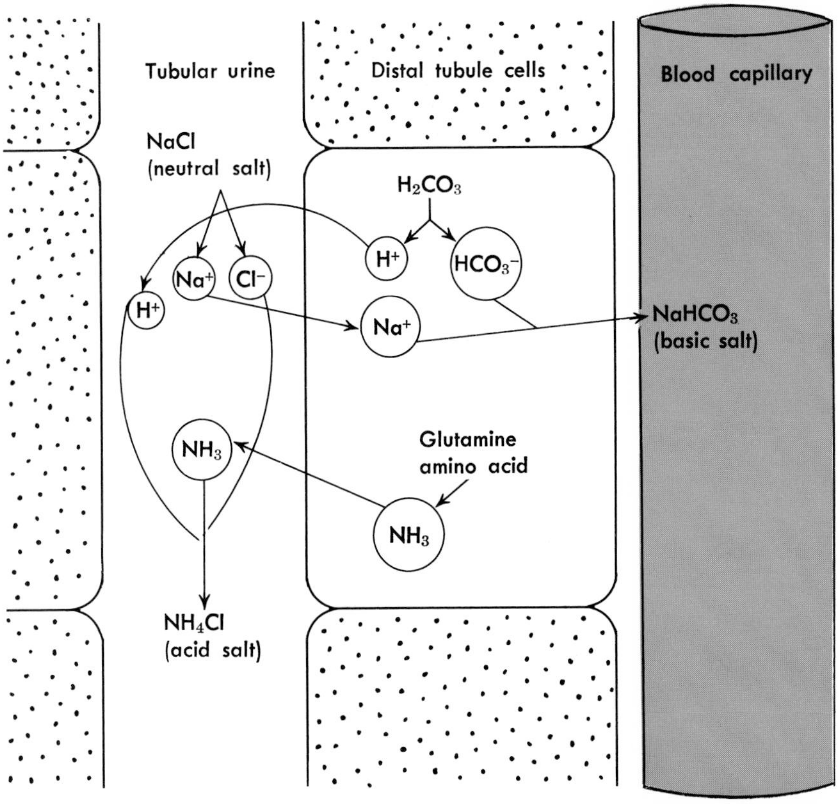

Fig. 16-6 Acidification of urine by tubule excretion of ammonia (NH_3). An amino acid (glutamine) leaves blood, enters a tubule cell, and is deaminized to form ammonia which is excreted into urine. In exchange, the tubule cell reabsorbs a basic salt (mainly $NaHCO_3$) into blood from urine.

increase blood pH. Quantitatively, however, ammonium ion excretion is more important than hydrogen ion excretion.

Renal tubule excretion of hydrogen and ammonia is controlled at least in part by the blood pH level. As indicated in Fig. 16-7, a decrease in blood pH accelerates tubule excretion of both hydrogen and ammonia. An increase in blood pH produces the opposite effects.

■ Acid-base imbalances

Acidosis

During the course of certain diseases, such as diabetic ketosis or starvation, for instance, abnormally large amounts of nonvolatile acids enter the blood. All three types of pH control mechanisms – buffers, respiratory, and urinary – are enlisted in

an all-out effort to compensate for the excess acid and to maintain acid-base balance. Basic bicarbonate buffers immediately react with the acids (Fig. 16-2). This decreases the ratio of base bicarbonate–carbonic acid of blood and therefore decreases blood pH. The decreased blood pH stimulates the respiratory centers. Respirations become faster and deeper (hyperventilation). Hence more carbon dioxide leaves the blood and less carbonic acid remains in it. This increases the ratio

$$\frac{B \cdot HCO_3}{H \cdot HCO_3}$$

and thereby increases blood pH back up toward normal. The kidney tubules increase their excretion of H^+ and NH_3 in exchange for reabsorbed Na^+ (Figs. 16-5 and 16-6).

519

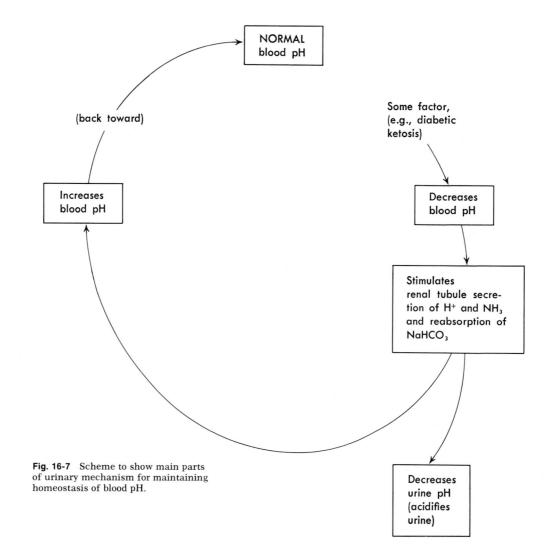

Fig. 16-7 Scheme to show main parts
of urinary mechanism for maintaining
homeostasis of blood pH.

And these actions also tend to increase
blood's base bicarbonate–carbonic acid
ratio and its pH.

If hyperventilation and urine acidification together succeed in preventing a decrease in the base bicarbonate–carbonic
acid ratio, blood pH remains normal and
a state of *compensated acidosis* exists. But
if, despite these homeostatic devices, the
ratio and pH decrease, *uncompensated
acidosis* develops.

Increased blood hydrogen ion concentration (that is, decreased blood pH), as
we have noted, stimulates the respiratory
center. For this reason, hyperventilation
is an outstanding clinical sign of acidosis.
Increases in hydrogen ion concentration

above a certain level depress the central
nervous system and, therefore, produce
such symptoms as disorientation and coma.

Alkalosis

Alkalosis develops less often than acidosis. Some circumstances, however, such as
ingestion of an excessive amount of an alkaline drug, or hyperventilation, or excessive vomiting can produce alkalosis. Base
bicarbonate increases above normal in
alkalosis. In compensated alkalosis, carbonic acid also increases—enough to maintain a normal base bicarbonate–carbonic
acid ratio and pH. In uncompensated alkalosis, the ratio and, therefore, the pH
increase.

outline summary

Mechanisms that control pH of body fluids

1 Meaning of term pH—negative logarithm of H ion concentration of solution
2 Types of pH control mechanisms
   a Buffers
   b Respirations
   c Secretion of urine of varying pH
3 Effectiveness of pH control mechanisms; range of pH—extremely effective, normally maintain pH within very narrow range of 7.35 to 7.45

Buffer mechanism for controlling pH of body fluids

1 Buffers defined
   a Substances that prevent marked change in pH of solution when acid or base added to it
   b Consist of weak acid (or its acid salt) and basic salt of that acid
2 Buffer pairs present in body fluids—mainly carbonic acid, proteins, hemoglobin, acid phosphate, and sodium and potassium salts of these weak acids
3 Action of buffers to prevent marked changes in pH of body fluids
   a Volatile acids, chiefly carbonic acid, buffered mainly by potassium salts of hemoglobin and oxyhemoglobin
   b Nonvolatile acids, such as lactic acid and ketone bodies, buffered mainly by sodium bicarbonate
   c Bases buffered mainly by carbonic acid

$$\left(\text{Ratio } \frac{B \cdot HCO_3}{H_2CO_3} = \frac{20}{1} \right.$$

   when homeostasis of pH at 7.4 exists)
,4 Evaluation of role of buffers in pH control—cannot maintain normal pH without adequate functioning of respiratory and urinary pH control mechanisms

Respiratory mechanism of pH control

1 Explanation of mechanism
   a Amount of blood carbon dioxide directly related to amount of carbonic acid and, therefore, to concentration of H ions
   b With increased respirations, less carbon dioxide remains in blood, hence less carbonic acid and fewer H ions; with decreased respirations, more carbon dioxide remains in blood, hence more carbonic acid and more H ions
2 Adjustment of respirations to pH of arterial blood—see Fig. 16-4

3 Some principles relating respirations and pH of body fluids
   a Acidosis ⟶ hyperventilation
      ↓
      increases elimination of CO_2
      ↓
      decreases blood CO_2
      ↓
      decreases blood H_2CO_3
      ↓
      decreases blood H ions (i.e., increases blood pH)
      ↓
      tends to correct acidosis (i.e., to restore normal pH)
   b Prolonged hyperventilation, by decreasing blood H ions excessively, may produce alkalosis
   c Alkalosis causes hypoventilation, which tends to correct alkalosis by increasing blood CO_2 and, therefore, blood H_2CO_3 and H ions
   d Prolonged hypoventilation, by eliminating too little CO_2, causes increase in blood H_2CO_3 and, consequently, in blood H ions, thereby may produce acidosis

Urinary mechanism of pH control

1 General principles about mechanism—plays vital role in acid-base balance because kidneys can eliminate more H ions from body while reabsorbing more base when pH tends toward acid side and eliminate fewer H ions while reabsorbing less base when pH tends toward alkaline side
2 Mechanisms that control urine pH
   a Secretion of H ions into urine—when blood CO_2, H_2CO_3, and H ions increase above normal, distal tubules secrete more H ions into urine to displace basic ion (mainly sodium) from a urine salt and then reabsorb sodium into blood in exchange for the H ions excreted
   b Secretion of NH_3—when blood hydrogen ion concentration increases, distal tubules secrete more NH_3, which combines with H ion of urine to form NH_4 ion, which displaces basic ion (mainly sodium) from a salt; basic ion then reabsorbed back into blood in exchange for ammonium ion excreted

Acid-base imbalances

1 Acidosis

    a Compensated metabolic acidosis—decreased alkaline reserve (mainly $NaHCO_3$), but ratio

$$\frac{B \cdot HCO_3}{H_2CO_3}$$

    maintained at normal 20/1 by proportionately decreasing blood carbonic acid by hyperventilation

    b Uncompensated acidosis—alkaline reserve

$$\frac{B \cdot HCO_3}{H_2CO_3}$$

    ratio and blood pH all decrease below normal

2 Alkalosis

    a Compensated alkalosis—opposite to compensated acidosis

    b Uncompensated alkalosis—opposite to uncompensated acidosis

review questions

1 Explain, in your own words, what the term pH means.
2 What is the normal range for pH of body fluids?
3 Explain what a buffer is in terms of its chemical composition and in terms of its function. Cite specific equations to illustrate your explanation.
4 What is the numerical value of the ratio of $B \cdot HCO_3/H_2CO_3$ when blood pH is 7.4 and the body is in a state of acid-base balance?
5 Is the ratio $B \cdot HCO_3/H_2CO_3$ necessarily abnormal when an acid-base disturbance is present? Give reasons to support your answer.
6 Is blood pH always abnormal when an acid-base disturbance is present? Give reasons for your answer.
7 Explain how the "respiratory mechanism of pH control" operates.
8 Why is hyperventilation a characteristic clinical sign in acidosis?

Situation: A patient has suffered a severe head injury. Respirations are markedly depressed.

9 Which, if either, do you think is a potential danger—acidosis or alkalosis? State your reasons.
10 Describe the homeostatic mechanisms that would operate to try to maintain acid-base balance in this patient.

Situation: A mother brings her baby to the hospital and reports that he has seemed very sick for the past twenty-four hours and that he has not eaten during that time and has passed no urine.

11 Do you think acidosis or alkalosis may be present in this baby?
12 Why?

Stress

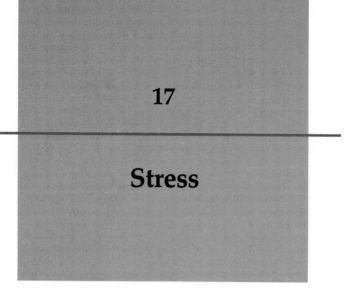

17

Stress

Physiological stress
History
Meaning
Causes
Results
Measurement

Psychological stress

Stress and disease

Not once in a lifetime, but many times, each one of us experiences stress. Most of us can usually recognize when we are in a state of stress but cannot define nor describe it adequately. Our plan for this chapter is to relate a brief history of the development of the concept of human stress and then attempt to answer, at least in part, the following questions: What is physiological stress? What causes stress? What responses does stress cause? What factors affect stress responses? How is stress measured? How, if at all, does phys-

iological stress relate to disease? What is psychological stress and how, if at all, does it relate to physiological stress? Part of our aim shall be to help you understand the intangibility of stress and the reasons for the lack of concise general statements regarding it. We hope, too, that you will gain a sufficiently accurate understanding of stress not to be misled when you read accounts of it in newspapers, magazines, and textbooks and even in professional journals.

■ Physiological stress
History

The concept of physiological stress is generally traced to Cannon's theory of homeostasis. As explained in Chapter 1, Cannon held that the body maintains homeostasis—a relative constancy of its internal environment—by means of numerous homeostatic mechanisms. Basically, these are stimulus-response systems. The stimulus is some specific change in the external or internal environment of the body. It triggers or initiates specific responses which tend to reverse the change and reestablish the former state. Thus, homeostatic responses tend to maintain homeostasis. Homeostatic responses are adaptative in nature – that is, they tend to maintain healthy survival. In almost every chapter of this book we have described several homeostatic mechanisms. You can find examples of these in Figs. 12-22, 12-24, and 10-27 (pp. 420, 422, and 335). Figs. 12-22 and 12-24 illustrate two of the several mechanisms that tend to maintain homeostasis of blood glucose concentration. Fig. 10-27 diagrams one of the mechanisms for maintaining homeostasis of arterial blood pressure. In each of these examples and, in fact, in all homeostatic mechanisms, a specific stimulus induces a series of specific responses and these responses tend to maintain homeostasis and achieve adaptation.

Hans Selye of McGill University in Montreal laid the cornerstone for the study of stress as distinct from homeostasis during the late 1940's and early 1950's. He conducted numerous experiments on animals in which he exposed them to different kinds of intense stimuli and observed their responses. He found that no matter what type of intense stimulus he used – whether it was physical trauma, infection, fatigue, poisons, x-ray, or whatever – the same group (syndrome) of responses always occurred. He called them *stress responses*. He also coined the term *general adaptation syndrome* to mean the entire group of stress responses he observed. Note that this term suggests that stress responses are adaptive in nature. Or, we might substitute the word protective for adaptive. The stress responses seemed to protect the animals from serious damage by extreme stimuli and to promote their healthy survival. Therefore, Selye looked upon the *general adaptation syndrome* as a crucial part of the body's arsenal against its potential enemies. He emphasized that not one particular kind of stimulus but any kind of extreme stimuli induces the same general adaptation syndrome of responses. Compare this with what you know about homeostatic mechanisms. A homeostatic mechanism, unlike the stress mechanism, consists of a specific response triggered by a specific stimulus (example: the specific response of accelerated liver glycogenolysis, triggered by the specific stimulus of low blood sugar). Homeostatic responses resemble stress responses in that both tend to achieve adaptation or healthy survival.

Meaning

Physiological stress is a state of the body. Extreme environmental stimuli acting on the body produce this state. A specific syndrome of responses manifest or make known its presence. Stress, like health or any other state or condition, is an intangible phenomenon. It cannot be seen, heard, tasted, smelled, felt, or measured. How, then, can we know that it exists? Only by the visible, tangible, measurable responses the body makes to stress. Most frequently studied of these "stress responses" are changes in the heart and blood vessels, respirations, blood chemistry, and urine chemistry. Changes probably occur, however, in every tissue and organ and fluid when the body is in a state of general stress.

Causes

What causes stress? The briefest answer we can give to that question is that stressors cause stress. A *stressor*, according to Selye, is "the agent that causes physiologic stress."* A stressor is any kind of extreme environmental stimuli that causes stress; mild stimuli usually are not stressors. Thus, extreme cold, extreme heat, loud noise, and pain are stressors. But coolness, warmth, soft music, and discomfort are not. Stressors usually are extreme stimuli, and frequently they are injurious or unpleasant—but not always. "A painful blow and a passionate kiss," Selye wrote, "can be equally stressful."* Even stressors that are extreme stimuli

*From Selye, H.: The stress syndrome, Amer. J. Nurs. **65**:97-99 (March), 1965; © American Journal of Nursing Company.

may not be extreme excesses. Some are extreme deficiencies of an essential kind of stimuli such as social contact. Solitary confinement in a prison, a long solitary ocean voyage, travel in space, withdrawal and social isolation of an aged person, blindness, and deafness have all been identified as stressors. But overcrowding, the opposite extreme of isolation, also acts as a stressor.

It is impossible to say definitely that certain stimuli are stressors and that certain other stimuli are not stressors. Why? Because individuals differ in their reactions to the same stimuli. Stimuli that are stressors for you may not be stressors for your friends. Or, stimuli that are stressors for you today may not have been stressors for you yesterday or may not be stressors for you tomorrow. About as definite as we can be about stressors is to say that they are generally extreme stimuli, that they are frequently unpleasant, and that many factors—especially the state of one's physical and mental health—affect individual reactions to such stimuli.

Results

Profound changes take place in the body during stress. Somehow, the state of stress inaugurates a series of events that bring about a syndrome of stress responses. We shall call this series of events and responses the stress mechanism. Some investigators suggest that the stress mechanism begins with stimulation of the hypothalamus. But just how a state of stress in the body stimulates this part of the brain remains a matter of conjecture. There are various theories about it but as yet no firm facts. The stimulated hypothalamus activates two different mechanisms: the sympathetic–adrenal medullary mechanism and the anterior pituitary–adrenocortical mechanism. Neurons whose cell bodies lie in the hypothalamus conduct impulses via various tracts to sympathetic centers in the thoracolumbar segments of the cord (Fig. 7-35, p. 226). With stimulation of these lower sympathetic centers, impulses go out from them to widely scattered visceral effectors, stimulating them to respond—to make stress responses, that is. Sympathetic stimulation causes immediate, short-lived responses which mobilize the body for maximal physical exertion. In classic terms, sympathetic activity prepares the body for "fight or flight." Stress responses induced by sympathetic impulses are summarized in Table 7-11 (p. 231). They function to quickly increase circulation, decrease digestion, and increase catabolism. Specific sympathetic stress reponses include a faster heart rate and larger stroke volume, constriction of blood vessels in the skin, kidneys, and most other viscera, and dilatation of blood vessels in the skeletal muscles. By causing these vasomotor responses, sympathetic activity increases arterial blood pressure and redistributes the blood. It increases blood flow to more active organs and decreases it to less active ones. Other stress responses caused by sympathetic stimulation are decreased secretion by digestive glands but increased secretion by sweat glands and the adrenal medulla. The increase in epinephrine secretion by the adrenal medulla is rapid, marked, and significant—significant because the resulting high blood level of epinephrine augments and prolongs sympathetic responses.

In addition to sending nerve impulses to sympathetic centers and thereby activating the sympathetic–adrenal medullary mechanism during stress, the hypothalamus also activates the anterior pituitary–adrenocortical mechanism. It does this not by nerve impulses, but by means of a chemical, a neurosecretion which it releases into the pituitary portal veins. The name of the neurosecretion is corticotropin-releasing factor (CRF). Transported to the anterior pituitary gland by the pituitary portal vein, CRF stimulates this all-important endocrine gland to increase its secretion of adrenocorticotropic hormone (ACTH). ACTH travels via the circulation to the adrenal gland. There it stimulates the adrenal cortex to secrete more glucocorticoids. In general, glucocorticoids (discussed in Chapter 9) are essential for food, water, and electrolyte metabolism and for enabling the body to cope successfully with stress. How they accomplish this later function, however, is not clear. Many investigators suggest that the most important role of glucocorticoids is a "permissive" or "sensitizing" one. What they mean by this is that glucocorticoids permit tissues to respond to the demands imposed by stress by making them more sensitive and responsive to other agents. For example, glucocorticoids increase the vasoconstriction response to norepinephrine. Conversely, animals with subnormal adrenocortical function show decreased responsiveness to vasoconstrictors such as norepinephrine. And as a result, these animals cannot maintain normal blood pressure nor a normal distribution of body fluids if stressed by any means. (See pp. 277 and 278 for other glucocorticoid-induced stress responses.)

Numerous factors have been shown to influence the occurrence, time of onset, pattern, and extent of the stress response. Because so many factors can influence stress responses, they vary widely in different individuals and in a single individual at different times. In fact, the variation in stress responses is so great that different investigators carrying out comparable studies often arrive at highly diverse conclusions. Perhaps you have already come across such discrepancies in your reading and have wondered who was right and who was wrong. Actually, it is quite possible that both were right and that the apparent discrepancy in their results and conclusions stemmed from some slight difference in one or more aspects of their studies.

Here is a list of factors that determine how your body will respond to major stressors, whether they be actual or anticipated: your basic health status and personality, your age and sex, your socioeconomic status, your nationality, ethnic origin, and religious affiliation, your marital status, your heredity, and last, though probably not least, your previous experience with a similar stressor. Whether stress results in healthy survival and increased resistance against future stressors or in exhaustion and death depends upon these and probably many other factors.

Measurement

The state of physiological stress can be detected and its magnitude can be measured indirectly by determining the extent of sympathetic or adrenocortical responses. Indicators of sympathetic activity most frequently used to evaluate stress are the following: the increases in the rate and force of heart contraction, the rise in systolic blood pressure, the increase in metabolic rate, the decrease in stomach and intestinal motility, the increase in epinephrine or norepinephrine levels in blood and urine, the activation of palmar sweat glands, and the dilatation of pupils.

The heart rate has been shown to increase in response to such varied stressors as anesthesia, annoying sounds, and the entrance of an attractive nurse into a patient's room. Even anticipation by patients in a coronary care unit of their transferral to a less closely supervised convalescent unit has been identified as a stressor that causes heart rate to speed up.

In normal persons, urinary epinephrine and norepinephrine levels have been found to increase in response to sensory deprivation (a stressor) and to decline after reestablishment of sensory stimulation.

Indicators of adrenocortical activity that are most frequently used to measure stress are eosinopenia and lymphocytopenia (a decrease in the number of circulating eosinophils and lymphocytes) and an increase in the levels of glucocorticoids (or their breakdown products) in blood and urine. Soldiers stressed by prolonged marching have seen found to have fewer circulating eosinophils than normal. This same stress indicator has been observed in college oarsmen when they were anticipating performance on exhibition days and in heart patients when they were anticipating transfer from a coronary care unit to a convalescent unit.

The amount of urinary adrenocorticoids is often used as a measure of stress. It has been found to increase in depressed persons who feel hopeless and doomed, in test pilots, and in college students taking examinations or attending arousing movies. In contrast, urinary corticoids were found to drop precipitously during attendance at showings of nature study films.

The level of adrenocorticoids in the blood plasma of disturbed patients having acute psychotic episodes has been found to be 70% higher than that in normal individuals or in calm patients. Another study showed that the plasma corticoid levels of chronically depressed patients were significantly lower than those of acutely anxious patients. Smoking or exposure to nicotine have also been shown to be stressors that cause a marked rise in plasma adrenocorticoids—by as much as 77% in both human beings and experimental animals.

Psychological stress

Recently, several investigators from the fields of psychology, psychiatry, and sociology have attempted to relate physiological stress to psychological (mental or emotional) phenomena. Although a concept of psychological stress has developed, it has not advanced so far as has the concept of physiological stress. Many analogies, however, exist between the two. For instance, many kinds of stimuli act as physiological stressors; they create a state of physiological stress in the body and trigger the same syndrome of physiological stress responses. So, too, do various types of stimuli act as psychological stressors; they create a state of psychological stress in the body and trigger the same syndrome of psychological stress responses. What are psychological stressors? Briefly, anything that you perceive as a threat—whether real or imagined—to your survival or self-image or doing anything that you consider wrong is for you a psychological stressor.

Psychological stress responses include such emotional reactions as anxiety, fear, guilt, hate, and anger. They also include such overt behavior as restlessness, fidgeting, hesitant speech, stammering, quarreling, criticizing, and crying. Ancient peoples intuitively recognized a relationship between psychological stress and physiological responses. For example, the Chinese in ancient times are said to have required a person suspected of lying to chew rice powder and spit it out. If the powder came out dry, not moistened by saliva, the suspect was judged guilty. Perhaps they reasoned, or intuitively knew, that lying makes a person "nervous" and that, in turn, makes his mouth dry. We reason similarly but use different terms. Lying induces psychological stress, we might say, and psychological stress acts in some way to cause the sympathetic response of decreased salivation. Both common observation and scientific studies have established that psychological stimuli can result in physiological responses. Undoubtedly you have observed this principle operating in yourself. Can you remember ever having been badly

frightened and noticing that your heart seemed to race and pound, or that the palms of your hands became clammy with sweat, or that your tongue felt so dry that it literally stuck to the roof of your mouth? In addition to such common observations, there now exists a body of scientific data substantiating the principle that psychological stressors can and often do produce physiological stress responses. A relatively new scientific discipline called psychophysiology has furnished much of this evidence. Using accepted research methods and sophisticated instruments (including polygraphs designed especially for this type research), scientists have investigated a wide variety of physiological responses made by individuals being subjected to psychological stressors. Their findings amply confirm the principle that psychological stressors can produce numerous physiological responses. They also established that identical stressors do not induce identical physiological responses in different individuals. In the words of one experimentor, "for one person the cardiovascular system may quite regularly mirror emotion most sensitively, whereas another person may be primarily a 'pulmonary reactor.' Moreover, some organ systems 'adapt'—become less responsive—after a number of stimulations. In any individual, therefore, certain autonomic responses are better indexes of emotion than others."*

◼ Stress and disease

Stress, as we have observed several times, produces different results in different individuals and in the same individual at different times. In one person, a certain amount of stress may induce responses that maintain or even enhance

*From Smith, B. M.: The polygraph, Sci. Amer. **216:** 25-31 (Jan.), 1967; © 1967 by Scientific American, Inc.; all rights reserved.

his health. But in another person that same amount of stress appears to "make him sick." Whether stress is "good" or "bad" for you seems to depend more upon your own body's responses to it than upon the severity of the stressors inducing it. You may recall that Selye emphasized the adaptive nature of stress responses. He coined the term *general adaptation syndrome* because he believed that stress responses usually enable the body to adapt successfully to the many stressors that assail it. He held that the state of stress alerts physiological mechanisms to meet the challenge imposed by stressors. But a challenge issued does not necessarily mean a challenge successfully met. Selye proposed that sometimes the body's adaptive mechanisms fail to meet the challenge issued by stressors and that, when they fail, disease results—diseases of adaptation, he called them. He and his co-workers succeeded in inducing a variety of pathologic changes by exposing sensitized animals to intense stimuli of various kinds. Their experimental animals developed a number of different diseases, among them hypertension, arthritis, arteriosclerosis, nephrosclerosis, and gastrointestinal ulcers. As a result of these experiments, many researchers have attempted to establish stress as a cause of disease in man. Their findings, however, have not been widely accepted as convincing evidence that stress, induced in a normal individual, can produce disease.

Around the middle of this century, one of the problems studied was the relationship of blood glucocorticoid concentration to disease. If stress is adaptive, the investigators reasoned, and helps the body combat the effects of many kinds of stressors (for example, infection, injury, burns, etc.), then possibly various diseases might be treated by adding to the body's natural output of glucocorticoids. In 1949, Hench and his colleagues[*] at the Mayo Clinic cautiously suggested that it might be helpful to give glucocorticoids to patients afflicted with various illnesses. Subsequently, they acted on their own suggestion and reported finding that a hormone of the adrenal cortex improved some of the clinical symptoms of rheumatoid arthritis. Since that time, cortisone has been administered to thousands of patients with widely different ailments. So often has it been used and so numerous have been the articles written about cortisone that today it would be almost impossible to find an adult who has never heard of this hormone.

[*]Hench, P. S., Kendall, E. C., Slocumb, C. H., and Polley, H. F.: The effect of a hormone of the adrenal cortex (17-hydroxy-11-dehydrocorticosterone; compound E) and of pituitary adrenocorticotropic hormone on rheumatoid arthritis: preliminary report, Proc. Staff Meet. Mayo Clin. **24**:181-197 (April), 1949.

outline summary

Physiological stress
  History
1 Cannon's theory
  a Body normally maintains relative constancy (homeostasis) of its internal environment by means of homeostatic mechanisms
  b Homeostatic mechanism consists of series of specific responses to specific stimulus
  c Stimulus that activates homeostatic mechanism is some specific change away from homeostasis in body's internal environment

  d Homeostatic responses reverse stimulus change that initiated them and thereby tend to maintain or restore homeostasis; homeostatic responses constitute adaptive mechanisms—that is, enable body to adapt to change in ways that promote healthy survival
2 Selye's theory
  a Any extreme change in body's external environment produces condition of stress in body, and this condition operates in some way to activate "stress mechanism"

b Stress mechanism consists of group (syndrome) of responses to internal condition of stress; stress responses are nonspecific in that same syndrome of responses occurs regardless of kind of extreme change that produced stress

c Stimulus that produces stress and thereby activates stress mechanism is nonspecific in that it can be any kind of extreme change in environment

d Stress responses are adaptive; tend to enable body to adapt to and survive extreme change. Selye referred to this syndrome of stress responses as *general adaptation syndrome*

Meaning

Physiological stress is internal state or condition of body produced by extreme environmental stimuli and identified by nonspecific syndrome of responses which it initiates

Causes

1 Stressors—that is, extreme environmental stimuli of any kind—cause physiological stress
2 Stressors are frequently but not always injurious or unpleasant stimuli
3 Stressors are frequently extreme excesses but may be extreme deficiencies of some essential kind of stimulus
4 Whether or not stimulus acts as stressor depends not only upon intensity of stimulus but also upon various characteristics of individual, notably his physical and mental condition

Results

1 Immediate result of physiological stress is occurrence of stress responses (*general adaptation syndrome* or stress mechanism)
2 According to one hypothesis, physiological stress produces stress responses by means of following series of events:
  a Physiological stress operates in some unknown way to stimulate hypothalamus
  b Stimulated hypothalamus activates both sympathetic–adrenal medullary mechanism and anterior pituitary–adrenocortical mechanism as follows:
    1 Various tracts conduct impulses from hypothalamus to sympathetic centers in cord, stimulating them to send increased impulses out to visceral effectors, causing them to make stress responses (e.g., increased epinephrine secretion by adrenal medulla; faster, stronger heart beat; constriction of blood vessels in skin, kidneys, and most viscera; dilatation of blood vessels in skeletal muscles, etc.)
    2 Certain neurons in the hypothalamus release corticotropin-releasing factor (CRF) into pituitary portal vein, which transports it to anterior pituitary gland where it

stimulates increased secretion of ACTH; ACTH stimulates adrenal cortex to increase its secretion of glucocorticoids; in some way not yet determined, glucocorticoids sensitize or permit various effectors to make appropriate stress responses
3 Numerous factors influence stress responses—individual's physical and mental condition, age, sex, socioeconomic status, heredity, religious affiliation, previous experience with similar stressor, etc.
4 Physiological stress most often results successfully—i.e., results in adaptation, healthy survival, and increased resistance; on other hand, physiological stress sometimes produces exhaustion and death; complex of factors listed in preceding item plus many others determine ultimate results of physiological stress

Measurement

1 Magnitude of physiological stress measured indirectly by determining:
  a Extent of sympathetic responses
  b Extent of adrenocortical responses
2 Indicators of sympathetic activity most often used to measure stress:
  a Increase in rate and force of heartbeat
  b Increase in systolic blood pressure
  c Increase in metabolic rate
  d Increase in blood and urine concentrations of catecholamines—viz., epinephrine and norepinephrine
  e Dilatation of pupils
3 Indicators of adrenocortical activity most often used to measure stress:
  a Decrease in number of circulating eosinophils and lymphocytes
  b Increase in blood and urine levels of glucocorticoids (or their breakdown products)

Psychological stress

1 Many different kinds of stimuli constitute psychological stressors; in general, any stimulus which individual perceives as threat to his well-being or to his self-image acts as psychological stressor
2 Psychological stressors produce state of psychological stress
3 State of psychological stress leads to psychological stress responses
4 Psychological stress responses include:
  a Emotional reactions such as anxiety, fear, guilt, hate, and anger
  b Behavioral responses such as restlessness, stammering, quarreling, criticizing, and crying
5 Psychological stressors frequently result in physiological stress responses—e.g., hearing "this is a stick-up" shouted at you creates state of psychological stress which, in turn, produces not only psychological stress response

of fear but also various physiological stress responses such as racing, pounding heartbeat, dry mouth, dilated pupils, etc.

Stress and disease

1 Selye held that stress could result in disease instead of adaptation
2 To date, research evidence that stress, induced in normal individual, can produce disease has not gained wide acceptance

review questions

1 Explain the meaning of the term physiological stress.
2 What causes physiological stress?
3 What responses does physiological stress cause? Construct a flow chart diagram summarizing the stress mechanism (use information in item 2 under Results of physiological stress in outline summary).
4 What factors affect stress responses?
5 How is physiological stress measured?
6 What is psychological stress?
7 How, if at all, does psychological stress relate to physiological stress?
8 How, if at all, does physiological stress relate to disease?

Supplementary readings

Chapter 1

1 Anthony, C. P.: Basic concepts in anatomy and physiology – a programmed presentation, ed. 2, St. Louis, 1970, The C. V. Mosby Co., pp. 1–7.
2 Cannon, W. B.: The wisdom of the body, ed. 2, New York, 1963, W. W. Norton & Co., Inc.

Chapter 2

1 Allfrey, V. G., and Mirsky, A. E.: How cells make molecules, Sci. Amer. **205**: 74–82 (Sept.), 1961.
2 Allison, A.: Lysosomes and disease, Sci. Amer. **217**:62–72 (Nov.), 1967.
3 Anthony, C. P.: Basic concepts in anatomy and physiology – a programmed presentation, ed. 2, St. Louis, 1970, The C. V. Mosby Co., pp. 8–21.
4 Baserga, R., and Kisieleski, W. E.: Autobiographies of cells, Sci. Amer. **209**: 103–110 (Aug.), 1963.
5 Beermann, W., and Clever, U.: Chromosome puffs, Sci. Amer. **210**:50–58 (April), 1964.
6 Brachet, J.: The living cell, Sci. Amer. **205**:50–61 (Sept.), 1961.
7 Crick, F. H. C.: The genetic code, Sci. Amer. **207**:66–74 (Oct.), 1962.
8 Green, D. E.: The mitochondrion, Sci. Amer. **210**:67–74 (Jan.), 1964.
9 Ham, A. W., and Leeson, T. S.: Histology, ed. 6, Philadelphia, 1969, J. B. Lippincott Co.

10 Hayashi, T.: How cells move, Sci. Amer. **205**:184–204 (Sept.), 1961.

11 Hayflick, L.: Human cells and aging, Sci. Amer. **218**:32–37 (March), 1968.

12 Hokin, M., and Hokin, L.: The chemistry of cell membranes, Sci. Amer. **213**: 78–86 (Oct.), 1965.

13 Holter, H.: How things get into cells, Sci. Amer. **205**:167–180 (Sept.), 1961.

14 Hurwitz, J., and Furth, J. J.: Messenger RNA, Sci. Amer. **206**:41–49 (Feb.), 1962.

15 Kornberg, A.: The synthesis of DNA, Sci. Amer. **219**:64–78 (Oct.), 1968.

16 Lehninger, A. L.: How cells transform energy, Sci. Amer. **205**:62–73 (Sept.), 1961.

17 Neutra, M., and Lebland, C. P.: The Golgi apparatus, Sci. Amer. **220**:100–107 (Feb.), 1969.

18 Nirenberg, M. W.: The genetic code II, Sci. Amer. **208**:80–94 (March), 1963.

19 Nomura, M.: Ribosomes, Sci. Amer. **221**: 28–35 (Oct.), 1969.

20 Racker, E.: The membrane of the mitochondrion, Sci. Amer. **218**:32–39 (Feb.), 1968.

21 Rich, A.: Polyribosomes, Sci. Amer. **209**: 44–53 (Dec.), 1963.

22 Solomon, A. K.: Pumps in the living cell, Sci. Amer. **207**:100–108 (Aug.), 1962.

23 Yanofsky, C.: Gene structure and protein structure, Sci. Amer. **216**:80–94 (May), 1967.

Chapter 3

1 Bevelander, G.: Essentials of histology, ed. 6, St. Louis, 1970, The C. V. Mosby Co.

2 Ham, A. W., and Leeson, T. S.: Histology, ed. 6, Philadelphia, 1969, J. B. Lippincott Co.

Chapter 6

1 Chapman, C. B., and Mitchell, J. H.: The physiology of exercise, Sci. Amer. **212**: 88–96 (May), 1965.

2 Hoyle, G.: How is muscle turned on and off? Sci. Amer. **222**:84–93 (April), 1970.

3 Huxley, H. E.: The mechanism of muscular contraction, Sci. Amer. **213**:18–27 (Dec.), 1965.

4 Morehouse, L. E., and Miller, A. T., Jr.: Physiology of exercise, ed. 5, St. Louis, 1967, The C. V. Mosby Co.

5 Porter, K. R., and Franzine-Armstrong, C.: The sarcoplasmic reticulum, Sci. Amer. **212**:72–81 (March), 1965.

Chapter 7

1 Agranoff, B. W.: Memory and protein synthesis, Sci. Amer. **216**:115–122 (June), 1967.

2 Dicara, L. V.: Learning in the autonomic nervous system, Sci. Amer. **222**:30–39 (Jan.), 1970.

3 Eccles, J.: The synapse, Sci. Amer. **212**: 56–66 (Jan.), 1965.

4 Hyden, H.: Satellite cells in the nervous system, Sci. Amer. **205**:62–70 (Dec.), 1961.

5 Luria, A. R.: The functional organization of the brain, Sci. Amer. **222**:66–79 (March), 1970.

6 Mountcastle, V. B., editor: Medical physiology, ed. 12, St. Louis, 1968, The C. V. Mosby Co., pp. 1057-1858.

7 Penfield, W., and Rasmussen, T.: The cerebral cortex of man, New York, 1968, Hafner Publishing Co., Inc.

8 Pribram, K. H.: The neurophysiology of remembering, Sci. Amer. **220**:73-86 (Jan.), 1969.

9 Wilson, V. J.: Inhibition in the central nervous system, Sci. Amer. **214**:102–110 (May), 1966.

Chapter 8

1 Botelho, S. Y.: Tears and the lacrimal gland, Sci. Amer. **211**:78–86 (Oct.), 1964.

2 Hubbard, R., and Kropf, A.: Molecular isomers in vision, Sci. Amer. **216**:64–70 (June), 1967.

3 Michael, C. R.: Retinal processing of visual images, Sci. Amer. **220**:105–114 (May), 1969.

4 Neisser, U.: The processes of vision, Sci. Amer. **219**:204–214 (Sept.), 1968.

5 Rock, I., and Harris, C. S.: Vision and touch, Sci. Amer. **216**:96–104 (May), 1967.

6 Rosenzweig, M. R.: Auditory localization, Sci. Amer. **205**:132–142 (Oct.), 1961.

7 Thomas, E. L.: Movements of the eye, Sci. Amer. **219**:88–95 (Aug.), 1968.

Chapter 9

1 Hawken, P.: Hypophysectomy with yttrium 90, Amer. J. Nurs. **65**:122–125 (Oct.), 1965.

2 McKusick, V. A., and Rimoin, D. L.: General tom thumb and other midgets, Sci. Amer. **217**:102–106 (July), 1967.

3 Mountcastle, V. B., editor: Medical phys-

iology, ed. 12, St. Louis, 1968, The C. V. Mosby Co., pp. 871–1050.

4 Shea, K. M., O'Connor, C. P., Karafelis, E. G., Thorn, G. W., and Kozak, G. P.: Teaching a patient to live with adrenal insufficiency, Amer. J. Nurs. **65**:80–85 (Dec.), 1965.

5 Wurtman, R. J., and Axelrod, J.: The pineal gland, Sci. Amer. **213**:50–60 (July), 1965.

Chapter 10

1 Adolph, E. F.: The heart's pacemakers, Sci. Amer. **213**:32–37 (March), 1967.

2 Anthony, C. P.: Basic concepts in anatomy and physiology – a programmed presentation, ed. 2, St. Louis, 1970, The C. V. Mosby Co., pp. 108–144.

3 Clarke, C. A.: The prevention of "rhesus" babies, Sci. Amer. **219**:46–52 (Nov.), 1968.

4 Effler, D. B.: Surgery for coronary disease, Sci. Amer. **219**:36–43 (Oct.), 1968.

5 Mayerson, H. S.: The lymphatic system, Sci. Amer. **208**:80–90 (June), 1963.

6 Mountcastle, V. B., editor: Medical physiology, ed. 12, St. Louis, 1968, The C. V. Mosby Co., pp. 1–280.

7 Porter, R. R.: The structure of antibodies, Sci. Amer. **217**:81–87 (Oct.), 1967.

8 Wood, E. J.: The venous system, Sci. Amer. **218**:86–96 (Jan.), 1968.

Chapter 11

1 Comroe, J. H., Jr.: The lung, Sci. Amer. **214**:57–66 (Feb.), 1966.

2 Mountcastle, V. B., editor: Medical physiology, ed. 12, St. Louis, 1968, The C. V. Mosby Co., pp. 613–862.

3 Winter, P. M., and Lowenstein, E.: Acute respiratory failure, Sci. Amer. **221**:23–29 (Nov.), 1969.

Chapter 12

1 Benzinger, T. H.: The human thermostat, Sci. Amer. **204**:134–147 (Jan.), 1961.

2 Hickey, M. C.: Hypothermia, Amer. J. Nurs. **65**:116–122 (Jan.), 1965.

3 Martin, M. M.: Diabetes mellitus: current concepts, Amer. J. Nurs. **66**:510–514 (March), 1966.

4 Morehouse, L. E., and Miller, A. T., Jr.: Physiology of exercise, ed. 5, St. Louis, 1967, The C. V. Mosby Co.

5 Programmed instruction: potassium imbalance, Amer. J. Nurs. **67**:343–366 (Feb.), 1967.

6 Rich, A.: Polyribosomes, Sci. Amer. **209**:44–53 (Dec.), 1963.

Chapter 13

1 Anthony, C. P.: Basic concepts in anatomy and physiology – a programmed presentation, ed. 2, St. Louis, 1970, The C. V. Mosby Co., pp. 145–149.

2 Grollman, A.: Diuretics, Amer. J. Nurs. **65**:84–89 (Jan.), 1965.

3 Trusk, C. W.: Hemodialysis for acute renal failure, Amer. J. Nurs. **65**:80–85 (Feb.), 1965.

Chapter 14

1 Behrman, S. J.: Management of infertility, Amer. J. Nurs. **66**:552–555 (March), 1966.

2 Eichner, E.: Progestins, Amer. J. Nurs. **65**:78–81 (Sept.), 1965.

3 Kormondy, E. J.: Introduction to genetics, New York, 1964, McGraw-Hill Book Co., pp. 23–45.

Chapter 15

1 Anthony, C. P.: Fluid imbalances – formidable foes to survival, Amer. J. Nurs. **63**:75–77 (Dec.), 1963.

2 Anthony, C. P.: Basic concepts in anatomy and physiology – a programmed presentation, ed. 2, St. Louis, 1970, The C. V. Mosby Co., pp. 150–154.

3 Burgess, R. E.: Fluids and electrolytes, Amer. J. Nurs. **65**:90–94 (Oct.), 1965.

4 Gamble, J. L.: Chemical anatomy, physiology, and pathology of extracellular fluid, ed. 6, Cambridge, Mass., 1958, Harvard University Press.

5 Mountcastle, V. B., editor: Medical physiology, ed. 12, St. Louis, 1968, The C. V. Mosby Co., pp. 287-366.

Chapter 16

1 Anthony, C. P.: Basic concepts in anatomy and physiology – a programmed presentation, ed. 2, St. Louis, 1970, The C. V. Mosby Co., pp. 154–157.

2 Frisell, W. R.: Acid-base chemistry in medicine, New York, 1968, The Macmillan Co.

Chapter 17

1 Hench, P. S., Kendall, E. C., Slocumb, C. H., and Polley, H. F.: The effect of a hormone of the adrenal cortex (17-hydroxy-11-dehydrocorticosterone; compound E) and of pituitary adrenocorticotropic hormone on rheumatoid arthritis: preliminary report, Proc. Staff Meet. Mayo Clin. **24**:181–197 (April), 1949.

2 Ingle, D. J.: Permissibility of hormone action; a review, Acta Endocr. (Kobenhavn) **17**:172–186 (Sept.–Dec.), 1954.

3 Pitts, F. N., Jr.: The biochemistry of anxiety, Sci. Amer. **220**:69–75 (Feb.), 1969.

4 Ross, R.: Wound healing, Sci. Amer. **220**:40–50 (June), 1969.

5 Selye, H.: The stress of life, New York, 1956, McGraw-Hill Book Co.

6 Selye, H.: The stress syndrome, Amer. J. Nurs. **65**:97–99 (March), 1965.

7 Smith, B. M.: The polygraph, Sci. Amer. **216**:25–31 (Jan.), 1967.

Additional references
Biochemistry

1 Orten, J. M., and Neuhaus, O. W.: Biochemistry, ed. 8, St. Louis, 1970, The C. V. Mosby Co.

Gross anatomy

1 Goss, C. M., editor: Gray's anatomy of the human body, ed. 28, Philadelphia, 1966, Lea & Febiger.

2 Hamilton, W. J., editor: Textbook of human anatomy, New York, 1957, The Macmillan Co.

3 Sobotta, J., and Figge, F. H. J.: Atlas of human anatomy, ed. 8, New York, 1963, Hafner Publishing Co., Inc. (3 vols.).

4 Spalteholz, W.: Atlas of human anatomy, ed. 16 (revised and re-edited by R. Spanner), New York, 1967, F. A. Davis Co.

Microscopic anatomy

1 Bloom, W., and Fawcett, D. W.: A textbook of histology, ed. 9, Philadelphia, 1968, W. B. Saunders Co.

2 Ham, A. W., and Leeson, T. S.: Histology, ed. 6, Philadelphia, 1969, J. B. Lippincott Co.

Developmental anatomy

1 Arey, L. B.: Developmental anatomy – a textbook and laboratory manual of embryology, ed. 7, Philadelphia, 1965, W. B. Saunders Co.

Physiology

1 Cannon, W. B.: The wisdom of the body, rev. ed., New York, 1963, W. W. Norton & Co., Inc.

2 Guyton, A. C.: Textbook of medical physiology, ed. 3, Philadelphia, 1966, W. B. Saunders Co.

3 Mountcastle, V. B., editor: Medical physiology, ed. 12, St. Louis, 1968, The C. V. Mosby Co.

4 Tuttle, W. W., and Schottelius, B. A.: Textbook of physiology, ed. 16, St. Louis, 1969, The C. V. Mosby Co.

Periodicals

American Journal of Medical Sciences
American Journal of Nursing
American Journal of Physiology
Annual Review of Physiology
Harvey Lectures
Journal of the American Medical Association
Journal of Anatomy
Journal of Neurophysiology
Science Newsletter
Scientific American

Abbreviations and prefixes

◼Abbreviations

A Angstrom unit
A Ch acetylcholine
A Ch E acetylcholinesterase
ACTH adrenocorticotropic hormone
ADP adenosine diphosphate
ATP adenosine triphosphate
BMR basal metabolic rate
BNA Basle Nomina Anatomica (see Glossary)
C centigrade
CA catecholamines
Cal large calorie
cm centimeter
CNS central nervous system
COMT catechol-O-methyl transferase
CP creatine phosphate
CRF corticotropin-releasing factor
CVA cardiovascular accident
DNA deoxyribonucleic acid
DPN diphosphopyridine nucleotide
ECF extracellular fluid
EFP effective filtration pressure
EPSP excitatory postsynaptic potential
ER endoplasmic reticulum
F Fahrenheit
FSH follicle-stimulating hormone
GH growth hormone
Hb hemoglobin
HbO₂ oxyhemoglobin
HP hydrostatic pressure
ICF intracellular fluid
ICSH interstitial cell–stimulating hormone
IF interstitial or intercellular fluid

IPSP inhibitory postsynaptic potential
kcal kilocalorie
kg kilogram
LH luteinizing hormone
MAO monamine oxidase
mEq milliequivalents
mg milligram
μ micron
ml milliliter
mm millimeter
mm Hg pres millimeter mercury pressure
MSH melanocyte-stimulating hormone
mv millivolt
MW molecular weight
NE norepinephrine
OP osmotic pressure
PAH para-aminohippuric acid
PBI protein-bound iodine
P_{CO_2} partial pressure of carbon dioxide
pH hydrogen-ion concentration; negative
    logarithm of hydrogen-ion concentration
P_{O_2} partial pressure of oxygen
PNS peripheral nervous system
RBC, rbc red blood cells
RNA ribonucleic acid
SD systolic discharge
SDA specific dynamic action
STH somatotropic hormone
SV stroke volume
TH thyrotropic hormone
TMR total metabolic rate
TSH thyroid-stimulating hormone
WBC, wbc white blood cells

■ Prefixes

ab- away from
ad- to, toward
adeno- glandular
amphi- on both sides
ante- before, forward
anti- against
bi- two, double, twice
circum- around, about
contra- opposite, against
de- away from, from
dys- difficult
ecto- outside
endo- in, within
ento- inside, within
epi- on, upon
eu- well
ex- from out of, from
extra- outside, beyond, in addition
hemi- half
hyper- over, excessive, above
hypo- under, deficient
infra- underneath, below
inter- between, among
intra- within, on the side
para- beside, to side of
peri- round about, beyond
post- after, behind
pre- before, in front of
pro- before, in front of
retro- backward, back
semi- half
sub- under, beneath
super- above, over
supra- above, on upper side
syn- with, together
trans- across, beyond

Glossary

abdomen body area between the diaphragm and pelvis.

abduct to move away from the midline; opposite of adduct.

absorption passage of a substance through a membrane (e.g., skin or mucosa) into blood.

acapnia marked decrease in blood carbon dioxide content.

acetabulum socket in the hip bone (os coxae or innominate bone) into which the head of the femur fits.

acetone bodies ketone bodies, acids formed during the first part of fat catabolism— viz., acetoacetic acid, beta-hydroxybutyric acid, and acetone.

Achilles tendon tendon inserted on calcaneus; so-called because of the Greek myth that Achilles' mother held him by the heels when she dipped him in the river Styx, thereby making him invulnerable except in this area.

acidosis condition in which there is an excessive proportion of acid in the blood.

acromion bony projection of the scapula; forms point of the shoulder.

adduct to move toward the midline; opposite of abduct.

adenohypophysis anterior pituitary gland.

adenoids glandlike; adenoids or pharyngeal tonsils are paired lymphoid structures in the nasopharynx.

adolescence period between puberty and adulthood.

adrenergic fibers axons whose terminals release norepinephrine and epinephrine.

adventitia, externa outer coat of a tube-shaped structure such as blood vessels.

aerobic requiring free oxygen; opposite of anaerobic.

afferent neuron transmitting impulses to the central nervous system.

albuminuria albumin in the urine.

aldosterone a hormone secreted by the adrenal cortex.

alkali reserve bicarbonate salts present in body fluids; mainly sodium bicarbonate.

alkalosis condition in which there is an excessive proportion of alkali in the blood; opposite of acidosis.

alveolus literally a small cavity; alveoli of lungs are microscopic saclike dilatations of terminal bronchioles.

ameboid movement movement characteristic of amebae – i.e., by projections of protoplasm (pseudopodia) toward which the rest of the cell's protoplasm flows.

amenorrhea absence of the menses.

amino acid organic compound having an NH_3 and a COOH group in its molecule; has both acid and basic properties; amino acids are the structural units from which proteins are built.

amphiarthrosis slightly movable joint.

ampulla saclike dilatation of a tube or duct.

anabolism synthesis by cells of complex compounds (e.g., protoplasm, hormones) from simpler compounds (amino acids, simple sugars, fats, minerals); opposite of catabolism, the other phase of metabolism.

anaerobic not requiring free oxygen; opposite of aerobic.

anastomosis connection between vessels – e.g., the circle of Willis is an anastomosis of certain cerebral arteries.

anemia deficient number of red blood cells or deficient hemoglobin.

anesthesia loss of sensation.

aneurysm blood-filled saclike dilatation of the wall of an artery.

angina any disease characterized by spasmodic suffocative attacks – e.g., angina pectoris, paroxysmal thoracic pain with feeling of suffocation.

Angstrom unit 1/10 of a millimicron or 1/10 millionth of a meter or about 1/250 millionth of an inch.

anorexia loss of appetite.

anoxemia deficient blood oxygen content.

anoxia deficient oxygen supply to tissues.

antagonistic muscles those having opposing action – e.g., muscles that flex the upper arm are antagonistic to muscles that extend it.

anterior front or ventral; opposite of posterior or dorsal.

antibody, immune body substance produced by the body that destroys or inactivates a specific substance (antigen) that has entered the body – e.g., diphtheria antitoxin is the antibody against diphtheria toxin.

antigen substance which, when introduced into the body, causes formation of antibodies against it.

antrum cavity – e.g., the antrum of Highmore, the space in each maxillary bone, or the maxillary sinus.

anus distal end or outlet of the rectum.

apex pointed end of a conical structure.

aphasia loss of a language faculty such as the ability to use words or to understand them.

apnea temporary cessation of breathing.

aponeurosis flat sheet of white fibrous tissue that serves as a muscle attachment.

aqueduct tube for conduction of liquid – e.g., the cerebral aqueduct conducts cerebrospinal fluid from the third to the fourth ventricle.

arachnoid delicate, weblike middle membrane of the meninges.

areola small space; the pigmented ring around the nipple.

arteriole small branch of an artery.

artery vessel carrying blood away from the heart.

arthrosis joint or articulation.

articular referring to a joint.

articulation joint.

arytenoid ladle-shaped; two small cartilages of the larynx.

ascites accumulation of serous fluid in the abdominal cavity.

asphyxia loss of consciousness due to deficient oxygen supply.

aspirate to remove by suction.

asthenia bodily weakness.

ataxia loss of power of muscle coordination.

atrium chamber or cavity – e.g., atrium of each side of the heart.

astrocytes star-shaped neuroglia, connective tissue cells in brain and cord.

atrophy wasting away of tissue; decrease in size of a part.

auricle part of the ear attached to the side of the head; earlike appendage of each atrium of heart.

autonomic self-governing, independent.

axilla armpit.

axon nerve cell process that transmits impulses away from the cell body.

baroreceptor, baroceptor receptor stimulated by change in pressure.

Bartholin seventeenth century Danish anatomist.

basophil white blood cell that stains readily with basic dyes.

biceps two headed.

bilirubin red pigment in the bile.

biliverdin green pigment in the bile.

BNA (Basle Nomina Anatomica) anatomic terminology accepted at Basle by the Anatomical Society in 1895.

Bowman nineteenth century English physician.

brachial pertaining to the arm.

bronchiectasis dilatation of the bronchi.

bronchiole small branch of a bronchus.

bronchus one of the two branches of the trachea.

buccal pertaining to the cheek.

buffer compound that combines with an acid or with a base to form a weaker acid or base, thereby lessening the change in hydrogen ion concentration that would occur without the buffer.

bursa fluid-containing sac or pouch lined with synovial membrane.

buttock prominence over the gluteal muscles.

calcitonin a hormone secreted by the parathyroid glands.

calculus stone formed in various parts of the body; may consist of different substances.

calorie heat unit; a large Calorie is the amount of heat needed to raise the temperature of 1 kg of water 1°C.

calyx cup-shaped division of the renal pelvis.

canaliculus little canal.

capillary microscopic blood vessel; capillaries connect arterioles with venules; also, microscopic lymphatic vessels.

carbhemoglobin, carbaminohemoglobin compound formed by union of carbon dioxide with hemoglobin.

carbohydrate organic compounds containing carbon, hydrogen, and oxygen in certain specific proportions—e.g., sugars, starches, cellulose.

carboxyhemoglobin compound formed by union of carbon monoxide with hemoglobin.

carcinoma cancer, a malignant tumor.

caries decay of teeth or of bone.

carotid from Greek word meaning to plunge into deep sleep; carotid arteries of the neck so called because pressure on them may produce unconsciousness.

carpal pertaining to the wrist.

casein protein in milk.

cast mold—e.g., formed in renal tubules.

castration removal of testes or ovaries.

catabolism breakdown of food compounds or of protoplasm into simpler compounds; opposite of anabolism, the other phase of metabolism.

catalyst substance that alters the speed of a chemical reaction.

cataract opacity of the lens of the eye.

catecholamines norepinephrine and epinephrine.

caudal pertaining to the tail of an animal; opposite of cephalic.

cecum blind pouch; the pouch at the proximal end of the large intestine.

celiac pertaining to the abdomen.

cellulose polysaccharide, the main plant carbohydrate.

centimeter 1/100 of a meter, about 2/5 of an inch.

centrioles two dots seen (with light microscope) in centrosphere; active during mitosis.

centromere structure that joins each pair of chromatids produced by chromosome duplication.

centrosphere, centrosome spherical area or body near center of cell.

cephalic pertaining to the head; opposite of caudal.

cerumen earwax.

cervix neck; any necklike structure.

chemoreceptor distal end of sensory dendrites especially adapted for chemical stimulation.

chiasm crossing; specifically, a crossing of the optic nerves.

cholecystectomy removal of the gallbladder.

cholesterol organic alcohol present in bile, blood, and various tissues.

cholinergic fibers axons whose terminals release acetylcholine.

cholinesterase enzyme; catalyzes breakdown of acetylcholine.

choroid, chorioid skinlike.

chromatids newly formed chromosomes.

chromatin deep-staining substance in the nucleus of cells; divides into chromosomes during mitosis.

chromosomes deep-staining, rod-shaped bodies in cell nucleus; composed of genes.

chyle milky fluid; the fat-containing lymph in the lymphatics of the intestine.

chyme partially digested food mixture leaving the stomach.

cilia hairlike projections of protoplasm.

circadian daily.

cochlea snail shell or structure of similar shape.

coenzyme nonprotein substance which activates an enzyme.

collagen principle organic constituent of connective tissue.

colloid solute particles with diameters of 1 to 100 millimicrons.

colostrum first milk secreted after childbirth.

commissure bundle of nerve fibers passing from one side to the other of the brain or cord.

concha shell-shaped structure—e.g., bony projections into the nasal cavity.

condyle rounded projection at the end of a bone.

congenital present at birth.

contralateral on the opposite side.

coracoid like a raven's beak in form.

corium true skin or derma.

coronal of or like a crown.

coronary encircling; in the form of a crown.

corpus body.

corpuscles very small body or particle.

cortex outer part of an internal organ—e.g., of the cerebrum and of the kidneys.

costal pertaining to the ribs.

crenation, plasmolysis shriveling of a cell due to water withdrawal.

cretinism dwarfism due to hypofunction of the thyroid gland.

cribriform sievelike.

cricoid ring shaped; a cartilage of this shape in the larynx.

cruciate cross shaped.

crystalloid solute particle less than 1 millimicron in diameter.

cubital pertaining to the forearm.

cutaneous pertaining to the skin.

cyanosis bluish appearance of the skin due to deficient oxygenation of blood.

cytokinesis dividing of cytoplasm to form two cells.

cytology study of cells.

cytoplasm the protoplasm of a cell exclusive of the nucleus.

deamination chemical reaction by which the amino group NH_3 is split from an amino acid.

deciduous temporary; shedding at a certain stage of growth—e.g., deciduous teeth.

decussation crossing over like an X.

defecation elimination of waste matter from the intestines.

deferens carrying away.

deglutition swallowing.

deltoid triangular—e.g., deltoid muscle.

dendrite, dendron branching or treelike; a nerve cell process that transmits impulses toward the cell body.

dens tooth.

dentate having toothlike projections.

dentine main part of a tooth, under the enamel.

dentition teething; also, number, shape, and arrangement of the teeth.

dermatome area of skin supplied by sensory fibers of a single dorsal root.

dermis, corium true skin.

dextrose glucose, a monosaccharide, the principal blood sugar.

dialysis separation; the separation of crystalloids from colloids by the faster diffusion of the former through a membrane.

diapedesis passage of blood cells through intact blood vessel walls.

diaphragm membrane or partition that separates one thing from another; the muscular partition between the thorax and abdomen; the midriff.

diaphysis shaft of a long bone.

diarthrosis freely movable joint.

diastole relaxation of the heart interposed between its contractions; opposite of systole.

diencephalon "tween" brain; parts of the brain between the cerebral hemispheres and the mesencephalon or midbrain.

diffusion spreading—e.g., scattering of solute particles.

digestion conversion of food into assimilable compounds.

diplopia double vision; seeing one object as two.

disaccharide sugar formed by the union of two monosaccharides; contains twelve carbon atoms.

542

distal toward the end of a structure; opposite of proximal.

diverticulum outpocketing from a tubular organ such as the intestine.

dorsal, posterior pertaining to the back; opposite of ventral.

Douglas Scottish anatomist of the late seventeenth and early eighteenth centuries.

dropsy accumulation of serous fluid in a body cavity, in tissues; edema.

duct canal or passage.

dura mater literally strong or hard mother; outermost layer of the meninges.

dyspnea difficult or labored breathing.

dystrophy faulty nutrition.

ectopic displaced; not in the normal place— e.g., extrauterine pregnancy.

edema excessive fluid in tissues; dropsy.

effector responding organ—e.g., voluntary and involuntary muscle, the heart, and glands.

efferent carrying from, as neurons that transmit impulses from the central nervous system to the periphery; opposite of afferent.

electrocardiogram graphic record of heart's action potentials.

electroencephalogram graphic record of brain's action potentials.

electrolyte substance that ionizes in solution rendering the solution capable of conducting an electric current.

electron minute, negatively charged particle.

elimination expulsion of wastes from the body.

embolism obstruction of a blood vessel by foreign matter carried in the bloodstream.

embryo animal in early stages of intrauterine development; the human fetus the first three months after conception.

emesis vomiting.

emphysema dilatation of pulmonary alveoli.

empyema pus in a cavity—e.g., in the chest cavity.

encephalon brain.

endocrine secreting into the blood or tissue fluid rather than into a duct; opposite of exocrine.

endoplasm cytoplasm located toward center of cell, as distinguished from ectoplasm located nearer periphery of cell.

endoplasmic reticulum network of tubules and vesicles in cytoplasm.

energy power or ability to do work; *kinetic* or *active*, due to moving particles (e.g., mechanical energy, heat); *potential* or *stored*, due to attraction between particles (e.g., chemical energy).

enteron intestine.

enzyme catalytic agent formed in living cells.

eosinophil, acidophil white blood cell readily stained by eosin.

epidermis "false" skin; outermost layer of the skin.

epinephrine secretion of the adrenal medulla.

epiphyses ends of a long bone.

erythrocyte red blood cell.

ethmoid sievelike.

eupnea normal respiration.

Eustachio Italian anatomist of the sixteenth century.

exocrine secreting into a duct; opposite of endocrine.

exophthalmos abnormal protrusion of the eyes.

extrinsic coming from the outside; opposite of intrinsic.

facilitation decrease in a neuron's resting potential to a point above its threshold of stimulation.

Fallopius sixteenth century Italian anatomist.

fascia sheet of connective tissue.

fasciculus little bundle.

fetus unborn young, especially in the later stages; in human beings, from third month of intrauterine period until birth.

fiber threadlike structure.

fibrin insoluble protein in clotted blood.

fibrinogen soluble blood protein that is converted to insoluble fibrin during clotting.

fibroblasts connective tissue cells that synthesize interstitial fibers and gels.

fibrocytes old fibroblasts.

filtration passage of water and solutes through a membrane due to hydrostatic pressure gradient.

fimbria fringe.

fissure groove.

flaccid soft, limp.

follicle small sac or gland.

fontanel "soft spots" of the infant's head; unossified areas in the infant's skull.

foramen small opening.

fossa cavity or hollow.

fovea small pit or depression.

fundus base of a hollow organ—e.g., the part farthest from its outlet.

ganglion cluster of nerve cell bodies outside the central nervous system.

gasserian named for Gasser, a sixteenth century Austrian surgeon.

gastric pertaining to the stomach.

gene part of the chromosome that transmits a given hereditary trait.

genitalia reproductive organs.

gestation pregnancy.

gland secreting structure.

gomerulus compact cluster—e.g., of capillaries in the kidneys.

glossal of the tongue.

glucagon hormone secreted by alpha cells of the islands of Langerhans.

glucocorticoids hormones that influence food metabolism; secreted by adrenal cortex.

glucokinase enzyme; catalyzes conversion of glucose to glucose-6-phosphate.

gluconeogenesis formation of glucose from protein or fat compounds.

glucose monosaccharide or simple sugar; the principal blood sugar.

gluteal of or near the buttocks.

glycerin, glycerol product of fat digestion.

glycogen "animal starch"; main polysaccharide stored in animal cells.

glycogenesis formation of glycogen from glucose or from other monosaccharides, fructose or galactose.

glycogenolysis hydrolysis of glycogen to glucose-6-phosphate or to glucose.

glyconeogenesis the formation of glycogen from protein or fat compounds.

gonad sex gland in which reproductive cells are formed.

graafian named for Graaf, a seventeenth century Dutch anatomist.

gradient a slope or difference between two levels—e.g., concentration gradient; a difference between the concentrations of two substances.

gustatory pertaining to taste.

gyrus convoluted ridge.

haversian named for Hevers, English anatomist of the late seventeenth century.

helix spiral; coil.

hemiplegia paralysis of one side of the body.

hemoglobin iron-containing protein in red blood cells.

hemolysis destruction of red blood cells with escape of hemoglobin from them into surrounding medium.

hemopoiesis blood cell formation.

hemorrhage bleeding.

hepar liver.

heparin substance obtained from the liver that inhibits blood clotting.

heredity transmission of characteristics from a parent to a child.

hernia, "rupture" protrusion of a loop of an organ through an abnormal opening.

hilus, hilum depression where vessels enter an organ.

His German anatomist of the late nineteenth century.

histology science of minute structure of tissues.

homeostasis relative uniformity of the normal body's internal environment.

hormone substance secreted by an endocrine gland.

hyaline glasslike.

hydrocortisone a hormone secreted by the adrenal cortex; compound F.

hydrolysis literally "split by water"; chemical reaction in which a compound reacts with water.

hymen Greek for skin; mucous membrane that may partially or entirely occlude the vaginal outlet.

hyoid shaped like the letter U; bone of this shape at the base of the tongue.

hypercapnia abnormally high blood CO_2 concentration.

hyperemia increased blood in a part.

hyperkalemia higher than normal concentration of potassium in the blood.

hypernatremia higher than normal concentration of sodium in the blood.

hyperopia farsightedness.

hyperplasia increase in the size of a part due to an increase in the number of its cells.

hyperpnea abnormally rapid breathing; panting.

hypertension abnormally high blood pressure.

hyperthermia fever; body temperature above 37° C.

hypertrophy increased size of a part due to an increase in the size of its cells.

hypervolemia larger volume of blood than normal.

hypokalemia lower than normal concentration of potassium in the blood.

hyponatremia lower than normal concentration of sodium in the blood.

hypophysis Greek for undergrowth; hence

the pituitary gland, which grows out from the undersurface of the brain.

hypothalamus part of the diencephalon; gray matter in the floor and walls of the third ventricle.

hypothermia subnormal body temperature; below 37° C.

hypovolemia subnormal volume blood.

hypoxia oxygen deficiency.

inclusions any foreign or heterogenous substance contained in a cell or in any tissue or organ that was not introduced as a result of trauma.

incus anvil; the middle ear bone that is shaped like an anvil.

inferior lower; opposite of superior.

inguinal of the groin.

inhalation inspiration or breathing in; opposite of exhalation or expiration.

inhibition an increase in a neuron's resting potential above its usual level.

innominate not named, anonymous—e.g., ossa coxae (hip bones) formerly known as innominate bones.

insulin hormone secreted by beta cells of the islands of Langerhans in the pancreas.

intercellular between cells; interstitial.

interneurons (*internuncial* or *intercalated neurons*) conduct impulses from sensory to motor neurons; lie entirely within the central nervous system.

interstitial of or forming small spaces between things; intercellular.

intima innermost.

intrinsic not dependent upon externals; located within something; opposite of extrinsic.

involuntary not willed; opposite of voluntary.

involution return of an organ to its normal size after enlargement; also retrograde or degenerative change.

ion electrically charged atom or group of atoms.

ipsilateral on the same side; opposite of contralateral.

irritability excitability; ability to react to a stimulus.

ischemia local anemia; temporary lack of blood supply to an area.

isotonic of the same tension or pressure.

keratin protein compound present in the human body, chiefly in hair and nails.

ketones acids (acetoacetic, beta-hydroxybutyric, and acetone) produced during fat catabolism.

ketosis excess amount of ketone bodies in the blood.

kilogram 1,000 gm, approximately 2.2 lb.

kinesthesia "muscle sense"—i.e., sense of position and movement of body parts.

labia lips.

lacrimal pertaining to tears.

lactation secretion of milk.

lactose milk sugar, a disaccharide

lacuna space of cavity—e.g., lacunae in bone contain bone cells.

lamella thin layer, as of bone.

lateral of or toward the side; opposite of medial.

lemniscus, medial a flat band of sensory fibers extending up from the medulla, through the pons and midbrain to the thalamus.

leukocyte white blood cell.

ligament bond or band connecting two objects; in anatomy, a band of white fibrous tissue connecting bones.

limbic lobe or *system* cerebral cortex on the medial surface of the brain that forms a border around the corpus callosum; older name rhinencephalon.

lipid fats and fatlike compounds.

loin part of the back between the ribs and hip bones.

lumbar of or near the loins.

lumen passageway or space within a tubular structure.

luteum golden yellow.

lymph watery fluid in the lymphatic vessels.

lymphocyte one type of white blood cells.

lysis the destruction of cells and other antigens by a specific lysin antibody.

lysosomes membranous organelles containing various enzymes that can dissolve most cellular compounds; hence called "digestive bags" or "suicide bags" of cells.

malleolus small hammer; projections at the distal ends of the tibia and fibula.

malleus hammer; the tiny middle ear bone that is shaped like a hammer.

Malpighii seventeenth century Italian anatomist.

maltose disaccharide or "double" sugar.

mammary pertaining to the breast.

mamillary like a nipple.

manometer instrument used for measuring the pressure of fluids.

manubrium handle; upper part of the sternum.

mastication chewing.

matrix ground substance in which cells are embedded.

meatus passageway.

medial of or toward the middle; opposite of lateral.

mediastinum middle section of the thorax — i.e., between the two lungs.

medulla Latin for marrow; hence, the inner portion of an organ in contrast to the outer portion or cortex.

meiosis nuclear division in which the number of chromosomes are reduced to half their original number before the cell divides in two.

membrane thin layer or sheet.

menstruation monthly discharge of blood from the uterus.

mesencephalon midbrain.

mesentery fold of peritoneum that attaches the intestine to the posterior abdominal wall.

mesial situated in the middle; median.

metabolism complex process by which food is utilized by a living organism.

metabolite any substance produced by metabolism.

metacarpus "after" the wrist; hence, the part of the hand between the wrist and fingers.

metatarsus "after" the instep; hence, the part of the foot between the tarsal bones and toes.

meter about 39.5 in.

microglia one type connective tissue cell found in the brain and cord.

micron 1/1,000 of a millimeter; about 1/25,000 of an inch.

micturition urination, voiding.

milliequivalent a unit of chemical equivalence computed by the following formula:

$$\frac{mg}{\text{atomic weight}} \times \text{Valence} = mEq$$

$$\text{e.g.,} \frac{3{,}266 \text{ mg Na}^+}{23 \text{ (At wt Na)}} \times 1 = 142 \text{ mEq Na}^+$$

millimeter 1/1,000 of a meter; about 1/25 of an inch.

mineralocorticoids hormones that influence mineral salt metabolism; secreted by adrenal cortex.

mitochondria threadlike structures.

mitosis indirect cell division involving complex changes in the nucleus.

mitral shaped like a miter.

molar concentration number of grams solute per liter of solution divided by the solute's molecular weight.

mole molecular weight of a compound in grams; also called gram molecular weight.

monosaccharide simple sugar.

Monro eighteenth century English surgeon.

morphology study of shape and structure of living organisms.

motoneurons (*motor* or *efferent neurons*) transmit nerve impulses away from the brain or spinal cord.

myelin lipoid substance found in the myelin sheath around some nerve fibers.

myocardium muscle of the heart.

myopia nearsightedness.

nares nostrils.

neurilemma nerve sheath.

neurohypophysis posterior pituitary gland.

neuron nerve cell, including its processes.

neurovesicles microscopic sacs in axon terminals; contain transmitter substance.

neutrophil white blood cell that stains readily with neutral dyes.

nuchal pertaining to the nape of the neck.

nucleotide component of DNA and RNA; a nucleotide is composed of a sugar (deoxyribose or ribose), a nitrogenous base (adenine, thymine, cytosine, or guanine), and a phosphate group.

nucleus spherical structure within a cell; a group of neuron cell bodies in the brain or cord.

occiput back of the head.

olecranon elbow.

olfactory pertaining to the sense of smell.

oligodendroglia a type of connective tissue cell found in the brain and cord.

ophthalmic pertaining to the eyes.

organelle cell organ; one of the specialized parts of a single-celled organism (protozoon), serving for the performance of some individual function.

os Latin for mouth and for bone.

osmosis movement of a fluid through a semipermeable membrane.

ossicle little bone.

oxidation loss of hydrogen or electrons from a compound or element.

oxyhemoglobin a compound formed by union of oxygen with hemoglobin.

palate roof of the mouth.

palpebrae eyelids.

papilla small nipple-shaped elevation.

paralysis loss of the power of motion or sensation, especially voluntary motion.

parenchyma the distinguishing, functional cells of an organ.

parietal of the walls of an organ or cavity.

parotid located near the ear.

parturition act of giving birth to an infant.

patella small, shallow pan; the kneecap.

Pavlov Russian physiologist of the late nineteenth and early twentieth centuries.

pectineal pertaining to the pubic bone.

pectoral pertaining to the chest or breast.

pelvis basin or funnel-shaped structure.

Peyer Swiss anatomist of the late seventeenth and early eighteenth centuries.

peripheral pertaining to an outside surface.

peroneus, peroneal of or near the fibula.

petrous rocklike.

pH hydrogen ion concentration; the negative logarithm of hydrogen ion concentration.

phagocytosis process by which a segment of cell membrane forms a small pocket around a bit of solid outside the cell, breaks off from rest of the membrane, and moves into the cell; briefly, ingestion and digestion of particles by a cell.

phalanges finger or toe bones.

phrenic pertaining to the diaphragm.

pia mater gentle mother; the vascular innermost covering (meninges) of the brain and cord.

pilomotor mover of a hair.

pineal shaped like a pine cone.

pinocytosis a process by which a segment of cell membrane forms a small pocket around a bit of fluid outside the cell, breaks off from rest of membrane, and moves into the cell.

piriformis pear shaped.

pisiform pea shaped.

plantar pertaining to the sole of the foot.

plasma liquid part of the blood.

plasmolysis shrinking of a cell due to water loss by osmosis.

plexus network.

plica fold.

polymorphonuclear having many-shaped nuclei.

polyribosomes a group of ribosomes working together to synthesize proteins.

polysaccharide a carbohydrate containing a large number of saccharide groups, $(C_6H_{10}O_5)$ — e.g., starch, glycogen.

pons bridge.

popliteal behind the knee.

posterior following after; hence, located behind; opposite of anterior.

potential, action difference in electrical charges on inner and outer surfaces of the cell membrane during impulse conduction; outer surface negative to inner.

potential or *potential difference* difference in electrical charges — e.g., on outer and inner surfaces of cell membrane.

potential, resting difference in electrical charges on inner and outer surfaces of the cell membrane when it is not conducting impulses; outer surface positive to inner.

Poupart seventeenth century French anatomist.

presbyopia "oldsightedness"; farsightedness of old age.

pressoreceptors receptors stimulated by a change in pressure; baroreceptors.

pronate to turn palm downward.

proprioreceptors receptors located in the muscles, tendons, and joints.

protoplasm living substance.

proximal next or nearest; located nearest the center of the body or the point of attachment of a structure.

psoas pertaining to the loin, the part of the back between the ribs and hip bones.

psychosomatic pertaining to the influence of the mind (notably the emotions) on body functions.

pterygoid wing shaped.

puberty age at which the reproductive organs become functional.

racemose like a cluster of grapes.

ramus branch.

Ranvier French pathologist of the late nineteenth and early twentieth centuries.

receptor peripheral ending of a sensory neuron.

reflex involuntary action.

refraction bending of a ray of light as it passes from a medium of one density to one of a different density.

refractory resisting stimulation.

renal pertaining to the kidney.

reticular netlike.

reticulum a network.

rhinencephalon see limbic lobe.

ribosomes organelles in cytoplasm of cells; synthesize proteins so nicknamed "protein factories."

rugae wrinkles or folds.

sagittal like an arrow; longitudinal.

salpinx tube; oviduct.

sartorius tailor; hence, the thigh muscle used to sit cross-legged like a tailor.

sciatic pertaining to the ischium.

sclera from Greek for hard.

scrotum bag.

sebum Latin for tallow; secretion of sebaceous glands.

sella turcica Turkish saddle; saddle-shaped depression in the sphenoid bone.

semen Latin for seed; male reproductive fluid.

semilunar half-moon shaped.

senescence old age.

serratus saw toothed.

serum any watery animal fluid; clear, yellowish liquid that separates from a clot of blood.

sesamoid shaped like a sesame seed.

sigmoid S shaped.

sinus cavity.

soleus pertaining to a sole; a muscle in the leg shaped like the sole of a shoe.

somatic of the body framework or walls, as distinguished from the viscera or internal organs.

sphenoid wedge shaped.

sphincter ring-shaped muscle.

splanchnic visceral.

squamous scalelike.

stapes stirrup; tiny stirrup-shaped bone in the middle ear.

Starling English physiologist of the late nineteenth and early twentieth centuries.

stereognosis awareness of the shape of an object by means of touch.

stimulus agent that causes a change in the activity of a structure.

stratum layer.

stress, physiological according to Selye, a condition in the body produced by all kinds of injurious factors that he calls "stressors" and manifested by a syndrome.

stressor any injurious factor that produces biological stress—e.g., emotional trauma, infections, severe exercise, etc.

striated marked with parallel lines.

stroma Greek for mattress or bed; hence, the framework or matrix of a structure.

sudoriferous secreting sweat.

sulcus furrow or groove.

superior higher; opposite of inferior.

supinate to turn the palm of the hand upward; opposite of pronate.

Sylvius seventeenth century anatomist.

symphysis Greek for a growing together.

synapse joining; point of contact between adjacent neurons.

synovia literally "with egg"; secretion of the synovial membrane resembles egg white.

synthesis putting together of parts to form a more complex whole.

systole contraction of the heart muscle.

talus ankle; one of the bones of the ankle.

tarsus instep.

tegmentum part of the brainstem that covers the posterior surface of the cerebral peduncles and the pons.

tendon band or cord of fibrous connective tissue that attaches a muscle to a bone or other structure.

thorax chest.

thrombosis formation of a clot in a blood vessel.

tibia Latin for shin bone.

tonus continued, partial contraction of muscle.

tract bundle of axons located within the central nervous system.

trauma injury.

trochlear pertaining to a pulley.

trophic having to do with nutrition.

tropic having to do with a turning or a change.

tunica covering.

turbinate shaped like a cone or like a scroll or spiral.

tympanum drum.

umbilicus navel.

utricle little sac.

uvula Latin for a little grape; a projection hanging from the soft palate.

vagina sheath.

vagus Latin for wandering.

valve structure that permits flow of a fluid in one direction only.

vas vessel or duct.

vastus wide, of great size.

Vater German anatomist of the late seventeenth and early eighteenth centuries.

vein vessel carrying blood to the heart.

ventral of or near the belly; in man, front or anterior; opposite of dorsal or posterior.

ventricle small cavity.

vermiform worm shaped.

villus hairlike projection.

viscera internal organs.

volar pertaining to the palm of the hand or the sole of the foot; palmar; plantar.

vomer ploughshare.

Willis seventeenth century English anatomist.

xiphoid sword shaped.

zygoma yoke.

Index

A

A band of striated muscle, 114–115
AB blood types, 302
Abdominal:
  aorta, descending, 318–319
  breathing, 376
  cavity, 3
    dropsy, cause, 326
  inflammation, function of greater
    omentum, 392
  organs, veins, 322–323
  reflex, somatic reflex, 223
  regions, 5, 7
  wall:
    horizontal section, 154
    motion, muscles, 138
    muscles, 128, 142
      deep, 153
    weak places in, 127
  viscera, front, 394
Abdominopelvic cavity, 3
Abducens nerve, summary, 206, 208
Abduction, definition, 105
Abductor muscles, 127, 129
Abnormal constituents, elimination in
  urine, 461
Abnormalities, hormone imbalance in,
  265
Absorption:
  end products of digestion, function of
    small intestine, 401
  food, 414–416
  prevention, mechanism, 416
  salt, active transport mechanisms, 416
  water:
    active transport mechanisms, 416
    function of large intestine, 402
Acceleration, sensation, mechanism, 257
Accelerator nerves, sympathetic nerves
  to heart, 314
Accessory:
  nerve, summary, 207, 210
  organs:
    mouth, 394
    skin, 56–58
  reproductive glands, male, 470–472

Accessory—cont'd
  structures of eye, 247, 249
Accommodation of lens, mechanism, 252
Acetabulum:
  direct view into, 89
  marking of os coxae, 89, 100
Acetylcholine:
  release by cholinergic fibers, 229
  role in muscle contraction, 121
  transmitter substance, 176, 178
Achilles reflex, somatic reflex, 223
Achlorhydria in pernicious anemia, 400
Acid(s):
  amino (see Amino acid)
  fixed or nonvolatile, buffering, 515
Acid-base:
  balance, 512–522
    dependence of kidney function, 451
    maintenance, role of sodium and
      potassium, 429
    role of hormones, 265
  imbalances, 519–520
Acidosis, 519–520
  hyperkalemia in, 518
  and hyperventilation, relationship, 517
Acoustic:
  meatus, external, 254, 255
  nerve, summary, 206, 210
Acromegaly resulting from excess growth
  hormone, 267, 268
Acromioclavicular joint, description, 108
Acromion process of scapula, marking,
  84, 85, 99
ACTH:
  effect on glucocorticoid secretion, 428
  increase in blood glucose concentration
    caused by, mechanism, 419
  protein catabolic hormone, 428
  role in stress response, 527
  secretion, control, 270
Actin:
  muscle cells, 115, 116
  role in muscle fiber contraction, 117
Action(s):
  muscle, skeletal, hints on deducing, 126
  potential, 171–178

Activating system, reticular, 215
Active transport:
  chemistry, 28, 29
  mechanism(s), 27–29
    and tubular reabsorption, 453, 454
    water and salt absorption, 416
  role in tubular secretion, 456
Acuity, visual, fraction indicating, 252
Adam's apple prominence in male, reasons, 366
Adaptation:
  definition, 6
  diseases of, 530
  syndrome, general, of Selye, 525, 530
Adaptive responses, 6
Adduction, definition, 105
Adductor:
  muscle group, 127, 128, 129, 134, 147,
    149
  tubercle, marking of femur, 102
Adenohypophysis (see Anterior pituitary
  gland)
Adenoids, location, 365
Adipose:
  cell, fat mobilization, 425
  tissue, 45, 50, 51
    calories to synthesize one pound, 434
    principal fat storage, 425
Adrenal:
  cortex, 276–281
    target gland of adrenocorticotropin,
      269
    virilizing tumor, 279
  gland, 276
  medulla, 281–282
    epinephrine secretion during stress,
      419
Adrenalin, effect on reticular activating
  system, 215
Adrenergic:
  effects of visceral effectors, 231
  fibers, norepinephrine release, 229
Adrenocorticoids, amount, measure of
  stress, 528
Adrenocorticotropin, function, 269

Adrenoglomerulotropin, 281, 285
Adulthood, skeletal changes, 103
Afferent neurons, description, 167, 168
Age:
  changes:
    body:
      functions, 8
      structures, 4
    facial bones, 103
    skeletal muscle, 121
    skeleton, 103
  factor influencing basal metabolic rate, 430
  fluid volume variation, 503
  old:
    atrophy of spleen, 354
    bone:
      changes, 103
      degeneration, 68
    skeletal changes from adulthood to, 103
  and oxygen consumption, 119
  and sex, basal metabolic rates, 431
Agglutinins, definition, 302
Air:
  amount exchanged in respirations, 374–376
  and blood gas exchange and alveolar ventilation, direct relationship, 376
  cells, mastoid, marking of temporal bone, 94
  inspired, volume, inverse relation to resistance opposing air flow, 374
  minimal, 374
  residual:
    amount, 374
    increase in emphysema, 375
  sinuses, special feature of skull, 97
  tidal, amount exhaled after inspiration, 374
  volume exchanged during normal and forced respiration, 375
Airway:
  obstruction, slowing of air flow, 374
  protection by cartilaginous rings in wall of trachea, 368
Albumins, blood proteins synthesized by liver cells, 405
Aldosterone, 278, 280, 281
  mechanism for decreasing fluid output, 498
  role in fluid and electrolyte balance, 507
Aldosteronism and hypertension, relationship, 280
Alerting mechanism, 215
Alimentary:
  canal, 390
  tract, blood vessels embedded in coat, 390
Alkalosis, 520
  and hypoventilation, relationship, 517
All-or-none law of muscle cell contraction, 118
Alpha and beta cells of islands of Langerhans, 406
Alveolar:
  and capillary membranes, permeability to gases, 377
  ducts, 368
  process, marking of mandible and maxilla, 96
  sacs, 368, 369

Alveolar—cont'd
  ventilation, 376
  walls, 368
    loss of elasticity in emphysema, 375
Alveolar-capillary membrane, surface area, oxygen diffusion affected by, 380, 381
Alveoli:
  and alveolar sacs, 369
  clusters, metal cast, 370
  lung, and atmosphere, gas pressure gradient between, 374
  tremendous number in two lungs, 368
Amino acid(s):
  absorption, role of transport mechanism, 414
  chain, protein composition, 426
  concentration, in portal and hepatic veins, 416
  molecules of protein compounds composed of, 411
Ammonia:
  excretion:
    effect on blood pH, 518
    tubule, urine acidification, 519
  tubular secretion, 456
Amphetamines, effect on reticular activating system, 215
Ampulla:
  of semicircular canal, 257, 259
  of Vater, opening of common bile duct in, 403
Amylase:
  (amylopsin), digestive enzyme of pancreatic juice, 409
  (ptyalin), digestive enzyme of saliva, 411
Amylopsin, digestive enzyme of pancreatic juice, 409
Anabolic hormone, insulin, 283
Anabolism, 29
  and catabolism, 416
    protein, concurrent, 427
    relationship between, 417
  fat, 425
  primary in protein metabolism, 426
  protein:
    essential part performed by liver cells, 405
    functions, 426
    promotion by estrogens, 283
    stimulation:
      by growth hormone, 268
      by testosterone, 468
    synopsis of process, 426–427
  synthesis of hormones by, 265
Anal:
  canal, 398, 402
  sphincter:
    muscles, 402
    stretching, effect on respirations, 384
Anastomosis(es):
  arterial, artery-to-artery connection, 320
  between coronary arteries or branches, 312
  important protective function, 320
Anatomical:
  dead space, 376
  neck of humerus, marking, 99
  position, definition, 6
Androgen secretion by adrenal cortex, 280

Androsterone in urine of men and women, 284
Anemia:
  causes, 298
  criterion for diagnosis, 295
  decrease in blood viscosity caused by, 335
  hemoglobin deficiency, cellular hypoxia produced by, 377
  hemolytic, formation of erythrocytes in spleen during, 355
  pernicious:
    atrophy of gastric mucosa, 400
    development, 297
    production by absence of intrinsic factor, mechanism, 400
  size of red cell indicative, 295
Anesthesia, arterial blood oxygen tension during, 383
Anesthetics, effect on reticular activating system, 215
Aneurysm:
  blood vessel abnormality, 316
  middle cerebral arteries, symptoms, 211
Anger, vasoconstriction and rise in arterial blood pressure, 340
Angiotensin II, role in hypertension, 280, 282
Angiotonin, function, 458
Angle, marking of mandible, 96
Ankle:
  jerk, somatic reflex, 223
  joint, description, 109
Anorexia, accompaniment of fever, 435
Antagonism and summation, autonomic, principle, 229
Antagonists, skeletal muscle group, 125
Anterior:
  ligament, 477
  median fissure, spinal cord groove, 179
  pituitary:
    gland, 265–272
      master gland, 265
      and neurohypophysis, differences, 265
      and ovaries, negative feedback mechanism between, 487
      secretion, regulation during stress, 271
      and testes, negative feedback mechanism between, 468–469
    hormones and target organs, 266
Antiallergic effects of glucocorticoids, 278
Antibodies:
  anti-Rh, two possible sources, 302
  blood plasma, 302
  cellular:
    definition, 299
    or tissue-rejecting, synthesis by lymphocytes, 348, 354
  circulating:
    definition, 299
    produced by plasma cells, 354
Antidiuretic hormone, 272
  effects, 455
  role in fluid and electrolyte balance, 507
Antigens:
  blood, 300–302
  cellular, definition, 299
  red blood cell, reaction with plasma antibodies, 302

Antigens—cont'd
  response of leukocytes, 299
Anti-immunity effect of glucocorticoids, 278
Anti-insulin effect of growth hormone, 268
Antilymphocyte serum to prevent transplant rejection, 299
Anti-Rh antibodies, two possible sources, 302
Antithrombins, factors opposed to blood clotting, 305
Antrum of Highmore, 76
  marking of maxilla, 96
Anus, opening of anal canal, 402
Aorta:
  abdominal, descending, 318–319
  ascending, 318–319
  injections of blood from heart, cause of pulse, 347
  thoracic, descending, 318–319
Aortic:
  arch, 318–319
  and carotid:
    cardiac pressoreflexes, regulation of heart rate, 335
    vasomotor pressoreflex mechanism, 337
  valve insufficiency, pulse pressure increased, 346
Apex:
  heart, location, 306
  lung, pointed upper margin, 371
Aphasia, 188
Apnea, 376, 384
Aponeurosis:
  connective tissue, 120
  internal oblique muscle, 154
  linea alba, 138
Appendicitis, 407
  function of greater omentum, 392
Appendicular skeleton, 84–93
  bones, 70–71
Appendix, vermiform, 398, 401, 406
Appetite:
  center, factors stimulating, 435
  regulation, role of hypothalamus, 190
Aqueduct of Sylvius, cerebral, 196, 197
Aqueous humor, 242, 245, 246
Arachnoid membrane of meninges, 195
Arbor vitae of cerebellum, 184, 192
Arc reflex:
  course of nerve impulse, 172–175
  intersegmental, 177
  monosynaptic, 172, 173
  principle, in autonomic nervous system, 223
  three-neuron, 167, 172, 175
    correlation of structures, 222
  two-neuron, 172, 173
    conduction over, in knee jerk, 221
Arch(es):
  aortic, 318–319
  deciduous, 396
  foot, 91, 92
    dorsal venous, 322–324
    fallen, 92
  longitudinal, markings of tarsals, 92, 102
  metatarsal, markings of tarsals, 92, 102

Arch(es)—cont'd
  neural, of vertebrae, marking, 98
  palmar:
    branches of radial and ulnar arteries, 318–319
    and plantar, arterial anastomoses, 321
    vein of upper extremity, 321
  plantar:
    branches of anterior and posterior tibial arteries, 318
    vein of lower extremities, 322
  pubic, marking of os coxae, 101
  superciliary, marking of frontal bone, 94
  transverse, marking of tarsals, 92, 102
  volar, vein of upper extremity, 321
Areola, color change in, 481
Areolar tissue, 44, 48, 50, 51
Arm (*see* Extremity, upper)
Arousal mechanism, 215
  hypothalamus, role, 190
  thalamus, role, 188
Arterial:
  anastomosis, artery-to-artery connection, 320
  bleeding, pressure points to stop, 347
  blood:
    conversion of venous blood to, 380
    gases, influence on respirations, 382–383
    oxygen saturation, explanation, 379
    pH, 515
      adjustment of respiration to, 516
    pressure (*see* Blood pressure, arterial)
    volume, 331, 332
  and venous bleeding, relation of blood pressure to, 346
  walls, elasticity, cause of pulse, 347
Arteries:
  base of brain, 321
  and capillaries, relationship, 320
  composing circle of Willis, 321
  coronary, 310, 311, 312
  face and head, 320
  function, 315
  main, 318–321
  principal, of body, 319
  umbilical, 326, 327
Arteriole(s), 315
  and capillaries, small diameter, cause of peripheral resistance, 333
  cross-sectional area, 346
  diameter, factors regulating, 336
  resistance vessels, important function, 315
  runoff, rate controlled by peripheral resistance, 333
  vasodilatation, local control in localized areas, 340
Arteriosclerosis:
  blood vessel abnormality, 316
  with hypertension, illustrative of Poiseuille's law, 342
  pulse pressure increased in, 346
Artery:
  branchial, use for blood pressure measurement, 345
  comparative thickness of three coats, 317
  definition, 314
Arthrodial joint, summary, 106
Articular cartilage of long bones, 66

Articulating processes, of vertebrae, markings, 98
Articulations, 104–110
Ascending:
  aorta, 318–319
  colon, in vertical position on right side of abdomen, 398, 402
Aschheim-Zondek pregnancy test, 285
Ascites, cause, 326
Asphyxiation:
  cause, 408
  death from, result of obstruction of trachea, 368
Association tracts of cerebrum, 184
Astigmatism, explanation, 250, 252
Astrocytes, 165, 166
Atherosclerosis, increased blood-clotting tendency, 305
Atlantoepistropheal joint, description, 108
Atlas:
  description, 98
  viewed from below, 83
Atmosphere and lung alveoli, gas pressure gradient between, 372–374
Atoms, rearrangement within molecule, enzymes causing, 411
ATP:
  breakdown, 116, 118
  molecules, storage of energy in, 417
  synthesis, 33
Atria of heart, 308, 309
Atrioventricular:
  bundle, 312, 313
  node, 312, 313
  valves, 308–311
    action, 310
Auditory:
  apparatus:
    anatomy, 254–259
    physiology, 259–260
  area, primary, of cortex, 186, 191
  canal, external, description, 254, 255
  meatus, marking of temporal bone, 94
  nerve, summary, 206, 210
  ossicles, 70
    description, 255, 256
  reflex centers, 194
  tubes:
    location and functions, 255, 256
    opening into nasopharynx, 365
Auricle of external ear, 254, 255
Automatic movements, production, 219
Autonomic:
  antagonism and summation, principle, 229
  centers, higher, function of hypothalamus, 189
  chemical transmitters, principle, 228
  functioning and homeostasis, principle, 228
  functions, 231
  innervation, principles, 228
  nerve fibers, innervation of blood vessels, 316
  nervous system, 223–232
    divisions, 224
    general principles, 228–232
    innervation of heart, 313
    microscopic structure, 224
    preganglionic and postganglionic fibers, 226
  neurons, locations, 227

Autonomic – cont'd
  reflex(es):
    centers, higher, location and function, 232
    mechanism, 221
Axial skeleton, 68–84
  bones, 69–70
Axillary:
  artery, 318–319
  cluster of lymph nodes, 352
  vein of upper extremity, 321
Axis, description, 83, 98
Axon:
  description, 167, 168, 169
  diameter and rate of conduction, 168, 174

B

Babinski reflex, 221, 223
Back muscles, 128, 142–143
Bacteria (*see* Microorganisms)
Bainbridge reflex, 334
Balance:
  acid-base (*see* Acid-base balance)
  electrolyte, dependence on kidney function, 451
  energy, relationship to body weight, 433
  fluid:
    dependence on kidney function, 451
    and electrolyte, 496–511
  mechanism of maintenance, 257
  nitrogen, 427
    simultaneous with protein balance, 427
  protein, normal condition, 427
  water, maintenance, role of sodium and potassium, 429
Ball and socket joint, 106, 107
Bands, A and I, of striated muscle, 114–115
Baroreceptors, role in heart rate regulation, 334
Bartholin's glands, 480
Basal:
  conditions, description, 430
  ganglia of cerebrum, 185, 187, 189
  metabolic rate(s), 430–432
    age and sex, 431
    computation, 432
    and total metabolic rates, factors determining, 431
Base of lung resting on diaphragm, 371
Basilar membrane, floor of cochlear duct, 259
Basilic vein, median, at elbow, site of intravenous injections, 322, 323, 324
Basophil, 294
Bending movement, flexion, 105
Benzedrine, influence on basal metabolic rate, 431
Betz cells of upper motoneuron, 216
Bicarbonate:
  ions, carbon dioxide in plasma carried as, 379
  pairs in body fluids, 514
Biceps:
  brachii muscle, 128, 129, 132, 143–144
  femoris muscle, 135, 151
Bicipital groove of humerus, marking, 86, 99

Bicuspid valve, location and function, 308–310
Bile, 403, 404, 405
  ducts, 403
  emulsification of fats in small intestine, 411, 412
  secretion, chemical control, 414
  summary of action, 409
Bilirubin and biliverdin, breakdown products of hemoglobin, 296
Biological clock, pineal body, 285
Bipolar neurons, location, 169
Birth, skull at, 80
Bishydroxycoumarin, anticoagulant, clinical value, 306
Bladder, 459–460
  urinary, and male reproductive organs, 471
Blastocyst, 489
Bleeding:
  arterial:
    pressure points to stop, 347
    and venous, relation of blood pressure to, 346
  excessive, treatment, 306
Blind spot:
  location, 241
  reason for, 245
Blood, 293–306
  accumulation in leg veins while standing still, 345
  amount in heart, determinant of myocardial fiber stretching, 333
  arterial (*see* Arterial blood)
  in capillaries of digestive organs, flow through portal system, 329
  cells, 294–300
    classifications, 294, 301
    human, 294
    red (*see* Erythrocyte)
    summary of basic facts, 301
    white (*see* Leukocyte)
  circulation, definition, 328
  clotting, 303–306
    hastening, clinical methods, 306
    mechanism, 303
      role of platelets, 300
    pharmaceutical preparations that retard, 306
    role of prothrombin and fibrinogen, 405
  coagulation (*see* Blood clotting)
  as connective tissue, 45, 50
  constituents, union of oxygen and carbon dioxide with, 377
  deoxygenated, carried by all veins but pulmonary, 315
  fetal, mixture of oxygenated and deoxygenated, 328
  flow:
    abnormally slow, clotting factor, 305
    course, 329, 330
    direction to and from heart, 309
    volume per minute, factors determining, 342
  glucocorticoid concentration, relationship to disease, 530
  glucose concentration:
    decrease by insulin and thyroid hormones, 419
    homeostatic mechanism regulating, 420, 422
    increase, role of growth hormone and ACTH, 419

Blood – cont'd
  groups, 300–303
  injections, from heart into aorta, cause of pulse, 347
  and interstitial fluid:
    filtration pressure between, calculation, 505
    fluid exchange, 351
  layer in capillaries, 380
  oxygen, 382
  oxygenated, carried by all arteries but pulmonary, 315
  pH:
    effect of ammonia excretion, 518
    normal, urinary mechanism for maintaining, 520
  plasma, 303
    antibodies, 302
    and blood serum, difference, 305
    carbon dioxide dissolved in, method of transport, 379
    electrolyte composition, 503
    as extracellular fluid, 8
    major body fluid, 293
    oxygen and carbon dioxide transport, 377
    proteins, importance, 303
    solutes, 303
  platelets, 296, 300
  pressure, 345–346
    arterial:
      average, 345
      clinical measurement, 345
      determinants, 331
      factor in respiration control mechanism, 383
      homeostasis, maintenance, 335, 337
    and circulation, normal, proteins essential for, 405
    diastolic, definition and clinical importance, 345
    gradient, 330, 331
      factors determining, 343
      and peripheral resistance, magnitude, 342
    influence of kidneys, 458
    normal, maintenance by glucocorticoids, 278
    sounds through stethoscope, 345
    systolic, definition, 345
  protein:
    concentration, determinant of viscosity, 335
    synthesis, 405
  reservoir(s):
    function of spleen, 355
    venous plexus and sinuses of skin and abdominal organs, 340
  return from chest organs, 322
  serum, difference from plasma, 304, 305
  sinuses, 97
  storage, function of spleen, 355
  sugar homeostasis, maintenance, function of liver, 404
  supply:
    blood vessels, 316–317
    heart, 311–312
  temperature:
    factor influencing respirations, 384
    in hypothalamus, influence on appetite centers, 435
    increased, cardiac accelerator, 335

Blood — cont'd
  transfusion:
    danger, 302
    source of anti-Rh antibodies, 302
  transport of gases, method, 377
  types, 300–303
  universal:
    donor, 302
    recipient, 303
  velocity, inversely proportional to cross-
      sectional area of vessels, 346
  venous:
    conversion to arterial blood, 380
    pH, 515
  vessels, 314–328
    abnormalities, 316
    additional, in fetus, to reach ma-
        ternal blood, 326
    alimentary tract, embedded in sub-
        mucous coat, 390
    blood supply, 316–317
    coats, 316–317
    endothelial lining, clotting factors,
        305
    functions, 318
    main, 318-328
    nephron unit, 449
    nerve supply, 316
    resistance, revealed by diastolic pres-
        sure, 345
    small, influence of medullary isch-
        emic reflex on, 340
    structure, 316–317
    wettable surface, 304
  viscosity, 333, 335
  volume:
    arterial, determinant of arterial blood
        pressure, 331
    inverse variation with excess body
        fat, 293
    normal, maintenance, renin–angio-
        tensin II mechanism, 281
Bloodstream, return of lymph to, via
    thoracic duct, 348
Body(ies):
  anterior view, descriptive terms, 5
  cavities, 3
  changes during stress, 525
  ciliary, description, 242
  control, 163–289
  erect and moving, 61–161
  fluids:
    buffer pairs in, 514
    chemical structure and function,
        relationship, 503
    pH, control mechanisms, 512
  function(s):
    changes with age, 8
    emotions and role of hypothalamus,
        190
    environment and, 6
    essential part played by water, 429
    generalizations, 6–8
    interdependence, 158
      physiological principle, 330
  gallbladder, 405
  heat production, muscle function, 114
  integration, 163–289
  ketone:
    fat catabolism, buffering, 515
    oxidation via citric acid cycle, 425
  lateral view, descriptive terms, 5
  main motor tracts, 217
  mandible, marking, 96

Body(ies) — cont'd
  Nissl, description, 168
  parts, movement, 114
  pineal, 284–285
  planes, 4, 5
  response to major stressors, factors
      determining, 528
  rib, marking, 99
  sphenoid bone marking, 95
  sternum, 84
    marking, 99
  stomach, central part, 399
  as structural unit, 2
  structure(s):
    changes with age, 4
    generalizations, 2–4
    terminology, 4–6
  temperature, homeostasis, 435–439
  uterus, upper portion, 475
  vertebrae, 82
    marking, 97
  weight, relationship of energy balance,
      433
Bone(s), 65–103
  age changes, 68
  appendicular skeleton, 70–71
  arm, 70–71
  attachment to, by skeletal muscles, 124
  axial skeleton, 69–70
  carpal, forming wrist, 88
  changes in old age, 103
  compact, section showing haversian
      system, 65
  cuboid, 71, 91
  depressions and openings, definitions,
      93
  ethmoid, 81
  face, 76–79
    age changes, 103
  finger, 71
  foot, 71
    right, view from above, 91
  function in movement, 125
  hand, 71
    and wrist, right, 88
  hip, right, side view, 89
  hyoid, 70, 79, 82
  leg, 71
  lower extremities, 71
  markings, 93–102
    definitions, 93
    identification, 94–102
  marrow:
    damage, causes, 298
    red:
      erythropoiesis, site, 295
      platelet formation, site, 300
      and yellow, description, 300
    tissue, 45
  matrix, formation, 67
  metacarpal, framework of hand, 88
  microscopic structure, 63–64
  nasal septum, 97
  orbits, 79, 97
  palatine, location and function, 363,
      364
  pelvis, 71
  processes with muscle attachments, 93
  projections or processes, definitions, 93
  shoulder girdle, 70–71
  skeleton, 69–71
  skull, 68–79
    front view, 72
    viewed from right side, 73

Bone(s) — cont'd
  tarsal, view from above, 91
  temporal, mastoid cells, middle ear
      connection, 255, 256
  thorax, 84
  tissue, 45
  toe, 71
  upper extremities, 70–71
  vertebral column, 70
Bony:
  labyrinth, components, 255, 256–257
  and membranous labyrinths, relation-
      ships, 257
  pelvis:
    description, 101
    and diaphragm, position of stomach
        and large intestine in relation
        to, 398
  sinuses, special feature of skull, 97
Borders of scapula, markings, 84, 85, 99
Bowel, small, circular folds and layers of
    wall, 390
Bowman's capsule, description, 448
Boyle's law, operation in respiration, 372
Brachia:
  conjunctivum cerebelli, 192
  pontis, cerebellar, 192
Brachial:
  artery, 318–319
    use for blood pressure measurement,
        345
  plexus:
    clinical significance, 200
    description, 199
    intermixing fibers, 202
  vein of upper extremity, 321
Brachialis muscle, 128, 129, 132, 143, 145
Brachioradialis muscle, 128, 129, 143,
    146
Brain, 182–195
  arteries at base, 321
  centers higher, vasomotor control by,
      340
  and cord:
    coverings, 195–196
    fluid spaces, 196
    meninges, 195
  neurons, function, 170
  size, 182
  structures, 183–184
  two-way contact with all body tissues,
      272
  venous sinuses in relation to, 325
Brainstem, 193–195
  attachment of cranial nerves, 185
Breast(s), 480–482
  anterior view, 481
  development and functioning, role of
      estrogen, 283
  lactating, milk ejection, oxytocin func-
      tion, 272
Breastbone, description, 70
Breath-holding and carbon dioxide level
    in arterial blood, 384
Breathing:
  effect on circulation, 344
  movements, factor in lymph pressure
      gradient, 351
Bridges, cross, of myosin filaments, 115,
    116, 117
Brim, pelvic, 101
Broad ligaments, 474, 476
Broca's motor speech area, 188, 191
Bronchi, 368–369, 371

Bronchial tree, 368
Bronchioles, 368
Brunner, glands of, and goblet cells, secretion of mucus, 401
Buccal cavity, 393–394
Buffer(s):
    action, 514–515
    mechanism for controlling pH of body fluids, 513–515
    pairs in body fluids, 514
    role in pH control, effectiveness, 515
Bulbar:
    conjunctiva, location, 243
    inhibitory area, location, 220
Bulbourethral glands, 470–472
Bundle, atrioventricular, 312, 313
Burns, edema formation, mechanism, 504
Bursae, definition, 156
Bursitis, locations, 157
Buttocks muscles, 128, 147, 150

C

Cachexia, pituitary, 268
Caffeine, influence on basal metabolic rate, 431
Calcaneus, 71, 91, 102
Calcium:
    blood, homeostasis, maintenance, 274, 276
    heart and intestinal contractions, role, 429
    muscle contraction and relaxation, role, 117
Calculi, renal, 459
Calyces, renal, 448, 458
Canal(s):
    alimentary, 390
    anal, 398, 402
    auditory, external, marking of temporal bone, 94
    carotid, marking of temporal bone, 95
    ear, description, 254, 255
    haversian, central canal of bone, 63
    inguinal, herniation through, 156
    lacrimal, location, 249, 250
    of Schlemm, location, 242
    semicircular, of inner ear, 255, 257
    Volkmann's, of bone, 64
Cancellous bone, spongy type, 64
Cancer cells and lymph nodes, 353
Cannon's theory of homeostasis, 524
Canthus, description, 249
Capillary(ies):
    and alveolar membranes, permeability to gases, 377
    arteries and, relationship, 320
    cross-sectional area, 346
    function, 315
    lung, gas exchanges, 371, 380, 381
    lymphatic, 348
    membranes, water movement across, 505
    "missing link" in Harvey's theory of circulation, 315
    Starling's law, 505
    tissue, pressures at arterial and venous ends, 506
    walls, structure, 317
Capitulum of humerus marking, 86, 99
Capsular:
    colloidal osmotic pressure, 452
    hydrostatic pressure, 451

Capsule:
    Bowman's, description, 448
    of Glisson, 403
    internal, of cerebrum, 184, 189
    renal, 448
Carbamino compounds, 379
Carbohydrate(s):
    absorption, 415
    anabolism and catabolism, 429
    digestion, 411, 412
    "fat-sparing" effect, 426
    metabolism (*see* Metabolism, carbohydrate)
Carbon dioxide:
    enzymes that add or remove, 411
    exchange, 376, 377, 382
    and oxygen:
        blood content, homeostasis, mechanism, 339
        pressure gradients, 377
        transport by red blood cells, 295
    regulatory chemical, 265
    respiratory stimulant, 384
    tension of arterial blood, influence on respirations, 382–383
    transport in blood, methods, 379
Carbonic acid inside red blood cell, buffering, 514
Cardiac: (*see also* Heart)
    center of medulla, 194
    cycle, 314, 316
    minute output, 332
    muscle, 46, 47, 49
        contractions, autonomic reflexes, 221
        effectors, parasympathetic effects, 231
        layer of heart wall, 307, 308
        modified, of connection system of heart, 312
        rabbit, electron photomicrograph, 49
        tissue, 44, 45
    nerves, 314
    pressoreflexes, 334, 335
    sphincter, 400
Carotid:
    and aortic:
        cardiac pressoreflexes, 335
        vasomotor pressoreflex mechanism, 337
    arteries, 318–319
    canal, marking of temporal bone, 95
    chemoreflexes, adjustment of blood pH, 516
Carpal:
    bones, 71, 88
    joints, 108
Cartilage:
    articular, of long bones, 66
    and bone, similarities and differences, 64
    classification, 45
    costal:
        function, 84
        origin of pectoralis major muscle, 131
        rib marking, 99
    cricoid, lowest of laryngeal cartilages, 366, 367
    epiphyseal, role in long bone growth, 68
    laryngeal, 366
    microscopic structure, 64
    nasal, 77
    rings:
        bronchi, 368

Cartilage – cont'd
    rings – cont'd
        trachea, 367, 368
            lacking in esophagus, 398
        thyroid, Adam's apple, 364, 366
Cartilaginous joints, 106, 107
Caruncle, location, 249, 250
Catabolic:
    effect of insulin, 283
    rate, meaning of metabolic rate, 430
Catabolism:
    and anabolism:
        metabolic processes, 416
        protein, concurrent, 427
        relationship between, 417
    carbohydrate by cells, 417
    carbon dioxide production inside cells, 382
    energy release, 32, 33–34
    fat:
        ketogenesis and oxidation, 425
        ketone bodies, buffering, 515
        processes, 424
        promotion by growth hormone, 268
    foods:
        only means of heat production, 436
        preferential order, 434
    glucose, 29–34
        stimulation by thyrotropin, 419
    heat produced by muscle cells, 114
    primary in carbohydrate and fat metabolism, 426
    protein, 427–428
        and fat, first step, function of liver, 404
        promotion by progesterone, 284
    regulatory chemicals product of, 265
Cauda equina of spinal cord, 198
Caudate:
    lobe, part of right lobe of liver, 402
    nucleus of cerebrum, 185, 187, 189
Cavity(ies):
    abdominal, 3
        dropsy, cause, 326
    body, 3
    buccal, 393–394
    chest, 3
    cranial, 3
        floor, 72, 76
        venous sinuses, 321, 322, 325
    dorsal, 3
    glenoid, of scapula, marking, 84, 85, 99
    heart, 308, 309
    and humors of eye, 245–247
    long bones, 67
    nasal, lateral wall exposed, 364
    pelvic, 3
    spinal, 3
        contents, 168
    thoracic, 3
    tympanic, 255, 256
    uterus, 476
    ventral, 3
Cecum, 398, 402
    and ascending colon, place of attachment like letter T, 401, 402
Cell(s), 11–41
    air, mastoid, marking of temporal bone, 94
    alpha and beta, of islands of Langerhans, 406
    Betz, of upper motoneuron, 216

Cell(s) — cont'd
blood, 294-300
classifications, 294, 301
human, 294
red (*see* Erythrocyte)
summary of basic facts, 301
white (*see* Leukocyte)
breathing, internal respiration, 372
cancer, and lymph nodes, 353
chief, secretion of gastric juice enzymes, 400
cytoplasm, 14
definition, 3
different parts of renal tubule, 450
division, reproductive process, 34
epithelial:
anterior pituitary gland, types, 266
proliferation, acceleration by estrogens, 283
functions, 6
role of hypothalamus in controlling, 190
gastric mucosa, secretions, 400
goblet, secretion of mucus, 401
graft-rejecting, in rejection mechanism, 299
hair:
hearing and equilibrium sense organs, 257, 259
olfactory sense organs, 260
liver, diversity of function, 403
mastoid, of temporal bone, middle ear connections, 255, 256
membrane, 12, 14, 15
movement of substances through, 20-29
polarization by sodium pump, 17
structure, triple-layer hypothesis, 15
water movement through, 506
metabolism, 29-34
muscle (*see* Muscle cell)
nervous system, 165-171
nucleus, description, 19
physiology, 19-41
plasma, formation, 354, 355
processes and cell body, relation, 170
replacement, function of protein anabolism, 426
reproduction, 34-39
reproductive, 469
reticuloendothelial, phagocytosis, 298, 355
Schwann, description, 168
secreting of pancreas, exocrine glands, 405
structural units of body, 2
structure, 11-19
typical, seen under electron microscope, 13
water important constituent, 429
Cellular metabolism, vital part of circulatory system, 293
Centers, reflex, autonomic, higher, location and function, 232
Central nervous system, composition, 178
Centrioles of centrosome, 19
Centromere, 38
Centrosome, 19
Centrosphere, 19
Cephalic phase of gastric secretion, reflex and psychic control, 413
Cerebellar disease, symptoms, 193
Cerebellum, 184, 185, 190-193

Cerebral:
aqueduct of Sylvius, 196, 197
cortex, 183-184
consciousness and, 211
control over all cells by way of hypothalamus, 271
functions, 186
human, map, 191
impulses from, role in control of vasoconstriction, 340
and lower centers, hypothalamus as link, 190
motor pathways from, classification, 216
pyramidal pathway from, 216
respiration control mechanism, factor, 384
sensory:
areas, 186-187, 191
neural pathways to, 211
source of nerve impulses, 174
visual area, conduction to, 254
nucleus(i), 185, 187, 189
peduncles of midbrain, 194
tracts, 184
ventricles, 196, 197
Cerebrospinal fluid, 196-198
Cerebrum, 183-188
basal ganglia, 189
and cerebellum, common characteristics, 192
Cerumen, secretion, 256
Ceruminous glands of skin, 58
Cervical:
canal, uterine, 476
glands, superficial, group of lymph nodes, 352
plexus, spinal nerve branches from, 200
vertebra(ae), 79, 82, 83
enumeration, 70
parts, 98
Cervix, uterine, 475
Chambers:
anterior cavity of eyeball, 242, 245
heart, 308, 309
Cheekbones, description, 69
Chemical(s):
changes:
blood clotting mechanism, 304
glycolysis, 30
muscle, during exercise, 118-119
constituents of fluid compartments, 502, 503
control of bile secretion, 414
digestion, 409, 410-414
photosensitive, in rods and cones, 254
reactions:
citric acid cycle, 31, 32-33
glycogenesis, 417
and reflex control of gastric secretion, 413
regulatory, comparison with hormones, 265
structure of enzymes, 410-411
transmitters, autonomic, principle, 228
Chemistry of active transport, 28, 29
Chemoreflexes:
carotid, adjustment of blood pH, 516
vasomotor, 336, 338, 340
Chest, 371-372
cavity, 3
or costal breathing, 376
muscles, 128, 142

Chest — cont'd
organs, veins, 322
parietal and visceral pleura, 55
wall motion, muscles, 139
Cheyne-Stokes respirations, 376
Chiasma, optic, description, 208
Chief cells, secretion of enzymes of gastric juice, 400
Chloride shift in blood, mechanism, 378
Choanae, 363
Choking reflex, protective device, 384
Cholecystectomy, 405
Cholecystitis, 405
Cholecystokinin, 413, 414
Cholinergic:
effects of visceral effectors, 231
fibers, release of acetylcholine, 229
Cholinesterase, role in synaptic conduction, 176
Chordae tendineae, description, 308, 310
Chorioid, middle coat of eye, 241, 242
Chorionic gonadotropins, secretion of placenta, 285
Choroid plexus, 196
Chromatids, formation, 38
Chromatin granules, 19, 38
Chromosome(s):
composition, 19
reduction in sex cells, 488
replication, 34, 38
Ciliary:
body, description, 242
muscle of eye, function, 247
Ciliated columnar epithelium, pseudo-stratified, 43
Circle of Willis, arterial anastomosis, 321
Circular:
folds, small bowel, 390
theory of mind and body relations, 272
Circulation, 328-345
cerebrospinal fluid, 196-197
collateral, of heart, 312
control mechanisms:
functions, 330
parts, some of many, 341, 343
fetal, 326-328
function, one essential, 315
hepatic cells, 403
increase or decrease, causes, 342
lymph, 351
portal, 322, 326, 416
principles, 330-340
systemic, 320-322
and pulmonary, relationship, 329
varied according to activity of different tissues, 330
veins in, role, 315
Circulatory:
mechanism, essential function, 315
system, 292-361
lymphatic system part, 348
and respiratory system, interdependence, 362
Circumduction, definition, 105
Cisterna chyli, 350, 351
Citrate, blood clotting retarding agent, 306
Citric acid cycle, 30, 31-34
ketogenesis and, 425
second step in protein catabolism, 427
Clavicle, 70, 84
insertion of trapezius muscle, 130
muscle origins, 131, 137

Clavicular joints, description, 108
Cleft:
  palate, description, 363
  synaptic, description, 174
Clitoris, location, 480
Clot, blood, 306
Clotting, blood (*see* Blood clotting)
Coagulation, blood (*see* Blood clotting)
Coats:
  blood vessels, 316–317
  digestive:
    organ walls, 390
    tract, modifications, 391
  eyeball, 241–245
  fallopian tubes, 477
  gallbladder, 405
  large intestine, 391, 402
  small intestine, 390, 391
  stomach, 391, 399
  ureter walls, 458
  wall of uterus, 475
Coccyx:
  description, 70
  location in vertebral column, 79
Cochlea, location and description, 255, 257–259
Cochlear:
  duct, shape and location, 257, 258
  nerve, composition, 210
Coenzyme, form of prosthetic group of enzyme molecule, 410
Cold stimuli to skin, temporary apnea caused by, 384
Collar bones, 70
Collecting tubule, renal, 449, 450
Colliculi of midbrain, 184, 194
Colloids in blood plasma, 303
Colon, 398, 402
  digestive process performed, 408
Color:
  skin, 56, 57
  vision, function of cones, 254
Colostrum, antecedent of milk, 482
Column(s):
  renal, 448
  spinal cord, 180
  vertebral, 79, 82–84
    age changes in skeleton, 103
    bones, 70
    markings, 97–98
    shape, 83
Columnar epithelium, 43, 44
Common:
  bile duct, 403
  iliac arteries, 318–319
  path, final, principle, 215
Communication:
  essential to control of body activities, 164
  systems of body, 165
Compact bone, 64, 65
Components of connective tissue, 119–120
Compound(s):
  B, 277
  complex, built from food by anabolism, 416
  F, 277
Conchae:
  division of nasal cavities, 363, 365
  inferior:
    description, 69
    middle, and superior, 78
  superior and middle, ethmoid bone markings, 81, 96

Conduction:
  heat regulation, 437
  nerve impulse, 172–174
    function of neurons, 169–170
    across neuroeffector junctions, 178
    rate, 174
    stimulus, 170
    across synapses, 174, 176, 178
    to visual cortex, 254
  paths, central, of autonomic nervous system, 226
  sole function of modified cardiac muscle, 313
  sound wave, in hearing mechanism, 259
  system of heart, 312–313
Condyles, bone markings, 90, 95, 96, 102
Condyloid joint, summary, 106
Cones, rods and, photoreceptor neurons, 243, 244
Congestive heart disease, decreased vital capacity, 375
Conjunctiva, description, 243, 249
Connective tissue, 44–45, 47–48, 50–51
  components, 119–120
  loose ordinary, 48, 50
  submucous coat, second layer of alimentary canal walls, 390
Conscious:
  mechanism to regulate skin temperature, 439
  proprioception:
    neural pathway, 214
    neurons, 212
Consciousness:
  cerebral cortex and, 211
  mechanism, 215
Constriction of pupil, mechanism, 253
Contraceptive pills, rationale, 283
Contractions, muscle (*see* Muscle contraction)
Contracture, muscle, definition, 123
Contralateral reflex arc, three-neuron, 175
Control of body (*see* Body control)
Convection, heat regulation, 437
Convergence:
  definition, 253
  principle, 214, 215
Convoluted tubules, function, 453
Convolutions of cerebral cortex, 184
Convulsions, 123
Coracobrachialis muscle, 145
Coracoid process of scapula, marking, 84, 85, 99
Cord:
  spermatic, internal structures, 472–473
  spinal (*see* Spinal cord)
  umbilical, formation by umbilical arteries and veins, 327
Cornea:
  description, 242
  irregular, cause of astigmatism, 252
Corneal reflex, somatic reflex, 223
Coronal suture, 72, 97
Coronary arteries, 310, 311, 312, 318, 319
Coronoid:
  fossa of humerus, marking, 86, 99
  process, marking:
    mandible, 96
    ulna, 87, 100
Corpora:
  cavernosa penis, 472
  quadrigemina of midbrain, 194

Corpus:
  callosum of cerebrum, 183
  cavernosum urethrae, 472
  luteum:
    development, 483
    progesterone secretion, 284
    target of luteinizing hormone, 269
  striatum, composition, 185
Corpuscle:
  Malpighian, 448
  Meissner's, nerve ending in skin, 240
  renal, 448
  tactile, nerve ending in skin, 240
Cortex:
  adrenal (*see* Adrenal cortex)
  cerebral (*see* Cerebral cortex)
  kidney, 448
Corti, organ of, 258, 259
Corticospinal tracts, 180–182
  lateral, 217
    motor pathways, 216
    ventral, 218
Corticosterone, glucocorticoid, 277
Corticotropin-releasing factor:
  neurosecretion, 271
  role in stress response, 527
Cortisol, 277
  secretion, stimulation by adrenocorticotropin, 269
Cortisone, rationale for use, 530
Costal:
  breathing, shallow or chest breathing, 376
  cartilage(s):
    function, 84
    origin of pectoralis major muscle, 131
    rib marking, 99
Cowper's glands, 470–472
Cranial:
  cavity, 3
    floor, 72, 76
    venous sinuses, 321, 322, 325
  nerves, 204–211
    attachment to brainstem, 185
    reflex centers, 195
Craniosacral division of autonomic nervous system, 224
Cranium, 72–73, 76
  age changes in skeleton, 103
  bones, 69
Creatine phosphate, role in muscle contraction, 116, 118
Crest:
  iliac, markings of os coxae, 89, 100
  marking of tibia, 102
  pubic, marking of os coxae, 89, 101
Cretinism symptoms, 274
Cribriform plate, ethmoid bone marking, 81, 95
Cricoid cartilage, lowest of laryngeal cartilages, 366, 367
Crista(e):
  ampullaris, 259
  galli, ethmoid bone marking, 81, 96
  of mitochondria, 16, 17
Cross bridges of myosin filament, 115, 116, 117
Cross-eye, definition, 253
Crown, neck, and root, parts of tooth, 397
Crystalloids in blood plasma, 303
Cubital superficial or supratrochlear lymph nodes, 352

Cuboid bones, 71, 91
Cul-de-sac of Douglas, 477
Cuneatus tract of spinal cord, 180–182
Cuneiform bones, 71, 91
Cupula, location, 259
Curves of vertebral column, 98
Cushing's syndrome, 278, 279
Cuspid valves, 308–311
    action, 310
Cutaneous membrane (*see* Skin)
Cyanosis of skin, 56
Cycle(s):
    cardiac sounds during, 314, 316
    citric acid, 30, 31–34
        ketogenesis and, 425
        second step in protein catabolism,
            427
    endometrial, 483–484
    female sexual, 482–488
    gonadotropic, 484
    menstrual, 483–486
    myometrial, 484
    ovarian, 483
Cystic duct, 403, 405
Cytokinesis, 34
Cytology, study of cell, 11
Cytoplasm:
    cell, 14–19
    neuron, 166
Cytoplasmic granules, characteristic of
    granular leukocytes, 298

D

Dalton's law of partial pressures, 376
Daylight vision, function of cones, 254
Dead space, anatomic, 376
Deamination, first step in protein catabo-
    lism, 427
Debt, oxygen, in exercise, 119
Deciduous:
    arch, 396
    teeth, 395, 396
Decussation of crossed corticospinal
    tracts, 194
Deep:
    fascia, 120
    reflex, knee jerk, 222
    veins, 321
Defecation, 408, 410
    promotion by parasympathetic im-
        pulses, 230
Defense:
    function:
        lymph nodes, 352
        spleen, 355
    mechanism, importance of leukocytes,
        298
Deformities, congenital, embryology in
    study, 488
Deglutition, definition, 408
Deltoid:
    muscle, 128, 129, 131, 142–143
    tuberosity of humerus, marking, 86, 99
Dendrites, description, 167
Dens:
    axis, marking, 83, 98
    second cervical vertebra, 82
Dense fibrous tissue, 50, 120
Dentate nuclei of cerebellum, 192
Dentine, entrance of bacteria through
    broken enamel to, 397
Dentition, deciduous and permanent sets,
    397

Deoxygenated blood:
    carried by all veins but pulmonary, 315
    oxygenated and, fetal blood mixture,
        328
Depressions in bone, definition, 9
Depressor:
    muscle group, 127
    nerves, vagus fibers to heart, 314
Dermis, epidermis and, 54, 56
Descending:
    abdominal aorta, branches, 318
    colon, location, 398, 402
    thoracic aorta, branches, 318
Desoxycorticosterone, mineralocorticoid,
    278
Development, hormones regulators, 265
Diabetes:
    insipidus, role of antidiuretic hormone,
        272
    mellitus:
        insulin deficiency, 419
        untreated, carbohydrate utilization,
            417
    renal, 455
Diabetogenic effects:
    excess glucocorticoids, 278
    growth hormone, 268
Dialysis, description, 22
Diapedesis of leukocytes, aid in defense
    mechanism, 298
Diaphragm:
    descent during inspiration, role in
        establishing lymph pressure gra-
        dient, 351
    front, 155
    liver and pancreas, position relative to,
        403
    muscle of chest wall, motion, 139
    stomach and large intestine, position
        relative to, 398
Diaphragmatic or abdominal breathing,
    376
Diaphysis of long bone, 66
Diarthroses, freely movable joints, 104,
    106, 107
Diarthrotic joints:
    movements, 105–110
    structure, 104
Diastole, relaxation of atria, 314
Diastolic blood pressure, definition and
    importance, 345
Dicumarol, anticoagulant, clinical value,
    306
Diencephalon, structures, 188
Diet(s):
    high-protein, rationale, 433
    protein-poor, cause of negative nitro-
        gen balance, 427
Differential count, percentage count of
    white blood cells, 300, 301
Diffusion, 20–33
    mechanism in tubular secretion, 456
Digestion, 407–414 (*see also* Food)
    absorption of end products, function of
        small intestine, 401
    definition, 389
    promotion by parasympathetic im-
        pulses, 230
    residues, 412
    wastes, elimination, function of large
        intestine, 402
Digestive:
    enzymes:
        extracellular action, 411

Digestive—cont'd
    enzymes—cont'd
        secreted by pancreas, 406
    gland secretion, control, 412–414
    hormones, action, summary, 413
    juice(s):
        action, summary, 409
        flow, control, 412
        small intestine, 401
    organs:
        blood from, detour through liver,
            322
        walls, 390–393
    system, 389–445
    tract:
        glands and smooth muscle, parasym-
            pathetic dominance, 230
        modification of coats, 391
        and respiratory tract, pharynx hall-
            way for, 365, 407
Dinitrophenol, influence on basal meta-
    bolic rate, 431
Diplopia, double vision, 253
Disease(s):
    of adaptation, 530
    cerebellar, symptoms, 193
    coronary, 311
    Graves', 274, 275
    heart, congestive, decreased vital ca-
        pacity, 375
    infectious, hypertrophy of spleen, 354
    Simmonds', result of growth hormone
        deficiency, 268
    and stress, 529–530
Disk, optic, description, 241, 245
Distal:
    and collecting renal tubules, functions
        in urine formation, 457
    convoluted tubule, renal, 449, 450
Divergence:
    phenomenon, result of intersegmental
        reflex arcs, 177
    principle, 214
Division:
    cell, reproductive process, 34
    nuclear, 34–39
DNA:
    genes and, 36, 37–38
    molecule, Watson-Crick structure, 36,
        37
    replication, biological importance, 34,
        38
    structure, 36–38
Doan, Dr. Charles Austin, performed first
    splenectomy, 355
Donnan equilibrium principle, 503
Donor, universal, blood type O, 302
Dorsal:
    cavity, 3
    flexion, definition, 105
    venous arch of foot, 322–324
Double helix, DNA molecule, 36
Douglas, cul-de-sac of, 477
Dropsy of abdominal cavity, cause, 326
Drugs, influence on basal metabolic rate,
    431
Dual autonomic innervation, principle,
    228
Duct(s):
    alveolar, 368
    bile, 403
    cochlear, shape and location, 257, 258
    cystic and hepatic, 405
    ejaculatory, 470, 471

Duct(s)—cont'd
  lymphatic and thoracic, right, main lymphatic trunks, 348
  nasolacrimal, description, 249, 250
  pancreatic, 406
  of Santorini, 406
  seminal, 470
  Stensen's, 395
  of testes, 467, 468, 470
  thoracic:
    description, 348
    and tributaries, position, 350
  Wharton's, 395
  of Wirsung, 406
Ductless glands, definition, 263
Ductus:
  arteriosus, 327, 328
  deferens, 470
  venosus, 326, 327
Duodenojejunal recess, 393
Duodenum, upper part of small intestine, 400
Dura mater of meninges, 195
Dural sinuses of head, projected on skull, 325
Dwarf and giant, pituitary, 267
Dwarfism, hormone imbalance in, 265
Dyspnea, difficult or labored breathing, 376

E

Ear, 254–259
  bones, 70
  inner, organs of balance, 255, 257
Eardrum, location, 255, 256
Ectoderm, 490
Ectopic pregnancy, 489
Edema formation, mechanism, 504, 505, 507
Effective filtration pressure:
  between glomeruli and capsules, 451
  glomerular, calculation, 452, 453
Effectors:
  role in impulse conduction, 172
  visceral, definitions, 223
Efferent neurons, description, 167, 168
Ejaculatory duct, 470, 471
Elastic cartilage, 45
  structure, 64
Elbow joints, description, 108
Electrical currents in heart, clinical importance, 313
Electrocardiograph, visible record of heart conduction, 313
Electroencephalograms, use in study of cerebral function, 186
Electrolyte(s):
  balance:
    dependence on kidney function, 451
    role of mineralocorticoids in maintenance, 280
  blood plasma, 303
  composition of blood plasma, 503
  concentrations and osmotic pressures of body fluids, 507
  elimination in urine, 461
  and fluid:
    balance, 496–511
      role of hormones, 265
    imbalance, relationship, 508
    metabolism, effect of estrogens on, 283

Electrolyte(s)—cont'd
  metabolism, function of mineralocorticoids, 278
  reabsorption:
    mechanism, 455
    and secretion, 456
Electromyography, newer method of muscle function study, 121
Electron-transport particles, 33
Elimination of wastes of digestion, function of large intestine, 402
Ellipsoidal joint, 106, 107
Embryo, human, stages in development, 489
Embryology, 488–490
Embryonic disc, 490
Eminence, intercondylar, marking of tibia, 90, 102
Emissary veins between superficial and deep veins, projected on skull, 325
Emotional expressions, production, 219
Emotions:
  and bodily functions, role of hypothalamus, 190
  effect on heart rate, mechanism, 334
  influence on basal metabolic rate, 431
  role of thalamus in mechanism, 188
Emphysema:
  increase in dead air space, 376
  reduced functional alveolar-capillary membrane area, 380
  stretching of alveolar walls, 375
Enamel, intact, bacterial resistance, 397
End buttons or end feet, 174, 177
Endarthroses, summary, 106
Endocardium, 307, 308
Endocrine:
  glands, 58
    and hormones involved in glucose homeostatic mechanism, 418
    importance, 265
    location in female, 264
    main, 264
  glandular cells of islands of Langerhans, 406
  system, 263–285
    and nervous system:
      hypothalamus link between, 271
      similarity and differences, 263
Endolymph, cochlear duct fluid, 257, 258, 259
Endometrial or menstrual cycle, 483–484
Endometrium:
  layers composing, 475
  preparation for gestation by progesterone, 284
Endomysium of connective tissue, 119
Endoplasmic reticulum, 14–15
Endosteum of long bones, 67
Endothelial lining of blood vessels, clotting factors, 305
Enema, anatomical reason for placing patient on left side, 398, 402
Energy:
  balance:
    relationship to body weight, 433
    role of hormones, 265
  expended by specific dynamic action of foods and total metabolic rate, 432
  generated in mitochondria, 16, 18, 34
  heat, measurement, 430
  muscle cell, ATP breakdown, 118

Energy—cont'd
  release:
    by catabolism, 32, 33–34
    rate, meaning of metabolic rate, 430
  source(s):
    for muscle contraction, 118–119
    reserve, body's largest, stored fats, 426
    used in skeletal muscle work, determinant of total metabolic rate, 432
    for work of anabolism, provision by catabolism, 417
Energy-supplying foods, carbohydrates and fats, 426
Enterocrinin, stimulation of intestinal juice secretion, 414
Enterogastric reflex, control of stomach emptying process, 409
Enterogastrone, lessening of gastric secretion and motility, 413
Entoderm, 490
Environment, influence on body function, 6
Enzyme(s):
  built from food by anabolism, 416
  chemical structure, 410–411
  digestive:
    juices, hydrolysis of foods catalyzed by, 410
    and mucus contained in intestinal digestive juice, 401
    secreted by pancreas, 406
  gastric juice, secretion of chief cells, 400
  properties, 411
  systems for naming, 410–411
Eosinopenia induced by glucocorticoids, 278
Eosinophil, 294
Epicardium, 307, 308
Epicondyles of humerus, markings, 86, 99
Epidermis and dermis, 54, 56
Epididymis, 468
Epiglottis, 366
Epimysium of connective tissue, 119
Epinephrine:
  abundance during stress, 419, 527
  blood:
    concentration, modifier of heart rate, 313
    glucose increased by, 419
  cardiac accelerator, 335
  secretion, control mechanism, 282
Epiphyseal cartilage, role in long bone growth, 68
Epiphysis(es):
  cerebri, 284–285
  long bones, 66
    age changes, 103
Episiotomy, reasons for, 480
Epistropheus, description, 83, 98
Epithelial:
  cells:
    anterior pituitary gland, types, 266
    proliferation, acceleration by estrogens, 283
    surface of gastric mucosa, mucus secreted, 400
    squamous, tissue, walls of alveoli, 368
  tissue, 42–45
Epithelium:
  columnar, 43

Epithelium—cont'd
  pigment, of retina, 244
  squamous, 43
Epsom salts, nonabsorbability, 416
Equilibration, result of net diffusion, 22
Equilibrium:
  maintenance, role of cerebellum, 193
  principle, Donnan, 503
  sense organ:
    crista ampullaris, 259
    macula in utricle, 257
Eruption of teeth, 395–397
Erythrocyte(s), 295–298
  antigens, reaction with plasma anti-
    bodies, 302
  count, determinant of viscosity, 335
  determination of number, 298
  and platelet destruction, function of
    spleen, 355
  replacement, 426
Erythropoiesis, red blood cell formation,
  295
Erythropoietin, kidney hormone, 297
Esophageal peristalsis, mechanism, 408
Esophagus:
  description and location, 398
  digestive process performed, 408
  modifications of muscle and fibrose-
    rous coats, 391
Esophoria, description, 253
Estrogen(s):
  deficiency, cause of menopause, 271
  functions, 283
  secretion:
    adrenal cortex, 280
    cyclical changes, 486
    ovaries, 283, 478
    placenta, 285
    stimulation:
      follicle-stimulating hormone, 269
      luteinizing hormone, 269
Ethmoid:
  bone, 73, 81
    description, 69
    markings, 81, 95–96
  cells, 365
  sinuses:
    ethmoid bone markings, 81, 96
    drainage, 363
Eupnea, normal quiet breathing, 376
Eustachian tubes:
  location and functions, 255, 256
  opening into nasopharynx, 365
Evaporation of water, method of heat
  loss, 436
Eversion, definition, 110
Evertor muscles, abductors of foot, 129
Excitatory postsynaptic potential, 176
Excretory organs and substances excreted, 447
Exercise:
  application of Starling's law of heart,
    334
  chemical changes in muscle, 118–119
  effect on heart rate, 335
  mechanism for shifting blood from
    reservoirs to active structures,
    340
  oxygen debt, 119
  respiration control, 384
Exocrine glands, 58, 59
  cells around small pancreatic ducts,
    406
  definition, 263

Exophoria, description, 253
Exophthalmic goiter, symptoms, 274
Expiration:
  mechanism, 373
  stimulation from pneumotaxic center,
    384
Expiratory reserve volume, definition,
  374
Expressions, emotional, production, 219
Extension, straightening movement, 105
Extensor:
  muscles, 127, 129
  reflex, knee jerk, 222
External:
  auditory meatus, marking of temporal
    bone, 94
  genitals of male, organs constituting,
    472
  iliac:
    arteries, 318–319
    veins of lower extremities, 322
  oblique muscle, 128, 138, 142
Exteroreceptors, surface receptors, 239
Extracellular fluid:
  components, 348
  description, 8
  functions, 501
  increased volume induced by gluco-
    corticoids, 278
  role in control of antidiuretic hormone
    secretion, 272
Extrapyramidal tracts, complexity, 219
Extremity(ies):
  lower, 89–93
    age changes in skeleton, 103
    bones, 71
    muscles, 128, 129, 135, 147
    nerves, 203
    veins, 322–324
  upper, 84–88
    bones, 70–71
    and hand, innervation, 200
    muscles, 128–129, 130–131, 143–146
    veins, 321–322, 324
Extrinsic:
  factor, role in erythropoiesis, 297
  muscles of eye, 284
Eye(s):
  anatomy, 241–249
  accessory structures, 247, 249
  convergence, definition, 253
  extrinsic muscles, 248
  muscles, classification and summary,
    247
  refracting media, 250
Eyeball:
  cavities and humors, 245–247
  coats, 241–245
  left, horizontal section, 242
  and lids, longitudinal section, 243
Eyebrows, function, 247
Eyeground, right, 241
Eyelashes, function, 247
Eyelids:
  description, 249
  longitudinal section, 243

F

Face:
  bone, 69, 76–79
    age changes, 103

Face—cont'd
  expressions:
    head muscles causing, 156
    muscles, 141
  frontal section, 365
  and head, main arteries, 320
  and neck, sagittal section, 364
  nerve, summary, 206, 210
Facilitated diffusion, 23
Facilitatory impulses, 219, 220
Factor, intrinsic, 400
Falciform ligament, division of liver into
  two main lobes, 402
Fallen arches, 92
Fallopian tubes, 474, 477
Falx cerebelli and cerebri, functions, 196
Farsightedness, explanation, 250, 251
Fascia, deep, 120
Fasciculus gracilis tract of spinal cord,
  180–182
Fat(s):
  absorption, 414, 415
  anabolism and catabolism, 429
  body, excess, inverse variation of blood
    volume with, 293
  catabolism:
    first step, function of liver, 405
    ketone bodies, buffering, 515
    processes, 424
    promotion by growth hormone, 268
  concentration in portal and hepatic
    veins, 416
  content of body, fluid volume variation
    with, 502
  digestion, 411, 412
  glucose storage as, principle, 424
  metabolism, 424–426
    effect of glucocorticoids, 278
    hormone control, 426
  mobilization, promotion by growth hor-
    mone, 268
  and protein, gluconeogenesis by liver,
    418
  storage, 425
"Fat-sparing" effect of carbohydrates,
  426
Fatigue, muscle, 123
Fauces, 365, 393
Faucial tonsils, location, 365
Fear, intense vasoconstriction and rise in
  arterial blood pressure, 340
Feedback:
  control of hormone secretion, 270
  mechanism, negative:
    between anterior pituitary and ova-
      ries, 487
    between anterior pituitary and testes,
      468–469
    erythrocyte homeostasis, 297
Female:
  genitals, external, 479
  and male skeletons, differences, 103
  reproductive organs, 474–482
  sexual cycles, 482–488
  urethra, structure and function, 460
Femoral:
  artery, 318–319
  rings, herniation through, 156
  veins of lower extremities, 322–324
Femur:
  description, 71, 89
  insertion of muscles, 134
  markings, 90, 101–102
  origin of muscles, 135, 136

Femur—cont'd
  right, anterior surface, 90
Fenestra ovalis and rotunda of middle
    ear, 255, 256, 257
Fertility, male, 473
Fertilization of ovum, fallopian tubes
    usual site, 477
Fetal circulation, 326–328
Fetus, Rh-positive, source of anti-Rh anti-
    bodies, 302
Fever:
  anorexia accompaniment, 435
  basal metabolic rate increased, 431
  production, mechanism, 439
Fiber(s):
  adrenergic norepinephrine release, 229
  arrangement of skeletal muscles, 119
  cholinergic, acetylcholine release, 229
  muscle, 114
    relaxation, mechanism, 117
  myocardial, length, regulator of heart-
      beat strength, 333
  nerve:
    diameter and rate of conduction, 168,
        174
    myelinated, white matter, 178
  Purkinje, 312, 313
  Sharpey's, 67
Fibrillation, 123
Fibrin, role in blood clotting, 304
Fibrinogen:
  blood protein synthesized by liver cells,
      405
  role in blood clotting, 304
Fibrinolysis, mechanism, in clot dissolu-
    tion, 306
Fibroserous coat of alimentary canal,
    outer layer, 390
Fibrosis of muscle, age change, 121
Fibrous:
  cartilage, 45
    structure, 64
  connective tissue, 45
  joints, 106, 107
  tissue, dense, 50, 120
Fibula:
  articulation, 91
  description, 71
  insertion of biceps femoris muscle, 135
  markings, 90, 102
  origin of soleus muscle, 136
  tibia and, right, anterior surface, 90
Filaments:
  myofibril, 115–116
  myosin, cross bridges, 115, 116, 117
Filtration:
  air, to and from lungs, function of nose,
      365
  defense function of lymph nodes, 352
  glomerular, 451–454
    volume, 452
  and phagocytosis of microorganisms
      by spleen, 355
  pressure between blood and interstitial
      fluid, calculation, 505
Filum terminale, formation, 196
Fimbriae of fallopian tubes, 477
Final common path, principle, 215
Fingers, bones, 71
Fissure(s):
  anterior median, spinal cord groove,
      179
  cerebrum, 183

Fissure(s)—cont'd
  orbital, superior, sphenoid bone mark-
      ing, 95
  palpebral, variation in size, 249
Flaccid muscles, definition, 121
Flat bones, 65, 67
Flatfoot, 92
Flavors, combination of causes, 260
Flexion, definitions, 105
Flexor muscles, 127, 129
Floating ribs, 70
Floor of cranial cavity, 72, 76
Fluid(s):
  balance:
    dependence on kidney function, 451
    role of mineralocorticoids in main-
        tenance, 280
  body:
    bicarbonate buffer pairs, 514
    chemical structure and function, re-
        lationship, 503
    pH, mechanisms of control, 512
  cerebrospinal, 196–198
  compartments:
    body, 496
    chemical constituents, 502, 503
    comparison, 501
  conduction of sound waves, 259
  distribution, homeostasis, 501
  and electrolyte:
    balance, 496–511
      role of hormones, 265
    imbalance, relationship, 508
    metabolism, effect of estrogens, 283
    therapy, rationale, 498
  exchange between blood and intersti-
      tial fluid, 351
  extracellular:
    description, 8
    increased volume induced by gluco-
        corticoids, 278
    role in control of antidiuretic hor-
        mone secretion, 272
  intercellular, definition, 8
  interstitial:
    and blood:
      filtration pressure between, calcu-
          lation, 505
      fluid exchange between, 351
    definition, 8
  pericardial, secretion by serous mem-
      brane, 307
  seminal, composition and course, 472–
      473
  shift, principles concerning, 506
  tissue, interstitial fluid, 348
Folding movement, flexion, 105
Follicle(s):
  graafian:
    ovary, 478
    secretion of estrogens, 283
    target of follicle-stimulating hor-
        mone, 269
  ovarian, and ovum development, suc-
      cessive stages, 478
Follicle-stimulating hormone:
  cyclical changes in amounts, 486
  function, 269
  ovarian effects, 485
  secretion, inhibition by estrogens, 283
Fontanels, special feature of skull, 97
Food(s): (see also Digestion)
  absorption, 414–416

Food(s)—cont'd
  catabolism:
    only method of heat production, 436
    preferential order, 434
  changes brought about by mechanical
      digestion, 407
  churning and breaking up, function of
      stomach, 400
  complex compounds built from, by
      anabolism, 416
  digestion, function of small intestine,
      401
  energy-supplying, 426
  intake, mechanisms for regulating, 435
  metabolism, function of liver, 404, 428
  modification necessary to reach cells,
      389
  one-way route, mechanism controlling,
      407
  specific dynamic action and metabolic
      rate, 432
  storage:
    as glycogen, fat, and tissue proteins,
        434
    until partial digestion, function of
        stomach, 400
  tissue-building, proteins, 426
  and water, comparative importance in
      maintaining life, 429
Foot:
  arches, 91, 92
  bones, 71, 91
  dorsal venous arch, 322–324
  motion, muscles, 129, 136
  structure, 91
Foramen:
  carotid, 95
  infraorbital, 79, 96
  intervertebral, 98
  jugular, 94
  magnum, 95
  mandibular, 96
  mental, 96
  obturator, 89, 101
  optic, 95
  ovale:
    description and function, 328
    fetal structure, 327
    sphenoid bone marking, 95
  rotundum, 95
  spinal, of vertebrae, 98
  stylomastoid, 94
  supraorbital, 94
  vertebral, 82
Forearm (see Extremity, upper)
Forehead bone, description, 69
Foreign objects, aspirated, right bron-
    chus frequent site, 368
Foresight, location, 187
Fossa:
  coronoid, of humerus, 86, 99
  iliac, 101
  jugular, 94
  mandibular, 94
  olecranon, of humerus, 100
  pituitary, of sella turcica, 265
Fovea centralis, description, 241, 242,
    243
Frenulum, 394
Frenum, 394
Frontal:
  bone:
    description, 69, 72
    markings, 94

Frontal—cont'd
  lobe of cerebrum, 183
  sinuses, 72
    drainage, 363, 365
    marking of frontal bone, 94
    tuberosities, marking of frontal bone,
      94
Fuel, preferred:
  of body, carbohydrates, 417
  energy, principle, 421
Function (*see also* under respective part)
  body (*see* Body function)
  and structure, relationship, 8
Fundus:
  eye, right, 241
  gallbladder, 404
  stomach, location, 399
  uterus, 475
Funiculi of spinal cord, 180

G

Gallbladder, 404, 405
Ganglion(a):
  basal, of cerebrum, 185, 187, 189
  definition, 178
  gasserian, removal for tic douloureux,
    208
  parasympathetic, location, 224
  spinal, formation and location, 198–
    199
  spiral, of auditory relay, 257, 258
  sympathetic, description, 224
Gas(es):
  blood transport, method, 377
  exchange:
    between blood and air, 362, 371
    in lungs, tissues, and capillaries,
      378–382
    pressure gradient, cause of air move-
      ment in lungs, 372
    transportation and exchange between
      blood and tissue, circulation, 362
Gasserian ganglion, removal for tic dou-
    loureux, 208
Gastric:
  juice:
    composition, 400
    secretion, reflex and chemical con-
      trol, 413
    summary of action, 409
  mucosa, role in erythropoiesis, 297
  phase of gastric juice secretion, chem-
    ical control, 413
Gastrin, source and action, 413
Gastrocnemius muscle, 128, 129, 136, 147
Gastrointestinal tract, 390
General adaptation syndrome of Selye,
    525, 530
Genes:
  determinants of heredity, 37
  and DNA, 36, 37–38
  and other factors determining skin
    color, 57
  role in protein anabolism, 426
  zygote, 489
Genitalia, external:
  female, 479
  male, organs constituting, 472
Germ layers, primary, 490
Giant and dwarf, pituitary, 267
Gigantism, hormone imbalance in, 265
Gingivae and periodontal membrane, in-
    flammation, 397

Ginglymus, summary, 106
Girdle:
  pelvic, description, 89, 101
  shoulder:
    bones, 70–71
    description, 89
Glabella, marking of frontal bone, 94
Gland(s):
  Bartholin's, 480
  of Brunner, and goblet cells, secretion
    of mucus, 401
  bulbourethral, 470–472
  ceruminous, of skin, 58
  cervical, superficial, 352
  classification, 58
  Cowper's, 470–472
  digestive:
    secretion, control, 412–414
    tract, parasympathetic dominance,
      230
  ductless, definition, 263
  endocrine, 58
    definition, 263
    and hormones involved in glucose
      homeostatic mechanism, 418
    importance, 265
    location in female, 264
    main, 263, 264
  exocrine, 58, 59
    definition, 263
  intestinal mucosa showing, section,
    392
  lacrimal, location, 249, 250
  largest, of body, liver, 402
  of Lieberkühn, secretion of digestive
    enzymes, 401
  lymph, 351–354
  parasympathetic effects, 231
  parathyroid, 276
  parotid, location and duct opening, 395
  pineal, 284–285
  pituitary, 265–273
    anterior, 265–272
      secretion, regulation during stress,
        271
    location, 184, 185
    posterior, 265, 272–273
    stalk, 184, 189
  prostate, 470–471
  reproductive, accessory, male, 470–472
  salivary, 394–395
  Skene's, lesser vestibular, 480
  skin, 58
  stomach, embedded in mucosa, 400
  sublingual, location and duct opening,
    395
  submaxillary, location and duct open-
    ing, 395
  sweat, substances excreted, 447
  sympathetic effects, 231
  thyroid, 273–275
    and parathyroid, location, 274
  tubuloacinar, compound, pancreas
    classified as, 405
  vestibular, greater and lesser, 480
Glandular secretion, form of reflex, 221
Glans penis, 472
Glaucoma, mechanism, 246, 247
Glenoid cavity of scapula, marking, 84,
    85, 99
Gliding joint, 106, 107
Glisson, capsule of, 403
Globulins, blood proteins synthesized by
    liver cells, 405

Globus pallidus of cerebrum, 185
Glomerular filtration, 451–454
Glomerulocapsular pressures, 453
Glomerulus:
  Bowman's capsule, 448
  function in urine formation, 457
Glossopharyngeal nerve, summary, 207,
    210
Glottis, 366
Glucagon:
  blood glucose increased by, 419
  insulin:
    antagonist, 283
    mechanisms, blood glucose homeo-
      stasis result, 422
  secretion:
    alpha endocrine cells of pancreas,
      406
    islands of Langerhans, 282
  sugar-regulating hormone, 419
Glucocorticoid(s), 277–278
  blood glucose increased by, 419
  concentration, blood, relationship to
    disease, 530
  protein catabolic hormones, 427, 428
  secretion:
    effect of ACTH, 428
    stimulation by adrenocorticotropin,
      269
  stress response function, 527
Glucokinase, enzyme, activity increased
    by insulin, 418
Gluconeogenesis:
  from fats or proteins, 418
  principle, 424
Gluconeogenic effect of excess gluco-
    corticoids, 278
Glucose:
  absorption, postulated mechanism, 414,
    415
  blood, concentration, factors regulat-
    ing, 419–420, 422
  catabolism, 29–34
    stimulation by thyrotropin, 419
  concentration:
    portal and hepatic veins, 416
    and rate of utilization, influence on
      appetite control, 435
  homeostatic mechanism, glands and
    hormones involved, 404, 418
  metabolism, hormonal and neural con-
    trol, 418
  phosphorylation, reaction with ATP,
    417
  preferred energy fuel, 421
  storage as fat, principle, 424
  transport:
    mechanism, 417
    through cell membranes, accelera-
      tion by insulin, 418
"Glucostat theory" of appetite control,
    435
Gluteal:
  group, muscles of thigh motion, 134
  lines, markings of os coxae, 101
  tubercle, marking of femur, 102
Gluteus muscles, 128, 129, 134, 147, 150
Glycogen:
  fats, and tissue proteins, order of catab-
    olism, 434
  formation from glucose, 417
Glycogenesis:
  carbohydrate metabolism process by
    liver, 404

Glycogenesis—cont'd
  formation of glycogen from glucose, 417
  principle, 421
  promotion by insulin, 283
Glycogenolysis:
  carbohydrate metabolism process by liver, 404
  gluconeogenesis, and glycogenesis, mechanism, 421
  liver:
    acceleration by glucagon, 283
    and muscle, accelerated by epinephrine, 419
  meanings in different cells, 417
  principle, 421
Glycolysis, 29–31
  chemical changes, 30
Glycosuria, renal, 455
Goblet cells, secretion of mucus, 401
Goiter, exophthalmic, symptoms, 274
Golgi apparatus:
  description, 15–16, 17
  mucus-secreting goblet cell, 17
  neuron, 166
Gonadotropic:
  cycles, 484
  substances, excretion in urine during pregnancy, 285
Gonadotropin:
  chorionic, secretion by placenta, 285
  secretion during childhood, 269
Gonads, female (ovaries), 478
Gonococci, infection of Bartholin's and Skene's glands, 480
Graafian follicles:
  ovary, 478
  secretion of estrogens, 283
  target of follicle-stimulating hormone, 269
Gracilis:
  muscle, 128, 147, 149
  tract of spinal cord, 180–182
Graded strength principle of skeletal muscle contraction, 124
Graft-rejecting cells, role in rejection mechanism, 299
Granular leukocytes, 294
Granules, chromatin, 19, 38
Graves' disease, 274, 275
Gravity, effect on posture, 157
Gray matter:
  cerebral cortex, 183
  distribution in spinal cord, 180
  neuron cell bodies, 178
Groin, site of inguinal lymph nodes, 352
Groove:
  bicipital, of humerus, marking, 99
  intertubercular, of humerus, 86, 99
  lacrimal, marking of maxilla, 96
  radial, of humerus, marking, 99
Growth:
  cause of positive nitrogen balance, 428
  cells and body, function of protein anabolism, 426
  effect of estrogens, 283
  hormone, 266–268
    blood glucose increased by, 419
    excess, acromegaly resulting from, 267, 268
    shift from carbohydrate to fat utilization caused by, 419
  normal, role of calcium, 429
  regulation by thyroid hormone, 273

Growth—cont'd
  regulators, hormones as, 265
  years, oversecretion of growth hormone during, result, 268
Gums, pyorrhea, 397
Gustatory:
  area, primary, of cortex, 187
  sense organs, 260
Gyrus of cerebral cortex, 184

H

H zone of striated muscle, 114–115
Hair:
  accessory skin organ, 56–57
  cells:
    hearing and equilibrium sense organs, 257, 259
    olfactory sense organ, 260
Hammer, anvil, and stirrup, auditory ossicles, 255, 256
Hamstring muscles, 128, 147, 151
  flexors of leg, 129
  lower leg motion, 135
  thigh, 151
Hand:
  and arm, innervation, 200
  bones, 71, 88
  joints, description, 109
  motion, muscles, 129, 133
Hard palate, bones composing, 393
Haversian:
  canal, 63
  system, 64, 65
Head:
  age changes in skeleton, 103
  and face, arteries, main, 320
  femur, marking, 90, 101
  humerus, marking, 86, 99
  mandible, marking, 96
  motion, muscles, 129, 137
  muscles, 156
  and neck, veins, 322, 325
  radius, 85
    marking, 87, 100
  rib, marking, 99
  ulna, marking, 87, 100
Healing of wounds:
  function of protein anabolism, 426
  slower, induced by glucocorticoids, 278
Health, maximal, and good posture, reciprocally related, 158
Hearing, mechanism, 259–260
Heart, 306–314 (see also Cardiac)
  atria, 308, 309
  blood supply, 311–312
  boundaries, clinical importance, 307
  chambers, 308, 309
  collateral circulation, 312
  conduction system, 312–313
  covering, pericardium, 307
  disease, congestive, decreased vital capacity, 375
  and intestinal contractions, role of calcium, 429
  nerve supply, 313
  pacemaker, mechanism, 313
  physiology, 314
  rate:
    factors regulating, 334–336
    modifiers, 313
    regulation, 230
    and stroke volume, determinants of cardiac minute output, 332

Heart—cont'd
  sounds during cycle, 314
  Starling's law, 124, 333, 334
  valves, 308–311
  ventricles, 308, 309
  wall, component layers, 307, 308
Heartbeat:
  mechanism, 314
  strength, factors affecting, 333
Heat:
  body, production, muscle function, 114
  energy, measurement, 430
  production and loss, processes causing, 436–437
  receptors, skin, stimulation, cardiac accelerator, 335
  transfer through mucous membranes and skin, 436
Heels, high, effect on weight bearing, 93
Helix, double, DNA molecule, 36
Helmholtz' theory of lens accommodation, 252
Hematocrit, volume percentage of red cells in whole blood, 298
Hemispheres of cerebrum, 183
Hemoglobin:
  deficiency anemia, cellular hypoxia produced by, 377
  levels, normal, 295
  main protein to combine with carbon dioxide, 379
  molecules in each red cell, 295
  and oxygen, 377
  pairs in body fluid, 514
  salvage by macrophages of spleen, 355
Hemolysis, 27
Hemolytic anemia, 355
Hemopoiesis:
  lymph nodes, function, 354
  red marrow, function, 67
  spleen, function, 355
Hemopoietic connective tissue, 45
Hemorrhage, decrease in blood viscosity caused by, 335
Hemorrhoids, enlargements of veins in anal canal, 402
Hemosiderin, breakdown product of hemoglobin, 296
Henle, loop of:
  function, 453
    urine formation, 457
  limbs, 449, 450
Heparin, antithrombin, 305
Hepatic: (see also Liver)
  cells, circulation, 403
  or common bile duct, obstruction, results, 405
  and cystic ducts, function, 405
  lobules, anatomical units of liver, 403
  vein of abdomen, 322
Heredity:
  determined by genes, 37
  and skin color, 56, 57
Hering-Breuer inspiratory reflexes, 383
Hernia of abdominal wall, 127
Heterophoria, description, 253
High-protein diets, rationale, 433
Hilum:
  concave notch in mesial surface of kidney, 448
  lung, slit through which root enters, 371
Hinge joint, 106, 107

Hip:
  bone(s):
    description, 71
    right, 89
  joint, description, 109
Histogenesis and organogenesis, 490
Homeostasis, 6–8
  adaptive responses and, 6, 7, 8
  arterial blood pressure, maintenance, 335, 337
  autonomic functioning and, principle, 228
  blood:
    glucose, importance of liver, 418
    oxygen and carbon dioxide content, mechanism to restore, 339
    sodium, primary mineralocorticoid function, 280
    sugar, maintenance, function of liver, 404
  body temperature, 435–439
    role of skeletal muscle contraction, 114
  Cannon's theory, 524
  hormones, role in maintenance, 265
  and movement, 62
  receptors, role, 239
  total body water, 499
  urine excretion and elimination, vital mechanisms, 446
  vital part of circulatory system in maintenance, 293
Homeostatic:
  mechanism:
    blood glucose concentration, 420, 422
    control of tropic hormone secretion, 270
    erythrocyte, 296
    restore normal blood flow, local vasodilatation, 342
  and stress responses, resemblance, 526
Horizontal plate:
  ethmoid bone marking, 81, 95
  palatine marking, 96
Hormonal control:
  body activities, 414
  and neural control of glucose metabolism, 418
Hormone(s):
  adrenal medulla, 281
  anabolic, insulin, 283
  anterior pituitary, 266
    and carbohydrate metabolism, 419
  antidiuretic, 272, 455
    role in fluid and electrolyte balance, 507
  blood concentrations, modifiers of heart rate, 313
  built from food by anabolism, 416
  control:
    fat metabolism, 426
    protein metabolism, 428
    testes, 468
  digestive, action, 413
  endocrine gland secretions, 265
  follicle-stimulating, function, 269
  growth, 266–268, 427, 428
    shift from carbohydrate to fat utilization caused by, 419
  interstitial cell–stimulating, in male, function, 269
  lactogenic (prolactin), lactation stimulated by, 268, 482
  luteinizing, functions, 269

Hormone(s)–cont'd
  luteotropic, of anterior pituitary, 268
  melanocyte-stimulating, site of production, 269
  and nerve impulses, comparison, 265
  ovarian, role:
    breast development, 480
    lactation, 481
  posterior pituitary, synthesis by hypothalamus, 272
  replacement therapy after hypophysectomy, 270
  role in cyclic changes in female, 484
  secretion:
    anterior pituitary during stress, 271
    feedback control, 270
    during ovarian cycles, 483
    ovaries, function, 478
    regulation by hypothalamus, 190
    small intestine, function, 401
    testes, function, 468
  sex, secretion by adrenal cortex, 280
  somatotropic, deficiency and excess, results, 267, 268
  thyroid:
    anabolic or catabolic, 428
    components, 273
    hypersecretion and hyposecretion, effects, 274
    influence on basal metabolism, 431
  thyroid-stimulating:
    function, 269
    and glucose catabolism, 419
  translation of nerve impulses into, by neuroendocrine transducer, 271
  tropic, secretion, control by negative feedback mechanism, 269
Horns of spinal cord, 178, 180
Housemaid's knee, prepatellar bursitis, 157
Humerus:
  description, 71
  markings, 85, 86, 99–100
  muscle:
    insertions, 131
    origins, 132
  right, anterior and posterior views, 86
Humid atmosphere, retardation of evaporation, 436
Humor(s):
  aqueous, 242, 245, 246
  and cavities of eye, 245–247
  vitreous, 242, 245
Hyaline cartilage, 45, 64
Hydrocephalus, definition, 197
Hydrochloric acid:
  absence in stomach in pernicious anemia, 400
  gastric juice, 400, 409, 410
Hydrocortisone, glucocorticoid, 277
Hydrogen ion(s):
  concentration:
    arterial blood, influence on respirations, 382, 383
    body fluids, 512
  loss, effect of aldosterone, 280
  tubular secretion, 456
Hydrolysis, chemical digestion of food result of, 410, 411
Hymen, location, 479
Hyoid bone, 79
  description, 70
  with muscle attachments, 82
Hypercalcemia, 274, 276

Hypercapnia:
  or hypoxia, severe, vasomotor chemoreflex response, 338
  stimulator of chemoreceptors, 340
Hyperemia, reactive, local vasodilatation, 342
Hyperextension, definition, 105
Hyperglycemia in stress, principle, 424
Hyperglycemic:
  effect:
    excess glucocorticoids, 278
    growth hormone, 268
  mechanism of ACTH and growth hormone, 419
Hyperkalemia in acidosis, 518
Hypermetropia, explanation, 250, 251
Hypertension:
  and aldosteronism, relationship, 280
  important cause, 280
  postulated mechanism, 282
  renal, theories, 458
Hypertonic solution, 27
Hyperventilation and acidosis, relationship, 517, 520
Hypocalcemia, cause of tetany, 276
Hypoglossal nerve, summary, 207, 211
Hypoglycemia, effect of glucocorticoid deficiency, 278
Hypoglycemic effect of insulin, 268
Hypophysectomy, hormone replacement therapy after, 270
Hypophysis cerebri, 265–273
  location, 184, 185, 189
Hypothalamic–anterior pituitary mechanism, operation, 271
Hypothalamus:
  attachment of pituitary body, 265
  control:
    body, implications, 271
    vasoconstriction, role, 340
  epinephrine secretion, role, 282
  functions, 189–190
  site:
    appetite-regulating centers, 435
    heat-dissipating stimulus, 437
  structure and location, 188
  synthesis of posterior pituitary hormones, 272
Hypotonic solutions, 27
Hypoventilation and alkalosis, relationship, 517
Hypoxia:
  or hypercapnia, severe, vasomotor chemoreflex response, 338
  stimulator of chemoreceptors, 340
Hypoxic neurons of respiratory centers, fewer impulses to respiratory muscles from, 383

I

I bands of striated muscle, 114–115
Idiocy, hormone imbalance in, 265
Ileocecal:
  region and vermiform appendix, 398, 401
  sphincter, papillary form, 401
  valve at opening of ileum into large intestine, 402
Ileum, lower part of small intestine, 401
Iliac:
  arteries, 318–319
  crests, 89, 100
  fossa, 101

Iliac—cont'd
spines, 89, 100
veins, external, of lower extremities, 322–323
Iliacus muscle, 134, 152
Iliopectineal line, marking of os coxae, 101
Iliopsoas muscle, 129, 134, 140, 152
Iliotibial tract, insertion of gluteus muscles, 134
Ilium:
description, 89
marking of os coxae, 89, 100
origin of muscles, 131, 134, 135
Illness, wasting, and nitrogen balance, 427, 428
Image, retinal, formation, 249–253
Immobility, increased tendency to blood clotting caused by, 305
Immune reaction, attempt to prevent transplant rejection, 299
Immunity:
mechanism, 299
role of thymus, 284
role of lymphocytes in development, 299
Imperforate hymen, 479
Impulse(s):
conduction through heart, course, 313
nerve, 171–178
conduction, 172–178
function of neurons, 169–170
to visual cortex, 254
facilitatory and inhibitory, 220
and hormones, comparison, 265
from various receptors, route to cardiac centers, 336
Incontinence, cause, 460
Incus:
description, 70
location in ear, 255, 256
Indirect calorimetry, method of determining basal metabolic rate, 432
Infancy to adulthood, skeletal changes from, 103
Infarction, myocardial, cause and result, 312
Infection(s):
middle ear:
route, 256
and transverse sinuses, correlation, 322
result of solids or liquids in respiratory tract, 408
tendency to spread, induced by glucocorticoids, 278
thymus, role in defense mechanisms against, 284
Infectious diseases, hypertrophy of spleen during, 354
Inflammation:
abdominal, function of greater omentum, 392
decrease induced by glucocorticoids, 278
edema formation, mechanism, 504
Infraorbital foramen, marking of maxilla, 79, 96
Infundibulum:
fallopian tube, 477
gallbladder, 405
Inguinal:
canals, herniation through, 156
lymph nodes in groin, 352

Inhibitory:
area, bulbar, location, 220
impulses to lower motoneuron, 220
nerves, vagus fibers to heart, 314
postsynaptic potential, 176
tracts, impulse conduction, 219
Injections, intravenous, 322, 323, 324
Injuries, cranial nerve, symptoms, 211
Injurious particles, removal by lymph nodes, 353
Inlet, pelvic, 101
Inner ear, 256–259
organs of balance, 255, 257
Innervation, autonomic, single and dual, principles, 228
Innominate:
artery, 318–319
bones (*see* Os coxae)
veins of head and neck, 322–323
Insemination, 488
Insertion bone, moved by muscle contraction, 126
Inspiration, 372, 373, 374
Inspiratory reserve volume, definition, 374
Insula of cerebrum, 183
Insulin:
anabolic and catabolic hormone, 283
deficiency in diabetes mellitus, 419
effects, 418
function, 282
with glucagon, maintenance of blood glucose homeostasis, 420
hypoglycemic effect, 268
secretion, 282, 406
Integration of body, 163–289
Intercellular fluid, definition, 8
**Intercondylar eminence, marking of tibia, 90, 102
Intercondyloid notch, marking of femur, 102
Intermedin, melanocyte-stimulating hormone, 269
Internal:
auditory meatus, marking of temporal bone, 94
capsule of cerebrum, 184
jugular and subclavian veins, relation of thoracic duct, 350
oblique muscle:
abdominal wall motion, 138
aponeurosis, 154
Interneurons, description, 167, 168
Intersegmental reflex arcs, 177
Interstitial:
cell-stimulating hormone in male, function, 269
fluid:
and blood:
filtration pressure between, calculation, 505
fluid exchange between, 351
definition, 8, 348
major body fluid, 293
Intertubercular groove of humerus marking, 86, 99
Intervertebral foramina of vertebral column, markings, 98
Intestinal:
contractions, role of calcium, 429
juice:
secretion, control, 414
summary of action, 411

Intestinal—cont'd
mucosa, section, villi, central lacteal, and glands, 392
phase of gastric juice secretion, chemical control, 413
Intestine(s):
large, 398, 402
digestive process in, 408
and stomach in relation to diaphragm and bony pelvis, 398
modifications of mucous, muscle, and fibroserous coats, 391
small:
digestive process performed, 408
functions, 401
section, circular folds and layers of wall, 390
substances excreted, 447
Intracellular fluid, 293, 501
Intravenous injections, 322, 323, 324
Intrinsic factor:
absence in stomach in pernicious anemia, 400
necessary for erythropoiesis, 297
secretion, function of stomach, 400
Intubation, emergency measure to open trachea, 368
Inversion, definition, 110
Invertor muscle, adductor of foot, 129
Involuntary structures innervated by autonomic nervous system, 223
Iodine:
protein-bound, 273
in thyroid hormones, 273
Ipsilateral reflex:
arcs, 173, 175
knee jerk, 222
Iris:
description, 242, 243
muscle of eye, function, 247
Iron atoms in hemoglobin molecule, 295
Irregular bones, 65, 67
Ischemic reflex, medullary:
factor in control of arteriole diameter, 336
mechanism, 339, 340
Ischial:
spine, marking of os coxae, 89, 101
tuberosity, marking of os coxae, 89, 101
Ischium:
description, 89
marking of os coxae, 89, 100
muscle origins, 135
Island(s):
of Langerhans, 282–283
endocrine glandular cells, 406
site of insulin and glucagon secretion, 419
of Reil of cerebrum, 183
Isometric contraction of skeletal muscle, 122
Isosmotic solutions, 26
Isotonic:
contraction of skeletal muscle, 121
solutions, 26
Isthmus of thyroid gland, location, 273

J

Jaundice, result of obstruction of hepatic or common bile duct, 405
Jaw bones, 69
Jejunum, 401

Jelly roll hypothesis of myelin formation, 168
Joint(s), 104–110 (*see also* Bones)
  bone projections fitting into, 93
  bones and, relation of muscle tissue, 113
  classification, 104–109
  diarthrotic, movements, 105–110
  function in movement, 125
  individual, 108–109
  movable, 104
Jugular:
  foramen, marking of temporal bone, 94
  fossa, marking of temporal bone, 94
  vein, internal, 322–323, 325
    ligation for transverse sinus thrombus, 322
    and subclavian vein, relation of thoracic duct, 350
Juice(s):
  digestive:
    action, summary, 409
    flow, control, 412
  gastric, 409, 413
  intestinal, 409, 414
  pancreatic, 409, 414
Junctions:
  neuroeffector, conduction across, 178
  neuromuscular, description, 120

K

Keith-Flack node of heart, 313
Ketogenesis:
  and citric acid cycle, 425
  mechanism, 425
Ketone bodies:
  from fat catabolism, buffering, 515
  oxidation via citric acid cycle, 425
17-Ketosteroids, urinary, derivation, 284
Kidneys, 446–458 (*see also* Renal)
  coronal section, 448
  function, 451
  influence on blood pressure, 458
  medulla, 448
Kinesthesia, neural pathway, 214
Knee:
  jerk, somatic reflex, 221
  joint, description, 109
Kneecap, description, 71
Knobs, synaptic, description and function, 174, 176–177
Krebs' cycle (*see* Citric acid cycle)
Kyphosis of vertebral column, 84, 98

L

Labia:
  majora, 480
  minora, 480
Labor:
  function of uterus, 477
  initiation, role of oxytocin, 272
Labyrinth(s):
  components, 256–259
  membranous and bony, relationship, 257
Lacrimal:
  apparatus, description, 249, 250
  bones, 69
    description, 78
  groove, marking of maxilla, 96
Lactating breast, milk ejection from, oxytocin function, 272

Lactation:
  estrogens, role, 283
  and milk ejection, mechanism of control, 481–482
Lacteal, central, section of intestinal mucosa showing, 392
Lactogenic hormone (prolactin):
  anterior pituitary gland, 268, 482
  lactation stimulated by, 482
Lactose, enzyme of intestinal juice, 409
Lacunae of bone, 64
Lambdoidal suture(s), 72
  special feature of skull, 97
Lamella, layer of bone matrix, 63
Laminae of vertebrae, marking, 97
Langerhans, islands of, 282–283, 419
  endocrine glandular cells, 406
Language, use and understanding, site, 188
Large intestine (*see* Intestine, large)
Laryngeal cartilages, 366
Laryngopharynx, 365
Larynx, 366–367
  or pharynx, stimulation, temporary apnea caused by, 384
  posterior view, showing muscles, 366
  relation of thyroid and parathyroid glands to, 274
Latissimus dorsi muscle, 128, 129, 131, 142
Law:
  Boyle's, operation in respiration, 372
  Dalton's, of partial pressures, 376
  Poiseuille's, application to flow of gases, 374
Layers of skin, 54, 56
Leg (*see* Extremity, lower)
Lemniscus, medial, description, 214
Lens, accommodation, mechanism, 252
Lenticular nucleus of cerebrum, 185, 187, 189
Leptomeninges, definition, 196
Leukemia, white blood cells, 300
Leukocyte(s), 294, 298–300
  classification, 294, 301
  count:
    clinical significance, 300
    differential, 301
  leukemia, 300
  phagocytic, 298
  stages of development, 296, 300
Leukocytosis, 300
Leukopenia, 300
Levator:
  ani muscle, 128, 157
  coccygeus, 128, 157
  muscle group, 127
Libido, estrogen contribution, 283
Lieberkühn, glands of, secretion of digestive enzymes, 401
Ligament(s):
  suspensory, function, 242
  uterine, 476–477
Light rays, refraction, 249–252
"Lights," slaughterhouse term for lungs, reason, 374
Limbic system, location and function, 232
Line(s):
  gluteal, marking of os coxae, 101
  iliopectineal, marking of os coxae, 101
  nuchal, occipital bone markings, 95
  popliteal, marking of tibia, 102
  Z, of striated muscle, 114–115

Linea:
  alba, muscle insertion, 138
  aspera, marking of femur, 102
Lingual tonsils, location, 365
Lipase, enzyme of pancreatic juice, 409
Lipogenesis, 283, 425
Liver, 402–405 (*see also* Hepatic)
  cell(s):
    deamination, site, 427
    glycogenolysis, 418, 419
    ketogenesis, site, 425
  food deposition, 416
  gluconeogenesis, 418, 427
  importance:
    blood glucose homeostasis, 418
    food metabolism, 428
  mouse, with mitochondria, electron photomicrograph, 424
  and pancreas in normal positions, 403
  protein synthesis, 304, 427
  venous return from, obstruction, 326
Lobes:
  cerebrum, 183
  liver, 402
  lung, 371
  prefrontal, site of foresight and personality, 187
Lobules, hepatic, anatomical units of liver, 403
Locomotion, form of body movement, 114
Long bones, 65–67
Longitudinal:
  fissure of cerebrum, 183
  sinus, 322–323
Loop of Henle:
  function, 453, 457
  limbs, 449, 450
Loose, ordinary connective tissue, 48, 50
Lordosis of vertebral column, 84, 98
Lower extremity (*see* Extremity, lower)
Lumbar:
  plexus, formation and branches, 201
  puncture:
    definition, 198
    optimum site, 196
  vertebrae, 79
    enumeration, 70
    markings, 83, 98
Lumbodorsal fascia, muscle origins, 138
Lung(s):
  alveoli and atmosphere, gas pressure gradient between, 374
  breathing, external respiration, 372
  collapse caused by pneumothorax, 374
  dog, metal cast of air spaces, 370
  gas exchange, 380–381
  pharynx, trachea, and, 369
  structure and function, 371
  substances excreted, 447
  and trachea, projection in relation to rib cage and clavicles, 367
Luteal phase of menstrual cycle, 484
Luteinizing hormone, 269, 485
Luteotropic hormone of anterior pituitary gland, 268
Lymph:
  circulation, 351
  drainage, 348, 350
  glands, 351–354
  nodes, 351–354
    front of trunk, 349
    skin section, eliminating bacteria, 353
  pressure gradient, mechanisms, 351

Lymphatic(s):
  capillaries, 348
  connective tissue, 45
  function, 351
  superficial, of arm and trunk, 349
  system, 348–355
  tissue:
    derivation of nongranular leuko-
      cytes, 299
    involution, induced by glucocorti-
      coids, 278
    spleen, 354
    and veins, similarity and differences,
      351
  vessels, 348–351
Lymphocyte(s), 294
  formation, 354, 355
  rejection of transplanted organ, role,
    348
  tissue-attacking, role in graft rejection,
    299
Lymphocytopenia induced by glucocorti-
    coids, 278
Lysosomes, description and function, 18

M

Macrophages of spleen, function, 355
Macula:
  lutea, location, 241, 243
  in utricle, 257, 259
Magnesium sulfate, nonabsorbability, 416
Malar bones, description, 69, 77
Male:
  and female skeletons, differences, 103
  fertility, 473
  neck, prominence of Adam's apple, 366
  reproductive organs, 467–474
    and urinary bladder, 471
  urethra, structure and functions, 460
Malleolus:
  lateral, marking of fibula, 90, 102
  medial, marking of tibia, 90, 102
Malleus, 70, 255, 256
Malpighian corpuscle, 448
Maltase, enzyme of intestinal juice, 409
Mamillary bodies, location, 184, 185, 189
Mammalian ovary, 474, 478
Mandible:
  description, 69, 77
  markings, 96
Mandibular:
  foramen, marking, 96
  fossa, marking of temporal bone, 94
Manubrium of sternum, 84
  marking, 99
Margin, supraorbital, marking of frontal
    bone, 94
Markings, bone, 93–102
Marrow:
  bone:
    damage, causes, 298
    red:
      erythropoiesis, site, 295
      platelet formation, site, 300
    tissue, 45
    yellow, 300
  cavity of long bones, 67
Mastication:
  definition, 408
  muscles, 141
Mastoid:
  air cells, 73, 94

Mastoid – cont'd
  cells of temporal bone, middle ear con-
    nection, 255, 256
Matrix, bone:
  calcification, 63
  formation, 67
Maxilla(ae):
  keystone of face bones, 76
  markings, 96
Maxillary:
  bones, description, 69
  sinus, 76, 96
    drainage, 363, 365
Meatus(i):
  acoustic, external, 254, 255
  auditory, marking of temporal bone, 94
  nasal, 78, 363, 365
  urinary, location, 480
Mechanical digestion, 407–410
Mediastinum, division of thoracic cavity,
    371
Medulla:
  adrenal, 281–282
    epinephrine secretion by during
      stress, 419
  kidney, 448
  oblongata, 193–194
Medullary:
  cavity of long bones, 67
  ischemic reflex:
    influence on small blood vessels, 340
    mechanism, 339
Megakaryocytes, forerunners of plate-
    lets, 300
Meiosis of sex cells, 488
Meissner's corpuscle, nerve ending in
    skin, 240
Melanin, determinant of skin color, 56, 57
Melanocyte-stimulating hormone, 269
Melatonin, hormone of pineal gland, 285
Membrane(s), 53–58
  alveolar and capillary:
    permeability to gases, 377
    surface area, influence on oxygen
      diffusion, 380
  arachnoid, of meninges, 195
  basilar, floor of cochlear duct, 259
  cell (*see* Cell membrane)
  cutaneous (*see* Skin)
  inner limiting of retina, 244
  mucous, 53
    ciliated, lining of nose and respira-
      tory tract, 363
    heat transfer through, 436
  periodontal, pyorrhea and, 397
  Reissner's, roof of cochlear duct, 259
  selectively permeable, 23
  serous and synovial, 53–54
  tectorial, or organ of Corti, 259
  tympanic, 255, 256
Membranous labyrinth, components, 255,
    256–259
Memory, complex neural process, 187
Menarche, 486
Meninges of brain, 195
  location, 209
Meningitis, 196
Menopause, 486
  estrogen deficiency cause, 271
Menstrual cycle, 483–486
Menstruation, function of uterus, 477
Mental:
  foramen, marking of mandible, 96
  functions of cerebral cortex, 186–188

Mesentery, 392–393
Mesoderm, 490
Metabolic:
  defects, resulting from insulin defi-
    ciency, 419
  rate(s), 430–435
    regulation by thyroid hormone, 273
Metabolism, 416–435
  body, maintenance, 291–463
  carbohydrate, 417–424
    control:
      by pancreatic secretions, 406
      under stress, 423
    corticoid control, 277–278
    influence of growth hormone, 268
    processes carried on by liver, 404
  cell, 29–34
  cellular, role of circulatory system, 293
  electrolyte and fluid, effect of estro-
    gens, 283
  fat, 424–426
    effect of glucocorticoids, 278
    hormone control, 426
  food:
    function of liver, 404
    importance of liver, 428
  hormones, regulators, 265
  mineral salts, 278, 429
  protein, 426
    effect of estrogens, 283
    hormone control, 428
    promotion by glucocorticoids, 277
  vitamins, 428
Metacarpal bones, 71, 88
Metatarsal:
  arch, marking of tarsals, 92, 102
  bones, 71
Microglia, 165, 166
Microorganisms:
  defense against:
    circulatory system, role, 293
    spleen, role, 355
  immunity against, function of thymus,
    284
  removal by lymph nodes, 253
Microvilli of cell membrane, 14
Midbrain, 194–195
Middle ear cavity, description, 255, 256
Milieu interne, Claude Bernard's term, 8
Milk:
  ejection:
    from lactating breast, function of
      oxytocin, 272
    mechanism for control, 482
  secretion:
    function of lactogenic hormone, 268
    promotion by progesterone, 284
Mind and body relations, circular theory,
    272
Mineral(s):
  salts, importance, 428
  as control mechanisms, examples, 429
Mineralocorticoids, 27, 280
Minimal:
  air, 374
  and basic metabolic rates, difference,
    430
Mitochondria(on), 16–18
  energy generated in, 16, 18, 34
  mouse liver, electron photomicrograph,
    424
  neuron, 166
  structure, 17
  and synaptic knob, 174

Mitosis, 34–39
Mitral:
　stenosis, description, 311
　valve, location and function, 308, 309
Mittelschmerz, cause, 483
Modiolus, description, 257
Molar tooth, section, 397
Molecular structure of skeletal muscle
　　cells, 115–116
Molecules, large, of hormones, 265
Monocyte(s), 294
　formation:
　　lymph nodes, 354
　　spleen, 355
Monosynaptic reflex arc, 172, 173
Mons veneris, 480
Morphine, effects, 380
Morula, 489
Motility of leukocytes, aid in defense
　　mechanism, 298
Motion, Newton's laws of, application to
　　circulation, 330
Motoneurons:
　description, 167, 168
　lower, 220, 221
　pyramidal pathway, 216
　somatic:
　　definition, 203
　　impulse-transmitting nerve cell, 120
　upper, injury, symptoms, 221
Motor:
　aphasia, site, 188
　area, primary, of cortex, 187, 191
　end plate, description, 120
　fibers of cranial nerves, 206–207
　functions:
　　cerebral cortex, 186
　　spinal cord, 182
　nerve essential to skeletal muscle,
　　121
　neural pathways to skeletal muscles,
　　215–221
　pathways from cerebral cortex, classi-
　　fication, 216
　tracts:
　　body, main, 217
　　cerebral, 184
　　descending, in spinal cord, 180–182
　unit, 120
Mouse liver, electron photomicrograph,
　　424
Mouth, 393–394
　accessory organs, 394
　digestive process performed, 408
Movable joints, 104
Movement(s):
　automatic, production, 219
　diarthrotic joints, 105–110
　function:
　　bones, 125
　　skeletal system, 63
　and homeostasis, 62
　importance in survival, 113
　neuromusculoskeletal unit, 125
　reflex, complex, role of thalamus, 188
　relation of motor units to, 120
　substances through cell membranes,
　　20–29
　voluntary, complex mechanism, 187
Mucosa:
　gastric, role in erythropoiesis, 297
　intestinal, section, 392
Mucous:
　lining of alimentary canal, 390

Mucous – cont'd
　membrane, 53
　　ciliated, lining nose and respiratory
　　　tract, 363
　　heat transfer through, 436
Multipolar neurons, location, 169
Mumps, 395
Murmur, heart, abnormal heart sound,
　　314
Muscle(s): (*see also* Skeletal muscles)
　abdominal wall motion, 138
　actions:
　　hints on deducing, 126
　　synergic control function of cere-
　　　bellum, 192
　anal sphincter, 402
　arm motion, lower, 132
　attachments:
　　bone processes, 93
　　hyoid bone, 82
　cardiac, 46, 47, 49
　　contractions, autonomic reflexes, 221
　　effectors, 231
　　layer of heart wall, 307, 308
　　rabbit, electron photomicrograph, 49
　cell(s):
　　chemical changes during exercise,
　　　118–119
　　glycogenolysis, 417
　　skeletal, 114–119
　　terminology, 114
　chest wall motion, 139
　ciliary, location, 242
　contraction(s):
　　acetylcholine, role, 121
　　all-or-none law, 118
　　ATP breakdown, role, 116, 118
　　classification, 121–123
　　creatine phosphate, role, 116, 118
　　energy sources, 118–119
　　factor in lymph pressure gradient,
　　　351
　　graded strength principle, 124
　　mechanism, 116
　　myosin and actin, role, 117
　　nerve impulses, role, 117
　　and relaxation, calcium, role, 117
　　somatic reflex, 221
　　strength, factors, 124, 125
　　temperature homeostasis, role, 114
　　venous blood return to heart influ-
　　　enced by, 342, 344
　　voluntary, heat-gaining process, 437
　　walking, movement of venous blood,
　　　345
　　warm-up principle, 122, 124
　deep, abdominal wall, 153
　energy, ATP breakdown, chemical re-
　　action, 118
　eye, 247–248
　facial expression and mastication, 141
　features, names descriptive of, 127
　fibers:
　　or cells, 114
　　relaxation, mechanism, 117
　fibrosis, age change, 121
　flaccid, 121
　foot motion, 136
　forearm, right, anterior, 146
　function:
　　importance to normal life, 113
　　methods of study, 121
　hand motion, 133

Muscle(s) – cont'd
　head, 156
　　motion, 137
　layers and interior of stomach, 399
　leg motion, lower, 135
　and liver glycogenolysis, acceleration
　　by epinephrine, 419
　location concerning part moved, 125
　motor neural pathways, 215–221
　names, reasons, 126
　and nerves, role of potassium in func-
　　tion, 429
　pelvic floor, 129
　pull against gravity, 158
　respiratory, changes in size of thorax
　　produced by, 373
　shoulder motion, 130
　skeletal:
　　age changes, 121
　　attachment to bones, 124
　　function, 121–126
　　group action, 125
　　grouped:
　　　by function, 127, 129
　　　by location, 128
　　organs, 119–127
　　origins, insertions, and actions, hints
　　　on deducing, 126
　　structure, 119–121
　　voluntary, 46, 47
　smooth:
　　contractions, autonomic reflexes, 221
　　digestive tract, parasympathetic
　　　dominance, 230
　　effectors, sympathetic and parasym-
　　　pathetic effects, 231
　　tissue, 44, 45
　spastic, definition, 121
　sphincter, 127, 400, 401, 402
　striated, 114–115
　thigh, 128, 147–151
　　motion, 134
　tissue, 44, 46–47
　　relation to bones and joints, 113
　tone, 219, 220
　trunk, 140, 142
　upper extremity, 143
　　motion, 131
　visceral, 46, 47
　walls of esophagus, 398
Muscular:
　coat of alimentary canal, 390
　effort, excessive, necessary in emphy-
　　sema, 375
　system, 113–161
　work, determinant of body heat produc-
　　tion, 436
Myelin:
　formation, jelly roll hypothesis, 168
　nerve fiber, and rate of impulse conduc-
　　tion, 174
　sheath description, 168, 169
Myelinated nerve fibers, white matter,
　　178
Myeloid:
　hemopoietic connective tissue, 45
　and lymphatic tissues, hemopoietic,
　　299
　tissue derivation of granular leuko-
　　cytes, 299
Myocardial:
　fibers, length, regulator of heartbeat
　　strength, 333
　infarction, cause and result, 312

Myocardium, cardiac muscle, 307, 308, 309
Myofibrils, 114–116
Myoglobin of skeletal muscle, 115
Myometrial:
 contractions, role of estrogen and progesterone, 283, 284
 cycle, 484
Myometrium, muscular strength, 475
Myopia, explanation, 250, 251
Myosin of muscle cells, 115–117
Myotatic reflex, knee jerk, 222
Myxedema, 274, 275

N

Nails, accessory skin organ, 57
Nares, anterior and posterior, 363
Nasal:
 bones, description, 69, 77
 cartilage, description, 77
 cavity, 364
 conchae, 78
 meati, 78
 septum, bones and cartilages forming, 97, 364
Nasolacrimal ducts, description, 249, 250
Nasopharynx, 365
Navicular bones, 71, 91
Near reflex of pupil, 253
Nearsightedness, explanation, 250, 251
Neck:
 anatomical, of humerus, 99
 and face, sagittal section, 364
 femur, 90, 101
 gallbladder, 404
 and head, veins, 322, 325
 male, prominence of Adam's apple, 366
 mandible, 96
 muscles, 128, 142
 rib, marking, 99
 root, and crown, parts of tooth, 397
 superficial cervical glands, site, 352
 surgical, of humerus, 86, 99
Negative feedback mechanism:
 control of tropic hormone secretion, 270
 erythrocyte homeostasis, 297
Nephron:
 constituent parts, 451
  functions in urine formation, 457
 unit, 450
  blood vessels, 499
Nephrosis, edema formation, mechanism, 504
Nerve(s):
 abducens, summary, 206, 208
 accelerator, sympathetic nerves to heart, 314
 accessory, summary, 207, 210
 acoustic, summary, 206, 210
 auditory, summary, 206, 210
 cardiac, sympathetic fibers, component of cardiac plexuses, 314
 cochlear, composition, 210
 cranial, 204–211
  attachment to brainstem, 185
  contrasted with spinal, 204
  injury, symptoms, 211
  reflex centers, 195
 definition, 178
 depressor, vagus fibers to heart, 314
 facial, summary, 206, 210

Nerve(s)–cont'd
 fibers:
  diameter and rate of conduction, 168, 174
  myelinated, white matter, 178
 glossopharyngeal, summary, 207, 210
 hypoglossal, summary, 207, 211
 impulse, 171–178
  conduction, 172–174, 176, 178
   function of neurons, 169–170
   sound, 260
   stimulus, 170
  definition, 171
  and hormones, comparison, 265
  mechanism, 171, 173
  skeletal muscle contraction, role, 117
  translation into hormones by neuroendocrine transducer, 271
 inhibitory, vagus fibers to heart, 314
 lower extremity, 203
 mixed:
  cranial, 204
  sensory and motor, 203
 motor, essential to skeletal muscle, 121
 oculomotor, summary, 206, 208
 olfactory, summary, 204, 206
 optic, 244, 245
  summary, 206, 208
 pneumogastric, summary, 207, 210
 sciatic, formation, 202
 spinal, 198–204
  accessory, summary, 207, 210
  cranial contrasted with, 204
  formation, 209
  segmental arrangement, 202, 205
 supply:
  blood vessels, 316
  heart, 313
  skeletal muscles, 120
 tissue, 44–45
 and tracts, comparison, 184
 trifacial, summary, 206, 208
 trigeminal:
  main divisions, 209
  summary, 206, 208
 trochlear, summary, 206, 208
 trunk, cross section, 199
 vagus:
  parasympathetic fibers, component of cardiac plexuses, 314
  summary, 207, 210
 vestibular, composition, 210
Nervous system, 164–238
 autonomic, 223–232
  divisions, 224
  general principles, 228–232
  innervation of heart, 313
  microscopic structure, 224
  preganglionic and postganglionic fibers, 226
 cells, 165–171
 central, composition, 178
 and endocrine system:
  hypothalamus link between, 271
  similarity and differences, 263
 maintenance of posture, role, 158
 organs, 178–211
 parasympathetic, 226
 peripheral, composition, 178
 sympathetic, 226
 tumors, neuroglia and, 165
Net diffusion, 21–22
Neural:
 arch of vertebrae, marking, 98

Neural–cont'd
 and hormonal control of glucose metabolism, 418
 pathways:
  course, 172, 174
  motor, to skeletal muscles, 215–221
  sensory, 211–215
Neurilemma, description, 168, 169
Neuroeffector junctions, conduction across, 178
Neuroendocrine transducer:
 adrenal medulla, 282
 nerve impulses translated into hormones, 271
 pineal body, 285
Neurofibrils, description, 168
Neuroglia, 165–166
Neurohypophysis, 184, 189, 265, 272–273
Neuromuscular:
 irritability, result of calcium deficiency, 276
 junction, description, 120
Neuromusculoskeletal unit, physiological unit for movement, 125
Neuron(s), 166–171
 anterior horn, function, 215
 autonomic, locations, 277
 brain, function, 170
 cell bodies, gray matter, 178
 classification, 168
 component of motor unit, 120
 conscious proprioception, 212
 function, 169–171
 nonconducting, role in nerve impulse, 170
 pain, 212
 photoreceptor, rods and cones, 243, 244
 preganglionic and postganglionic, of autonomic system, 224–228
 sensory, 212
Neurosecretions, release into pituitary portal system, 271
Neurovesicle and synaptic knob, 174
Neutrophil, 294
 multilobular nuclei, 298
Newton's laws of motion, application to blood circulation, 330
Night vision, role of rhodopsin, 254
Nipples, stimulation, oxytocin secretion increased by, 273
Nissl bodies, description, 168
Nitrogen balance, 427
  negative, from excess of glucocorticoids, 277
Nitrogenous wastes, elimination in urine, 460
Nociceptors, pain receptors, 240
Nodes:
 conduction system of heart, 312–313
 heart, atrioventricular and sinoatrial, 312
 lymph (*see* Lymph nodes)
 of Ranvier:
  description, 168, 169
  and rate of impulse conduction, 174
Nonautonomy, principle, 230
Nonconducting neuron, role in nerve impulse, 170
Nonelectrolytes in blood plasma, 303
Nongranular leukocytes, 294
Norepinephrine:
 hormone of adrenal medulla, 281
 release by adrenergic fibers, 229
 transmitter substance, 176, 178

Nose, 363–365
Notch:
    intercondyloid, marking of femur, 102
    radial, marking of ulna, 87, 100
    sciatic, marking of os coxae, 89, 100
    semilunar, marking of ulna, 87, 100
    supraorbital, marking of frontal bone, 94
Nuchal lines, occipital bone markings, 95
Nuclear division, 34–39
Nucleoli, description and function, 19
Nucleus(i):
    cell, description, 19
    central nervous system, 178
    cerebral, 185, 187
    dentate, of cerebellum, 192
    neuron, 166
    prominent:
        hypothalamus, 189
        medulla oblongata, 193
    red, of midbrain, 194
    steroid compounds, 277
Nursing, oxytocin secretion increase by, 273
Nutrients, reabsorption and secretion, 456

O

Obstetrical importance of sacral promontory, 98
Obturator foramen, marking of os coxae, 89, 101
Occipital:
    bone, 69, 73, 95
        origin of trapezius muscle, 130
    lobe of cerebrum, 183
    protuberances, occipital bone markings, 95
Oculomotor nerve, summary, 206, 208
Odontoid process of axis, marking, 83, 98
Odors, role in taste sensation, 260
Olecranon:
    fossa of humerus, marking, 86, 100
    process, marking of ulna, 87, 100
Olfactory:
    area, primary, of cortex, 186
    nerves, summary, 204, 206
    receptors in nasal mucosa, 365
    sense organs, 260
Oligodendroglia, 165, 166
Olivary nucleus, inferior, 193
Olive of medulla oblongata, 185, 193
Omentum, greater and lesser, 392
Openings:
    bone, definitions, 93
    change in size, form of movement, 114
Optic:
    chiasma, description, 208
    disk, description, 241, 245
    foramen, sphenoid bone marking, 95
    nerves, 224, 245
        summary, 206, 208
    papilla, description, 245
    tract, summary, 208
Orbital fissure, superior, sphenoid bone marking, 95
Orbits:
    bones forming, 79, 97
    special feature of skull, 97
Ordinary connective tissue, loose, 48, 50
Organ(s):
    abdominal, veins, 322–323

Organ(s) – cont'd
    accessory, of skin, 56–58
    chest, veins of, 322
    of corti, 258, 259
    definition, 3
    digestive:
        blood from, detour through liver, 322
        system, 389–407
    nervous system, 178–211
    respiratory system, 363–372
    skeletal muscle, 119–127
    structural units of body, 2
    thoracic, veins of, 322
    transplant, mechanism of rejection, 299
Organelles:
    cytoplasm, 14–19
    neuron, 166
Organization:
    body, 2–10
    characteristic of life, 4, 16, 32
Organogenesis, 490
Orifice, vaginal, 480
Origin bone of muscle, 126
Oropharynx, location behind mouth, 365
Orthopnea, inability to breathe in horizontal position, 376
Os:
    coxae:
        description, 71
        insertion of external oblique muscle, 138
        markings, 89, 100–101
        muscle origins, 138
        pelvic girdle, 89
    external, of cervix, 476
    innominatum:
        origin of sartorius muscle, 135
        pelvic girdle, 89
    internal, of uterus, 476
Oscillograph, visible record of heart conduction, 313
Osmosis, 23–27
    in urine excretion processes, 454, 455, 456
Osmotic pressure(s), 24–27, 452, 453
    electrolyte concentrations of body fluids, 507
Ossicles, auditory, 70
Ossification of prebone structures, 67
Osteitis fibrosa generalisata, result of hypercalcemia, 276
Osteoporosis, senile, factors in development, 68
Otoliths, function, 257
Outlet, pelvic, description, 101
Oval window of middle ear, 255, 256, 257
Ovarian:
    changes and follicle-stimulating hormone, 485
    cycles, 483
    follicle and ovum development, successive stages, 478
    hormones:
        breast development, role, 480
        lactation, role, 481
    vein of abdomen, 322
Ovary(ies), 283–284
    and anterior pituitary, negative feedback mechanism between, 487
    cyclical changes, 485
    female gonads, 478
    mammalian, 474, 478

Overeating, distention of stomach, symptoms, 399
Oviducts, 474, 477
Ovoid joint, summary, 106
Ovulation, 485, 488
    function of ovaries, 478
    mechanism, 483
    promotion by follicle-stimulating hormone, 269
    time of occurrence, 484
Ovum:
    development, 478
    parts, 469
Oxidation, ketogenesis and, in fat catabolism, 425
Oxidation-reduction enzymes, function, 411
Oxygen:
    from air, exchange for carbon dioxide from blood in alveoli, 375
    association curves of human blood, 378
    blood, 382
    and carbon dioxide:
        blood content, homeostasis, mechanism to restore, 339
        pressure gradients, 377
        transport, red blood cell function, 295
    combination with hemoglobin, factors influencing, 377
    consumption:
        and age, 119
        in given time, basis of basal metabolic rate calculation, 432
    continual supply to cells, essential to life, 372
    debt in exercise, 119
    diffusion:
        from plasma into red blood cells, 377
        volume, factors determining, 380, 381
    partial pressure in atmosphere, formula, 376
    saturation of arterial blood, explanation, 379
    tension of arterial blood, influence on respirations, 382, 383
    utilization by cells, increase with activity, 382
Oxygenated blood:
    carried by all arteries but pulmonary, 315
    and deoxygenated blood, fetal blood mixture, 328
Oxyhemoglobin, compound of oxygen and hemoglobin, 377
Oxytocin:
    function, 272
    milk ejection, role, 482

P

Pacemakers, 313
Pain, 240–241
    receptors:
        sensory impulses from, influence on respirations, 384
        stimulation, cardiac accelerator, 335
    sensations, neural pathway, 214
    sensory impulse relays, 213
    and temperature neurons, 212
Painful stimuli, 384
Pairs, buffer, in body fluids, 514

Palate:
  cleft, 363
  hard, bones composing, 393
  soft, arch-shaped muscle, 393
Palatine:
  bones:
    description, 69, 78
    location and function, 363, 364
    markings, 96
  process, marking of maxilla, 96
  tonsils, location, 365
Pallidum of cerebrum, 185, 187, 189
Palmar arches, branches of radial and
    ulnar arteries, 318–319, 321
Palmaris longus, flexor of hand, 129
Palpebral:
  conjunctiva, location, 243
  fissure, variation in size, 249
Pancreas, 405–406
  and liver in normal positions, 403
Pancreatic:
  duct, 406
  juice:
    secretion, 406, 413
    summary of action, 411
Pancreozymin, source and action, 413
Papilla(ae):
  optic, description, 245
  renal, 448
  tongue, description and location, 394–
    395
  of Vater, opening of common bile duct,
    403
Paralysis:
  flaccid, "lower motoneuron sign," 221
  mechanism, 215, 219
  spastic, cause, 221
Paranasal sinuses, 363, 365
Parasympathetic:
  division of autonomic nervous system,
    224
  dominance of digestive glands and
    smooth muscle, principle, 230
  effects of visceral effectors, 231
  fibers of vagus nerve, component of
    cardiac plexuses, 314
  ganglia, location, 224
  nervous system, 226, 228
Parathormone, 274, 276
Parathyroid glands, 274, 276
Paraventricular nuclei of hypothalamus,
    189
Parietal:
  bones, 69, 72
  cells, hydrochloric acid of gastric juice
    secreted by, 400
  layer:
    pleura, lining of thoracic cavity, 371
    serous membrane, 53, 55
  lobe of cerebrum, 183
  peritoneum, 476
  pleura, chest section, 55
Parotid gland, location and duct opening,
    395
Partial pressure, 376, 377
Passive transport mechanisms, 20–27
Patella, 71, 91
Patellar reflex, somatic reflex, 221
Path, final common, principle, 215
Pathologic changes induced by intense
    stimuli, 530
Pathway(s):
  conduction, central, of autonomic ner-
    vous system, 226

Pathway(s)–cont'd
  motor:
    from cerebral cortex, classification,
      216
    neural, to skeletal muscles, 215–221
  pyramidal, from cerebral cortex, 216
  sensory neural, 211–215
  spinothalamic, 214
  sympathetic, 225
Pectineus muscle, adductor of thigh, 149
Pectoralis:
  major muscle, 128, 129, 131, 142
  minor muscle, 130
Pedicles of vertebrae, markings, 97
Peduncles:
  cerebellar, 192
  cerebral, of midbrain, 194
Pelvic:
  bones, 71
  brim, 101
  cavity, 3
  contents, female, 474
  floor, 128, 139, 157
  girdle, description, 89, 101
Pelvis:
  age changes in skeleton, 103
  bony, description, 101, 398
  renal, 448, 458
Penis, external structure, 472
Pepsin, digestive enzyme of gastric juice
    409
Peptic ulcer, parasympathetic excess,
    230
Peptidases, digestive enzymes of intes-
    tinal juice, 409
Pericardectomy, 308
Pericardial:
  fluid, secretion by serous membrane,
    307
  space, location, 307
Pericardium:
  covering of heart, 307–308
  division of pleural cavity, 371
  fibrous, 307
  serous:
    component layers, 307
    membrane, 53
Perilymph, vestibular fluid, 257, 258, 259
Perimysium of connective tissue, 119
Perineum, clinical importance, 480
Periodontal membrane, pyorrhea of gums
    and, 397
Periosteum of long bones, 67
Peripheral:
  nervous system, composition, 178
  resistance:
    determinant of arterial blood volume,
      and pressure, 332, 342
    factors determining, 335, 344
Peristalsis, 408, 410
  promotion by parasympathetic im-
    pulses, 230
Peritoneum:
  parietal, 473
  serous membrane, 53
Permanent teeth, 396
Pernicious anemia, atrophy of gastric
    mucosa, 400
Peroneus muscles, abductors of foot, 129
Perpendicular plate, ethmoid bone mark-
    ing, 81, 96
Personality traits, location, 187
Petrous portion, marking of temporal
    bone, 94

pH:
  blood:
    arterial:
      decreased, stimulator of chemore-
        ceptors, 340
      and venous, 515
    effect of ammonia excretion, 518
    urinary mechanism for maintaining,
      520
  body fluids, mechanisms of control, 512
  control:
    effectiveness of buffers, 515
    respiratory mechanism, 515–517
    urinary mechanism, 517
  homeostatic control mechanism, effec-
    tiveness, 513
  meaning of symbol, 512
  specific, optimal for enzymes, 410, 411
  urine, control mechanisms, 517
Phagocytic:
  leukocytes, 298
  reticuloendothelial cells, 298
Phagocytosis, 28, 298
  bacteria by lymph nodes, 352, 353
  microorganisms by spleen, 355
Phalanges, description, 71
Pharyngeal tonsils, adenoids, 365
Pharynx:
  description and functions, 365
  digestive process performed, 408
  or larynx, stimulation by chemicals or
    touch, temporary apnea caused
    by, 384
  trachea, and lungs, 369
Phases of mitosis, 34–35
Phenomenon, treppe (staircase), 122, 123
Phlebitis, 316
Phonation, aid in, function of nose, 365
Phosphate:
  blood, regulation by parathormone, 276
  buffer pair in body fluid, 514
Phosphorylating enzymes, function, 411
Photopupil reflex, function, 253
Photoreceptor neurons, rods and cones,
    243, 244
Photosensitive chemicals in rods and
    cones, 254
Phrenic nerves, clinical importance, 201
Physical and chemical agents, enzyme
    action inhibited, 411
Physiological:
  responses to psychological stressors,
    529
  stress, 524, 528
Physiology, cell, 19–41
Pia mater of meninges, 196
Pigmentation of skin, function of me-
    lanocyte-stimulating hormone,
    269
Pigments, elimination in urine, 461
Pineal body (pineal gland), 284–285
Pinkeye, 249
Pinna of external ear, 254
Pinocytosis, 28
Pisiform bone of wrist, 88
Pitressin, use in diabetes insipidus, 272
Pituitary:
  body, 265–273
  cachexia, deficiency of growth hor-
    mone cause, 268
  gland:
    anterior, 265–272
    hormones and target organs, 266
    location, 184, 185

Pituitary—cont'd
  gland—cont'd
    posterior, 265, 272–273
    stalk, 184, 189
    portal system, release of neurosecretions into, 271
Pivot joint, 106, 107
Placenta:
  function, 327
  estrogen secretion, 283
  lactation, role, 482
  progesterone secretion, 284
  temporary endocrine gland, 285
Planes of body, 4, 5
Plantar:
  arches, branches of anterior and posterior tibial arteries, 318
  flexion, definition, 105
  reflex, somatic reflex, 223
Plasma:
  blood (*see* Blood plasma)
  cells, formation:
    lymphatic tissue of lymph nodes, 354
    spleen, 355
  protein(s):
    importance, 303
    pair in body fluid, 514
Plate:
  horizontal:
    ethmoid bone marking, 95
    palatine marking, 96
  perpendicular, ethmoid bone marking, 81, 96
Platelet(s):
  destruction, function of spleen, 355
  factors, trigger for blood clotting, 304
  size and appearance, 294, 300
Pleura:
  parietal and visceral, chest section, 55
  serous membrane, 53, 55
  visceral, covering of lung, 371
Pleural division of thoracic cávity, 371
Pleurisy, inflammation of pleura, 371
Plexus(es):
  brachial:
    clinical significance, 200
    intermixing fibers, 202
  cardiac, course of fibers from, 314
  choroid, 196
  formed from anterior rami of spinal nerves, 200–201
  lumbar, formation and branches, 201
  sacral, formation and branches, 201
  and sinuses, venous, in skin and abdominal organs, blood reservoirs, 340
Pneumogastric nerve, summary, 207, 210
Pneumonia, mechanism in lung, 371
Pneumotaxic center:
  function, 1194
  maintenance of rhythmicity of respirations, 384
Pneumothorax, presence of air in thoracic cavity, 374
Poiseuille's law, application:
  circulation, 342
  flow of gases, 374
Polarization of cell membrane by sodium pump, 171
Polarized membrane of nonconducting neuron, 170
Polycythemia, criterion for diagnosis, 295
Polymorphonuclear leukocytes, neutrophils, 298

Polysomes, role in protein synthesis, 18
Pons varolii, 194, 195
Popliteal:
  artery, 318–319
  line, marking of tibia, 102
  vein, 322–323
Portal:
  circulation, 322, 326
    function, 416
  system, pituitary, release of neurosecretions into, 271
  vein:
    obstruction, cause of ascites, 326
    route of blood from digestive organs to liver, 322
    tributaries, 326
Position, anatomical, 6
Posterior:
  ligament, 477
  median sulcus, spinal cord groove, 179
  pituitary gland, 265, 272–273
Postganglionic:
  fibers of autonomic nervous system, 226
  neurons, location and course, 225, 227, 228
Postural reflexes, role of cerebellum, 193
Posture, 157, 158
  maintenance, muscle function, 114, 121, 157
Potassium:
  concentration, effect on fluid exchange, 507
  loss, effect of aldosterone, 280
  maintenance of acid-base and water balances, role, 429
  muscle and nerve function, importance, 429
Potential:
  action, 171–178
  difference across neuron membrane, 170
  osmotic pressure, 24–27
  resting, mechanism, 170
"Preferred energy fuel," principle, 421
Prefrontal lobes, site of foresight and personality, 187
Preganglionic:
  fibers of autonomic nervous system, 226
  neurons, location and course, 224–225, 227, 228
Pregnancy:
  cause of positive nitrogen balance, 428
  ectopic, 489
  excretion of gonadotropic substances in urine, 285
  function of uterus, 477
  influence on basal metabolic rate, 431
  Rh-negative woman with Rh-positive fetus, source of anti-Rh antibodies, 302
  test, Aschheim-Zondek, 285
Pregnant uterus, contractions, stimulation by oxytocin, 272
Premenstrual phase of cycle, 484
Prepuce:
  female, 480
  male, 472
Presbyopia, definition, 252
Pressoreflexes:
  cardiac, regulation of heart rate, 334, 335
  vasomotor, mechanism, 336, 337

Pressure(s):
  blood (*see* Blood pressure)
  glomerular filtration, effective, calculation, 452, 453
  glomerulocapsular, 453
  gradient:
    blood (*see* Blood pressure gradient)
    between Bowman's capsules and glomeruli, 451
  intraocular, normal, 245, 246
  intrathoracic, decrease during inspiration, 372
  osmotic, 25–27
    and electrolyte concentrations of body fluids, 507
  partial:
    definition, 376
    gas in liquid, determination, 377
  points, 347
  pulse, definition, 346
  sensation(s):
    neural pathway, 214
    neurons, 212
  tympanic membrane, equalization, 256
Prime movers, skeletal muscle group, 125
Process(es):
  bone, 93
    markings, 83–85, 87, 94–96, 98–100
  cell, and cell body, relation, 170
  number, classification of nerve fibers, 169
  spinous, of vertebrae, 82
Procoagulants, role in blood clotting, 304
Progesterone:
  effects, 284
  phase of menstrual cycle, 484
  secretion:
    adrenal cortex, 280
    ovaries, 478
    placenta, 285
    stimulation by luteinizing hormone, 269
Projection(s):
  bone, 93
  tracts:
    cerebrum, 184
    spinal cord, 180
Prolactin, hormone of anterior pituitary gland, 268
Proliferative phase of menstrual cycle, 484
Promontory, sacral, 98
Pronation, definition, 110
Pronator(s):
  muscle group, 127
  teres muscle, 128, 143, 145
Proprioception, conscious, neural pathway, 214
Proprioceptors, location, 239
Propulsion through tubes, form of movement, 114
Prostate gland, 470–471
Prosthetic group, nonprotein part of enzymes, 410
Protease, digestive enzyme of gastric juice, 409
Protection, function of skeletal system, 63
Protein(s):
  absorption, 415
  anabolism:
    and catabolism, 429

Protein(s)—cont'd
anabolism—cont'd
essential part performed by liver cells, 405
functions, 426
promotion by estrogens, 283
stimulation:
growth hormone, 268
testosterone, 468
synopsis of process, 426
balance, normal condition, 427
blood:
plasma, 303
synthesized by liver, 405
catabolism, 427–428
promotion by progesterone, 284
composition of enzymes, 410
compounds in muscle, 115
concentration, blood, determinant of viscosity, 335
digestion, 411, 412
and fat(s):
catabolism:
first step, function of liver, 405
shift from carbohydrate to, 417
gluconeogenesis by liver, 418
foreign, immunity against, function of thymus, 284
metabolism, 426–428
effect of estrogens, 283
hormone control, 428
promotion by glucocorticoids, 277
return from interstitial fluid to blood by lymphatics, 351
specific dynamic action, 432
steroid compounds, hormones, 265
synthesis:
polysomes, role, 18
promotion by insulin, 283
Protein-bound iodine, 273
Protein-poor diet, cause of negative nitrogen balance, 427
Prothrombin:
blood:
clotting, role, 304
protein synthesized by liver cells, 405
Protoplasm, 11–12
built from food by anabolism, 417
and complex compounds, synthesis, use of fats, 426
Protraction, definition, 110
Protuberances, occipital, bone markings, 95
Proximal convoluted tubule, renal, 449, 450
functions in urine formation, 457
Pseudostratified ciliated columnar epithelium, 43
Psoas:
major, muscle of thigh motion, 134
muscles, 152
Psyche and soma, hypothalamus as link, 190
Psychic functions, cortical activity responsible for, 187
Psychological:
stress, 528–529
stressors, physiological responses to, 529
Psychosomatic disorders, sympathetic dominance, 230

Pterygoid processes, sphenoid bone markings, 94
Ptosis of kidney, 448
Ptyalin, digestive enzyme of saliva, 409
Puberty, increased secretion of gonadotropins, stimulus, 269
Pubic:
arch, marking of os coxae, 101
bone:
insertion of internal oblique muscles, 138
marking of os coxae, 89, 100
origin of adductor muscles, 134
crest, marking of os coxae, 89, 101
rami, markings of os coxae, 89, 101
tubercle, marking of os coxae, 101
Pubis:
description, 89
marking of os coxae, 89, 100
Pulmonary circulation, 329
Pulp of tooth, nerves and blood vessels contained, 397
Pulse, 346–348
Pump, sodium, 28
and cell membrane polarization, 171
Punctae, location, 249, 250
Puncture, lumbar, 196, 197
Pupil of eye, 242, 243
constriction, mechanism, 253
Pupillary light reflex, function, 253
Purkinje fibers of heart, 312, 313
Putamen of cerebrum, 185, 187, 189
Pyloric sphincter, 400
Pylorus, 399
Pyorrhea, 397
Pyramidal:
pathway from cerebral cortex, 216
signs, causes, 221
tracts:
motor pathways, 216–219
spinal cord, 181
Pyramids:
medulla oblongata, 185, 193
renal, 448

Q

Quadrate lobe, part of right lobe of liver, 402
Quadratus lumborum muscle, 152
Quadriceps femoris muscle group, 128, 129, 135, 147–148

R

Radial:
artery, 318–319
groove of humerus, marking, 99
notch, marking of ulna, 87, 100
tuberosity, 85
marking, 87, 100
vein of upper extremity, 321
Radiation, method of heat loss, 436
Radioautography in study of cells, 12
Radius:
articulations, 85
description, 71
insertion of biceps brachii muscle, 132
and ulna, right, anterior and posterior surfaces, 87

Ramus(i):
marking of mandible, 96
pubic, markings of os coxae, 89, 101
spinal nerves, 198
Ranvier, nodes:
description, 168, 169
rate of impulse conduction, 174
Rates, metabolic, 430–435
Reabsorption:
and secretion mechanisms, 456
tubular, 453–455
Reactions, chemical, of citric acid cycle, 31, 32–33
Reactive hyperemia, local vasodilation, 342
Receptors:
impulse conduction, role, 172
specificity, principle, 240
Recipient, universal, blood, 303
Rectal columns, 402
Rectococcygeus muscle, 128, 157
Rectum, 398, 402
digestive process performed, 408
Rectus:
abdominis muscle, 129, 138, 154
femoris muscle, 129, 134, 135, 148
Recurring female cycles, 482–484
Red:
blood cell (*see* Erythrocyte)
bone marrow:
erythropoiesis, site, 295
location and importance, 67
nucleus of midbrain, 194
Reducing, weight, cardinal principle, 434
Referred pain, 241
Reflex(es), 221–223
arc(s):
course of nerve impulses, 172–175
intersegmental, 177
principle in autonomic nervous system, 223
three-neuron, 167, 172, 175
two-neuron, 172, 173
conduction over, knee jerk, 221
Bainbridge, mechanism controversial, 334
centers:
autonomic, higher, location and function, 232
central nervous system, 178
cord gray matter, 182
and chemical control of gastric secretion, 413
enterogastric, 409
functions of spinal cord, 182
Hering-Breuer, inspiratory, 383
medullary ischemic, mechanism, 339, 340
movement of food from pharynx to stomach, 407
pupil, 253
righting, impulses causing, 257
and sensations, functions of receptors, 239
spinal, arcs producing, 175
test results, clinical significance, 221
vasoconstrictor mechanism, 340
Refraction:
light rays, 249–252
normal eye, 251
Regions of abdomen, 5, 7
Regulatory chemicals, comparison with hormones, 265

Reil, island of, of cerebrum, 183
Reissner's membrane, roof of cochlear duct, 258, 259
Relaxation of muscle fiber, mechanism, 117
Renal: (*see also* Kidneys)
  calculi, 459
  capsule, 448
  columns, 448
  corpuscle, 448
  diabetes, 455
  glycosuria, 455
  hypertension, theories, 458
  papillae, 448
  pelvis, 448, 458
  pyramids, 448
  suppression under stress, 453
  tubule(s):
    role in urine pH control, 517–518
    sections, 450
  vein of abdomen, 322
Renin:
  function, 458
  secretion, excess, role in hypertension, 280
Repair, protein anabolism chief process, 426
Replication, DNA, 34, 38
Reproduction:
  cell(s), 34–39
    and body, function of protein anabolism, 426
  hormones regulators, 265
  human being, 465–490
Reproductive:
  cells, 469
  functions, role of hypothalamus in control, 190
  glands, accessory, male, 470–472
  organs:
    female, 474–482
    male, and urinary bladder, 471
  system, 466–490
    female, epithelial cell proliferation, 283
Residual air:
  amount, 374
  increase in emphysema, 375
Resistance:
  peripheral (*see* Peripheral resistance)
  vessels, arterioles, 315
Respiration(s):
  abnormal, 376
  adjustment to pH of arterial blood, 516
  control:
    during exercise, 384
    pneumotaxic center, 194
  controlling mechanism, 382–385
  depth and rhythmicity, regulation by Hering-Breuer reflexes, 383
  effect:
    circulation, 344
    position of stomach, 399
  external and internal, 372
  inhibition during swallowing, 408
  mechanisms, 372–374
  physiology, 372–385
  types, 376
  venous blood return to heart influenced by, 342
Respiratory:
  center:
    medulla, 194

Respiratory—cont'd
  center—cont'd
    stimulation by decreased blood pH, 520
  and circulatory systems, interdependence, 362
  control mechanism, 383
  mechanism of pH control, 515–517
  system, 362–385
Responses:
  adaptive, 6
  sympathetic, widespread area involved, 225
Restiform bodies, cerebellar, 192
"Resting" neuron, role in nerve impulse, 170
Resting potential, mechanism, 170
Retention of urine, catheterization for, 460
Reticular:
  activating system, 215
  formation in medulla oblongata, 193
Reticulocyte count, indicator of rate of erythropoiesis, 296
Reticuloendothelial:
  cells, phagocytic, 298, 353, 355
  connective tissue, 45, 50
Reticulospinal tracts, motor pathways, 180, 181, 182, 219
Reticulum:
  endoplasmic, 14–15
  sarcoplasmic, 114, 115
Retina, 241, 242, 243
  layers, 244
  stimulation, 253–254
  three-neuron relay, 244
Retinal image, formation, 249–253
Retraction, definition, 110
Rh antigen, blood, 302
Rhinencephalon, limbic system, 232
Rhodopsin, function in vision, 253–254
Rib(s), 70
  articulation, 84
  cage, position of liver and pancreas relative to, 403
  insertion of muscles, 138
  markings, 99
  origin of muscles, 130, 138
Ribosomes, description and function, 18
Ridges, supracondylar, markings of femur, 102
Righting reflexes, impulses causing, 257
Rigidity, mechanism of development, 221
Rings, femoral, herniation through, 156
RNA, role in protein anabolism, 426
Rods and cones, photoreceptor neurons, 243, 244
Rolando, fissure of, 183
Root:
  lung, components, 371
  neck, and crown, parts of tooth, 397
Rotation, pivoting movement, 105
Rotators, muscle group, 127
Round:
  ligaments, 474, 477
  window of middle ear, 255, 256, 257
Rupture of abdominal wall, 127

S

Saccharide compounds, digestion, 411, 412
Saccule, location, 256, 257

Sacral:
  plexus, formation, location, and branches, 201–202
  promontory, 98
Sacroiliac joint, description, 109
Sacrospinalis muscle, extensor of trunk, 129
Sacrum:
  description, 70
  location in vertebral column, 79
Sacs:
  alveolar, 368, 369
  lacrimal, location, 249, 250
Saddle joint, 106, 107
Sagittal:
  sinuses of cranial cavity, 322, 323
  suture, special feature of skull, 96
Saliva:
  secretion, reflex control only, 412
  summary of action, 409
Salivary glands, 394–395
Salt(s):
  absorption, active transport mechanisms in, 416
  mineral, metabolism, 429
Salt-retaining and water-retaining effects of progesterone, 284
Santorini, duct of, 406
Saphenous veins of lower extremities, 322, 324
Sarcolemma, muscle cell membrane, 114
Sarcomere of myofibril, 114–115
Sarcoplasm, muscle cell cytoplasm, 114
Sarcoplasmic reticulum:
  muscle cells, 114
  triad, 115
Sartorius, muscle of lower leg motion, 135
Satiety center, factors inhibiting, 435
Scala:
  tympani, 257–258
  vestibuli, 257–258
Scapula:
  description, 71
  insertion of muscles, 130
  markings, 84, 85, 99
  origin, 131, 132
  right, 84, 85
Scar tissue formation, function of protein anabolism, 426
Schlemm, canal, location, 242
Schwann cells, description, 168
Sciatic:
  nerve, formation, 202
  notch, marking of os coxae, 89, 100
Sclera, outer layer of eyeball, 241, 242
Scoliosis of vertebral column, 84, 98
Scrotum:
  external structure, 472
  testes contained in, 467
Sebaceous glands of skin, 58
Secretin, source and action, 413
Secreting cells of pancreas, exocrine glands, 405
Secretion:
  bile, chemical control, 414
  digestive glands, control, 412–414
  endocrine glands, circulation, 265
  gastric juice, 400, 413
  glandular, form of reflex, 221
  glucocorticoid, stimulating effect of ACTH, 428

Secretion – cont'd
  hormone(s):
    feedback control, 270
    function of ovaries, 478
    during ovarian cycles, 483
  intestinal juice, control obscure, 414
  pancreas, 406, 414
  saliva, reflex control only, 412
  tubular, 455–457
Segmental:
  distribution of spinal nerves, 202, 205
  reflex, knee jerk, 222
Sella turcica:
  pituitary body, site, 265
  sphenoid bone marking, 76, 95
Selye's general adaptation syndrome, 525, 530
Semicircular canals of inner ear, 255, 257, 259
Semilunar:
  notch, marking of ulna, 87, 100
  valves:
    description and function, 309–311
    frequency in thoracic duct, 351
    venous blood movement toward heart, role, 344
Semimembranosus muscle:
  lower leg motion, 135
  thigh, 151
Seminal:
  duct, 470
  fluid, composition and course, 472–473
  vesicles, 470–471
Seminiferous tubules, 467
Semispinalis capitis muscle, extensor of head, 129
Semitendinosus muscle:
  lower leg motion, 135
  thigh, 151
Sensation(s):
  consciousness and cerebral cortex, 211
  crude awareness and thalamus, 211
  and receptors by which mediated, 240
  structures producing, 186
  thalamus, role in mechanism, 188
Sense organs, 239–260
Sensory:
  area, general, of cerebral cortex, 186, 191
  cerebral tracts, 184
  fibers of cranial nerves, 206–207
  functions:
    cerebral cortex, 186
    spinal cord, 182
  impulses from thermal or pain receptors, influence on respiration, 384
  neural pathways, 211–215
  neurons, description, 167, 168
  tracts, ascending, in spinal cord, 180–182
Septum, nasal:
  bones forming, 97
  and cartilages forming, 364
  partition, 363
Serotonin, postulated secretion of pineal body, 285
Serous and synovial membranes, 53–54
Serratus anterior muscle, 128, 130, 142
Serum:
  antilymphocyte, to prevent transplant rejection, 299
  blood, difference from plasma, 304
Sesamoid bones, location, 71

Sex:
  and age, basal metabolic rates, 431
  cells, meiosis, 488
  fluid volume variation with, 502
  hormones, secretion by adrenal cortex, 280
  organs:
    female, 475–482
    male, 467–474
    and characteristics, secondary, promotion by testosterone, 284
Sexual:
  behavior, normal, testosterone contribution to, 284
  cycles, female, 482–488
Shaft of rib, marking, 99
Shape of skeletal muscles, 119
Sharpey's fibers of periosteum, 67
Sheath(s):
  myelin, description, 168, 169
  tendon, of connective tissue, 120
Shin bone, tibia, 71
Shivering, heat-gaining process, 437
Short bones, 65, 67
Shoulder:
  blades, 71
  girdle, 70–71, 84
  joint, 108
  motion, muscles, 130
  muscle, 128, 142–143
Sigmoid colon, description, 398, 402
Signs, pyramidal, causes, 221
Simmonds' disease, result of growth hormone deficiency, 268
Single autonomic innervation, principle, 228
Sinoatrial node of heart, 312, 313–314
Sinus(es):
  dural, of head, projected on skull, 325
  ethmoid bone markings, 81, 96
  frontal, 72
    marking of frontal bone, 94
  longitudinal, site of intravenous injections in infants, 322
  maxillary, 76
    marking, 96
  paranasal:
    drainage, 363
    frontal section through face, 365
    special feature of skull, 97
  sphenoid, 73
    bone markings, 78, 95
  transverse, 322
  venous:
    cranial cavity, 321, 322, 325
    relation to brain and skull, 325
    spleen, 354
Size:
  factor influencing basal metabolic rate, 430
  skeletal muscles, 119
Skeletal:
  muscle(s), 46, 47, 113–158 (*see also* Muscle)
    age changes, 121
    attachment of bones, 124
    cells, 114–119
      molecular structure, 115–116
    contractions, in temperature homeostasis, 114
    function, 121–126
    functioning, role of calcium, 429
    group action, 125

Skeletal – cont'd
  muscle(s) – cont'd
    grouped:
      by function, 127, 129
      by location, 128
    motor nerve essential to, 121
    nerve supply, 120
    organs, 119–127
    origins, insertions, and actions, hints on deducing, 126
    structure, 119–121
    tissue, 44, 46
    work, energy used in, determinant of total metabolic rate, 432
  system, 62–112
    definition, 63
Skeleton, 68–103
  age changes, 103
  anterior, 74
  appendicular, 84–93
    bones, 70
  axial, 68–84
    bones, 69–70
  bones, 69–71
  male and female, differences, 103
  posterior, 75
Skene's glands, lesser vestibular, 480
Skin, 54–58
  accessory organs, 56–58
    nails as, 57
  color:
    factors determining, 57
    heredity and, 56, 57
  glands, 58
  heat:
    receptors, stimulation, cardiac accelerator, 335
    transfer through, 436
  layers, 54, 56
  microscopic section, 55
  substances excreted, 447
  temperature, conscious mechanism to regulate, 439
  terminology, 58
  thermal receptors, 437
    impulses from, influence on respirations, 384
Skull:
  at birth, 80
  bones, 68–79
  special features, 96–97
Small intestine:
  digestive process performed, 408
  functions, 401
  section, circular folds and layers of wall, 390
Smell, organ of, function of nose, 365
Smooth muscle:
  contractions:
    autonomic reflexes, 221
    digestive tract, parasympathetic dominance, principle, 230
  effectors, parasympathetic and sympathetic effects, 231
  tissue, 44, 45
Sodium:
  blood, homeostasis, primary function of mineralocorticoids, 280
  concentration, effect on fluid exchange, 507
  hydroxide, buffering by carbonic acid, 515
  maintenance of acid-base and water balances, role, 429

Sodium – cont'd
    pump, 28
        and cell membrane polarization, 17
        and water retention, effect of aldo-
            sterone, 280
Soft palate, arch-shaped muscle, 393
Soleus muscle, 128, 129, 136, 147
Solutes:
    blood plasma, 303
    reabsorption, mechanism, 453
Solutions:
    hypertonic and hypotonic, 27
    isosmotic and isotonic, 26
Soma and psyche, hypothalamus as link,
    190
Somatic:
    motoneurons, 120, 203
    pain, 241
    reflexes, mechanism, 221
Somatopsychosomatic theory of mind
    and body relations, 272
Somatotropic hormone, abnormalities
    and results, 267, 268
Somatotropin, mechanism of function,
    266
Somesthetic sensory area of cortex, 186,
    191
Sound waves, course, in hearing mecha-
    nism, 259
Space, pericardial, location, 307
Spastic muscles, definition, 121
Spasticity, mechanism of development,
    221
Specific dynamic action of foods, 432
Specificity of receptors, principle, 240
Speech:
    area, Broca's motor, 188, 191
    functions, site, 188
Sperm:
    count and sterility, relationship, 473
    course, 472
Spermatic:
    cords, internal structures, 472–473
    or ovarian vein of abdomen, 322
Spermatogenesis, function of testes, 468
Spermatozoon, parts, 469
Sphenoid:
    bone, 73
        description, 69
        markings, 95
    sinuses, 73, 78, 95
        drainage, 363, 364
Spheroidal joint, summary, 106
Sphincter muscles, 127, 400, 401, 402
Sphygmomanometer, description, 345
Spinal:
    cavity, 3
    contents, 178
    cord, 178–182
        and brain, coverings and fluid spaces,
            195–196
        major tracts, 181
        with meninges, 209
        reflex, knee jerk, 222
        relation:
            to surrounding structures, 179
            to sympathetic trunk, 209
            to vertebra, 209
    foramen of vertebrae, marking, 98
    ganglion, formation and location, 198,
        199
    nerve(s), 198–204
        accessory, summary, 207, 210
        cranial contrasted with, 204

Spinal – cont'd
    nerves(s) – cont'd
        distribution, 199–203
        formation, 209
        segmental arrangement, 202, 205
        reflexes, arcs producing, 175
Spine(s): (see also Spinal; vertebral col-
    umn)
    iliac, markings of os coxae, 89, 100
    ischial, marking of os coxae, 89, 101
    scapula, marking, 85, 99
Spinocerebellar tracts of spinal cord,
    181–182
Spinothalamic:
    pathway, 214
    tract(s):
        lateral, 213
        spinal cord, 180–182
Spinous process of vertebrae, 82
    marking, 98
Spiral ganglion of auditory relay, 257, 258
Spirometer, measurement of air ex-
    changes in breathing, 374
Spleen, 354–355
    and liver, exceptional sites of erythro-
        poiesis, 295
Splenectomy, first, landmark (Dr. Charles
    Austin Doan), 355
Splenic vein and superior mesenteric
    vein, union to form portal vein,
    322, 326
Squamous:
    epithelial tissue, walls of alveoli, 368
    epithelium, 43, 44
    portion, marking of temporal bone, 94
    suture, 72
Staircase phenomenon of skeletal mus-
    cle contraction, 122
Stalk, pituitary, 184, 189, 265
Standing and walking, comparative dis-
    comfort, reasons, 345
Stapes:
    description, 70
    location in ear, 255, 256
Starling's law:
    capillaries, 505
    heart, 124, 333, 334
Starvation, cause of negative nitrogen
    balance, 427
Steapsin, digestive enzyme of pancreatic
    juice, 409
Stenosis, mitral, 311
Stensen's duct, 395
Stereognosis, definition, 214
Sterility:
    hormone imbalance, 265
    male, and sperm count, relationship,
        473
Sternoclavicular joint, 108
Sternocleidomastoid muscle, 128, 129,
    137, 142
Sternum:
    description, 70
    markings, 99
    muscle origins, 131, 137
    parts, 84
Steroid compounds, nucleus, 277
Stimulation:
    retina, 253–254
    threshold, in impulse conduction, 172
Stimulus:
    definition, 172
    nerve impulse conduction, 170
    skeletal muscle contraction, 121

Stimulus – cont'd
    strength, variation, and muscle con-
        traction, 124
Stomach:
    curves, 400
    digestive process performed, 408
    emptying process, 409
    functions, 400
    modifications of mucous, muscle, and
        fibroserous coats, 391
    position:
        frequent alteration, 399
        of liver and pancreas in relation to,
            403
    sphincter muscles, 400
Strabismus, definition, 253
Straight sinus of cranial cavity, 322–323
Straightening movement, extension, 105
Strata, skin, 54, 56
Stratified squamous epithelium, 43, 44
Strength of contraction, skeletal muscle,
    124, 125
Stress, 523–532
    adrenal medulla function, 281
    antidiuretic hormone secretion stim-
        ulus, 273
    carbohydrate metabolism under, con-
        trol mechanism, 419, 423
    and disease, 529–530
    and homeostatic responses, resem-
        blance, 526
    hyperglycemia, principle, 424
    hypothalamus function, 282
    magnitude, indicators, 528
    regulation of anterior pituitary secre-
        tion, 271
    renal suppression, 453
    resistance, function of glucocorticoids,
        277
    response:
        hormones regulators, 265
        induced by glucocorticoids, 278
    sympathetic dominance, principle, 230
Stressor(s):
    individual differences in reaction, 525
    major, body response, factors deter-
        mining, 528
    psychological, physiological responses,
        529
Stretch reflex, knee jerk, 222
Striated muscle, 114–115
    tissue, 44, 46
Stroke volume of heart, 332, 333
Structure:
    body (see Body structure)
    cell, 11–19
    DNA, 36–38
    and function, relationship, 8
    skeletal muscles, 119–121
    Watson-Crick, DNA molecule, 36, 37
Student's elbow, olecranon bursitis, 157
Styloid process:
    marking:
        radius, 87, 100
        temporal bone, 94
        ulna, 87, 100
    radius, 85
Stylomastoid foramen, marking of tem-
    poral bone, 94
Subarachnoid space, location, 196
Sublingual gland, location and duct open-
    ing, 395
Subclavian:
    artery, 318–319

Subclavian – cont'd
  vein of upper extremity, 321
Subdural space, location, 196
Submaxillary gland, location and duct opening, 395
Submental and submaxillary lymph nodes, 352
Submucous coat of connective tissue, 390
Substrate, specific, action of enzymes, 411
Succus entericus, 409
Sucrase, digestive enzyme of intestinal juice, 409
Sugar (*see* Glucose)
Sulcus(i):
  cerebrum, 183
  posterior median, spinal cord groove, 179
Summation, antagonism and, autonomic, principle, 229
Superciliary arches, marking of frontal bone, 94
Superficial veins situated near surface of body, 321
Supination, definition, 105
Supinators, muscle group, 127
Support, function of skeletal system, 63
Supporting structures of male reproductive organs, 472
Suppression, urine, seriousness, 460
Supracondylar ridges, markings of femur, 102
Supraoptic nuclei of hypothalamus, 189
Supraorbital notch, marking of frontal bone, 94
Suprarenal vein of abdomen, 322
Surface area:
  body's, factor influencing basal metabolic rate, 430
  man, chart for determining, 433
Surgical neck of humerus, marking, 86, 99
Survival:
  body and species, 6
  chief function of body, 164
Suspensory ligament, function, 242
Suture(s):
  coronal, 72
  lambdoidal, 72
  special feature of skull, 96–97
  squamous, 72
Swallowing, 367, 407–409
Sweat glands:
  skin, 58
  substances excreted, 447
Sylvius:
  aqueduct of, cerebral, 196, 197
  fissure of, cerebral, 183
Sympathetic:
  division of autonomic nervous system, 224
  dominance under stress, principle, 230
  effects of visceral effectors, 231
  fibers of cardiac nerves, component of cardiac plexuses, 314
  ganglia, description, 224
  impulses:
    conducted to sinoatrial node, modifier of heart rate, 313
    stress responses induced, 527
  nervous system, 226
    preganglionic and postganglionic neurons, 224–225, 227

Sympathetic – cont'd
  pathways, 225
  responses, widespread area involved, 225
  stress responses, specific, 527
  trunk, spinal cord relation to, 209
Symphysis pubis:
  description, 109
  marking of os coxae, 101
Synapses:
  definition, 172
  nerve impulse conduction across, 174, 176, 178
Synaptic:
  cleft, description, 174
  knobs, 174, 176–177
Synarthroses, 105, 106, 107
Synchondroses, 106
Syndesmoses, 106
Syndrome:
  Cushing's, 278, 279
  general adaptation, of Selye, 525, 530
Synergic control of muscle action, function of cerebellum, 192
Synergists, skeletal muscle group, 126
Synovial membranes, serous and, 53–54
Synthesis:
  continual, of enzymes, necessity, 411
  protein (*see* Protein anabolism)
System(s):
  autonomic nervous, 223–232
    innervation of heart, 313
  circulatory, 292–361
  communication, of body, 165, 169
  conduction, of heart, 312–313
  definition, 3
  digestive, 389–445
  endocrine, 263–285
  haversian, of bone, 64, 65
  limbic, location and function, 232
  lymphatic, 348–355
  muscular, 113–161
  nervous, 164–238
    autonomic, 223–232
    and endocrine system, hypothalamus link between, 271
    maintenance of posture, role, 158
    parasympathetic, 224–232
    sympathetic, 224–232
  peripheral nervous, composition, 178
  pituitary portal, release of neurosecretions into, 271
  portal, function, 416
  reproductive, 466–490
  respiratory, 362–385
  reticular activating, 215
  reticuloendothelial, 50
  skeletal, 62–112
  structural units of body, 2
  T, of skeletal muscle cells, 114–115
  urinary, 446–461
Systemic:
  blood pressure gradient, factors determining, 343
  circulation, 320–322, 328
  and pulmonary circulation, relationship, 329
Systole:
  contraction of atria, 314
  ventricular, start of new pulse wave, 347
Systolic:
  blood pressure, definition, 345

Systolic – cont'd
  discharge, volume of blood pumped by one contraction, 332

T

T system of skeletal muscle cells, 114
Tactile corpuscle, nerve ending in skin, 240
Talus, 71, 91
  marking of tarsals, 91, 102
Target:
  gland stimulation, function of tropic hormone, 269
  organs of anterior pituitary hormone, 266
Tarsal bones:
  description, 71
  markings, 91, 92, 102
  muscle insertions, 136
Taste(s):
  area, primary, of cortex, 187
  buds:
    location, 260
    papillae of tongue, 394
  classification, 260
  corpuscles, distribution over tongue, 260
  zones, overlapping, location on tongue, 395
Tectorial membrane of organ of Corti, 259
Teeth, 395–397
Telephone system and nervous system, similarity, 169
Temperature:
  blood:
    factor influencing respirations, 384
    increased, cardiac accelerator, 335
    hypothalamus, influence on appetite centers, 435
  body:
    homeostasis, 435–439
    hypothalamus, role in maintaining, 190
  homeostasis, role of skeletal muscle contraction, 114
  and pain neurons, 212
  receptors, sensory impulse relays from, 213
  sensations, neural pathway, 214
  skin, conscious mechanism to regulate, 439
Temporal:
  bone(s), 72
    description, 69
    markings, 94–95
    mastoid cells, middle ear connection, 255, 256
    muscle insertion, 137
  lobe of cerebrum, 183
Tendon:
  connective tissue, 119
  reflex, knee jerk, 222
  sheaths of connective tissue, 120
Tension, synonym for partial pressure, 377
Tensor(s):
  fasciae latae muscle, 128, 134, 147
  muscle group, 127
Tentorium cerebelli, function, 196
Terms descriptive of body, 4–6

Testis(es), 467–468
  and anterior pituitary, negative feed-
    back mechanism between, 468–
    469
  ducts, 467, 468, 470
  and epididymis, tubules, 467
  and left spermatic cord, lateral view,
    473
  secretions, 284
  undescended, 473
Testosterone, 284, 427, 428, 468
  secretion stimulation by interstitial
    cell–stimulating hormone, 269
Tetanic contraction, skeletal muscle, 122
Tetanus, muscle, 123
Tetany, result of hypocalcemia, 276
Thalamus:
  crude awareness and, 211
  structure, location, and functions, 188,
    189
Thermal receptors, skin, 384, 437
"Thermostat theory" of appetite control,
    435
Thermostatic control of heat production
    and loss, 437
Thigh:
  bone, description, 71
  motion, muscles, 129, 134
  muscles, 128, 147–149, 151
Thoracic:
  aorta, descending, 318–319
  cavity, 3
    air pressure gradient, 372
  duct, 348, 350
  joints, description, 108
  organs, veins of, 322
  vertebrae, 79
    enumeration, 70
    parts, 97–98
Thoracolumbar division of autonomic
    nervous system, 224
Thorax, 371–372
  age changes in skeleton, 103
  bones forming, 70, 84
  origin of diaphragm, 139
Three-neuron:
  reflex arc, 167, 172, 175
    correlation of structures, 222
  relay in retina, 244
Throat, description, 365
Thrombin, role in blood clotting, 304
Thrombus of transverse sinus, internal
    jugular vein ligation, 322
Thumb, mobility, importance, 88
Thymus, 284
Thyrocalcitonin, function, 274
Thyroglobulin, 273
Thyroid:
  cartilage, Adam's apple, 364, 366
  function test, protein-bound iodine
    blood level, 273
  gland, 273–275
  hormone:
    anabolic or catabolic, 428
    blood concentration, modifier of
      heart rate, 313
    components, 273
    hypersecretion, and hyposecretion,
      effects, 274
    influence on basal metabolism, 431
    secretion, stimulation by thyrotropin,
      269
Thyroid-stimulating hormone, function,
    269

Thyrotropin:
  function, 269
  glucose catabolism stimulated by, 419
Thyrotropin-releasing factor, neurosecre-
    tion, 271
Thyroxine, component of thyroid hor-
    mone, 273
Tibia:
  articulation, 91
  description, 71
  and fibula, right, anterior surface, 90
  insertion of muscles, 134, 135
  markings, 90, 102
  origin of muscles, 136
Tibial:
  arteries, anterior and posterior, 318–
    319
  tuberosity, marking, 90, 102
  veins of lower extremities, 322–323
Tibialis anterior muscle, 128, 129, 136,
    147
Tibiofibular joint, description, 109
Tic douloureux, treatment, 208
Tidal air, amount exhaled after inspira-
    tion, 374
Time components of simple muscle
    twitch, 122
Tissue(s), 42–51
  adipose:
    calories to synthesize one pound, 434
    principal fat storage, 425
  capillaries:
    gas exchange between arterial blood
      and cells, 381
    pressures at arterial and venous
      ends, 506
  classification, 44–45
  coats of alimentary canal, modifica-
    tions in different organs, 390–
    391
  connective (*see* Connective tissue)
  definition, 3
  differentiation, regulation by thyroid
    hormone, 273
  fluid, interstitial fluid, 348
  gas exchange, 382
  lymphatic:
    derivation of nongranular leuko-
      cytes, 299
    involution, induced by glucocorti-
      coids, 278
    myeloid, derivation of granular leuko-
      cytes from, 299
  protein:
    mobilization, 427
    synthesis, 427
  scar, formation, function of protein
    anabolism, 426
  site of phagocytosis by leukocytes, 298
  structural units of body, 2
  transplant, mechanism of rejection,
    299
  wasting, excess of glucocorticoids
    cause, 277
Tissue-attacking lymphocytes, role in
    graft rejection, 299
Toes, bones, 71
Tone, muscle, 219, 220
Tongue, description, 394, 395
Tonic contraction of skeletal muscle, 121
Tonicity of muscle, allowing pull against
    gravity, 158
Tonsils, 365
Total metabolic rate, 431–432

Touch sensations, neural pathways, 214
Toxic wastes, dilution, importance of
    water, 429
Toxins, elimination in urine, 461
Trabeculae, bone processes, 64
Trachea, 367–368
  obstruction, cause of death from as-
    phyxiation, 368
  relation of thyroid and parathyroid
    glands to, 274
Tracheotomy, method, 368
Tract(s):
  cerebellar, 192
  cerebral, 184
  corticospinal, 216, 217, 218
  definition, 178
  digestive, parasympathetic dominance,
    230
  extrapyramidal, complexity, 219
  motor, 217
  and nerves, comparison, 184
  optic, summary, 208
  pyramidal, motor pathways, 216–219
  reticulospinal, motor pathways, 219
  spinal cord, major ascending and de-
    scending, 181–182
  spinothalamic, lateral, 213
Transducer:
  endocrine, nerve impulses translated
    into hormones, 271
  neuroendocrine, 282, 285
Transfusion, blood, 302
Transmitter(s):
  chemical autonomic, principle, 228
  substance in synaptic conduction, 174,
    176
Transplanted organ or tissue, mechanism
    of rejection, 299
Transport:
  active chemistry, 28, 29
  mechanisms:
    active, 27–29
      role in food absorption, 414
      and tubular reabsorption, 453, 454
    passive, 20–27
Transportation, primary function of cir-
    culatory system, 293
Transversalis, muscle of abdominal wall
    motion, 138
Transverse:
  colon, location, 398, 402
  processes of vertebrae, marking, 98
  sinus(es) of cranial cavity, 322–323
Trapezius muscle, 128, 129, 130, 142–143
Tree, bronchial, 368
Treppe phenomenon of skeletal muscle
    contraction, 122, 123
Triad of sarcoplasmic reticulum, 115
Triceps brachii muscle, 128, 129, 132,
    143–144
Tricuspid valve, 308–310
Trifacial nerve, summary, 206, 208
Trigeminal nerve, 206, 208, 209
Triiodothyronine, component of thyroid
    hormone, 273
"Triple-layer hypothesis" of cell mem-
    brane molecular structure, 15
Trochanters, markings of femur, 90, 101–
    102
Trochlea, marking:
  femur, 102
  humerus, 86, 100
Trochlear nerve, summary, 206–208
Trochoid joint, summary, 106

Tropic hormones of anterior pituitary gland, 268, 269
Tropomyosin of muscle cells, 115–116
Troponin of muscle cells, 115–116
Trunk:
    age changes in skeleton, 103
    muscles, 129, 140, 142
    sympathetic, spinal cord relation to, 209
Trypsin, digestive enzyme of pancreatic juice, 409
Tubercle(s):
    adductor, marking of femur, 102
    gluteal, marking of femur, 102
    humerus, 86, 99
    pubic, marking of os coxae, 101
    rib, marking, 99
Tuberosity(ies):
    deltoid, of humerus, marking, 86, 99
    frontal, marking of frontal bone, 94
    ischial, marking of os coxae, 89, 101
    radial, 85
        marking, 87, 100
    tibial, marking, 90, 102
Tubes:
    auditory (Eustachian), 255, 256, 365
    fallopian, 474, 477
    propulsion through, form of movement, 114
    uterine, 474, 477
Tubular:
    reabsorption, 452, 455
        water, regulation, 499
    secretion, 454–457
Tubule(s):
    convoluted, 453
    renal:
        role in urine pH control, 518
        sections, 449, 450
    seminiferous, 467
    T system, 114–115
    testis and epididymis, 467
Tubuloacinar gland, compound, pancreas classified as, 405
Tumor(s):
    nervous system, neuroglia and, 165
    virilizing, of adrenal cortex, case, 279
Turbinate(s):
    bones, description, 69
    division of nasal cavities by, 363, 365
    superior and middle, ethmoid bone markings, 81, 96
Turk's saddle, sphenoid bone marking, 76, 95
Twitch:
    contraction of skeletal muscle, 122
    and tetanus, 123
Two-neuron reflex arc, 172, 173
    conduction over, knee jerk, 221
Tympanic:
    cavity (middle ear), 255, 256
    membrane (eardrum), 255, 256
Type O blood, universal donor blood, 302

U

Ulcer, peptic, parasympathetic excess, 230
Ulna:
    articulation, 85
    description, 71
    insertion of muscles, 132
    markings, 87, 100

Ulnar:
    artery, 318–319
    vein of upper extremity, 321
Umbilical:
    arteries:
        description, 627
        and vein, fetal structures, 326
    cord, formation by umbilical arteries and vein, 327
    vein, description, 327
Umbilicus, herniation through, 156
Underarm and upper chest, site of axillary lymph nodes, 352
Unipolar neurons, location, 169
Universal:
    donor blood, type O, 302
    recipient blood, 303
Upper extremity (*see* Extremity, upper)
Ureters, 458–459
Urethra, 460
    relationship to prostate gland, clinical significance, 470
Urinary:
    bladder and male reproductive organs, 471
    meatus, 460, 480
    mechanism of pH control, 517, 520
    system, 446–461
Urination, promotion by parasympathetic impulses, 230
Urine:
    acidification, 518, 519
    chemical composition, 460
    excretion, 451–458
        and elimination, vital functions, 446
    pH, control mechanisms, 517
    retention, catheterization, 460
    suppression, seriousness, 460
    volume:
        control mechanisms, 458
        regulation, 499
Uterine tubes, 474, 477
Uterosacral ligaments, 477
Uterus, 474, 477
    cyclical changes, 486
    pregnant, contractions, stimulation by oxytocin, 272
Utricle, 256–257
Uvula, 393

V

Vagina, 479
Vaginal orifice, 480
Vagus nerve:
    parasympathetic fibers, component of cardiac plexuses, 314
    summary, 207–210
Valve(s):
    aortic, insufficiency, pulse pressure increased, 346
    atrioventricular, 308–311
    cuspid, 308–311
    heart, 308–311
    ileocecal, 402
    semilunar, 309–311
        frequent in thoracic duct, 351
    venous, 317
        action, 344
Valvulae conniventes, section of small bowel, 390
Valvular:
    disease, indicated by heart murmur, 314

Valvular – cont'd
    insufficiency, description, 311
Varicose veins, blood vessel abnormality, 316
Vas deferens, 470
Vasa vasorum, definition, 316
Vasoconstriction, skin, heat-gaining process, 437
Vasodilatation:
    localized areas, mechanism, 340, 342
    major mechanism, 336
    and vasoconstriction, regulation of heat radiation, 436
Vasomotor:
    center of medulla, 194
    chemoreflex(es), mechanism, 340
    control by higher brain centers, 340
    pressoreflexes, mechanism, 336, 337
Vasopressin, use in diabetes insipidus, 272
Vastus muscles, 135, 148
Vater:
    ampulla of, 403, 406
    papilla of, 403
Vein(s):
    abdominal organs, 322–323
    corresponding to and near main arteries, 321
    deep and superficial, 321
    definition, 315
    emissary, 325
    head and neck, 322, 325
    lower extremities, 322–324
    and lymphatics, similarities and differences, 351
    portal, 322, 326
    principal, 321–322
        of body, 323
    saphenous, of lower extremity, 322, 324
    showing thickness of three coats, 317
    thoracic organs, 322
    umbilical, 326, 327
    upper extremities, 321–322, 324
    valves, 317
    varicose, blood vessel abnormality, 316
Venous:
    arch, dorsal, of foot, 322–324
    and arterial bleeding, relation of blood pressure to, 346
    blood:
        conversion to arterial blood, 380
        pH, 515
    networks, blood reservoirs, 340
    plexuses in skin and abdominal organs, blood reservoirs, 340
    pulse, in large veins only, 348
    return:
        to heart, factors influencing, 342
        from liver, obstruction, cause of ascites, 326
    sinuses:
        cranial cavity, 321, 322, 325
        spleen, 354
    valves, action, 344
Ventilation, alveolar, 376
Ventral cavity, 3
Ventricles:
    brain, 196–197
    heart, 308, 309
    systole, start of new pulse wave, 347
Venules:
    small veins, 315
    smaller cross-sectional area than capillaries, 346

Vermiform:
  appendix, size, shape, and location, 406
  process and ileocecal region, 398, 401
Vermis of cerebellum, 192
Vertebra(ae):
  cervical, 82, 83
    parts, 98
  lumbar, markings, 83, 98
  muscle origins, 130, 131, 134
  parts, 82, 97
  spinal cord relation to, 209
  thoracic, parts, 97–98
Vertebral:
  artery, location, 320
  column, 78, 82–84
    age changes in skeleton, 103
    bones, 70
    markings, 97–98
    shape, 82
  foramen, 82
  joints, description, 108
Vesicles, seminal, 470–471
Vessels:
  blood (*see* Blood vessels)
  lymphatic, 348–351
Vestibular:
  glands, greater and lesser, 480
  nerve, composition, 210
Vestibule, location and description, 255, 256, 480
Vibrations, conduction of sound waves, 259
Villi, section of intestinal mucosa, 392
Virilizing tumor of adrenal cortex, case, 279
Viscera:
  abdominal, front, 394
  contained in body cavities, 3
Visceral:
  effectors:
    definitions, 223
    parasympathetic and sympathetic effects, 231
  functions of cerebral cortex, 186
  layer of serous membrane, 53, 55
  muscle, 46, 47
  pain, 241
  pleura:
    chest section, 55
    covering of outer surface of lung, 371
  reflexes, mechanism, 221
Visceroreceptors, internal receptors, 239
Viscosity, blood, 333, 335
Vision:
  daylight and color, 254
  night, role of rhodopsin, 254
  physiology, 249–254
Visual:
  acuity, fraction indicating, 252

Visual — cont'd
  area, primary of cortex, 186, 191
  cortex, impulse conduction to, 254
  reflex centers, 194
Vital:
  capacity, 374–375
  centers of medulla, 194
Vitamin:
  A in rhodopsin cycle, 254
  B$_{12}$, 400
  K, role in blood clotting, 304
  metabolism, 428
Vitreous humor, description, 242, 245
Vocal cords, true and false, 364, 366
Voice:
  box, larynx, 366
  pitch, determined by vocal cords, 367
Volar arch, vein of upper extremity, 321
Volkmann's canals of bone, 64
Voluntary:
  movements, complex mechanism, 187
  muscle:
    contractions, heat gaining process, 437
    tissue, 44, 45
Vomer, 69, 79
Vulva, 479–480

W

Waking state, role of hypothalamus, 190
Walking and standing, comparative discomfort, reasons, 345
Walls of digestive organs, 390–393
Warm-up principle of muscle contraction, 122, 124
Wastes:
  nitrogenous, elimination in urine, 460
  reabsorption and secretion, 456
Wasting illness, cause of negative nitrogen balance, 427
Water:
  absorption:
    active transport mechanisms, 416
    function of large intestine, 402
  avenues of entrance and exit in body, 498, 500
  balance:
    blood proteins, important for maintenance, 405
    maintenance:
      hypothalamus, role, 190
      sodium and potassium, role, 429
  evaporation, method of heat loss, 436
  intake and output, normal values, 498
  movement:
    through capillary membranes, 505
    through cell membranes, 506

Water — cont'd
  reabsorption:
    mechanism, 455
    and secretion, 456
  retention, role of antidiuretic hormone, 273
  return from interstitial fluid to blood by lymphatics, 351
  and sodium retention, effect of aldosterone, 280
  total body, homeostasis, 499
Watson-Crick structure of DNA molecule, 36, 37
Wave, pulse, new with each ventricular systole, 347
Weight:
  body, relationship of energy balance to, 433
  gain, calculation, 433
  reduction, cardinal principle, 434
Wettable surface of blood vessels, 304
Wharton's duct, 395
White:
  blood cells (*see* Leukocyte)
  matter:
    cerebrum, 184
    division of nervous tissue, 178
    spinal cord, 180
Willis, circle of, 321
Windows of middle ear, 255, 256, 257
Windpipe, trachea, 367–368
Wings, great and lesser, sphenoid bone markings, 95
Wirsung, duct of, 406
Wormian bones:
  location, 71
  special feature of skull, 97
Wound healing:
  function of protein anabolism, 426
  slower, induced by glucocorticoids, 278
Wrist:
  bones, 88
  joint, description, 108

X

X and Y chromosomes, 488
Xiphoid process of sternum, 84
  marking, 99

Z

Z line of striated muscle, 114–115
Zone, H, of striated muscle, 114–115
Zygomatic:
  bones, description, 69, 77
  process, marking of temporal bone, 94
Zygote, 489
Zymogen cells, 400